OPTIMIZATION OF CHEMICAL PROCESSES

BUILDING THE LITERATURE OF A PROFESSION

Fifteen prominent chemical engineers first met in New York more than 60 years ago to plan a continuing literature for their rapidly growing profession. From industry came such pioneer practitioners as Leo Baekeland, Arthur D. Little, Charles L. Reese, John V. N. Dorr, M. C. Whitaker, and R. S. McBride. From the universities came such eminent educators as William H. Walker, Alfred H. White, D. D. Jackson, J. H. James, Warren K. Lewis, and Harry A. Curtis. H. C. Parmelee, then editor of *Chemical and Metallurgical Engineering*, served as chairman and was joined subsequently by S. D. Kirkpatrick as consulting editor.

After several meetings, this committee submitted its report to the McGraw-Hill Book Company in September 1925. In the report were detailed specifications for a correlated series of more than a dozen texts and reference books which have since become the McGraw-Hill Series in Chemical Engineering and which became the cornerstone of the chemical engineering curriculum.

From this beginning there has evolved a series of texts surpassing by far the scope and longevity envisioned by the founding Editorial Board. The McGraw-Hill Series in Chemical Engineering stands as a unique historical record of the development of chemical engineering education and practice. In the series one finds the milestones of the subject's evolution: industrial chemistry, stoichiometry, unit operations and processes, thermodynamics, kinetics, and transfer operations.

Chemical engineering is a dynamic profession, and its literature continues to evolve. McGraw-Hill and its consulting editors remain committed to a publishing policy that will serve, and indeed lead, the needs of the chemical engineering profession during the years to come.

THE SERIES

Bailey and Ollis: *Biochemical Engineering Fundamentals*
Bennett and Myers: *Momentum, Heat, and Mass Transfer*
Beveridge and Schechter: *Optimization: Theory and Practice*
Brodkey and Hershey: *Transport Phenomena: A Unified Approach*
Carberry: *Chemical and Catalytic Reaction Engineering*
Constantinides: *Applied Numerical Methods with Personal Computers*
Coughanowr and Koppel: *Process Systems Analysis and Control*
Douglas: *Conceptual Design of Chemical Processes*
Edgar and Himmelblau: *Optimization of Chemical Processes*
Fahien: *Fundamentals of Transport Phenomena*
Finlayson: *Nonlinear Analysis in Chemical Engineering*
Gates, Katzer, and Schuit: *Chemistry of Catalytic Processes*
Holland: *Fundamentals of Multicomponent Distillation*
Holland and Liapis: *Computer Methods for Solving Dynamic Separation Problems*
Katz, Cornell, Kobayashi, Poettmann, Vary, Elenbaas, and Weinaug: *Handbook of Natural Gas Engineering*
King: *Separation Processes*
Luyben: *Process Modeling, Simulation, and Control for Chemical Engineers*
McCabe, Smith, J. C., and Harriott: *Unit Operations of Chemical Engineering*
Mickley, Sherwood, and Reed: *Applied Mathematics in Chemical Engineering*
Nelson: *Petroleum Refinery Engineering*
Perry and Chilton (Editors): *Chemical Engineers' Handbook*
Peters: *Elementary Chemical Engineering*
Peters and Timmerhaus: *Plant Design and Economics for Chemical Engineers*
Probstein and Hicks: *Synthetic Fuels*
Reid, Prausnitz, and Sherwood: *The Properties of Gases and Liquids*
Resnick: *Process Analysis and Design for Chemical Engineers*
Satterfield: *Heterogeneous Catalysis in Practice*
Sherwood, Pigford, and Wilke: *Mass Transfer*
Smith, B. D.: *Design of Equilibrium Stage Processes*
Smith, J. M.: *Chemical Engineering Kinetics*
Smith, J. M., and Van Ness: *Introduction to Chemical Engineering Thermodynamics*
Treybal: *Mass Transfer Operations*
Valle-Riestra: *Project Evolution in the Chemical Process Industries*
Van Ness and Abbott: *Classical Thermodynamics of Nonelectrolyte Solutions: With Applications to Phase Equilibria*
Van Winkle: *Distillation*
Volk: *Applied Statistics for Engineers*
Walas: *Reaction Kinetics for Chemical Engineers*
Wei, Russell, and Swartzlander: *The Structure of the Chemical Processing Industries*
Whitwell and Toner: *Conservation of Mass and Energy*

OPTIMIZATION OF CHEMICAL PROCESSES

T. F. Edgar and D. M. Himmelblau

Department of Chemical Engineering
University of Texas

McGRAW-HILL, INC.
New York St. Louis San Francisco Auckland
Bogotá Caracas Lisbon London Madrid
Mexico Milan Montreal New Delhi Paris
San Juan Singapore Sydney Tokyo Toronto

This book was set in Times Roman.
The editors were B. J. Clark and John M. Morriss;
the production supervisor was Friederich W. Schulte.
The drawings were done by Oxford Illustrators, Ltd.
Project supervision was done by Harley Editorial Services.

Printed and bound by Impresora Donneco Internacional
S.A. de C.V., a division of R. R. Donnelley & Sons Company.

Manufactured in Mexico

OPTIMIZATION OF CHEMICAL PROCESSES

4 5 6 7 8 9 0 DOR DOR 9 9 8 7 6 5 4 3 2 1

ISBN 0-07-018991-9

Library of Congress Cataloging-in-Publication Data

Edgar, Thomas F.
Optimization of chemical processes.
1. Chemical Processes. 2. Mathematical optimization.
I. Himmelblau, David Mautner. II. Title.
TP155.7.E34 1988 660.2′8 86-34269
ISBN 0-07-018991-9

ABOUT THE AUTHORS

Thomas F. Edgar is professor and chairman of chemical engineering at the University of Texas at Austin. He earned his B.S. in chemical engineering from the University of Kansas and received his Ph.D. from Princeton University. Before receiving his doctorate, he was employed by Continental Oil Company. He is General Editor of the technical journal *In Situ* and a consultant to a number of companies in the field of process control. Professional honors include selection as the 1980 winner of the AIChE Colburn Award, Outstanding AIChE Student Chapter Counselor in 1974, and Katz Lecturer at the University of Michigan in 1981. He is listed in *Who's Who in America.*

He has published over 100 papers in the fields of process control, optimization, and mathematical modeling of processes such as separations and combustion. He is coauthor of the book *Chemical Process Control* (John Wiley, 1988) and author of *Coal Processing and Pollution Control Technology*, published by Gulf Publishing Company in 1983. Dr. Edgar has also been Chairman of the CAST Division of AIChE in 1986 and President of the CACHE Corporation from 1981 to 1984.

David M. Himmelblau is the Bob R. Dorsey Professor of Chemical Engineering at the University of Texas at Austin. He received his B.S. degree from the Massachusetts Institute of Technology and M.S. and Ph.D. degrees from Washington University. He has taught at the University of Texas at Austin for over 25 years. Prior to that time he worked for companies such as the International Harvester Company, Simpson Logging Company and Excel Battery Company.

Among his more than 160 publications are 11 books including one of the most widely used introductory books in chemical engineering, books on process analysis and simulation, decomposition, fault detection in chemical processes, and applied nonlinear programming. He has served in a number of positions in the AIChE including Chairman of the Educational Projects Committee, the Student Chapters Committee, and the CAST Division. From 1974 to 1976 he was a Director of the AIChE. His principal areas of research are in fault detection and diagnosis, process analysis and simulation, and optimization, and he also has been extensively involved in the development of computer-aided instructional materials for chemical engineers.

CONTENTS

PREFACE

The chemical industry has undergone significant changes during the past 15 years due to the increased cost of energy and increasingly stringent environmental regulations. Modification of both plant design procedures and plant operating conditions have been implemented in order to reduce costs and meet constraints. Most industry observers believe that the emphasis in the near future will be on improving efficiency and increasing profitability of existing plants rather than on plant expansion. One of the most important engineering tools that can be employed in such activities is optimization. As computers have become more powerful, the size and complexity of problems which can be solved by optimization techniques have correspondingly expanded.

To be able to make the most effective applications of optimization in the chemical process industries, you must understand both the theory *and* the practice of optimization, both of which are explained in this book. Optimization algorithms have reached a degree of maturity in recent years due to the extensive evaluation of proposed optimization techniques on a wide range of problems. We have chosen to focus on only the few superior techniques that offer the most potential for success and give reliable results, rather than on an extensive coverage of algorithms proposed in the optimization literature.

The book is intended to be introductory in nature in presenting the necessary tools for problem solving. We have placed more emphasis on how to formulate optimization problems appropriately than on the theory underlying optimization algorithms because many engineers and scientists find this phase of their decision-making process the most exasperating and difficult. In the book rigorous proofs are omitted, being replaced by plausibility arguments without sacrificing correctness. Ample references are cited for those who wish to explore the theoretical concepts in more detail.

The book contains three main sections. Because of the importance of problem formulation, Part I of the book presents the methodology of how to specify the three key components of an optimization problem, namely,

1. The objective function
2. The process model
3. The constraints

Part I, comprised of three chapters, motivates the study of optimization by giving examples of different types of problems that may be encountered in chemical engineering. After discussing the three essential features of every optimization problem, we provide a list of six steps that must be used to varying degrees in solving an optimization problem. It is important that a potential user of optimization techniques be able to translate a verbal description of the problem into the appropriate mathematical description. It is also important for a user to understand how the problem formulation has a major impact on the ease of problem solution. We show how problem simplification and sensitivity analysis are important steps in analysis of model-building and estimating the unknown parameters in models. Chapter 3 discusses how the objective function should be developed. We focus on economic factors in this chapter, and present several alternative methods of evaluating profitability.

Part II covers the theoretical and computational basis for proven techniques in optimization. The choice of a specific technique must mesh with the three components listed above. Part II begins with a chapter (Chap. 4) that provides the essential conceptual background for optimization, namely the concepts of convexity, concavity, necessary and sufficient conditions for an extremum, and the Hessian matrix. Chapter 5 follows with a brief explanation of one-dimensional search methods. We choose to describe only a few effective ones. Chapter 6 presents reliable unconstrained optimization and methods. Chapter 7 treats linear programming theory and applications using both graphical and matrix methods to illustrate and emphasize the concepts involved. Chapter 8 covers modern nonlinear programming methods, mainly the generalized reduced gradient method, augmented Lagrange method, and successive quadratic programming. We conclude Part II with a chapter on optimization of staged and discrete processes, highlighting mixed-integer programming problems and methods. Only deterministic optimization problems are treated throughout the book as lack of space precludes stochastic variables, constraints, and coefficients.

Although we include many simple applications in Parts I and II to illustrate the optimization techniques and algorithms, Part III of the book is exclusively devoted to illustrations and examples of optimization procedures and examples classified according to their applications: heat transfer and energy conservation (Chap. 10), separations (Chap. 11), fluid flow (Chap. 12), reactor design (Chap. 13), and plant design (Chap. 14). Many students and professionals learn by example or analogy and often find out how to solve a problem by examining the solution to similar problems. By organizing applications of optimization in this manner, you can focus on a single class of applications of particular interest without requiring extensive review of the book. We present a spectrum of different methods in each of these chapters rather than overemphasize a few selected techniques. The Introduction to Part III lists each application classified by the technique employed. In some cases the optimization method may be analytical, leading to simple design rules; other examples illustrate numerical methods. In some applications the problem statement may be so complex that it cannot be explicitly written out, as in plant design, and thus requires the use of a process simulator. No exercises are included in Part III, but an instructor can (*a*) modify the variables, parameters, conditions, or constraints in an example, and (*b*) suggest a different technique of solution, to obtain exercises for solution by students.

An understanding of optimization techniques does not require complex mathematics. We require as background only a few tools from calculus and linear algebra to explain the theory and computational techniques and provide you with an understanding of how optimization techniques work (or, in some cases, fail to work). We have included two appendices in the book as supplementary reading for those without the necessary mathematical background.

Presentation of each optimization technique is followed by examples to illustrate an application. We also have included many practically-oriented homework problems. In university courses, this book could be used at the upper-division or the first-year graduate levels, and the material has been used for such courses at the University of Texas. The book contains more than enough material for a 15-week course on optimization, and, because of its emphasis on applications, it may also serve as one of the supplementary texts in a senior design course.

In addition to use as a textbook, we believe the book is also suitable for use in individual study, industrial practice, industrial short courses, and other continuing education programs.

We wish to acknowledge the helpful suggestions of several colleagues in developing this book, especially Yaman Arkun, Georgia Institute of Technology; Lorenz T. Biegler, Carnegie-Mellon University; James R. Couper, University of Arkansas; James Fair, University of Texas-Austin; Ignacio Grossman, Carnegie-Mellon University; K. Jayaraman, Michigan State University; I. Lefkowitz, Case Western Reserve University; Leon Lasdon, University of Texas; and Mark Stadtherr, University of Illinois-Urbana Champaign. Several of the examples in Chap. 10 through 14 were provided by friends in industry and in universities and are acknowledged there. We also recognize the help of many graduate students in developing solutions to the examples.

T. F. Edgar
D. M. Himmelblau

OPTIMIZATION OF CHEMICAL PROCESSES

PART I

PROBLEM FORMULATION

Problem formulation is perhaps the most crucial step in resolving a problem that involves optimization. Problem formulation requires identifying the essential elements of a conceptual or verbal statement of a given application, and organizing them into a prescribed mathematical form, namely

1. The objective function (economic criterion)
2. The process model (constraints)

The objective function represents profit, cost, energy, yield, etc., in terms of the key variables of the process being analyzed. The process model and constraints describe the interrelationships of the key variables. It is important to learn a systematic approach for assembling the physical and empirical relations and data involved in an optimization problem, and Chaps. 1, 2, and 3 cover the recommended procedures. Chapter 1 presents six steps for optimization that can serve as a general guide for problem solving in design and operations analysis. Numerous examples of problem formulation in chemical engineering are presented to illustrate the steps.

Chapter 2 summarizes the characteristics of process models and explains how to build a process model. Special attention is focused on developing mathematical models, particularly empirical models, by fitting empirical data by least squares, which itself is an optimization procedure.

Chapter 3 treats the most common type of objective function, the cost or revenue function. Historically, the majority of optimization applications have involved trade-offs between capital costs and operating costs. The nature of the trade-off depends on a number of assumptions such as the desired rate of return on investment, service life, depreciation method, and so on. While an objective function based on net present value is preferred for the purposes of optimization, discounted cash flow based on spreadsheet analysis can be employed as well.

It is important to recognize that many possible mathematical problem formulations can result from an engineering analysis, depending on the assumptions made and the desired accuracy of the model. To solve an optimization problem, the mathematical formulation of the model must mesh satisfactorily with the computational algorithm to be used. Therefore, you should recognize that a certain amount of artistry, judgment, and experience is required during the problem formulation phase of optimization.

CHAPTER 1

THE NATURE AND ORGANIZATION OF OPTIMIZATION PROBLEMS

Optimization is one of the major quantitative tools in the machinery of decision-making. A wide variety of problems in the design, construction, operation, and analysis of chemical plants (as well as many other industrial processes) can be resolved by optimization. In this chapter we examine the basic characteristics of optimization problems and their solution techniques, and describe some typical benefits and applications in the chemical and petroleum industries.

1.1 WHAT OPTIMIZATION IS ALL ABOUT

A well-known approach to the principle of optimization was first scribbled centuries ago on the walls of an ancient Roman bathhouse in connection with a choice between two aspirants for emperor of Rome. It read—"De doubus malis, minus est semper aligendum"—of two evils, always choose the lesser.

Optimization pervades the fields of science, engineering, and business. In physics, many different optimal principles have been enunciated, describing natural phenomena in the fields of optics and classical mechanics. The field of statistics treats various principles termed "maximum likelihood," "minimum loss," and "least squares," while business makes use of "maximum profit," "minimum cost," "maximum use of resources," "minimum effort," in the endeavor to increase profits. A typical engineering problem can be posed as follows: You have a process that can be represented by some equations or perhaps solely by experimental data. You also have a single performance criterion in mind such as minimum cost. The goal of optimization is to find the values of the variables in the process that yield the best value of the performance criterion. What is usually involved is a trade-off between capital and operating costs. The ingredients described above—process or model and the performance criterion—comprise the optimization "problem."

Typical problems in chemical engineering process design or plant operation have many, and possibly an infinite number of, solutions. Optimization is concerned with selecting the best among the entire set by efficient quantitative methods. Computers and associated software make the computations involved in the selection feasible and cost-effective. But to employ them requires (1) critical analysis of the process or design, (2) insight as to what the appropriate performance objectives are, i.e., what is to be accomplished, and (3) use of past experience, sometimes called "engineering judgment," before useful information results.

1.2 WHY OPTIMIZE?

Why are engineers interested in optimization? What benefits result from optimization versus intuitive decision-making? Engineers work to improve the initial design of equipment, and strive for enhancements in the operation of the equipment once it is installed in order to realize the largest production, the greatest profit, the maximum cost, the least energy usage, and so on. The use of monetary value provides a convenient measure of different but otherwise incompatible objectives, but not all problems have to be considered in a monetary (cost versus revenue) framework.

In plant operations, benefits arise from improved plant performance, such as improved yields of valuable products (or reduced yields of contaminants), reduced energy consumption, higher processing rates, and longer times between shutdowns. Optimization can also lead to reduced maintenance costs, less equipment wear, and better staff utilization. In addition, intangible benefits arise from the interactions among plant operators, engineers, and management. It is extremely helpful to systematically identify the objective, constraints, and degrees of freedom in a process or a plant, leading to such benefits as improved quality of design, faster and more reliable trouble-shooting, and faster decision-making.

Predicting benefits must be done with care. Design and operating variables in most plants are always coupled in some way. If the fuel bill for a distillation column is $3000 per day, a 5 percent savings may justify an energy conservation project. However, in a unit operation such as distillation, it is incorrect to simply sum the heat exchanger duties and claim a percentage reduction in total heat required. A reduction in the reboiler heat duty may influence both the product purity, which can translate to a change in profits, and the condenser cooling requirements. Hence, it may be misleading to ignore the indirect and coupled effects that process variables have on costs.

What about the argument that the formal application of optimization is really not warranted because of the uncertainty that exists in the mathematical representation of the process and/or the data used in the model of the process? Certainly such an argument has some merit. Engineers have to use judgment in applying optimization techniques to problems that have considerable uncertainty associated with them, both from the standpoint of accuracy and the fact that the plant operating parameters and environs are not always static. In some cases it may be possible to carry out an analysis via deterministic optimization and then add on stochastic features to the analysis to yield quantitative predictions of the degree of uncertainty. Whenever the model of a process is idealized and the input and parameter data only known approximately, the optimization results must be treated judiciously. They can provide upper limits on expectations. Another way to evaluate the influence of uncertain parameters in optimal design is to perform a sensitivity analysis. It is possible that the optimum value of a process variable is unaffected by certain parameters (low sensitivity), therefore having precise values for these parameters will not be crucial to finding the true optimum. We discuss how a sensitivity analysis is performed later on in this chapter.

1.3 SCOPE AND HIERARCHY OF OPTIMIZATION

Optimization can take place at many levels in a company ranging from a complex combination of plants and distribution facilities down through individual plants, combinations of units, individual pieces of equipment, subsystems in a piece of equipment, or even smaller entities (Beveridge and Schechter, 1970). Optimization problems can be found at all these levels. Thus the scope of an optimization problem can be the entire company, a plant, a process, a single unit operation, a single piece of equipment in that operation, or any intermediate system between these. The complexity of analysis may involve only gross features, or may examine minute detail, depending upon the use to which the results will be put, the availability of accurate data, and the time available in which to carry out the optimization. In a typical industrial company there are three areas (levels) in which optimization is used: (1) management, (2) process design and equipment specification, and (3) plant operations (see Fig. 1.1).

Management makes decisions concerning project evaluation, product selection, corporate budget, investment in sales versus research and development, new plant construction, i.e., when and where should new plants be constructed, and so forth. At this level much of the information that is available is at best qualitative or has a high degree of uncertainty. Therefore many management decisions for optimizing some feature(s) of a large company have the potential to be significantly in error when put into practice, especially if the timing is wrong. In general, the magnitude of the objective function, as measured in dollars, is much larger at the management level than at the other two levels.

Individuals engaged in process design and equipment specification are concerned with the choice of a process and nominal operating conditions. They answer questions such as: Do we design a batch process or a continuous process? How many reactors do we use in producing a petrochemical? What should the configurations of the plant be and how do we arrange the processes so that the operating efficiency of the plant is at a maximum? What is the optimum size

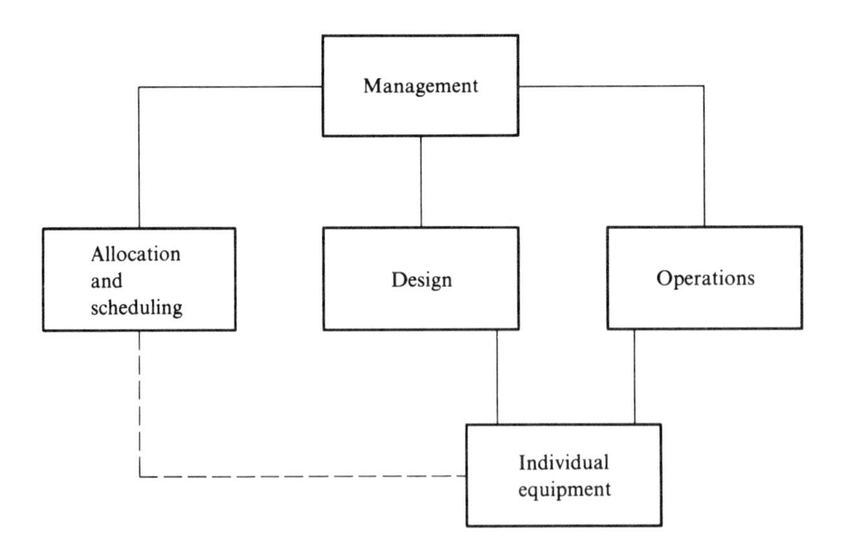

Figure 1.1 Hierarchy of levels of optimization.

of a unit or combination of units? Such questions can be resolved with the aid of so-called process design simulators or flowsheeting programs. These are large computer programs that carry out the material and energy balances for individual pieces of equipment and combine them into an overall production unit. Iterative use of such a simulator is often required in order to arrive at a desirable process flowsheet.

Other decisions which are more specific in nature are made in process design including the actual choice of equipment (for example, in the selection of heat exchangers one has over ten different types of exchangers from which to choose) and the materials of construction of various process units.

The third constituency employing optimization operates on a totally different time scale than the other two. Process design and equipment specification is usually performed prior to the implementation of the process, and management decisions to implement designs are usually made far in advance of the process design step. On the other hand, optimization of operating conditions is carried out monthly, weekly, daily, hourly, or even at the extreme every minute. Plant operations are concerned with operating controls for a given unit at certain temperatures, pressures, flowrates, etc., that are the best in some sense. For example, the selection of the percentage of excess air in a process heater is quite critical and involves a balance on the fuel-air ratio to assure complete combustion and at the same time make the maximum use of the heating potential of the fuel.

Plant operations deal with the allocation of raw materials on a daily or weekly basis. One classical optimization problem, which is discussed later in this text, is the allocation of raw materials in a refinery. Typical day-to-day optimization in a plant minimizes steam consumption or cooling water consumption.

Plant operations are also concerned with the overall picture of shipping, transportation, and distribution of products to engender minimal costs. For example, the frequency of ordering, the method of scheduling production, and scheduling delivery are very critical to maintaining a low-cost operation.

There are a number of attributes of processes affecting costs or profits that make them attractive for the application of optimization:

1. Sales limited by production: If additional products can be sold beyond current capacity, then economic justification of design modifications is relatively easy. Often, increased production can be attained with only slight changes in operating costs (raw materials, utilities, etc.) and with no change in investment costs. This situation implies a higher profit margin on the incremental sales.
2. Sales limited by market: This situation is susceptible to optimization only if improvements in efficiency or productivity can be obtained; hence, in this case there may be less economic incentive for implementation than in case 1 above, because no additional products are made. Reductions in per-unit manufacturing costs (via optimizing usage of utilities and feedstocks) are generally the main targets.

3. Large unit throughputs: High volume of production offers great potential for increased profits, because small savings in production costs per unit are greatly magnified. Most large chemical and petroleum processes fall into this classification.
4. High raw material or energy consumption: Significant savings can be made by reducing consumption of those items if they have high unit costs.
5. Product quality exceeds product specifications: If the product quality is significantly better than that required by the customer, higher than necessary production costs and wasted capacity may occur. By operating close to customer specification (constraints), cost savings can be obtained.
6. Losses of valuable components through waste streams: The chemical analysis of various plant exit streams, both to the air and water, should indicate if valuable materials are being lost. Adjustment of air/fuel ratios in furnaces to minimize hydrocarbon emissions and hence fuel consumption is one such example. Pollution regulations also influence permissible air and water emissions.
7. High labor costs: In processes where excessive handling is required, such as in batch operation, often bulk quantities can be handled at lower cost and with less manpower. Revised layouts of facilities can reduce costs. Sometimes no direct reduction in the labor force results, but the intangible benefits of a lessened work load can allow the operator to assume greater responsibility.

Two valuable sources of data for identifying opportunities for optimization include (1) profit and loss statements for the plant or the unit and (2) the periodic operating records for the plant. The profit and loss statement contains much valuable information on sales, prices, manufacturing costs, and profits, while the operating records present information on material and energy balances, unit efficiencies, production levels, and feedstock usage.

Because of the complexity of chemical plants, complete optimization of a given plant can be quite an extensive undertaking. In the absence of complete optimization we often rely on "incomplete optimization" of which a special variety is termed *suboptimization*. Suboptimization involves optimization for one phase of an operation or a problem ignoring some factors which have an effect, either obvious or indirect, on other systems or processes in the plant. Suboptimization is often necessary because of economic and practical considerations, limitations on time or manpower, and the difficulty of obtaining answers in a hurry. Suboptimization is useful when neither the problem formulation nor the available techniques permit obtaining a reasonable solution to the full problem. In most practical cases, suboptimization at least provides a rational technique for approaching an optimum.

However, you must recognize a major principle: suboptimization of all elements does *not* necessarily ensure attainment of an overall optimum for the entire system. Subsystem objectives may not be compatible nor mesh with overall objectives.

1.4 EXAMPLES OF APPLICATIONS OF OPTIMIZATION

Optimization can be applied in numerous ways to chemical processes and plants. Typical projects where optimization has been used include:

1. Determination of best sites for plant location
2. Routing of tankers for the distribution of crude and refined products
3. Pipeline sizing and layout
4. Equipment and entire plant design
5. Maintenance and equipment replacement scheduling
6. Operation of equipment, such as tubular reactors, columns, absorbers, etc.
7. Evaluation of plant data to construct a model of a process
8. Minimization of inventory charges
9. Allocation of resources or services among several processes
10. Planning and scheduling of construction

These examples provide an introduction to the types of variables, objective functions, and constraints that will be encountered in subsequent chapters.

In this section we provide four illustrations of "optimization in practice," i.e., optimization of process operations and design. These examples will help illustrate the general features of optimization problems, a topic treated in more detail in Sec. 1.5.

EXAMPLE 1.1 ECONOMIC INSULATION THICKNESS

Insulation design is a classic example of overall cost-saving that is especially pertinent when fuel costs are high. The addition of insulation should save money through reduced heat losses; on the other hand, the insulation material can be expensive. How much insulation should be added is a question that can be answered by optimization.

Assume that the bare surface of a vessel is at 700°F with an ambient temperature of 70°F. The surface heat loss is 4000 Btu/(h)(ft^2). Add one inch of calcium silicate insulation and the loss will drop to 250 Btu/(h)(ft^2). At an installed cost of \$4.00/ft^2 and a cost of energy at \$5.00/10^6 Btu, a savings of \$164 per year (8760 hours of operation) per square foot would be realized. A simplified payback calculation shows a payback period of

$$\frac{\$4.00/(\text{ft}^2)}{\$164/(\text{ft}^2)(\text{year})} = 0.0244 \text{ year or 9 days!}$$

As additional inches of insulation are added, the increments must be justified by the savings obtained. Figure E1.1 shows the way to picture the outcome of adding additional layers of insulation. Since insulation can only be added in 0.5-in increments, the possible capital costs are shown as a series of dots; these costs are

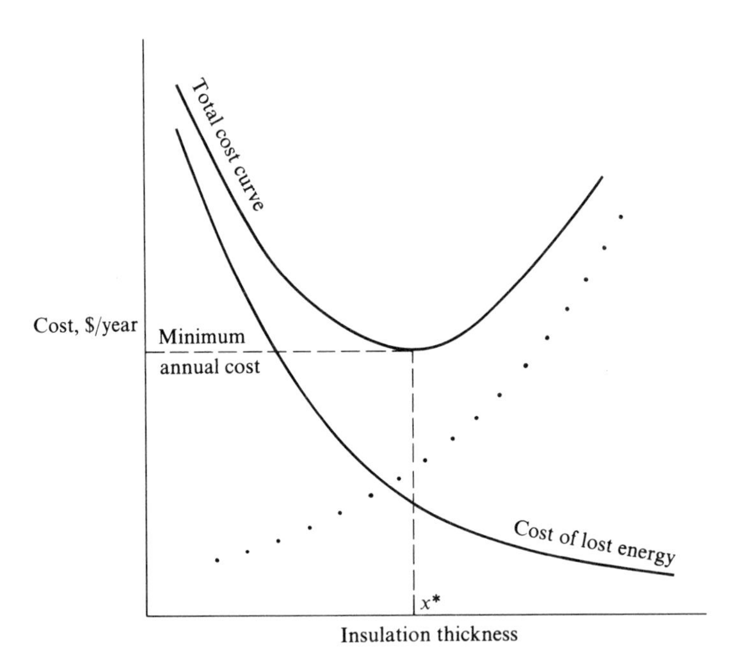

Figure E1.1 The effect of insulation thickness on total cost (x^* = optimum thickness). Insulation can be purchased in 0.5-in increments. (The total cost function is shown as a smooth curve for convenience, although the sum of the two costs would not actually be smooth.)

prorated because the insulation lasts for several years before having to be replaced. In Fig. E1.1 the energy loss cost is a continuous curve since it can be calculated directly from heat transfer principles. The total cost is also shown as a continuous function. Note that at some point total costs begin increasing as the insulation thickness increases because little or no benefit in heat conservation results. The trade-off between energy cost and capital cost, and the optimum insulation thickness, can be determined by optimization. Further discussion of capital versus operating costs appears in Chap. 3; in particular, see Example 3.3.

EXAMPLE 1.2 OPTIMAL BOILER OPERATING CONDITIONS

Another example of optimization can be encountered in the operation of a boiler. Engineers focus attention on utilities and powerhouse operations within refineries and process plants because of the large amounts of energy consumed by these plants, and the potential for significant reduction in the energy required for utilities generation and distribution. Control of environmental emissions adds complexity and constraints in optimizing boiler operations. In a boiler it is desirable to optimize the air/fuel ratio so that the thermal efficiency is maximized; however, environmental regulations encourage operation under "fuel-rich" conditions and lower combustion temperatures in order to reduce the emissions of nitrogen oxides (NO_x). Unfortunately such operating conditions also decrease efficiency due to some unburned fuel escaping through the stacks, resulting in an increase in undesirable hydrocarbon (HC) emissions. Thus, a conflict in operating criteria arises.

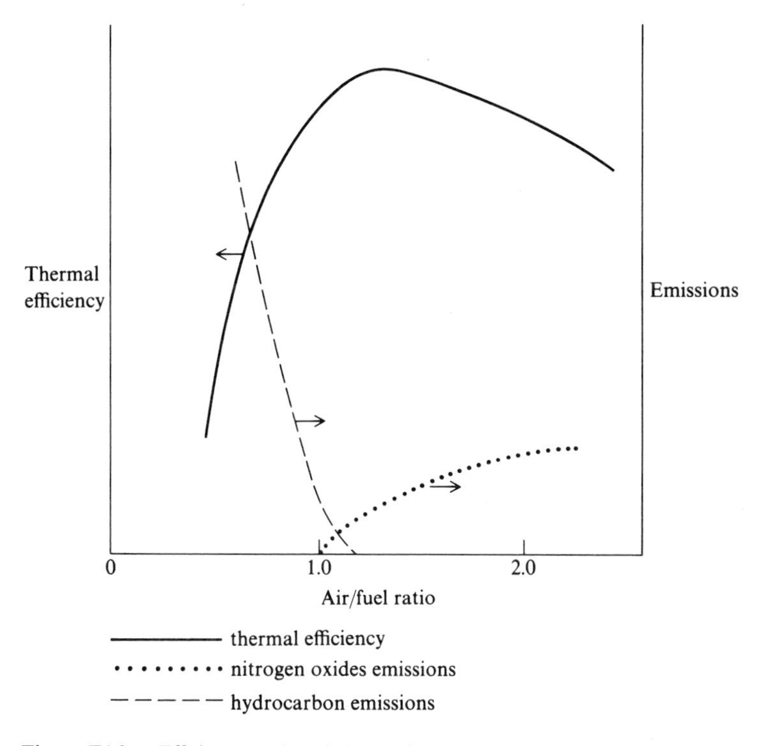

Figure E1.2*a* Efficiency and emissions of a boiler as a function of air/fuel ratio. (1.0 = stoichiometric air/fuel ratio.)

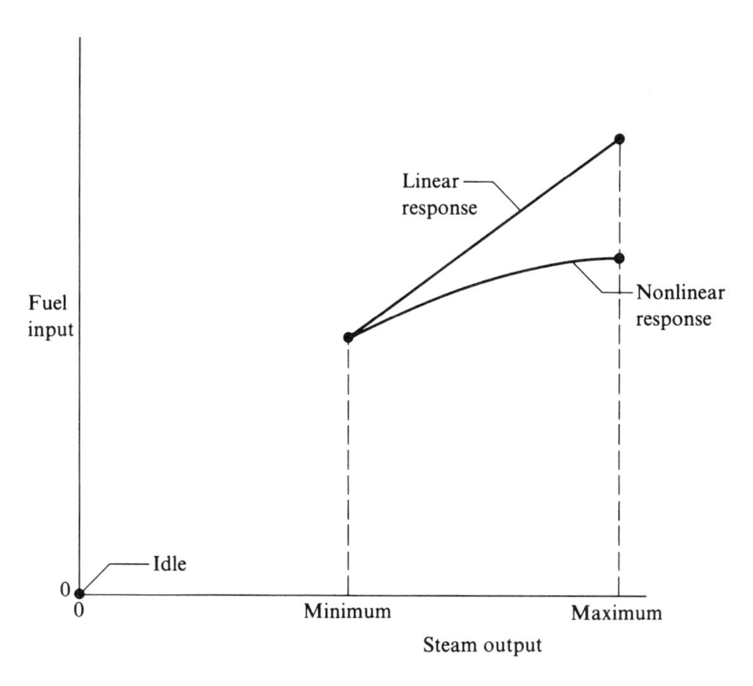

Figure E1.2*b* Discontinuity in operating regime.

Figure E1.2*a* illustrates the trade-offs between efficiency and emissions, suggesting that in this case more than one performance criterion may exist; we are forced to consider maximizing efficiency versus minimizing emissions, resulting in some compromise of the two objectives.

Another feature of boiler operations is the widely varying demands caused by changes in process operations, plant unit start-ups and shutdowns, and daily and seasonal cycles. Because utility equipment is often operated in parallel, a common operating question raised by demand swings is when to idle or to bring on-line another boiler, turbine, or other piece of equipment, and which one it should be.

Answering this question is complicated by the feature that most powerhouse equipment cannot be operated continuously all the way down to the idle state as illustrated by Fig. E1.2*b* for boilers and turbines. Instead, a range of continuous operation may exist for certain conditions, but a discrete jump to a different set of conditions (here idling conditions) may be required if demand changes. In formulating many optimization problems, discrete variables (on-off, high-low, integer 1, 2, 3, 4, etc.) must be accommodated (Robnett, 1979).

EXAMPLE 1.3 OPTIMUM DISTILLATION REFLUX

Prior to 1974 when fuel costs were low, operation of distillation column trains used a strategy involving the substantial consumption of utilities such as steam and cooling water in order to maximize separation (i.e., product purity) for a given tower. However, the operation of any one tower involves certain limitations or

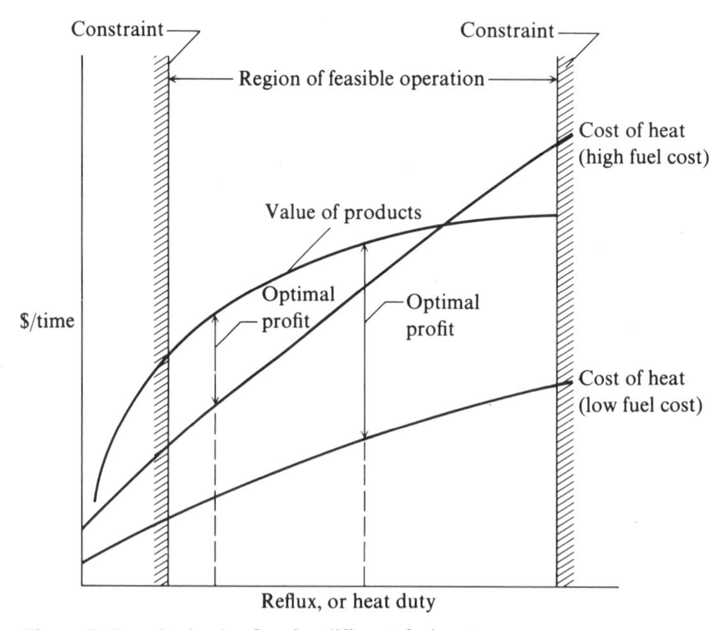

Figure E1.3*a* Optimal reflux for different fuel costs.

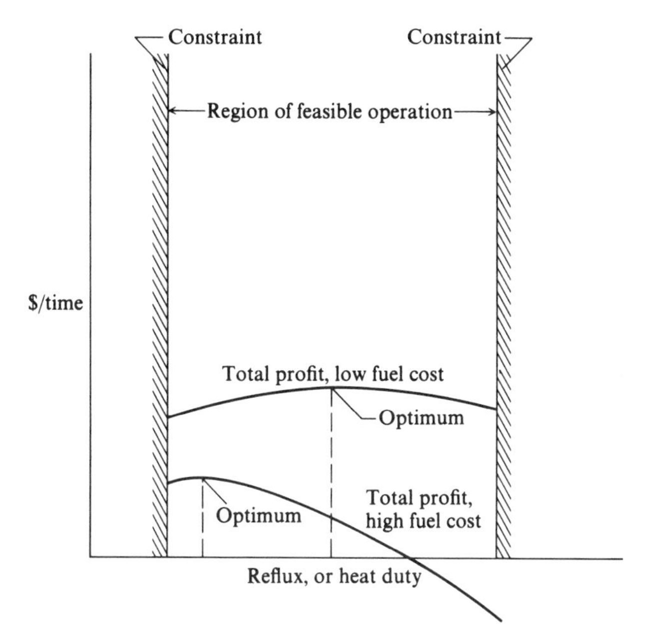

Figure E1.3*b* Total profit for different fuel costs.

constraints on the process such as the condenser duty, tower tray flooding, or reboiler duty.

The need for energy conservation suggests a different objective, namely minimizing the reflux ratio. Then one can ask: How low can the reflux ratio be set? From the viewpoint of optimization, there is an economic minimum value below which the energy savings are less than the cost of product quality degradation (Latour, 1981). Figures E1.3*a* and E1.3*b* show such a point. Operators tend to overreflux a column because this strategy makes it easier to stay well within the product specifications. Often columns are operated with a fixed flow control for reflux so that the reflux ratio is higher than needed when feed rates drop off. This application is discussed in more detail in Chap. 11.

EXAMPLE 1.4 MULTIPLANT PRODUCT DISTRIBUTION

A common problem encountered in large chemical companies involves the distribution of a single product (Y) manufactured at several plant locations. Generally, the product needs to be delivered to several customers located at various distances from each plant. Therefore, it is desirable to determine how much Y must be produced at each of m plants ($Y_1, Y_2, \ldots, Y_m$) and how, for example, Y_m should be allocated to each of n demand points ($Y_{m1}, Y_{m2}, \ldots, Y_{mn}$). The cost-minimizing solution to this problem not only involves the transportation costs between each supply and demand point but also the production cost versus capacity curves for each plant. The individual plants probably vary with respect to design production rate and some plants may be more efficient than others, having been constructed at a later date. Both of these factors contribute to a unique functionality between production cost and production rate. Because of the particular distribution of transportation costs, it may be desirable to manufacture more product from an old, inefficient plant (at higher cost) than from a new, efficient one because new customers may be located very close to the old plant. On the other hand, if the old plant is operated far above its design rate, costs could become exorbitant, forcing a reallocation by other plants in spite of high transportation costs. In addition, no doubt constraints exist on production levels from each plant that also affect the product distribution plan.

1.5 THE ESSENTIAL FEATURES OF OPTIMIZATION PROBLEMS

The solution of optimization problems as we shall see involves the use of various features of mathematics. Consequently, the formulation of an optimization problem must be via mathematical expressions, although this does not necessarily imply great complexity. Not all problems are susceptible to quantitative statement and analysis, but we will restrict our coverage to quantitative methods. From a practical viewpoint, it is important to mesh properly the problem statement with the anticipated solution technique.

A wide variety of optimization problems have amazingly similar structure. Indeed, it is this similarity that has enabled the recent progress in optimization techniques. We note a common interest on the part of chemical engineers, petroleum engineers, physicists, chemists, traffic engineers, etc., in precisely the same mathematical problem structures, each with a different application in the real world. We can make use of this structural similarity to develop a framework or methodology within which any problem can be studied. This section describes how any process problem, complex or simple, for which one desires the optimal solution should be organized. To do so, as mentioned previously, you must (*a*) consider the model representing the process and (*b*) choose a suitable objective criterion to guide the decision-making.

You will find every optimization problem contains three essential categories:

1. At least one objective function to be optimized (profit function, cost function, etc.)
2. Equality constraints (equations)
3. Inequality constraints (inequalities)

Categories 2 and 3 comprise the model of the process or equipment; category 1 is sometimes called the *economic model*.

By a *feasible solution* of the optimization problem we mean a set of variables that satisfy categories 2 and 3 to the desired degree of precision. Figure 1.2 illustrates the *feasible region*, that is the region of feasible solutions, defined by categories 2 and 3. In this case the feasible region consists of a line bounded by

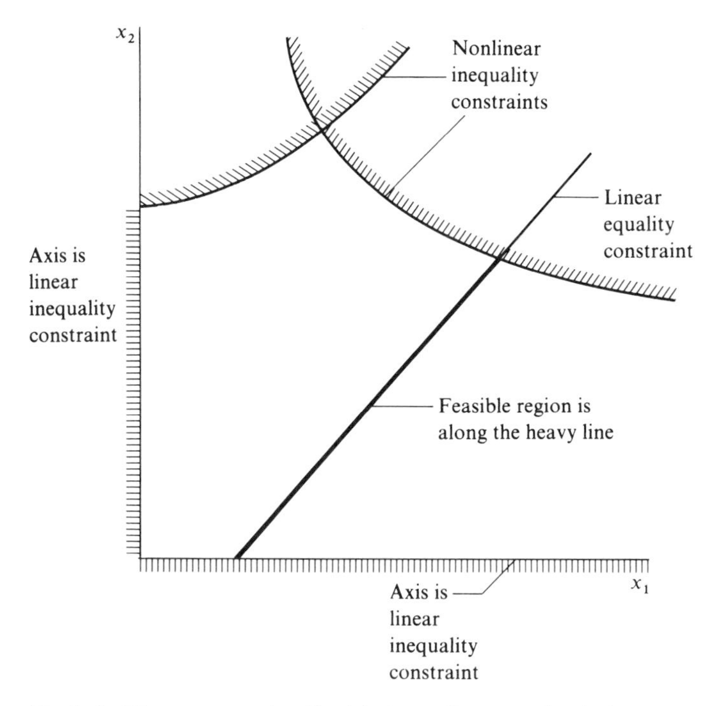

The dashed lines represent the side of the inequality constraints in the plane that form part of the infeasible region. The heavy line shows the feasible region.

Figure 1.2 Feasible region for an optimization problem involving two independent variables.

two inequality constraints. An *optimal solution* is a set of values of the variables that satisfy the components of categories 2 and 3 and also provides an optimal value for the function in category 1. In most cases the optimal solution is a unique one; in some it is not. If you formulate the optimization problem so that there are no residual degrees of freedom among the variables in categories 2 and 3, optimization is not needed to obtain a solution for a problem. More specifically, if m_e equals the number of independent consistent equality constraints and m_i equals the number of independent inequality constraints that are satisfied as equalities (equal to zero), and if the number of variables whose values are unknown is equal to $m_e + m_i$, then at least one solution exists of the relations in components 2 and 3 regardless of the optimization criterion. (Multiple solutions may exist when models in categories 2 and 3 are comprised of nonlinear relations.) If a unique solution exists, no optimization is needed to obtain a solution—one just solves a set of equations—and need not worry about optimization methods because the unique feasible solution is by definition the optimal one.

On the other hand, if more process variables whose values are unknown exist in category 2 than there are independent equations, the process model is called *underdetermined*, that is the model has an infinite number of feasible solutions so that the objective function in category 1 is the additional criterion used to reduce the number of solutions to just one (or a few) by specifying what is the "best" solution. Finally, if the equations in category 2 contain more independent equations than variables whose values are unknown, then the process model is *overdetermined* and there is no solution which satisfies all the constraints exactly. To reach a resolution of the difficulty, sometimes we may choose to relax some or all of the constraints. A typical example of an overdetermined model might be the reconciliation of process measurements for a material balance. One approach to yield the desired material balance would be to resolve the set of inconsistent equations by minimizing the sum of the errors of the set of equations (usually by a procedure termed *least squares*).

In this text the following notation will be used for each category of the optimization problem:

$$\text{Minimize:} \quad f(\mathbf{x}) \qquad \text{objective function} \qquad (a)$$

$$\text{Subject to:} \quad \mathbf{h}(\mathbf{x}) = \mathbf{0} \qquad \text{equality constraints} \qquad (b)$$

$$\mathbf{g}(\mathbf{x}) \geq \mathbf{0} \qquad \text{inequality constants} \qquad (c)$$

where $\mathbf{x}$ is a vector of n variables $(x_1, x_2, \ldots, x_n)$, $\mathbf{h}(\mathbf{x})$ is a vector of equations of dimension m_1, and $\mathbf{g}(\mathbf{x})$ is a vector of inequalities of dimension m_2. The total number of constraints is $m = (m_1 + m_2)$.

EXAMPLE 1.5 OPTIMAL SCHEDULING: FORMULATION OF THE OPTIMATION PROBLEM

In this example we illustrate the formulation of the components of an optimization problem.

We want to schedule the production in two plants, A and B, each of which can manufacture two products, no. 1 and no. 2. How should the scheduling take place to maximize profits while meeting the market requirements based on the following data:

Plant	Material processed, lb/day 1	2	Profit, $/lb 1	2
A	M_{A1}	M_{A2}	S_{A1}	S_{A2}
B	M_{B1}	M_{B2}	S_{B1}	S_{B2}

How many days per year (365 days) should each plant operate processing each kind of material? *Hints*: Does the table contain the variables to be optimized? How do you use the above information to mathematically formulate the opitimization problem? What other factors must you consider?

Solution. How should we start to convert the words of the problem into mathematical statements? First, let us define the variables. There will be four of them (t_{A1}, t_{A2}, t_{B1}, and t_{B2}, designated as a set by the vector $\mathbf{t}$) representing, respectively, the number of days per year each plant operates on each material as indicated by the subscripts.

What will the objective function be? We will select the annual profit so that

$$f(\mathbf{t}) = t_{A1}M_{A1}S_{A1} + t_{A2}M_{A2}S_{A2} + t_{B1}M_{B1}S_{B1} + t_{B2}M_{B2}S_{B2} \qquad (a)$$

Next, do any equality constraints evolve from the problem statement or from implicit assumptions? If each plant runs 365 days per year there are two equality constraints

$$t_{A1} + t_{A2} = 365 \qquad (b)$$

$$t_{B1} + t_{B2} = 365 \qquad (c)$$

Finally, are there any inequality constraints that evolve from the problem statement or implicit assumptions? On first glance it may appear there are none, but further thought indicates t must be nonnegative since negative values of t have no physical meaning:

$$t_{Ai} \geq 0 \qquad i = 1, 2 \qquad (d)$$

$$t_{Bi} \geq 0 \qquad i = 1, 2 \qquad (e)$$

Do negative values of the coefficients S have physical meaning?

Other inequality constraints might be added after further analysis, such as a limitation on the total amount of material 2 that can be sold (L_1):

$$t_{A2}M_{A2} + t_{B2}M_{B2} \leq L_1 \qquad (f)$$

or a limitation on production rate for each product at each plant, namely

$$\left.\begin{aligned} M_{A1} &\leq L_2 \\ M_{A2} &\leq L_3 \\ M_{B1} &\leq L_4 \\ M_{B2} &\leq L_5 \end{aligned}\right\} \qquad (g)$$

To find the optimal $\mathbf{t}$, we need to optimize (a) subject to constraints (b) to (g).

EXAMPLE 1.6 MATERIAL BALANCE RECONCILIATION

Suppose the flow rates entering and leaving a process are measured periodically. Determine the best value for stream A in pounds for the process shown from the three hourly measurements indicated of B and C in Fig. E1.6, assuming steady state operation at a fixed operating point. The process model is:

$$M_A + M_C = M_B \qquad (a)$$

where M is the mass per unit time of throughput.

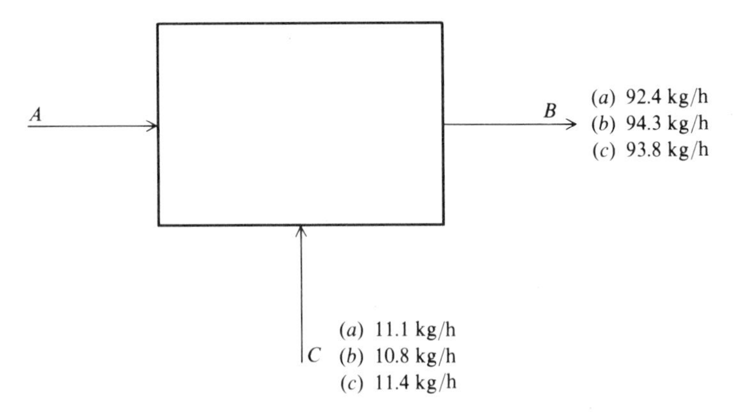

Figure E1.6

Solution. We need to set up the objective function first. Let us minimize the sum of the squares of the deviations between input and output as the criterion so that the objective function becomes

$$f(M_A) = (M_A + 11.1 - 92.4)^2 + (M_A + 10.8 - 94.3)^2 + (M_A + 11.4 - 93.8)^2 \quad (b)$$

A sum of squares is used since this guarantees that $f > 0$ for all values of M_A; a minimum at $f = 0$ implies no error.

No equality constraints remain in the problem. Are there any inequality constraints? (*Hint:* What about M_A?) The optimum value of M_A can be found by differentiating f with respect to M_A; this leads to an optimum value for M_A of 82.4, and is the same result as that obtained by computing from the averaged measured values, $M_A = \bar{M}_B - \bar{M}_C$. Other methods of reconciling material (and energy) balances are discussed by Tamhane and Mah (1985).

1.6 GENERAL PROCEDURE FOR SOLVING OPTIMIZATION PROBLEMS

No single method or algorithm of optimization exists that can be applied efficiently to all problems. The method chosen for any particular case will depend primarily on (1) the character of the objective function and whether it is known explicitly, (2) the nature of the constraints, (3) the number of independent and dependent variables.

Table 1.1 lists the six general steps for the analysis and solution of optimization problems. You do not have to follow the cited order exactly, but you should cover all of the steps eventually. Shortcuts in the procedure are allowable and the easy steps can be performed first. Each of the steps will be examined in more detail in subsequent chapters.

Remember, the general objective in optimization is to choose a set of values of the variables subject to the various constraints that will produce the desired optimum response for the chosen objective function.

Table 1.1 The six steps used to solve optimization problems

1. Analyze the process itself so that the process variables and specific characteristics of interest are defined, i.e., make a list of all of the variables.
2. Determine the criterion for optimization and specify the objective function in terms of the above variables together with coefficients. This step provides the performance model (sometimes called the economic model when appropriate).
3. Develop via mathematical expressions a valid process or equipment model that relates the input-output variables of the process and associated coefficients. Include both equality and inequality constraints. Use well-known physical principles (mass balances, energy balances), empirical relations, implicit concepts, and external restrictions. Identify the independent and dependent variables to get the number of degrees of freedom.
4. If the problem formulation is too large in scope:
   (*a*) break it up into manageable parts and/or
   (*b*) simplify the objective function and model
5. Apply a suitable optimization technique to the mathematical statement of the problem.
6. Check the answers, and examine the sensitivity of the result to changes in the coefficients in the problem and the assumptions.

Steps 1, 2, and 3 deal with the mathematical definition of the problem, i.e., identification of variables, specification of the objective function, and statement of the constraints. We devote considerable attention to problem-formulation in the remainder of this chapter, as well as in Chaps. 2 and 3. If the process to be optimized is very complex, it may be necessary to reformulate the problem so that it can be solved with reasonable effort.

Step 4 suggests that the mathematical statement of the problem be simplified as much as possible without losing the essence of the problem. First, you might decide to neglect those variables which have an insignificant effect on the objective function. This step can be done either in an ad hoc fashion, based on engineering judgment, or by performing a mathematical analysis and determining the weights which should be assigned to each variable via simulation. Second, a variable which appears in a simple form within an equation can be eliminated, i.e., can be solved for explicitly and then eliminated from other equations, the inequalities, and the objective function. Such variables are then deemed to be dependent variables.

As an example, in heat exchanger design, you might initially include in the problem the following variables: heat transfer surface, flow rates, number of shell passes, number of tube passes, number and spacing of the baffles, length of the exchanger, diameter of the tubes and shell, the approach temperature, and the pressure drop. Which of the variables are independent and which are not? This question can become quite complicated in a problem with many variables. You will find that each problem has to be analyzed and treated as an individual case;

generalizations are difficult. Often the decision is quite arbitrary although instinct indicates that the controllable variables be initially selected as the independent ones.

If an engineer were familiar with a particular heat exchanger system, he or she might decide that certain variables can be neglected based on the notion of the controlling or dominant heat transfer coefficient. In such a case only one of the flowing streams is important in terms of calculating the heat transfer in the system, and the engineer might decide at least initially to eliminate from consideration those variables related to the other stream.

A third strategy can be carried out when there are many constraints and many variables in the problem. We assume that some variables are fixed and let the remainder of the variables represent degrees of freedom (independent variables) in the optimization procedure. For example, the pressure in a distillation column optimization might be fixed at the minimum pressure (as limited by condenser cooling).

Finally, analysis of the objective function may permit some simplification of the problem. For example, if one product (A) from a plant is worth $30 per pound and all other products from the plant are worth $5 or less per pound, then we might initially decide to maximize the production of A only.

Step 5 in Table 1.1 involves the computation of the optimum point. Quite a few techniques exist to obtain the optimal solution for a problem. We will describe several methods in detail later on. In general, the solution of most optimization problems involves the use of a digital computer to obtain numerical answers. It is fair to state that over the past 15 years, substantial progress has been made in developing efficient and robust digital methods for optimization calculations. Much is known about which methods are most successful, although comparisons of candidate methods often are of an ad hoc nature based on test cases of simple problems. Virtually all numerical optimization methods involve iteration, and the effectiveness of a given technique often depends on a good first guess as to the values of the variables at the optimal solution.

The last entry in Table 1.1 involves checking the candidate solution to determine that it is indeed optimal. In some problems you can check that the sufficient conditions for an optimum are satisfied. More often an optimal solution may exist, yet you cannot demonstrate that the sufficient conditions are satisfied. All you can do is show by repetitive numerical calculations that the value of the objective function is superior to all known alternatives. A second consideration is how sensitive is the optimum to changes in parameters in the problem statement. A sensitivity analysis for the objective function value is important, and is illustrated as part of the next example.

EXAMPLE 1.7 THE SIX STEPS OF OPTIMIZATION FOR A MANUFACTURING PROBLEM

This example examines a simple problem in detail so that you can gain some understanding of how to execute the steps for optimization listed in Table 1.1. You also will see in this example that optimization can give insight into the nature of optimal operations and how optimal results might compare with simple or arbitrary rules of thumb so often used in practice.

Suppose you are a chemical distributor who wishes to optimize the inventory of a specialty chemical. You expect to sell Q barrels of this chemical over a given year at a fixed price with demand spread evenly over the year. If $Q = 100{,}000$ barrels (units) per year, you must decide on a production schedule. Unsold production is kept in inventory. In order to determine the optimal production schedule you must quantify those aspects of the problem which are important from a cost viewpoint [Baumol (1972)].

Step 1. One option is to produce 100,000 units in one run at the beginning of the year, and allow the inventory to be reduced to zero at the end of the year (at which time another 100,000 units are manufactured). Another option you have is to make 10 runs of 10,000 apiece. It is clear that there will be much more money tied up in inventory in the former case than in the latter case. Funds tied up in inventory are funds that could be invested in other areas or placed in a savings account. Therefore you might conclude that it would be cheaper to make the product ten times a year.

However, if you extend this notion to an extreme and make 100,000 production runs of one unit each (actually one unit every 315 seconds), the decision obviously is impractical, since the cost of producing 100,000 units, one unit at a time, will be exorbitant. Therefore, it appears that the desired operating procedure lies somewhere in between the two extremes. In order to arrive at some quantitative answer to this problem, first define the three operating variables that appear to be important: number of units of each run (D), the number of runs per year (n), and the total number of units produced per year (Q). Then you must obtain details about the costs of operations. In so doing, a cost (objective) function and a mathematical model will be developed, as discussed below. After obtaining a cost model, any constraints on the variables will be indentified, which will allow selection of independent and dependent variables.

Step 2. Let the business costs be split up into two factors: (1) the carrying cost or the cost of inventory, and (2) the cost of production. Let D be the number of units produced in one run. Q (annual production level) is assigned to be a known value. If the problem were posed so that a minimum level of inventory is specified, it would not change the structure of the problem.

The cost of the inventory not only includes the cost of the money which is tied up in the inventory, but also a storage cost which is a function of the inventory size. Warehouse space must exist to store all the units produced in one run. In the objective function let the cost of carrying the inventory be $K_1 D$, where the parameter K_1 essentially lumps together cost of working capital for the inventory itself and the storage costs.

Assume that the annual production cost in the objective function is proportional to the number of production runs required. The cost per run is assumed to be a linear function of D, given by the following equation:

$$\text{Cost per run} = K_2 + K_3 D \qquad (a)$$

The cost parameter K_2 is a setup cost and denotes a fixed cost of production—equipment must be made ready, cleaned, etc. The parameter K_3 is an operating cost parameter. The operating cost is assumed to be proportional to the number of

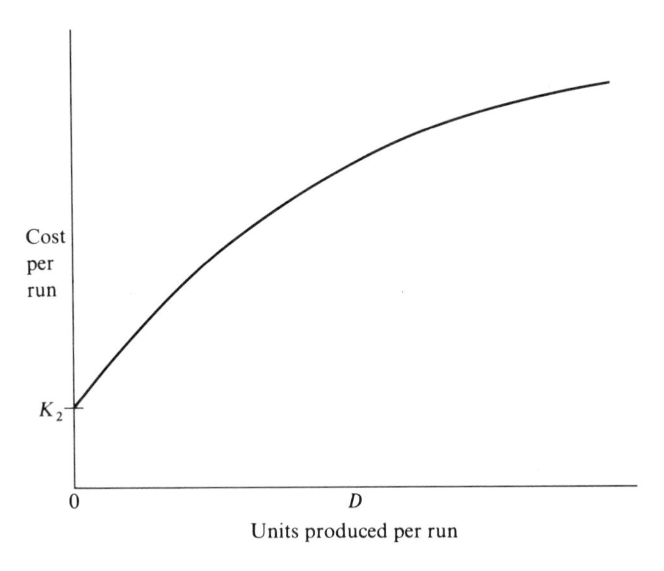

Figure E1.7 Nonlinear cost function for manufacturing.

units manufactured. Equation (*a*) may be an unrealistic assumption because the incremental cost of manufacturing could decrease somewhat for large runs; consequently, instead of a linear function, you might choose a nonlinear cost function of the form:

$$\text{Cost per run} = K_2 + K_4 D^{1/2} \tag{b}$$

as is shown in Fig. E1.7. The effect of this alternate assumption will be discussed later. The annual production cost can be found by multiplying either Eq. (*a*) or (*b*) by the number (n) of production runs per year.

The total annual manufacturing cost (C) for the product is the sum of the carrying costs and the production costs, namely

$$C = K_1 D + n(K_2 + K_3 D) \tag{c}$$

Step 3. The objective function in (*c*) is a function of two variables, D and n. However, D and n are directly related, namely $n = Q/D$. Therefore only one independent variable exists for this problem, which we will select to be D. The dependent variable is therefore n. Eliminating n from the objective function in (*c*) gives

$$C = K_1 D + \frac{K_2 Q}{D} + K_3 Q \tag{d}$$

What other constraints exist in this problem? None are stated explicitly, but several implicit constraints exist. One of the assumptions made in arriving at Eq. (*c*) is that over the course of one year production runs of integer quantities may be involved. Can D be treated as a continuous variable? Such a question is crucial prior to using differential calculus to solve the problem. The occurrence of integer variables in principle prevents the direct calculation of derivatives of functions of integer variables. In the simple example here, with D being the only variable and large, you can treat D as being continuous. After obtaining the optimal D, the practical value for D is obtained by rounding up or down. However, there is no guarantee that $n = Q/D$ will be an integer, but as long as you operate from year to year, there should be no restriction on n.

What other constraints exist? You know that D must be positive. Are there any equality constraints that relate D to the other known parameters of the model? If there were, then the sole degree of freedom in the process model could be eliminated and optimization would not be needed!

Step 4. Not needed.

Step 5. Look at the total cost function, Eq. (*c*). Observe that the cost function includes a constant term, K_3Q. If the total cost function is differentiated, the term K_3Q will vanish and thus K_3 will not enter into the determination of the optimal value for D. K_3 will, however, contribute to the total cost.

Two approaches can be employed to solve for the optimal value of D: analytical or numerical. A simple problem has been formulated so that an analytical solution can be obtained. Recall from calculus that if you differentiate the cost function with respect to D and equate the total derivative to zero

$$\frac{dC}{dD} = K_1 - \frac{K_2 Q}{D^2} = 0 \tag{e}$$

you can obtain the optimal solution for D

$$D^{\text{opt}} = \sqrt{\frac{K_2 Q}{K_1}} \tag{f}$$

Equation (*f*) was obtained without knowing specific numerical values for the parameters. If K_1, K_2, or Q change for one reason or another, then the calculation of the new value of D^{opt} is straightforward. Thus, the virtue of an analytical solution (versus a numerical one) is apparent.

Suppose you are given values of $K_1 = 1.0$, $K_2 = 10{,}000$, $K_3 = 4.0$, and $Q = 100{,}000$. Then D^{opt} from Eq. (*f*) is 31,622.

You can also quickly verify for this problem that D^{opt} from Eq. (*f*) minimizes the objective function by taking the second derivative of C and showing that it is positive. Equation (*g*) helps demonstrate the sufficient conditions for a minimum.

$$\frac{d^2C}{dD^2} = \frac{2K_2 Q}{D^3} > 0 \tag{g}$$

Details concerning the necessary and sufficient conditions for minimization are presented in Chap. 4.

Another benefit of obtaining an analytical solution is that you can gain some insight into how production should be scheduled. For example, suppose the optimum number of production runs per year were 4.0 (25,000 units per run), and the projected demand for the product were doubled ($Q = 200{,}000$) for the next year. Using intuition you might decide to double the number of units produced (50,000 units) with 4.0 runs per year. However, as can be seen from the analytical solution, the new value of D^{opt} should be selected according to the square root of Q rather than the first power of Q. This relationship is known as the *economic order quantity* in inventory control and demonstrates some of the pitfalls which may result from making decisions by simple analogies or intuition.

We mentioned earlier that this problem was purposely designed so that an analytical solution could be obtained. Suppose now that the cost per run follows a nonlinear function such as shown earlier in Fig. E1.7. Let the cost vary as given by Eq. (*b*) thus allowing for some economy of scale. Then the total cost function becomes

$$C = K_1 D + \frac{K_2 Q}{D} + \frac{K_4 Q}{D^{1/2}} \tag{h}$$

After differentiation and equating the derivative to zero, you get

$$\frac{dC}{dD} = K_1 - \frac{K_2 Q}{D^2} - \frac{K_4 Q}{2D^{3/2}} = 0 \tag{i}$$

Note that Eq. (*i*) is a rather complicated polynomial equation that cannot explicitly be solved for D^{opt}; you have to resort to a numerical solution as discussed below.

A dichotomy arises in attempting to minimize function (*h*). You can either (1) minimize the cost function (*h*) directly, or (2) find the roots of Eq. (*i*). Which is the best procedure? In general it is easier to minimize C directly by a numerical method rather than take the derivative of C, equate it to zero, and solve the resulting nonlinear equation. This guideline also applies to functions of several variables.

The second derivative of Eq. (*h*) is

$$\frac{d^2C}{dD^2} = \frac{2K_2 Q}{D^3} + \frac{3K_4 Q}{4D^{5/2}} \tag{j}$$

A numerical procedure to obtain D^{opt} directly from Eq. (*d*) could also have been carried out by simply choosing values of D and computing the corresponding values of C from Eq. (*d*) ($K_1 = 1.0$; $K_2 = 10{,}000$; $K_3 = 4.0$; $Q = 100{,}000$).

$D \times 10^{-3}$	10	20	30	40	50	60	70	80	90	100
$C \times 10^{-3}$	510	470	463	465	470	477	484	492	501	510

From the listed numerical data you can see that the function has a single minimum in the vicinity of $D = 20{,}000$ to $40{,}000$. Subsequent calculations in this range (on a finer scale) for D will yield a more precise value for D^{opt}.

Observe that the objective function value for $20 \le D \le 60$ does not vary significantly. However, not all functions behave like C in Eq. (*d*)—some exhibit sharp changes in the objective function near the optimum.

Step 6. You should always be aware of the sensitivity of the optimal answer, i.e., how much the optimal value of C changes when a variable such as D changes or a coefficient in the objective function changes. Parameter values usually contain errors or uncertainties. Therefore, information concerning the sensitivity of the optimum to changes or variations in a parameter is very important in optimal process design. For some problems a sensitivity analysis can be carried out analytically, but in others the sensitivity coefficients must be determined numerically.

In this example problem, we can analytically calculate the changes in C^{opt} in Eq. (*d*) with respect to changes in the various cost parameters. Substitute D^{opt} from Eq. (*f*) into the total cost function

$$C^{opt} = 2\sqrt{K_1 K_2 Q} + K_3 Q \tag{k}$$

Next, take the partial derivatives of C^{opt} with respect to K_1, K_2, K_3, and Q

$$\frac{\partial C^{opt}}{\partial K_1} = \sqrt{\frac{K_2 Q}{K_1}} \tag{l1}$$

$$\frac{\partial C^{opt}}{\partial K_2} = \sqrt{\frac{K_1 Q}{K_2}} \tag{l2}$$

$$\frac{\partial C^{opt}}{\partial K_3} = Q \tag{l3}$$

$$\frac{\partial C^{opt}}{\partial Q} = \sqrt{\frac{K_1 K_2}{Q}} + K_3 \tag{l4}$$

Equations (*l*1) through (*l*4) are absolute sensitivity coefficients.

Similarly, we can develop expressions for the sensitivity of D^{opt}:

$$D^{opt} = \sqrt{\frac{K_2 Q}{K_1}} \tag{f}$$

$$\frac{\partial D^{opt}}{\partial K_1} = \frac{-1}{2K_1}\sqrt{\frac{K_2 Q}{K_1}} \tag{m1}$$

$$\frac{\partial D^{opt}}{\partial K_2} = \frac{1}{2K_2}\sqrt{\frac{K_2 Q}{K_1}} \tag{m2}$$

$$\frac{\partial D^{opt}}{\partial K_3} = 0 \tag{m3}$$

$$\frac{\partial D^{opt}}{\partial Q} = \frac{1}{2Q}\sqrt{\frac{K_2 Q}{K_1}} \tag{m4}$$

Suppose we now substitute numerical values for the constants in order to clarify how these sensitivity functions might be used. For

$$Q = 100{,}000 \qquad K_1 = 1.0 \qquad K_2 = 10{,}000 \qquad K_3 = 4.0$$

then

$$D^{opt} = 31{,}622$$

$$C^{opt} = D^{opt} + \frac{10^9}{D^{opt}} + 400{,}000 = \$463{,}240$$

$$\frac{\partial C^{opt}}{\partial K_1} = 31{,}620 \qquad \frac{\partial D^{opt}}{\partial K_1} = -15{,}810$$

$$\frac{\partial C^{opt}}{\partial K_2} = 3.162 \qquad \frac{\partial D^{opt}}{\partial K_2} = 1.581$$

$$\frac{\partial C^{opt}}{\partial K_3} = 100{,}000 \qquad \frac{\partial D^{opt}}{\partial K_3} = 0$$

$$\frac{\partial C^{opt}}{\partial Q} = 4.316 \qquad \frac{\partial D^{opt}}{\partial Q} = 0.158$$

What can we conclude from the above numerical values? It appears that D^{opt} is extremely sensitive to K_1, but not to Q. However, you must realize that a one unit change in Q (100,000) is quite different from a one unit change in K_1 (0.5). Therefore, in order to put the sensitivities on a more meaningful basis, you should compute the relative sensitivities: for example, the relative sensitivity of C^{opt} to K_1 is

$$S_{K_1}^C = \frac{\partial C^{\text{opt}}/C^{\text{opt}}}{\partial K_1/K_1} = \frac{\partial \ln C^{\text{opt}}}{\partial \ln K_1} = \sqrt{\frac{K_2 Q}{K_1}} \cdot \frac{K_1}{C^{\text{opt}}} = \frac{31{,}622(1.0)}{463{,}240} = 0.0683 \qquad (n)$$

Application of the above idea for the other variables yields the other relative sensitivities for C^{opt}. Numerical values are

$$S_{K_3}^C = 0.863$$

$$S_{K_2}^C = 0.0683 \qquad S_Q^C = 0.932$$

Changes in the parameters Q and K_3 have the largest relative influence on C^{opt}, significantly more than K_1 or K_2. The relative sensitivities for D^{opt} are

$$S_{K_1}^D = -0.5 \qquad S_{K_2}^D = S_Q^D = 0.5 \qquad S_{K_3}^D = 0$$

so that all the parameters except for K_3 have the same influence (in terms of absolute value of fractional changes) on the optimum value of D.

For a problem for which we cannot obtain an analytical solution, you would have to determine sensitivities numerically. You would compute (1) the cost for the base case, i.e., for a specified value of a parameter; (2) change each parameter separately (one at a time) by some arbitrarily small value, such as plus 1 percent or 10 percent and then calculate the new cost. You might repeat the procedure for minus 1 percent or 10 percent. The variation of the parameter, of course, can be made arbitrarily small to approximate a differential; however, when the change approaches an infinitesimal value, the numerical error engendered may confound the calculations.

1.7 OBSTACLES TO OPTIMIZATION

If the objective function and constraints in an optimization problem are "nicely behaved," optimization presents no great difficulty. In particular, if the objective function and constraints are all linear, there is a powerful method known as linear programming for solving the optimization problem (refer to Chap. 7). For this specific type of problem it is known that a unique solution exists if any solution exists. However, most optimization problems in their natural formulation are not linear.

To make it possible to work with the relative simplicity of a linear problem, we often modify the mathematical description of the physical process so that it fits the available method of solution. Many persons employing computer codes for optimization do not fully appreciate the relation between the original problem and the problem being solved; the computer shows its neatly printed output with an authority which the reader feels unwilling, or unable, to question.

In this text we will discuss optimization problems based on behavior of physical systems that have a complicated objective function and/or constraints; for these problems some optimization procedures may be inappropriate and sometimes misleading. Often optimization problems exhibit one or more of the following characteristics causing difficulty and/or failure to calculate the desired optimal solution:

1. The objective function and/or the constraint functions may have finite discontinuities in the continuous parameter values. For example, the price of a compressor or heat exchanger may not change continuously as a function of variables such as size, pressure, temperature, and so on. Consequently, increasing the level of a parameter in some ranges has no effect on cost whereas in other ranges a jump in cost occurs.
2. The objective function and/or the constraint functions may be nonlinear functions of the variables. When one considers real process equipment, the existence of truly linear behavior and system behavior is somewhat of a rarity. This does not preclude the use of linear approximations, but one must interpret the results of such approximations with considerable care.
3. The objective function and/or the constraint functions may be defined in terms of complicated interactions of the variables. A familiar case of interaction is the temperature and pressure dependence in the design of pressure vessels. For example, if the objective function is given as $f = 15.5x_1x_2^{1/2}$, the interaction between x_1 and x_2 precludes the determination of unique values of x_1 and x_2. Many other more complicated and subtle interactions are common in engineering systems. The interaction prevents calculation of unique values of the variables at the optimum.
4. The objective function and/or the constraint functions may exhibit nearly "flat" behavior for some ranges of variables or exponential behavior for other ranges. What this means is that the value of the objective function or a constraint is not sensitive, or is very sensitive, respectively, to changes in the value of the variables.
5. The objective function may exhibit many local optima whereas the global optimum is sought. A solution to the optimization problem may be obtained that is less satisfactory than another solution elsewhere in the region. The better solution may be reached only by initiating the search for the optimum from a different starting point.

In subsequent chapters we will examine these obstacles and discuss some ways of mitigating such difficulties in performing optimization, but you should be aware these difficulties cannot always be alleviated.

REFERENCES

Baumol, W. J., *Economic Theory and Operations Analysis*, Third Edition, Prentice-Hall, Englewood Cliffs, New Jersey (1972).

Beveridge, G. S., and R. S. Schechter, *Optimization: Theory and Practice*, McGraw-Hill, New York (1970).

Latour, P. R., "Requirements for Successful Closed Loop Optimization of Petroleum Refining Process," in *Digital Computer Applications to Process Control* (*Proceed. 6th IFAC/IFIP Conf., Dusseldorf, Oct. 1980*), Oxford (1980).

Robnett, J. D., "Engineering Approaches to Energy Conservation," *Chem. Eng. Prog.*, **59** (March 1979).

Tamhane, A. C., and R. S. H. Mah, "Data Reconciliation and Gross Error Detection in a Chemical Process Network," *Technometrics*, **27**: 409 (1985).

SUPPLEMENTARY REFERENCES

Ackoff, R. L., and M. W. Sasieni, *Fundamentals of Operations Research*, John Wiley, New York (1968).

Beale, E. M. L., "The Evolution of Mathematical Programming Systems," *J. Oper. Res. Soc.*, **36**: 357 (1985).

Noyes, J. L., "A Survey of Mathematical Programming in the Soviet Union," (AD-A116 969/7) Foreign Technology Div., Wright Patterson AFB, OH, 1982.

Peters, M. S., and K. D. Timmerhaus, *Plant Design and Economics for Chemical Engineers*, McGraw-Hill, New York (1980).

Reklaitis, G. V., A. Ravindran, and K. M. Ragsdell, *Engineering Optimization*, John Wiley, New York (1983).

Rudd, D. F., and C. C. Watson, *Strategy of Process Engineering*, John Wiley, New York (1968).

PROBLEMS

For each of the following six problems, formulate the objective function, the equality constraints (if any), and the inequality constraints (if any). Specify and list the independent variable (s), the number of degrees of freedom, and the coefficients in the optimization problem. Solve the problem and state the complete optimal solution values.

1.1. A poster is to contain 300 cm^2 of printed matter with margins of 6 cm at the top and bottom and 4 cm at each side. Find the overall dimensions that minimize the total area of the poster.

1.2. A box with a square base and open top is to hold 1000 cm^3. Find the dimensions that require the least material (assume uniform thickness of material) to construct the box.

1.3. Find the area of the largest rectangle with its lower base on the x-axis and whose corners are bounded at the top by the curve $y = 10 - x^2$.

1.4. Three points x are selected a distance h apart (x_0, $x_0 + h, x_0 + 2h$), with corresponding values f_0, f_1, and f_2. Find the maximum or minimum attained by a quadratic function passing through all three points.

1.5. Find the point on the curve $f = 2x^2 + 3x + 1$ nearest the origin.

1.6. Find the volume of the largest right circular cylinder that can be inscribed inside a sphere of radius R.

1.7. An organic chemical is produced in a batch reactor. The time required to successfully complete one batch of product depends on the amount charged to (and produced from) the reactor, and it has been correlated to be $t = 2P^{0.4}$, where P is the amount of product in pounds per batch and t in hours. A certain amount of nonproduction time is associated with each batch for charging, discharging, and minor maintenance, namely 14 h/batch. The operating cost for the batch system is \$50/h while operating. The capital costs including storage depend on the size of each batch and have been prorated on an annual basis to be $C_I = \$800\ P^{0.7}$.

The required annual production is 300,000 lb/year, and the process can be operated 320 days/year (24 h/day). Total raw material cost at this production level is \$400,000/year.

(*a*) Formulate an objective function using P as the only variable.

(*b*) Are there any constraints on P? (Give relations if applicable).

(*c*) Solve for the optimum value of P. Check that it is a minimum. Also check applicable constraints.

(*d*) Is this a "flat optimum," i.e., is the objective function insensitive to P? Draw an approximate plot of the respective capital and operating cost components as a function of P.

1.8. For a two-stage adiabatic compressor where the gas is cooled to the inlet gas temperature between stages, the theoretical work is given by:

$$W = \frac{kp_1V_1}{k-1}\left[\left(\frac{p_2}{p_1}\right)^{(k-1)/k} - 2 + \left(\frac{p_3}{p_2}\right)^{(k-1)/k}\right]$$

where $k = C_p/C_v$
p_1 = inlet pressure
p_2 = intermediate stage pressure
p_3 = outlet pressure
V_1 = inlet volume

We wish to optimize the intermediate pressure p_2 so that the work is a minimum. Show that if $p_1 = 1$ atm and $p_3 = 4$ atm, $p_2^{\text{opt}} = 2$ atm.

1.9. You are the manufacturer of PCl_3, which you sell in barrels at a rate of P barrels per day. The cost per barrel produced is:

$$C = 50 + 0.1P + 9000/P \qquad \text{dollars}$$

For example, for $P = 100$ barrels/day, $C = \$150$/barrel. The selling price per barrel is \$300. Determine:

(*a*) The production level giving the minimum cost per barrel

(*b*) The production level which maximizes the profit per day

(*c*) The production level at zero profit

(*d*) Why are the answers in (*a*) and (*b*) different?

1.10. It is desired to cool a gas [$C_p = 0.3$ Btu/(lb)(°F)] from 195 to 90°F, using cooling water at 80°F. Water costs \$0.20/1000 ft^3, and the annual fixed charges for the exchanger are \$0.50/ft^2 of inside surface, with a diameter of 0.0875 ft. The heat transfer coefficient is $U = 8$ Btu/(h)(ft^2)(°F) for a gas rate of 3000 lb/h. Plot the annual cost of cooling water and fixed charges for the exchanger as a function of the outlet water temperature. What is the minimum total cost? How would you formulate the problem to obtain a more meaningful result?

1.11. The total cost (dollars per year) for pipeline installation/operation for an incompressible fluid can be expressed as follows:

$$C = C_1 D^{1.5} \cdot L + C_2 m \Delta p / \rho$$

where C_1 is the installed cost of the pipe per foot of length computed on an annual basis ($C_1 D^{1.5}$ is expressed in dollars per year per foot length), C_2 is based on \$0.05/kWh, 365 days/year and 60 percent pump efficiency.

D = diameter (to be optimized)
L = pipeline length = 100 miles
m = mass flow rate = 200,000 lb/h
$\Delta p = 2\rho v^2 L/(D g_c) \cdot f$ = pressure drop, psi
ρ = density = 60 lb/ft^3
v = velocity = $(4m)/(\rho \pi D^2)$
f = friction factor = $(0.046\mu^{0.2})/(D^{0.2} V^{0.2} \rho^{0.2})$
μ = viscosity = 1 cP

(*a*) Find general expressions for D^{opt}, v^{opt}, and C^{opt}.
(*b*) For $C_1 = 0.3$ (D expressed in inches for installed cost), calculate D^{opt} and v^{opt} for the following values of μ and ρ (watch your units!)

$$\mu = 0.2 \text{ cP}, 1 \text{ cP}, 10 \text{ cP}$$

$$\rho = 50 \text{ lb/ft}^3, 60 \text{ lb/ft}^3, 80 \text{ lb/ft}^3$$

1.12. Calculate a new expression for D^{opt} if $f = 0.005$ (rough pipe), independent of the Reynolds number. Compare your results with Prob. 1.11 for $\mu = 1$ cP and $\rho = 60$ lb/ft^3.

1.13. Calculate the relative sensitivities of D^{opt} and C^{opt} in Prob. 1.11 to changes in ρ, μ, m, and C_2 (cost of electricity). Use the base case parameters as given in Prob. 1.11, with $C_1 = 0.3$.

Pose each of the following problems as an optimization problem. Include all of the features mentioned in connection with the first four steps of Table 1.1 but do not solve the problem.

1.14. A chemical manufacturing firm has discontinued production of a certain unprofitable product line. This has created considerable excess production capacity on the three existing batch production facilities. Management is considering devoting this excess capacity to one or more of three new products: call them products 1, 2, and 3. The available capacity on the existing units which might limit output is summarized in the following table:

Unit	Available time, hours/week
A	20
B	10
C	5

Each of the three new produces requires the following processing time for completion:

	Productivity, hours/batch		
Unit	**Product 1**	**Product 2**	**Product 3**
A	0.8	0.2	0.3
B	0.4	0.3	
C	0.2	....	0.1

The sales department indicates that the sales potential for products 1 and 2 exceeds the maximum production rate and that the sales potential for product 3 is 20 batches per week.

The profit per batch would be \$20, \$6, and \$8, respectively, on products 1, 2, and 3.

How much of each product should be produced to maximize profits of the company?

1.15. You have been asked to design an efficient treatment system for runoff from rainfall in an ethylene plant. The accompanying figure gives the general scheme to be used.

The rainfall frequency data for each recurrence interval will fit an empirical equation in the form of

$$R = a + b(t)^2$$

where R = cumulative inches of rain during time t
t = time, h
a and b = constants that have to be determined by fitting the observed rainfall data

Four assumptions should be made:

1. The basin is empty at the beginning of the maximum intensity rain.
2. As soon as water starts to accumulate in the basin the treatment system is started and water is pumped out of the basin.
3. Stormwater is assumed to enter the basin as soon as it falls. (This is normally a good assumption since the rate at which water enters the basin is small compared to the rate at which it leaves the basin during a maximum intensity rain.)
4. All the rainfall becomes runoff.

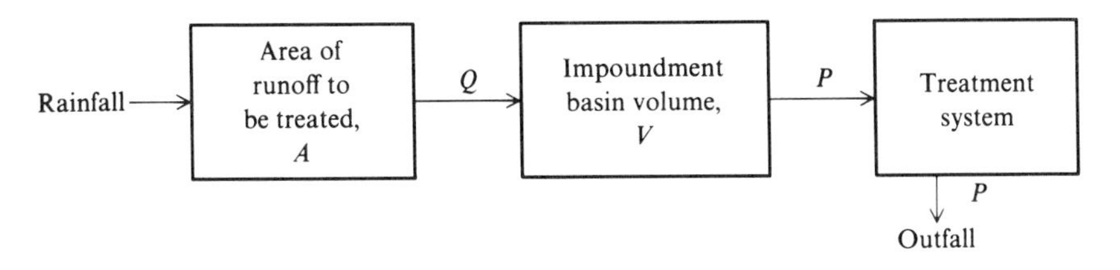

Figure P1.15

The basin must not overflow so that any amount of water that would cause the basin to overflow must be pumped out and treated. What is the minimum pumping rate, P_{max}, required?

Other notation: Q = volumetric flow rate of water entering basin
P = volumetric treatment rate in processing plant

1.16. Optimization of a distributed parameter system can be posed in various ways. An example is a packed, tubular reactor with radial diffusion. Assume a single reversible reaction takes place. To set up the problem as a nonlinear programming problem, write the appropriate balances (constraints) including initial and boundary conditions using the following notation:

x = extent of reaction $\quad$ t = time
T = dimensions temperature $\quad$ r = dimensionless radial coordinate

Do the differential equations have to be expressed in the form of analytical solutions?

The objective function is to maximize the total conversion in the effluent from the reactor over the cross-sectional area at any instant of time. Keep in mind that the heat flux through the wall is subject to physical bounds.

1.17. A waste heat boiler (see Fig. P1.17) is to be designed for steady state operation under the following specifications.

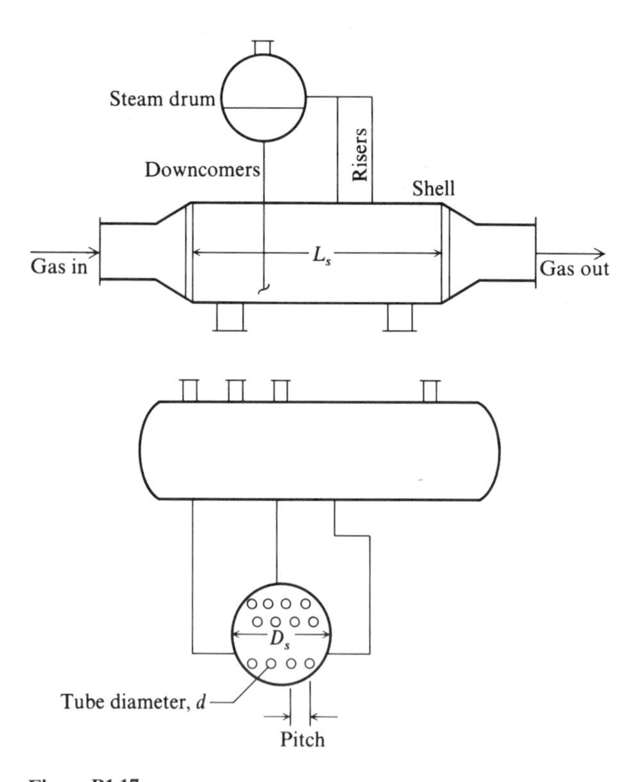

Figure P1.17

Total gas flow:	25,000 kg/h
Gas composition:	SO_2(9%), O_2(12%), N_2(79%)
Gas temperatures:	in = 1200°C; out = 350°C
Stream pressure outside tubes:	250 kPa
Gas properties:	$C_p = 0.24$ kcal/(g)(°C)
	$\mu = 0.14$ kg/(m)(h)
	$k = 0.053$ kcal/(m)(h)(°C)

Cost data are:

Shell:	\$2.50/kg
Tubes:	\$150/m^2
Electricity:	\$0.60/kWh
Interest rate:	14%

Base the optimization on just the cost of the shell, tubes, and pumping costs for the gas. Ignore maintenance and repairs.

Formulate the optimization problem using only the following notation (as needed):

A	surface area of tubes, m^2
C_s	cost of shell, \$
C_t	cost of tubes, \$
C_{pi}	heat capacity of gas, kcal/(kg)(°C)
D	diameter of shell, m
d_o, d_i	tube outer and inner diameters, m
f	friction factor
g	acceleration due to gravity, m/s^2
h_i	gas side heat transfer coefficient inside the tubes, kcal/(m^2)(h)(°C)
i	interest rate, fraction
k	gas thermal conductivity, kcal/(m)(h)(°C)
L_s	length of shell, m
MW	molecular weight of gas
n	number of tubes
N	life of equipment, years
Q	duty of the boiler, kcal/h
$T1$, $T2$	gas temperature entering and leaving the boiler, °C
T	temperature in general
ρ_g	density of gas, kg/m^3
μ_g	viscosity of gas, kg/(m)(h)
V	gas velocity, m/s
W_g	gas flow, kg/h
W_s	weight of shell, tons
η	efficiency of blower
ΔP_g	gas pressure drop, kPa
Z	shell thickness, m

How many degrees of freedom are in the problem you have formulated?

CHAPTER 2

FITTING MODELS TO DATA

Constraints in optimization problems arise from physical bounds on the variables, empirical relations, physical laws, and so on as mentioned in Sec. 1.4. The mathematical relations describing the process also comprise constraints. How to estimate the values of the coefficients in the models and develop empirical relations using the methods of optimization is the topic of this chapter. Mathematical models are employed in all areas of science, engineering, and business to solve problems, design equipment, interpret data, and communicate information. Eykhoff (1974) has defined a mathematical model as "a representation of the essential aspects of an existing system (or a system to be constructed) which presents knowledge of that system in a usable form." For the purpose of optimization, we shall be concerned with developing quantitative (mathematical) expressions for the system model so that we can use the rules of mathematics and computer calculations to extract useful concepts.

Modeling can be valuable because it is an abstraction and helps avoid repetitive experimentation and observations. However, the potential cost and time savings offered by using a mathematical model must be weighed against the fact that the model only imitates reality and does not incorporate all features of the real system being modeled. Hence, the process of interest contains information not readily available or perhaps not even valid in the model. In the development of a model, the user must decide what factors are relevant and how complex the model should be. For example, as a modeler you should consider the following questions:

1. Should the process be modeled on a macroscopic or microscopic level and what level of effort will be required for either approach?
2. Can the process be described adequately using principles of chemistry and physics?
3. What is the desired accuracy of the model and how does its accuracy influence its ultimate use?
4. What measurements are available and what data are available for model verification?
5. Is the process actually composed of smaller, simpler subsystems which are more easily analyzed?

The answers to these questions depend on the usage of the model. As the model of the process becomes more complex, optimization involving the model usually becomes more difficult.

In this chapter we will discuss a number of the factors that have to be taken into consideration in constructing a process model. In addition, we will examine the use of optimization to assist in developing mathematical models, i.e., to estimate the values of unknown coefficients in models, yielding a compact and reasonable representation of process data. Additional information can be found in textbooks specializing in mathematical modeling (Eykhoff, 1974; Himmelblau, 1970; Seinfeld and Lapidus, 1974; Box, Hunter, and Hunter, 1978; Luyben, 1973; Friedly, 1972).

2.1 CLASSIFICATION OF MODELS

Two general categories of models exist:

1. Those based on physical theory
2. Those based on strictly empirical descriptions (so-called black-box models)

Mathematical models based on physical and chemical laws (e.g., mass and energy balances, thermodynamics, chemical reaction kinetics) are frequently employed in optimization applications (see Chaps. 10 through 14). These models are conceptually attractive because a general model for any system size can be developed, even before the system is constructed. On the other hand, an empirical model can be devised which simply correlates input-output data without any physicochemical analysis of the process. The following examples illustrate different categories of models that might be employed in association with the optimization of chemical processes.

Models of a Reactor

An isothermal tubular reactor operates with axial dispersion of concentration. In order to optimize the reactor operations, it is important to be able to predict the concentration of a valuable component A at the reactor exit. Wen and Fan (1975) give a transient model for such a reactor, obtained by physical analysis of the system.

Reactor Model 1

$$\frac{\partial c}{\partial t} = E_z \frac{\partial^2 c}{\partial z^2} - \bar{u} \frac{\partial c}{\partial z} + r$$

where c = concentration of A
t = time
E_z = axial dispersion coefficient
z = distance along the reactor
$\bar{u}$ = average fluid velocity
r = rate of production of $A = f(c_A)$

You can solve the differential equation to obtain $c(z, t)$ for appropriate boundary conditions. An alternate model evolves from dividing the reactor volume into p completely mixed compartments of equal size, V/p, arranged in series, as shown in Fig. 2.1. This arrangement is known as a mixing-cell model. Because of the dispersion, a back-flow term, fv, exists as well as a forward flow for each compartment. The material balance equation for the nth compartment is

Reactor Model 2

$$\left(\frac{V}{p}\right) \frac{\partial c_n}{\partial t} = fv(c_{n+1} - 2c_n + c_{n-1}) + V(c_{n-1} - c_n) + r_n \left(\frac{V}{p}\right)$$

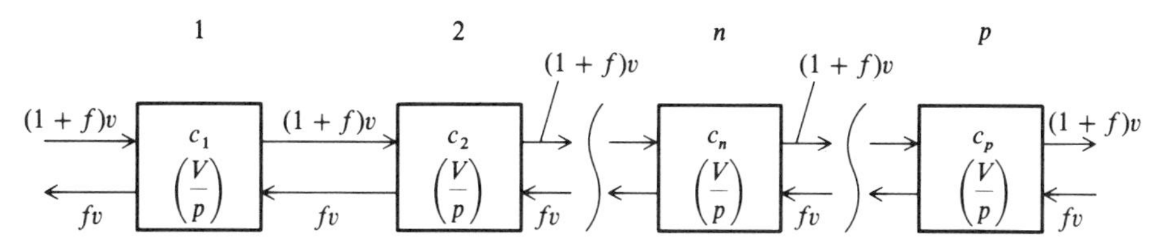

Figure 2.1 Compartments in series with back-flow model (p stages).

The effect of using the compartment model in lieu of the differential model is that for model 2 the concentration of A exists only at discrete points (with respect to distance), rather than continuously (as a continuous function of z) as for the analytical solution of the differential equation. If the reactor operates in the steady state most of the time, $\partial c/\partial t = \partial c_n/\partial t = 0$ and model 2 reduces to a set of p linear algebraic equations (in p unknowns; $c_1, c_2, \ldots, c_p$) that can be solved for the reactor outlet concentration, c_p. You could use such a model as a constraint in calculating the optimal volume of the reactor, given the selling price of the product A and the costs of fabricating and operating the reactor.

Models of an Electrostatic Precipitator

A coal combustion pilot plant is used to obtain efficiency data on the collection of particulate matter by an electrostatic precipitator (ESP). The ESP performance has been varied by changing the surface area of the collecting plates. Figure 2.2 shows the efficiency data (η) as a function of the specific collection area (A), measured as plate area/volumetric flow rate.

Two models of different complexity have been proposed to fit the performance data:

$$\text{Precipitator Model 1:} \qquad \eta = b_1 A + b_2$$

$$\text{Precipitator Model 2:} \qquad \eta = 100\left[1 - \frac{\gamma_1 e^{-\gamma_2 A}}{\gamma_1 + \gamma_3 A}\right]$$

Model 1 is linear in the coefficients, and Model 2 is semiempirical. The solid line in Fig. 2.2 was drawn using $b_1 = 0.156$ and $b_2 = 83.8$. Model 2 was derived taking into account the physical characteristics of the particulate matter, including particle size and electrical properties, hence its complex form. The parameter γ_1 can be obtained from characterization of the particle size distribution (and was equal to 0.04 for the data shown in Fig. 2.2). The parameters γ_2 and γ_3 were estimated to be 0.011 and 0.008, respectively, leading to the dashed line in Fig. 2.2. As can be seen in Fig. 2.2, model 2 provides a better fit of the data than model 1 over the range of A considered. Once a suitable performance model is selected, it could be used as a constraint in calculating the optimal cost of the collection system.

Although we have illustrated some process models above, the major difficulty that generally exists in developing such models is your ability to describe processes by quantitative mathematical relations. Most process equipment is so

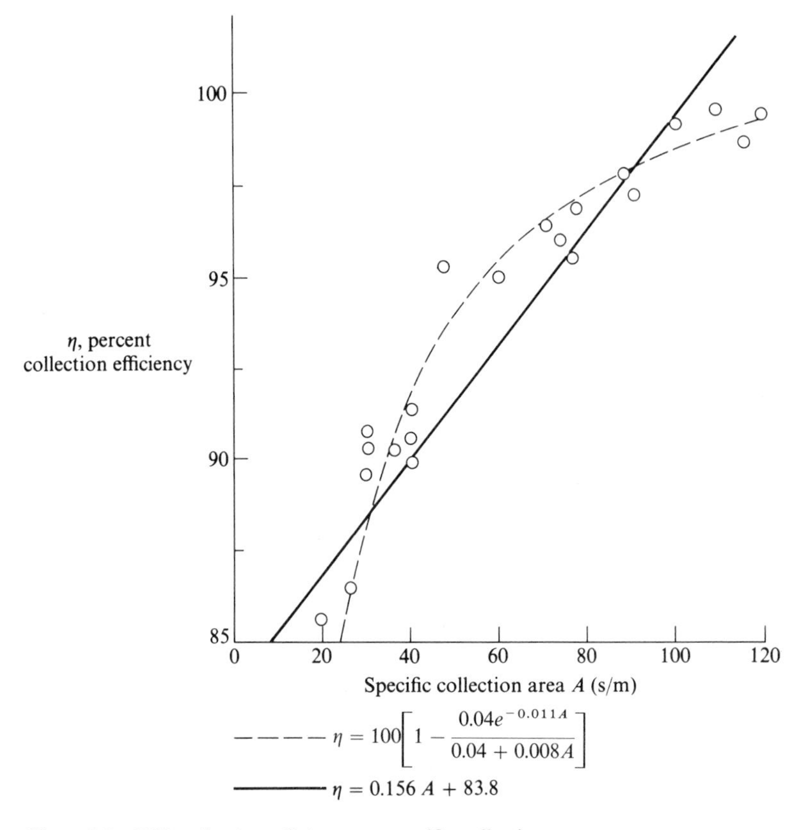

Figure 2.2 ESP collection efficiency vs. specific collection area.

complicated in operation that application of conservation laws leads to innumerable questions in modeling, many of which cannot be answered. For example, thermodynamic or chemical kinetics data which are not available may be required in such a model, increasing the level of effort necessary to formulate the model. On the other hand, while the development of black-box models may require less effort and the resulting models may be simpler in form, empirical models are usually only relevant for restricted ranges of operation and scale-up. Thus, a model such as the ESP model 2 might have to be completely reformulated for a different size range of particulate matter or for a different coal type. You might have to use several black-box models to achieve suitable accuracy for different operating conditions.

In addition to the classification of models as theoretically based versus empirical, we can generally group models according to the following types:

Linear	vs. nonlinear
Steady state	vs. unsteady state
Lumped parameter	vs. distributed parameter
Continuous	vs. discrete variables

1 LINEAR VS. NONLINEAR. Linear models exhibit the important property of superposition; nonlinear ones do not. Equations (and hence models) are linear if the dependent variables or their derivatives appear only to the first power; otherwise they are nonlinear. In practice the ability to use a linear model for a process is of great significance, since the manipulation and solution of linear models which may be required during optimization calculations is an order of magnitude easier than for nonlinear ones.

To test for the linearity versus nonlinearity of a model, examine the equation(s) that represents the process. If any one term in the equations is nonlinear, then the model itself is nonlinear. By implication, the process is nonlinear.

Examine models 1 and 2 for the electrostatic precipitator. Is model 1 linear in A? Model 2? The superposition test in each case is: Does

$$J(ax_1 + bx_2) = aJ(x_1) + bJ(x_2) \tag{2.1a}$$

and

$$J(kx) = kJ(x) \tag{2.1b}$$

where J is any operator contained in the model such as square, differentiation, and so on, k is a constant, and x_1 and x_2 are variables. ESP model 1 is linear in A

$$J(b_1 A + b_2) = b_1 J(A) + b_2$$

but ESP model 2 is nonlinear because

$$\left(\frac{\gamma_1 e^{-\gamma_2(A_1 + A_2)}}{\gamma_1 + \gamma_3(A_1 + A_2)}\right) \neq \left(\frac{\gamma_1 e^{-\gamma_2 A_1}}{\gamma_1 + \gamma_3 A_2}\right) + \left(\frac{\gamma_1 e^{-\gamma_2 A_2}}{\gamma_1 + \gamma_3 A_2}\right)$$

For model 1 of the reactor, the differential equation is linear if r is a linear function of c because differentiation is a linear operation. For example, a second derivative is linear

$$\frac{\partial^2}{\partial z^2}(ac_1 + bc_2) = a\frac{\partial^2}{\partial z^2}c_1 + b\frac{\partial^2}{\partial z^2}c_2$$

Likewise $\partial/\partial t$ and $\partial/\partial z$ are linear operators. However, if the term r is nonlinear in c, then the model is also nonlinear. The same comments apply to the mixing cell approximation, model 2.

2 STEADY STATE VS. UNSTEADY STATE. Other synonyms for steady state are time-invariant, static, or stationary. These terms refer to a process in which the values of the dependent variables remain constant with respect to time. Unsteady state processes are also called nonsteady state, transient, or dynamic, and represent the situation in which the process-dependent variables change with time. A typical example of an unsteady state process is the operation of a batch distillation column, which would exhibit a time-varying product composition. In the reactor example given above we mentioned that a transient model reduces to a steady state model when $\partial/\partial t = 0$. Virtually all optimization problems treated in this book that involve models are based on steady state models. Optimization problems involving dynamic models usually pertain to "optimal control" problems (see Sec. 8.10).

3 DISTRIBUTED VS. LUMPED PARAMETERS. Briefly, a lumped-parameter representation means that spatial variations are ignored, and that the various properties and the state of the system can be considered homogeneous throughout the entire volume. A distributed-parameter representation, on the other hand, takes into account detailed variations in behavior from point to point throughout the system. The differential equation for the reactor (model 1) above illustrated a distributed parameter model in which c was dependent on z. In contrast, in the mixing cell approximation (model 2) each compartment represented a lumped-parameter section of the reactor model because the concentration in the nth mixing cell is "lumped" into an average concentration c_n (even though in the process c_n will vary axially across the distance represented by the cell). All real systems are, of course, distributed in that there are some variations of states throughout them. Because the spatial variations often are relatively small, they may be ignored, leading to a lumped approximation. If both spatial and transient characteristics are to be included in a model, a partial differential equation or a series of stages is required to describe the process behavior.

It is not easy to determine whether lumping in a single compartment in a process model is valid for representing the process. A good rule of thumb is that if the response of the process is essentially the same at all points in the process, then the process model can be lumped as a single unit. If the response shows significant instantaneous differences in any direction along the vessel, then it should be treated using an appropriate differential equation or series of compartments. In an optimization problem it is desirable to simplify a distributed model by using an equivalent lumped-parameter system, although you must be careful to avoid masking the salient features of the distributed element (hence building an inadequate model). In this text, we will mainly consider optimization techniques applied to lumped systems.

4 CONTINUOUS VS. DISCRETE VARIABLES. Continuous means that a variable can assume any value within an interval; discrete means the variable can take on only distinct values in the interval. An example of a discrete variable would be one that assumes integer values only. Often in chemical engineering discrete variables and continuous variables occur simultaneously in a problem. If you wish to optimize a compressor system, for example, you must select the number of compressor stages (an integer) in addition to the suction and production pressure of each stage (positive continuous variables). Optimization problems without discrete variables are far easier to solve than those with even one discrete variable.

An engineer typically strives to treat discrete variables as continuous ones even at the cost of achieving a suboptimal solution when the continuous variable is rounded off. Consider the variation of the cost of insulation with thickness shown in Fig. E1.1. Although insulation is only available in 0.5-inch increments, it is helpful in an optimization problem to use a continuous approximation for the thickness as illustrated further on in Example 3.3.

2.2 HOW TO BUILD A MODEL

Model-building can be divided for convenience of presentation into four phases: problem definition and formulation, preliminary and detailed analysis, evaluation, and interpretation/application. You should keep in mind that model-building is an iterative procedure. Figure 2.3 summarizes the activities to be carried out, that are discussed below. The content of this section is quite limited in scope; before actually embarking on a comprehensive model development program, you should consult other textbooks, such as those mentioned in the introduction to this chapter.

1 PROBLEM DEFINITION AND FORMULATION PHASE. In this phase you must define the problem to be solved and to identify the important elements

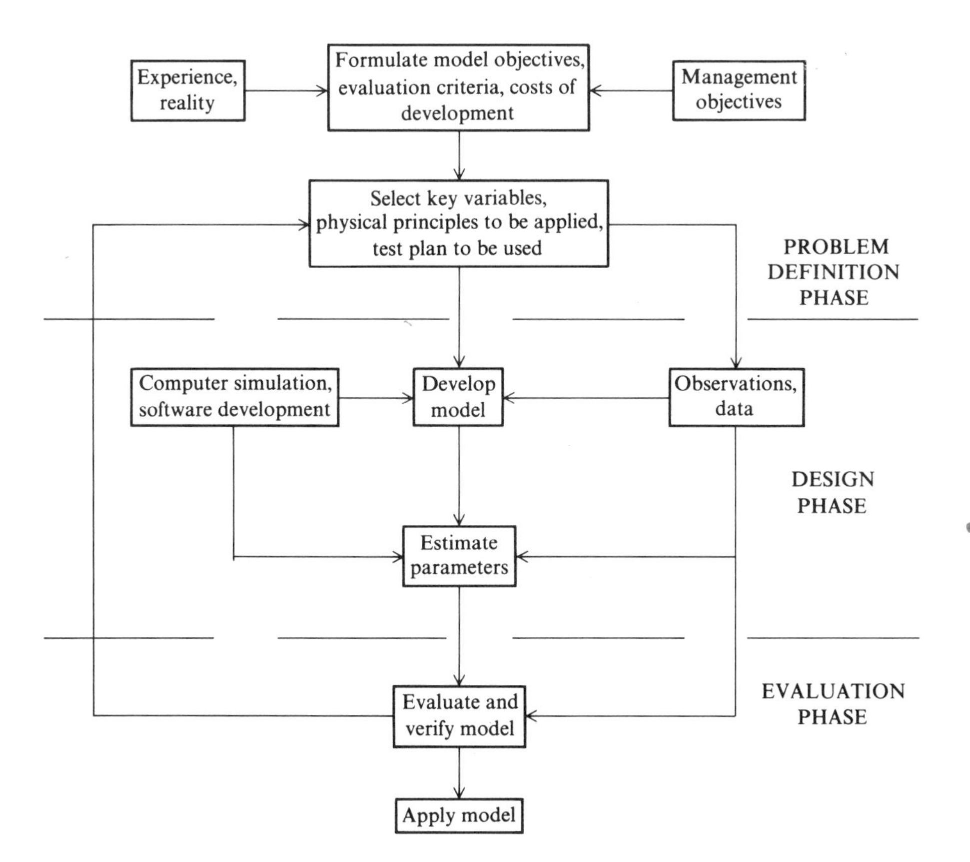

Figure 2.3 Major activities in model-building prior to application.

that pertain to the problem and its solution. The degree of accuracy needed in the model and the potential uses of the model must be determined. You must also evaluate the structure and complexity of the model; ascertain

1. the number of independent variables to be included in the model
2. the number of independent equations required to describe the system (sometimes called the "order" of the model)
3. the number of unknown parameters in the model

In the previous section we have addressed some of these issues in the context of physical vs. empirical models. These issues are also intertwined with the question of model verification; what kinds of data are available for determining that the model is a valid description of the process? Model-building is an iterative process, as shown by the recycle of information in Fig. 2.3.

Before carrying out the actual modeling work, it is important to evaluate the economic justification for (and benefits of) the modeling effort, and the capability of support staff for carrying out such a project. Primarily you should determine that a successfully developed model will indeed help solve the optimization problem.

2 **DESIGN PHASE.** Activities in the design phase include specification of the information content, general description of the programming logic and algorithms necessary to develop and employ a useful model, formulation of the mathematical description of such a model, and simulation of the model. First, define the input and output variables and determine what are the "system" and the "environment." Also, select the specific mathematical representation(s) to be used in the model, as well as the assumptions and limitations of the model resulting from its translation into actual computer code. Computer implementation of the model requires that you verify the availability and adequacy of computer hardware and software, specify computer input/output media, develop program logic and flowsheets, and define program modules and their structural relationships. Use of existing subroutines and data bases saves your time but can add complexity to an optimization problem for the reasons explained in Chap. 14.

3 **EVALUATION PHASE.** This phase is intended as a final check of the model as a whole. Testing of individual model elements should be conducted during earlier phases. Evaluation of the model is carried out according to the evaluation criteria and test plan established in the problem definition phase. Next, carry out sensitivity testing of the model inputs and parameters and determine if the apparent relationships are physically meaningful. Use actual data in the model when possible. This step is also referred to as diagnostic checking, and may entail statistical analysis of the fitted parameters (Himmelblau, 1970; Box, Hunter, and Hunter, 1978).

Finger and Naylor (1967) have stated that model validation consists of three parts: validation of logic, validation of model assumptions, and validation of model behavior. These tasks involve comparison with historical input-output data, or data in the literature, comparison with pilot plant performance, and simulation. In general, data used in formulating a model should not be used to validate it if at all possible. Because model evaluation involves multiple criteria, it is helpful to obtain expert opinion in the verification of models, i.e., what is the impression of the model when reviewed by people who know the process being modeled?

No single validation procedure is appropriate for all models. Nevertheless, it is appropriate to ask the question: What would you really like the model to do? In the best of all possible worlds, you would like the model to predict the desired features of the process performance with suitable accuracy, but this is often an elusive goal.

2.3 FITTING FUNCTIONS TO EMPIRICAL DATA

A model relates the output, i.e., the dependent variable(s), to the independent variable(s). Each equation in the model usually includes one or more coefficients that are presumed constant. The term *parameter* as used here will mean coefficient and possibly input or initial condition. With the help of experimental data, we can determine the *form* of the model, and subsequently (or simultaneously) estimate the value of some or all of the parameters in the model.

2.3.1 How to Determine the Form of a Model

Models can be written in a variety of mathematical forms. Figure 2.4 shows a few of the possibilities, some of which have already been illustrated in Sec. 2.1. This section focuses on the simplest case, namely models comprised of algebraic equations. Emphasis here will be on estimating the coefficients in simple models and not on the complexity of the model.

Selection of the form of an empirical model requires judgment as well as some skill in recognizing how response patterns match possible algebraic functions. Optimization methods can help in the selection of the model structure as well as in the estimation of the unknown coefficients. If you can specify a quantitative criterion which defines what is "best" in representing the data, then the model representation can be improved by adjustment of the form of the model to improve the value of the criterion. The best model presumably exhibits the least error between actual data and the predicted response in some sense. Typical relations for empirical models might be

$$y = a_0 + a_1x_1 + a_2x_2 + \cdots \qquad \text{linear in the variables and coefficients}$$

$$y = a_0 + a_{11}x_1^2 + a_{12}x_1x_2 + \cdots \qquad \text{linear in the coefficients, nonlinear in the variables } (x_1, x_2)$$

$$G(s) = \frac{1}{a_0 + a_1 s + a_2 s^2} \qquad \text{nonlinear in all the coefficients}$$

$$\text{Nu} = a(\text{Re})^b \qquad \text{nonlinear in the coefficient } b$$

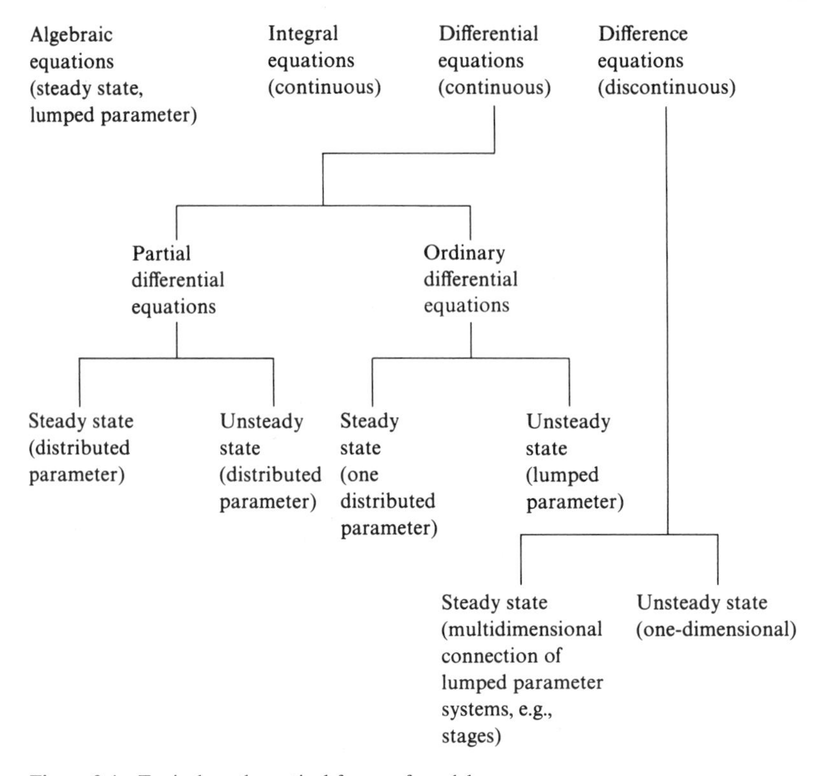

Figure 2.4 Typical mathematical forms of models.

When the model is linear in the coefficients, the coefficients can be estimated by a procedure called *linear regression.* If the coefficients appear in the function in a nonlinear fashion, the estimation of the coefficients is referred to as *nonlinear regression.* We will describe the linear regression problem in more detail in the next section (Sec. 2.4).

Graphical methods for determining the form of the function of a single variable (or two variables) can save considerable time. The response (y) versus the independent variable (x) can be plotted and the resulting form of the model evaluated visually. Figure 2.5 shows experimental heat transfer data which have been plotted on log-log coordinates. Since the plot appears to be approximately linear over most of the range of Re, then a selection of the model to represent Nu vs. Re is much easier than if the relation between the Nu and the Re were extremely nonlinear. A straight line in Fig. 2.5 would correspond to $\log \text{Nu} = \log a + b \log \text{Re}$ or $\text{Nu} = a(\text{Re})^b$. Observe the scatter of experimental data in Fig. 2.5, especially for large values of the Re.

If two independent variables are involved in the model, a plot such as Fig. 2.6 can be of assistance; in this case the second independent variable becomes a

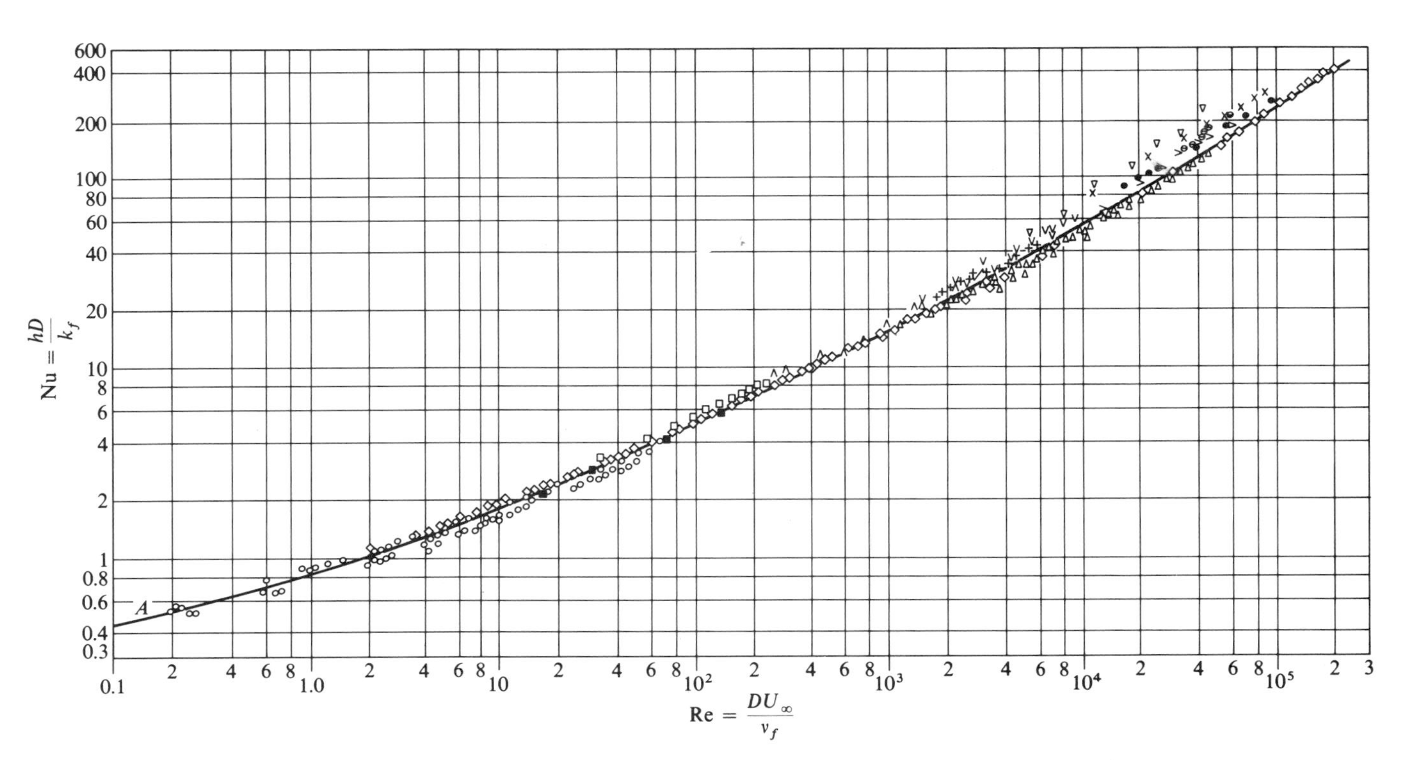

Figure 2.5 Average Nusselt number vs. Reynolds number for a circular cylinder in air, placed normal to the flow (Gebhart, 1971).

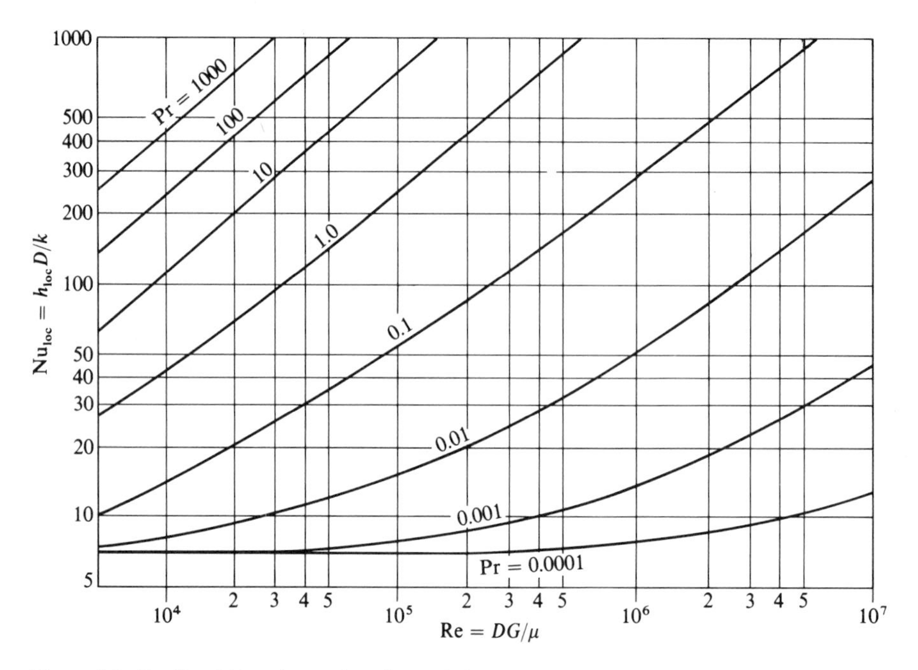

Figure 2.6 Predicted Nusselt numbers for turbulent flow with constant well heat flux (*adapted from Bird, et al., 1964*).

parameter that is held constant at various levels. Figure 2.7 shows a variety of nonlinear functions and their associated plots (Himmelblau, 1970). These plots can assist in selecting relations for nonlinear functions of y versus x.

Now let us examine an application of analyzing data with a model.

EXAMPLE 2.1 ANALYSIS OF THE HEAT TRANSFER COEFFICIENT

Suppose the overall heat transfer coefficient of a shell and tube heat exchanger is monitored daily as a function of the flow rates in both the shell and tube sides (w_s and w_t, respectively). U has the units of Btu/(h)(°F)(ft^2) and w_s and w_t are in lb/h. Figures E2.1*a* and E2.1*b* illustrate measured data patterns. Determine the form of a model based on physical analysis.

Solution. You could elect to simply fit U as a function of w_s and w_t; there appears to be very little effect of w_s on U while U appears to vary linearly with w_t (except at the upper range of w_t where it begins to level off). A single line could be drawn through the data using visual evaluation, or a more quantitative approach can be used based on a physical analysis of the exchanger. First determine why w_s has no effect on U. This result can be explained by the formula for the overall heat transfer coefficient

$$\frac{1}{U} = \frac{1}{h_s} + \frac{1}{h_t} + \frac{1}{h_f} \tag{a}$$

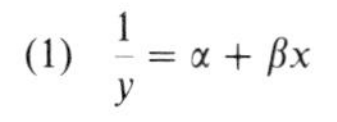

(1) $\dfrac{1}{y} = \alpha + \beta x$

A. $\dfrac{1}{y} = -0.1 + 0.3x$

B. $\dfrac{1}{y} = 0.1 + 0.3x$

C. $\dfrac{1}{y} = -0.5 + 0.3x$

D. $\dfrac{1}{y} = 0.5 + 0.3x$

Equation (1)

(2) $y = \alpha + \dfrac{\beta}{x}$

A. $y = -0.1 + \dfrac{0.3}{x}$

B. $y = 2 + \dfrac{0.3}{x}$

C. $y = 4 + \dfrac{0.3}{x}$

D. $y = 6 + \dfrac{0.3}{x}$

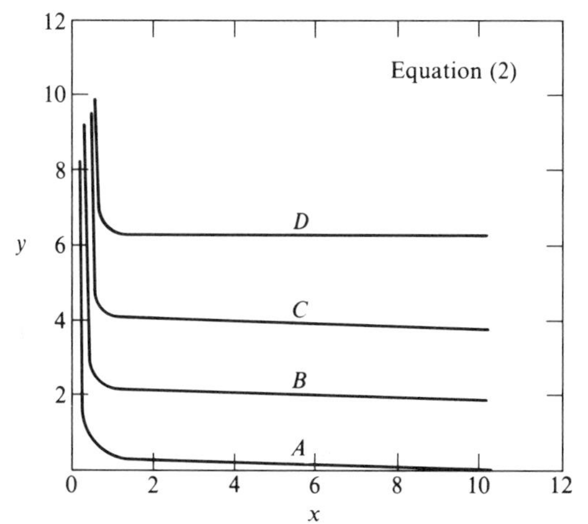

(3) $\dfrac{x}{y} = \alpha + \beta x$

A. $\dfrac{x}{y} = -0.1 + 0.3x$

B. $\dfrac{x}{y} = 0.1 + 0.3x$

C. $\dfrac{x}{y} = -0.4 + 0.3x$

D. $\dfrac{x}{y} = 4 + 0.3x$

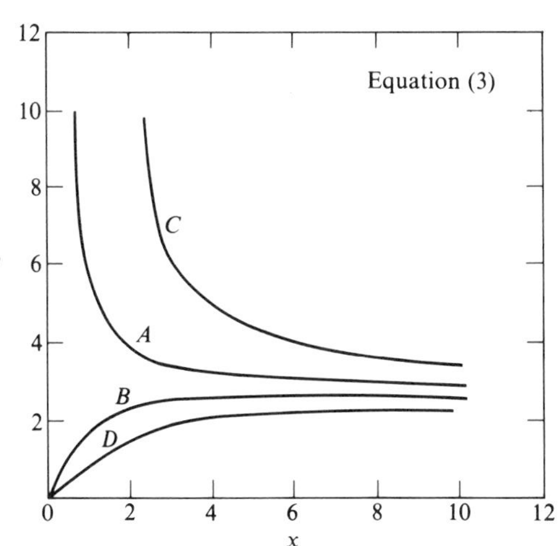

Figure 2.7 (*Continued*)

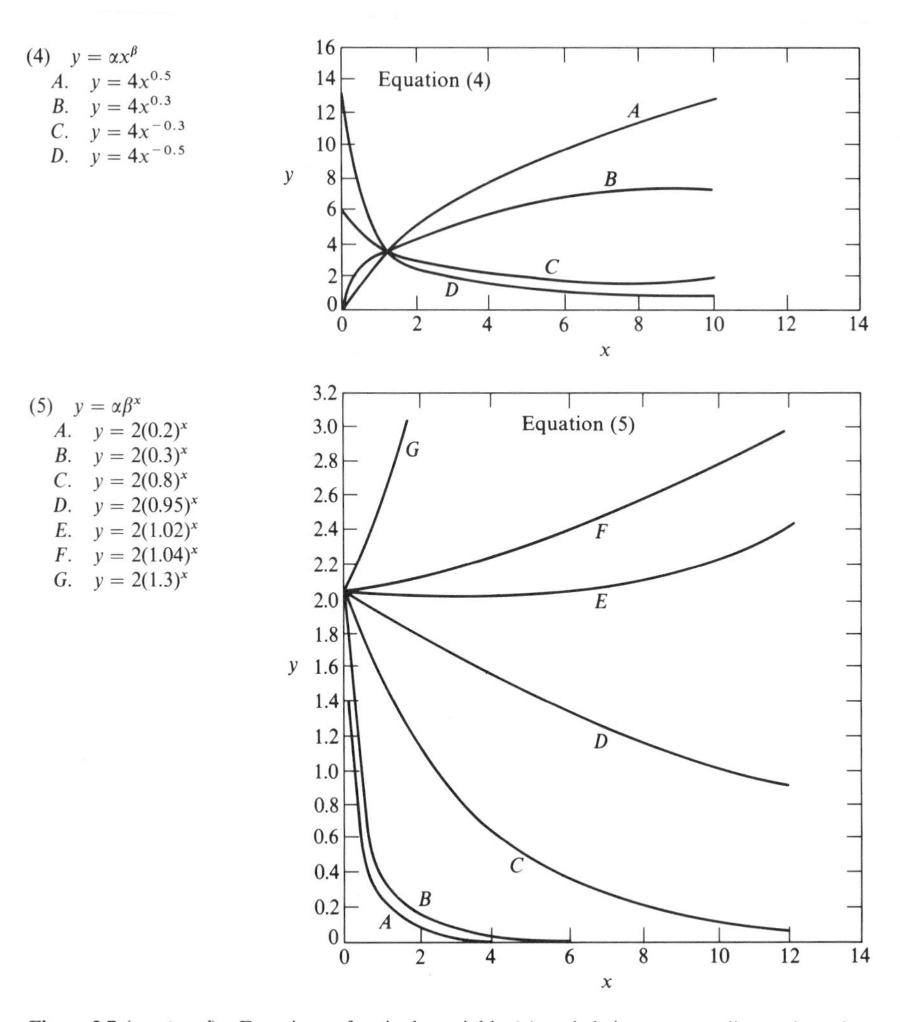

Figure 2.7 (*continued*) Functions of a single variable (x) and their corresponding trajectories.

in which h_s and h_t are the shell and tube side heat transfer coefficients, respectively, and h_f is the fouling coefficient. If h_t is small and h_s is large, U is dominated by h_t, hence changes in w_s have little effect. Next examine the data for U vs. w_t in the context of Fig. 2.7. For a reasonable range of w_t the pattern is similar to curve D in case 3, where

$$\frac{x}{y} = \alpha + Bx \tag{b}$$

which can also be written as

$$\frac{1}{y} = \frac{\alpha}{x} + B \tag{c}$$

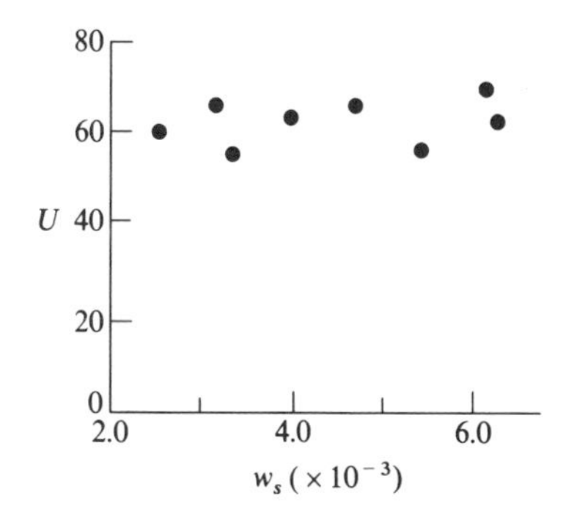

Figure E2.1*a* Variation of overall heat transfer coefficient with shell-side flow rate (w_s) for $w_t = 8000$.

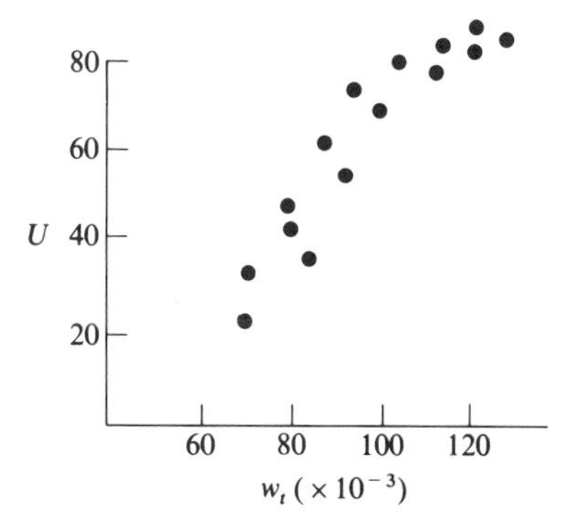

Figure E2.1*b* Variation of overall heat transfer coefficient with tube-side flow rate (w_t) for $w_t = 4000$.

Note the similarity of (c) to Eq. (a), where $x = h_t$ and $y = U$. From a standard heat transfer coefficient correlation (Bird et al., 1964), you could find that h_t also varies according to $K_t w_t^{0.8}$ (K_t is a coefficient that depends on the fluid physical properties and exchanger geometry). Thus, the data in Fig. E2.1b could be interpreted in terms of a semitheoretical model. Equation (a) could be used as the model for heat exchanger performance, estimating the unknowns K_t and h_f (h_s could be ignored or "lumped" into h_f) from plant data.

In fitting either an empirical or theoretically based model, keep in mind that the number of data sets must be equal to or greater than the number of coefficients in the model, the values of which are to be estimated. For example, with three data points of y versus x, you can estimate at most the values of three coefficients. Examine Fig. 2.8. A straight-line correlation for the three points might be adequate, but the data can be fitted exactly via a quadratic model

$$y = c_0 + c_1 x + c_2 x^2 \tag{2.1}$$

Each data point, a pair (y, x), would correspond to one equation of $y(x)$ with three unknown coefficients. The set of three data points therefore yields three linear equations in three unknowns (the coefficients). The solution of these equations can be obtained by using Gaussian elimination (see App. B).

Note that you can postulate a wide range of functional forms for $y(x)$, such as

$$y = c_0 g_0(x) + c_1 g_1(x) + c_2 g_2(x) \tag{2.2}$$

(for example, $g_0 = 1$; $g_1 = \log x$; $g_2 = 1/x$). In all cases three linear equations in three unknowns would still result.

You should use physical insight and common sense in fitting data. In Fig. 2.8, a quadratic function fitted to the three points exhibits a maximum. Intuition, experience, or physical reasoning can indicate if a maximum (or minimum)

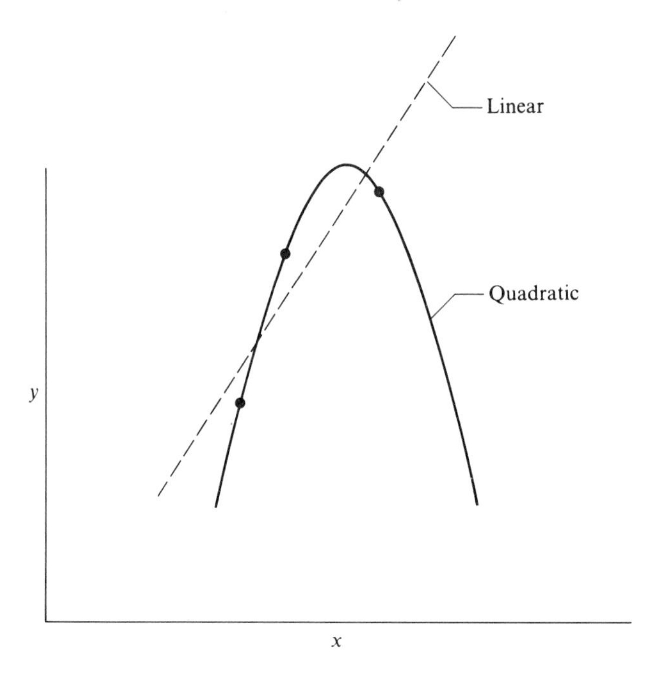

Figure 2.8 Linear vs. quadratic fit for three data points.

should occur, which may validate the model structure. In the absence of such information, the simplest adequate model (with the fewest number of coefficients) should be used.

If p is the number of data sets and n is the number of undetermined coefficients in the model, then you should collect enough data sets such that $p > n$ rather than $p = n$. Although you now have an inconsistent set of equations, optimization can be used to obtain the best solution of the p equations, i.e., the "best" values of the unknown coefficients, according to some selected criterion. We describe a method in the next section called *least squares* that minimizes the sum of the squares of the errors between the predicted and the experimental values of the dependent variable y for each data point x. A quadratic objective function is minimized with respect to the unknown coefficients; for models in which the coefficients appear linearly, this procedure leads to a set of linear equations that can be solved uniquely.

2.4 THE METHOD OF LEAST SQUARES

Various criteria might be used to estimate the coefficients in a model from experimental data. For each of p data points, we can define the error ε_j as the difference between the observation Y_j, $j = 1, 2, \ldots, p$, and the predicted model response $y_j(\mathbf{x})$

$$Y_j - y_j = \varepsilon_j \qquad j = 1, \ldots, p$$

The independent variables in the vector **x** can be different variables or different functions of the same variable, such as x, x^2, x^3, etc. The independent variables are assumed to be known exactly, or at least the error involved in each element of **x** is substantially less than that involved in Y. You might think that the overall sum of the errors could be of utility as an objective function; however, this idea is not appropriate because such an objective function allows positive and negative errors to cancel. A second criterion would be to sum the absolute values of the errors

$$f_1 = \sum_{j=1}^{p} |\varepsilon_j| \tag{2.3}$$

Another would be to minimize the absolute value of the maximum error. Both of these latter criteria can be used via library computer codes.

However, the classical error criterion is the quadratic error summation:

$$f_2 = \sum_{j=1}^{p} \varepsilon_j^2 \tag{2.4}$$

Criterion f_2 is different from criterion f_1 in that it weights large errors much more than extremely small ones in estimating the coefficients. In addition f_2, when minimized, leads to an analytical solution for the unknown coefficients. In some cases, if the error in ε_j is known or can be estimated, it is appropriate to use weighting factors for the ε_j that are inversely proportional to the known error. By this procedure you reduce the confidence limits for the estimated dependent variable (Hald, 1952), but note that the confidence limits really are based on the degree of the experimental error as described in most texts on statistics. If weights are included in the summation, you would use

$$f_3 = \sum_{j=1}^{p} w_j \varepsilon_j^2 \tag{2.5}$$

Let us use the linear model $y = \beta_0 + \beta_1 x$ to illustrate the principal features of the least squares method to estimate the model coefficients with $w_j = 1$. The objective function is

$$f_2 = \sum_{j=1}^{p} (Y_j - y_j)^2 = \sum_{j=1}^{p} (Y_j - \beta_0 - \beta_1 x_j)^2 \tag{2.6}$$

There are two unknown coefficients, β_0 and β_1, and p known pairs of experimental values of Y_j and x_j. We want to minimize f_2 with respect to β_0 and β_1. Recall from calculus that you take the first partial derivatives of f_2 and equate them to zero, to get the necessary conditions for a minimum (the rationale is described in more detail in Sec. 4.5)

$$\frac{\partial f_2}{\partial \beta_0} = 0 = 2 \sum_{j=1}^{p} (Y_j - \beta_0 - \beta_1 x_j)(-1) \tag{2.7a}$$

$$\frac{\partial f_2}{\partial \beta_1} = 0 = 2 \sum_{j=1}^{p} (Y_j - \beta_0 - \beta_1 x_j)(-x_j) \tag{2.7b}$$

Let b_0 and b_1 be the estimates of β_0 and β_1, respectively, obtained by solving Eqs. (2.7). (We use different symbols to distinguish between true values and estimated values.) Rearrangement yields a set of linear equations in two unknowns, b_0 and b_1

$$\sum_{j=1}^{p} b_0 + \sum_{j=1}^{p} b_1 x_j = \sum_{j=1}^{p} Y_j$$

$$\sum_{j=1}^{p} b_0 x_j + \sum_{j=1}^{p} b_1 x_j^2 = \sum_{j=1}^{p} x_j Y_j$$

The summation $\sum_{j=1}^{p} b_0$ is $(p)(b_0)$ and in the other summations the constants b_0 and b_1 can be removed from within the summation signs

$$b_0(p) + b_1 \sum_{j=1}^{p} x_j = \sum_{j=1}^{p} Y_j \tag{2.8a}$$

$$b_0 \sum_{j=1}^{p} x_j + b_1 \sum_{j=1}^{p} x_j^2 = \sum_{j=1}^{p} x_j Y_j \tag{2.8b}$$

The above two linear equations in two unknowns, b_0 and b_1, can be solved quite easily. The predicted value of y, $\hat{Y}$, is $\hat{Y} = b_0 + b_1 x$.

EXAMPLE 2.2 APPLICATION OF LEAST SQUARES

Fit the model $y = \beta_0 + \beta_1 x$ to the following data (Y is the measured response and x the independent variable)

x	Y
0	0
1	2
2	4
3	6
4	8
5	10

Solution. The computations needed to solve Eq. (2.8) are

$$\Sigma x_j = 15 \qquad \Sigma x_j Y_j = 110$$

$$\Sigma Y_j = 30 \qquad \Sigma x_j^2 = 55$$

Then

$$6b_0 + 15b_1 = 30$$

$$15b_0 + 55b_1 = 110$$

Solution of these two equations yields

$$b_0 = 0 \qquad b_1 = 2$$

and the model becomes $\hat{Y} = 2x$ where $\hat{Y}$ is the predicted value for a given x.

Now that we have presented the basic ideas of fitting a model using simple example, we will extend the procedure to a general model which is linear in the coefficients.

$$y = \sum_{i=0}^{n} \beta_i x_i \qquad x_0 = 1 \tag{2.9}$$

In Eq. (2.9) $x_0 = 1$, a constant, so that an intercept is included in the equation. There are n independent variables x_i, $i = 1, \ldots, n$. Independent here means controllable, or adjustable, not functionally independent. Equation (2.9) is linear with respect to the β_i but x_i can be nonlinear. However, keep in mind that the values of x_i will be substituted for x_i prior to solving for the b_i, the estimates of β_i, hence the functional form of x_i even if nonlinear is not a matter of concern. For example, if the model is quadratic

$$y = \beta_0 + \beta_1 x + \beta_1 x^2$$

we specify

$$x_0 = 1$$
$$x_1 = x$$
$$x_2 = x^2$$

and the general structure of Eq. (2.9) is satisfied.

Introduction of Eq. (2.9) into the objective function (2.4) gives

$$\begin{aligned} f_2 &= \sum_{j=0}^{p} (Y_j - y_j)^2 \\ &= \sum_{j=0}^{p} \left(Y_j - \sum_{i=0}^{n} \beta_i x_{ij} \right)^2 \end{aligned} \tag{2.10}$$

The independent variables are now identified by a double subscript, the first index designating the independent variables $(i = 0, \ldots, n)$ and the second the sequence of p data points $(j = 1, \ldots, p)$.

Next, differentiate f_2 with respect to $\beta_0, \beta_1, \ldots, \beta_n$, and equate the $(n + 1)$ partial derivatives to zero, obtaining $(n + 1)$ equations in $(n + 1)$ unknown values of the estimated coefficients $(b_0, \ldots, b_n)$:

$$\begin{aligned}
b_0 \sum_{j=1}^{p} x_{0j}^2 \quad + b_1 \sum_{j=1}^{p} x_{0j}x_{1j} + b_2 \sum_{j=1}^{p} x_{0j}x_{2j} + \cdots + b_n \sum_{j=1}^{p} x_{0j}x_{nj} &= \sum_{j=1}^{p} Y_j x_{0j} \\
b_0 \sum_{j=1}^{p} x_{1j}x_{0j} + b_1 \sum_{j=1}^{p} (x_{1j})^2 + b_2 \sum_{j=1}^{p} x_{1j}x_{2j} + \cdots + b_n \sum_{j=1}^{p} x_{1j}x_{nj} &= \sum_{j=1}^{p} Y_j x_{1j} \\
b_0 \sum_{j=1}^{p} x_{2j}x_{0j} + b_1 \sum_{j=1}^{p} x_{2j}x_{1j} + b_2 \sum_{j=1}^{p} (x_{2j})^2 + \cdots + b_n \sum_{j=1}^{p} x_{2j}x_{nj} &= \sum_{j=1}^{p} Y_j x_{2j} \\
&\vdots \\
b_0 \sum_{j=1}^{p} x_{nj}x_{0j} + b_1 \sum_{j=1}^{p} x_{nj}x_{ij} + b_2 \sum_{j=1}^{p} x_{nj}x_{2j} + \cdots + b_n \sum_{j=1}^{p} (x_{nj})^2 &= \sum_{j=1}^{p} Y_j x_{nj}
\end{aligned} \tag{2.11}$$

Note the symmetry of the summation terms in x_{ij}. This set of $(n+1)$ equations in $(n+1)$ unknowns can be solved on a computer using one of the many readily available routines for solving simultaneous linear equations.

Equations (2.11) can be expressed in more compact form if matrix notation is employed (see App. B). Let

$$\mathbf{b} = \begin{bmatrix} b_0 \\ b_1 \\ \vdots \\ b_n \end{bmatrix} \qquad \mathbf{Y} = \begin{bmatrix} Y_1 \\ Y_2 \\ \vdots \\ Y_p \end{bmatrix}$$

$$\mathbf{X} = \begin{bmatrix} 1 & x_{11} & x_{12} & \cdots & x_{1n} \\ 1 & x_{21} & x_{22} & \cdots & x_{2n} \\ \vdots & \vdots & \vdots & & \vdots \\ 1 & x_{p1} & x_{p2} & \cdots & x_{pn} \end{bmatrix}$$

Equations (2.11) can then be expressed as

$$\mathbf{X}^T\mathbf{X}\mathbf{b} = \mathbf{X}^T\mathbf{Y} \tag{2.12}$$

which has the formal solution via matrix algebra

$$\mathbf{b} = (\mathbf{X}^T\mathbf{X})^{-1}\mathbf{X}^T\mathbf{Y} \tag{2.13}$$

EXAMPLE 2.3 APPLICATION OF LEAST SQUARES

The analysis of labor costs involved in the fabrication of heat exchangers can be used to predict the cost of a new exchanger of the same class. Let the cost be expressed as a linear equation

$$C = \beta_0 + \beta_1 N + \beta_2 A$$

where β_0, β_1, and β_2 are constants, N = number of tubes, and A = shell-surface area. Estimate the values of the constants β_0, β_1, and β_2 from the data in Table E2.3.

Solution. The matrices to be used in calculating **b** are as follows (the weights on the errors are all unity):

$$(\mathbf{X}^T\mathbf{X}) = \begin{bmatrix} 10 & 891 & 3{,}430 \\ 891 & 86{,}241 & 349{,}120 \\ 3{,}430 & 349{,}120 & 1{,}472{,}700 \end{bmatrix}$$

$$(\mathbf{X}^T\mathbf{Y}) = \begin{bmatrix} 2{,}135 \\ 207{,}290 \\ 844{,}800 \end{bmatrix}$$

Table E2.3 Data for mild-steel floating-head exchangers (0–500 psig) working pressure (Shahbenderian, 1961)

Y	x_1	x_2
Labor cost, $	**Area, *A***	**Number of tubes, *N***
310	120	550
300	130	600
275	108	520
250	110	420
220	84	400
200	90	300
190	80	230
150	55	120
140	64	190
100	50	100

The solution of (2.12) or (2.13) gives the estimates of β_0, β_1, and β_2

$$b_0 = 38.177$$

$$b_1 = 1.164$$

$$b_2 = 0.209$$

You should check to see if these coefficients yield a reasonable fit to the data in Table E2.3.

EXAMPLE 2.4 APPLICATION OF LEAST SQUARES

Ten data points were taken in an experiment in which the independent variable x is the mole percentage of a reactant and the dependent variable Y is the yield (in percent):

x	Y
20	73
20	78
30	85
40	90
40	91
50	87
50	86
50	91
60	75
70	65

Fit a quadratic model with these data and determine the value of x that maximizes the yield.

Solution. The model is $y = \beta_0 + \beta_1 x + \beta_2 x^2$ or $y = \beta_0 + \beta_1 x_1 + \beta_2 x_2$ where $x = x_1$ and $x^2 = x_2$. First construct the $\mathbf{X}$, $\mathbf{X}^T\mathbf{X}$, and $\mathbf{X}^T\mathbf{Y}$ matrices and vectors.

$$\mathbf{X} = \begin{bmatrix} 1 & 20 & 400 \\ 1 & 20 & 400 \\ 1 & 30 & 900 \\ 1 & 40 & 1600 \\ 1 & 40 & 1600 \\ 1 & 50 & 2500 \\ 1 & 50 & 2500 \\ 1 & 50 & 2500 \\ 1 & 60 & 3600 \\ 1 & 70 & 4900 \end{bmatrix} \quad (\text{columns } x_0,\ x,\ x^2)$$

$$\mathbf{X}^T\mathbf{X} = \begin{bmatrix} 10 & 430 & 20{,}900 \\ 430 & 20{,}900 & 1{,}105{,}000 \\ 20{,}900 & 1{,}105{,}000 & 61{,}970{,}000 \end{bmatrix}$$

$$\mathbf{X}^T\mathbf{Y} = \begin{bmatrix} 821 \\ 35{,}060 \\ 1{,}675{,}000 \end{bmatrix}$$

Note that $\mathbf{X}^T\mathbf{X}$ is

$$\begin{bmatrix} p & \Sigma x & \Sigma x^2 \\ \Sigma x & \Sigma x^2 & \Sigma x^3 \\ \Sigma x^2 & \Sigma x^3 & \Sigma x^4 \end{bmatrix}$$

and $\mathbf{X}^T\mathbf{Y}$ is

$$\begin{bmatrix} \Sigma Y \\ \Sigma x Y \\ \Sigma x^2 Y \end{bmatrix}$$

The vector $\mathbf{b} = [b_0 \quad b_1 \quad b_2]^T$ is obtained by solving Eqs. (2.12) or (2.13)

$$b_0 = 35.66$$

$$b_1 = 2.63$$

$$b_2 = -0.032$$

The predicted optimum can be formed by differentiating

$$\hat{Y} = b_0 + b_1 x + b_2 x^2$$

with respect to x and setting the derivative to zero to get

$$x^{\text{opt}} = \frac{-b_1}{2b_2} = 41.09$$

This result agrees with that estimated from a plot of the data. The predicted yield at the optimum ($\hat{Y}$) is 88.8.

Additional useful information can be extracted from a least squares analysis if four basic assumptions are made in addition to the presumed linearity of y in $\mathbf{x}$:

1. The x's are deterministic variables (not random).
2. The variance of ε_j is constant or varies only with the x's.

3. The observations Y_j are mutually statistically independent.
4. The distribution of Y_j about y_j given x_j is a normal distribution.

Then you could make use of the experimental data at replicate values of x_i (such as 20, 40, and 50 in Example 2.4) to determine if the model represented the data adequately. For further details see Himmelblau (1970); Box, Hunter, and Hunter (1978); Draper and Hunter (1967); Box and Hill (1967).

How to estimate the best values of coefficients in nonlinear models (nonlinear regression) will be deferred until we take up numerical methods of unconstrained optimization in Chap. 6. What you can do is to directly minimize f in Eqs. (2.3), (2.4), or (2.5) with respect to the coefficients using a computer code. However, the calculations are iterative rather than noniterative as with linear regression.

2.5 FACTORIAL EXPERIMENTAL DESIGNS

The calculation of the values of estimated coefficients in a model can be simplified and made more precise if you collect the experimental data to be used at suitable values of $\mathbf{x}$. For example, so-called orthogonal (factorial) experimental designs permit considerable simplification in Eq. (2.11). Suppose in such a series of experiments the values of x_i take on the integer values 0, 1, and -1, and you want to fit the linear model $y = \beta_0 + \beta_1 x_1 + \beta_2 x_2$. Suppose that the values of x_1 and x_2 in the experiment are deliberately chosen to be as follows (an *experimental design*)

Experiment no.	Response Y	Experimental design x_1	x_2
1	Y_1	-1	-1
2	Y_2	1	-1
3	Y_3	-1	1
4	Y_4	1	1
5	Y_5	0	0

Such a series of experiments corresponds to the four corners of a square in the $x_1 - x_2$ space as in Fig. E2.5. You can show that the summations in Eq. (2.11) simplify in this case to

$$\sum_{j=1}^{5} x_{1j} = 0 \qquad \sum_{j=1}^{5} x_{2j} = 0 \qquad \sum_{j=1}^{5} x_{1j}x_{2j} = 0$$

$$\sum_{j=1}^{5} x_{1j}^2 = 4 \qquad \sum_{j=1}^{5} x_{2j}^2 = 4$$

For the above experimental design

$$\mathbf{X}^T\mathbf{X} = \begin{bmatrix} 5 & 0 & 0 \\ 0 & 4 & 0 \\ 0 & 0 & 4 \end{bmatrix}$$

$$\mathbf{X}^T\mathbf{X} = \begin{bmatrix} \Sigma Y_j \\ -Y_1 + Y_2 - Y_3 + Y_4 \\ -Y_1 - Y_2 + Y_3 + Y_4 \end{bmatrix}$$

It is quite easy to solve Eqs. (2.11) now because they are uncoupled; the inverse of ($\mathbf{X}^T\mathbf{X}$) for Eq. (2.13) can be obtained by merely taking the reciprocal of the diagonal elements. One other even more important advantage of an orthogonal design is that such a design minimizes the covariances of the parameter estimates **b**, that is, you can have more confidence in the values calculated for b_i than would occur by a nonorthogonal design (Box, Hunter, and Hunter, 1978; Biles and Swain, 1980).

EXAMPLE 2.5 ESTIMATION USING AN ORTHOGONAL DESIGN

Assume a reactor is operating at the reference state of 220°F and 3 atm pressure. We can set up an orthogonal factorial design for this process so that the coded values of the variables are 1, −1, and 0. Examine Fig. E2.5. Suppose we select

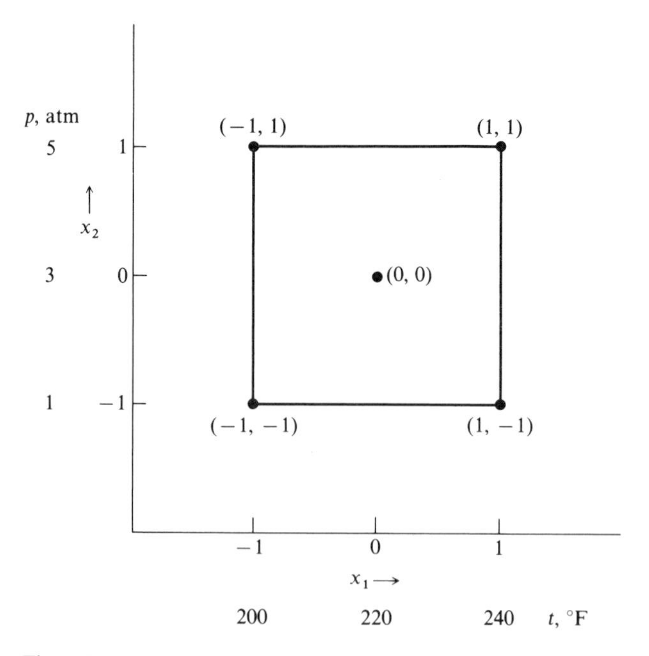

Figure E2.5 Two-level factorial design for the variables temperature and pressure.

perturbations in the operating conditions of $\pm 20°F$ for the temperature and ± 2 atm for the pressure. Then x_1 and x_2, the coded variables, are calculated in terms of the actual operating conditions as follows (check Fig. E2.5):

$$x_1 = \frac{t(°\mathrm{F}) - 220}{20}$$

$$x_2 = \frac{p(\mathrm{atm}) - 3}{2}$$

The following data are collected:

Y (yield)	x_1	x_2
20.500	−1	−1
60.141	1	−1
58.890	−1	1
67.712	1	1
77.870	0	0
78.933	0	0
70.100	0	0

The extra data at the (0, 0) point are needed to obtain a measure of the error involved in the experiment. We first use a linear model to fit the data, i.e.,

$$Y = \beta_0 + \beta_1 x_1 + \beta_2 x_2$$

Solution. Simple calculations show that

$$(\mathbf{X}^T\mathbf{X}) = \begin{bmatrix} 7 & 0 & 0 \\ 0 & 4 & 0 \\ 0 & 0 & 4 \end{bmatrix} \qquad (\mathbf{X}^T\mathbf{X})^{-1} = \begin{bmatrix} 1/7 & 0 & 0 \\ 0 & 1/4 & 0 \\ 0 & 0 & 1/4 \end{bmatrix}$$

$$\mathbf{X}^T\mathbf{Y} = \begin{bmatrix} 434.1 \\ 48.46 \\ 37.96 \end{bmatrix}$$

With these matrices you can compute the estimates of β_i by solving (2.13) and get the regression line

$$\hat{Y} = 62.020 + 12.116x_1 + 9.490x_2$$

Hence

$$\hat{Y} = 62.020 + 12.116\left(\frac{t(°\mathrm{F}) - 220}{20}\right) + 9.490\left(\frac{p(\mathrm{atm}) - 3}{2}\right)$$

2.6 FITTING A MODEL TO DATA SUBJECT TO CONSTRAINTS

Suppose you want to fit some experimental data, but insist that the model response pass exactly through one particular point (called the "base point"). How does this requirement affect the optimization procedure? You now must solve a constrained optimization problem rather than an unconstrained one.

Consider a linear model with two independent variables

$$y = \beta_0 + \beta_1 x_1 + \beta_2 x_2$$

If you specify the value of y at a known point (y^0, x_1^0, x_2^0), this specification is tantamount to generating one equality constraint relating β_0, β_1, and β_2, that is linear in the coefficients. In order to fit the data by the least-squares procedure described in Sec. 2.4, you might use substitution to eliminate the constraint. In other words, solve the constraint for one coefficient and eliminate that coefficient from the model by direct substitution. A more attractive alternative is to use a model of the form

$$(y - y^0) = c_0 + c_1(x_1 - x_1^0) + c_2(x_2 - x_2^0) \tag{2.14}$$

The origin for (2.14) is (y^0, x_1^0, x_2^0), and when the intercept is $c_0 = 0$, the model is forced through the desired point.

If Eq. (2.14) is to be used as the model with $c_0 = 0$, it is first necessary to transform all the experimental data (Y, x_1, x_2) into deviations from the base point (Y^0, x_1^0, x_2^0). Therefore the general matrix developed earlier for Eq. (2.11) must be modified because of the absence of a constant term (corresponding to x_0) in the new model. What is done is to delete the first column of the $\mathbf{X}$ matrix to get a reduced system of n equations and n unknowns, and also define the following deviations

$$\Delta\mathbf{y} = \mathbf{y} - \mathbf{y}^0 \qquad \Delta\mathbf{Y} = \mathbf{Y} - \mathbf{y}^0 \qquad \Delta\mathbf{x}_1 = \mathbf{x}_1 - \mathbf{x}_1^0 \qquad \Delta\mathbf{x}_2 = \mathbf{x}_2 - \mathbf{x}_2^0$$

to replace the independent and dependent variables. Then, to solve for b_1 and b_2 requires the calculations of (with $w_j = 1$)

$$(\Delta\mathbf{x})^T\Delta\mathbf{x} = \begin{bmatrix} \sum_{j=1}^{p} \Delta x_{ij}^2 & \sum_{j=1}^{p} \Delta x_{ij}\Delta x_{2j} \\ \sum_{j=1}^{p} \Delta x_{ij}\Delta x_{2j} & \sum_{j=1}^{p} \Delta x_{2j}^2 \end{bmatrix} \qquad \text{replaces } \mathbf{X}^T\mathbf{X}$$

$$(\Delta\mathbf{x})^T\Delta Y = \begin{bmatrix} \sum_{j=1}^{p} \Delta x_{ij}\Delta Y_j \\ \sum_{j=1}^{p} \Delta x_{2j}\Delta Y_j \end{bmatrix} \qquad \text{replaces } \mathbf{X}^T\mathbf{Y}$$

Then you solve the equivalent of either Eq. (2.11) or (2.13) for c_1 and c_2.

Suppose now a quadratic function, $y = \beta_0 + \beta_1 x + \beta_2 x^2$, has to pass through a base point, (y^0, x^0), and the coefficients are to be obtained by least squares for a number of data points. We use the same procedure as before:

$$y - y^0 = \Delta y = \beta_1(x - x^0) + \beta_2(x - x^0)^2 = \beta_1 \Delta x + \beta_2 (\Delta x)^2 \tag{2.15}$$

Let $x_1 = (x - x^0) = \Delta x$ and $x_2 = (x - x^0)^2 = (\Delta x)^2$. Also, $\Delta Y = (Y - y^0)$. Equation (2.11) becomes

$$\begin{bmatrix} \sum_{i=1}^{p} (\Delta x_i)^2 & \sum_{i=1}^{p} (\Delta x_i)^3 \\ \sum_{i=1}^{p} (\Delta x_i)^3 & \sum_{i=1}^{p} (\Delta x_i)^4 \end{bmatrix} \cdot \begin{bmatrix} b_1 \\ b_2 \end{bmatrix} = \begin{bmatrix} \sum_{j=1}^{p} \Delta x_i \Delta Y_i \\ \sum_{i=1}^{p} (\Delta x_i)^2 \Delta Y_i \end{bmatrix} \tag{2.16}$$

which can be solved for the coefficients in **b**.

More details on nonlinear curve-fitting with base points can be found in Beveridge and Schechter (1967). In Chap. 8 you will learn about methods which can be used to minimize F_1 or F_2 subject to sets of equations and inequality constraints.

EXAMPLE 2.6 FITTING A QUADRATIC FUNCTION PASSING THROUGH A BASE POINT

Fit Eq. (2.15) using the data below with the base point at (2, 2).

x	Y	Δx	ΔY	
0.5	0.6	−1.5	−1.4	
1.0	1.4	−1.0	−0.6	$\Delta x = x - 2.0$
2.1	2.0	0.1	0	$\Delta Y = Y - 2.0$
3.4	3.6	1.4	1.6	

Solution. The summation terms for Eq. (2.15) are

$$\sum_{i=1}^{4} (\Delta x_i)^2 = 5.22 \qquad \sum_{i=1}^{4} (\Delta x_i)^4 = 9.90$$

$$\sum_{i=1}^{4} (\Delta x_i)^3 = -1.63 \qquad \sum_{i=1}^{4} (\Delta x_i \Delta Y_i) = 4.94$$

$$\sum_{i=1}^{4} (\Delta x_i^2 \Delta Y_i) = -0.61$$

The solution of Eq. (2.16) is

$$\Delta y = 0.993\Delta x + 0.149(\Delta x)^2$$

or

$$\hat{Y} - 2.0 = 0.993(x - 2.0) + 0.149(x - 2.0)^2$$

REFERENCES

Beveridge, G. S., and R. S. Schechter, *Optimization: Theory and Practice*, McGraw-Hill, New York (1967).

Biles, W. E., and J. J. Swain, *Optimization and Industrial Experimentation*, Wiley-Interscience, New York (1980).

Bird, R. B., W. E. Stewart, and E. N. Lightfoot, *Transport Phenomena*, Wiley, New York (1964).

Box, G. E. P., and J. W. Hill, "Discrimination Among Mechanistic Models," *Technometrics*, **9**: 57 (1967).

Box, G. E. P., W. G. Hunter, and J. S. Hunter, *Statistics for Experimenters*, Wiley-Interscience, New York (1978).

Draper, N. R., and W. G. Hunter, "The Use of Prior Distribution in the Design of Experiments for Parameter Estimation in Nonlinear Situations," *Biometrika*, **54**: 147 (1967).

Eykhoff, P., *System Identification*, Wiley-Interscience, New York (1974).

Finger, G. S., and T. H. Naylor, "Verification of Computer Simulation Models," *Manage. Sci.*, **14**: 92 (1967).

Friedly, J. C., *Dynamic Behavior of Processes*, Prentice-Hall, Englewood Cliffs, New Jersey (1972).

Gebhart, B., *Heat Transfer*, McGraw-Hill, New York (1971).

Hald, A., *Statistical Theory with Engineering Applications*, Wiley, New York (1952).

Himmelblau, D. M., *Process Analysis*, Wiley, New York (1970).

Luyben, W. L., *Process Modeling, Simulation, and Control for Chemical Engineers*, McGraw-Hill, New York (1973).

Seinfeld, J. H., and L. Lapidus, *Process Modeling, Estimation, and Identification*, Prentice-Hall, Englewood Cliffs, New Jersey (1974).

Shahbenderian, A. P., "The Application of Statistics to Cost Estimating," *Brit. Chem. Eng.*, p. 16 (January 1961).

Wen, C. Y., and L. T. Fan, *Models for Flow Systems and Chemical Reactors*, Marcel Dekker, New York (1975).

SUPPLEMENTARY REFERENCES

Bendor, E. A., *An Introduction to Mathematical Modeling*, John Wiley, New York (1978).

Churchill, S. W., *The Interpretation and Use of Rate Data*, Scripta Publishing Company, Washington, D.C. (1974).

Davis, M. E., *Numerical Methods and Modeling for Chemical Engineers*, John Wiley, New York (1983).

Innis, G. S., and E. A. Rexstad, "Simulation Model Simplification Techniques," *Simulation* **42**(7): 7 (1983).
Zeigler, B. P., *Theory of Modeling and Simulation*, John Wiley, New York (1976).

PROBLEMS

2.1. Classify the following models as linear or nonlinear

(*a*) Two-pipe heat exchanger (streams 1 and 2)

$$\frac{\partial T_1}{\partial t} + v\frac{\partial T_1}{\partial z} = \frac{2h_1}{S_1\rho_1 C_{p_1}}(T_2 - T_1)$$

$$\frac{\partial T_2}{\partial t} = \frac{2h_1}{\rho_2 C_{p_2} S_2}(T_2 - T_1)$$

BC: $T_1(t, 0) = a$ $\quad IC$: $T_1(0, z) = 0$

$T_2(t, 0) = b$ $\quad T_2(0, z) = T_0$

T = temperature $\quad \rho$ = density

t = time $\quad C_p$ = heat capacity

S = area factor

(*b*) Diffusion in a cylinder

$$\frac{\partial C}{\partial t} = D\left(\frac{\partial^2 C}{\partial r^2} + \frac{1}{r}\frac{\partial C}{\partial r}\right)$$

$$C(0, r) = C_0$$

$$\frac{\partial C(t, 0)}{\partial r} = 0$$

$$C(t, R) = C_0$$

C = concentration $\quad r$ = radial direction

t = time $\quad D$ = constant

2.2. Classify the following equations as linear or nonlinear (y = dependent variable: x, z = independent variables)

(*a*) $y_1^2 + y_2^2 = a^2$

(*b*) $v_x \dfrac{\partial v_y}{\partial x} = \mu \dfrac{\partial^2 v_y}{\partial z^2}$

2.3. Classify the models in Probs. 2.1 and 2.2 as steady state or unsteady state.

2.4. Classify the models in Probs. 2.1 and 2.2 as lumped or distributed.

2.5. By what type of model would you represent the following process?

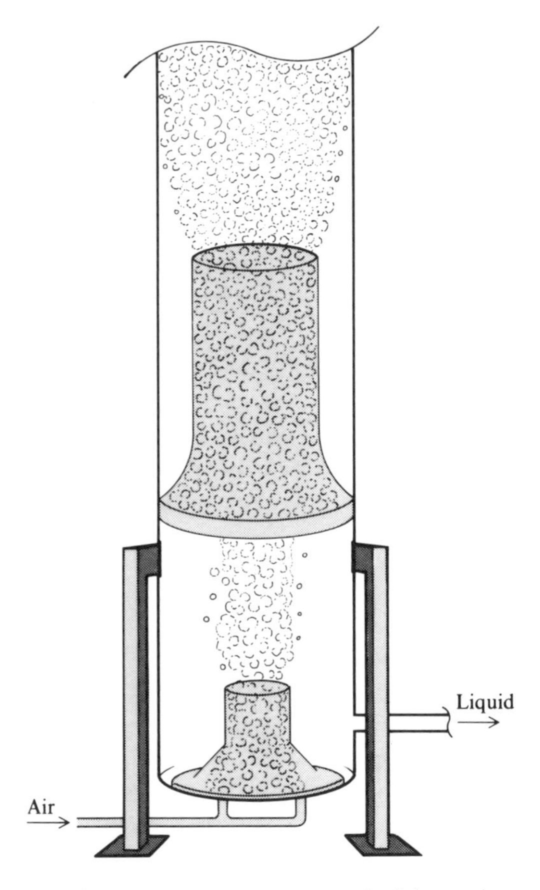

A waste water treatment system for laboratories uses five stacked venturi sections to assure maximum oxygenation efficiency.

Figure P2.5

2.6. Determine the number of independent variables, the number of independent equations, and the number of degrees of freedom for the reboiler shown below. What variables should be specified to make the solution of the material and energy balances determinate? (Q = heat transferred)

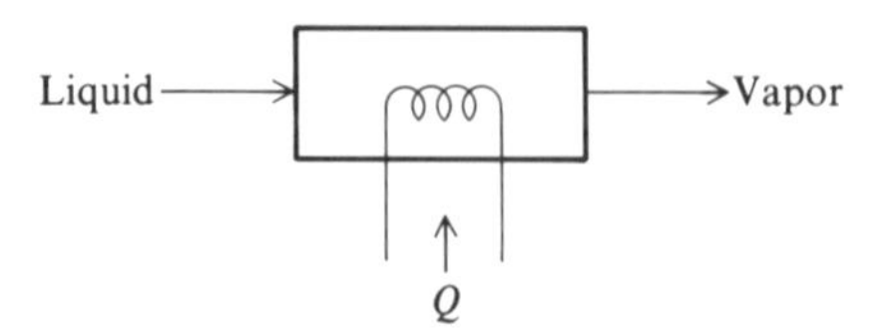

Figure P2.6

2.7. If to the equilibrium stage shown below you add a feed stream, determine the number of degrees of freedom for a binary mixture. (Q = heat transferred)

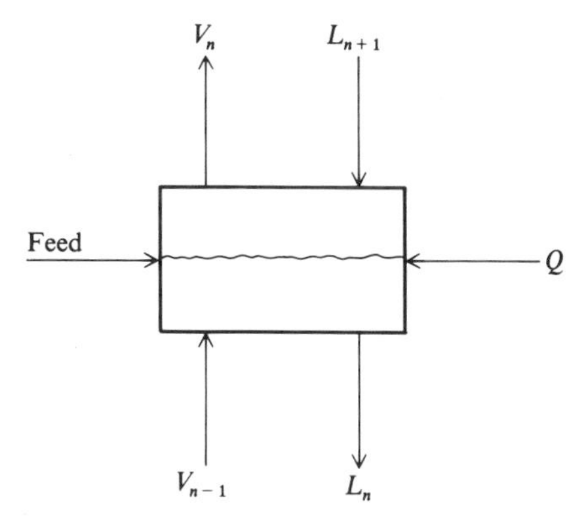

Figure P2.7

2.8. How many variables should be selected as independent variables for the furnace shown below?

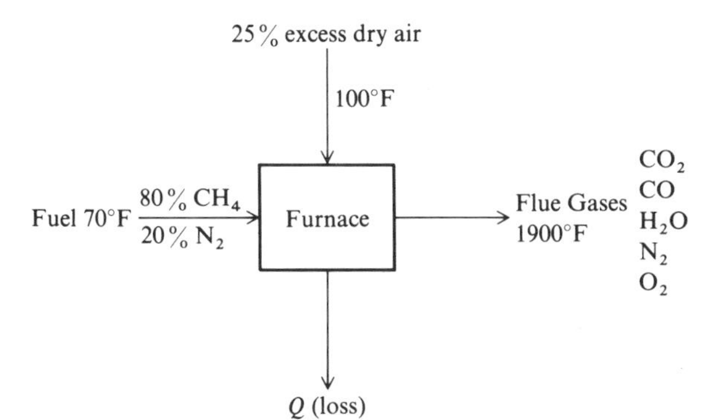

Figure P2.8

2.9. Determine the number of independent variables, the number of independent equations, and the number of degrees of freedom in the following process (A, B, and D are chemical species):

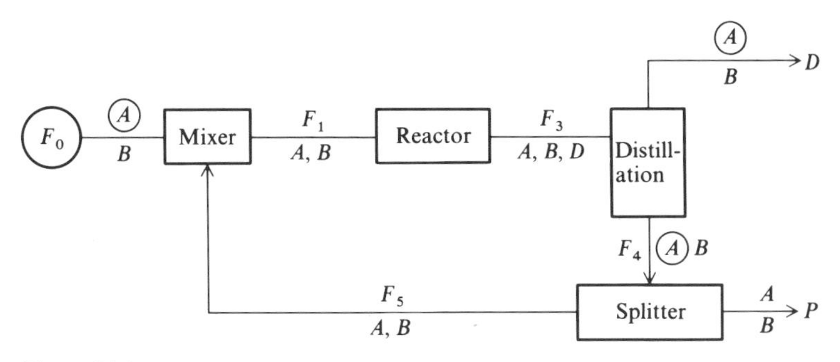

Figure P2.9

(Problem 2.9 is continued overleaf)

The encircled variables have known values. The reaction parameters in the reactor are known as is the fraction split at the splitter between F_4 and F_5. Each stream is a single phase.

2.10. (*a*) Suppose that you wished to curve-fit a set of data (shown below) with the equation

$$y = c_0 + c_1 e^{3x} + c_2 e^{-3x}$$

x	y
0	1
1	2
2	2
3	1

Calculate c_0, c_1, and c_2 (show what summations need to be calculated). How do you find c_1 and c_2 if c_0 is set equal to zero?

(*b*) If the desired equation were $y = a_1 x e^{-a_2 x}$, how could you use least-squares to find a_1 and a_2?

2.11. Fit the following data using the least-squares method with the equation:

$$y = c_0 + c_1 x$$

x	Y
0.5	0.6
1.0	1.4
2.1	2.0
3.4	3.6

Compare the results with a graphical (visual) estimate.

2.12. Fit the same data in Prob. 2.11 using a quadratic fit. Repeat for a cubic model ($y = c_0 + c_1 x + c_2 x^2 + c_3 x^3$)? Plot the data and the curves.

2.13. Determine the best functional relation to fit the following data:

(*a*)

x	Y
1	5
2	7
3	9
4	11

(*b*)

x	Y
2	94.8
5	87.9
8	81.3
11	74.9
14	68.7
17	64.0

(*c*)

x	Y
2	0.0245
4	0.0370
8	0.0570
16	0.0855
32	0.1295
64	0.2000
128	0.3035

(*d*)

x	Y
0	8290
20	8253
40	8215
60	8176
80	8136
100	8093

2.14. The following data have been collected:

x	Y
10	1.0
20	1.26
30	1.86
40	3.31
50	7.08

Which of the following three models best represents the relationship between Y and x?

$$y = e^{\alpha + \beta x}$$

$$y = e^{\alpha + \beta_1 x + \beta_2 x^2}$$

$$y = \alpha x^{\beta}$$

2.15. Given the following equilibrium data for the distribution of SO_3 in hexane, determine a suitable linear (in the parameters) empirical model to represent the data.

x pressure (psia)	Y weight fraction hexane
200	0.846
400	0.573
600	0.401
800	0.288
1000	0.209
1200	0.153
1400	0.111
1600	0.078

2.16. You are given data for Y versus x and asked to fit an empirical model of the form:

$$y = \alpha + \beta x$$

where β is a *known* value. Give an equation to calculate the best estimate of α.

2.17. A replicated two-level factorial experiment is carried out as follows (the dependent variables are yields):

Time, h	Temperature, °C	Yield, %
1	240	24
5	240	42
1	280	3
5	280	19
1	240	24
5	240	46
1	280	5
5	280	21

Find the coefficients in a first-order model, $Y = \beta_0 + \beta_1 x_1 + \beta_2 x_2$. ($Y$ = yield, x_1 = time, x_2 = temperature.)

2.18. An experiment based on a hexagon design was carried out with four replications at the origin, giving the data below:

Factor levels			Design levels	
Yield, %	Temperature, °C	Time, h	x_1	x_2
96.0	75	2.0	1.000	0
78.7	60	2.866	0.500	0.866
76.7	30	2.866	−0.500	0.866
54.6	15	2.0	−1.000	0
64.8	30	1.134	−0.500	−0.866
78.9	60	1.134	0.500	−0.866
97.4	45	2.0	0	0
90.5	45	2.0	0	0
93.0	45	2.0	0	0
86.3	45	2.0	0	0

Coding: $x_1 = \dfrac{\text{temperature} - 45}{30}$ $\quad x_3 = \dfrac{\text{time} - 2}{1.000}$

Fit the full second-order (quadratic) model to the data.

2.19. In Probs. 2.11 and 2.12, the data were fitted with various polynomials. Using the base point concept, fit linear and quadratic equations through the point (0, 0). Repeat for a base point of (0.5, 0.5). What value of y do you predict in each case for $x = 4.0$?

2.20. A reactor converts an organic compound to product P by heating the material in the presence of an additive A. The additive can be injected into the reactor, while steam can be injected into a heating coil inside the reactor to provide heat. Some conversion can be obtained by heating without addition of A, and vice versa. In order to predict the yield of P, Y_p (lb mole product per lb mole feed), as a function of the mole fraction of A, X_A, and the steam addition S (in lb/lb mole feed), the following data have been obtained.

Y_p	X_A	S
0.2	0.3	0
0.3	0	30
0.5	0	60

(*a*) Fit a linear model

$$Y_p = c_0 + c_1 X_A + c_2 S$$

that provides a "least-squares" fit to the data.

(*b*) If we require that the model always must fit the point $Y_p = 0$ for $X_A = S = 0$, obtain c_0, c_1, and c_2 so that a least-squares fit is obtained.

CHAPTER

3

FORMULATION OF OBJECTIVE FUNCTIONS

The formulation of objective functions is one of the crucial steps in the application of optimization to a practical problem. As discussed in Chap. 1, you must be able to translate a verbal statement or concept of the desired objective into mathematical terms. In the chemical process industries, the objective function often is expressed in units of currency (for example, U.S. dollars) because the goal of the enterprise is to minimize costs or maximize profits subject to a variety of constraints. Therefore, we omit from consideration such subjective and philosophical goals as "building a better world" or "developing a more humane society," and treat only goals which are readily measurable. Problems involving multiple objective functions (vector-valued performance objectives), problems that are of interest in several fields, have been discussed by Chankong and Haimes (1983), Carlsson and Kochetkov (1983), Zeleny (1982), and Hansen (1983); we do not cover this topic in this book because of the complexity of the methodology involved.

In this chapter we have chosen the simplest, most direct approach—to formulate an objective function in terms of cost or profit; other measures of merit such as yield are easier to quantify and can be found in several examples in Chaps. 10 to 14. However, if income or expenses are to be optimized, the formulation of the objective function becomes complex. You must draw on the ideas of economic analysis and project evaluation. We shall concentrate on those specific aspects of process economics which are important in optimization, namely the evaluation of the profitability of a given project or design.

The term "profitability" is used to measure the amount of profit generated, but this measure can be expressed in different ways. A variety of quantitative choices exist, and the considerations that must be included in economic evaluation will depend on the specific corporate environment in which you find yourself and the problem to be solved. Such variables as the current and future interest rate, use of borrowed versus equity capital, depreciation of capital, and tax structure all affect the profitability of a given project.

In our discussion of how to develop objective functions, one complicating factor in most real problems is that cash flows (income and expenses) may occur in a nonuniform fashion over a period of years. Some costs are *capital costs*, namely the cost of equipment or structures with a useful life of more than one year, while other costs occur on a yearly basis and are termed *operating costs*. In order to handle the dynamic nature of costs, income, and profits, the concept of discounted cash flow (DCF) will be introduced.

3.1 INVESTMENT COSTS AND OPERATING COSTS IN OBJECTIVE FUNCTIONS

In this section we consider three categories of objective functions that include operating and capital costs. The first category of objective functions involves no capital costs at all but just operating costs and revenues. Such cases are often referred to as "supervisory control" problems and arise when capital costs are a fixed sum (the equipment is already in place). We cannot influence these costs by optimizing the operating variables. See Example 3.1 below for one such application. A second category is optimization of capital equipment in circumstances where no operating costs are involved. Many mechanical design problems fall in this category; Example 3.2 presents the optimum sizing of a cylindrical pressure vessel.

The third category of objective functions includes both capital costs and operating costs. Such problems usually involve some capital expenditure in order to reduce operating costs or manufacture additional product. Capital costs here include the purchase price of the equipment plus a term for a return on the investment (Happel and Jordan, 1975). Both design variables and operating conditions can be optimized in such a problem, but you will find that this type of objective function poses the greatest conceptual difficulty in formulation. The principal difficulty is how to mesh the two types of costs in a sensible way so that the objective function is comprised of terms with common units (note that the units of capital cost are dollars vs. dollars per year for operating costs). Example 3.3 gives a typical application in this third category.

EXAMPLE 3.1 OPERATING PROFITS AS THE OBJECTIVE FUNCTION

A chemical plant makes three products (E, F, G) and utilizes three raw materials (A, B, C) in limited supply. Each of the three products is produced in a separate process (1, 2, 3); a schematic of the plant is shown in Fig. E3.1. The available

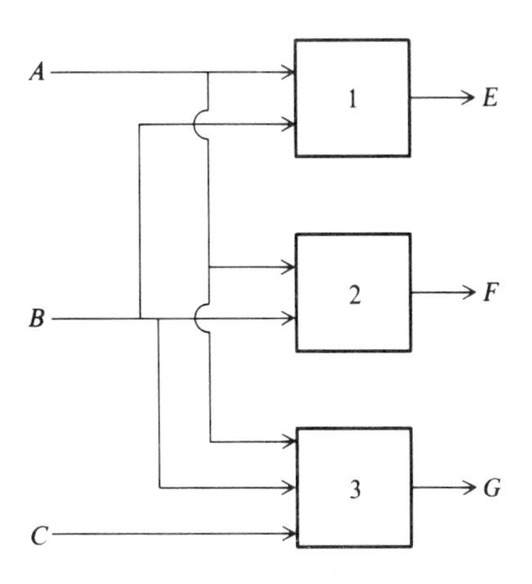

Figure E3.1 Flow diagram for multiproduct plant.

materials A, B, and C do not have to be totally consumed. The reactions involving A, B, and C are as follows:

Process data

Process 1: $A + B \rightarrow E$

Process 2: $A + B \rightarrow F$

Process 3: $3A + 2B + C \rightarrow G$

Raw material	Maximum available, lb/day	Cost ¢/lb
A	40,000	1.5
B	30,000	2.0
C	25,000	2.5

Process	Product	Reactant requirements (lb) per lb product	Processing cost	Selling price of product
1	E	$\frac{2}{3}A$, $\frac{1}{3}B$	1.5 ¢/lb E	4.0 ¢/lb E
2	F	$\frac{2}{3}A$, $\frac{1}{3}B$	0.5 ¢/lb F	3.3 ¢/lb F
3	G	$\frac{1}{2}A$, $\frac{1}{6}B$, $\frac{1}{3}C$	1.0 ¢/lb G	3.8 ¢/lb G
		(mass is conserved)		

¢ = \$/100.

Formulate the objective function to maximize the total operating profit per day in the units of \$/day.

Solution. The notation for the mass flowrates of reactants and products is:

$$x_1 = \frac{\text{lb}}{\text{day}} A$$

$$x_2 = \frac{\text{lb}}{\text{day}} B$$

$$x_3 = \frac{\text{lb}}{\text{day}} C$$

$$x_4 = \frac{\text{lb}}{\text{day}} E$$

$$x_5 = \frac{\text{lb}}{\text{day}} F$$

$$x_6 = \frac{\text{lb}}{\text{day}} G$$

The income in dollars per day from the plant is found from the selling prices $(0.04x_4 + 0.033x_5 + 0.038x_6)$. The operating costs in dollars per day include

(*a*) Raw material costs: $0.015x_1 + 0.02x_2 + 0.025x_3$
(*b*) Processing costs: $0.015x_4 + 0.005x_5 + 0.01x_6$

$$\text{Total costs in dollars per day} = 0.015x_1 + 0.02x_2 + 0.025x_3 + 0.015x_4 + 0.005x_5 + 0.01x_6$$

The daily profit is found by subtracting daily operating costs from the daily income:

$$f(\mathbf{x}) = 0.025x_4 + 0.028x_5 + 0.028x_6 - 0.015x_1 - 0.02x_2 - 0.025x_3$$

Note that the six variables in the objective function (x_1 through x_6) are constrained through material balances, namely

$$x_1 = 0.667x_4 + 0.667x_5 + 0.5x_6$$

$$x_2 = 0.333x_4 + 0.333x_5 + 0.167x_6$$

$$x_3 = 0.333x_6$$

The variables also satisfy the overall material balance $x_1 + x_2 + x_3 = x_4 + x_5 + x_6$, but the overall balance is not an independent equation. There are also bounds on the amounts of A, B, and C processed:

$$0 \le x_1 \le 40{,}000$$

$$0 \le x_2 \le 30{,}000$$

$$0 \le x_3 \le 25{,}000$$

The optimization problem in this example is comprised of a linear objective function and linear constraints, hence linear programming would be the best technique to use in solving it (see Chap. 7).

The next example treats a case in which only capital costs are to be optimized.

EXAMPLE 3.2 CAPITAL COSTS AS THE OBJECTIVE FUNCTION

Suppose you wanted to find the configuration that minimizes the capital costs of a cylindrical pressure vessel. In order to select the best dimensions (length, L, and diameter, D) of the vessel, formulate a suitable objective function for the capital costs and find the optimal (L/D) that minimizes the cost function. Let the tank volume be V, which is fixed. Compare your result with the design rule-of-thumb, $(L/D)^{\text{opt}} = 3.0$.

Solution. Let us begin with a simplified geometry for the tank based on the following assumptions:

1. Both ends are closed and flat.
2. The vessel walls (sides and ends) are of constant thickness (t) with density ρ, and the wall thickness is not a function of pressure.
3. The cost of fabrication and material is the same for both the sides and ends, and is S (dollars per unit weight).
4. There is no wasted material during fabrication due to the available width of metal plate.

The surface area of the tank using the above assumptions is equal to

$$\underset{\text{(ends)}}{2\left(\frac{\pi D^2}{4}\right)} + \underset{\text{(cylinder)}}{\pi D L} = \frac{\pi D^2}{2} + \pi D L \qquad (a)$$

From assumptions (2) and (3), you might set up several different objective functions:

$$f_1 = \frac{\pi D^2}{2} + \pi D L \qquad \text{(units of area)} \qquad (b)$$

$$f_2 = \rho\left(\frac{\pi D^2}{2} + \pi D L\right)\cdot t \qquad \text{(units of weight)} \qquad (c)$$

$$f_3 = S\cdot\rho\cdot\left(\frac{\pi D^2}{2} + \pi D L\right)\cdot t \qquad \text{(units of cost in dollars)} \qquad (d)$$

Note that all of these objective functions differ from each other only by a multiplicative constant; this constant has no effect on the values of the independent variables at the optimum. Therefore, for simplicity, we will use f_1 to determine the optimal values of D and L. Implicit in the problem statement is that a relation exists between volume and length, namely the constraint

$$V = \frac{\pi D^2}{4}\cdot L \qquad (e)$$

Hence there is only one independent variable in the problem.

Next use (e) to remove L from (b) to obtain the objective function

$$f_4 = \frac{\pi D^2}{2} + \frac{4V}{D} \qquad (f)$$

Differentiation of f_4 with respect to D for constant V, equating the derivative to zero, and solving the resulting equation gives

$$D^{\text{opt}} = \left(\frac{4V}{\pi}\right)^{1/3} \qquad (g)$$

This result implies that $f_4 \sim V^{2/3}$, a relationship close to the familiar "six-tenths" rule used in cost estimating. From (*e*), $L^{opt} = (4V/\pi)^{1/3}$; this yields a rather surprising result, namely

$$\left(\frac{L}{D}\right)^{opt} = 1 \qquad (h)$$

The $(L/D)^{opt}$ ratio is significantly different from that stated earlier in the example, namely, $L/D = 3$; this difference must be due to the assumptions (perhaps erroneous) made earlier regarding vessel geometry and fabrication costs.

Brummerstedt (1944) and Happel and Jordan (1975) have discussed a somewhat more realistic formulation of the problem of optimizing a vessel size, making the following modifications in the original assumptions:

1. The ends of the vessel are 2:1 ellipsoidal heads, with an area for the two ends of $2(1.16D^2) = 2.32D^2$.
2. The cost of fabrication for the ends is higher than the sides; Happel suggested a factor of 1.5.
3. The thickness t is a function of the vessel diameter, allowable steel stress, pressure rating of the vessel, and a corrosion allowance. For example, a design pressure of 250 psi and a corrosion allowance of $\frac{1}{8}$ inch gives the following formula for t in inches (in which D is expressed in feet):

$$t = 0.0108D + 0.125 \qquad (i)$$

The above assumptions require that the objective function be expressed in dollars since area and weight are no longer directly proportional to cost

$$f_5 = \rho[\pi DLS + (1.5S)(2.32D^2)]t \qquad (j)$$

The unit conversion of t from inches to feet does not affect the optimum (L/D), nor do the values of ρ and S, which are multiplicative constants. Therefore, the modified objective function, substituting Eq. (*i*) in Eq. (*j*) is

$$f_6 = 0.0339D^2L + 0.435D^2 + 0.3927DL + 0.0376D^3 \qquad (k)$$

The volume constraint is also different than previously used because of the dished heads:

$$V = \frac{\pi D^2}{4}\left(L + \frac{D}{3}\right) \qquad (l)$$

Equation (*l*) can be solved for L and substituted into Eq. (*k*). However, no analytical solution for D^{opt} by direct differentiation of the objective function is possible now because the expression for f_6, when L is eliminated, leads to a complicated polynomial equation for the objective function:

$$f_7 = 0.0432V + 0.5000\frac{V}{D} + 0.3041D^2 + 0.0263D^3 \qquad (m)$$

When f_7 is differentiated, a fourth-order polynomial in D results; no simple analytical solution is possible to obtain the optimum value of D. Therefore, it is better to use a numerical search for D^{opt} based on f_7 (rather than examining $df_7/dD = 0$). However, such a search will need to be performed for different values of V and the design pressure, parameters which are embedded in Eq. (*i*). Recall that Eq. (*m*) is based on a design pressure of 250 psia. Happel (1975) has presented a solution for $(L/D)^{\text{opt}}$ as follows:

Optimum (L/D)

	Design pressure, psi		
Capacity, gal.	**100**	**250**	**400**
2,500	1.7	2.4	2.9
25,000	2.2	2.9	4.3

In Chap. 5 you will learn how to obtain such a solution. Note that for small capacities and low pressures, the optimum (L/D) approaches the ideal case, Eq. (h), considered above. It is clear from the above table that the rule of thumb for $(L/D)^{\text{opt}}$ can be in error by as much as ± 50 percent from the actual optimum. You should exercise some caution with strict interpretation of these results, since the optimum does not take into account materials wasted during fabrication.

Next we consider an example in which both operating costs and capital costs are included in the objective function. The solution of this example will require that the two types of costs be put on some common basis, namely, dollars per year.

EXAMPLE 3.3 OPTIMUM THICKNESS OF INSULATION

In specifying the insulation thickness for a cylindrical vessel or pipe, it is necessary to consider both the costs of the insulation as well as the value of the energy saved by adding the insulation. In this example we determine the optimum thickness of insulation for a large pipe which contains a hot liquid. The insulation is added to reduce heat losses from the pipe. Rubin (1982) has presented tables which show the economic insulation thickness as a function of pipe size, fuel cost, and pipe temperature, based on a 7.5 mile-per-hour wind and 60°F air. Below we develop an analytical expression for insulation thickness based on a mathematical model.

The rate of heat loss from a large insulated cylinder (see Fig. E3.3), for which the insulation thickness is much smaller than the cylinder diameter and the inside heat transfer coefficient is very large, can be approximated by the formula:

$$Q = \frac{A\Delta T}{x/k + 1/h_c} \tag{a}$$

where ΔT = average temperature difference between pipe fluid and ambient surroundings, °F
A = surface area of pipe
x = thickness of insulation, ft
h_c = outside convective heat transfer coeffient, Btu/(h)(ft^2)(°F)
k = thermal conductivity of insulation, Btu/(h)(ft)(°F)
Q = heat loss, Btu/h

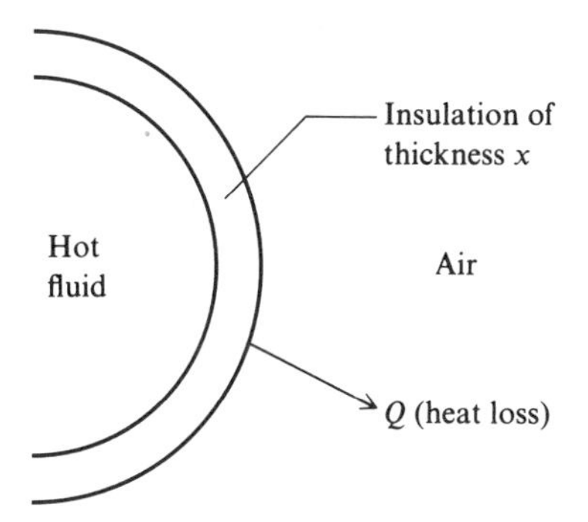

Figure E3.3 Heat loss from insulated pipe.

All of the above parameters in (*a*) are fixed values except for x, the variable to be optimized. Assume the cost of installed insulation per unit area can be represented by the relation $F_0 + F_1x$, where F_0 and F_1 are constants (F_0 = fixed installation cost and F_1 = incremental cost per foot of thickness). The insulation has a lifetime of five years and must be replaced at that time. The funds to purchase and install the insulation can be borrowed from a bank and paid back in five annual installments. Define r as the fraction of the installed cost to be paid each year to the bank. The value of r selected depends on the interest rate of the funds borrowed and will be discussed in Sec. 3.3. Alternatively, for use of internal funds, a "hurdle" rate of return comparable to a bank interest rate can be specified. The hurdle rate depends on a specified annual rate of return and annual depreciation charges (see Sec. 3.5).

Let the replacement value of the heat lost from the pipe be H_t ($\$/10^6$ Btu). Let Y be the number of hours per year operated.

1. Formulate an objective function to maximize the savings in operating cost due to heat conserved less the annualized cost of the insulation.
2. Obtain an analytical solution for x^{opt}.

Solution. If operating costs are stated in terms of dollars per year, then the capital costs must be stated in equivalent units. Since the funds required for the insulation are to be paid back in equal installments over a period of five years, the payment per year is $r(F_0 + F_1x)A$. The energy savings due to insulation can be calculated from the difference between $Q(x = 0)$, or Q_0, and Q:

$$Q_0 - Q = h_c \Delta T A - \frac{\Delta T A}{x/k + 1/h_c} \tag{b}$$

The objective function to be maximized is the value of heat conserved per year less the annualized capital cost (also in dollars per year):

$$f = (Q_0 - Q)\left(\frac{\text{Btu}}{\text{h}}\right) \cdot Y\left(\frac{\text{h}}{\text{year}}\right) \cdot H_t\left(\frac{\text{dollars}}{\text{Btu}}\right)$$
$$- (F_0 + F_1x)Ar\left(\frac{\text{dollars}}{\text{year}}\right) \tag{c}$$

Substitute Eq. (*b*) into (*c*), differentiate f with respect to x, and solve for the optimum ($df/dx = 0$):

$$x^{\text{opt}} = k\left\{\left(\frac{H_t Y \Delta T}{10^6 k F_1 r}\right)^{1/2} - \frac{1}{h_c}\right\} \tag{d}$$

You should examine how x^{opt} varies with the different parameters in (*d*) and confirm that the trends are physically meaningful. Note that the heat transfer area A does not appear in (*d*). Why? Could you formulate f as a cost minimization problem, i.e., the sum of the value of heat lost plus insulation cost? Does it change the result for x^{opt}? How do you use this result to select the correct commercial insulation size (see Example 1.1)?

3.2 CONSIDERATION OF THE TIME VALUE OF MONEY

In Example 3.3, the funds to purchase and install the insulation were to be repaid over a prescribed period of time. We used the concept of an annualization factor, or yearly capital charge, to obtain the capital costs on a yearly basis. In fact, if the interest rate and number of years for repayment are specified, then the annualization factor and a uniform payment schedule (equal installment payments) can be computed rather easily. However, money has a value which varies with respect to time. A payment of \$5000 now is not equivalent to a similar payment of \$5000 five years later. Payment in equal yearly installments (principal plus interest) is but one way the loan repayment in Example 3.3 might be structured. Another method might be annual payments of interest but no principal for several years, followed by a second period of higher payments consisting of principal plus interest. Given a large variety of ways in which to repay a loan, some method to analyze the effect of time on allowable payments is needed. For the general case in which income and costs vary each year, we must first define two terms: future worth and present worth.

Future Worth

Future worth is the value of an investment after a prescribed period of time, given some rate of appreciation (or depreciation). Consider the case of a savings account, in which the funds invested are allowed to increase each year by the addition of interest. Suppose the account earns a 6 percent annual interest rate and the interest is paid at the end of the year. We want to compute how the account value grows over time. If \$100 is invested initially, 6 percent of that \$100 will be paid as interest at the end of the year. Therefore, \$6 is added to the original \$100. During the second year no money is withdrawn, therefore, the \$106 is retained. An additional 6 percent interested computed on the \$106 makes the total value of the account at the end of the second year \$106 plus

0.06 times \$106 or a total of \$112.36. This compounding effect continues annually, and at the end of five years \$133.86 would be in the savings account.

It is easy to develop a general formula for investment growth for the case when fractional interest (i) is compounded once per year. (*Note:* On most occasions we will cite i in percent, as is the common practice, even though i is treated as a fraction.) If P is the original investment, then $P(1 + i)$ is the amount accumulated after one year. Using the same reasoning, the value of the investment in successive years for discrete interest payments is

$$t = 2 \text{ years} \qquad F_2 = P(1 + i) + iP(1 + i) = P(1 + i)^2 \tag{3.1a}$$

$$t = 3 \text{ years} \qquad F_3 = P(1 + i)^2 + iP(1 + i)^2 = P(1 + i)^3 \tag{3.1b}$$

$$t = n \text{ years} \qquad F_n = P(1 + i)^n \tag{3.1c}$$

The symbol F_i is called the *future worth* of the investment after year i, that is, the future value of a current investment P based on a specific interest rate.

Can you extend this formula to the case in which interest is compounded q times per year? In this case the multiplier for computing the future worth of an investment after n years is given by

$$F_n = P\left(1 + \frac{i}{q}\right)^{nq} \tag{3.2}$$

If a monthly interest rate is given ($= i/q$), then set $q = 12$ in the exponent in Eq. (3.2). For continuous compounding, let $q \to \infty$, and Eq. (3.2) reduces to

$$F_n = Pe^{in} \tag{3.3}$$

Present Worth

We can reverse the above analysis to determine the current value of a "cash flow" which occurs at some time in the future. Cash flow can result from income (positive cash flow) or an expense (negative cash flow). The current value is called the *present worth* of a cash flow. Suppose the value of the savings account will be \$150 five years from now. What is its corresponding present worth? To account for the influence of interest rate, the present value of an income available one year from now is reduced by factor of $(1 + i)$, where i is the annual interest rate; in a corporate setting, i is the average rate of return a company earns on its capital. Therefore, for a cash flow F_n occurring n years in the future, the present worth P corresponding to F_n can be calculated using the following formula, assuming interest is calculated annually:

$$P = \frac{F_n}{(1 + i)^n} \tag{3.4}$$

Note that Eq. (3.4) follows directly from Eq. (3.1c).

Suppose that an investment yields a constant cash flow equal to F every year for n years. Then, the total present worth of all of the cash flows is found by summing the present worths for each year, namely

$$P = \frac{F_1}{1+i} + \frac{F_2}{(1+i)^2} + \cdots + \frac{F_{n-1}}{(1+i)^{n-1}} + \frac{F_n}{(1+i)^n} \tag{3.5}$$

Let $F_i = F$, multiply (3.5) by $(1 + i)$, and subtract the resulting equation from (3.5) to get

$$-iP = -F + \frac{F}{(1+i)^n} \tag{3.6}$$

or

$$F = \frac{i(1+i)^n}{(1+i)^n - 1} \cdot P \tag{3.7}$$

Equation (3.7) can also be used to calculate the constant annual payment for repaying a loan. Given the prevailing rate of interest i, the number of years n, and the loan amount P, the annual repayment F can be computed. The fraction $(0 \le r \le 1)$

$$r = \frac{i(1+i)^n}{(1+i)^n - 1} \tag{3.8}$$

is called the repayment multiplier, and can be used to determine the annual capital charge $(=rP)$. The reciprocal of r is called the *series present worth factor.* The repayment multiplier is compiled in handbooks and some textbooks [e.g., Perry (1984), Jelen and Black (1983)]; a shortened version of the table is given in Table 3.1. Various calculators perform the analysis automatically.

Table 3.1 Values for the fraction $r = \dfrac{i(1+i)^n}{(1+i)^n - 1}$

n = number of years i = interest rate, %

Interest rate

n	$i \to 1$	2	4	6	8	10	12	14	16	18
1	1.010	1.020	1.040	1.060	1.080	1.100	1.120	1.140	1.160	1.180
2	0.507	0.515	0.530	0.545	0.561	0.576	0.592	0.607	0.623	0.639
3	0.340	0.347	0.360	0.374	0.388	0.402	0.416	0.431	0.445	0.460
5	0.206	0.212	0.225	0.237	0.251	0.264	0.277	0.291	0.305	0.320
10	0.106	0.111	0.123	0.136	0.149	0.163	0.177	0.192	0.207	0.222
15	0.072	0.078	0.090	0.103	0.117	0.132	0.147	0.163	0.179	0.196
20	0.055	0.061	0.074	0.087	0.102	0.117	0.134	0.151	0.169	0.187
25	0.045	0.051	0.064	0.078	0.094	0.110	0.128	0.145	0.164	0.183
30	0.039	0.045	0.058	0.073	0.089	0.106	0.124	0.143	0.162	0.181
40	0.030	0.037	0.051	0.067	0.084	0.102	0.121	0.141	0.160	0.180
50	0.026	0.032	0.047	0.063	0.082	0.101	0.120	0.140	0.160	0.180
75	0.019	0.026	0.042	0.061	0.080	0.100	0.120	0.140	0.160	0.180
100	0.016	0.023	0.041	0.060	0.080	0.100	0.120	0.140	0.160	0.180

If the interest is calculated continuously, rather than annually, merely substitute e^i for $1 + i$ in Eq. (3.7) [note that $(1 + i)$ is a first-order Taylor series approximation of e^i]. The resulting equation for F is

$$F = \frac{i(e^{in})}{e^{in} - 1} \cdot P \tag{3.9}$$

EXAMPLE 3.4 PAYMENT SCHEDULE

Suppose you obtain a \$100,000 mortgage on a house, agreeing to pay 10 percent interest and to repay the loan over 25 years. What is the annual payment you must make if the interest is compounded just once per year?

Solution. From Table 3.1, the repayment multiplier is 0.110, hence, the yearly payment is

$$F = (0.110)(\$100{,}000) = \$11{,}000/\text{year}$$

Check this value by using Eq. (3.7).

You should understand the relationship between the savings account and the loan repayment. The growth of principal and interest in a savings account can be thought of as a "one-time" repayment after n years of an initial investment. For the example of the house mortgage, a one-time repayment of an initial \$100,000 after 25 years would be given by Eq. (3.1c): $F_n = (1 + i)^n \cdot (\$100{,}000)$. For $i = 0.10$ and $n = 25$, $F_n = \$1{,}083{,}471$! Large numbers always result from compounding over a long period of time with interest rates of 10 percent or higher.

Rather than fixed annual cash flows, in the chemical industry you are more likely to experience a variable annual return from an investment. Suppose \$20 million is invested in a new plant facility. The incremental profit from that facility will tend to be lower in the early stages of its operating life, due to higher start-up expenses. Because of evolutionary improvements in operating conditions, annual profits should reach some peak after the initial start-up period is completed. However, in the later stages of the plant life, you might expect earnings to begin to decrease because of equipment replacement, equipment failures, extended shutdowns, lower sales, and so on. Therefore, the company's net cash flow profile will be variable with time although the profile may follow a predictable pattern.

Let us show how you might analyze a variable cash return over several years by means of a simple example. Suppose that you were asked to loan \$100 to an acquaintance. The acquaintance wants to repay the loan over a period of five years and is offering you the following payback schedule:

First year: \$50 Second year: \$25 Third year: \$25
Fourth year: \$15 Fifth year: \$10

Now you must determine if this loan is in your best interest, i.e., whether it will be more profitable to place the $100 in a rather secure savings account from which you might be able to earn, say, 8 percent interest over the five years. The concept of present worth can be employed to compare the two investment alternatives. (We shall ignore the risk of nonpayment from your acquaintance!)

The present worth for a series of unequal annual returns (at the end of each year) can be found from Eq. (3.5)

$$P = \frac{F_1}{1+i} + \frac{F_2}{(1+i)^2} + \frac{F_3}{(1+i)^3} + \frac{F_4}{(1+i)^4} + \frac{F_5}{(1+i)^5} \tag{3.10}$$

Assuming an 8 percent interest rate, you can compute the present worth for each year of the nonuniform payback schedule given above:

$$t = 1 \text{ year} \qquad P_1 = \frac{F_1}{(1+i)} = \frac{50}{1.08} = \$46.29$$

$$t = 2 \text{ years} \qquad P_3 = \frac{F_2}{(1+i)^2} = \frac{25}{(1.08)^2} = \$21.43$$

$$t = 3 \text{ years} \qquad P_3 = \frac{25}{(1.08)^3} = \$19.85$$

$$t = 4 \text{ years} \qquad P_4 = \frac{15}{(1.08)^4} = \$11.03$$

$$t = 5 \text{ years} \qquad P_5 = \frac{10}{(1.08)^5} = \$6.81$$

$$P = \sum_{k=1}^{5} P_k = \$105.41$$

The total present worth is the sum of the individual present worths. Compare this number with the amount which was originally invested ($100). The above calculation shows that the payback schedule proposed by your friend is satisfactory, because the present worth of the return on the investment is greate than the $100 originally invested. Would the same outcome be true for an interest rate of 10 percent?

An alternative method to evaluate the proposed loan versus investment is to find the value of i such that Eq. (3.10) is satisfied when $P = \$100$. This value of i is called the *internal rate of return* and is discussed in more detail in the next section, Sec. 3.3. The computed value of i can be compared with the prevailing institutional interest rate (say 8 percent) or a company's "hurdle" rate. If the computed rate of return is higher than other alternatives (in this case, 8 percent) and the investment risk is acceptable, then the loan under consideration is an attractive one.

The value of i that satisfies Eq. (3.10) when $P = \$100$ must be found by iterative calculations, because this equation cannot be solved explicitly for i.

Equation (3.10) has a fairly smooth behavior so that interpolation and extrapolation are feasible; consider three trial values of i and calculate P for each value:

i	P
0.06	109.76
0.08	105.41
0.10	101.36

Based on the trend, we can project i to be approximately 0.106, which yields $P = \$100.18$. This means that the payback schedule is equivalent to an interest rate of 10.6 percent. To solve for i in Eq. (3.10) a scheme based on interpolation-extrapolation can be carried out on a calculator or computer; alternatively, the Newton-Raphson iterative method can be employed to solve (3.10).

3.3 MEASURES OF PROFITABILITY

In the chemical process industries, potential projects are evaluated using some measure of profitability. The projects with highest priorities are the ones with the highest expected profitability; "expected" implies that probabilistic considerations must be taken into account (Hecty, 1968). In this section, however, we are concerned with a deterministic approach for evaluating profitability, keeping in mind that different definitions of profitability can lead to different priority rankings. In the context of optimization, the objective function is the chosen profitability measure. We shall use the term *cash flow* in these measures; a cash flow in a given year is defined as the sum of the net profit after taxes are deducted plus depreciation (see Sec. 3.5). The cash flow is presumed to be available for various uses.

The simplest profitability criteria do not take into account the "time value" of money, such as present worth. Rather, they take an integrated view of profits over a period of several years. The *payback period* PBP (in years) is one criterion, defined as

$$\text{PBP} = \frac{\text{depreciable fixed investment}}{\text{average annual cash flow}} \tag{3.11}$$

PBP is the minimum length of time necessary to recover the original fixed capital investment in the form of cash flow from the project. Note that there is no weighting of earlier cash flows versus later cash flows, and there is no consideration of project earnings after the initial investment has been recovered. On the other hand, no assumptions regarding equipment life or the appropriate interest rate for capital use need be made. A related profitability criterion is the return on investment (ROI) or the "profit rate," which is the ratio, expressed as a percentage, of the net income after taxes to the sum of fixed capital investment plus working capital. In project selection we seek to maximize ROI.

More popular profitability measures take into account variable cash flows and the time value of money (Woods, 1975; DeGarmo and Canada, 1973). Such measures include (1) the internal rate of return (IRR) discussed in the previous section, also referred to as discounted cash flow, and (2) the net present value (NPV). Alternatively, we may ignore taxes and compute a before-tax cash flow. The advantages of using the internal rate of return are:

1. No assumption as to the interest rate to be applied to capital is required—it is the unknown to be calculated.
2. The internal rate of return takes into account the time value of cash flows occurring at different points in time and for different equipment lives as well as for investments at different times.

The disadvantage of using the internal rate of return is the need to make assumptions about equipment life and (perhaps) the iterative calculations required to determine i [see Eq. (3.10)]. Keep in mind that in calculating i you assume that the cash generated by the project will continue to earn at the same rate of return as the project did originally. According to Brigham (1982), this may be an unrealistic assumption.

The net present value (NPV) of a project is defined as

$$\text{NPV} = \sum_{j=1}^{n} \frac{F_j}{(1+i)^j} - I_0 \tag{3.12}$$

where i is the stated interest rate for capital, F_j are the yearly cash flows over n years (assumed to occur in one "lump" at the end of each year), and I_0 is the initial investment (made at $j = 0$). NPV differs from IRR in that i is specified a priori and not calculated; when NPV $= 0$, you solve Eq. (3.12) to find i, which is the IRR. An appropriate index for ranking the profitability of various alternative projects is $\Pi = \text{NPV}/I_0$. If several investments I_j are made over several years, then

$$\text{NPV} = \sum_{j=1}^{n} \frac{F_j}{(1+i)^j} - \sum_{j=0}^{n-1} \frac{I_j}{(1+i)^j} \tag{3.13}$$

In this case, the internal rate of return is the value of i that makes NPV $= 0$ in (3.13), i.e., the sum of the discounted cash inflows equals the sum of the discounted cash outflows. We use the convention that I_j is an investment made at the end of year j. Given n, F_j, and I_j, Eq. (3.13) must be solved iteratively for the IRR.

In Eq. (3.13) you should note that there is no provision for accumulating funds to replace equipment during the cash flow generating period. Furthermore, if you simply generate cash flows which are merely equal to the investment, then the internal rate of return is zero, i.e., for an initial investment only, $\sum_{j=1}^{n} F_j = I_0$, then $i = 0$. The question of equipment depreciation and its effect on yearly cash flows is discussed later on in this chapter.

Table 3.2 Formulas for evaluating profitability

	Lumped initial investment, I_0 and constant cash flow, F	Distributed investments, I_j, and variable cash flows, F_j
Payback period, PBP, years	$\text{PBP} = \frac{I_0}{F}$	$\text{PBP} = \frac{\sum_{j=0}^{n-1} I_j}{\sum_{j=1}^{n} F_j}$
Return on investment (ROI)	$\text{ROI} = \frac{\text{NI}}{I_0}$	
Internal rate of return (IRR), discounted cash flow	$i = \frac{F}{I_0} - \frac{i}{(1+i)^n - 1}$ (*Solve for* i = IRR)	$\sum_{j=1}^{n} \frac{F_j}{(1+i)^j} - \sum_{j=0}^{n-1} \frac{I_j}{(1+i)^j} = 0$
Net present value (NPV), \$	$\text{NPV} = F\left[\frac{(1+i')^n - 1}{i'(1+i')^n}\right] - I_0$	$\text{NPV} = \sum_{j=1}^{n} \frac{F_j}{(1+i')^j} - \sum_{j=0}^{n-1} \frac{I_j}{(1+i')^j}$

Nomenclature

1. i' is the interest value of money in NPV, generally taken as the opportunity interest that the company must forgo by not investing in the next best alternative. i is the internal rate of return. ROI, i', and i are fractions; to obtain %, multiply by 100.
2. n is the total number of time periods (normally years) between start-up (which is time zero) and end of operation of the equipment; j is a particular time period in the project life.
3. F_j = cash flow (\$/year) taken at the end of time period j; I_j = investment (\$) taken at the end of time period j; I_0 is initial investment; NI = net income after taxes (\$/year).

Tables 3.2 and 3.3 summarize the formulas, advantages, and disadvantages of the different methods. There is an active computer software market for spreadsheets to calculate profitability, especially for multiyear projects; perhaps the best known program is LOTUS 1-2-3.

EXAMPLE 3.5 CALCULATION OF PAYOUT TIME, RATE OF RETURN, AND NET PRESENT VALUE

Two alternative projects are under consideration. Project A has a project life of 10 years and requires an initial investment of \$100,000 with an annual cash flow after taxes of \$20,000 per year for each of five years followed by \$10,000 per year for years 6 through 10. Project B has a life of 10 years and requires the same investment but has cash flows of \$15,000 per year for each year. Evaluate projects A and B using (a) payout period, (b) internal rate of return, and (c) net present value for an interest rate of 10 percent ($i = 0.10$).

Solution. The payout period for project A is (\$100,000)/(\$20,000) = 5 years, while it is (\$100,000)/(\$15,000) = 6.67 years for project B. Applying Eq. (3.12) for NPV = 0 and $n = 10$, we find that the IRR (project A) is 10 percent, while for B the IRR is 8 percent. This comparison is predictable because project A generates more profit earlier in its lifetime. For $i = 0.10$, the NPV of project A is $-\$640$, while that for

Table 3.3 Comparisons of various methods used in economic analyses

Payback period (PBP)	Net present value (NPV)	Internal rate of return (IRR)
	Definition	
Number of years for the net after-tax income to recover the net investment without considering time value of money	Present worth of receipts less the present worth of disbursements	IRR = interest rate i such that NPV of receipts less NPV of disbursements equals zero
	Advantages	
1. Measure of fluidity of an investment 2. Commonly used and well understood	1. Works with all cash flow patterns 2. Easy to compute 3. Gives correct ranking in most project evaluations	1. Gives rate of return which is a familiar measure and indicates relative merits of a proposed investment 2. Treats variable cash flows 3. Does not require reinvestment rate assumption
	Disadvantages	
1. Does not measure profitability 2. Neglects life of assets 3. Does not properly consider the time value of money and distributed investments or cash flows	1. It is not always possible to specify a reinvestment rate for capital recovered 2. Size of NPV ($) sometimes fails to indicate relative profitability	1. Implicitly assumes that capital recovered can be reinvested at the same rate. 2. Requires trial-and-error calculation 3. Can give multiple answers for distributed investments

project B is $-\$7840$, hence NPV($A$) > NPV($B$), which is consistent with the other two profitability criteria examined above. In fact, we can show that for all reasonable values of i, NPV(A) > NPV(B). However, the negative values for NPV for $i > 0.10$ would discourage investment in either project A or B.

If the project lives of A and B differed, the analysis would become somewhat more complicated for IRR. Suppose project A has a 10-year life and B now has a 20-year life. Some textbooks recommend that in the rate-of-return analysis, you can operate two successive 10-year projects (A) and compare profitability with the 20-year project (B) so that the total project lives are equivalent. However, project A may only have predictable profits for 10 years; after that time lower profits may result, hence merely implementing it twice may not be meaningful. NPV does not require that the total lives (or multiples thereof) be equal for comparison. Thus, ambiguous and sometimes contradictory results can arise in using IRR vs. NPV [see Brigham (1982)]. Jelen and Black (1983) have suggested a comparison based on uniform annual cost, called *unacost*.

EXAMPLE 3.6 CALCULATION OF IRR AND NPV FOR PROJECTS WITH DIFFERENT LIFETIMES

Suppose project C has a 20-year life and a yearly after-tax cash flow of \$48,000 for an initial investment of \$300,000. Project D has a 5-year life, with a yearly cash flow of \$110,000 for an initial investment of \$300,000. Compare the internal rate of return and net present value ($i = 0.08$) for each option.

Solution. Because the annual cash flows are all equal for C as well as for D, we can apply Eq. (3.7). The internal rates of return are $i_C = 0.15$ for project C and $i_D = 0.25$ for project D. The advantage of project D is a more concentrated period of

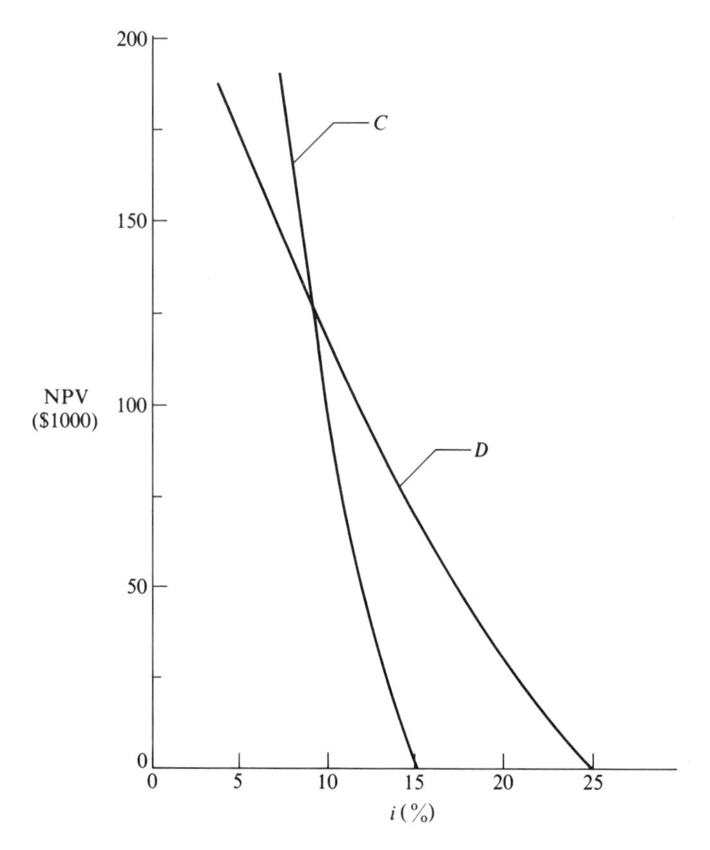

Figure E3.6 Comparison of the net present value (NPV) for two projects as a function of i.

cash generation at a higher level. For a value of $i = 0.08$, the NPV of each project is as follows:

$$\text{NPV}(C) = \left(\sum_{j=1}^{20} \frac{48{,}000}{(1+i)^j} \right) - 300{,}000$$

$$= 470{,}600 - 300{,}000 = \$170{,}600$$

$$\text{NPV}(D) = \left(\sum_{j=1}^{5} \frac{110{,}000}{(1+i)^j} \right) - 300{,}000$$

$$= 438{,}200 - 300{,}000 = \$138{,}200$$

Therefore, based strictly on this calculation, project C would be favored over D because over its lifetime (20 years versus 5 years), it would generate more (discounted) cash flow. However, this conclusion is in conflict with that obtained by comparing the IRR's of the two projects. The ranking based on NPV may change if a different interest rate is assumed. Figure E3.6 shows how NPV varies for each project as a function of i (note the crossover point). Brigham (1982) has concluded that the use of NPV is preferable to IRR, because NPV gives more realistic results for a wide variety of cases, especially when cash flows vary greatly from year to year.

3.4 OPTIMIZING PROFITABILITY

In order to optimize profitability, you first need to determine the independent variables which influence

(*a*) Capital investment
(*b*) Income
(*c*) Operating costs
(*d*) Cash flow

Assuming there are p independent variables, denote (*a*) as $I(x_1, x_2, \ldots, x_p)$ and the cash flow $F(x_1, x_2, \ldots, x_p)$ is obtained from (*a*), (*b*), and (*c*).

Next you need to define the specific profitability measure, such as payback time, return on investment, net present value, or internal rate of return. The first three objective functions are *explicit* functions of F and I. However, IRR is an *implicit* function of F and I; once F and I are specified, an iterative solution must be used to find i [see Eq. (3.7)]. In order to maximize the IRR, you must vary the independent unknowns and compute I and F for each trial set of $(x_1, x_2, \ldots, x_p)$. Next solve Eq. (3.7) for i and continue the procedure until you find the value of $\mathbf{x}$ which maximizes the IRR. Hence the solution of Eq. (3.7) is imbedded in the iterative optimization procedure. For example, Fig. 3.1 shows a plot of IRR vs. plant capacity, where the optimum occurs at 125 million pounds per year. For the more general case when F_i is not the same for each year, Eq. (3.5) must be used in conjunction with the optimization procedure.

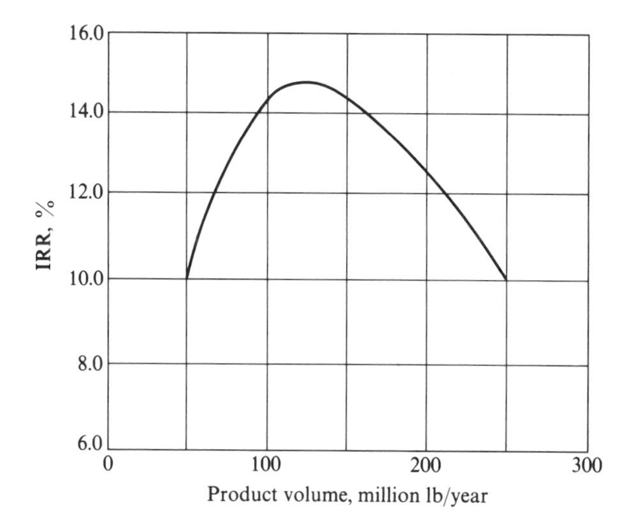

Figure 3.1 Variation of IRR with plant capacity; maximum return in this case is 14.8% at a volume of 125 million pounds per year.

Maximization of the rate of return on investment is one alternative to maximizing IRR, and it has the advantage of being an explicit performance criterion. Use of this profitability measure, ROI $= F/I_0$, is feasible only when F is constant each year. The use of ROI is a reasonable approximation to IRR, at least for the purposes of optimization, since the formula ROI $= F/I_0 = [i(1 + i)^n]/[(1 + i)^n - 1]$ is well approximated by a relation mentioned by Happel and Jordan (1975): ROI $\approx 1/n + 0.65i$. The approximation is most accurate for $5 \leq n \leq 20$ and implies that ROI is proportional to i (or IRR). Therefore maximizing ROI appears to be a reasonable alternative. However, as shown in the next example, an objective function based on the return on investment may lead to unexpected optima.

EXAMPLE 3.7 OPTIMIZATION OF HEAT TRANSFER SURFACE TAKING INTO ACCOUNT THE TIME VALUE OF MONEY

We want to optimize the heat transfer area of a steam generator. A hot oil stream from a reactor needs to be cooled down, providing a source of heat for steam production. As shown in Fig. E3.7, the hot oil enters the generator at 400°F and leaves at an unspecified temperature T_2; the hot oil transfers heat to a saturated

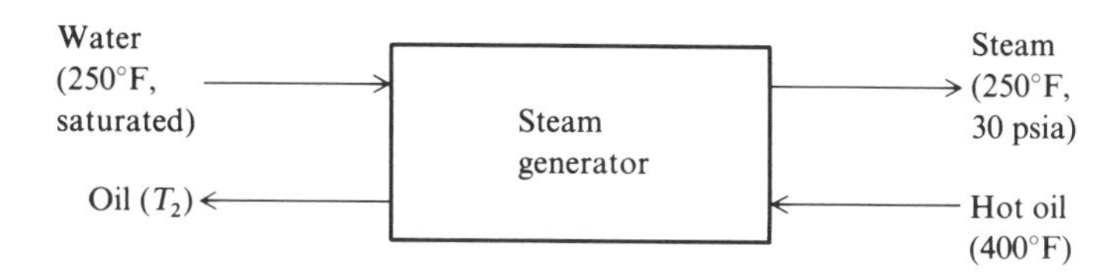

Figure E3.7 Steam generator flow diagram, Example 3.7.

liquid water stream at 250°F, yielding steam (30 psia, 250°F). The other operating conditions of the exchanger are

$$U = 100 \text{ Btu/(h)(ft}^2\text{)(°F)} \qquad \text{overall heat transfer coefficient}$$

$$w_{\text{oil}} C_{p_{\text{oil}}} = 7.5 \times 10^4 \text{ Btu/(°F)(h)}$$

We neglect the cost of the energy of pumping and the cost of water, and only consider the investment cost of the heat transfer area. The heat exchanger cost is $25 per square foot of heat transfer surface. You can expect a credit of $2/10^6 Btu for the steam produced. Assume the exchanger will be in service 8000 h/year. Find the outlet temperature T_2 and heat exchanger area A which maximize the return on investment ROI, as measured by F/I_0.

Solution. In this problem recognize that an exchanger of infinitely large area will maximize the energy recovery in the stream but at an exorbitant cost. Hence we expect there to be a trade-off between capital cost and energy savings. The variables to be optimized include T_2 and A as well as the amount of steam generated, w_{steam}. First determine if any equality constraints exist in the problem. The energy balance for the steam generator is

$$w_{\text{oil}} C_{p_{\text{oil}}}(400 - T_2) = \frac{UA(400 - T_2)}{\ln [150/(T_2 - 250)]} \tag{a}$$

or

$$w_{\text{oil}} C_{p_{\text{oil}}} = \frac{UA}{\ln [150/(T_2 - 250)]} \tag{b}$$

The water converted to steam is obtained from

$$w_{\text{oil}} C_{p_{\text{oil}}}(400 - T_2) = \Delta H_v w_{\text{steam}} \tag{c}$$

where $\Delta H_v = 950$ Btu/lb and $w_{\text{steam}} =$ lb/h. Recognize that Eq. (*b*) relates the variables T_2 and A, hence they are not independent. In addition, Eq. (*c*) relates T_2 and w_{steam}. Therefore we can express all costs in terms of T_2 with the aid of Eqs. (*b*) and (*c*). The capital cost is (dropping the "oil" subscript):

$$I_0 = (25)(A) = \frac{25wC_p}{100} \ln \left(\frac{150}{T_2 - 250} \right) \tag{d}$$

The annual credit for the value of the steam is

$$F = \left\{ 2\text{x}10^{-6} \left[\frac{\$}{\text{Btu}} \right] \right\} \left\{ wC_p(400 - T_2) \left[\frac{\text{Btu}}{\text{h}} \right] \right\} \left\{ 8000 \frac{\text{h}}{\text{year}} \right\}$$

$$F = [0.016] \left[wC_p(400 - T_2) \frac{\$}{\text{year}} \right] \tag{e}$$

Note that wC_p for the oil appears in the expressions for both F and I_0 and thus cancels. The profitability ratio is therefore

$$\text{ROI} = \frac{F}{I_0} = \frac{0.064(400 - T_2)}{\ln [150/(T_2 - 250)]} \tag{f}$$

The maximum value of ROI must be found numerically because of the complicated expressions appearing on the right-hand side of (f). The optimum is at $T_2 = 400°F$, which is the same temperature as at the inlet, corresponding to $A \to 0$. At the optimum an extremely high rate of return occurs ($r = 9.6$), which can be found by applying L'Hopital's rule to the above expression for ROI when $T_2 = 400°F$. This outcome, of course, is an unrealistic answer, since it suggests the optimum return consists of an exchanger with infinitesimal area! Why does this result occur? The difficulty with using ROI as an objective function is that nothing in Eq. (f) constrains the area to be above a minimum size; in fact, as $T_2 \to 400°$, the investment I_0 is decreasing faster than is the numerator, leading to a maximum value at $T_2 = 400°$. If $T_2 > 400°$, the rate of return becomes negative.

From the above example, you can see that the ratio of F/I_0 may yield unrealistic results for an optimum. This occurs here because $I_0 \to 0$ for $T_2 \to 400$. Consider reformulating the problem using the net present value (NPV) of before-tax profits as an alternative objective function. Use of NPV means that a rate of return on the capital is specified. We follow the same approach used to derive Eq. (3.7); Eq. (3.12) for a constant annual cash return ($F_j = F$) over n years gives

$$\text{NPV} = \left\{\frac{(1+i)^n - 1}{i(1+i)^n}\right\}F - I_0 \tag{3.14}$$

or

$$\text{NPV} = \frac{F}{r} - I_0 \tag{3.15}$$

where r is given by Eq. (3.8). Since r is fixed by the assumptions about i and n, an equivalent criterion is

$$r \cdot \text{NPV} = F - rI_0 \tag{3.16}$$

Note that this modified objective function ($r \cdot \text{NPV}$) is equivalent to the use of the annualization factor (repayment multiplier) to obtain the capitalization charge as applied in Example 3.6. In problems in which you seek to minimize only costs rather than maximizing profit because there is no stated income, then F in Eq. (3.16) is negative. An example arises in optimizing pipe size to minimize pump operating costs and pipe investment costs. Instead of maximizing $r \cdot \text{NPV}$, you can minimize ($-r \cdot \text{NPV}$).

EXAMPLE 3.8

Let us return to Example 3.7, and use the net present value analysis to determine the optimum value of T_2. Assume an interest rate for capital of 15 percent and a period of 10 years.

Solution. The objective function for net present value (to be maximized with respect to T_2) is

$$f = F - rI_0$$

$$= 2\text{x}10^{-6}\, wC_p(400 - T_2)(8000) - r \cdot 25 \cdot A \text{ (\$/year)} \qquad (a)$$

Elimination of A in terms of T_2 [Eq. (b) of Example 3.7] gives:

$$f = (0.016)wC_p(400 - T_2) - 25r\frac{wC_p}{U}\ln\frac{150}{T_2 - 250} \qquad (b)$$

Note that wC_p is a common factor of the two terms and will not be involved in calculating T_2. We can differentiate Eq. (b) and set $df/dT_2 = 0$:

$$\frac{df}{dT} = 0 = \frac{25r}{100}\frac{1}{T_2 - 250} - 0.016 \qquad (c)$$

$$T_2 = 250 + 15.62r \qquad (d)$$

If $r = 0.2$ ($n = 10$, $i = 15$ percent in Table 3.1), then $T_2 = 253.1°\text{F}$, a 3.1° approach (somewhat lower than the normal approach temperatures of 5 to 10°F recommended in design manuals). The optimal approach temperature, according to the analysis in this example, depends on U, r, and the ratio of the value of steam to the cost-per-unit area for the heat exchanger.

To calculate the annual profit before taxes, we compute the value of $F = (2 \times 10^{-6})(wC_p)(400 - T_2)(8000)$, which would be \$176,280. The optimum value of A is 2905 ft^2, so the original investment is \$72,625. The payout is, therefore, less than one year. Remember that while higher values of ROI can be obtained by selecting T_2 closer to 250°F, maximization of ROI leads to the meaningless solution obtained previously.

While the rate of return on investment (F/I_0) did not lead to meaningful results in Example 3.7, there are some conditions under which this criterion can be employed effectively to obtain a reasonable value for the optimum. For example, if the heat transfer area costs were assumed to be $I_0 = I'_0 + 25A$ (I'_0 is the fixed installation cost for the exchanger), then maximizing F/I_0 would yield a more realistic result for T_2. Note that at $T_2 = 400°\text{F}$, ROI = 0.0, rather than 9.6 obtained earlier for Eq. (f) in Example 3.7. Another case which gives a meaningful answer for ROI occurs when several projects are considered simultaneously with a constraint on total capital expenditure. If \$100 million is to be invested among three projects, then an overall rate of return for the three projects, defined as $(F^1 + F^2 + F^3)/(I^1 + I^2 + I^3)$, can be formulated. The optimum, when calculated, should be meaningful because it is guaranteed that \$100 million will be committed to plant investment. In fact, $I^1 + I^2 + I^3$ in this case is a constant value (\$100 million), hence we simply optimize $F^1 + F^2 + F^3$.

Decisions made on the basis of the internal rate of return often favor investment in smaller facilities rather than large plants because the ratio of profit to investment is optimized. The heat exchanger example considered above is a case in point. Adelson (1968), in comparing the NPV-vs.-IRR analysis, stated that plant capacities obtained by optimizing NPV typically are larger than those obtained by optimizing IRR. Low capital, high maintenance projects are favored by maximizing IRR. However, most of the differences between IRR and NPV analysis occur for cash-flow profiles which vary radically over time. For most industrial optimization problems, fairly uniform cash flows occur (e.g., the same operating costs per unit of production each year), so that optimizing either NPV or IRR should be acceptable. The choice in each analysis will be based on company policy and/or the need for an explicit rather than implicit objective function.

3.5 PROJECT FINANCIAL EVALUATION

In the previous section we only briefly mentioned the financial assumptions used in profitability analysis. Any detailed evaluation of a project requires specifying the following parameters:

1. Initial investment
2. Future cash flows
3. Salvage value
4. Economic life
5. Depreciation
6. Depletion
7. Tax credit
8. Taxes
9. Inflation
10. Debt/equity ratio

The earlier parts of this chapter have focused on items 1 and 2, using before-tax returns or cash flows to measure profitability. However, factors 3 to 10 above can have a significant influence on profitability. They are discussed below, along with comments on how the objective function (profitability) is affected. You should recognize that many of these factors are subject to government regulation via the tax code; hence they are subject to change. You are strongly advised to obtain most recent tax information prior to carrying out project evaluations.

Salvage Value

Salvage value is the price that can be actually obtained or is imputed to be obtained from the sale of used property if, at the end of its usage, the equipment (property) still has some utility. Salvage value is influenced by the current cost of equivalent equipment, its commercial value, whether the equipment must be dismantled and relocated to have utility for others, and the (projected) physical condition of the equipment. Salvage value can be thought of as a cash flow that may occur several years in the future and thus would be heavily discounted (in terms of present worth). However, salvage value is not a consideration in several depreciation methods, such as ACRS or declining balance (see below).

Economic Life

The economic life of equipment is also known as the service life (Jelen and Black, 1983), and data on economic lives have been tabulated for various types of equipment by manufacturing associations as well as the U.S. Internal Revenue Service (IRS). The estimate of the economic life of equipment has a major impact on depreciation calculations, which in turn influence profits and taxes. The IRS has issued several publications over the years detailing average lives as well as lower and upper limits for manufacturing equipment; see Peters and Timmerhaus (1980) and Internal Revenue Service Publication 456 (1971). The shorter the economic life, the faster a piece of equipment can be depreciated, which generally improves cash flow by reducing income taxes paid (see discussion on depreciation and taxes below). For economic lives in the 10- to 20-year range, profitability is not very sensitive to the specific value of the equipment or property life. However, when $n \leq 7$, the economic life becomes a more important variable in profitability analysis.

Depreciation

Depreciation is a measure of the decrease in the value of equipment and improvements over a period of time. This decrease in value results from wear and deterioration, and inadequacy of the equipment relative to newer designs or technological advances (so-called functional depreciation or obsolescence). Because all of these factors are difficult to quantify, standard formulas for depreciation as a function of time have been developed. Depreciation reduces operating profits by deducting the costs associated with the recovery of capital that has been invested in physical equipment. Depreciation is prorated or assigned in various ways and makes up part of the total cost of production of a given product or products as do material and labor costs.

Depreciation is an allowable cost in the computation of income taxes, and hence affects after-tax profits. For example, for a $10,000 investment for equipment with a life of five years, you could choose to deduct $2000 each year for five years from the operating income. Note, however, that such a deduction does not imply a deposit of cash into an equipment-replacement fund in the amount of the depreciation. Depreciation is merely a noncash charge that reduces operating income, and hence income taxes, through the imputed annualized cost of the capital investment. The federal government has definite rules and regulations concerning how depreciation may be determined. Because these regulations change somewhat from year to year, new project evaluations should be based on the most recent regulations. Revisions in the income tax laws are often instituted with the express purpose of making capital investment more attractive by yielding a higher rate of return.

Historically, the commonly used depreciation methods include:

1. Straight line
2. Declining balance
3. Sum of year's digits
4. Accelerated Cost Recovery System

plus combinations of the above methods. We discuss the basic methods below. More details on depreciation methods can be found in the books by Happel and Jordan (1975), Jelen and Black (1983), and Peters and Timmerhaus (1980).

Straight-Line Depreciation

In the straight-line (SL) method for determining depreciation, it is assumed that the equipment value declines linearly with respect to time. The annual depreciation cost ($\bar{d}$) is

$$\bar{d} = \frac{I_0 - S_v}{n} \tag{3.17}$$

where I_0 = capital investment (in dollars), S_v = salvage value (in dollars), and n = economic life (years). The book value of the equipment can be found for any year j as $I_0 - j\bar{d}$. For example, if the investment I_0 = \$10,000 and the salvage value S_v = \$1000, the annual depreciation for an asset with a 5-year life is 9000/5 = 1800.

Declining-Balance Depreciation

The declining-balance (DB) method allows you to take more depreciation early in a project than in its later stages; however, no salvage value is allowed ($S_v = 0$). The number of years over which the item is to be depreciated is computed. The reciprocal of the number of years is called the *depreciation factor*. In the first year, you take the fraction of depreciation equal to the depreciation factor. In the second year, however, you multiply the remaining balance to be depreciated by the depreciation factor. For example, if the depreciation of an asset is estimated to be \$10,000, and the life is 5 years, the first-year depreciation would be $\frac{1}{5}$(\$10,000) = \$2000 and the second year of depreciation would be $\frac{1}{5}$(\$10,000 − \$2000) = \$1600. In succeeding years you again compute the remaining amount to be depreciated and then calculate the specific year's depreciation based on that figure. Double declining balance (DDB) interpreted literally means that the depreciation factor described above is multiplied by two. Therefore, you are able to take twice as much depreciation as would be available with a standard declining balance method. In either method (DB or DDB), no salvage value is used ($S_v = 0$).

For a depreciation factor of d_f and initial investment I_0, the book value V_j of the asset after j years is

$$V_j = I_0(1 - d_f)^j \tag{3.18}$$

The cumulative amount of depreciation taken after j years is, therefore,

$$D_j = I_0[1 - (1 - d_f)^j] \tag{3.19}$$

and the incremental amount of depreciation taken in year j is $I_0 d_f(1 - d_f)^{j-1}$. General formulas which allow you to calculate the depreciation schedule as a function of the lifetime of an asset for both declining balance and double declining balance have been given by Happel and Jordan (1975).

Increased depreciation in the beginning of a project translates into reduced income taxes at that time. In order to maximize the present worth of the profits, income tax reductions should be taken as early as possible. You should recognize that less depreciation can be taken in later years, but future incremental income is discounted more heavily in the present worth formula. Of course, too much depreciation can change profits into losses, which may not be desirable.

The strict use of DDB will not allow 100 percent of the possible depreciation to be taken, hence, it is permissible to modify the declining-balance method to eliminate this disadvantage by switching the depreciation method to straight-line after a period of time, say one-half of the project life [Happel and Jordan (1975)]. In this case, you may use $S_v \neq 0$.

Sum of the Years Digits

Another possible depreciation method is sum-of-the-years-digits (SYD) method. In this method you set the number of years (n) over which the asset is to be depreciated. You then calculate the sum of the digits running from 1 to n, $1 + 2 + \cdots + n$, which is equal to $[n(n + 1)]/2$. Therefore, the fraction of depreciation taken in any year j will be given by $[n - j + 1]/[n(n + 1)/2]$ and the cumulative depreciation fraction remaining after j years is $(n - j)(n - j + 1)/[(n)(n + 1)]$. For a five-year life, the sum of the digits is 15. The first year's depreciation is 5/15 of $(I_0 - S_v)$, the second year's depreciation is 4/15, etc. One advantage of SYD over the declining-balance methods is that this method permits taking all of the possible depreciation during the lifetime of the asset. However, as mentioned above, one variant of the double-declining-balance method allows a 100 percent write-off by utilizing DDB depreciation over part of the life of the asset and straight-line depreciation for the remainder of the life.

ACRS Depreciation

The Economic Recovery Act of 1981 made major changes to the income tax laws and depreciation rules. In this Act, the IRS set up a new system of depreciation to be used for all assets acquired after January 1, 1981, known as the "Accelerated Cost Recovery System" (ACRS). This act was modified in 1986, returning to straight-line and declining-balance calculations in 1987. Effectively all assets were placed in categories, according to their economic life.

For the period of 1981–1985, the depreciation schedule was as follows for equipment pertinent to manufacturing:

Year	3-year	5-year	10-year
1	25%	15%	8%
2	38%	22%	14%
3	37%	21%	12%
4	....	21%	10%
5	....	21%	10%
6	....	....	10%
7	....	....	9%
8	....	....	9%
9	....	....	9%
10	....	....	9%

You will note that the above figures are not much different from straight-line depreciation. An asset acquired at any time during the year qualifies for a full year's depreciation allowance. The rates for the 15-year class of properties are a bit more complicated and can be found in the IRS Publication 534.

The 1986 revision to ACRS involves 8 cost recovery classes ranging from, 3 to 31.5 years. R and D equipment was classified in the 5-year class, and chemical process equipment in the 7-year class, with straight-line depreciation applicable to both types of equipment.

Figure 3.2 illustrates the rate of depreciation taken for the four methods discussed here. Table 3.4 lists the fractional depreciation schedule for each of the methods and compares the present worth of depreciation taken during any given year by each method (the standard DB method is not presented here since

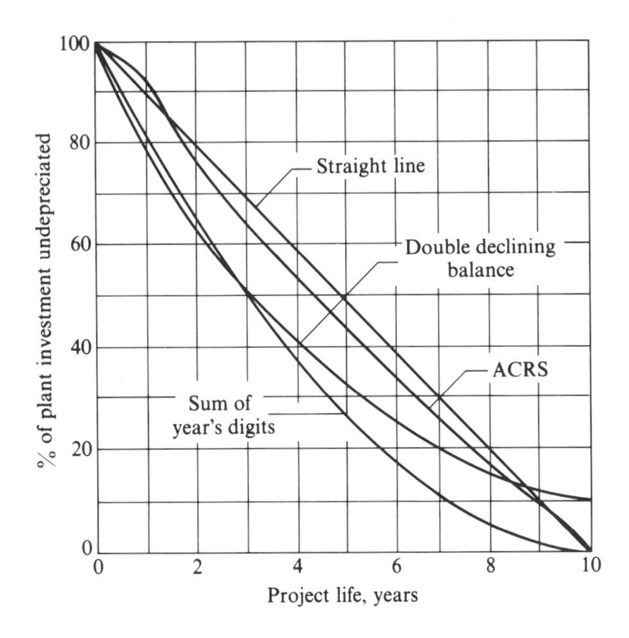

Figure 3.2 Comparison of depreciation methods with zero salvage value.

Table 3.4 Comparison of depreciation deductions as fraction of investment

($n = 10$; $i = 0.10$; $S_v = 0$)

Year	Straight line		Double declining balance		Combination straight line and declining balance		Sum of the digits		ACRS	
	Annual deduction	Present worth	Annual deduction	Present worth	Annual deduction	Present worth	Annual deduction	Present worth	Annual deduction	Present worth
1	0.10	0.0909	0.2000	0.1820	0.2000	0.1820	0.1820	0.1650	0.08	0.0727
2	0.10	0.0826	0.1600	0.1320	0.1600	0.1320	0.1634	0.1350	0.14	0.1157
3	0.10	0.0751	0.1280	0.0961	0.1280	0.0961	0.1454	0.1095	0.12	0.0902
4	0.10	0.0683	0.1020	0.0696	0.1020	0.0696	0.1272	0.0870	0.10	0.0683
5	0.10	0.0621	0.0820	0.0509	0.0820	0.0509	0.1090	0.0677	0.10	0.0621
6	0.10	0.0565	0.0656	0.0370	0.0656	0.0370	0.0910	0.0513	0.10	0.0564
7	0.10	0.0513	0.0524	0.0269	0.0656	0.0336	0.0728	0.0372	0.09	0.0462
8	0.10	0.0467	0.0420	0.0196	0.0656	0.0306	0.0546	0.0255	0.09	0.0420
9	0.10	0.0424	0.0336	0.0142	0.0656	0.0278	0.0364	0.0154	0.09	0.0382
10	0.10	0.0386	0.0268	0.0104	0.0656	0.0254	0.0182	0.0070	0.09	0.0347
Total	1.00	0.6145	0.8924	0.6387	1.000	0.6850	1.000	0.7006	1.000	0.6265

it obviously is inferior to the DDB method). Table 3.4 also shows the present worth ($i = 0.10$) of the cumulative sum of the depreciation deductions for each year. A higher present worth corresponds to lower income taxes in the earlier part of a project, which in turn improves the cash flow. Table 3.4 shows that for a 10-year life and the interest rate of 10 percent the SYD method is superior to even the combined double-declining balance plus straight-line (taken during the last four years of asset life). ACRS is slightly better than SL.

It is important that you do not generalize upon the results in Table 3.4, because for other interest rates and for short project lives, the double-declining-balance-plus-straight-line method may be superior in present worth to the sum-of-the-years-digits method. The combined-depreciation method presents an optimization problem in itself, namely, at what point in the calculations do you switch to the straight-line method from the double-declining-balance method? Intuitively, it would seem that if during the first year a straight-line approach yielded more depreciation than a declining-balance method, the methods should be changed. However, this transition requires more thorough examination. See Happel and Jordan (1975) for a discussion on the optimization of the change-over point.

EXAMPLE 3.9 COMPARISON OF DEPRECIATION METHODS

A piece of capital equipment costs \$6000 and has a service life of three years, and no salvage value. Compute the depreciation schedules using the following methods: SL, DDB, SYD, and ACRS. Compare the present worths of each schedule for $i = 0.07$ and $i = 0.20$.

Solution. The depreciation schedules are as follows:

Year	SL	DDB	SYD	ACRS
1	2000	4000	3000	1500
2	2000	1333	2000	2280
3	2000	444	1000	2220

For $i = 7$ percent, the present worths are

$$\text{SL:} \quad \frac{2000}{1.07} + \frac{2000}{(1.07)^2} + \frac{2000}{(1.07)^3} = \$5249$$

$$\text{DDB:} \quad \frac{4000}{1.07} + \frac{1333}{(1.07)^2} + \frac{444}{(1.07)^3} = \$5266$$

$$\text{SYD:} \quad \frac{3000}{1.07} + \frac{2000}{(1.07)^2} + \frac{1000}{(1.07)^3} = \$5367$$

$$\text{ACRS:} \quad \frac{1500}{1.07} + \frac{2280}{(1.07)^2} + \frac{2220}{(1.07)^3} = \$5205$$

SYD gives the largest value but all are quite close to each other.

For $i = 20$ percent, the present worths are

SL:	\$4213
DDB:	\$4516
SYD:	\$4468
ACRS:	\$4118

For the higher rate of interest, DDB is now favored. Therefore it is a more attractive depreciation method, i.e., it would improve cash flow.

Depletion

When natural resources are consumed in producing goods or services, depletion of these resources measures the decrease in value that has occurred. Therefore, depletion is kindred to depreciation. Since these resources usually cannot be replaced, the Internal Revenue Service allows a percentage of the yearly income to be deducted as a cost in order to encourage further discovery and development of natural resources, because such ventures often involve considerable uncertainty and risk (e.g., such as a dry hole in oil drilling). The percentage depletion (really a fraction to be multiplied by yearly income) allowed by IRS varies according to the industry (e.g., oil and gas—22 percent; brick and tile clay—5 percent) (Jelen and Black, 1983). From a project evaluation standpoint, depletion involves a constant charge to be applied to reduce annual income, so the formulation of an objective function involving depletion is not difficult.

Tax Credit

Periodically Congress has permitted the use of tax credits as a direct reduction from income taxes themselves. One example is tax credits for energy conservation devices. Tax credits have been used historically to stimulate capital investment in the United States. Such deductions are more valuable than depreciation because they represent direct deductions from the tax bill after taxes are computed on income. Investment tax credits are usually taken in the first year in which the equipment is put into operation and therefore might be viewed as a reduction in capital requirements. If I_0 is the investment and i is the interest value of capital, a tax credit fraction equal to t_c $(0 \le t_c \le 1)$ at the end of the first year yields an effective or net capital investment of

$$I_0 - \frac{t_c I_0}{1+i} = I_0\left(1 - \frac{t_c}{1+i}\right) \tag{3.20}$$

Income Taxes

The federal income tax on profits from corporations is based on income after all costs, including depreciation, have been deducted. Since depreciation affects taxable income, it is an important consideration in estimating profitability. The federal income tax rate for large corporations (profit greater than $100,000) was recently 34 percent. State income taxes may push the total tax rate to about 40 percent. Therefore a depreciation amount of $1 reduces taxes about $0.40. At this level of taxation, the before-tax rate of return will be roughly 1.67 the after-tax rate of return.

Inflation

Inflation can be a significant factor in analysis of profitability. High inflation rates frequently occur in many countries. In computing the rate of return or net present value, you would like to obtain a measure of profitability which is independent of the inflation rate. If you inflate projections of future annual income, the computed rate of return may largely result from the effects of inflation. Most companies strive for an internal rate of return (after taxes) of 10 to 20 percent in the absence of inflation; this figure would rise if projected future income is increased to include the effects of inflation (i.e., selling prices are raised yearly). Furthermore, costs will also rise because of inflation.

Griest (1979) has discussed the effects of inflation on profitability analysis and has pointed out that the percentage change in profits after income taxes rarely increases at the rate of inflation, largely due to the effects of taxation. Assumptions about inflation can change the relative ranking of project alternatives based on net present value; special techniques based on probability may be required because the inflation is difficult to predict.

Debt-Equity Ratio

Debt to equity ratio quantifies the sources of funds used for capital investment, and is generally expressed as percent/percent, for example, 75/25 means 75 percent debt, 25 percent equity. Debt financing involves borrowing funds (either from banks, insurance companies, or other lenders, or by selling bonds) based on fixed or adjustable interest rates and specified lengths of time until the loan is due. Equity financing involves selling shares of stock or partnership shares to raise investment funds and/or the expenditure of retained earnings of the company. Both debt and equity financing can be used on the same project. Compared to 100 percent equity financing, the rate of return on an investment can be increased if the interest rate for borrowed capital is appropriate because interest payments are considered to be an expense in computing income taxes. Suppose that the debt interest rate is 12 percent and the equity interest rate is 12 percent. Because interest payments are deductible, the effective debt interest rate after taxes for a tax rate of 40 percent is 7.2 percent. According to DeGarmo and Canada (1973), for a 50-50 debt-to-equity ratio, the equivalent interest rate is 9.6 percent ($0.5 \times 7.2\% + 0.5 \times 12\%$).

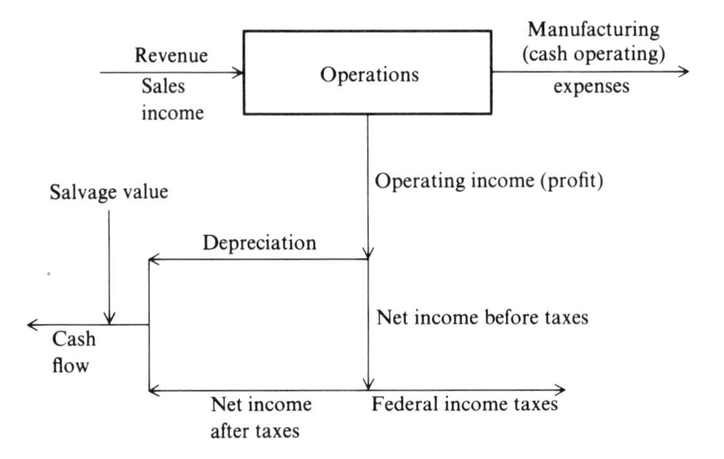

Figure 3.3 Computation of cash flow.

Below we present an example of profitability analysis which incorporates most of the factors mentioned in this section. Figure 3.3 summarizes the cash flow calculation which is to be carried out.

EXAMPLE 3.10 INTERNAL RATE OF RETURN FOR EQUITY VERSUS DEBT FINANCING

Suppose an extruder is to be purchased for manufacturing plastic widgets. The capital investment is \$24,000 and the equipment has a life of five years, after which the salvage value is \$4000. The income from the machine is expected to be \$12,000 the first year and \$15,000 per year thereafter. As the equipment ages, its operating and maintenance costs increase. Table E3.10-1 (line *B*) shows the expense profile. Assume a tax rate of 50 percent with no investment tax credit. Compute the internal rate of return for (*a*) 100 percent equity financing and (*b*) 50 percent debt/50 percent equity. Use straight-line depreciation; for debt financing, assume equal annual payments (principal plus interest) to the lender.

Solution. The cash flow calculations for case (*a*) are shown in Table E3.10-1. The straight-line depreciation is \$4000/year, based on $(I_o - S_v)/n = (24{,}000 - 4000)/5 =$ \$4000. In this table the salvage value is included as a nontaxable return in year 5 since it was never expensed or depreciated. The internal rate of return is 16.53 percent for this case. Note that the total of the after-tax cash flows exceeds the original investment. If $\sum_{i=1}^{5} F_k = 20{,}000$, then $i = 0$ percent, implying that the capital was returned dollar for dollar (but with no appreciation).

Suppose one-half the capital (\$12,000) is borrowed and to be repaid over five years at an interest rate of 10 percent. Allowing for a salvage value of \$4000, Table E3.10-2 can be constructed. In this case, interest expenses are subtracted under line *B*. Note that the taxes each year are lower than in the 100 percent equity case (Table E3.10-1), but also the net income after taxes is lower; however, the equity investment amount has been reduced. In computing the IRR, you must adjust line *G* to account for principal payments to the lender, which increase over time (line

Table E3.10-1 Cash flow calculations for purchase of extruder

	Year				
	1	2	3	4	5
A. Income	\$12,000	\$15,000	\$15,000	\$15,000	\$15,000
B. Expenses	2,000	3,000	5,000	7,000	8,000
C. Profit $(A - B)$	10,000	12,000	10,000	8,000	7,000
D. Depreciation (straight line)	4,000	4,000	4,000	4,000	4,000
E. Net income before taxes $(C - D)$	6,000	8,000	6,000	4,000	3,000
F. Taxes $(0.5E)$	3,000	4,000	3,000	2,000	1,500
G. Net income after taxes $(E - F)$	3,000	4,000	3,000	2,000	1,500
H. Salvage	0	0	0	0	4,000
I. Cash flow $(D + G + H)$	7,000	8,000	7,000	6,000	9,500

$$P = \sum_{k=1}^{5} \frac{S_k}{(1+i)^k} = \frac{7000}{1+i} + \frac{8000}{(1+i)^2} + \frac{7000}{(1+i)^3} + \frac{6000}{(1+i)^4} + \frac{9500}{(1+i)^5}$$

$$P = \$24{,}000$$

By iterative solution $i = 0.1653$ (16.53 %).

Table E3.10-2 50 percent debt/50 percent equity financing for Example 3.10

	Year				
	1	2	3	4	5
A. Income	12,000	15,000	15,000	15,000	15,000
B. Expenses	2,000	3,000	5,000	7,000	8,000
Interest	1,200	1,000	790	550	300
C. Profit	8,800	11,000	9,210	7,450	6,700
D. Depreciation	4,000	4,000	4,000	4,000	4,000
E. Net income before taxes	4,800	7,000	5,210	3,450	2,700
F. Taxes	2,400	3,500	2,600	1,720	1,350
G. Net income after taxes	2,400	3,500	2,610	7,730	1,350
H. Principal	1,970	2,170	2,380	2,620	2,860
I. Salvage					4,000
J. Principal $(D + G - H + I)$	4,430	5,330	4,230	3,110	6,490

Rate of return

$$12{,}000 = \frac{4430}{1+i} + \frac{5330}{(1+i)^2} + \frac{4230}{(1+i)^3} + \frac{3110}{(1+i)^4} + \frac{6490}{(1+i)^5}$$

$$i = 0.2707 \ (27.07\,\%)$$

H). The returns in line I are lower than for the first case (100 percent equity); however, the capital investment has been reduced by a factor of two. This leads to an internal rate of return on the equity of 27 percent for 50/50 D/E, compared to 16.53% for the 100 percent equity case.

The increase in IRR using debt financing of assets is known as the principle of "leverage." A similar result can often be obtained by leasing equipment because the lease payments are completely deductible as expenses for income tax purposes.

Keep in mind that a year-by-year analysis of the discounted cash flow of a given project may be a cumbersome task. When used in the context of optimization of one or more variables, you must use repetitive calculations (a "case-study" approach) in order to maximize the internal rate of return or net present value. As mentioned earlier, however, many optimization problems can be formulated using constant yearly income and costs. In such cases, the net present value criterion can easily be employed to develop an explicit expression for the objective function.

EXAMPLE 3.11

Formulate net present value for a project lasting n years with an initial investment I_0, interest rate of capital i, constant annual expense E, constant annual revenue A, salvage value S_v, yearly depreciation D_j, and a tax rate t. Compare NPV for tax rates of 0 and 50 percent, $S_v = 0$, and straight-line depreciation. Use $n = 10$ and $i = 0.15$ $(r = 0.2)$.

Solution. The net present value is

$$\text{NPV} = -I_0 + \left\{\sum_{j=1}^{n} \frac{A}{(1+i)^j} - \sum_{j=1}^{n} \frac{E}{(1+i)^j}\right\}(1-t) + t\sum_{j=1}^{n} \frac{D_j}{(1+i)^j} + \frac{S_v}{(1+i)^n} \tag{a}$$

Since R and E are constants (the same each year), the series can be simplified as

$$\text{NPV} = -I_0 + (A - E)\left(\frac{1-t}{r}\right) + t\sum_{j=1}^{n} \frac{D_j}{(1+i)^j} + \frac{S_v}{(1+i)^n} \tag{b}$$

where r is the repayment multiplier defined in Eq. (3.8). Multiply (b) by r to get an equivalent objective function

$$r(\text{NPV}) = -rI_0 + (A - E)(1 - t) + rt\sum_{j=1}^{n} \frac{D_j}{(1+i)^j} + \frac{(S_v)(r)}{(1+i)^n} \tag{c}$$

The summation term can be computed once the depreciation schedule is known; in general,

$$\sum_{j=1}^{n} \frac{D_j}{(1+i)^j} = \gamma I_0 \tag{d}$$

where $\gamma < 1$ due to the discounting effect. The relative size of γ depends upon the depreciation method selected. For $n = 10$ and $i = 0.15$, γ is 0.614 for SYD while $\gamma = 0.50$ for SL. A larger value of γ denotes a more favorable depreciation schedule and higher net present value. For $n = 3$ and $i = 0.10$, γ (DDB) is 0.87 while γ (SL) = 0.83. For the specific case of straight-line depreciation, $D_j = I_0/n$ and

$$rt \sum_{j=1}^{n} \frac{I_0}{n(1+i)^j} = rt\left(\frac{I_0}{rn}\right) = \frac{tI_0}{n} \tag{e}$$

Hence, Eq. (*c*) can be simplified to

$$r(\text{NPV}) = \left(\frac{t}{n} - r\right)I_0 + (1-t)(A-E) + \frac{S_v(r)}{(1+i)^n} \tag{f}$$

This is an explicit objective function which can be maximized. For $S_v = 0$, $n = 10$, and $r = 0.2$, you can compare before-tax profitability ($t = 0$) to after-tax profitability ($t = 0.5$) for NPV:

$$t = 0: \quad r(\text{NPV}) = A - E - 0.2I_0 \tag{g}$$

$$t = 0.5: \quad r(\text{NPV}) = 0.5(A - E) - 0.15I_0 \tag{h}$$

To facilitate comparison of (*g*) and (*h*), multiply (*h*) by 2:

$$2(r \cdot \text{NPV}) = A - E - 0.3I_0 \tag{i}$$

Equations (*g*) and (*i*) indicate that the inclusion of taxes and depreciation in the objective function increases the relative importance of I_0, the investment; in a maximization problem, I_0 would be forced to be smaller. Thus, in comparing before-tax and after-tax optimum present values, we expect the latter case will have a lower optimum capacity, i.e., require a smaller plant capital investment (but the difference may not be too significant). It is clear that for a before-tax analysis ($t = 0$), the objective function is easier to develop; on the other hand, if a computer is used to perform the optimization calculations, then the complexity of the objective function is not a serious restriction.

3.6 COST ESTIMATION

The estimation of future operating and capital costs is an important facet of chemical engineering design. Most modern methods of evaluating costs via a computer use cost correlations based on historical and projected data. Such correlations are usually expressed in terms of key operating or design variables, which in turn may correspond to independent variables that can be optimized. Cost estimation formulas generally yield the so-called preliminary estimates, which have a probable accuracy of -15 to $+30$ percent. For this degree of accuracy, the cost estimates are usually based on a flowsheet of the plant. Other types of estimates and their accuracy include order of magnitude (-30 to $+50$ percent) and definitive (-5 to $+15$ percent). The latter estimate involves the development of drawings and complete mechanical specifications; many of the factors involved are not subject to optimization. However, analysis at the preliminary level certainly can take advantage of optimization techniques.

Operating Cost Estimation

Operating costs include the costs of raw materials, direct operating labor, labor supervision, maintenance, plant supplies, utilities (steam, gas, electricity, fuel), property taxes, and insurance. Sometimes certain operating cost components are directly expressed as a fraction of the capital investment cost. Table 3.5 is a brief checklist for estimating operating costs; note that such items as property taxes, insurance, and maintenance are computed as fractions of total fixed capital investment. There are shortcut methods for estimating almost all of the components of operating costs (including labor costs) based on process complexity and capacity [see Jelen and Black (1983) and Ulrich (1984)]. For example, typical refinery maintenance costs are roughly $0.05P$ (P = plant capital investment) while a sulfuric acid alkylation unit has a maintenance cost of $0.07P$. One rule of thumb is that labor requirements vary according to the 0.2 power of the capacity ratio when plant capacities are scaled up or down. Peters and Timmerhaus (1980) have given typical labor requirements for process equipment such as for a batch reactor, one worker per shift. Another useful correlation for labor require-

Table 3.5 Preliminary operating cost estimates

A. Direct production cost
1. Materials
   a. Raw materials—estimate from price lists
   b. By-product and scrap credit—estimate from price lists
2. Utilities—from literature or similar operations
3. Labor—from manning tables, literature, or similar operations
4. Supervision—10 to 25% of labor
5. Payroll charges—30 to 45% of labor plus supervision
6. Maintenance—2 to 10% of investment per year
7. Operating supplies—0.5 to 1.0% of investment per year or 6 to 10% of operating labor
8. Laboratory—10 to 20% of labor per year
9. Waste disposal—from literature, similar operations, or separate estimate
10. Royalties—1 to 5% of sales
11. Contingencies—1 to 5% of direct costs

B. Indirect costs
1. Depreciation—10 to 20% of investment per year
2. Real estate taxes—1 to 2% of investment per year
3. Insurance—0.5 to 1.0% of investment per year
4. Interest—10 to 12% of investment per year
5. General plus overhead—50 to 70% of labor, supervision, and maintenance, or 6 to 10% of sales

C. Distribution costs
1. Packaging—estimate from container costs
2. Shipping—from carriers or 1 to 3% of sales

Source: Jelen and Black (1983), Couper (1986)

ments is the Wessel equation (1952), which is a log-log plot of operating labor man-hours required vs. capacity.

Jelen and Black (1983) and Peters and Timmerhaus (1980) discuss important considerations in developing detailed operating cost estimates. In recent times utility costs have been of particular importance because of the rising costs of fuel. Table 3.6 shows typical ranges for costs of utilities as of 1986, that can be applied to future years by employing an industrial price index. However, these costs can vary significantly according to plant location, so you should obtain local data when possible.

Capital Cost Estimation

The capital cost or fixed capital investment not only includes the delivered cost of equipment but also must be concerned with certain associated costs: installation and siting, piping, instrumentation, insulation, electrical, and engineering cost required for installation. The initial investment cost may well include other charges such as loan fees, legal fees, interest expense during construction, the cost of additional working capital to support operations of the new facility, and

Table 3.6 Rates for industrial utilities, 1986

Utility	Cost, $ (1986)	Unit
Steam		
500 psig	4.00 - 5.00	1000 lb
100 psig	3.00 - 4.00	1000 lb
Exhaust	2.00 - 3.00	1000 lb
Electricity		
Purchased	0.03 - 0.06	kWh
Self-generated	0.016- 0.045	kWh
Cooling water		
Well	0.065- 0.35	1000 gal
River or salt	0.045- 0.13	1000 gal
Tower	0.045- 0.175	1000 gal
Process water		
City	0.25 - 1.00	1000 gal
Filtered and softened	0.35 - 0.90	1000 gal
Distilled	1.70 - 2.70	1000 gal
Compressed air		
Process air	0.045- 0.135	1000 ft^3
Instrument	0.09 - 0.27	1000 ft^3
Natural gas	2.00 - 8.00	1000 ft^3
Manufactured gas	1.20 - 3.60	1000 ft^3
Fuel oil	0.45 - 1.05	gal
Coal	22.00 -50.00	ton
Refrigeration (ammonia) to 34°F	2.00	ton/day (288,000 Btu removed)

Source: Jelen and Black (1983), Couper (1986)

start-up costs. Therefore, several authors (Jelen and Black, 1983; Happel and Jordan, 1975; Ulrich, 1984) suggest using some multiplier or factor of purchased equipment cost in optimization calculations so that the realistic installed capital costs are used, especially when both capital and operating costs appear in an objective function.

Table 3.7 lists process variables which can be used to compute capital costs for different types of equipment. Cost curves for a wide variety of process equipment have been presented by Ulrich (1984) and Pikulik and Diaz (1977); Fig. 3.4 shows a typical example. Such cost curves generally appear as nearly straight lines on log-log plots, indicating a power-law relationship between capital cost and capacity, with exponents typically ranging from 0.5 to 0.8. Relations for preliminary or "study-grade" capital cost estimates have been included in flowsheet simulators such as FLOWTRAN (Seader, et al., 1977) and ASPEN (Evans, et al., 1979). Mulet, et al. (1981*a*, 1981*b*), Corripio, et al. (1982*a*, 1982*b*), and Kuri and Corripio (1984) have reviewed various cost correlations used in the ASPEN design program. Kuri and Corripio (1984) reported a comparison of ASPEN cost predictions with vendor quotes. The program was within ± 15 percent error for quotes on a fan, heat exchanger, pump, storage tank, and pressure vessel. Larger errors occurred for a cyclone and distillation column (± 40 percent). These errors should be considered only as representative of what might occur in practice because of the limited number of cases reported. As mentioned

Table 3.7 Process variables used in cost estimation for typical process equipment

Equipment type	Economic variables
1. Flashdrum	Diameter, height, material of construction, internal pressure
2. Distillation column, tray absorber	Diameter, height, internal pressure, material of construction, tray type, number of trays, condenser, reboiler (see 3)
3. Condenser, reboiler, heat exchanger (shell and tube)	Heat transfer surface area, type, shell design pressure, materials for shell and tube
4. Absorber (packed)	Diameter, height, internal pressure, material of construction, packing type, packing volume
5. Process furnace or direct-fired heater	Design type, absorbed heat duty, pressure, tube material, capacity
6. Pumps (centrifugal, reciprocating)	Fluid density, capacity, dynamic head, type, driver, operating condition limits, material of construction
7. Gas compressor	Brake horsepower, driver type
8. Storage tank	Tank capacity, type, and storage pressure
9. Boiler	Steam flow rate, design pressure, steam superheat
10. Reactor	Type, diameter, height, design pressure, material of construction, capacity

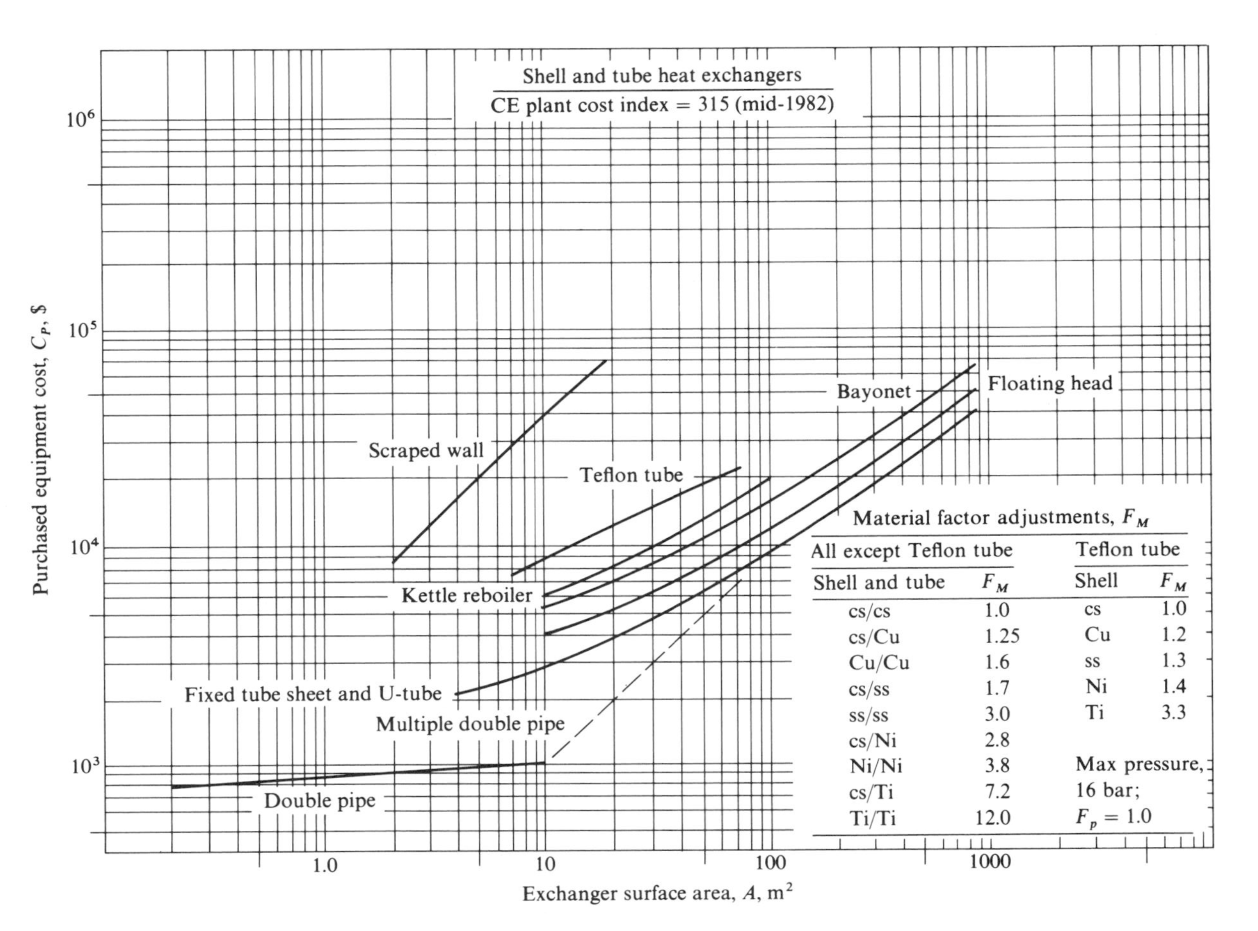

Material factor adjustments, F_M

All except Teflon tube		Teflon tube	
Shell and tube	F_M	Shell	F_M
cs/cs	1.0	cs	1.0
cs/Cu	1.25	Cu	1.2
Cu/Cu	1.6	ss	1.3
cs/ss	1.7	Ni	1.4
ss/ss	3.0	Ti	3.3
cs/Ni	2.8		
Ni/Ni	3.8	Max pressure,	
cs/Ti	7.2	16 bar;	
Ti/Ti	12.0	$F_p = 1.0$	

Figure 3.4 Purchased equipment costs for various types of heat exchangers (*adapted from Ulrich, 1984*).

earlier, such "study-grade" estimates are based on correlations of cost data reported in the literature. There are also resources available for more detailed cost estimates, e.g., COST by Icarus Corporation (1985) and Richardson (1979). While the process variables given in Table 3.7 may be used as input variables for the cost correlations, most software discussed above translates the process variables into weight of metal, using assumed mechanical design specifications, from which the cost is computed. Examples of this latter approach include the sizing of pressure vessels and towers. Heat transfer equipment cost correlations usually are expressed in terms of heat transfer surface area. Fluid-gas separator costs are generally functions of flowrate.

As an example of a typical formula for capital cost estimation, examine the following relation:

$$\log C_B = a_1 + a_2 \log S \tag{3.21}$$

where C_B is the base cost, S is the size parameter, and a_1 and a_2 are coefficients to be estimated from valid data. A base cost typically corresponds to carbon steel construction and pressure below 100 psig. Note that Eq. (3.21) is equivalent to

$$C_B = a_1' S^{a_2} \tag{3.22}$$

the familiar formula for scale-up, where a_2 is typically about 0.6. Ulrich (1984) has tabulated scale-up factors for a variety of process equipment types. A slightly different correlation provides a more accurate fit of cost data by using three coefficients:

$$\log C_B = a_1 + a_2 \log S + a_3 (\log S)^2 \tag{3.23}$$

The estimated capital cost can be found from base cost by

$$C_E = C_B f_D f_M f_P \tag{3.24}$$

where f_D = design type cost factor
f_M = material of construction cost factor
f_P = pressure rating factor

The design type refers to variations in equipment configuration (e.g., fixed-head versus floating-head in a heat exchanger). The adjustment for material of construction is used principally to account for the use of alloy steel instead of carbon steel. The pressure rating factor allows adjusting costs for pressures other than 100 psig. Obviously, higher pressure operation causes additional capital costs because of thicker vessel walls, etc.; f_p may be a discontinuous function.

EXAMPLE 3.12 CAPITAL COST ESTIMATION

Suppose the cost for a fixed-head heat exchanger constructed of 316 stainless steel operating at 300 to 600 psig is to be estimated. The base case is a carbon steel, floating head exchanger operating at 100 psig of area A. For such operation (Kuri and Corripio, 1984), the base cost is

$$C_B = \exp\left[8.551 - 0.30863\,(\ln A) + 0.06811\,(\ln A)^2\right] \qquad (a)$$

where A is the exchanger heat transfer area in square feet ($150 \leq A \leq 12{,}000$ ft^2) and C_B is in dollars. Multiply C_B by factors f_D, f_P, and f_M, calculated as follows:

For a fixed head (versus floating head)

$$f_D = \exp\left[-1.1156 + 0.0906\,(\ln A)\right] \qquad (b)$$

For 300 to 600 psig, the correction is

$$f_P = 1.0305 + 0.07140\,(\ln A) \qquad (c)$$

For 316 stainless steel, the correction is

$$f_M = 2.7 \qquad (d)$$

Eq. (3.24) can then be used to determine the actual capital cost for a specified A.

For equipment such as distillation columns, the costs of several components (trays, shell) must be calculated. We should also mention that many cost estimation programs include calculations of NPV or IRR as well as cost. Therefore, with such a code you can carry out profitability calculations iteratively with the process design calculations embedded in the program.

Accounting for the Effects of Inflation

Escalation or inflation must be applied to any cost calculation based on old cost data, i.e., referenced to a specific date, or designed to project into the future. This maxim applies to the computer programs discussed earlier (Seader, et al., 1977; Evans, et al., 1979). Inflation indices for capital costs may be kept within a company or taken from published indices based on data (Ulrich, 1984; Pikulik and Diaz, 1977) from:

1. ENR—*Engineering News-Record* Construction and Building Indexes
2. CE—*Chemical Engineering* Plant Cost Index
3. M & S—Marshall and Swift Equipment Index (also appears in *Chemical Engineering*)
4. NRC—Nelson Refinery Construction Index (appears in *Oil and Gas Journal*)

Figure 3.5 shows how the different indices have varied during the past 35 years; projection into the future is a subjective issue. Ulrich (1984) suggests that the

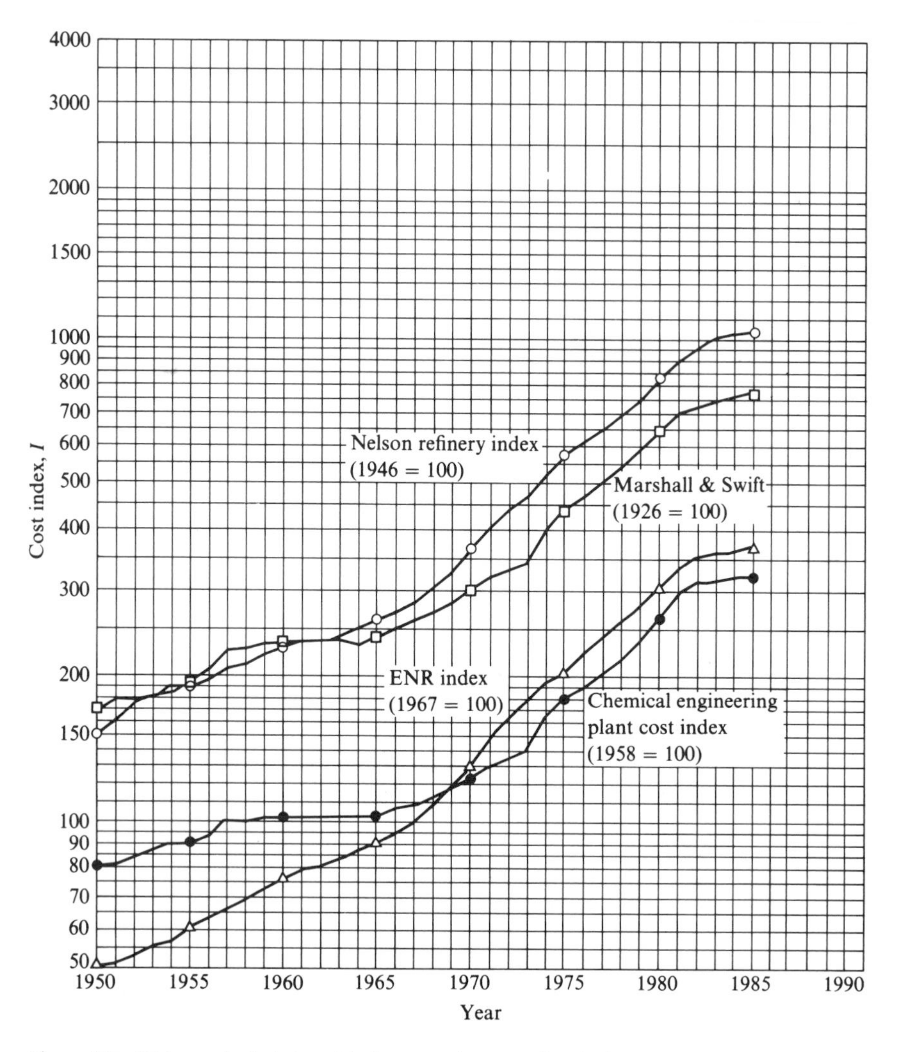

Figure 3.5 History of selected cost indexes pertinent to chemical process construction (1950–1985).

CE index is the most appropriate one for chemical engineering applications (see Table 3.8 for 12 years of data). To determine capital costs (C_x) in the year X in the future, given a known cost C_y in year Y, you simply multiply C_y by the ratio of the index (I_x/I_y):

$$C_x = C_y \cdot I_x/I_y \tag{3.25}$$

The Bureau of Labor statistics provide information which permits computation of estimated future labor and material costs [see Ulrich (1984)].

Table 3.8

CE plant cost index (1975–1986)		
1975 = 182.4	1979 = 238.7	1983 = 316.9
1976 = 192.1	1980 = 261.2	1984 = 322.7
1977 = 204.1	1981 = 297.0	1985 = 325.3
1978 = 218.8	1982 = 314.0	1986 = 318.5

REFERENCES

Adelson, R. M., "Criteria for Capital Investment: An Approach through Decision Theory," *Oper. Res. Q.*, **16** (1): 19–50 (1968).

Brigham, E. F., *Financial Management: Theory and Practice*, 3d ed., Dryden Press, Chicago (1982).

Brummerstedt, E. F., *Natl. Pet. News*, **36**, R282, R362, R497 (1944).

Carlsson, C., and Y. Kochetkov, eds., *Theory and Practice of Multiple Criteria Decision-Making*, Elsevier, New York (1983).

Chankong, V., and Y. Y. Haimes, *Multiobjective Decision-Making*, Elsevier Science Pub. Co., New York (1983).

Corripio, A. B., K. S. Chrien, and L. B. Evans, "Estimate Costs of Heat Exchangers and Storage Tanks Via Correlations," *Chem. Eng.*, p. 125 (Jan. 25, 1982).

Corripio, A. B., K. S. Chrien, and L. B. Evans, "Estimate Costs of Centrifugal Pumps and Electric Motors," *Chem. Eng.*, p. 115 (Feb. 22, 1982).

Couper, J. R., University of Arkansas, personal communication (1986).

DeGarmo, E. P., and J. R. Canada, *Engineering Economy*, Macmillan Publishing, New York (1973).

Evans, L. B., et al., "ASPEN: An Advanced System for Process Engineering," *Comp. Chem. Eng.*, **3**: 319 (1979).

Griest, W. H., "Making Decisions in an Inflationary Environment," *Chem. Eng. Prog.*, p. 13, (June, 1979).

Hansen, P., ed., *Essays and Surveys on Multiple Criteria Decision-Making*, Springer-Verlag (1983).

Happel, J., and D. G. Jordan, *Chemical Process Economics*, 2d ed., Marcel Dekker, New York (1975).

Hecty, D. B., "Risk Analysis in Capital Investment," *Harvard Business Review*, p. 96 (January–February), 1968.

Icarus Corporation, "COST System User's Manual," 7th ed., Version 14.1, Rockville, Maryland (1985).

Internal Revenue Service, "Depreciation—Guidelines and Rules," Publication no. 456, July, 1962; amended in 1971 as "Class Life Asset Depreciation Range System."

Jelen, F. C., and J. H. Black, *Cost and Optimization Engineering*, McGraw-Hill, New York (1983).

Kuri, C. J., and A. B. Corripio, "Two Computer Programs for Equipment Cost Estimation and Economic Evaluation of Chemical Processes," *Chem. Eng. Educ.*, p. 14 (Winter, 1984).

Mulet, A., A. B. Corripio, and L. B. Evans, "Estimate Costs of Pressure Vessels Via Correlations," *Chem. Eng.*, p. 145 (Oct. 5, 1981).

Mulet, A., A. B. Corripio, and L. B. Evans, "Estimate Costs of Distillation and Absorption Towers Via Correlations," *Chem. Eng.*, p. 77 (Dec. 28, 1981).

Perry's Chemical Engineers' Handbook, 6th ed., McGraw-Hill, New York (1984).

Peters, M. S., and K. D. Timmerhaus, *Plant Design and Economics for Chemical Engineers*, McGraw-Hill, New York (1980).

Pikulik, A., and H. E. Diaz, "Cost Estimating for Major Process Equipment," *Chem. Eng.*, p. 106 (Oct. 10, 1977).

Richardson Engineering Services, Inc., Solano Beach, California (1979).

Rubin, F. L., "Can you Justify more Piping Insulation," *Hydrocarbon Process*, p. 152 (July, 1982).

Seader, J. D., W. D. Seider, and A. Pauls, *FLOWTRAN Simulation—An Introduction*, CACHE Corporation, Ann Arbor, Michigan (1977).

Ulrich, G. D., *A Guide to Chemical Engineering Process Design and Economics*, Wiley, New York (1984).

Wessel, H. E., "New Graph Correlates Operating Labor Data for Chemical Processes," *Chem. Eng.*, 209 (July, 1952).

Woods, D. R., *Financial Decision-Making in the Process Industries*, Prentice-Hall, Englewood Cliffs, New Jersey (1975).

Zeleny, M., *Multiple Criteria Decision-Making*, McGraw-Hill, New York (1982).

SUPPLEMENTARY REFERENCES

Allen, D. H., *A Guide to the Evaluation of Projects*, Inst. Chem. Eng., Rugby, England (1972).

Axtell, O., and J. M. Robertson, *Economic Evaluation in the Chemical Process Industries*, John Wiley and Sons, New York (1986).

Happel, J., et al., *Process Economics*, Am. Inst. Chem. Eng., New York (1974).

Holland, F. A., et al., "Process Economics," *Perry's Chemical Engineers' Handbook*, 6th Ed., chap. 25, McGraw-Hill, New York (1984).

Kharbanda, O. P., *Process Plant and Equipment Cost Estimation*, Sevak Pub., Bombay (1977).

Park, W. R., and D. E. Jackson, *Cost Engineering Analysis*, "A Guide to Economic Evaluation of Engineering Projects," 2d ed., Wiley-Interscience, New York (1984).

Riggs, J. L., and T. West, *Essentials of Engineering Economics*, McGraw-Hill, New York (1986).

Weaver, J. B., "Project Selection in the 1980s," *Chem. Eng. News*, **34** (Nov. 2, 1981).

Woods, D. R., *Financial Decision-Making in the Process Industry*, Prentice-Hall (1975).

PROBLEMS

3.1. If you borrow $100,000 from a lending agency at 10 percent yearly interest and wish to pay it back in 10 years in equal installments paid annually at the end of the year, what will be the amount of each yearly payment? Compute the principal and interest for each year.

3.2. Compare the "present value" of various depreciation methods (straight-line, declining-balance, sum-of-the-digits, ACRS) as in Table 3.2 for $i = 0.12$ and $n = 10$ years.

3.3. Repeat Prob. 3.2 for $n = 3$ and $n = 5$.

3.4. It has been stated that the most beneficial depreciation method is the one which yields the largest present value for the depreciation summed over the life of the equipment. Why do we desire to maximize the present worth of depreciation when profits are actually more important?

3.5. An automobile costing $10,000 is to be used for business purposes. The auto has a service life of 3 years with no salvage value. Two alternatives are under consideration:

(*a*) Purchase the automobile by obtaining a bank loan ($i = 0.13$), to be repaid in 36 monthly installments.

(*b*) Lease the automobile (in this case all costs are deductible).

You are paying income taxes at a rate of 40 percent. What is the highest monthly lease cost equivalent to purchase of the auto, if you use ACRS depreciation in case *a*. Use net present value ($i = 0.12$).

3.6. Consideration is being given to two plans for supplying water to a plant. Plan *A* requires a pipeline costing $160,000 with annual operation and upkeep costs of $2200, and an estimated life of 30 years with no salvage. Plan *B* requires a flume costing $34,000 with a life of 10 years, a salvage value of $5600, and annual operation and upkeep of $4500 plus a ditch costing $58,000, with a life of 30 years and annual costs for upkeep of $2500. Using an interest rate of 12 percent, compare the net present values of the two alternatives.

3.7. A process requires pure water which can be produced satisfactorily either by distillation or by ion exchange using activated resins. The cost of installation for a distillation unit is $8000 with operating costs of $1.00 per 1000 gal of water produced. The ion-exchange installation costs $2.00 per gallon per hour feed rate, and suitable resins cost $1.00 per pound. After operating the ion-exchange system to produce pure water for 1 hour it is necessary to regenerate the resins for 20 minutes before pure water can be obtained again. Resin needs may be considered as 1 pound per 10,000 gallons of treated water, and operating costs are $0.40 per 1000 gallons. Based on present worth, determine which installation is more economical for a water duty of 1,000 gal/h. Assume 8400 h/year operation and 10-year life for both with interest at 12 percent.

3.8. Cost estimators have provided reliable cost data as given below for the chlorinators in the methyl-chloride plant addition. Analysis of the data and recommendations of the two alternatives are needed. Use present worth for $i = 0.10$ and $i = 0.20$.

	Chlorinators	
	Glass-lined	**Cast iron**
Installed cost	$24,000	$7200
Estimated useful life	10 years	4 years
Salvage value	4000	800
Miscellaneous annual costs as percent of original cost	10	20

Maintenance costs

Glass-lined. $230 at the end of the second year, $560 at the end of the fifth year, and $900 at the end of each year thereafter.
Cast iron. $730 each year.

The product from the glass-lined chlorinator will be essentially iron-free and it is estimated that this will yield a product quality premium of $1700 per year.

3.9. Preliminary consideration is being given to starting up a mixed fertilizer plant as an adjunct to manufacturing operations. It is now estimated that the annual fixed charges (amortization) for the new plant will amount to $40,000 and that the other costs (including labor, raw materials, fuel, sales, etc.) will be 10.5 cents per pound of mixed fertilizer product. We plan to sell direct to the consumer at an estimated minimum average sale price of 14.5 cents per pound. Market forecasts predict an annual sale at this price of 1000 pounds per square mile. Since the plant is in the midst of the potential sales area, delivery to the consumer will be by truck. A trucking contractor has agreed to deliver the fertilizer for $8 per ton for loading and unloading, plus 30 cents per ton-mile for delivery. For the time being, we can assume consumption to be uniformly distributed over the area.

We need to know the maximum distance we can afford to ship the fertilizer and what will be the average profit per ton.

3.10. Three projects (A, B, C) all earn a total of \$125,000 over a period of five years (after-tax earnings, nondiscounted). For the cash-flow patterns shown below, predict by inspection which project will have the largest rate of return. Why?

	Cash flow, $10^3		
Year	***A***	***B***	***C***
1	45	25	10
2	35	25	30
3	25	25	45
4	15	25	30
5	5	25	10

3.11. Resolve the Example 3.10 (Table E3.10-1) assuming a zero salvage value. Calculate the rate of return.

3.12. Repeat Example 3.10 using double-declining-balance, sum-of-the-years'-digits, and ACRS depreciation methods. Calculate the rate of return for each case (100 percent equity).

3.13. Suppose that an investment of \$100,000 will earn after-tax profits of \$10,000 per year over 20 years. However, due to uncertainties in forecasting, the projected after-tax profits may be in error by ± 20 percent. Discuss how you would determine the sensitivity of the rate of return to an error of this type. Would you expect the rate of return to increase by 20 percent of its computed value for a 20 percent increase in annual aftertax profits (i.e., to \$12,000)?

3.14. The installed capital cost of a pump is \$200/hp and the operating costs are 4¢/kWh. For 8000 h/year of operation, an efficiency of 70 percent, and a cost of capital $i = 0.10$, for $n = 5$ years, determine the relative importance of the capital vs. operating costs.

3.15. Formulate a single mathematical relationship that gives the present worth of a proposed investment including

(*a*) Sum of discounted incomes
(*b*) Initial investment
(*c*) Present worth of working capital required to operate
(*d*) Present worth of salvage values

Use the following notation:

i = interest rate (assumed constant)
I = initial investment
n = year in future
L = year in which equipment is dismantled
S_v = salvage value of investment
t = tax rate
x = dollars earned in year n
w = working capital

3.16. A technique of accommodating uncertainty in evaluating capital projects is to calculate (*a*) the IRR and (*b*) the NPV for a range of conditions rather than for a single set of conditions. It is proposed to install a cogeneration project with three options:

1. A single unit installation
2. Two units each having the same capacity as in option 1
3. Same option as 2 but additional power is to be generated during utility peak pricing periods.

Based on the following data, compare the IRR and NPV of the three options.

Condition 1—No cost escalation

Condition 2—Annual cost-escalation rates, as follows: electric energy, 9 percent natural gas, 9 percent; fuel oil, 12 percent; and operating costs, 5 percent

Condition 3—Annual cost-escalation rates are the same as in condition 2, except that the electric energy rate is reduced to 7 percent

Factors common to each alternative:

Business energy credit—20 percent of initial installation cost

Income tax rate—40 percent

Project economic life—20 years

Depreciation method—sum-of-years-digits

Depreciable life—11 years (no salvage)

Discount rate for NPV—10 percent

Per $1000 of project costs, it is estimated the costs are

	Option		
	1	2	3
Initial installation	652	1209	1209
First-year cash flows			
Reduced electrical costs	244	382	424
Added costs			
Gas (gas and oil)	90	158	192
Gas only	104	184	218
Oil (gas and oil)	26	56	56
Operation and maintenance	38	52	54

3.17. The longer it takes to build a facility, the lower its rate of return. Formulate the ratio of total investment I divided by annual cash flow C (profit after taxes plus depreciation) in terms of one-, two-, and three-year construction periods if i = interest rate, and n = life of facility (no salvage value).

Next, suppose that a $1-million facility will yield a cash flow of $150,000 per year for 15 years after construction. What will be the rate of return if the facility takes one, two, or three years to build?

3.18. McChesney and McChesney (*Chemical Engineering*, May 3, 1982, p. 70) described 20 factors that must be considered in a full analysis of the economics of insulation installation:

Insulation factors

1. Cost of installed insulation, of thickness t, per linear foot $C_{I(t)}$
2. Thermal conductivity of insulation, k_I
3. Thermal resistance of insulation surface, R_S
4. Pipe diameter (nominal), d_1
5. Ambient temperature, θ_3
6. Ambient wind speed
7. Pipe temperature, θ_1
8. Amortization period of insulation, n
9. Pipe-complexity factor
10. Maintenance and insurance costs

Fuel factors

11. Type of fuel and cost, C_F
12. Expected annual price rise of fuel expressed as a decimal, f

Heat-producing-plant factors

13. Efficiency of conversion of fuel to heat, E
14. Number of hours of operation per year, N
15. Capital investment in heat-producing plant
16. Amortization period of heat-producing plant

Economic factors

17. Cost of money, i.e., return on investment in insulation required, i
18. Tax rate
19. Cost of money to finance heat-producing plant
20. Economic model used for determining the economic thickness of insulation

For simplicity they assumed a pipe-complexity factor of unity, ignored maintenance and insurance costs, assumed zero wind speed, and ignored the economic aspects of the heat-producing plant; this represents an incremental positive cash flow.

Table P3.18 gives the results of solving the heat transfer equations for the specific case of a straight horizontal pipe of nominal 8-in. dia., carrying dry saturated steam at 500°F, located in still air, lagged with various thicknesses of calcium silicate insulation. In addition, the boiler was assumed to be 83% efficient, operating for a full 8760 hours per year, using fuel costing $2.84 per million Btu.

Given the data in the Table P3.18, how would you determine (*a*) the net present value of future cash flows, and (*b*) the effect of increases in prices of fuel over the amortization period?

Thickness of calcium silicate insulation, t, in	Surface temperature of insulation Θ_2, °F	Heat loss rate per linear foot $Q_{(t)}/L$ Btu h^{-1} ft^{-1}	Dollar value of heat lost per linear foot per year of operation $C_{H(t)}$ (\$) ft^{-1} $year^{-1}$	Installed cost of insulation per linear foot $C_{I(t)}$ (\$) ft^{-1}
0	500	3708	111	—
$\frac{1}{2}$	210	756	22.7	2.0
1	160	479	14.5	4.1
$1\frac{1}{2}$	136	358	10.7	6.2
2	122	292	8.74	8.4
$2\frac{1}{2}$	113	249	7.47	10.5
3	107	220	6.58	12.6
$3\frac{1}{2}$	102	198	5.93	14.7
4	99	181	5.42	16.8
$4\frac{1}{2}$	95	167	5.02	18.9
5	92	156	4.69	21.0
$5\frac{1}{2}$	90	147	4.42	23.1
6	89	140	4.18	25.2

If you are asked to determine the economic thickness of the insulation, how would you calculate this value. (*Hint:* Add up the cost of the heat lost and the insulation, both per linear foot). Assume $i = 0.12$.

3.19. A chemical valued at \$0.94/lb is currently being dried in a fluid-bed dryer that allows 0.1 percent of the 4-million-lb/year throughput to be carried out in the exhaust. An engineer is considering installing a \$10,000 cyclone that would recover the fines; extra pressure drop is no concern. What is the expected pay-back period for this investment? Maintenance costs are estimated to be \$300/year. The inflation rate is 8 percent and the interest rate 15 percent.

3.20. The following letter appeared in *Science* (vol. 208, Apr. 25, 1980, p. 349).

***When* Should the Gas Guzzler Go?**

Taking into account loan financing parameters and the income tax deductibility of interest payments, as well as the discount rate and various other parameters, the optimal amount of time to keep the guzzler can be found by the following algorithm.

1. Obtain the values of k, the discount rate in percentage per month; Z, the monthly cost of insurance and repairs for the guzzler; H, the same for the high-mileage car you are considering; C, the cost of the new car; S, the trade-in value of the guzzler; P, the price of gasoline; D, the annual miles driven; M, the guzzler's mileage in miles per gallon; L, the replacement period for the new car in years; f, the trade-in value of an L-year-old high-mileage car as a fraction of its purchase price; h, the amount by which the discount rate exceeds the escalation rate for repairs and insurance on the high-mileage car; z, the same for the guzzler; c, the amount by which the discount rate exceeds the escalation rate of new car purchases; g, the amount by which the discount rate exceeds the fuel escalation rate (this may be negative!); s, the amount by which it exceeds the guzzler's trade-in value escalation rate; i, the same for the loan interest rate; I, the loan interest rate itself; R, the ratio of mileage of the new car to the improvement in mileage that would result from dumping the guzzler; d, the down-payment fraction on new-car loans; N, the new-car loan period in months; and finally T, your marginal income tax rate.

2. Calculate the monthly gasoline cost savings by

$$A = DP/12\ MR$$

3. Calculate an upper bound on the optimum time to keep the guzzler. From my earlier letter, $Y_{\max}$ = the smaller of $(C/12A)$ and (years until the guzzler will drop dead).
4. The formula for the capital recovery function is

$$\text{crf}(r, n) = \begin{cases} r/[1 - (1 + r)^{-n}] & \text{if } r \neq 0 \\ 1/n & \text{if } r = 0 \end{cases}$$

5. Find the Y from the set $\{0, 1/12, 2/12, \ldots, Y_{\max}\}$ that maximizes V. I'll leave the search procedure up to you, the reader. V, the net present value of keeping the guzzler for Y more years before replacing it with a sequence of high-mileage cars instead of replacing it immediately, may be obtained as follows:

Expenses:

$$X = H/\text{crf}(h, 12Y) - Z/\text{crf}(z, 12Y) + A/\text{crf}(g, 12Y)$$

Capital required:

$$Q = \frac{C \times c}{\text{crf}(c, 12Y)}\left[1 + \frac{1 - f}{(1 + c)^{12L} - 1}\right] - \frac{S \times s}{\text{crf}(s, 12Y)}$$

Down payments;

$$D = dQ$$

Loan payments:

$$E = (1 - d)\ Q\ \text{crf}(I, N)/\text{crf}(k, N)$$

Tax reductions:

$$B = T\left\{E - \frac{(1 - d)\ Q[\text{crf}(I, N) - I]}{(1 + I)\ \text{crf}(i, N)}\right\}$$

Finally,

$$V = X + D + E - B$$

If none of the values of Y produces a positive value of V, get rid of the guzzler now—unless, of course, you *like* it better.

In closing, I might point out that the trade-off considered here is that of a continuation of high operating costs versus an initial investment followed by low operating costs.

Discuss whether or not analysis has been carried out correctly.

PART II

OPTIMIZATION THEORY AND METHODS

Part II describes modern techniques of optimization and translates these concepts into computational methods and algorithms. A vast literature on optimization techniques exists, hence we have focused solely on methods which have been proved effective for a wide range of problems. Optimization methods have matured sufficiently during the past twenty years so that fast and reliable methods are available to solve each important class of problem.

Six chapters make up Part II of this book, covering the following areas:

1. Mathematical concepts (Chapter 4)
2. One-dimensional search (Chapter 5)
3. Unconstrained multivariable optimization (Chapter 6)
4. Linear programming (Chapter 7)
5. Nonlinear programming (Chapter 8)
6. Optimization involving staged processes and discrete variables (Chapter 9)

The topics are grouped so that unconstrained methods are presented first, followed by constrained methods. The last chapter in Part II deals with discontinuous (integer) variables, a common category of problem in chemical engineering but one quite difficult to solve without great effort.

As optimization methods as well as computer hardware have been improved over the past two decades, the degree of difficulty of the problems that can be solved has expanded significantly. Continued improvements in optimization algorithms and computer hardware and software should enable optimization of large-scale nonlinear problems involving thousands of variables, both continuous and integer, some of which may be stochastic in nature.

CHAPTER 4

BASIC CONCEPTS OF OPTIMIZATION

In order to understand the strategy of optimization procedures, certain basic concepts must be described. In this chapter we will examine the properties of objective functions and constraints to establish a basis for analyzing optimization problems. Those features which are desirable (and also undesirable) in the formulation of an optimization problem are identified. Both qualitative and quantitative characteristics of functions will be described. In addition, we will present the necessary and sufficient conditions to guarantee that a supposed extremum is indeed a minimum or a maximum.

4.1 CONTINUITY OF FUNCTIONS

In carrying out analytical or numerical optimization you will find it preferable and more convenient to work with continuous functions of one or more variables than with functions containing discontinuities. Functions having continuous derivatives are also preferred. What does continuity mean? Examine Fig. 4.1. In case *A*, the function is clearly discontinuous. Is case *B* also discontinuous?

We define the property of continuity as follows. A function of a single variable x is continuous at a point x_0 if

(*a*) $f(x_0)$ exists

(*b*) $\lim_{x \to x_0} f(x)$ exists

(*c*) $\lim_{x \to x_0} f(x) = f(x_0)$

If $f(x)$ is continuous at every point in region R, then $f(x)$ is said to be continuous throughout R. For case B in Fig. 4.1, the function of x has a "kink" in it, but $f(x)$ does satisfy the property of continuity. However, $f'(x) \equiv df(x)/dx$ does not. Therefore, the function in case B is continuous but not continuously differentiable.

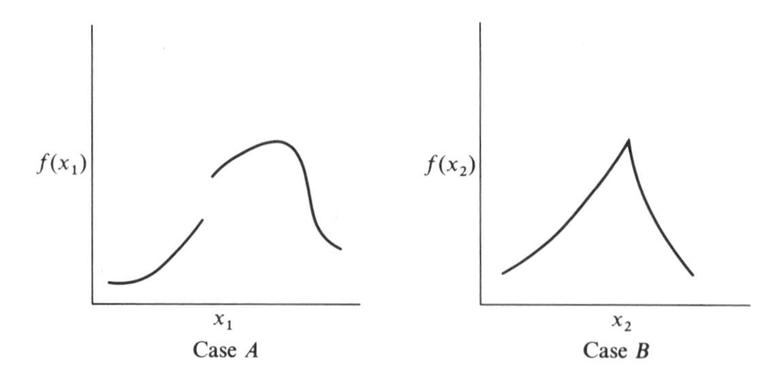

Figure 4.1 Functions with discontinuities in the function and/or derivatives.

EXAMPLE 4.1 ANALYSIS OF FUNCTIONS FOR CONTINUITY

Are the following functions continuous? (*a*) $f(x) = 1/x$; (*b*) $f(x) = \ln x$. In each case specify the range of x for which $f(x)$ and $f'(x)$ are continuous.

Solution

(*a*) $f(x) = 1/x$ is continuous except at $x = 0$; $f(0)$ is not defined. $f'(x) = -1/x^2$ is continuous except at $x = 0$.

(*b*) $f(x) = \ln x$ is continuous for $x > 0$. For $x \leq 0$, $\ln(x)$ is not defined. As to $f'(x) = 1/x$, see (*a*).

A discontinuity in a function may or may not cause difficulty in optimization. In case *A* in Fig. 4.1, the maximum occurs reasonably far from the discontinuity and may or may not be encountered in the search for the optimum. In case *B*, if a method of optimization that does not use derivatives is employed, then the "kink" in $f(x)$ will probably be unimportant, but methods employing derivatives might fail, because the derivative becomes undefined at the discontinuity and has different signs on each side of the discontinuity. Hence small changes in x do not lead to convergence.

One type of discontinuous objective function, one that allows only discrete values of the independent variable(s), occurs frequently in process design because the process variables assume only specific values rather than *continuous* values. Examples are the cost per unit diameter of pipe, the cost per unit area for heat exchanger surface, or the insulation cost considered in Example 1.1. For a pipe, we might represent the installed cost as a function of the pipe diameter as shown in Fig. 4.2. See also Noltie (1978). Although in fact discontinuous, the cost function can for most purposes be approximated as a continuous function because of the relatively small differences in available pipe diameters. You could

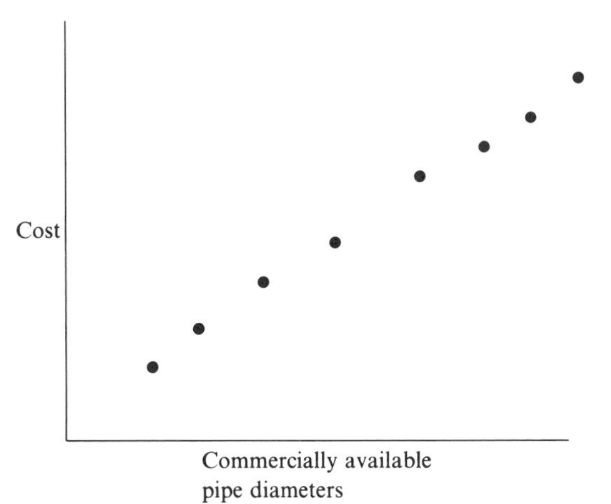

Figure 4.2 Installed pipe cost as a function of diameter.

then disregard the discrete nature of the function and optimize the cost as if the diameter were a continuous variable. Once the optimum value of the diameter is obtained for the continuous function, the discrete-valued diameter nearest to the optimum that is commercially available can be selected. A suboptimal value for installed cost will result, but such a solution should be sufficiently adequate for engineering purposes because of the narrow intervals between discrete values of the diameter.

EXAMPLE 4.2 OPTIMIZATION INVOLVING AN INTEGER-VALUED VARIABLE

Consider a catalytic regeneration cycle in which there is a simple trade-off between costs incurred during regeneration and the increased revenues due to the regenerated catalyst. Let x_1 be the number of days during which the catalyst is used in the reactor, and x_2 be the number of days for regeneration. The reactor start-up crew is only available in the morning shift, so $x_1 + x_2$ must be an integer.

We will assume that the reactor feed-flow rate q (kg/day) is constant as is the cost of the feed C_1 ($/kg), the value of the product C_2 ($/kg), and the regeneration cost C_3 ($/regeneration cycle). We will further assume that the catalyst deteriorates gradually according to the linear relation

$$d = 1.0 - kx_1$$

where 1.0 represents the weight fraction conversion of feed at the start of the operating cycle, and k is the deterioration factor in units of weight fraction per day. Define an objective function and find the optimal value of x_1.

Solution. For one complete cycle of operation and regeneration, the objective function for the total profit per day is comprised of

$$\frac{\text{Profit}}{\text{Day}} = \text{product value} - \text{feed cost} - (\text{regeneration cost per cycle})\cdot(\text{cycles per day})$$

or in the defined notation

$$f(\mathbf{x}) = \frac{qC_2 x_1 d_{\text{avg}} - qC_1 x_1 - C_3}{x_1 + x_2} \tag{a}$$

where $d_{\text{avg}} = 1.0 - (kx_1/2)$.

The maximum daily profit for an entire cycle would be obtained by maximizing Eq. (a) with respect to x_1. When the first derivative of Eq (a) is set equal to zero and the resulting equation solved for x_1, the optimum is

$$x_1^{\text{opt}} = -x_2 + \left[x_2^2 + \left(\frac{2}{k}\right)\left(x_2 - \frac{C_1 x_2}{C_2} + \frac{C_3}{qC_2}\right)\right]^{1/2}$$

Suppose $x_2 = 2$, $k_1 = 0.02$, $q = 1000$, $C_2 = 1.0$, $C_1 = 0.4$, and $C_3 = 1000$. Then $x_1^{\text{opt}} = 12.97$ (rounded to 13 days if x_1 is an integer).

Clearly, treating x_1 as a continuous variable may be improper if x_1 is 1, 2, 3, etc., but is probably satisfactory if x_1 is 15, 16, 17, etc. You might specify x_1 in terms of shifts of four or eight hours instead of days to obtain finer subdivisions of time.

You will find in real life that other problems involving discrete variables may not be so nicely posed. For example, if the cost is a function of the number of discrete pieces of equipment, such as compressors, the optimization procedure cannot ignore the integer character of the cost function because usually only a small number of pieces of equipment are involved. You cannot install 1.54 compressors, and rounding off to 1 or 2 compressors may be quite unsatisfactory. This subject will be discussed in more detail in Chap. 9.

4.2 UNIMODAL VERSUS MULTIMODAL FUNCTIONS

In formulating an objective function, it is far better, if possible, to choose a unimodal than a multimodal function as a performance criterion. Compare Fig. 4.3*a* with Fig. 4.3*b*. A unimodal function $f(x)$ (in the range specfied for x) has a single extremum (minimum or maximum) whereas a multimodal function has two or more extrema. If $f'(x) = 0$ at the extremum, the point is called a *stationary point* (and can be a maximum or a minimum). Multiple stationary points occur for multimodal functions. The distinction between the *global extremum*, the biggest or smallest among a set of extrema, and *local extrema* (any extremum) becomes significant in numerous practical optimization problems involving nonlinear functions. (Numerical procedures usually terminate at a local extremum, and that point may not be the point you are seeking.) More precisely, for a function of a single variable, if x^* is the point where $f(x)$ reaches a maximum (see Fig. 4.3*a*), unimodality is defined as

$$\begin{aligned} f(x_1) < f(x_2) < f(x^*) \qquad x_1 < x_2 < x^* \\ f(x_4) < f(x_3) < f(x^*) \qquad x^* < x_3 < x_4 \end{aligned} \tag{4.1}$$

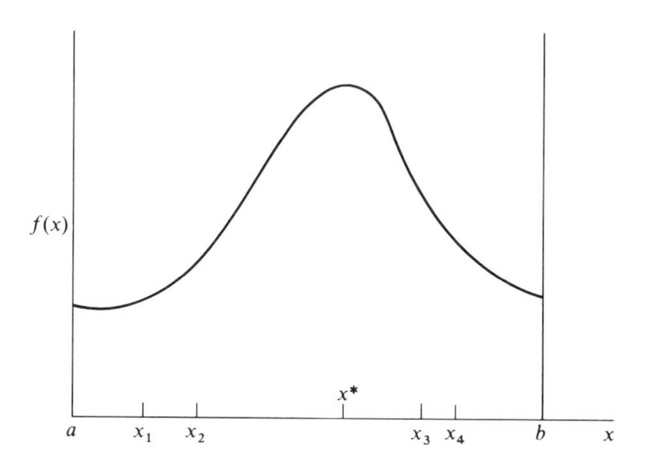

Figure 4.3*a* A unimodal function.

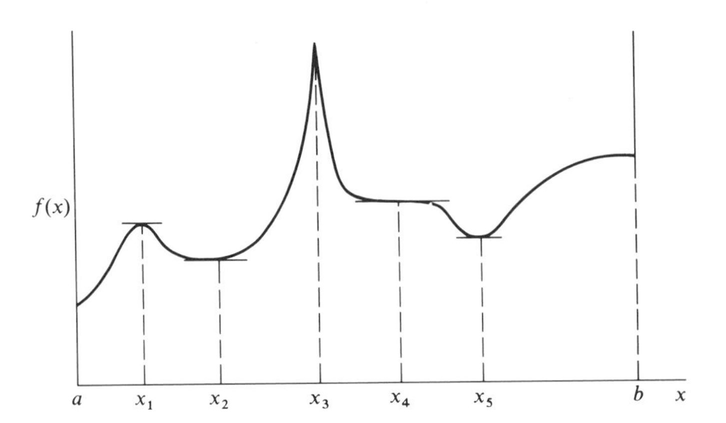

Figure 4.3*b* A multimodal function.

For a maximum, the function $f(x)$ monotonically increases from the left to the maximum, and monotonically decreases to the right of the maximum. An analogous set of relations to (4.1) can be written for a minimum. In Fig. 4.4, the point x_4 represents a saddle point (the concept of which is clear when you later on examine Fig. 4.14). We shall in Chap. 5 describe numerical techniques that are based on the underlying assumption that the function treated is unimodal. The property of unimodality is difficult to establish analytically, as demonstrated by Wilde and Beightler (1967). For functions of one or two variables, the function can be plotted, making it obvious whether or not the function is unimodal. For example, Fig. 4.4 illustrates the contours of a multimodal function of two variables with two minima.

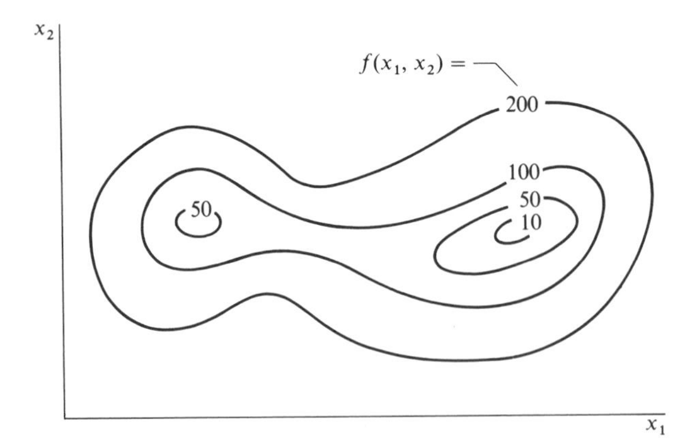

Figure 4.4 Contours of $f(x_1, x_2)$, a multimodal function.

4.3 CONVEX AND CONCAVE FUNCTIONS

Determination of convexity, or concavity, will help you establish whether a local optimal solution is also the global optimal solution (the best among all solutions), a matter of some concern in view of the remarks made in Sec. 4.2 and in Chap. 1 regarding multiple optima. When the objective function is known to have certain properties (defined below), computation of the optimum can be accelerated by using appropriate optimization algorithms. A function is called *concave* over the region R if the following relation holds. For any two different values of x (x may be a vector $\mathbf{x}$), x_a and x_b lying in the region R,

$$f[\theta x_a + (1 - \theta)x_b] \geq \theta f(x_a) + (1 - \theta) f(x_b) \tag{4.2}$$

where θ is a scalar having a value between 0 and 1. The function is *strictly concave* if the greater than or equal to sign of (4.2) is replaced with the greater than ($>$) sign.

Figure 4.5 illustrates the concepts involved in Eq. (4.2). If the values of $f(x)$ on each straight line (in the figure the dashed line) connecting pairs of function values $f(x_a)$ and $f(x_b)$ for all pairs x_a and x_b lie on or below the function values themselves, then the function is concave. If the values of $f(x)$ on each straight line between $f(x_a)$ and $f(x_b)$ lie above or on the function values themselves, then the function is said to be *convex*. For a convex function the inequality sign would be reversed in Eq. (4.2). What would *strictly convex* imply with respect to the $\leq$ sign? The expressions "strictly concave" or "strictly convex" imply that the dashed lines between $f(x_a)$ and $f(x_b)$ cannot fall on the function itself except at the end points of the line. The common example of a function that is convex but not strictly convex is a straight line; the line is also concave but not strictly concave.

Equation (4.2) is not a convenient equation to use in testing for convexity or concavity. Instead, we will make use of the second derivative of $f(x)$, or $\nabla^2 f(\mathbf{x})$ if x is a vector. $\nabla^2 f(\mathbf{x})$ is called the Hessian matrix of $f(\mathbf{x})$, often denoted by the symbol $\mathbf{H}(\mathbf{x})$, and is the symmetric matrix of second derivatives of

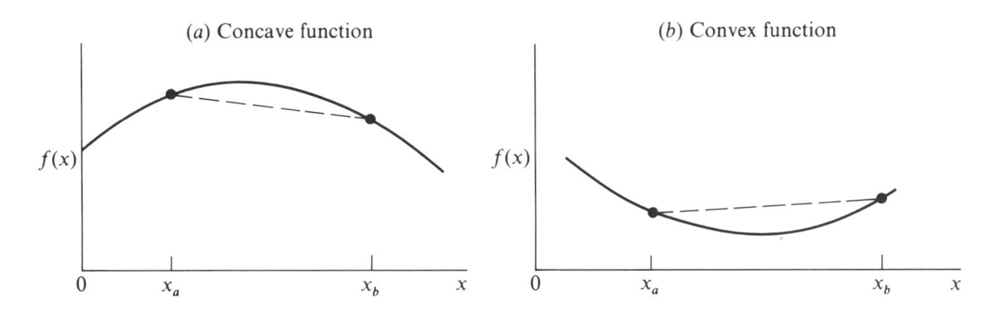

Figure 4.5 Comparison of concave (a) and convex (b) functions.

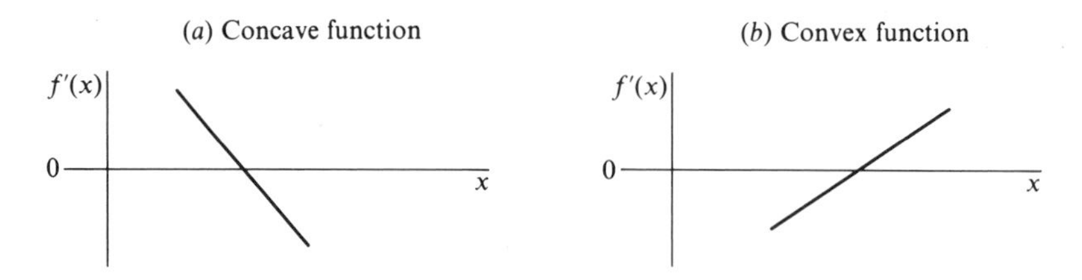

Figure 4.6 Plot of the first derivatives of a quadratic concave and convex function of a single variable.

$f(\mathbf{x})$ (see App. B). For example, if $f(\mathbf{x})$ is a function of two variables and is quadratic

$$f(\mathbf{x}) = h_{11}x_1^2 + h_{12}x_1x_2 + h_{22}x_2^2$$

$$\mathbf{H}(\mathbf{x}) \equiv \nabla^2 f(\mathbf{x}) = \begin{bmatrix} \dfrac{\partial^2 f(\mathbf{x})}{\partial x_1^2} & \dfrac{\partial^2 f(\mathbf{x})}{\partial x_1 \partial x_2} \\ \dfrac{\partial^2 f(\mathbf{x})}{\partial x_2\, \partial x_1} & \dfrac{\partial^2 f(\mathbf{x})}{\partial x_2^2} \end{bmatrix} = \begin{bmatrix} 2h_{11} & h_{12} \\ h_{12} & 2h_{22} \end{bmatrix}$$

Suppose we examine the simplest function of one variable that can have curvature, a quadratic function, such as shown in Figs. 4.5*a* and *b*. A plot of the first derivative of $f(x)$ would appear as indicated in Fig. 4.6. Figure 4.7 illustrates the value of the second derivative, $f''(x) = d^2f(x)/dx^2$.

From Fig. 4.7, we conclude that if the sign of the second derivative of $f(x)$ in the range of $a \leq x \leq b$ is always negative or zero, then $f(x)$ is concave; if the sign of $f''(x)$ is positive or zero (nonnegative) for $a \leq x \leq b$, then $f(x)$ will be convex. *Strictly concave* means that the sign of $f''(x)$ is always negative and *strictly convex* means that the sign of $f''(x)$ is always positive in the specified range of x. A general proof of these statements is available in Aoki (1971).

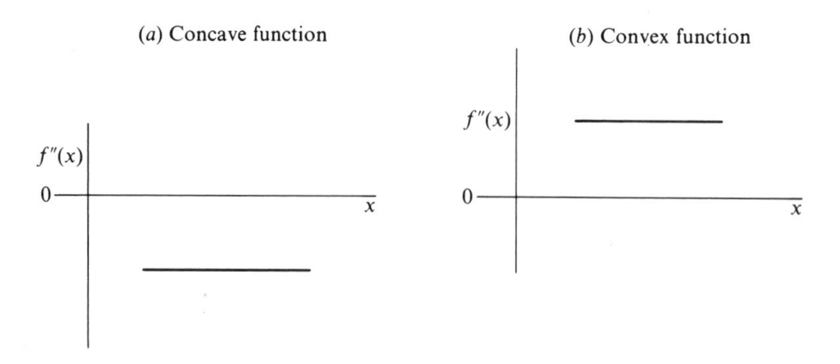

Figure 4.7 Second derivative of $f(x)$ for a quadratic function.

EXAMPLE 4.3 ANALYSIS FOR CONVEXITY AND CONCAVITY

For each of the functions below,

(*a*) $f(x) = 3x^2$
(*b*) $f(x) = 2x$
(*c*) $f(x) = -5x^2$
(*d*) $f(x) = 2x^2 - x^3$

determine if $f(x)$ is convex, concave, strictly convex, strictly concave, all, or none of these classes in the range $-\infty \leq x \leq \infty$.

Solution

(*a*) $f''(x) = 6$, always positive, hence $f(x)$ is both strictly convex and convex.

(*b*) $f''(x) = 0$ for all values of x, hence $f(x)$ is convex and concave. Note straight lines are both convex and concave simultaneously.

(*c*) $f''(x) = -10$, always negative, hence $f(x)$ is both strictly concave and concave.

(*d*) $f''(x) = 6 - 3x$; may be positive or negative depending on the value of x, hence $f(x)$ is not convex or concave over the entire range of x.

The concepts of concavity and convexity also apply to a multivariable function $f(\mathbf{x})$. For any objective functions, the Hessian matrix $\mathbf{H}(\mathbf{x})$ must be evaluated to determine the nature of $f(\mathbf{x})$. First let us summarize the definitions of matrix types:

1. $\mathbf{H}$ is *positive definite* if and only if $\mathbf{x}^T \mathbf{H} \mathbf{x}$ is >0 for all $\mathbf{x} \neq \mathbf{0}$.
2. $\mathbf{H}$ is *negative definite* if and only if $\mathbf{x}^T \mathbf{H} \mathbf{x} < 0$ for all $\mathbf{x} \neq \mathbf{0}$.
3. $\mathbf{H}$ is *indefinite* if $\mathbf{x}^T \mathbf{H} \mathbf{x} < 0$ for some $\mathbf{x}$ and >0 for other $\mathbf{x}$.

Definitions 1 and 2 can be extended to $\mathbf{H}$ being positive semidefinite ($\mathbf{x}^T \mathbf{H} \mathbf{x} \geq 0$) or negative semidefinite ($\mathbf{x}^T \mathbf{H} \mathbf{x} \leq 0$), for all $\mathbf{x}$. It can be shown from a Taylor series expansion that if $f(\mathbf{x})$ has continuous second partial derivatives, $f(\mathbf{x})$ is concave if and only if the Hessian matrix is negative semidefinite. For $f(\mathbf{x})$ to be strictly concave, $\mathbf{H}$ must be negative definite. For $f(\mathbf{x})$ to be convex $\mathbf{H}(\mathbf{x})$ must be positive semidefinite and for $f(\mathbf{x})$ to be strictly convex, $\mathbf{H}(\mathbf{x})$ must be positive definite.

Two convenient tests can be used to establish the status of $\mathbf{H}(\mathbf{x})$ for strict convexity:

(1) All diagonal elements must be positive and the determinants of all leading principal minors, det $\{M_i(\mathbf{H})\}$, and of $\mathbf{H}(\mathbf{x})$ itself, det $(\mathbf{H})$, are positive (>0). Keep in mind that $\mathbf{H}(\mathbf{x})$ *must be a symmetric matrix.*

Another test is

(2) All the eigenvalues of $\mathbf{H}(\mathbf{x})$ are positive (>0).

Table 4.1 Relationship between the character of $f(\mathbf{x})$ and the state of $\mathbf{H}(\mathbf{x})$

$f(\mathbf{x})$ is	$\mathbf{H}(\mathbf{x})$ is	All the eigenvalues of $\mathbf{H}(\mathbf{x})$ are	Determinants of the leading principal minors of $\mathbf{H}^*$ (Δ_i)
Strictly convex	Positive definite	>0	$\Delta_1 > 0, \Delta_2 > 0, \ldots$
Convex	Positive semi-definite	≥ 0	$\Delta_1 \geq 0, \Delta_2 \geq 0, \ldots$
Concave	Negative semi-definite	≤ 0	$\Delta_1 \leq 0, \Delta_2 \geq 0, \Delta_3 \leq 0, \ldots$ (alternating sign)
Strictly concave	Negative definite	<0	$\Delta_1 < 0, \Delta_2 > 0, \Delta_3 < 0$ (alternating sign)

Refer to App. B for details of how to calculate the leading principal minors and eigenvalues. For strict concavity, two alternative definitions similarly can be employed:

(1) All diagonal elements must be negative and $\det(\mathbf{H})$ and $\det\{M_i(\mathbf{H})\} > 0$ if i is even ($i = 2, 4, 6, \ldots$); $\det(\mathbf{H})$ and $\det\{M_i(\mathbf{H})\} < 0$ if i is odd ($i = 1, 3, 5, \ldots$), where M_i is the ith principal minor determinant.
(2) All the eigenvalues of $\mathbf{H}(\mathbf{x})$ are negative (<0).

To establish convexity and concavity, the strict inequalities $>$ or $<$, respectively, in the above tests are replaced by $\geq$ or $\leq$, respectively. If a function has a stationary point where the Hessian has eigenvalues of mixed signs, the function is neither convex nor concave.

Table 4.1 summarizes the relations between convexity, concavity, and the state of the Hessian matrix of $f(\mathbf{x})$. We have omitted the indefinite case for $\mathbf{H}$, that is, when $f(\mathbf{x})$ is neither convex or concave.

EXAMPLE 4.4 DETERMINATION OF POSITIVE DEFINITENESS

Classify the function $f(\mathbf{x}) = 2x_1^2 - 3x_1x_2 + 2x_2^2$ using the categories in Table 4.1, or state that it does not belong in any of the categories.

Solution

$$\frac{\partial f(x)}{\partial x_1} = 4x_1 - 3x_2 \qquad \frac{\partial^2 f(x)}{\partial x_1^2} = 4 \qquad \frac{\partial^2 f(x)}{\partial x_2^2} = 4$$

$$\frac{\partial f(x)}{\partial x_2} = -3x_1 + 4x_2 \qquad \frac{\partial^2 f(x)}{\partial x_1 \partial x_2} = \frac{\partial^2 f(x)}{\partial x_2 \partial x_1} = -3 \qquad \mathbf{H}(\mathbf{x}) = \begin{bmatrix} 4 & -3 \\ -3 & 4 \end{bmatrix}$$

Both diagonal elements are positive.

The leading principal minors are:

$$M_1(\text{order } 1) = 4 \qquad \det M_1 = 4$$

$$M_2(\text{order } 2) = \mathbf{H} \qquad \det M_2 = 7$$

hence $\mathbf{H(x)}$ is positive definite. Consequently, $f(\mathbf{x})$ is strictly convex (as well as convex).

EXAMPLE 4.5 DETERMINATION OF POSITIVE DEFINITENESS

Repeat the analysis of Example 4.4 for $f(\mathbf{x}) = x_1^2 + x_1x_2 + 2x_2 + 4$.

Solution

$$\mathbf{H(x)} = \begin{bmatrix} 2 & 1 \\ 1 & 0 \end{bmatrix}$$

$$\det(\mathbf{H(x)}) = -1$$

The principal minors are:

$$M_1 = 2 \qquad M_2 = \begin{bmatrix} 2 & 1 \\ 1 & 0 \end{bmatrix} \qquad \det M_2 = -1$$

$$\det M_1 = 2$$

Consequently, $f(\mathbf{x})$ does not fall into any of the categories in Table 4.1. We conclude that no unique extremum exists. Can you demonstrate the same result by calculating the eigenvalues of $\mathbf{H}$?

EXAMPLE 4.6 DETERMINATION OF CONVEXITY AND CONCAVITY

Repeat the analysis of Example 4.4 for $f(\mathbf{x}) = 2x_1 + 3x_2 + 6$.

Solution

$$\mathbf{H(x)} = \begin{bmatrix} 0 & 0 \\ 0 & 0 \end{bmatrix}$$

hence the function is both convex and concave.

EXAMPLE 4.7 DETERMINATION OF CONVEXITY

Consider the following objective function: is it convex? Use eigenvalues in the analysis.

$$f(\mathbf{x}) = 2x_1^2 + 2x_1x_2 + 1.5x_2^2 + 7x_1 + 8x_2 + 24$$

Solution

$$\frac{\partial^2 f(x)}{\partial x_1^2} = 4 \qquad \frac{\partial^2 f(x)}{\partial x_2^2} = 3 \qquad \frac{\partial^2 f(x)}{\partial x_1 \partial x_2} = \frac{\partial^2 f(x)}{\partial x_2 \partial x_1} = 2$$

Therefore the Hessian matrix is

$$\mathbf{H(x)} = \begin{bmatrix} 4 & 2 \\ 2 & 3 \end{bmatrix}$$

Next determine the eigenvalues of $\mathbf{H}(\mathbf{x})$.

$$\mathbf{H} - \alpha\mathbf{I} = \begin{bmatrix} 4-\alpha & 2 \\ 2 & 3-\alpha \end{bmatrix}$$

$$\det \begin{bmatrix} 4-\alpha & 2 \\ 2 & 3-\alpha \end{bmatrix} = (4-\alpha)(3-\alpha) - 4 = 0$$

$$\alpha^2 - 7\alpha + 8 = 0$$

$$\alpha_1 = 5.56$$

$$\alpha_2 = 1.44$$

Because both eigenvalues are positive, the function is strictly convex (and convex, of course) for all values of x_1 and x_2.

EXAMPLE 4.8 CLASSIFICATION OF STATIONARY POINTS

Find the stationary points and their classification for the nonlinear function

$$f(\mathbf{x}) = x_1^3 + x_2^2 - 3x_1 + 8x_2 + 2$$

Solution. The stationary points are found from solving simultaneously

$$\frac{\partial f}{\partial x} = 0 = 3x_1^2 - 3$$

and

$$\frac{\partial f}{\partial x_2} = 0 = 2x_2 + 8$$

and are $(1, -4)$ and $(-1, -4)$. Next find the Hessian:

$$\mathbf{H}(\mathbf{x}) = \begin{bmatrix} 6x_1 & 0 \\ 0 & 2 \end{bmatrix}$$

For $\mathbf{x}^* = (1, -4)$, $\mathbf{H}$ is positive definite, hence $f(\mathbf{x}^*)$ is convex at that point. For $\mathbf{x}^* = (-1, -4)$, $\mathbf{H}$ is indefinite. This point corresponds to a saddlepoint, a topic to be discussed later in Sec. 4.6.

In some instances we can use the concept that a sum of convex (concave) functions is also convex (concave). We can examine any $f(\mathbf{x})$ in terms of its component parts. If the function is separable into $f(\mathbf{x}) = f_1(\mathbf{x}) + f_2(\mathbf{x})$, and if $f_1(\mathbf{x})$ is convex and $f_2(\mathbf{x})$ is convex, then $f(\mathbf{x})$ is convex. For example

$$f(\mathbf{x}) = a_1(x_1 - c_1)^2 + a_2(x_2 - c_2)^2 \qquad a_i \geq 0$$

is convex.

4.4 CONVEX REGION

A convex region (set of points) plays a useful role in optimization involving constraints. Figure 4.8 illustrates the concept of a convex region and one that is

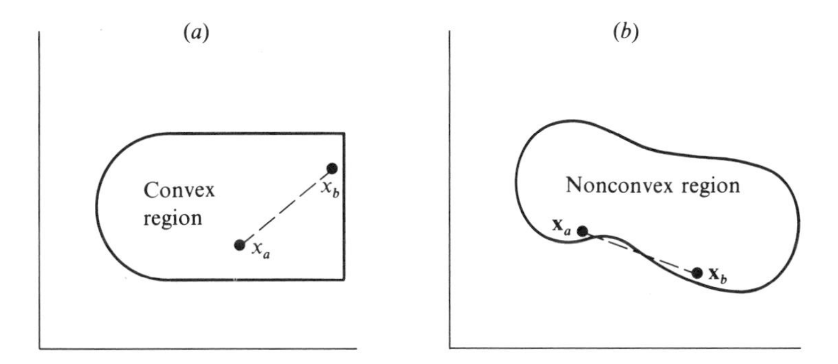

Figure 4.8 Convex and nonconvex regions.

not convex. A convex set of points exists if for any two points in a region, $\mathbf{x}_a$ and $\mathbf{x}_b$, all points $\mathbf{x} = \mu\mathbf{x}_a + (1 - \mu)\mathbf{x}_b$, where $0 \le \mu \le 1$, on the line joining $\mathbf{x}_a$ and $\mathbf{x}_b$ are in the set. In Fig. 4.8*b* note how this requirement is not satisfied along the dashed line. If a region is completely bounded by concave functions for the case in which all $g_i(\mathbf{x}) \ge 0$, then the functions form a closed convex region. For the case in which all the inequality constraints stated in the form $g_i(\mathbf{x}) \le 0$ are convex functions, the functions form a convex region. Keep in mind that straight lines are both concave and convex functions.

EXAMPLE 4.9 DETECTION OF A CONVEX REGION

Does the following set of constraints that form a closed region form a convex region

$$-x_1^2 + x_2 \ge 1$$

$$x_1 - x_2 \le -2$$

Solution. A plot of the two functions indicates the region circumscribed is closed. The arrows in Fig. E4.9 designate the directions in which the inequalities hold. Write the inequality constraints as $g_i \ge 0$. Therefore

$$g_1(\mathbf{x}) = -x_1^2 + x_2 - 1 \ge 0$$

$$g_2(\mathbf{x}) = -x_1 + x_2 - 2 \ge 0$$

That the enclosed region is convex can be demonstrated by showing that both $g_1(\mathbf{x})$ and $g_2(\mathbf{x})$ are concave functions:

$$\mathbf{H}[g_1(\mathbf{x})] = \begin{bmatrix} -2 & 0 \\ 0 & 0 \end{bmatrix}$$

$$\mathbf{H}[g_2(\mathbf{x})] = \begin{bmatrix} 0 & 0 \\ 0 & 0 \end{bmatrix}$$

Since all eigenvalues are zero or negative, according to Table 4.1 both g_1 and g_2 are concave and the region is convex.

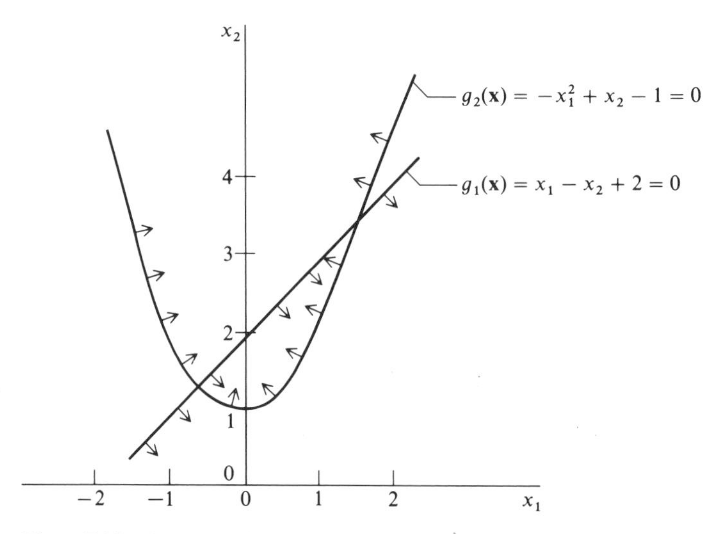

Figure E4.9 Convex region composed of two concave functions.

EXAMPLE 4.10 CONSTRUCTION OF A CONVEX REGION

Construct the region given by the following inequality constraints; is it convex? $x_1 \le 6$; $x_2 \le 6$; $x_1 \ge 0$; $x_1 + x_2 \le 6$; $x_2 \ge 0$

Solution. See Fig. E4.10 for the region delineated by the inequality constraints. By visual inspection, the region is convex. This set of linear inequality constraints

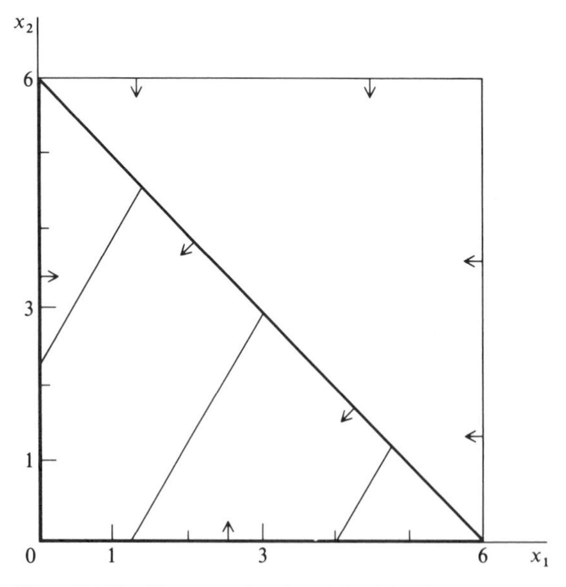

Figure E4.10 Diagram of region defined by linear inequality constraints.

forms a convex region since all the constraints are concave. In this case the convex region is closed.

As mentioned before, the existence of convex regions has an important bearing on optimization of functions subject to constraints. For example, examine Fig. 4.9. In optimization problems involving constraints, points that satisfy *all* the constraints are said to be *feasible* points; all other points are *nonfeasible*

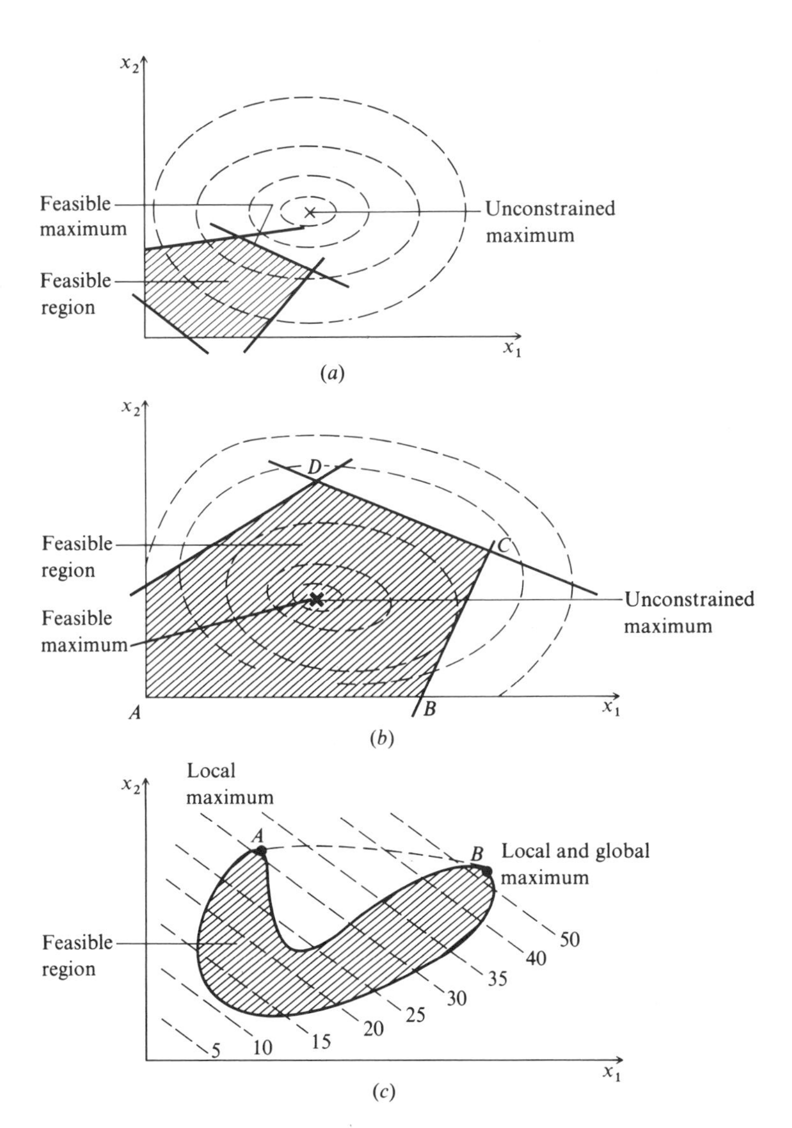

Figure 4.9 Effect of the character of the region of search in solving constrained optimization problems.

points. Inequality constraints specify a *feasible region* comprised of the set of points that are feasible whereas equality constraints limit the feasible set of points to hypersurfaces (multidimensional surfaces defined by $h_i(x_1, x_2, \ldots, x_n) = 0$), curves, or perhaps even a single point. Solid lines in Fig. 4.9 represent inequality constraint boundaries. Dashed lines are the contours of the objective function. All points that satisfy all the inequality constraints (*a*) as strict inequalities are termed *interior points*, and (*b*) as equalities are termed *boundary points*. All other points are *exterior* points.

In Fig. 4.9*a* the maximum of the unconstrained objective function lies outside the feasible region. It is clear that the given set of constraints requires that the search for the optimum terminate at a lesser value of the objective function (for the maximum) termed the constrained solution. In Fig. 4.9*a* one of the constraints at the optimal solution is "active," that is, the inequality constraint is satisfied at its boundary as an equality (=0). If no constraints are "active" at the optimal solution then the unconstrained maximum can be reached as shown in Fig. 4.9*b*. Figure 4.9*c* illustrates the importance of having the constraints comprise a convex region if a global extremum is to be located. The search for the maximum of $f(\mathbf{x})$, if initiated in the left-hand part of the feasible region, might well terminate at point A, whereas a search starting in the right-hand side of the region might terminate at B. Therefore, we conclude that the nature of the search region has an important bearing on the potential for obtaining suitable results in optimization. If the feasible region in Fig. 4.9*c* were extended, as shown by the dashed line, to make the region convex, then the search from any initial point would converge to the same answer.

4.5 NECESSARY AND SUFFICIENT CONDITIONS FOR AN EXTREMUM OF AN UNCONSTRAINED FUNCTION

In optimization of an unconstrained function we are concerned with finding the minimum or maximum of an objective function $f(\mathbf{x})$, a function of one or more variables. The problem can be interpreted geometrically as finding the point in an n-dimension space at which the function has an extremum. Examine Fig. 4.10 in which the contours of a function of two variables are displayed.

An optimal point $\mathbf{x}^*$ is completely specified by satisfying what are called the *necessary* and *sufficient conditions* for optimality. A condition N is necessary for a result R if R can be true only if the condition is true ($R \Rightarrow N$). However, the reverse is not true, that is if N is true, R is not necessarily true. A condition is sufficient for a result R if R is true if the condition is true ($S \Rightarrow R$). A condition T is necessary and sufficient for result R if R is true if and only if T is true ($T \Leftrightarrow R$).

The easiest way to develop the necessary and sufficient conditions for a minimum or maximum of $f(\mathbf{x})$ is to start with a Taylor series expansion about the presumed extremum $\mathbf{x}^*$

$$f(\mathbf{x}) = f(\mathbf{x}^*) + \nabla^T f(\mathbf{x}^*)\,\Delta\mathbf{x} + \tfrac{1}{2}(\Delta\mathbf{x}^T)\nabla^2 f(\mathbf{x}^*)\,\Delta\mathbf{x} + 0_3(\Delta\mathbf{x}) + \cdots \tag{4.3}$$

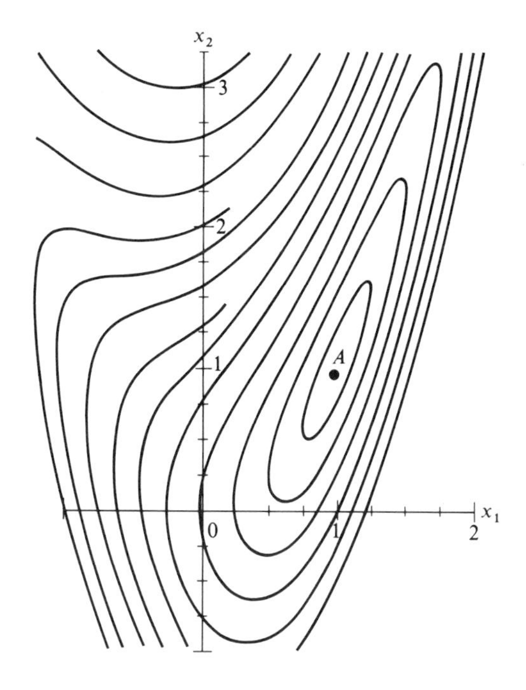

Figure 4.10*a* A function of two variables with a single stationary point, the extremum.

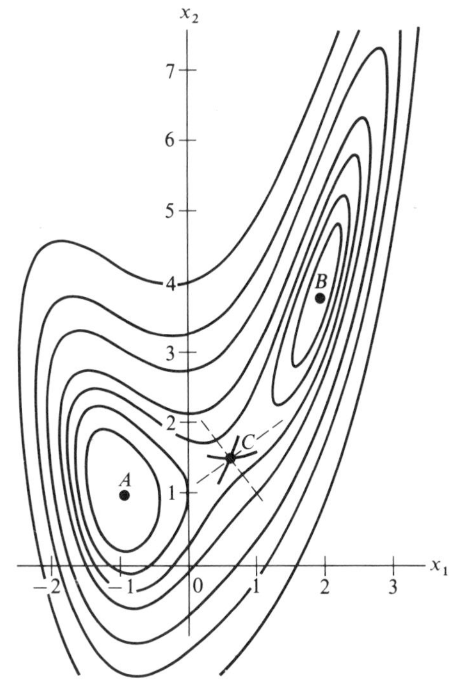

Figure 4.10*b* A function of two variables with three stationary points and two extrema, *A* and *B*.

where $\Delta\mathbf{x} = \mathbf{x} - \mathbf{x}^*$, the perturbation of $\mathbf{x}$ from $\mathbf{x}^*$. We assume all terms in Eq. (4.3) exist and are continuous, but will ignore the terms of order 3 or higher $[0_3(\Delta\mathbf{x})]$, and simply analyze what occurs for various cases involving just the terms through the second order.

By defining a local minimum is a point $\mathbf{x}^*$ such that no other point in the vicinity of $\mathbf{x}^*$ yields a value of $f(\mathbf{x})$ less than $f(\mathbf{x}^*)$, or

$$f(\mathbf{x}) - f(\mathbf{x}^*) \geq 0 \tag{4.4}$$

($\mathbf{x}^*$ is a global minimum if (4.4) holds for any $\mathbf{x}$ in the n-dimensional space of $\mathbf{x}$). Similarly, $\mathbf{x}^*$ is a local maximum if

$$f(\mathbf{x}) - f(\mathbf{x}^*) \leq 0 \tag{4.5}$$

Examine the second term on the right-hand side of Eq. (4.3): $\nabla^T f(\mathbf{x}^*)\,\Delta\mathbf{x}$. Because $\Delta\mathbf{x}$ is arbitrary and can have both plus and minus values of elements, we must insist that $\nabla f(\mathbf{x}^*) = \mathbf{0}$ or otherwise we could add a term to $f(\mathbf{x}^*)$ so that Eq. (4.4) for a minimum, or (4.5) for a maximum, would be violated. Hence, a *necessary condition* for a minimum or maximum of $f(\mathbf{x})$ is that the gradient of $f(\mathbf{x})$ vanishes at $\mathbf{x}^*$

$$\nabla f(\mathbf{x}^*) = \mathbf{0} \tag{4.6}$$

that is, $\mathbf{x}^*$ is a stationary point.

With the second term on the right-hand side of Eq. (4.3) forced to be zero, we next examine the third term: $\frac{1}{2}(\Delta\mathbf{x}^T)\nabla^2 f(\mathbf{x}^*)\,\Delta\mathbf{x}$. This term establishes the character of the stationary point (minimum, maximum, or saddle point). In Fig. 4.10*b*, *A* and *B* are minima while *C* is a saddle point. Note how movement along one of the perpendicular search directions (dashed lines) from point *C* increases $f(\mathbf{x})$ whereas movement in the other direction decreases $f(\mathbf{x})$. Thus, satisfaction of the necessary conditions does not guarantee a minimum or maximum. Figure 4.11 illustrates the character of $f(x)$ if the objective function is a function of a single variable.

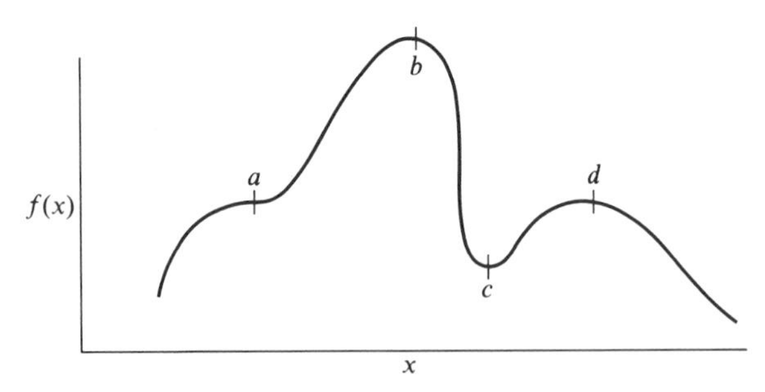

a—inflection point (scalar equivalent to a saddle point)
b—global maximum (and local maximum)
c—local minimum
d—local maximum

Figure 4.11 A function exhibiting different types of stationary points.

To establish the existence of a minimum or maximum at $\mathbf{x}^*$, we know from Eq. (4.3) with $\nabla f(\mathbf{x}^*) = \mathbf{0}$ and the conclusions reached in Sec. 4.3 concerning convexity that for $\Delta \mathbf{x} \neq 0$

$\nabla^2 f(\mathbf{x}^*)$	$\Delta \mathbf{x}^T \nabla^2 f(\mathbf{x}^*) \Delta \mathbf{x}$	Near $\mathbf{x}^*$, $f(\mathbf{x}) - f(\mathbf{x}^*)$
Positive definite	>0	Increases
Positive semidefinite	≥ 0	Possibly increases
Negative definite	<0	Decreases
Negative semidefinite	≤ 0	Possibly decreases
Indefinite	Both ≤ 0 and ≥ 0 depending on $\Delta \mathbf{x}$	Increases, decreases, neither

Consequently, $\mathbf{x}^*$ can be classified as

$\nabla^2 f(\mathbf{x}^*)$	$\mathbf{x}^*$
Positive definite	Unique ("isolated") minimum
Negative definite	Unique ("isolated") maximum

These two conditions are known as the *sufficiency conditions.*

In summary, the necessary conditions (1 and 2 below) and the sufficient condition (3) to guarantee that $\mathbf{x}^*$ is an extremum are as follows:

1. $f(\mathbf{x})$ is twice differentiable at $\mathbf{x}^*$.

2. $\nabla f(\mathbf{x}^*) = \mathbf{0}$, that is, a stationary point exists at $\mathbf{x}^*$.

3. $\mathbf{H}(\mathbf{x}^*)$ is positive definite for a minimum to exist at $\mathbf{x}^*$, and negative definite for a maximum to exist at $\mathbf{x}^*$.

Of course, a minimum or maximum may exist at $\mathbf{x}^*$ even though it is not possible to demonstrate the fact using the three conditions. For example, if $f(x) = x^{4/3}$, $x^* = 0$ is a minimum but $\mathbf{H}(0)$ is not defined at $x^* = 0$, hence condition 3 is not satisfied.

EXAMPLE 4.11 CALCULATION OF A MINIMUM OF $f(x)$

Does $f(x) = x^4$ have an extremum? If so, what is the value of x^* and $f(x^*)$ at the extremum?

Solution

$$f'(x) = 4x^3 \qquad f''(x) = 12x^2$$

Set $f'(x) = 0$ and solve for x; hence $x = 0$ is a stationary point. Also, $f''(0) = 0$, meaning that condition 3 is *not* satisfied. Figure E4.11 is a plot of $f(\mathbf{x}) = x^4$. Thus, a minimum exists for $f(x)$ but the sufficiency condition is not satisfied.

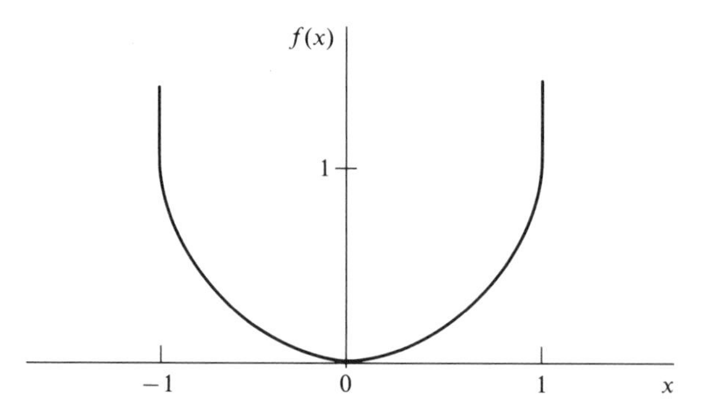

Figure E4.11

If both first and second derivatives vanish at the stationary point, then further analysis is required to evaluate the nature of the function. For functions of a single variable, take successively higher derivatives and evaluate them at the stationary point. Continue this procedure until one of the higher derivatives is not zero (the nth one); hence, $f'(x^*), f''(x^*), \ldots, f^{(n-1)}(x^*)$ all vanish. Two cases must be analyzed:

(*a*) If n is even, the function attains a maximum or a minimum; a positive sign of $f^{(n)}$ indicates a minimum, a negative sign a maximum.

(*b*) If n is odd, the function exhibits a saddle point.

For more details refer to Beveridge and Schechter (1970).

For application of these guidelines to $f(x) = x^4$, you will find $d^4f(x)/dx^4 = 24$ for which n is even and the derivative is positive, so that a minimum exists.

EXAMPLE 4.12 CALCULATION OF EXTREMA

Identify the stationary points of the following function (Fox, 1971), and determine if any extrema exist.

$$f(\mathbf{x}) = 4 + 4.5x_1 - 4x_2 + x_1^2 + 2x_2^2 - 2x_1x_2 + x_1^4 - 2x_1^2x_2$$

Solution. For this function, three stationary points can be located by setting $\nabla f(\mathbf{x}) = \mathbf{0}$:

$$\frac{\partial f(x)}{\partial x_1} = 4.5 + 2x_1 - 2x_2 + 4x_1^3 - 4x_1x_2 = 0 \qquad (a)$$

$$\frac{\partial f(x)}{\partial x_2} = -4 + 4x_2 - 2x_1 - 2x_1^2 = 0 \qquad (b)$$

The set of nonlinear equations (*a*) and (*b*) has to be solved, say by Newton's method, to get the pairs (x_1, x_2) as follows:

Stationary point (x_1, x_2)	$f(\mathbf{x})$	**Hessian matrix eigenvalues**		**Classification**
(1.941, 3.854)	0.9855	37.03	0.97	Local minimum
(−1.053, 1.028)	−0.5134	10.5	3.5	Local minimum (also the global minimum)
(0.6117, 1.4929)	2.83	7.0	−2.56	Saddle point

Figure 4.10*b* shows contours for the objective function in this example. Note that the global minimum can only be identified by evaluating $f(\mathbf{x})$ for all the local minima. For general nonlinear objective functions, it is usually difficult to ascertain the nature of the stationary points without detailed examination of each point.

EXAMPLE 4.13

In many types of processes such as batch constant-pressure filtration or fixed-bed ion exchange, the production rate decreases as a function of time. At some optimal time, t^{opt}, production is terminated (at P^{opt}) and the equipment is cleaned. Figure E4.13*a* illustrates the cumulative throughput $P(t)$ as a function of time (t) for such a process. For one cycle of production and cleaning, the overall production rate is

$$R(t) = \frac{P(t)}{t + t_c} \qquad (a)$$

where $R(t)$ is the overall production rate per cycle (mass/time) and t_c is the cleaning time (assumed to be constant).

Determine the maximum production rate and show that P^{opt} is indeed the maximum throughout.

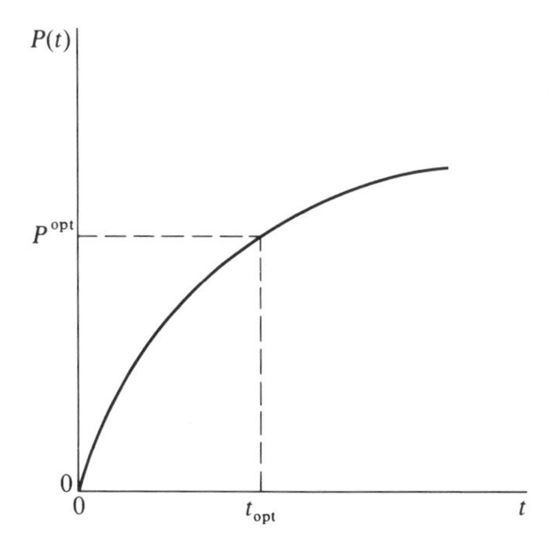

Figure E4.13*a*

Solution. Differentiate $R(t)$ with respect to t and equate the derivative to 0:

$$\frac{dR(t)}{dt} = \frac{-P(t) + [dP(t)/dt](t + t_c)}{(t + t_c)^2} = 0$$

$$P^{\text{opt}} = \left.\frac{dP(t)}{dt}\right|_{\text{opt}} (t + t_c) \tag{b}$$

The geometric interpretation of Eq. (*b*) is the classical result (Walker et al., 1937) that the tangent to $P(t)$ at P^{opt} intersects the time axis at $-t_c$. Examine Fig. E4.13*b*. The maximum overall production rate is

$$R^{\text{opt}} = \frac{P^{\text{opt}}}{t^{\text{opt}} + t_c} \tag{c}$$

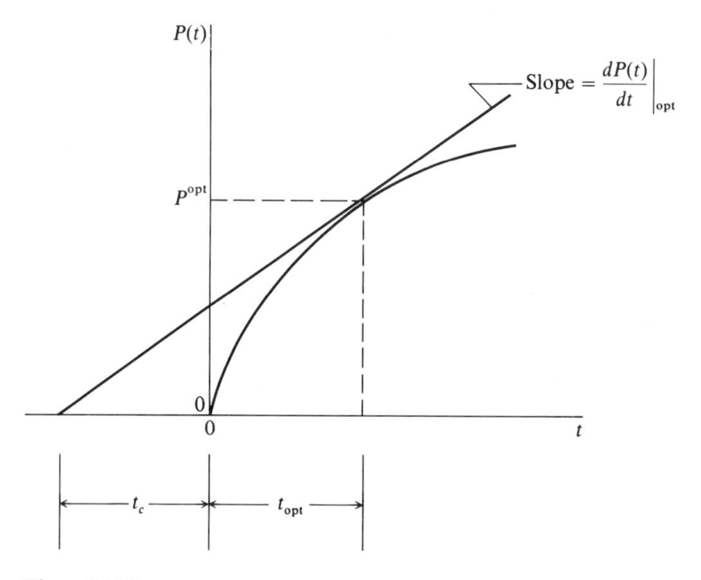

Figure E4.13*b*

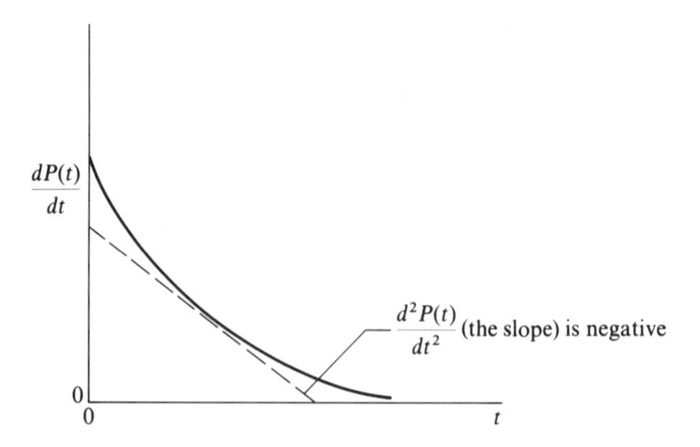

Figure E4.13*c*

Does P^{opt} meet the sufficiency condition to be a maximum? Is

$$\frac{d^2R(t)}{dt^2} = \frac{2P(t) - 2[dP(t)/dt](t + t_c) + [d^2P(t)/dt^2](t + t_c)^2}{(t + t_c)^3} < 0 \quad ? \qquad (d)$$

Rearrangement of (d) and introduction of (b) into (d) or the pair (P^{opt}, t^{opt}) gives

$$\frac{d^2P(t)}{dt^2}(t + t_c)^2 < 0$$

From Fig. E4.13*b* we note in the range $0 < t < t^{opt}$ that $dP(t)/dt$ is always positive and decreasing so that $d^2P(t)/dt^2$ is always negative (see Fig. E4.13*c*). Consequently, the sufficiency condition is met.

4.6 INTERPRETATION OF THE OBJECTIVE FUNCTION IN TERMS OF ITS QUADRATIC APPROXIMATION

If a function of two variables is quadratic or approximated by a quadratic function $f(\mathbf{x}) = b_0 + b_1x_1 + b_2x_2 + b_{11}x_1^2 + b_{22}x_2^2 + b_{12}x_1x_2$ then the eigenvalues of $\mathbf{H}(\mathbf{x})$ can be calculated and used to interpret the nature of $f(\mathbf{x})$ at $\mathbf{x}^*$. Table 4.2 lists some conclusions that can be reached by examining the eigenvalues of $\mathbf{H}(\mathbf{x})$ for a function of two variables, and Figs. 4.12 through 4.16 illustrate the

Table 4.2 Geometric interpretation of a quadratic function

Case	Eigenvalue Relations	Signs α_1	Signs α_2	Types of contours	Geometric interpretation	Character of center of contours	Figure
1	$\alpha_1 = \alpha_2$	−	−	Circles	Circular hill	Maximum	4.12
2	$\alpha_1 = \alpha_2$	+	+	Circles	Circular valley	Minimum	4.12
3	$\alpha_1 > \alpha_2$	−	−	Ellipses	Elliptical hill	Maximum	4.13
4	$\alpha_1 > \alpha_2$	+	+	Ellipses	Elliptical valley	Minimum	4.13
5	$\lvert\alpha_1\rvert = \lvert\alpha_2\rvert$	+	−	Hyperbolas	Symmetrical saddle	Saddle point	4.14
6	$\lvert\alpha_1\rvert = \lvert\alpha_2\rvert$	−	+	Hyperbolas	Symmetrical saddle	Saddle point	4.14
7	$\alpha_1 > \alpha_2$	+	−	Hyperbolas	Elongated saddle	Saddle point	4.14
8	$\alpha_2 = 0$	−		Straight lines	Stationary* ridge	None	4.15
9	$\alpha_2 = 0$	+		Straight lines	Stationary* valley	None	4.15
10	$\alpha_2 = 0$	−		Parabolas	Rising ridge*†	At ∞	4.16
11	$\alpha_2 = 0$	−		Parabolas	Falling valley*†	At ∞	4.16

* These are "degenerate" surfaces.

† The condition of rising or falling must be evaluated from the linear terms in $f(\mathbf{x})$.

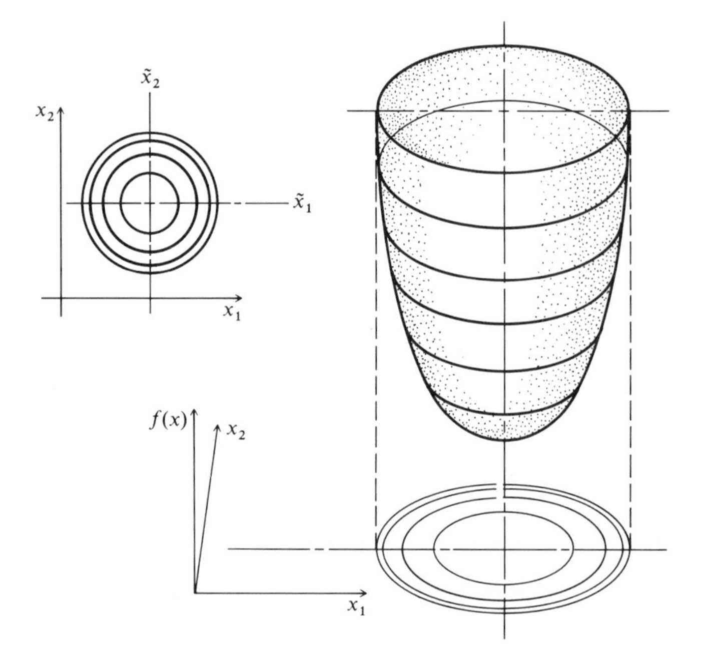

Figure 4.12 Geometry of second-order objective function of two independent variables—circular contours.

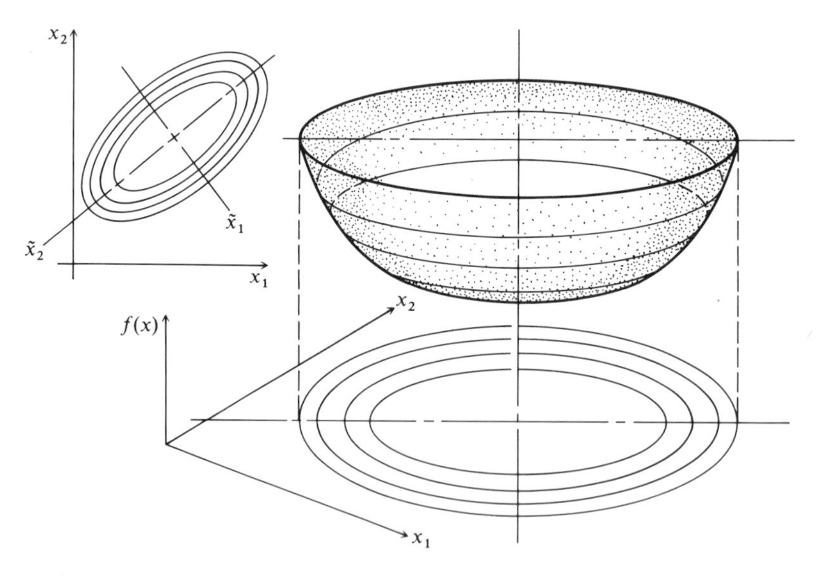

Figure 4.13 Geometry of second-order objective function of two independent variables—elliptical contours.

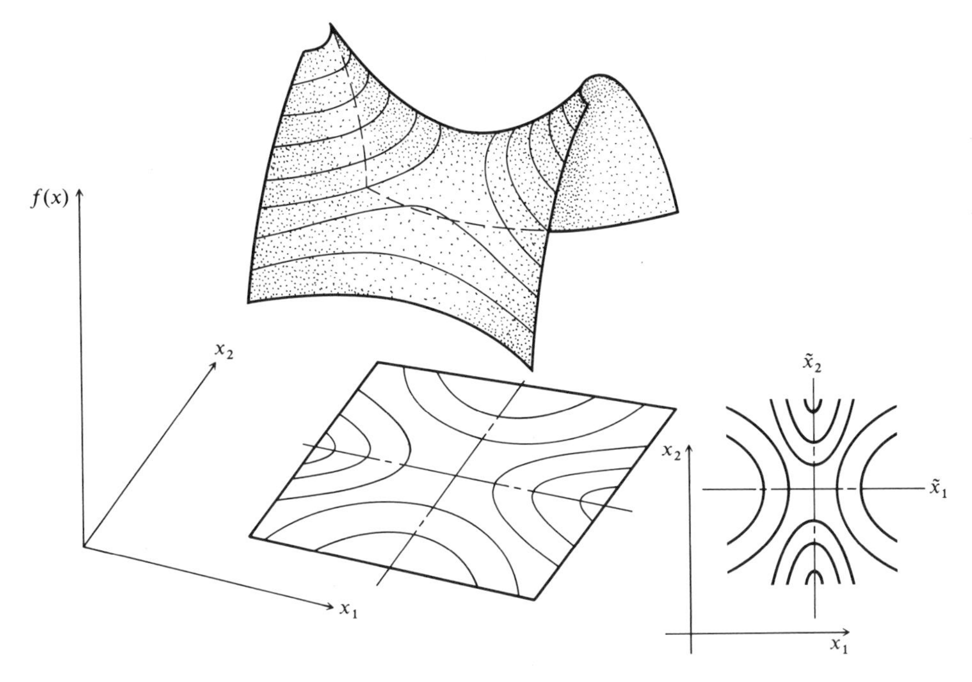

Figure 4.14 Geometry of second-order objective function of two independent variables—saddle point.

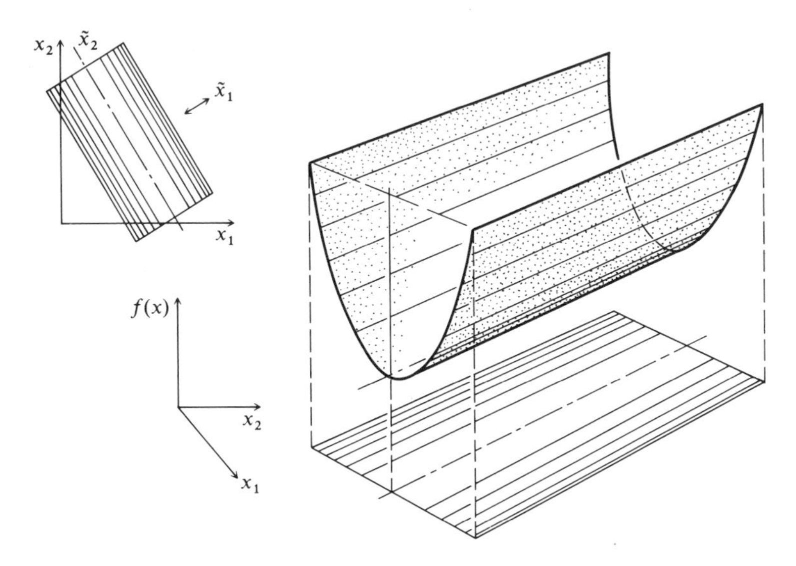

Figure 4.15 Geometry of second-order objective function of two independent variables—valley.

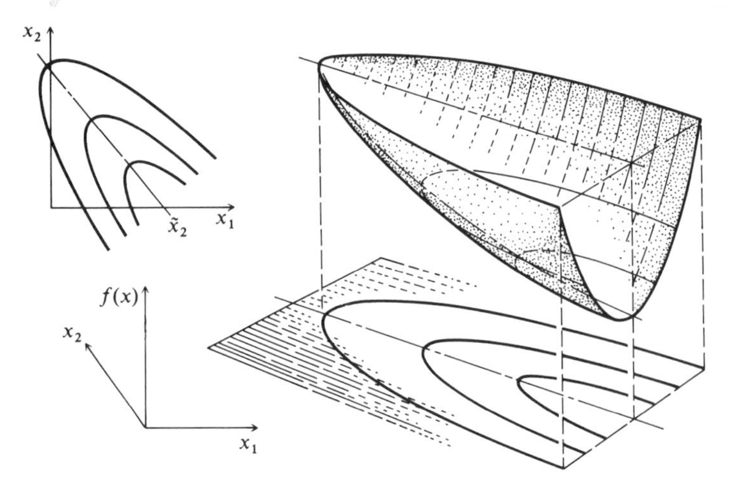

Figure 4.16 Geometry of second-order objective function of two independent variables—falling valley.

different types of surfaces corresponding to each case that arises for quadratic function. By implication, analysis of a function of many variables via examination of the eigenvalues can be conducted whereas contour plots are limited to functions of only two or three variables.

Figures 4.12 and 4.13 correspond to objective functions in well-posed optimization problems. In Table 4.2, cases 1 and 2 (Fig. 4.12) correspond to contours of $f(\mathbf{x})$ that are concentric circles, but such functions rarely occur in practice. Elliptical contours such as correspond to cases 3 and 4 are most likely for well-behaved functions. Cases 5 to 10 correspond to degenerate problems, those in which there is no finite maximum/minimum and/or perhaps nonunique optima appear.

EXAMPLE 4.14 INTERPRETATION OF AN OBJECTIVE FUNCTION IN TERMS OF ITS EIGENVALUES

Examine the function $f(\mathbf{x}) = 2x_1^2 + x_2^2 + 2x_1x_2$

$$\mathbf{H}(\mathbf{x}) = \begin{bmatrix} 4 & 2 \\ 2 & 2 \end{bmatrix} \qquad \det\begin{bmatrix} (4-\alpha) & 2 \\ 2 & (2-\alpha) \end{bmatrix} = 0$$

$$\alpha^2 - 6\alpha + 4 = 0$$

The eigenvalues are $\alpha = (6 \pm \sqrt{36 - 16})/2 = 3 \pm \sqrt{5}$, both positive, but not equal, corresponding to case 4 in Table 4.2.

If $f(\mathbf{x}) = x_1^2 + x_2^2 + 2x_1x_2$

$$\mathbf{H}(\mathbf{x}) = \begin{bmatrix} 2 & 2 \\ 2 & 2 \end{bmatrix} \qquad \det \begin{bmatrix} (2-\alpha) & 2 \\ 2 & (2-\alpha) \end{bmatrix} = 0$$

$$\alpha^2 - 4\alpha = 0 \qquad \text{so} \qquad \alpha_1 = 4 \qquad \text{and} \qquad \alpha_2 = 0$$

This case corresponds to case 9 in Table 4.2.

For well-posed quadratic objective functions the contours always form a convex region; for more general nonlinear functions, they do not (see for example Figs. 4.10*a* or 4.10*b*). It is helpful to construct contour plots to assist in analyzing the performance of multivariable optimization techniques when applied to problems of two or three dimensions. Most computer libraries have contour plotting routines to generate the desired figures.

As indicated in Table 4.2, the eigenvalues of the Hessian matrix of $f(\mathbf{x})$ indicate the shape of a function. For a positive definite symmetric matrix, the eigenvectors (refer to App. B) form an orthonormal set. For example, in two dimensions, if the eigenvectors are $\mathbf{v}_1$ and $\mathbf{v}_2$, $\mathbf{v}_1^T\mathbf{v}_2 = 0$ (the eigenvectors are perpendicular to each other). The eigenvectors also correspond to the directions of the principal axes of the contours of $f(\mathbf{x})$. This information can be used to make a variable transformation which eliminates the "tilt" in the contours, and yields search directions that are efficient. The next example illustrates such a calculation.

EXAMPLE 4.15

For $f(x) = 3x_1^2 + 2x_1x_2 + 1.5x_2^2$ find the equations for the principal axes and determine a transformation $\mathbf{x} = \mathbf{V}\mathbf{z}$ such that $f = a_{11}z_1^2 + a_{22}z_2^2$, thus eliminating the interaction term.

Solution. The optimum is at (0, 0). First, compute $\mathbf{H}(\mathbf{x})$ and the corresponding eigenvalues and eigenvectors.

$$\mathbf{H} = \begin{bmatrix} 6 & 2 \\ 2 & 3 \end{bmatrix}$$

The eigenvalues are 2 and 7 and the corresponding eigenvectors are

$$\mathbf{v}_1 = \begin{bmatrix} -1 \\ 2 \end{bmatrix} \qquad \text{and} \qquad \mathbf{v}_2 = \begin{bmatrix} 2 \\ 1 \end{bmatrix}$$

Refer to App. B for the procedure to calculate these quantities. Note that $\mathbf{v}_1^T\mathbf{v}_2 = 0$.

From linear algebra (Amundson, 1966), the transformation which eliminates the interaction (x_1x_2) term is

$$x = \mathbf{V}\mathbf{z} \qquad \text{where} \qquad \mathbf{V} = [\mathbf{v}_1 \quad \mathbf{v}_2]$$

Substituting for **V** and $\mathbf{z} = \begin{bmatrix} z_1 \\ z_2 \end{bmatrix}$

$$x_1 = -z_1 + 2z_2 \tag{a}$$

$$x_2 = 2z_1 + z_2 \tag{b}$$

If Eqs. (*a*) and (*b*) are substituted into the objective function, the result in terms of z_1 and z_2 is

$$f(z) = 5z_1^2 + 17.5z_2^2$$

The equations for the principal axes can be found from the eigenvectors, $[-1 \quad 2]$, and $[2 \quad 1]$

$$x_2 = 0.5x_1$$

$$x_2 = -2.0x_1$$

Figure E4.15 illustrates the contours of $f(\mathbf{x})$ and the principal axes for this Example. The variable transformation represents rotation of the variable axes from the $x_1 - x_2$ plane to the $z_1 - z_2$ plane.

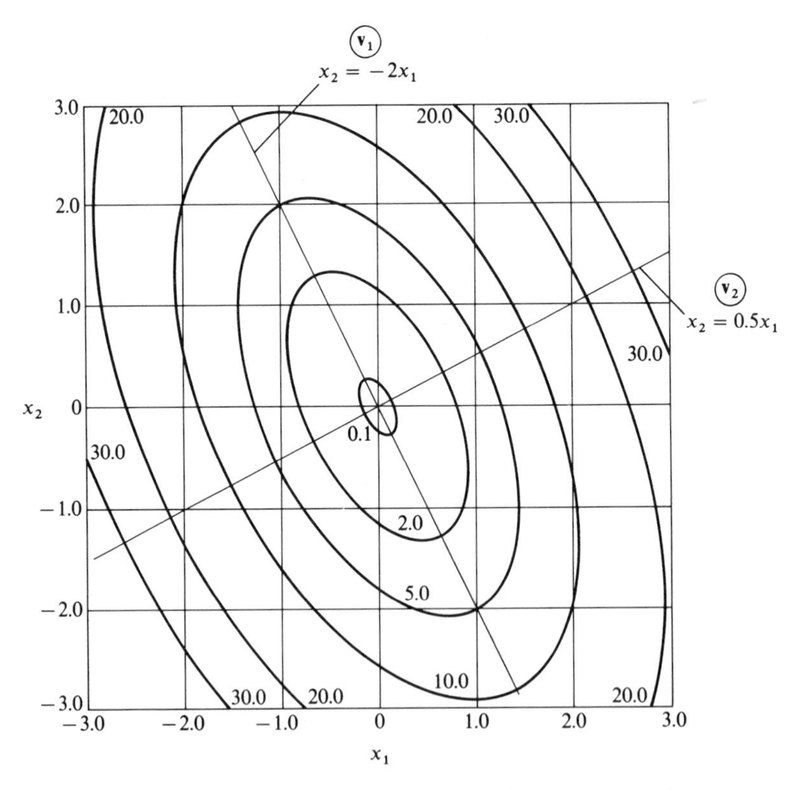

Figure E4.15 Contours and principal axes for $f = 3x_1^2 + 2x_1x_2 + 1.5x_2^2$.

This example was somewhat simplified because the origin of the principal axes and the origin for the original axes were the same. If they were not the same, then the origin for the principal axes would be at the extremum, and you would transform the variables so that the new variables in the principal axes, $\mathbf{y}$, would be related to $\mathbf{x}$ by $\mathbf{y} = \mathbf{x} - \mathbf{x}^*$, and thus eliminate linear terms in $f(\mathbf{x})$.

One of the primary requirements of any successful optimization technique is the ability to be able to move rapidly in a local region along a narrow valley (in minimization) toward the minimum of the objective function. In other words, an efficient algorithm will select a search direction which generally follows the axis of the valley rather than jumping back and forth across the valley. Valleys (ridges in maximization) occur quite frequently, at least locally, and these types of surfaces have the potential to slow down greatly the search for the optimum. A valley lies in the direction of the eigenvector associated with a small eigenvalue of the Hessian matrix of the objective function. For example, if the Hessian matrix of a quadratic function is

$$\mathbf{H} = \begin{bmatrix} 1 & 0 \\ 0 & 10 \end{bmatrix}$$

then the eigenvalues are $\alpha_1 = 1$ and $\alpha_2 = 10$. The eigenvector associated with $\alpha_1 = 1$, that is, the x_1 axis, is lined up with the valley in the ellipsoid. Transformation techniques such as those discussed above can be used to allow the problem to be more efficiently solved by a search technique. (See Chap. 6.)

Valleys and ridges corresponding to cases 1 through 4 can lead to a minimum or maximum, respectively, but not for other cases. Do you see why?

REFERENCES

Amundson, N. R., *Mathematical Methods in Chemical Engineering: Matrices and Their Application*, Prentice-Hall, Englewood Cliffs, New Jersey (1966).

Aoki, M., *Introduction to Optimization Techniques*, Macmillan Co., New York (1971), p. 24.

Beveridge, G. S. G., and R. S. Schechter, *Optimization: Theory and Practice*, McGraw-Hill, New York (1970), p. 126.

Fox, R. L., *Optimization Methods for Engineering Design*, Addison-Wesley, Reading. Massachusetts (1971), p. 42.

Noltie, C. B., *Optimum Pipe Size Selection*, Gulf Publ. Co., Houston, Texas (1978).

Walker, W. H., W. K. Lewis, W. H. McAdams, and E. R. Gilliland, *Principles of Chemical Engineering*, 3d ed., McGraw-Hill, New York (1937), p. 357.

Wilde, D. J., and C. S. Beightler, *Foundations of Optimization*, Prentice-Hall, Englewood Cliffs, New Jersey (1967), p. 220.

SUPPLEMENTARY REFERENCES

Avriel, M., *Nonlinear Programming*, Prentice-Hall, Englewood Cliffs, New Jersey, (1976).

Jeter, M. W., *Mathematical Programming*, Marcel Dekker, New York (1986).

Walsh, G. R., *Methods of Optimization*, John Wiley, New York (1975).

Wilde, D. J., *Optimum Seeking Methods*, Prentice-Hall, Englewood Cliffs, New Jersey (1964).

PROBLEMS

4.1. Classify the following functions as continuous (specify the range) or discrete:

(*a*) $f(x) = e^x$

(*b*) $f(x) = ax_{n-1} + b(x_0 - x_n)$ where x_n represents a stage in a distillation column

(*c*) $f(\mathbf{x}) = x_D - x_s/(1 + x_s)$ where x_D = concentration of vapor from a still and x_s is the concentration in the still

4.2. The future worth S of a series of n uniform payments each of amount P is

$$S = \frac{P}{i}[(1 + i)^n - 1]$$

where i is the interest rate per period. If i is considered to be the only variable, is it discrete of continuous? Explain. Repeat for n. Repeat for both n and i being variables.

4.3. In a plant the gross profit P in dollars is

$$P = nS - (nV + F)$$

where n is the number of units produced per year, S is the sales price in dollars per unit, V is the variable cost of production in dollars per unit, and F is the fixed charge in dollars. Suppose that the average unit cost is calculated as

$$\text{Average unit cost} = \frac{nV + F}{n}$$

Discuss under what circumstances n can be treated as a continuous variable.

4.4. One rate of return is the ratio of net profit P to total investment

$$R = 100\frac{P(1 - t)}{I} = 100(1 - t)\frac{[S - (V + F/n)]}{I/n}$$

where t is the fraction tax rate and I is the total investment in dollars. Find the maximum R as a function of n for a given I if n is a continuous variable. Repeat if n is discrete. (See Prob. 4.3 for other notation.)

4.5. Classify the following functions as unimodal, multimodal, or neither:

(*a*) $f(x) = 2x + 3$

(*b*) $f(x) = x^2$

(*c*) $f(x) = a \cos(x)$

4.6. Are the following functions unimodal, multimodal, or neither:

(*a*) $f(\mathbf{x}) = x_1^2 + x_2^2 + x_1x_2$

(*b*) $f(\mathbf{x}) = x_1^3 - 4x_1^2x_2 + x_1x_2^2$

(*c*) $f(\mathbf{x}) = e^{-(x_1^2 + x_2^2)}$

4.7. Determine the convexity or concavity of the following objective functions:

(*a*) $f(x_1, x_2) = (x_1 - x_2)^2 + x_2^2$
(*b*) $f(x_1, x_2, x_3) = x_1^2 + x_2^2 + x_3^2$
(*c*) $f(x_1, x_2) = e^{x_1} + e^{x_2}$

4.8. Given a linear objective function,

$$f = x_1 + x_2$$

explain why a nonconvex region such as region A in Fig. P4.8 will yield difficulties in the search for the maximum.

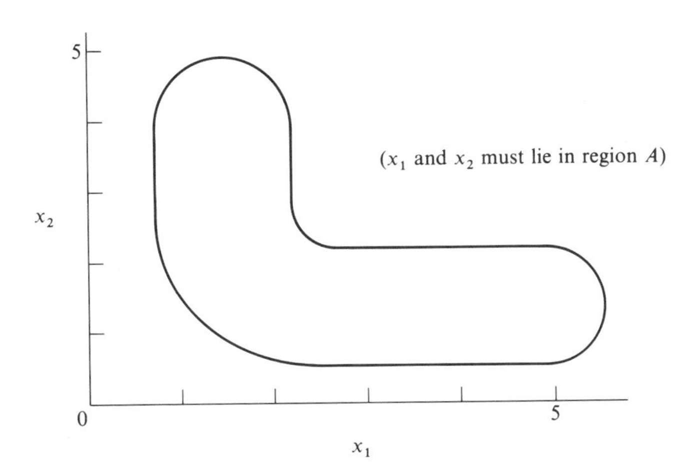

Figure P4.8

Why is region A not convex?

4.9. Are the following functions convex? Strictly convex? Why?

(*a*) $2x_1^2 + 2x_1x_2 + 3x_2^2 + 7x_1 + 8x_2 + 25$
What are the optimum values of x_1 and x_2?
(*b*) e^{5x}

4.10. A reactor converts an organic compound to product P by heating the material in the presence of an additive A (mole fraction $= x_A$). The additive can be injected into the reactor, while steam can be injected into a heating coil inside the reactor to provide heat. Some conversion can be obtained by heating without addition of A, and vice versa.

The product P can be sold for \$50 per lb-mol. For 1 lb-mol of feed, the cost of the additive (in dollars per lb-mol) as a function of x_A is given by the formula, $0.1 + 0.5x_A + x_A^2$. The cost of the steam (in dollars) as a function of S is $0.2 + 0.03S + 0.01S^2$. (S = lb steam/lb-mol feed). The yield equation is $y_p = 0.1 + 0.3x_A + 0.02S + 0.001x_AS$; y_p = lb-mol product P/lb-mol feed.

(*a*) Formulate the profit function (basis of 1.0 lb-mol feed) in terms of x_A and S.

$$f = \text{income} - \text{costs}$$

(*b*) Maximize f subject to the constraints

$$0 \leq x_A \leq 1 \qquad S \geq 0$$

by any method you choose.

(*c*) Is f a concave function? Demonstrate mathematically why it is or why it is not concave.

(*d*) Is the region of search convex? Why?

4.11. Show that $f = e^{x_1} + e^{x_2}$ is convex. Is it also strictly convex?

4.12. Show that $f = |x|$ is convex.

4.13. Is the following region constructed by the four constraints convex? Closed?

$$x_2 \geq 1 - x_1$$
$$x_2 \leq 1 + 0.5x_1$$
$$x_1 \leq 2$$
$$x_2 \geq 0$$

4.14. Does the following set of constraints form a convex region?

$$g_1(x) = -(x_1^2 + x_2^2) + 9 \geq 0$$
$$g_2(x) = -x_1 - x_2 + 1 \geq 0$$

4.15. Consider the following problem:

Minimize $\quad f(\mathbf{x}) = x_1^2 + x_2$

Subject to $\quad g_1(\mathbf{x}) = x_1^2 + x_2^2 - 9 \leq 0$

$$g_2(\mathbf{x}) = (x_1 + x_2^2) - 1 \leq 0$$
$$g_3(\mathbf{x}) = (x_1 + x_2) - 1 \leq 0$$

Does the constraint set form a convex region? Is it closed? (*Hint*: a plot will help decide.)

4.16. The objective function work requirement for a three-stage compressor can be expressed as

$$f = \left(\frac{p_2}{p_1}\right)^{0.286} + \left(\frac{p_3}{p_2}\right)^{0.286} + \left(\frac{p_4}{p_3}\right)^{0.286}$$

$p_1 = 1$ atm and $p_4 = 10$ atm. The minimum occurs at a pressure ratio for each stage of $\sqrt[3]{10}$. Is f convex for $1 \leq p_2 \leq 10$, $1 \leq p_3 \leq 10$?

4.17. Happel and Jordan [*Chemical Process Economics*, Marcel Dekker, New York, 1975, p. 178] reported an objective function (cost) for the design of a distillation column as follows:

$$\begin{aligned} f = {} & 14720(100 - P) + 6560R - 30.2PR + 6560 - 30.2P \\ & + 19.5n\,(5000R - 23PR + 5000 - 23P)^{0.5} \\ & + 23.2\,[5000R - 23PR + 5000 - 23P]^{0.62} \end{aligned}$$

where n = number of theoretical stages, R = reflux ratio, and P = percent recovery in bottoms stream. They reported the optimum occurs at $R = 8$, $n = 55$, and $P = 99$. Is f convex at this point? Are there nearby regions where f is not convex?

4.18. Consider the function

$$y = (x - a)^2$$

Note that $x = a$ minimizes y. Let $z = x^2 - 4x + 16$. Does the solution to $x^2 - 4x + 16 = 0$,

$$x = \frac{4 \pm \sqrt{-48}}{2} = 2 \pm j2\sqrt{3}$$

minimize z? ($j = \sqrt{-1}$).

4.19. The following objective function can be seen by inspection to have a minimum at $x = 0$:

$$f(x) = |x^3|$$

Can the criteria of Sec. 4.5 be applied to test this outcome?

4.20. (*a*) Consider the objective function,

$$f = 6x_1^2 + x_2^3 + 6x_1x_2 + 3x_2^2$$

Find the stationary points and classify them using the Hessian matrix.

(*b*) Repeat for

$$f = 3x_1^2 + 6x_1 + x_2^2 + 6x_1x_2 + x_3 + 2x_3^2 + x_2x_3 + x_2$$

(*c*) Repeat for

$$f = a_0x_1 + a_1x_2 + a_2x_1^2 + a_3x_2^2 + a_4x_1x_2$$

4.21. Classify the stationary points of

(*a*) $f = -x^4 + x^3 + 20$
(*b*) $f = x^3 + 3x^2 + x + 5$
(*c*) $f = x^4 - 2x^2 + 1$
(*d*) $f = x_1^2 - 8x_1x_2 + x_2^2$

according to Table 4.2

4.22. List stationary points and their classification (maximum, minimum, saddle point) of

(*a*) $f = x_1^2 + 2x_1 + 3x_2^2 + 6x_2 + 4$
(*b*) $f = x_1 + x_2 + x_1^2 - 4x_1x_2 + 2x_2^2$

4.23. Show that the eigenvectors of the Hessian for a two-variable quadratic optimization problem are orthogonal.

4.24. Figure P4.24 shows the contours of a quadratic optimization problem, which has a minimum at (2, 2) where $f(\mathbf{x}) = 5$. If the general quadratic function is described by

$$f = \tfrac{1}{2}\mathbf{x}^T A\mathbf{x} + \mathbf{b}^T\mathbf{x} + c$$

give characteristics of $\mathbf{A}$, $\mathbf{b}$, and c; be as quantitative as possible.

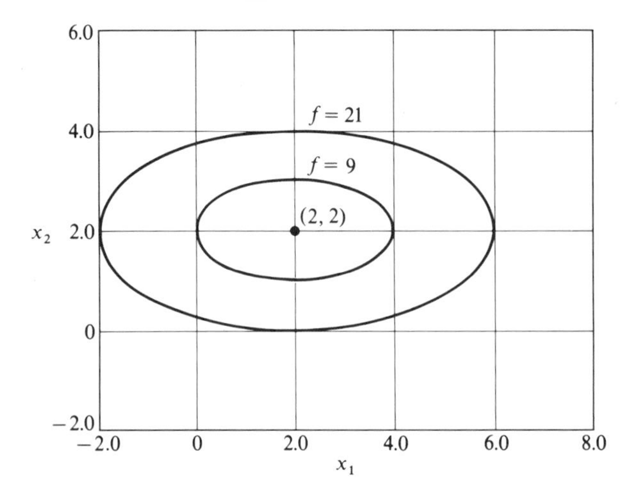

Figure P4.24

4.25. We wish to minimize

$$f(\mathbf{x}) = \underset{(a_1)}{-12x_1} + \underset{(a_2)}{2x_1^2} + \underset{(a_3)}{2x_2^2} - \underset{(a_4)}{2x_1x_2}$$

(*a*) Find the stationary point and determine if it is a maximum or minimum based on the Hessian matrix.

(*b*) The coefficients of each term (a_1, a_2, a_3, a_4) can affect whether a maximum or a minimum occurs. If three of the four coefficients are fixed at the above values and we vary the remaining one, find the individual values of a_1, a_2, a_3, and a_4 that change the character of the stationary point (e.g., from minimum to saddle point).

4.26. For

$$f = x_1^2 + x_1x_2 + x_2^2$$

find the transformation $\mathbf{x} = \mathbf{V}\mathbf{z}$ such that

$$f = a_{11}z_1^2 + a_{22}z_2^2$$

Illustrate the transformed coordinate system on a contour plot of $f(\mathbf{x})$.

CHAPTER 5

OPTIMIZATION OF UNCONSTRAINED FUNCTIONS: ONE-DIMENSIONAL SEARCH

A good technique for the optimization of a function of just one variable is essential for at least three reasons:

1. In many problems the constraints can be inserted into the objective function so that the dimensionality of the problem is reduced to one variable;
2. Some unconstrained problems inherently involve only one variable; and
3. Techniques for unconstrained and constrained optimization problems generally involve repeated use of a one-dimensional search as described in Chaps. 6 and 8.

Prior to the advent of high-speed computers, methods of optimization were limited primarily to the so-called *indirect methods*, that is methods of calculating a potential extremum by using the necessary conditions and analytical derivatives as well as values of the objective function. Modern computers have made possible direct methods, that is, search for an extremum by direct comparison of function values of $f(\mathbf{x})$ at a sequence of trial points $\mathbf{x}^{(1)}, \mathbf{x}^{(2)}, \ldots$ without involving analytical derivatives. Of course, as explained in Sec. 6.5, difference formulas can be substituted for analytical derivatives in which case the distinction between direct and indirect methods becomes blurred.

As an example consider the following function of a single variable x (see Fig. 5.1).

$$f(x) = x^2 - 2x + 1$$

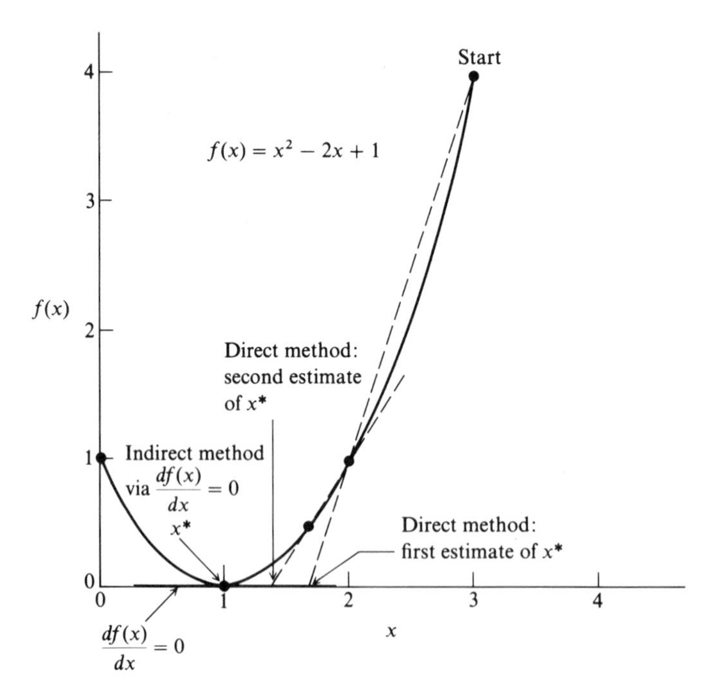

Figure 5.1 Direct versus indirect methods of finding a minimum.

An indirect analytical method of finding x^*, the minimum of $f(x)$, is to set the gradient of $f(x)$ equal to zero

$$\frac{df(x)}{dx} = 0 = 2x - 2$$

and solve the resulting equation to get $x^* = 1$; x^* can be tested for the sufficient conditions to ascertain that it is indeed a minimum:

$$\frac{d^2f(1)}{dx^2} = 2 > 0$$

To carry out a *direct method* of numerical minimization, you only use values of the objective function. You start with some initial value of x, say $x^0 = 0$, and calculate successive values of $f(x) = x^2 - 2x + 1$ for other values of x, values selected according to whatever strategy is to be employed. A number of different strategies are discussed in subsequent sections of this chapter. You stop when $f(x^{k+1}) - f(x^k) < \varepsilon$ where the superscript k designates the iteration number and ε is the prespecified tolerance or criterion of precision.

The indirect method has one inherent advantage even if numerical methods are used to compute function values, namely, convergence is normally faster. However, in engineering problems this advantage is often neutralized by lack of interest in quite precise determinations of an extremum in view of the uncertainty in the values of the coefficients in the objective function.

Direct numerical methods have the advantage that they can more easily treat problems involving functions with discontinuities, points of inflection, and end points. Also, the character of $f(\mathbf{x})$ in the vicinity of the extremum can be easily explored by numerical methods.

In passing we should point out that for nonlinear objective functions with more than one variable direct methods have been proved superior to indirect ones. For example, suppose that the nonlinear function $f(\mathbf{x}) = f(x_1, x_2, \ldots, x_n)$ is to be minimized. The necessary conditions that would be used in an indirect method are

$$\frac{\partial f(\mathbf{x})}{\partial x_1} = f_{x_1}(\mathbf{x}) = 0$$

$$\frac{\partial f(\mathbf{x})}{\partial x_2} = f_{x_2}(\mathbf{x}) = 0$$

$$\vdots$$

$$\frac{\partial f(\mathbf{x})}{\partial x_n} = f_{x_n}(\mathbf{x}) = 0$$

Each of the partial derivatives when equated to zero may well yield a nonlinear equation. Hence, the minimization of $f(\mathbf{x})$ is converted into a problem of solving a set of nonlinear equations in n variables, a problem that can be just as difficult to solve as the original problem. Also, the Hessian matrix of $f(\mathbf{x})$ is symmetric whereas the Jacobian of the equations of the necessary conditions is arbitrary in structure. Thus, most engineers prefer to attack the minimization problem directly by one of the numerical methods that are described in Chap. 6, rather than to use an indirect method. Even when minimizing a function of one variable by an indirect method, using the necessary conditions can lead to having to find the real roots of a nonlinear equation.

5.1 NUMERICAL METHODS FOR OPTIMIZING A FUNCTION OF ONE VARIABLE

Most algorithms for unconstrained and constrained optimization make use of an efficient unidimensional optimization technique to locate a local minimum of a function of one variable. Wilde (1964) and other general optimization books (such as Dennis and Schnabel, 1983) have reviewed one-dimensional search techniques which calculate the interval in which the minimum of a function lies. To apply these methods you need to know at the start an initial bracket Δ^0 that contains the minimum of the objective function $f(x)$, and that $f(x)$ is unimodal in the interval. It is very difficult to determine whether a general function is unimodal prior to optimization, but in many important cases objective functions do exhibit unimodality. There are various methods of varying the initial interval to reach a final interval Δ^n. In the next section we describe a few of the methods that in practice prove to be the most effective.

One method of optimization for a function of a single variable might be to set up as fine a grid as you wish for the values of x, and calculate the function value for every value on the grid. The optimum would be the best value of $f(x)$. While this procedure is not a very efficient method for finding the optimum, it can yield acceptable results. On the other hand, if we were to utilize this approach in optimizing a multivariable function, the computer time is quite likely to become prohibitive.

In selecting a search method to minimize or maximize a function of a single variable, you must be concerned with the trade-off between complexity of the procedure and number of function evaluations needed (which is directly related to the computer time used and hence cost). For example, in some problems a simulation may be required to generate the function values such as in determining the optimal number of trays in a distillation column. In other cases you have no functional description of the physical/chemical model of the process to be optimized, and are forced to operate the process at various input levels to evaluate the value of the process output. The generation of a new value of the objective function in such circumstances may be extremely costly, and no doubt the number of plant tests would be limited and have to be quite judiciously designed. In such circumstances, efficiency is a key criterion in selecting a minimization strategy.

5.2 SCANNING AND BRACKETING PROCEDURES

Almost every unidimensional search procedure requires that a bracket of the minimum be obtained as the first part of the strategy, and then the bracket is narrowed subsequently. Along with the statement of the objective function $f(x)$ there must be some statement of bounds on x or else the implicit assumption that x is unbounded ($-\infty < x < \infty$). For example, the problem

$$\text{Minimize} \quad f(x) = (x - 100)^2$$

has an optimal value of $x^* = 100$. Clearly you would not want to start at $-\infty$ (i.e., a large negative number) and try to bracket the minimum. Common sense would say to estimate the minimum x and set up a sufficiently wide bracket to contain the true minimum. Clearly, if you make a mistake and set up a bracket of $0 < x < 10$, you will find that the minimum occurs at one of the bounds, hence the bracket must be revised. In engineering and scientific work physical limits on temperature, pressure, concentration, and other physically meaningful variables place practical bounds on the region of search that might be used as an initial bracket.

Several strategies exist that we can use in order to scan the independent variable space and determine an acceptable range for search for the minimum of $f(x)$. As an example, in the function mentioned above, if we discretize the independent variable by a grid spacing of 0.01 and then initiate the search at zero, proceeding with consecutively higher values of x, much time and effort would be consumed in order to set up the initial bracket for x. Therefore, acceleration procedures are used in order to scan rapidly for a suitable range of x. One technique might involve using a functional transformation (e.g., $\log x$) in order to look at wide ranges of the independent variable. Another method might be to use a variable grid spacing. Consider a sequence in x given by the following formula:

$$x_{k+1} = x_k + \delta \cdot 2^{k-1} \tag{5.1}$$

Equation (5.1) allows for successively wider-spaced values given some base increment (delta). Table 5.1 lists the values of x and $f(x)$ for Eq. (5.1) with $\delta = 1$. Note that in nine calculations we have bounded the minimum of $f(x)$. Another scanning procedure could be initiated at $x = 63$, or $x = 255$, with δ reduced, and so on to find the minimum of $f(x)$. However, more efficient techniques are discussed in subsequent sections of this chapter.

In optimization of a function of a single variable, we recognize (as for general multivariable problems) that there is no substitute for a good first guess for the starting point in the search. Therefore, insight into the problem as well as

Table 5.1 Acceleration in fixing an initial bracket

x	0	1	3	7	15	31	63	127	255
$f(x)$	10^4	9801	9409	8649	7225	4761	1369	729	2325

previous experience are often very important factors influencing the amount of time and effort required to solve a given optimization problem.

The methods considered in the rest of this chapter are generally termed *sequential and logical* methods. They are called logical methods because a given step is pursued only if it yields an improved value for the objective function; they are referred to as sequential because only one step at a time is taken. First we cover indirect methods in Sec. 5.3, followed by a review of several direct methods in Sec. 5.4.

5.3 NEWTON, QUASI-NEWTON, AND SECANT METHODS OF UNIDIMENSIONAL SEARCH

Three basic procedures for finding an extremum of a function of one variable have evolved from applying the necessary optimality conditions to the function:

1. Newton's method
2. Finite difference approximation of Newton's method (sometimes called a quasi-Newton method)
3. Secant method

In comparing the effectiveness of these techniques, it is useful to examine the rate of convergence for each method. Rates of convergence can be expressed in various ways, but a common classification is as follows:†

Linear

$$\frac{\|\mathbf{x}^{k+1} - \mathbf{x}^*\|}{\|\mathbf{x}^k - \mathbf{x}^*\|} \le c \qquad 0 \le c \le 1 \tag{5.2}$$

(usually slow in practice)

Order p

$$\frac{\|\mathbf{x}^{k+1} - \mathbf{x}^*\|}{\|\mathbf{x}^k - \mathbf{x}^*\|^p} \le c \qquad c \ge 0, p \ge 1 \tag{5.3}$$

(fastest in practice)

If $p = 2$, the order of convergence is said to be quadratic.

† The symbols $\mathbf{x}^k$, $\mathbf{x}^{k+1}$, etc., refer to the kth or $(k + 1)$st stage of iteration and not to powers of $\mathbf{x}$.

Superlinear

$$\lim_{k \to \infty} \frac{\|\mathbf{x}^{k+1} - \mathbf{x}^*\|}{\|\mathbf{x}^k - \mathbf{x}^*\|} \to 0 \qquad (\text{or} < c_k \text{ and } c_k \to 0 \text{ as } k \to \infty) \tag{5.4}$$

(usually fast in practice)

For a function of a single variable $\|\mathbf{x}\| = \sqrt{\Sigma x_i^2} = x$ itself.

5.3.1 Newton's Method

Recall that the primary necessary condition for $f(x)$ to have a local minimum is that $f'(x) = 0$. Consequently, you can solve the equation $f'(x) = 0$ by Newton's method to get

$$x^{k+1} = x^k - \frac{f'(x^k)}{f''(x^k)} \tag{5.5}$$

making sure on each stage k that $f(x^{k+1}) < f(x^k)$ for a minimum. Examine Fig. 5.2.

To see what Newton's method implies about $f(x)$, suppose $f(x)$ is approximated by a quadratic function at x^k

$$f(x) = f(x^k) + f'(x^k)(x - x^k) + \tfrac{1}{2}f''(x^k)(x - x^k)^2 \tag{5.6}$$

Find $df(x)/dx = 0$, a stationary point of the quadratic model of the function. The result by differentiating Eq. (5.6) is

$$f'(x^k) + (\tfrac{1}{2})(2)f''(x^k)(x - x^k) = 0 \tag{5.7}$$

which can be rearranged to yield Eq. (5.5). Consequently, Newton's method is equivalent to using a quadratic model for a function in minimization (or maximization) and applying the necessary conditions.

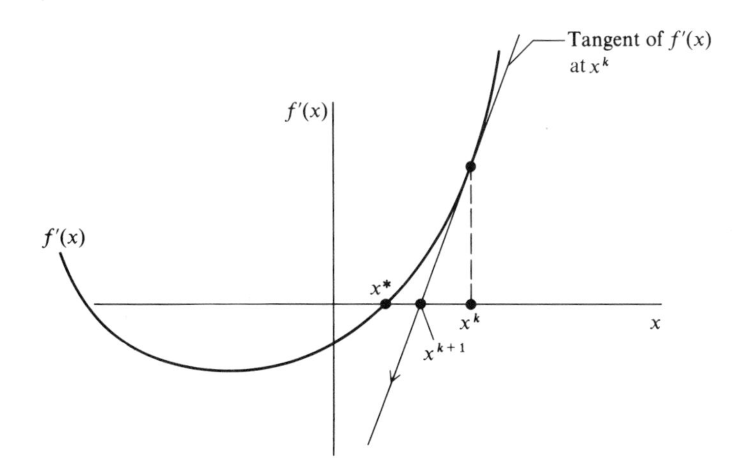

Figure 5.2 Newton's method applied to the solution of $f'(x) = 0$.

The advantages of Newton's method are

1. The procedure is locally quadratically convergent [$p = 2$ in Eq. (5.3)] to the extremum as long as $f''(x) \neq 0$.
2. For a quadratic function, the minimum is obtained in one iteration.

The disadvantages of the method are

1. You have to calculate both $f'(x)$ and $f''(x)$.
2. If $f''(x) \to 0$, the method converges slowly.
3. If more than one extremum exists, the method may not converge to the desired extremum (the global one) and might oscillate.

5.3.2 Quasi-Newton Method

A quasi-Newton method in general is one that imitates Newton's method. If $f(x)$ is not given by a formula, or the formula is so complicated that analytical derivatives cannot be formulated, you can replace Eq. (5.5) with a finite difference approximation

$$x^{k+1} = x^k - \frac{[f(x+h) - f(x-h)]/2h}{[f(x+h) - 2f(x) + f(x-h)]/h^2} \tag{5.8}$$

Central differences have been used in Eq. (5.8) but forward differences or any other difference scheme would suffice as long as the step size h is selected to match the difference formula and the computer (machine) precision for the computer on which the calculations are to be executed.

Other than the selection of the value of h, the only additional disadvantage of the quasi-Newton method is that additional function evaluations are needed on each iteration k [three for formula (5.8) versus two for formula (5.5)].

5.3.3 Secant Method

In the secant method the approximate model analogous to Eq. (5.7) to be solved is

$$f'(x^k) + m(x - x^k) = 0 \tag{5.9}$$

where m is the slope of the line connecting the point x^p and a second point x^q, given by

$$m = \frac{f'(x^q) - f'(x^p)}{x^q - x^p}$$

The secant approximates $f'(x)$ as a straight line (examine Fig. 5.3); as $x^q \to x^p$, m approaches the second derivative of $f(x)$. Thus Eq. (5.9) imitates Newton's

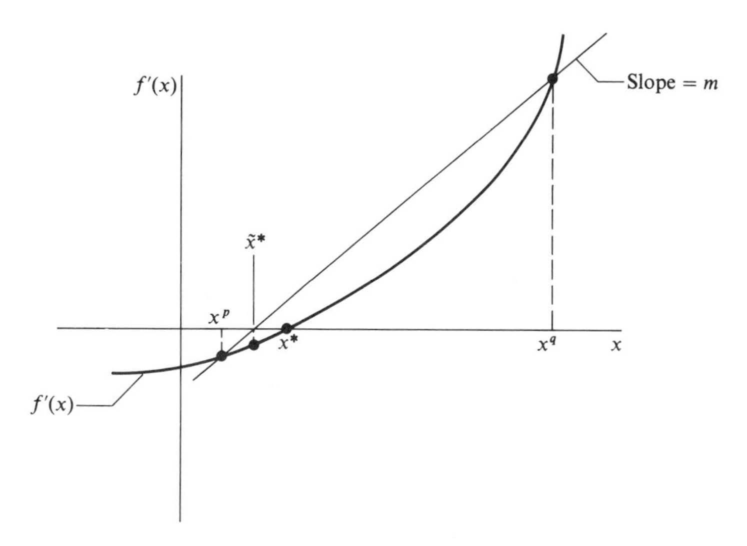

Figure 5.3 Secant method for solution of $f'(x) = 0$.

method (in this sense the secant method is also a quasi-Newton method; it might also be considered a region elimination method as described in Sec. 5.4)

$$\tilde{x}^* = x^q - \frac{f'(x^q)}{[f'(x^q) - f'(x^p)]/(x^q - x^p)} \tag{5.10}$$

where $\tilde{x}^*$ is the approximation to x^* achieved on one iteration k. Note that $f'(x)$ can itself be approximated by a finite difference substitute.

Secant methods start out by using two points x^p and x^q spanning the interval of x, points at which the first derivatives of $f(x)$ are of opposite sign. The zero of Eq. (5.9) is predicted by Eq. (5.10), and the derivative of the function is then evaluated at the new point. The two points retained for the next step are $\tilde{x}^*$ and either x^q or x^p, the choice being made so that the pair of derivatives $f'(\tilde{x}^*)$, and either $f'(x^p)$ or $f'(x^q)$, have opposite signs to maintain the bracket on x^*. (This variation is called "regula falsi" or the method of false position.) In Fig. 5.3, for the $(k + 1)$st search, $\tilde{x}^*$ and x^q would be selected as the end points of the secant line.

Secant methods may seem crude, but they work well in practice. The order of convergence is $(1 + \sqrt{5})/2 \approx 1.6$ for a single variable. Their convergence is slightly slower than a properly chosen finite difference quasi-Newton method, but they are usually more efficient in terms of total function evaluations to achieve a specified accuracy (see Dennis and Schnabel, 1983, Chap. 2).

For any of the three procedures outlined above, in minimization you assume the function is unimodal, bracket the minimum, pick a starting point, apply the iteration formula to get x^{k+1} (or $\tilde{x}^*$) from x^k (or x^p and x^q), and make sure that $f(x^{k+1}) < f(x^k)$ on each iteration so that progress is made toward the minimum. As long as $f''(x^k)$, or its approximation, is positive, $f(x)$ will decrease.

Of course, you must start in the correct direction to reduce $f(x)$ (for a minimum) by testing an initial perturbation in x. For maximization, minimize $-f(x)$.

EXAMPLE 5.1 COMPARISON OF NEWTON, FINITE-DIFFERENCE NEWTON, AND SECANT METHODS APPLIED TO A QUADRATIC FUNCTION

In this example, we will minimize a simple quadratic function $f(x) = x^2 - x$ that is illustrated in Fig. E5.1*a* using each of the three indirect methods presented in Sec. 5.3.

Solution. By inspection we can pick a bracket on the minimum, say $x = -3$ to $x = 3$. (Refer to Example 5.3 to see how a bracket can be determined by sequential calculations.) Let us assume $x^0 = 3$ is the starting point for the minimization.

Newton's method. For Newton's method sequentially apply Eq. (5.5). Examine Fig. 5.1*b* for $f(x) = x^2 - x$ and $f'(x) = 2x - 1$; $f''(x) = 2$. Note $f''(x)$ is always positive definite. For this example Eq. (5.5) is

$$x^1 = x^0 - \frac{f'(x^0)}{f''(x^0)} \qquad (a)$$

and

$$x^1 = 3 - \tfrac{5}{2} = 0.5$$

Because the function is quadratic and hence $f'(x)$ is linear, the minimum is obtained in one step. If the function were not quadratic, then additional iterations using Eq. (5.5) would take place.

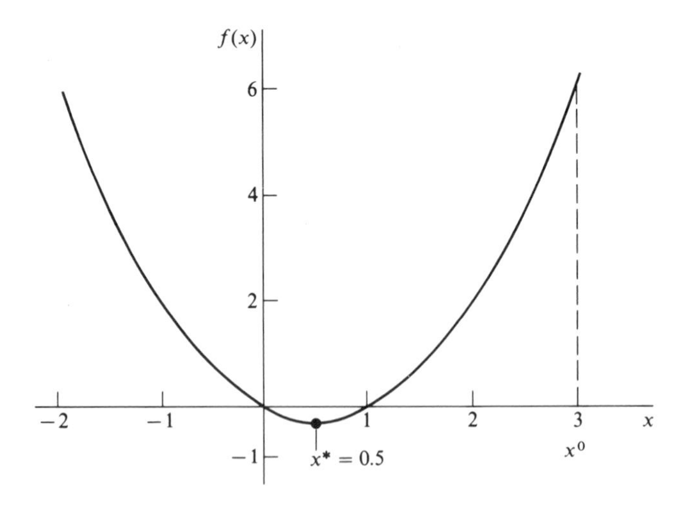

Figure E5.1*a*

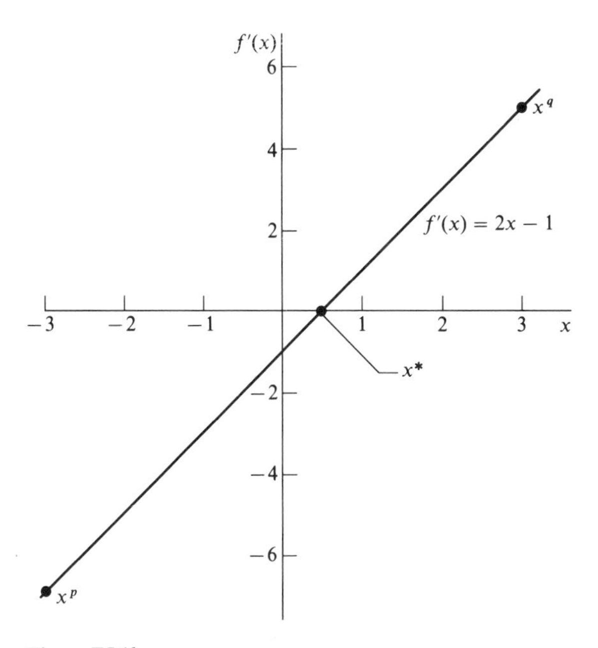

Figure E5.1*b*

Quasi-Newton method. Application of Eq. (5.8) to $f(x) = x^2 - x$ is illustrated below. However, here we will illustrate the use of a forward difference formula for $f'(x)$ and a three-point central difference formula for $f''(x)$

$$x^{k+1} = x^k - \frac{[f(x+h) - f(x)]/h}{[f(x+h) - 2f(x) + f(x-h)]/h^2} \tag{b}$$

with $h = 10^{-3}$:

$$x^1 = 3 - \frac{[f(3.001) - f(3.0)]/10^{-3}}{[f(3.001) - 2f(3.0) + f(2.999)]/(10^{-3})^2}$$

$$= 3 - (10^{-3})\frac{(6.005001 - 6.000000)}{(6.005001 - 12.000000 + 5.995001)}$$

$$= 3 - (10^{-3})\frac{0.005001}{0.000002} = 3 - 2.500500$$

$$= 0.499500$$

One more iteration could be taken to improve the estimate of x^*, perhaps with a smaller value of h (if desired).

Secant method. The application of Eq. (5.10) to $f(x) = x^2 - x$ starts with the two points $x = -3$ and $x = 3$ corresponding to the x^p and x^q, respectively, in Fig. 5.3:

$$f'(-3) = -7 \qquad f'(3) = 5$$

$$x^1 = 3 - \frac{5}{[5 - (-7)]/[3 - (-3)]} = 3 - 2.5 = 0.5$$

As before, the optimum is reached in one step because $f'(x)$ is linear and the linear extrapolation is valid.

EXAMPLE 5.2 MINIMIZING A MORE DIFFICULT FUNCTION

In this example we minimize a nonquadratic function $f(x) = x^4 - x + 1$ that is illustrated in Fig. E5.2*a*, using the same three methods as in Example 5.1. For a starting point of $x = 3$, minimize $f(x)$ until the change in x is less than 10^{-7}. Use $h = 0.1$ for the quasi-Newton method. For the secant method, use $x^q = 3$ and $x^p = -3$.

Solution.

Newton's method. For Newton's method, $f' = 4x^3 - 1$ and $f'' = 12x^2$, and the sequence of steps would be

$$x_1 = x_0 - \frac{4x_0^3 - 1}{12x_0^2} \tag{a}$$

$$= 3 - \frac{107}{108} = 2.009259$$

$$x_2 = 2.00926 - \frac{31.4465}{48.4454} = 1.36015$$

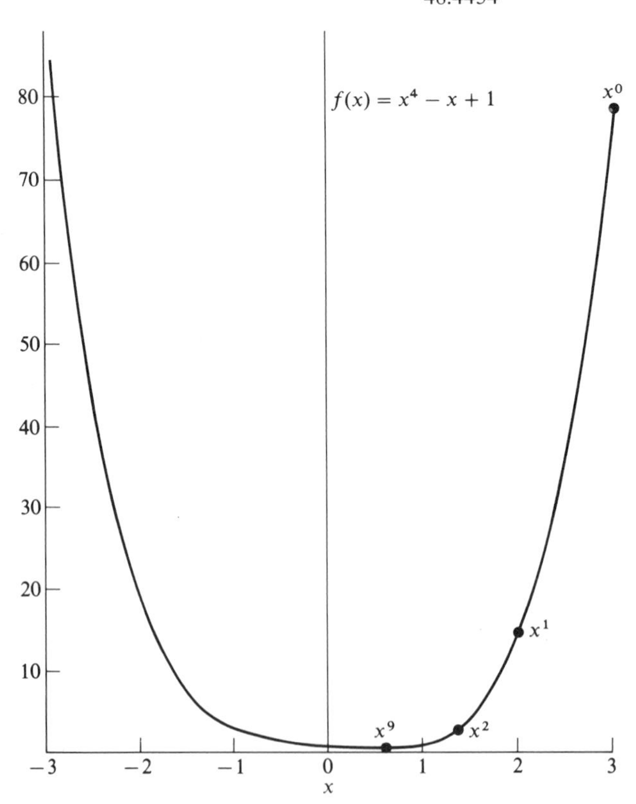

Figure E5.2*a*

Additional iterations yield the following values for x

k	x^k	$\dfrac{x^{k+1} - x^*}{x^k - x^*}$
0	3.00000	
1	2.009259	0.582
2	1.3601480	0.529
3	0.9518103	0.440
4	0.7265254	0.300
5	0.6422266	0.148
6	0.6301933	0.016
7	0.6299606	0.000
8	0.6299605	
9	0.6299605	

As you can see from the third column the rate of convergence of Newton's method is superlinear for this function.

Quasi-Newton. Equation (5.8) for this example is

$$x^{k+1} = x^k - \frac{h}{2} \frac{[f(x+h) - f(x-h)]}{[f(x+h) - 2f(x) + f(x-h)]} \tag{b}$$

For the same problem as used in Newton's method, the first iteration using (b) for $h = 10^{-4}$ is:

$$x^1 = 3 - \left[\frac{10^{-4}}{2}\right] \frac{[f(3.0001) - f(2.9999)]}{[f(3.0001) - 2f(3.000) + f(2.9999)]}$$

Other values of h give:

	x^k		
k	$h = 0.10$	$h = 10^{-4}$	$h = 10^{-7}$
0	3.00000	2.00926	3.00000
1	2.00833	1.36015	2.21568
2	1.35816	0.951811	1.46785
3	0.948531	0.726526	0.955459
4	0.721882	0.642227	0.736528
5	0.636823	0.630193	0.642986
6	0.624849	0.629960	0.631846
7	0.624668	0.6299605191	0.630035
8	0.624669	....	0.629964
9	0.624669313	....	0.629961
10	....	....	0.629961
11	....	....	0.629960525

For $h = 10^{-8}$, the procedure "blew up" after the second iteration.

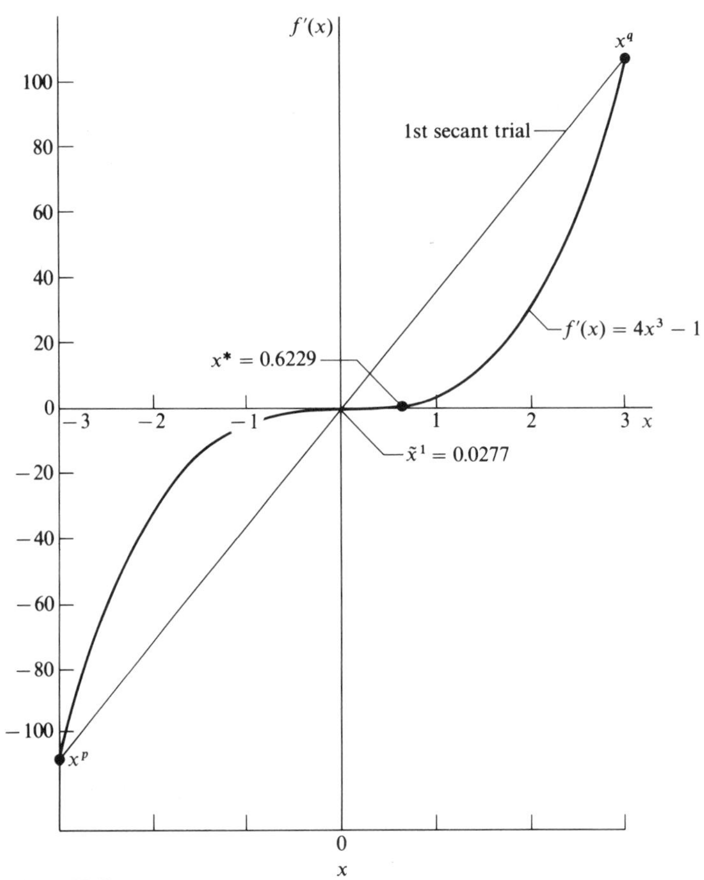

Figure E5.2*b*

Secant. The application of Eq. (5.10) yields the following results (examine Fig. E5.2*b*). Note how the shape of $f'(x)$ implies that a large number of iterations will be needed to reach x^*. Some of the values of $f'(x)$ and x during the search are shown below; notice that x^q remains unchanged in order to maintain the bracket with $f'(x) > 0$.

k	x^q	x^p	$f'(x^p)$
0	3.0	−3.0	−109.0000
1	3.0	0.0277778	−0.9991
2	3.0	0.055296	−0.9992
3	3.0	0.0825434	−0.9977
4	3.0	0.1094966	−0.9899
5	3.0	0.1361213	−0.9899
20	3.0	0.4593212	−0.6124
50	3.0	0.6223007	−0.0360
100	3.0	0.6299311	-1.399×10^{-4}
132	3.0	0.6299597	-3.952×10^{-6}

5.4 REGION ELIMINATION METHODS

Region elimination methods for a one-dimensional search make it possible to delete a calculated portion of the range of x from consideration on each successive stage of the search for an extremum of $f(x)$. When the remaining subinterval is sufficiently small, the search terminates.

The basic element underlying region elimination methods is the comparison of values of $f(x)$ at two or more points within the range of x. We must assume $f(x)$ is unimodal and has a minimum (is convex) within $a < x < b$. Figure 5.4 illustrates the three possible outcomes for values of $f(x)$ if two points are selected in the range a to b.

In Fig. 5.4*a*, the region to the left of x_1 must contain values of $f(x) > f(x_1)$ and hence can be eliminated from further consideration in searching for the minimum of $f(x)$. Similarly, in Fig. 5.4*c*, to the right of x_2, $f(x) > f(x_2)$ and the region can be eliminated. If the case of Fig. 5.4*b* arises, and the function is well-behaved (smooth), the optimum must lie between the points x_1 and x_2. In actual practice, case (*b*) is highly unlikely, simply because of the round-off error inherent in digital computers. However, we might allow some small tolerance for the equality of $f(x_1)$ and $f(x_2)$ and retain this case.

The location of the test points x_1 and x_2 should be chosen to make the search method efficient. If equal spacing is used, that is, $x_1 - a = b - x_2 = x_2 - x_1$, the search method is called *two-point equal interval search.* In this method, the interval of uncertainty is reduced by one-third at each iteration. Thus, if L^0 is the original interval $(b - a)$ and L^k is the interval after k iterations (here $(2/3)^k$ employs k as a power exponent)

$$L^k = (\tfrac{2}{3})^k L^0 \tag{5.11}$$

Note that two function evaluations are required per iteration.

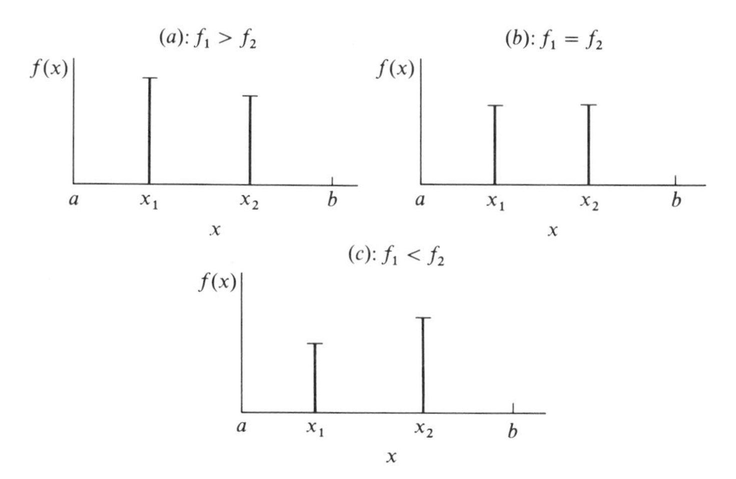

Figure 5.4 Three possible outcomes for $f(x)$ after dividing the interval (a, b) by two points, x_1 and x_2.

A more efficient method uses $x_1 - a = b - x_2$, but the points x_1 and x_2 are placed very close to each other. This method is called the *method of bisecting* or *dichotomous search.* As the two points x_1 and x_2 become an infinitesimal distance apart, you can see that the interval of uncertainty after k iterations approaches

$$L^k = (\tfrac{1}{2})^k L^0 \tag{5.12}$$

a distinct improvement for large values of k [compare Eqs. (5.12) and (5.11) for $k = 8$].

The most efficient region elimination methods such as the Fibonacci and Golden Section search methods (Reklaitis et al., 1983; Beveridge and Schechter, 1970) use a constant ratio for dividing the interval into segments. The golden section method which evolved from the Fibonacci numbers is discussed here.

The strategy employed in the golden section search is to locate the two interior points so that the interval eliminated on one iteration will be of the same proportion to the total interval regardless of the interval length. In addition, only *one* new point has to be calculated at each iteration (versus two in equal-interval and dichotomous search). How to suitably split the interval $(b - a)$ of a line into two segments x and y, was known in ancient times as the "golden section." The ratio of the whole line to the larger segment is the same as the ratio of the larger segment to the smaller (see Fig. 5.5). The ratio of x to y is computed as follows. For the golden section we let the whole interval be $x + y = 1$, and y be the bigger subinterval. Then $1/y = y/x$ or $1/(1 - x) = (1 - x)/x$. Rearrangement gives $x^2 - 3x + 1 = 0$ which has two roots

$$x = \frac{3 \pm \sqrt{5}}{2} = \begin{Bmatrix} 2.618 \\ 0.382 \end{Bmatrix}$$

of which only 0.382 is meaningful. Thus, the initial interval of 1 is divided into two segments

$$x = \frac{3 - \sqrt{5}}{2} = F_S \approx 0.382$$

$$y = 1 - \frac{3 - \sqrt{5}}{2} = F_B \approx 0.618$$

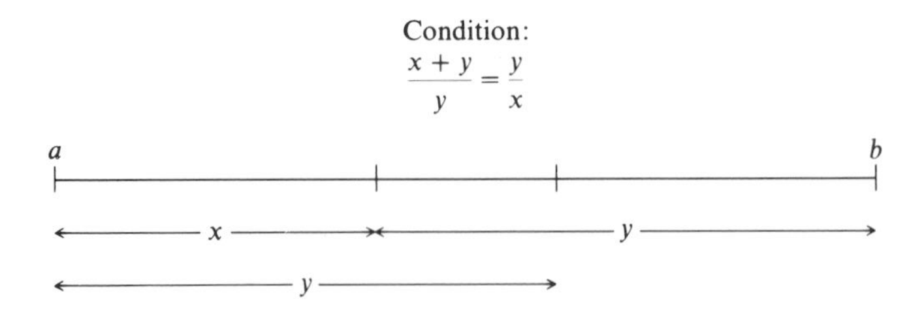

Figure 5.5 Two line segments (x, y) comprising the golden section.

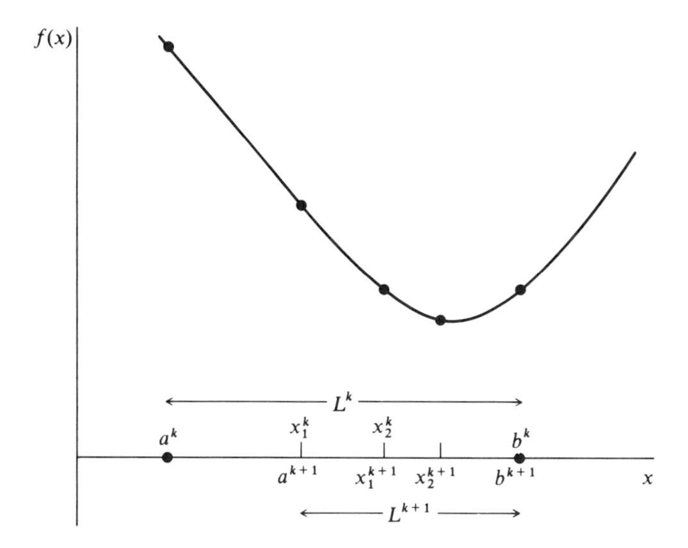

Figure 5.6 Selection of interior points in golden section search.

For any particular interval L^k, you apply the fractions F_S and F_B, to calculate the proper distances. If a^k and b^k are the current bounds of the interval at stage k in the search, then the two interior points are located at $x_1^k = a^k + F_S \cdot L^k$ and $x_2^k = b^k - F_S L^k = b^k + F_B L^k$. In Fig. 5.6, note that $f(x_1^k) > f(x_2^k)$, hence at the next step $a^{k+1} = x_1^k$ and $b^{k+1} = b^k$. The left interior point x_1^{k+1} replaces x_2^k, and we select the right interior point as $x_2^{k+1} = b^{k+1} - F_S(b^{k+1} - a^{k+1})$. Hence $x_1^{k+1} - a^{k+1} = b^{k+1} - x_2^{k+1}$ and the points are equidistant.

The final subinterval after k iterations can be determined as

$$L^k = (0.618)^k L^0$$

The number of function evaluations is $N + 1$ for N trials. This efficiency is slightly inferior to the Fibonacci method, as discussed by Beveridge and Schechter (1970). Therefore you can prespecify the final interval of uncertainty and determine the necessary number of test points required in the search. It should be pointed out that in practice the reduction of the interval of uncertainty to some arbitrary final value is meaningless unless it is put within the context of the value of the objective function. It may be more realistic in the iterative process to compare successive values of the objective function and determine at what point no significant decrease (for minimization) in the objective function is realizable.

EXAMPLE 5.3 APPLICATION OF THE GOLDEN SECTION SEARCH

The function being minimized is again $f(x) = x^2 - x$ as illustrated in Example 5.1, Fig. E5.1*a*. Use golden section search after bracketing the minimum ($x^0 = 3$, $\Delta x = 0.1$).

Solution. First, we bracket the minimum starting at $x^0 = 3$ and $\Delta x^0 = 0.1$. By a preliminary test (or by inspection), the search must start in the negative x direction.

Iteration	$x^{k-1} = 2^{k-1}\Delta x^0$	Value	$f(x^k)$
Start	x^0	3	6
1	$x^0 - \Delta x^0$	2.9	5.51
2	$x^1 - 2\Delta x^0$	2.7	4.59
3	$x^2 - 4\Delta x^0$	2.3	2.99
4	$x^3 - 8\Delta x^0$	1.5	0.75
5	$x^4 - 16\Delta x^0$	−0.1	0.11
6	$x^5 - 32\Delta x^0$	−3.3	14.19

Because the function increases so much between x^5 and x^6, we can try a half-step back from x^6 at $x = [-0.1 + (-3.3)]/2 = -1.7$ to reduce the range of x.

Table E5.3

Iteration	a^k	b^k	L^k	x_1^k	x_2^k	$f(x_1^k)$	$f(x_2^k)$
0	−1.7	1.5	3.2	−0.477708	0.277708	0.705912	−0.200586
1	−0.477708	1.5	1.977708	0.277708	0.744582	−0.200586	−0.190179
2	−0.477708	0.744582	1.222291	−0.010835	0.277708	0.010952	−0.200586
3	−0.010835	0.744582	0.755417	0.277708	0.456038	−0.200586	−0.248067
4	0.277708	0.744582	0.466873	0.456038	0.566252	−0.248067	−0.245610
5	0.277708	0.566252	0.288543	0.387922	0.456038	−0.237438	−0.248067
8	0.456039	0.524155	0.068116	0.482057	0.488137	−0.249678	−0.249997
11	0.491994	0.508074	0.016079	0.498136	0.501932	−0.249996	−0.249996
15	0.498136	0.500482	0.002346	0.499032	0.499586	−0.249999	−0.249999

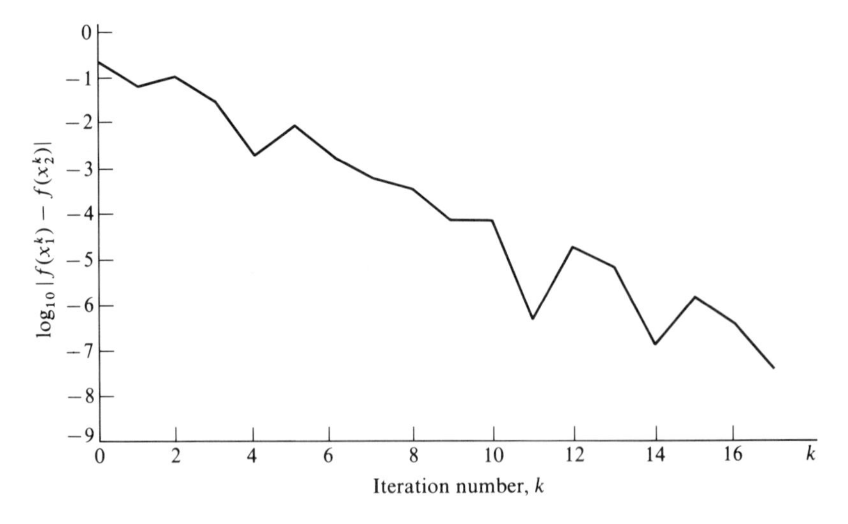

Figure E5.3

At $x = -1.7$, $f(x) = 4.59$ so the bracket is maintained by two of the three points to be used being

$$x_1^0 = -1.7 \qquad x_2^0 = 1.5$$

and $L^0 = 3.2$. Several stages of the search are listed in Table E5.3 (to six decimal places only). Here $F_s = 0.38197$ is used.

Figure E5.3 shows how the distance from the optimum changes from iteration to iteration.

5.5 POLYNOMIAL APPROXIMATION METHODS

Another class of methods of unidimensional minimization locates a point x near x^*, the value of the independent variable corresponding to the minimum of $f(x)$, by extrapolation and interpolation using polynomial approximates as models of $f(x)$. Both quadratic and cubic approximation have been proposed using function values only, and using both function and derivative values. Coggins (1964) pointed out that several of the techniques involving the fitting of a polynomial through selected points were slightly more efficient in practice in locating the minimum of $f(x)$ to within a specified precision than the golden section, and experience has substantiated this conclusion.

5.5.1 Quadratic Interpolation

We start with three points x_1, x_2, and x_3 in increasing order that may or may not be equally spaced, but the extreme points must bracket the minimum. From the analysis in Chap. 2, we know that a quadratic function $f(x) = a + bx + cx^2$ can be passed exactly through the three points, and that the function can be differentiated and the derivative set equal to 0 to yield the minimum of the approximating function

$$\tilde{x}^* = -\frac{b}{2c} \tag{5.13}$$

Suppose that $f(x)$ is evaluated at x_1, x_2, and x_3 to yield $f(x_1) \equiv f_1$, $f(x_2) \equiv f_2$, and $f(x_3) \equiv f_3$. The coefficients b and c can be evaluated from the solution of the three linear equations

$$f(x_1) = a + bx_1 + cx_1^2$$

$$f(x_2) = a + bx_2 + cx_2^2$$

$$f(x_3) = a + bx_3 + cx_3^2$$

via determinants or matrix algebra. Introduction of b and c expressed in terms of x_1, x_2, x_3, f_1, f_2, and f_3 into Eq. (5.13) gives

$$\tilde{x}^* = \frac{1}{2}\left[\frac{(x_2^2 - x_3^2)f_1 + (x_3^2 - x_1^2)f_2 + (x_1^2 - x_2^2)f_3}{(x_2 - x_3)f_1 + (x_3 - x_1)f_2 + (x_1 - x_2)f_3}\right] \tag{5.14}$$

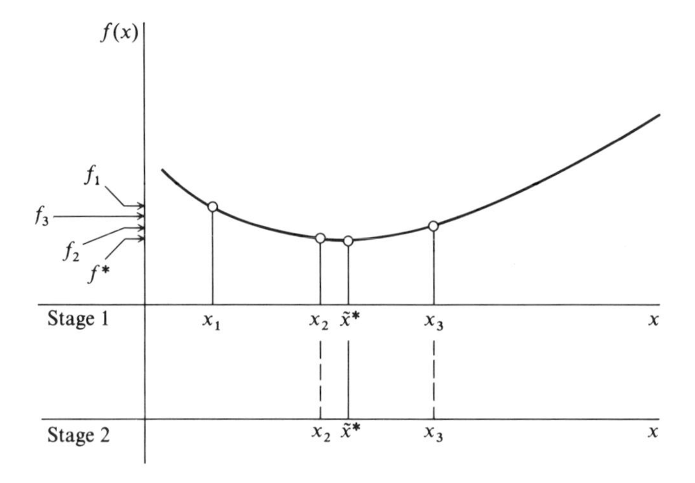

Figure 5.7 Two stages of quadratic interpolation.

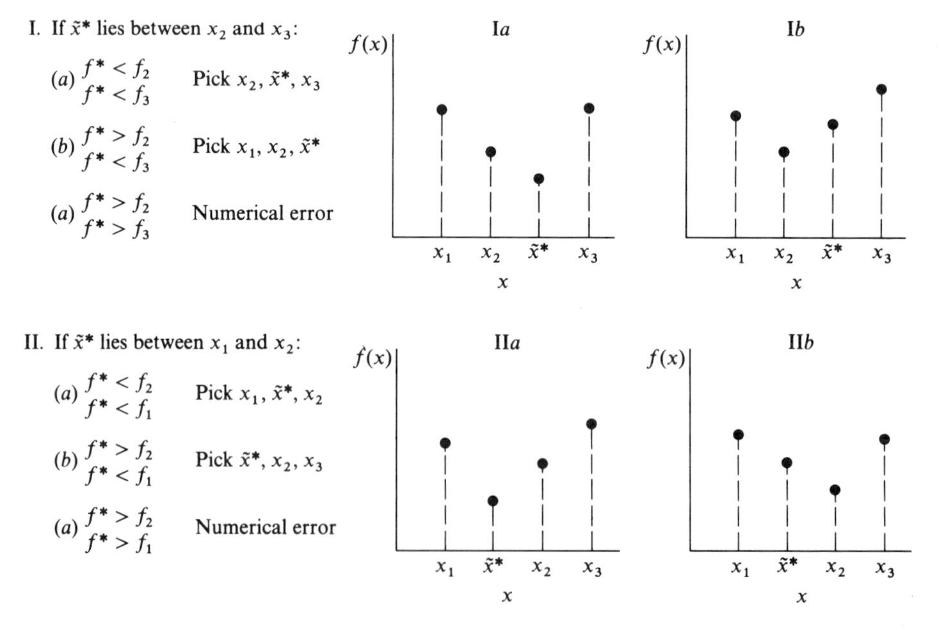

Figure 5.8 How to maintain a bracket on the minimum in quadratic interpolation.

To illustrate the first stage in the search procedure, examine the four points in Fig. 5.7 for stage 1. We want to reduce the initial interval $[x_1, x_3]$. By examining the values of $f(x)$ [with the assumptions that $f(x)$ is unimodal and has a minimum], we can discard the interval from x_1 to x_2 and use the region $(x_2, x_3]$ as the new interval. The new interval contains three points, $(x_2, \tilde{x}^*, x_3)$ that can be introduced into Eq. (5.14) to estimate a x^{opt}, and so on. In general, you evaluate $f(x^*)$ and discard from the set $\{x_1, x_2, x_3\}$ the point that corresponds to the greatest value of $f(x)$, unless a bracket on the minimum of $f(x)$ will be lost by so doing, in which case you discard the x so as to maintain the bracket. The specific tests and choices of x_i to maintain the bracket are illustrated in Fig. 5.8. In Fig. 5.8, $f^* \equiv f(\tilde{x}^*)$. If x^* and whichever of $\{x_1, x_2, x_3\}$ corresponding to the smallest $f(x)$ differ by less than the prescribed accuracy in x, or the prescribed accuracy in the corresponding values of $f(x)$ is achieved, terminate the search. Note that only function evaluations are used in the search, and that only one new function evaluation (for $\tilde{x}^*$) has to be carried out at each new iteration.

EXAMPLE 5.4 APPLICATION OF QUADRATIC INTERPOLATION

The function to be minimized is $f(x) = x^2 - x$ and is illustrated in Fig. E5.1*a*. The three points to bracket the minimum developed in Example 5.3 will be used to start the search for the minimum of $f(x)$; we use equally spaced points here but that is not a requirement of the method.

Solution

$$x_1 = -1.7 \qquad x_2 = -0.1 \qquad x_3 = 1.5$$
$$f(x_1) = 4.59 \qquad f(x_2) = 0.11 \qquad f(x_3) = 0.75$$
$$\Delta x = 1.6$$

Two different formulas for the quadratic interpolation can be examined, Eq. (5.8), the quasi-Newton method, and Eq. (5.13)

$$\tilde{x}^* = x_2 - \frac{\Delta x[f(x_3) - f(x_1)]}{2[f(x_3) - 2f(x_2) + f(x_1)]} \tag{5.8}$$

$$= -0.1 - \frac{1.6(0.75 - 4.59)}{2(0.75 - 2(0.11) + 4.59)} = 0.50 \tag{a}$$

$$\tilde{x}^* = \frac{1}{2}\frac{[x_2^2 - x_3^2]f(x_1) + [x_3^2 - x_1^2]f(x_2) + [x_1^2 - x_2^2]f(x_3)}{(x_2 - x_3)f(x_1) + (x_3 - x_1)f(x_2) + (x_1 - x_2)f(x_3)} \tag{5.14}$$

$$= \frac{1}{2}\frac{[(-0.1)^2 - (1.5)^2](4.59) + [(1.5)^2 - (-1.7)^2](0.11) + [(-1.7)^2 - (-0.1)^2](0.75)}{[(-0.1) - (1.5)](4.59) + [(1.5) - (-1.7)](0.11) + [-1.7) - (-0.1)](0.75)} \tag{b}$$

$$= 0.50$$

Note that a solution on the first iteration seems to be remarkable but keep in mind that the function is quadratic so that quadratic interpolation should be good even if approximate formulas are used for derivatives.

5.5.2 Cubic Interpolation

Cubic interpolation to find the minimum of $f(x)$ is based on approximating the objective function by a third-degree polynomial within the interval of interest, and then determining the associated stationary point of the polynomial

$$f(x) = a_1 x^3 + a_2 x^2 + a_3 x + a_4$$

Four points must be computed (that bracket the minimum) to estimate the minimum, either four values of $f(x)$, or the values of $f(x)$ and the derivative of $f(x)$, each at two points.

In the former case four linear equations are obtained with the four unknowns being the desired coefficients. Let the matrix $\mathbf{X}$ be

$$\mathbf{X} = \begin{bmatrix} x_1^3 & x_1^2 & x_1 & 1 \\ x_2^3 & x_2^2 & x_2 & 1 \\ x_3^3 & x_3^2 & x_3 & 1 \\ x_4^3 & x_4^2 & x_4 & 1 \end{bmatrix}$$

$$\mathbf{F}^T = [f(x_1) \quad f(x_2) \quad f(x_3) \quad f(x_4)]$$

$$\mathbf{A}^T = [a_1 \quad a_2 \quad a_3 \quad a_4]$$

$$\mathbf{F} = \mathbf{XA} \tag{5.15}$$

Then the extremum of $f(x)$ is obtained by setting the derivative of $f(x)$ equal to zero and solving for $\tilde{x}^*$

$$\frac{df(x)}{dx} = 3a_1 x^2 + 2a_2 x + a_3 = 0$$

so that

$$\tilde{x}^* = \frac{-2a_2 \pm \sqrt{4a_2^2 - 12a_1 a_3}}{6a_1} \tag{5.16}$$

The sign to use before the square root is governed by the sign of the second derivative of $f(\tilde{x}^*)$, that is, whether a minimum or maximum is sought. The vector $\mathbf{A}$ can be computed from $\mathbf{XA} = \mathbf{F}$ or

$$\mathbf{A} = \mathbf{X}^{-1}\mathbf{F} \tag{5.17}$$

After the optimum point $\tilde{x}^*$ is predicted, it is used as a new point in the next iteration and the point with the highest [lowest value of $f(x)$ for maximization] value of $f(x)$ is discarded.

If the first derivatives of $f(x)$ are available, only two points are needed and the cubic function can be fitted to the two pairs of the slope and function values. These four pieces of information can be uniquely related to the four coefficients in the cubic equation, which can be optimized for predicting the new, nearly optimal data point. If (x_1, f_1, f'_1) and (x_2, f_2, f'_2) are available, then the optimum $\tilde{x}^*$ (Fox, 1971) is

$$\tilde{x}^* = x_2 - \left[\frac{f'_2 + w - z}{f'_2 - f'_1 + 2w}\right](x_2 - x_1) \tag{5.18}$$

where $z = 3[f_1 - f_2]/[x_2 - x_1] + f'_1 + f'_2$
$w = [z^2 - f'_1 \cdot f'_2]^{1/2}$

In a minimization problem, you require $x_1 < x_2$, $f'_1 < 0$, and $f'_2 > 0$ (x_1 and x_2 bracket the minimum). For the new point ($\tilde{x}^*$), calculate $f'(\tilde{x}^*)$ to determine which of the previous two points to replace. The application of this method in nonlinear programming algorithms which use gradient information is straightforward and effective.

If the function being minimized is not unimodal locally, as has been assumed to be true in the above discussion, extra logic must be added to the unidimensional search code to ensure that the step size is adjusted to the neighborhood of the local optimum actually sought. For example, Figure 5.9 illustrates how a large initial step can lead to an unbounded solution to a problem when, in fact, a local minimum is sought.

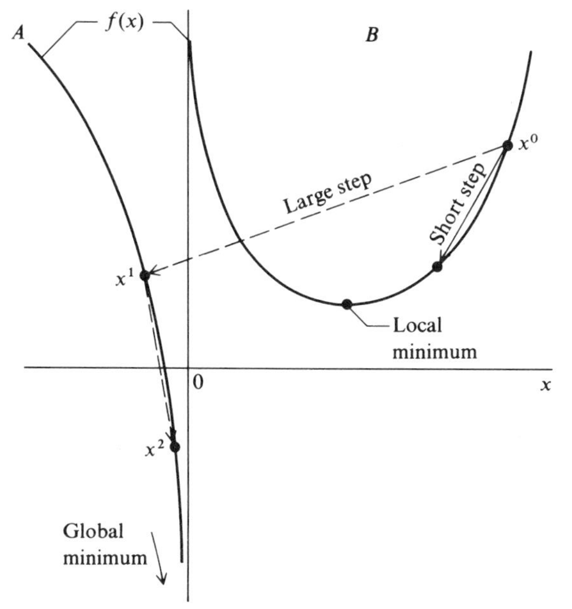

Figure 5.9 A unidimensional search for a local minimum of a multimodal objective function leads to an unbounded solution.

EXAMPLE 5.5 OPTIMIZATION OF A MICROELECTRONICS PRODUCTION LINE FOR LITHOGRAPHY

You are to optimize the thickness of resist used in a production lithographic process. There are a number of competing effects in lithography.

1. As the thickness t (measured in micrometers) grows smaller, the defect density grows larger. The number of defects per square centimeter of resist is given by

$$D_0 = 1.5t^{-3}$$

2. The chip yield in fraction of good chips for each layer is given by

$$\eta = \frac{1}{1 + \alpha D_0 a}$$

where a is the active area of the chip. Assume that 50 percent of the defects are "fatal" defects ($\alpha = 0.5$) detected after manufacturing the chip.

Assume four layers are required for the device. The overall yield is based on a series formula:

$$\eta = \frac{1}{(1 + \alpha D_0 a)^4}$$

3. Throughput decreases as resist thickness increases. A typical relationship is

$$V(\text{wafers/h}) = 125 - 50t + 5t^2$$

Each wafer has 100 chip sites with 0.25 cm^2 active area. The daily production level is to be 2500 finished wafers. Find the resist thickness to be used to maximize the number of good chips per hour. Assume $0.5 \le t \le 2.5$ as the expected range. Use cubic interpolation to find the optimal value of t, t^*. How many parallel production lines would be required for t^*, assuming 20 h/day operation each?

Solution. The objective function to be maximized is the number of good chips per hour, which is found by multiplying the yield, the throughput, and the number of chips per wafer (=100):

$$f = V\eta = (125 - 50t + 5t^2)\frac{100}{[1 + 0.5(1.5t^{-3})(0.25)]^4}$$

Using initial guesses of $t = 1.0$ and 2.0, cubic interpolation yielded the following values of f:

t	f	f′	
1.0	4023.05	−5611.10	
2.0	4101.73	2170.89	
1.414	4973.22	148.70	
1.395	4974.60	−3.68	(optimum)

Figure E5.5 is a plot of the objective function $f(t)$.

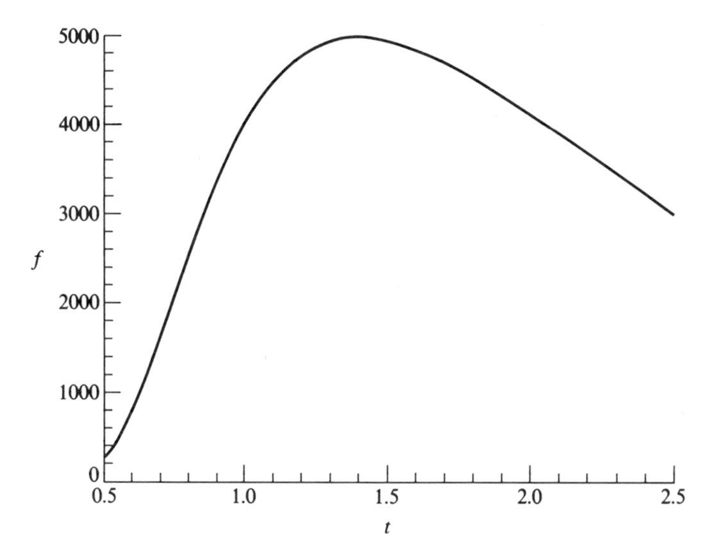

Figure E5.5 Plot of objective function (number of good chips per hour) versus resist thickness, t (μm).

The throughput for $t^* = 1.395$ is

$$V = 65.02 \text{ wafers/h}$$

If a production line is operated 20 h/day, two lines would be needed to achieve 2500 wafers/day.

5.6 HOW THE ONE-DIMENSIONAL SEARCH IS APPLIED IN A MULTIDIMENSIONAL PROBLEM

In minimizing a function $f(\mathbf{x})$ of several variables, the general procedure is to (a) calculate a search direction, and (b) reduce the value of $f(\mathbf{x})$ by taking one or more steps in that search direction. Chapter 6 describes in detail how to select search directions. Here we explain how to take steps in the search direction as a function of a single variable, the step length λ. Usually execution of this stepping is called a *unidimensional search* or *line search.*

Examine Fig. 5.10 in which contours of a function of two variables are displayed:

$$f(\mathbf{x}) = x_1^4 - 2x_2x_1^2 + x_2^2 + x_1^2 - 2x_1 + 5$$

Suppose that the negative gradient of $f(\mathbf{x})$, $-\nabla f(\mathbf{x})$, is selected as the search direction starting at the point $\mathbf{x}^T = [1 \quad 2]$. The negative gradient is the direc-

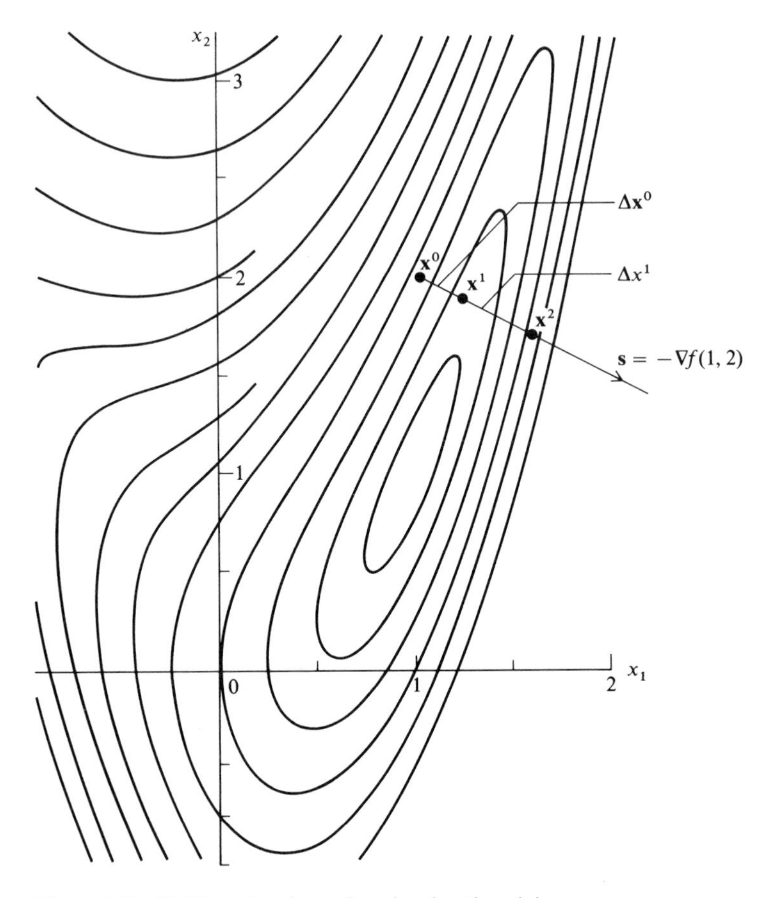

Figure 5.10 Unidimensional search to bracket the minimum.

tion which maximizes the rate of change of $f(\mathbf{x})$ in moving toward the minimum. To move in this direction we want to calculate a new $\mathbf{x}$

$$\mathbf{x}_{\text{new}} = \mathbf{x}_{\text{old}} + \lambda \mathbf{s}$$

where $\mathbf{s}$ is the search direction, a vector, and λ is a scalar denoting the distance moved along the search direction. Note $\lambda\mathbf{s} \equiv \Delta\mathbf{x}$, the vector for the step to be taken (encompassing both direction and distance).

Execution of a unidimensional search involves calculating a value of λ, and then taking steps in each of the coordinate directions as follows:

$$\text{In the } x_1 \text{ direction: } x_{1,\text{new}} = x_{1,\text{old}} + \lambda s_1$$

$$\text{In the } x_2 \text{ direction: } x_{2,\text{new}} = x_{2,\text{old}} + \lambda s_2$$

where s_1 and s_2 are the two components of $\mathbf{s}$ in the x_1 and x_2 directions, respectively. Repetition of this procedure accomplishes the unidimensional search.

EXAMPLE 5.6 EXECUTION OF A UNIDIMENSIONAL SEARCH

We illustrate two stages in bracketing the minimum in minimizing the function from Fox (1971)

$$f(\mathbf{x}) = x_1^4 - 2x_2x_1^2 + x_2^2 + x_1^2 - 2x_1 + 5$$

in the negative gradient direction

$$-\nabla f(\mathbf{x}) = -\begin{bmatrix} 4x_1^3 - 4x_2x_1 + 2x_1 - 2 \\ -2x_1^2 + 2x_2 \end{bmatrix}$$

starting at $\mathbf{x}^T = [1 \quad 2]$ where $f(\mathbf{x}) = 5$. Here

$$s = -\nabla f(1, 2) = -\begin{bmatrix} -4 \\ 2 \end{bmatrix}$$

We start to bracket the minimum by taking $\lambda^0 = 0.05$

$$x_1^1 = x_1^0 + (0.05)(4) = 1.2 \qquad (a)$$

$$x_2^1 = x_2^0 + (0.05)(-2) = 1.9 \qquad (b)$$

Steps (a) and (b) consist of one overall step in the direction $\mathbf{s} = [4 \quad -2]^T$, and yield $\Delta\mathbf{x}^T = [0.2 \quad -0.1]$. At $\mathbf{x}^1$, $f(1.2, 1.9) = 4.25$, an improvement.

For the next step, we let $\lambda^1 = 2\lambda^0 = 0.1$, and take another step in the same direction:

$$x_1^2 = x_1^1 + 0.1(4) = 1.6$$

$$x_2^2 = x_2^1 + 0.1(-2) = 1.7$$

$$\Delta\mathbf{x}^1 = [0.4 \quad -0.2]^T$$

At $\mathbf{x}^2$, $f(1.6, 1.7) = 5.10$, so that the minimum of $f(x)$ in direction $\mathbf{s}$ has been bracketed. Examine Fig. 5.10. The optimal value of λ along the search direction can be found to be $\tilde{\lambda}^* = 0.04875$ by one of the methods described in this chapter.

5.7 EVALUATION OF UNIDIMENSIONAL SEARCH METHODS

In this chapter we have described and illustrated only a few unidimensional search methods. Refer to Polak (1971) for many others including those of Armijo (1966) and Goldstein-Price (1967) that are well regarded. Naturally, you can ask which unidimensional search method is best to use, most robust, most efficient, etc. Unfortunately, the various algorithms are problem-dependent even if used alone, and if used as subroutines in optimization codes, also depend on how well they mesh with the particular code. Some codes call for almost exact minimization of a function of several variables by successive iterations to determine the value of the step length λ in the chosen search direction as described in the next chapter. Other codes simply take one or a few steps in the search direction, or in more than one direction, with no requirement for accuracy —only that $f(x)$ be reduced. Most computational evaluations on large sets of test problems show little difference among the interval reduction and Newton-type algorithms. Based on choices made by commercial and academic programmers, quadratic or cubic interpolation methods are favored. But a word of warning! Effective unidimensional codes contain many heuristic additions not described in this chapter to accommodate multiple extrema, round-off error, a high degree of curvature or essentially no curvature, and other undesirable features of functions of single variables found in practice.

REFERENCES

Armijo, L., "Minimization of Functions Having Continuous Partial Derivatives," *Pac. J. Math.*, **16**: 1 (1966).

Beveridge, G. S. G., and R. S. Schechter, *Optimization: Theory and Practice*, McGraw-Hill, New York (1970).

Coggins, G. F., Univariate Search Methods, Imperial Chemical Industries Ltd., Central Instr. Lab. Res. Note 64/11, 1964.

Dennis, J. E., and R. B. Schnabel, *Numerical Methods for Unconstrained Optimization and Nonlinear Equations*, Prentice-Hall, Englewood Cliffs, New Jersey (1983) chap. 2.

Fox, R. L., *Optimization Methods for Engineering Design*, Addison-Wesley, Reading, Massachusetts (1971) p. 42.

Goldstein, A. A., and J. F. Price, "An Effective Algorithm for Minimization," *Numerical Math.*, **10**: 184 (1967).

Polak, E. *Computational Methods in Optimization*, Academic Press, New York (1971).

Wilde, D. J., *Optimum Seeking Methods*, Prentice-Hall, Englewood Cliffs, New Jersey (1964).

SUPPLEMENTARY REFERENCES

Beightler, C. S., D. T. Phillips, and D. J. Wilde, *Foundations of Optimization*, 2d ed., Prentice-Hall, Englewood Cliffs, New Jersey (1979).

Brent, R. P., *Algorithms for Minimization Without Derivatives*, Prentice-Hall, Englewood Cliffs, New Jersey (1973).

Cooper, L., and D. Steinberg, *Introduction to Methods of Optimization*, W. B. Saunders Co., Philadelphia (1970).

Luenberger, D. G., *Introduction to Linear and Nonlinear Programming*, Addison-Wesley, Reading, Massachusetts (1973).

Reklaitis, G. V., R. A. Ravindran, and K. M. Ragsdell, *Engineering Optimization*, Wiley-Interscience, New York (1983).

Shoup, T. E., and F. Mistree, *Optimization Methods with Applications for Personal Computers*, Prentice-Hall, Englewood Cliffs, New Jersey (1987).

Weixnan, L., *Optimal Block Search*, Helderman, Berlin (1984).

PROBLEMS

5.1. Apply the following sequential one-dimensional search techniques to reduce the interval of uncertainty for the maximum of the function $f = 6.64 + 1.2x - x^2$ from $[0, 1]$ to less than 2 percent of its original size. Show all the iterations for the following search methods:

(*a*) Two-point equal interval
(*b*) Golden section search

5.2. You plan to use a one-dimensional search to find the operating pressure of the chemical reactor that corresponds to the maximum yield of the reactor. A possible range of pressures has been identified by laboratory work to be between 1 and 5 atm. You would like to operate within 0.03 atm of the optimum pressure. How many tests would be required in

(*a*) A two-point equal-interval search?
(*b*) Golden section search

5.3. Minimize the function $f = (x - 1)^4$. Use quadratic interpolation but no more than a maximum of ten function evaluations. The initial three points selected are $x_1 = 0$, $x_2 = 0.5$, and $x_3 = 2.0$.

5.4. Repeat Prob. 5.3 but use cubic interpolation via function and derivative evaluations. Use $x_1 = 0.5$ and $x_2 = 2.0$ for a first guess.

5.5. Repeat Prob. 5.3 for cubic interpolation with $x_1 = 1.5$, $x_2 = 3.0$, $x_3 = 4.0$, and $x_4 = 4.5$.

5.6. Repeat Prob. 5.3 for the quasi-Newton method, with $x_1 = -1$, $x_2 = -0.5$, $x_3 = 0.0$.

5.7. A one-dimensional search is to be carried out from the point (0, 3) using $\mathbf{s} = [1 \quad 1]^T$; $\mathbf{s}$ is the search direction. Evaluate

$$f = (x_1 - 1)^2 + 4(x_2 - 2)^2$$

at three points: (0, 3), (1, 4), and (−1, 2). Let $\mathbf{x} = \begin{bmatrix} 0 \\ 3 \end{bmatrix} + \lambda \begin{bmatrix} 1 \\ 1 \end{bmatrix}$. Use quadratic interpolation to find the minimum along this search direction (with respect to λ). Compare your answer with the optimum value (−0.6, 2.4). Comment on the advantages and disadvantages of quadratic interpolation vs. golden section search for this problem.

5.8. Determine whether the rate of convergence is (*a*) linear, (*b*) superlinear, or (*c*) quadratic from the values of x and $f(x)$ obtained on each iteration of the application of (1) Newton's method, (2) a quasi-Newton method, (3) secant method, (4) quadratic interpolation, (5) cubic interpolation, (6) two-point equal-interval search, and (7) golden section search to minimize the following functions:

(*a*) $x^2 - 6x + 3$
(*b*) $\sin(x)$ with $0 < x < 2\pi$
(*c*) $x^4 - 20x^3 + 0.1x$

5.9. In the golden section search, the interval that brackets the minimum is successively divided into two parts. If the initial interval is of length 0.1, how many consecutive subdivisions are needed to reduce the interval to at least 0.001?

5.10. The total annual cost of operating a pump and motor (C) in a particular piece of equipment is a function of x, the size (horsepower) of the motor, namely

$$C = \$500 + \$0.9x + \frac{\$0.03}{x}(150{,}000)$$

Find the motor size that minimizes the total annual cost.

5.11. A boiler house contains five coal-fired boilers, each with a nominal rating of 300 boiler horsepower (BHP). If economically justified, each boiler can be operated at a rating of 350 percent of nominal. Due to the growth of manufacturing departments, it has become necessary to install additional boilers. Refer to the data below. Determine the percent of nominal rating at which the present boilers should be operated. *Hint:* Minimize total costs per year per BHP output.

Data: The cost of fuel, coal, including the cost of handling coal and removing cinders, is $7 per ton, and the coal has a heating value of 14,000 Btu per pound. The overall efficiency of the boilers, from coal to steam, has been determined from tests of the present boilers operated at various ratings as:

Percent of nominal rating, R	Percent overall thermal efficiency, E
100	75
150	76
200	74
225	72
250	69
275	65
300	61

The annual fixed charges, C_F, in dollars per year on each boiler are given by the equation:

$$C_F = 14{,}000 + 0.04R^2$$

Assume 8550 hours operating per year.

Hint: You will find it helpful to first obtain a relation between R and E by least squares (refer to Chap. 2) to eliminate the variable E.

5.12. A laboratory filtration study is to be carried out at constant rate. The basic equation (Cook, L. N., "Laboratory approach optimizes filter-aid addition," *Chem. Eng.* July 23, 1984, p. 45) comes from the relation

$$\text{flow-rate} \propto \frac{\text{(pressure drop)(filter area)}}{\text{(fluid viscosity) (cake thickness)}}$$

Cook expressed filtration time as

$$t_f = \beta \frac{\Delta P_c A^2}{\mu M^2 c} x_c \exp(-a x_c + b)$$

where t_f = time to build up filter cake, min
ΔP_c = pressure drop across cake, psig (20)
A = filtration area, ft^2 (250)
μ = filtrate viscosity, centipoise (20)
M = mass flow of filtrate, lb$_m$/min (75)
c = solids concentration in feed to filter, lb$_m$/lb$_m$ filtrate (0.01)
x_c = mass fraction solids in dry cake
a = constant relating cake resistance to solids fraction (3.643)
b = constant relating cake resistance to solids fraction (2.680)
$\beta = 3.2 \times 10^{-8}$ (lb$_m$/ft)^2

Obtain the maximum time for filtration as a function of x_c by a numerical unidimensional search.

5.13. An industrial dryer for granular material can be modeled [Becker, H. A., P. L. Douglas, and S. Ilias, "Development of Optimization Strategies for Industrial Grain Dryer Systems," *Can. J. Chem. Eng.* **62**, 738 (1984)] with the total specific cost of drying $C(\$/m^3)$ being

$$C = [1.767 \ln (W_0/W_D)/\beta V_t] \frac{(F_A C_{pA} + UA)\Delta T C'_p}{\Delta H_C + PC'_E + C'_L}$$

where A = heat transfer area of dryer normal to the air flow, m^2 (153.84)
β = constant, function of air plenum temperature and initial moisture level
C'_E = unit cost of electricity, \$/kWh (0.0253)
C'_L = unit cost of labor, \$/h (15)
C'_p = unit cost of propane, \$/kg (0.18)
C_{pA} = specific heat of air, J/kg K (1046.75)
F_A = flow-rate of air, kg/h (3.38×10^5)
ΔH_c = heat combustion of propane, J/kg (4.64×10^7)
P = electrical power, kW (188)
ΔT = temperature difference $(T - T_1)$, K; the plenum air temperature T minus the inlet air temperature $T_1 (T_1 = 390 \text{ K})$
U = overall heat transfer coefficient from dryer to atmosphere, $W/(m^2)(K)$ (45)
V_t = total volume of the dryer, m^3 (56)
W_D = final grain moisture content (DB), kg/kg (0.1765)
W_0 = initial moisture content (DB), kg/kg (0.500)

Values for the coefficient

$$\beta = (-0.263\,1125 + 0.0028958T) W_0^{(-0.2368125 + 0.000966T)}$$

Find the minimum cost as a function of the plenum temperature T (in degrees Kelvin).

CHAPTER 6

UNCONSTRAINED MULTIVARIABLE OPTIMIZATION

The numerical optimization of general nonlinear multivariable objective functions requires that efficient and robust techniques be employed. Efficiency is important since these problems require an interative solution procedure. Trial and error becomes impractical for more than three or four variables. Robustness (ability to achieve a solution) is a desirable property since a general nonlinear function is unpredictable in its behavior; there may be relative maxima or minima, saddle points, regions of convexity/concavity, etc. In some regions the optimization algorithm may yield very slow progress toward the optimum, hence requiring excessive computer time. Fortunately, we can draw upon extensive numerical experience in testing nonlinear programming algorithms for unconstrained functions to evaluate various approaches proposed for the solution of such functions.

In this chapter we discuss the solution of the unconstrained optimization problem:

$$\text{Find } \mathbf{x}^* = [x_1 \quad x_2 \quad \cdots \quad x_n]^T \quad \text{that minimizes} \quad f(x_1, x_2, \ldots, x_n) \equiv f(\mathbf{x})$$

Most iterative procedures that are effective alternate between two phases in the optimization:

(*a*) choosing a search direction $\mathbf{s}^k$

(*b*) minimizing in that direction to some extent (or completely) to find a new point $\mathbf{x}^{k+1} = \mathbf{x}^k + \Delta\mathbf{x}^k$ where $\Delta\mathbf{x}^k \equiv \mathbf{x}^{k+1} - \mathbf{x}^k$ is known as the step size. For maximization, we minimize $-f(\mathbf{x})$.

In addition to (*a*) and (*b*), an algorithm must specify

(*c*) the initial starting vector $\mathbf{x}^0 = [x_1^0 \quad x_2^0 \quad \cdots \quad x_n^0]^T$ and

(*d*) the convergence criteria for termination

From a given starting point, a search direction is determined, and $f(\mathbf{x})$ is minimized in that direction. The search stops based on some criteria, and then a new search direction is determined, followed by another line search. The line search can be carried out to various degrees of precision. For example, we could use a simple successive doubling of the step size as a screening method until we detect the optimum has been bounded. At this point the screening search can be terminated and a more sophisticated method employed to yield a higher degree of accuracy. In any event, refer to the techniques discussed in Chap. 5 for ways to carry out the line search.

The NLP (nonlinear programming) methods which will be discussed in this chapter differ mainly in how they generate the search directions. Some nonlinear programming methods require information about derivative values, whereas other methods do not use derivatives and rely solely on function evaluations. Furthermore, finite difference substitutes can be used in lieu of derivatives as explained in Sec. 6.5. In most cases, methods that employ analytical derivatives use less computer time but more of your time for analysis. We shall discuss both. We first describe some simple nonderivative methods, and then present a series of methods which use derivative information. We also show how the nature of the objective function influences the effectiveness of the particular optimization algorithm.

6.1 DIRECT METHODS

Direct methods do not require the use of derivatives in determining the search direction. Under some circumstances the methods described in this section can be used effectively but they may be inefficient compared with methods discussed in subsequent sections. They do have the advantage of being simple to understand and execute.

6.1.1 Random Search

A random search method simply selects a starting vector $\mathbf{x}^0$, evaluates $f(\mathbf{x})$ at $\mathbf{x}^0$, and then randomly selects another vector $\mathbf{x}^1$, and evaluates $f(\mathbf{x})$ at $\mathbf{x}^1$. In effect, both a search direction and step length are chosen simultaneously. After one or more stages, the value of $f(\mathbf{x}^k)$ is compared with the best previous value of $f(\mathbf{x})$ from among the previous stages, and the decision is made to continue or terminate the procedure. Variations of this form of random search involve randomly selecting a search direction and then minimizing (possibly by random steps) in that search direction as a series of cycles. Clearly, *the* optimal solution can be obtained with a probability of 1 only as $k \to \infty$, but as a practical matter, if the objective function is quite flat, a suboptimal solution may be quite acceptable. Even though the method is inefficient insofar as function evaluations are concerned, it may provide a good starting point for another method. You might view random search as an extension of the case study method. Refer to Dixon and James (1980) for some practical algorithms.

6.1.2 Grid Search

Methods of experimental design discussed in most basic statistics books can equally well be applied to minimizing $f(\mathbf{x})$ (see Box and Hunter, 1962). You evaluate a series of points about a reference point selected according to some type of design such as the ones shown in Fig. 6.1 (for an objective function of two variables). Next you move to that point which improves the objective function the most, and repeat. For $n = 10$, we must examine $3^{10} - 1 = 59{,}048$ values of $f(\mathbf{x})$ if a three-level factorial design is to be used, obviously a prohibitive number of function evaluations.

6.1.3 Univariate Search

Another simple optimization technique would be to select n fixed search directions (usually the coordinate axes) for an objective function of n variables. Then

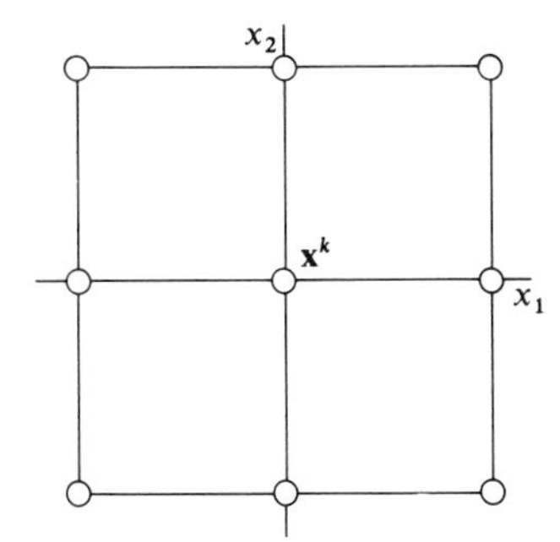

(a) Three-level factorial design ($3^2 - 1 = 8$ points plus center)

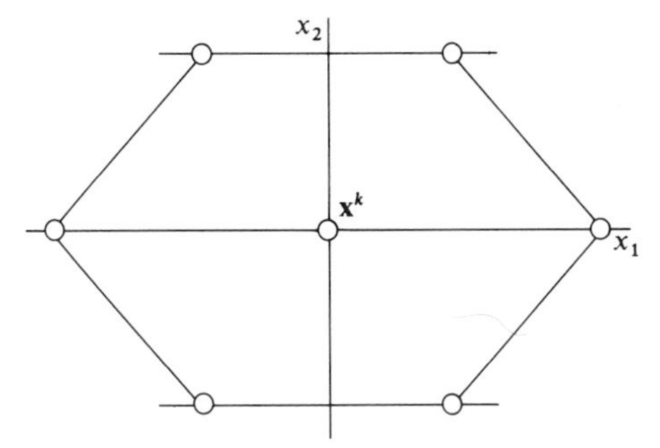

(b) Hexagon design (6 points + center)

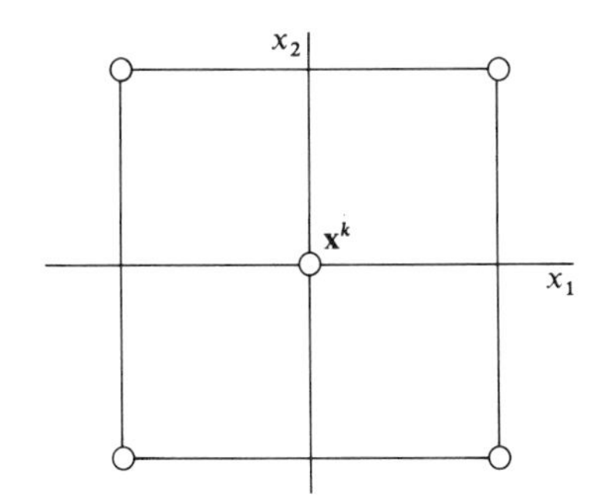

(c) Two-level factorial design ($2^2 = 4$ points plus center)

Figure 6.1 Various grid search designs to select vectors $\mathbf{x}$ to evaluate $f(\mathbf{x})$.

$f(\mathbf{x})$ is minimized in each search direction sequentially using a one-dimensional search. While this method is effective for a quadratic function of the form

$$f_1(\mathbf{x}) = \sum_{i=1}^{n} c_i x_i^2$$

because the search directions line up with the principal axes as indicated in Fig. 6.2*a*, it does not perform satisfactorily for more general quadratic objective functions of the form

$$f_2(\mathbf{x}) = \sum_{i=1}^{n} \sum_{j=1}^{n} d_{ij} x_i x_j$$

as illustrated in Fig. 6.2*b*. For the latter case, the step size decreases as the optimum is neared, thus slowing down the search procedure.

6.1.4 Simplex Method

The method of the "Sequential Simplex" formulated by Spendley, Hext, and Himsworth (1962) uses a regular geometric figure (a simplex) to select points at the vertices of the simplex at which to evaluate $f(\mathbf{x})$. In two dimensions the figure is an equilateral triangle. In three dimensions this figure becomes a regular tetrahedron, and so on. Each search direction points away from the vertex having the highest value of $f(\mathbf{x})$. Thus, the direction of search changes, but the step size is fixed for a given size simplex. Let us use a function of two variables to illustrate the procedure.

At each iteration, to minimize $f(\mathbf{x})$, $f(\mathbf{x})$ is evaluated at each of three vertices of the triangle. The direction of search is oriented away from the point with the highest value for the function through the centroid of the simplex. By making the search direction bisect the line between the other two points of the triangle, the direction will go through the centroid. A new point is selected in

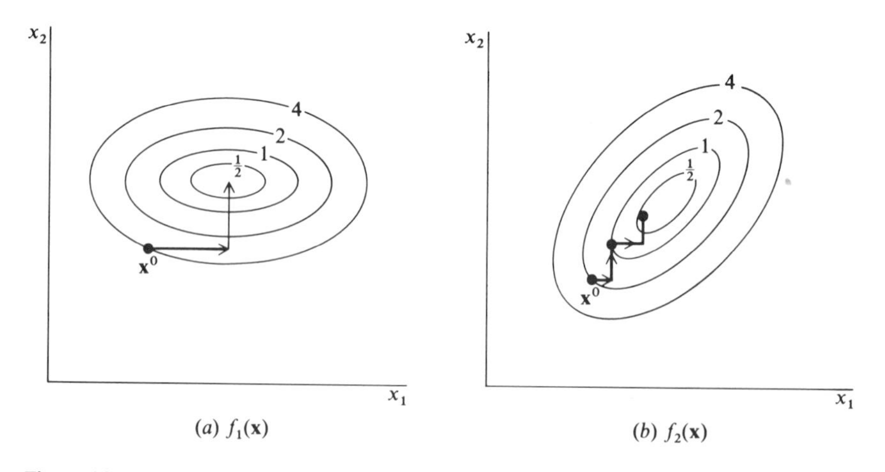

(*a*) $f_1(\mathbf{x})$ (*b*) $f_2(\mathbf{x})$

Figure 6.2 Execution of a univariate search on two different quadratic functions.

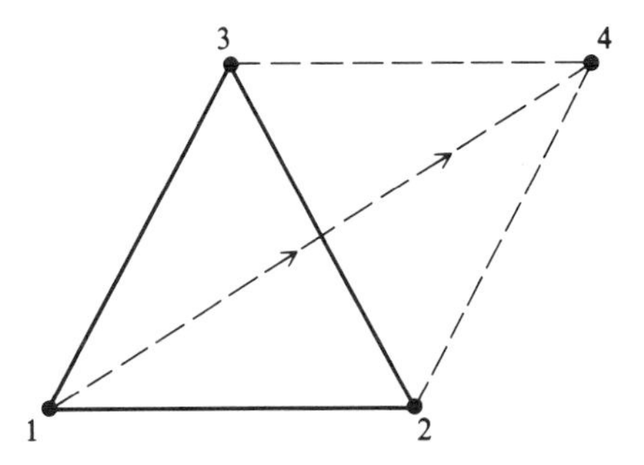

Figure 6.3 Reflection to a new point in the Simplex method.

this reflected direction (examine Fig. 6.3), preserving the geometric shape. The objective function is then evaluated at the new point, and a new search direction is calculated. The method proceeds rejecting one vertex at a time until the simplex straddles the optimum. Various rules are used to prevent excessive repeating of the same cycle or simplexes.

As the optimum is approached, the last equilateral triangle will straddle the optimum point or be within a distance of the order of its own size from the optimum (examine Fig. 6.4). Therefore, the procedure cannot get closer to the

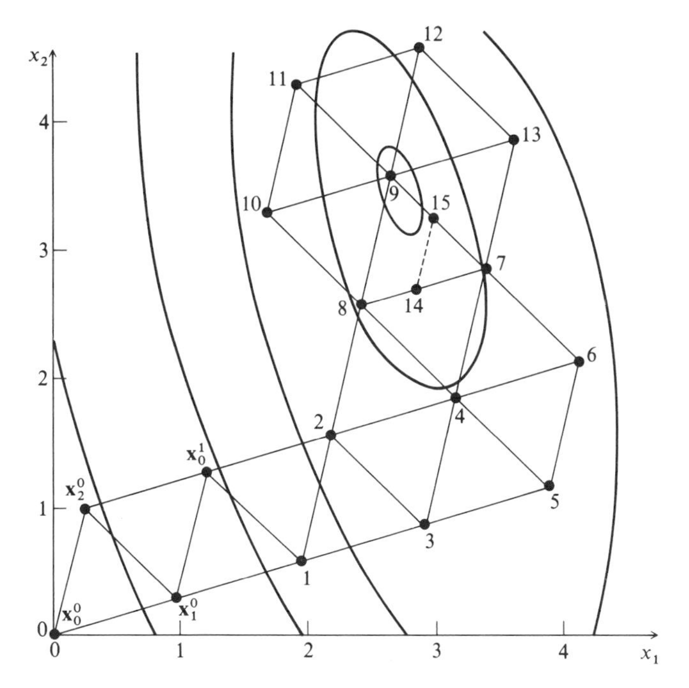

Figure 6.4 Progression to the vicinity of the optimum and oscillation around the optimum using the Simplex method of search. The original vertices are $\mathbf{x}_0^0$, $\mathbf{x}_1^0$, and $\mathbf{x}_2^0$. The next point (vertex) is $\mathbf{x}_0^1$. Succeeding new vertices are numbered starting with 1 and continuing to 13 at which point a cycle starts to repeat. The size of the Simplex is reduced to 7, 14, and 15, and then the procedure is continued (not shown).

optimum and would repeat itself so that the simplex size must be reduced, such as halving the length of all the sides of the simplex containing the vertex where the oscillation started. A new simplex composed of the midpoints of the ending simplex is constructed. When the simplex size is smaller than a prescribed tolerance, then the routine is stopped. Thus, the optimum position is determined to within a tolerance influenced by the size of the simplex.

Nelder and Mead (1965) described a more efficient (but more complex) version of the simplex method that permitted the geometric figures to expand and contract continuously during the search. Their method minimized a function of n variables using $(n + 1)$ vertices of a flexible polyhedron. Details of the method together with a computer code to execute the algorithm can be found in Himmelblau (1972).

6.1.5 Conjugate Search Directions

Experience has shown that directions called conjugate directions are much more effective as search directions than arbitrarily chosen search directions, such as in univariate search, or even orthogonal search directions. Two directions $\mathbf{s}^i$ and $\mathbf{s}^j$ are said to be *conjugate* (or *conjugated*) with respect to each other if

$$(\mathbf{s}^i)^T\mathbf{Q}(\mathbf{s}^j) = 0 \tag{6.1}$$

In general, a set of n linearly independent directions of search $\mathbf{s}^0, \mathbf{s}^1, \ldots, \mathbf{s}^{n-1}$ are said to be conjugate with respect to a positive definite square matrix $\mathbf{Q}$ if

$$(\mathbf{s}^i)^T\,\mathbf{Q}\mathbf{s}^j = 0 \qquad 0 \le i \ne j \le n - 1 \tag{6.2}$$

In optimization the matrix $\mathbf{Q}$ is the Hessian matrix of the objective function, $\mathbf{H}$. For a *quadratic function* $f(\mathbf{x})$ of n variables, in which $\mathbf{H}$ is a constant matrix, you are guaranteed to reach the minimum of $f(\mathbf{x})$ in n *stages* if you minimize *exactly* on each stage (Fletcher and Reeves, 1964). A conjugate direction in general is *not* a unique direction. However, in two dimensions, if you choose an initial direction $\mathbf{s}^1$ and $\mathbf{Q}$, $\mathbf{s}^2$ is fully specified as illustrated in Example 6.1 below.

Orthogonality is a special case of conjugacy because when $\mathbf{Q} = \mathbf{I}$, $(\mathbf{s}^i)^T\mathbf{s}^j = 0$ in Eq. (6.2). If the coordinates of $\mathbf{x}$ are translated and rotated by suitable transformations so as to align the new principal axes of $\mathbf{H}(\mathbf{x})$ with the eigenvectors of $\mathbf{H}(\mathbf{x})$ and to place the center of the coordinate system at the stationary point of $f(\mathbf{x})$ (refer to Figs. 4.12 to 4.16), then conjugacy can be interpreted as orthogonality in the space of the transformed coordinates.

Although authors and practitioners refer to a class of unconstrained optimization methods as "methods that use conjugate directions," for a general nonlinear function, the conjugate directions exist only for a quadratic approximation of the function at a single stage k. Once the objective function is modeled by a new approximation at stage $(k + 1)$, the directions on stage k are unlikely to be conjugate to any of the directions selected on stage $(k + 1)$.

In Example 6.1 below we will need to minimize a quadratic approximate in a given search direction. What this means is to compute the value for λ for the relation $\mathbf{x}^{k+1} = \mathbf{x}^k + \lambda \mathbf{s}^k$ that minimizes

$$f(\mathbf{x}) = f(\mathbf{x}^k + \lambda \mathbf{s}^k) = f(\mathbf{x}^k) + \nabla^T f(\mathbf{x}^k)\lambda \mathbf{s}^k + \tfrac{1}{2}(\lambda \mathbf{s}^k)^T \mathbf{H}(\mathbf{x}^k)(\lambda \mathbf{s}^k) \qquad (6.3)$$

where $\Delta \mathbf{x}^k = \lambda \mathbf{s}^k$. To get the minimum of $f(\mathbf{x}^k + \lambda \mathbf{s}^k)$, we differentiate (6.3) with respect to λ and equate the derivative to zero

$$\frac{df(\mathbf{x}^k + \lambda \mathbf{s}^k)}{d\lambda} = 0 = \nabla^T f(\mathbf{x}^k)\mathbf{s}^k + (\mathbf{s}^k)^T \mathbf{H}(\mathbf{x}^k)\lambda \mathbf{s}^k \qquad (6.4)$$

with the result

$$\lambda^{\text{opt}} = -\frac{\nabla^T f(\mathbf{x}^k)\mathbf{s}^k}{(\mathbf{s}^k)^T \mathbf{H}(\mathbf{x}^k)\mathbf{s}^k} \qquad (6.5)$$

EXAMPLE 6.1 CALCULATION OF CONJUGATE DIRECTIONS

Suppose we want to minimize $f(\mathbf{x}) = 2x_1^2 + x_2^2 - 3$ starting at $(\mathbf{x}^0)^T = [1 \quad 1]$ with the initial direction being $\mathbf{s}^0 = [-4 \quad -2]^T$. Find a conjugate direction to the initial direction $\mathbf{s}^0$.

Solution

$$\mathbf{s}^0 = -\begin{bmatrix} 4 \\ 2 \end{bmatrix} \qquad \mathbf{H}(\mathbf{x}) = \begin{bmatrix} 4 & 0 \\ 0 & 2 \end{bmatrix}$$

We need to solve Eq. (6.2) for $\mathbf{s}^1 = [s_1^1 \quad s_2^1]^T$ with $\mathbf{Q} = \mathbf{H}$ and $\mathbf{s}^0 = [-4 \quad -2]^T$.

$$(-1)[4 \quad 2]\begin{bmatrix} 4 & 0 \\ 0 & 2 \end{bmatrix}\begin{bmatrix} s_1^1 \\ s_2^1 \end{bmatrix} = 0$$

Because s_1^1 is not unique, we can pick $s_1^1 = 1$ and determine s_2^1

$$[-16 \quad -4]\begin{bmatrix} 1 \\ s_2^1 \end{bmatrix} = 0$$

Thus $\mathbf{s}^1 = [1 \quad -4]^T$ is a direction conjugate to $\mathbf{s}^0 = [-4 \quad -2]^T$.

Let us verify that we can reach the minimum of $f(\mathbf{x})$ in two stages using first $\mathbf{s}^0$ and then $\mathbf{s}^1$. Could we use the search directions in reverse order? From $\mathbf{x}^0 = [1 \quad 1]^T$ we could carry out a numerical search in the direction $\mathbf{s}^0 = [-4 \quad -2]^T$ to reach the point $\mathbf{x}^1$. But, because you must minimize exactly, Eq. (6.5) can be used to calculate analytically the overall step length because $f(\mathbf{x})$ is quadratic:

$$\lambda^0 = \frac{[4 \quad 2]\begin{bmatrix} -4 \\ -2 \end{bmatrix}}{[-4 \quad -2]\begin{bmatrix} 4 & 0 \\ 0 & 4 \end{bmatrix}\begin{bmatrix} -4 \\ -2 \end{bmatrix}} = \frac{20}{72}$$

Then

$$\mathbf{x}^1 = \mathbf{x}^0 + \lambda^0\mathbf{s}^0 = \begin{bmatrix}1\\1\end{bmatrix} + \frac{20}{72}\begin{bmatrix}-4\\-2\end{bmatrix} = \begin{bmatrix}-0.1111\\0.4444\end{bmatrix}$$

For the next stage, the search direction is $\mathbf{s}^1 = [1 \quad -4]^T$.

$$\lambda^1 = -\frac{[-0.4444 \quad 0.8888]\begin{bmatrix}1\\-4\end{bmatrix}}{[1 \quad -4]\begin{bmatrix}4 & 0\\0 & 2\end{bmatrix}\begin{bmatrix}1\\-4\end{bmatrix}} = 0.111$$

and

$$\mathbf{x}^2 = \mathbf{x}^1 + \lambda^1\mathbf{s}^1 = \begin{bmatrix}-0.1111\\0.4444\end{bmatrix} + 0.1111\begin{bmatrix}1\\-4\end{bmatrix} = \begin{bmatrix}0\\0\end{bmatrix}$$

as expected.

How can one calculate conjugate directions without using derivatives? Here is the basic concept. Refer to Fig. 6.5. Start at $\mathbf{x}^0$. Locate the point $\mathbf{x}^a$ in the direction $\mathbf{s}$ that is the minimum of $f(\mathbf{x})$. Then, start from another point $\mathbf{x}^1 \neq \mathbf{x}^0$, and locate the point $\mathbf{x}^b$, in the same direction $\mathbf{s}$, that is the minimum of $f(\mathbf{x})$. The vector (direction) $(\mathbf{x}^b - \mathbf{x}^a)$ is conjugate to $\mathbf{s}$. Examine Fig. 6.5.

Proof of the simple procedure is as follows. We will first show that the gradient of $f(\mathbf{x})$ at $\mathbf{x}^a$ is orthogonal to the search direction $\mathbf{s}$. The point $\mathbf{x}^a$ or $\mathbf{x}^b$ is to be obtained by minimizing $f(\mathbf{x}^0 + \lambda\mathbf{s})$ with respect to λ assuming $f(\mathbf{x})$ is a quadratic function

$$f(\mathbf{x}) = f(\mathbf{x}^0) + \nabla^T f(\mathbf{x}^0)\Delta\mathbf{x}^0 + \tfrac{1}{2}(\Delta\mathbf{x}^0)^T\mathbf{H}(\Delta\mathbf{x}^0)$$

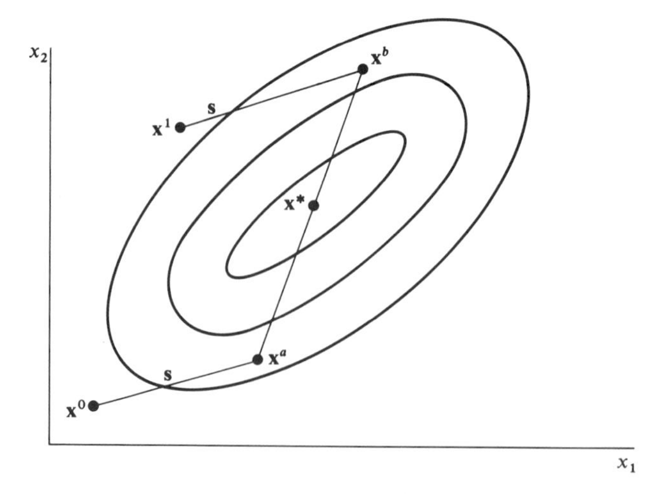

Figure 6.5 Determination of conjugate directions without calculating derivatives.

Then to get the exact minimum of $f(\mathbf{x})$ in direction $\mathbf{s}$ analytically we use Eq. (6.5)

$$\lambda = -\frac{\nabla^T f(\mathbf{x}^0)\mathbf{s}}{\mathbf{s}^T\mathbf{H}\mathbf{s}}$$

Suppose we write Eq. (6.3) in slightly different notation (note $\mathbf{x}^T\mathbf{b} = \mathbf{b}^T\mathbf{x}$)

$$f(\mathbf{x}) = a + \mathbf{x}^T\mathbf{b} + \tfrac{1}{2}\mathbf{x}^T\mathbf{H}\mathbf{x} \tag{6.3a}$$

Then

$$\nabla f(\mathbf{x}) = \mathbf{b} + \mathbf{H}\mathbf{x}$$

and

$$\nabla f(\mathbf{x}^0) = \mathbf{b} + \mathbf{H}\mathbf{x}^0$$

which when introduced into Eq. (6.4) gives (with $\lambda\mathbf{s} = \mathbf{x}^a - \mathbf{x}^0$)

$$(\mathbf{b} + \mathbf{H}\mathbf{x}^0)^T\mathbf{s} + \mathbf{s}^T\mathbf{H}\lambda\mathbf{s} = 0$$

$$\mathbf{s}^T(\mathbf{b} + \mathbf{H}\mathbf{x}^0) + \mathbf{s}^T\mathbf{H}(\mathbf{x}^a - \mathbf{x}^0) = 0$$

$$\mathbf{s}^T(\mathbf{b} + \mathbf{H}\mathbf{x}^a) = 0$$

so

$$\mathbf{s}^T\nabla f(\mathbf{x}^a) = 0 \tag{6.6}$$

Equation (6.6) demonstrates that if $f(\mathbf{x})$ is quadratic, the gradient of $f(\mathbf{x})$ at its minimum in a search direction $\mathbf{s}$ is orthogonal to $\mathbf{s}$.

Now, referring to Fig. 6.5, for the first search from $\mathbf{x}^0$

$$\mathbf{s}^T\nabla f(\mathbf{x}^a) = 0 = \mathbf{s}^T(\mathbf{H}x^a + \mathbf{b})$$

and for the second search from $\mathbf{x}^1$:

$$\mathbf{s}^T\nabla f(\mathbf{x}^b) = 0 = \mathbf{s}^T(\mathbf{H}\mathbf{x}^b + \mathbf{b})$$

By subtraction

$$\mathbf{s}^T\mathbf{H}(\mathbf{x}^b - \mathbf{x}^a) = 0 \tag{6.7}$$

hence $\mathbf{s}$ and $(\mathbf{x}^b - \mathbf{x}^a)$ are conjugate. The minimum of $f(\mathbf{x})$ will be found along the second search direction $(\mathbf{x}^b - \mathbf{x}^a)$ as illustrated in Fig. 6.5.

6.1.6 Powell's Method

Powell's method (1965) locates the minimum of a function f by sequential unidimensional searches from an initial point $(\mathbf{x}^0)$ along a set of conjugate directions generated by the algorithm. New search directions are introduced as the search progresses.

To clarify the beginning of Powell's method, let us use a function of just two variables. The search begins at an arbitrarily chosen point, (x_1^0, x_2^0). Two linearly independent directions of search ($\mathbf{s}_1^0, \mathbf{s}_2^0$) are chosen; for example, the coordinate axes ([1 0], [0 1]) are a convenient initial choice. A univariate search is carried out to find the optimal point in the x_1 coordinate direction

$$\mathbf{x}_1^0 = \mathbf{x}_0^0 + \lambda_1^0 \mathbf{s}_1^0$$

where λ_1^0 is the optimum step length. Then a univariate search is carried out from the point $\mathbf{x}_1^0$ in the x_2 coordinate direction to find the value of $\mathbf{x}$ that minimizes $f(\mathbf{x})$

$$\mathbf{x}_2^0 = \mathbf{x}_1^0 + \lambda_2^0 \mathbf{s}_2^0$$

Lastly, one more step from $\mathbf{x}_2^0$ of size $(\mathbf{x}_2^0 - \mathbf{x}_0^0)$ is taken corresponding to the overall step on the 0th stage. Examine Fig. 6.6.

In general, the kth stage of Powell's method employs n linearly independent search directions and can be considered to be initiated at $\mathbf{x}_0^k = \mathbf{x}_{n+1}^{k-1}$ as follows:

Step 1. From $\mathbf{x}_0^k$ determine λ_1^k by a unidimensional search in the $\mathbf{s}_1^k$ direction so that $f(\mathbf{x}_0^k + \lambda_1 \mathbf{s}_1^k)$ is a minimum. Let $\mathbf{x}_1^k = \mathbf{x}_0^k + \lambda_1^k \mathbf{s}_1^k$. From $\mathbf{x}_1^k$ determine λ_2^k so that $f(\mathbf{x}_1^k + \lambda_2 \mathbf{s}_2^k)$ is a minimum. Let $\mathbf{x}_2^k = \mathbf{x}_1^k + \lambda_2^k \mathbf{s}_2^k$. The search is continued sequentially in each direction, starting always from the last immediate point in the sequence until all the λ_i^k, $i = 1, \ldots, n$, are determined. The search for λ_0^k to minimize $f(\mathbf{x})$ in the direction $\mathbf{s}^{k-1}$ is taken into account in step 4 below.

Step 2. Powell pointed out that the search described in step 1 can lead to linearly dependent search directions, when, for example, one of the components of

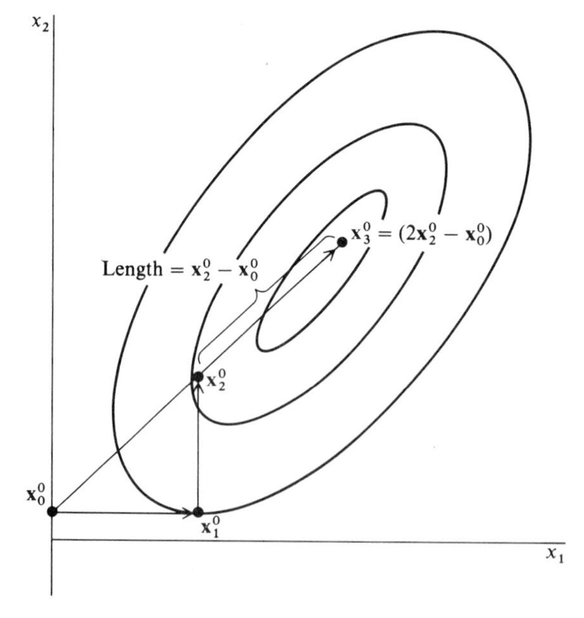

Figure 6.6 Initiation of Powell's method of optimization.

$\mathbf{s}^k$ becomes essentially zero because no progress was made in that particular direction. Two search directions thus might become essentially colinear. Hence it might be unwise to replace the old direction with a new one if by doing so the new set of directions became linearly dependent. Also, he demonstrated (for a quadratic function) that in scaling the search directions by

$$(\mathbf{s}_i^k)^T \mathbf{H} \mathbf{s}_i^k = 1 \qquad i = 1, \ldots, n$$

the determinant of the matrix whose columns comprise the search directions takes on its maximum value if and only if the $\mathbf{s}_i^k$ are mutually conjugate with respect to $\mathbf{H}$. He concluded that the overall direction of progress for the kth stage, $\mathbf{s}^k$, should replace an existing search direction only if the replacement vector increased the determinant of the matrix of the normalized search directions, for then the set of directions would be more effective. Hence, after minimizing $f(\mathbf{x})$ in each of the n directions as described in step 1, one additional step of size $(\mathbf{x}_n^k - \mathbf{x}_0^k)$ is taken corresponding to the total progress on the kth stage to yield the point $(2\mathbf{x}_n^k - \mathbf{x}_0^k)$. A test is then made—see step 3 below—to ascertain whether or not adding the new direction and dropping an old one decreases the determinant of the search directions.

Step 3. Let the largest reduction in $f(\mathbf{x})$ in any search direction on the kth stage be denoted by

$$\Delta^k = \max_{i=1,\ldots,n} \{f(\mathbf{x}_{i-1}^k) - f(\mathbf{x}_i^k)\}$$

The search direction corresponding to this maximum change in $f(\mathbf{x})$ will be designated $\mathbf{s}_m^k$. To make the notation more compact, let $f_1 = f(\mathbf{x}_0^k)$, $f_2 = f(\mathbf{x}_n^k)$, and $f_3 = f(2\mathbf{x}_n^k - \mathbf{x}_0^k)$, where $\mathbf{x}_0^k = \mathbf{x}_n^{k-1}$ and $\mathbf{x}_n^k = \mathbf{x}_{n-1}^k + \lambda_n^k \mathbf{s}_n^k = \mathbf{x}_0^k + \sum_{i=1}^n \lambda_i^k \mathbf{s}_i^k$.

If either $f_3 \geq f_1$ and/or $(f_1 - 2f_2 + f_3)(f_1 - f_2 - \Delta^k)^2 \geq 0.5\Delta^k\ (f_1 - f_3)^2$, use exactly the same directions $\mathbf{s}_1^k, \ldots, \mathbf{s}_n^k$ in the $(k+1)$st stage as in the kth stage, that is, $\mathbf{s}_i^{k+1} = \mathbf{s}_i^k$ for $i = 1, \ldots, n$, and start at $\mathbf{x}_0^{k+1} = \mathbf{x}_n^k$ [or from $\mathbf{x}_0^{k+1} = 2\mathbf{x}_n^k - \mathbf{x}_0^k = \mathbf{x}_{n+1}^k$, whichever $\mathbf{x}$ yields the lowest value of $f(\mathbf{x})$].

Step 4. If the test in step 3 is not satisfied, the direction $\mathbf{s}^k$ from $\mathbf{x}_0^k$ to $\mathbf{x}_n^k$ is searched for the minimum of $f(\mathbf{x})$, which will be used as the starting point for the next stage, $(k+1)$. The set of directions to be used in the $(k+1)$st stage are the same as the kth stage, with the exception of the direction $\mathbf{s}_m^k$, which is replaced by $\mathbf{s}^k$. However, $\mathbf{s}^k$ is placed in the last column of the matrix of directions instead of in the location of $\mathbf{s}_m^k$. Consequently, the directions to be used in the $(k+1)$st stage are

$$[\mathbf{s}_1^{k+1} \quad \mathbf{s}_2^{k+1} \quad \cdots \quad \mathbf{s}_n^{k+1}] = [\mathbf{s}_1^k \quad \mathbf{s}_2^k \quad \cdots \quad \mathbf{s}_{m-1}^k \quad \mathbf{s}_{m+1}^k \quad \cdots \quad \mathbf{s}_n^{(k)} \quad \mathbf{s}^{(k)}]$$

Step 5. A satisfactory convergence criterion for Powell's method is to terminate the search at the end of any stage in which the change in each independent variable is less than the required accuracy, ε_i, for $i = 1, \ldots, n$, or for $\|\mathbf{x}_n^k - \mathbf{x}_0^k\| \leq \varepsilon$.

EXAMPLE 6.2 APPLICATION OF POWELL'S METHOD

The application of Powell's algorithm to Rosenbrock's function

$$\text{Minimize } f(\mathbf{x}) = 100(x_2 - x_1^2)^2 + (1 - x_1)^2$$

illustrates what happens in the minimization of a nonquadratic function.

Step 0. We start at $\mathbf{x}^0 = [-1.2 \quad 1.0]^T$, where $f(\mathbf{x}^0) = 24.2$. The initial search directions are

$$\mathbf{s}_1^0 = \begin{bmatrix} 1 \\ 0 \end{bmatrix} \qquad \mathbf{s}_2^0 = \begin{bmatrix} 0 \\ 1 \end{bmatrix}$$

Step 1. An initial minimization of $f(\mathbf{x})$ is made using the search direction $\mathbf{s}_1^0$, that is, in the x_1 coordinate direction (the details of which are omitted) to reach the point $\mathbf{x}_1^0 = [-0.995 \quad 1.000]^T$, where $f(\mathbf{x}_1^0) = 3.990$. Then a search is made using the vector $\mathbf{s}_2^0$, that is, in the x_2 coordinate direction, yielding $\mathbf{x}_2^0 = [-0.995 \quad 0.990]^T$, where $f(\mathbf{x}_2^0) = 3.980$.

Step 2. Following a step of

$$\mathbf{x}_3^0 = 2\begin{bmatrix} -0.995 \\ 0.990 \end{bmatrix} - \begin{bmatrix} -1.200 \\ 1.000 \end{bmatrix} = \begin{bmatrix} -0.790 \\ 0.980 \end{bmatrix}$$

$$f(\mathbf{x}_3^0) = 15.872$$

New directions of search are computed as follows:

Steps 3 and 4

$$\mathbf{s}_1^1 = \mathbf{s}_2^0 = \begin{bmatrix} 0 \\ 1 \end{bmatrix}$$

$$\mathbf{s}_2^1 = \mathbf{x}_2^0 - \mathbf{x}_0^0 = \begin{bmatrix} -0.995 \\ 0.990 \end{bmatrix} - \begin{bmatrix} -1.200 \\ 1.000 \end{bmatrix} = \begin{bmatrix} 0.205 \\ -0.010 \end{bmatrix}$$

To normalize $\mathbf{s}_2^1$

$$\|\mathbf{s}_2^1\| = \sqrt{(\mathbf{s}_2^1)^T \mathbf{s}_2^1} = \sqrt{4.21 \times 10^{-2}} = 0.206$$

$$\mathbf{s}_2^1 = \begin{bmatrix} 0.999 \\ -0.0488 \end{bmatrix}$$

We use $\mathbf{s}_2^1$ because the criteria listed in step 3 are not satisfied ($\Delta^0 = 24.2 - 3.99 = 20.21$)

$$f(\mathbf{x}_3^0) = 15.872 < f(\mathbf{x}^0) = 24.2 \qquad (a)$$

and

$$[24.2 - 2(3.980) + 15.872][24.2 - 3.980 - 20.21]^2 < 0.5(20.21)(24.2 - 15.872)^2 \qquad (b)$$

The results of one additional stage of the search are given in Table E6.2, starting from $\mathbf{x}_0^1 = [-0.995 \quad 0.990]^T$:

Figure E6.2 illustrates the trajectory of the search. In total, the objective function was evaluated 1562 times to reach the minimum of $f(\mathbf{x})$ at $\mathbf{x}^* = [1.000 \quad 1.000]^T$ and $f(\mathbf{x}^*) = 1.338 \times 10^{-16} \approx 0$.

Table E6.2

Stage k	$(\mathbf{s}_1^k)^T$	$(\mathbf{s}_2^k)^T$	$\mathbf{x}^k$	$f(\mathbf{x}^k)$
1	[0 1]	[0.205 −0.010]	$x_0^1 = [-0.995 \quad 0.990]^T$	3.969
			$x_1^1 = [-0.990 \quad 0.990]^T$	3.959
			$x_2^1 = [-0.984 \quad 0.979]^T$	3.948
2	[0 1]	[0.453 −0.892]	$x_0^2 = [-0.761 \quad 0.540]^T$	3.257
			$x_1^2 = [-0.761 \quad 0.579]^T$	3.101
			$x_2^2 = [-0.702 \quad 0.462]^T$	2.986
3	[0.453 −0.892]	[0.608 −0.794]	$x_0^3 = [-0.503 \quad 0.203]^T$	2.510
			$x_1^3 = [-0.538 \quad 0.273]^T$	2.396
			$x_2^3 = [-0.466 \quad 0.178]^T$	2.301

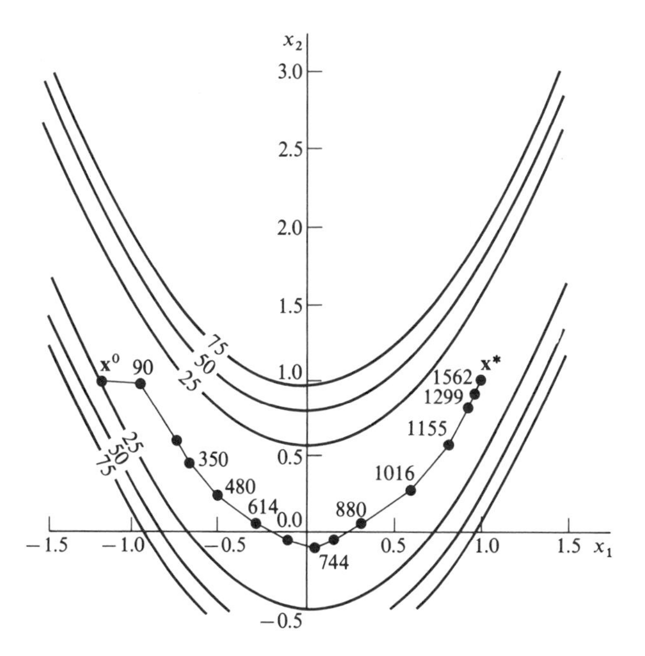

Figure E6.2 Search trajectory for the Powell algorithm (the numbers designate the function evaluation sequence number).

Brent (1973) proposed a modification to Powell's method to avoid linear dependence of the search directions and still retain quadratic convergence. He reset the search directions $\mathbf{s}_1, \ldots, \mathbf{s}_n$ periodically to be the columns of an orthogonal matrix selected so that if $f(\mathbf{x})$ were quadratic, the columns would remain conjugate. The columns of the orthogonal matrix were the normalized principal axes of $f(\mathbf{x})$. He also incorporated a random step prior to starting the unidimensional search, a heuristic that helps in poorly conditioned problems.

6.1.7 Summary

Direct methods were the earliest methods proposed for unconstrained optimization. They are currently used in the chemical process industries because they are simple to understand and execute. However, they are not as efficient and robust as many of the more modern indirect methods described in the sections that follow. But, for simple two-variable problems, direct methods are often quite satisfactory.

6.2 INDIRECT METHODS—FIRST ORDER

Indirect methods in contrast with the methods described in the previous section do make use of derivatives in determining the search direction for optimization. However, our classification of direct vs. indirect may not be as clear-cut as it seems because substitution of finite differences for derivatives will render the indirect methods "derivative-free." A good search direction should reduce (for minimization) the objective function so that if $\mathbf{x}^0$ is the original point and $\mathbf{x}^1$ is the new point

$$f(\mathbf{x}^1) < f(\mathbf{x}^0)$$

Such a direction $\mathbf{s}$ is called a descent direction and satisfies the following requirement at any point

$$\nabla^T f(\mathbf{x})\mathbf{s} < 0$$

To see why, examine the two vectors $\nabla f(\mathbf{x}^k)$ and $\mathbf{s}^k$ in Fig. 6.7. The angle between them is θ, hence

$$\nabla^T f(\mathbf{x})\mathbf{s}^k = |\nabla f(\mathbf{x}^k)||\mathbf{s}^k| \cos \theta$$

If $\theta = 90°$ as in Fig. 6.7, then steps along $\mathbf{s}^k$ will not reduce (improve) the value of $f(\mathbf{x})$. If $0 \leq \theta < 90°$, no improvement is possible and $f(\mathbf{x})$ increases. Only if $\theta > 90°$ will the search direction yield smaller values of $f(\mathbf{x})$, hence $\nabla^T f(\mathbf{x}^k)\mathbf{s}^k < 0$.

We first examine the classic method of using the gradient and then examine a conjugate gradient method.

6.2.1 Gradient Method

In this section we briefly set forth the strategy of the gradient (steepest-descent/ascent) method of unconstrained optimization. This method uses only the first derivatives of the objective function in the calculations.

You will recall that the gradient is the vector at a point $\mathbf{x}$ that gives the (local) direction of the greatest increase in $f(\mathbf{x})$ and is orthogonal to the contour of $f(\mathbf{x})$ at $\mathbf{x}$. For maximization, the search direction is simply the gradient (when used the algorithm is called "steepest ascent"); for minimization, the search direction is the negative of the gradient ("steepest descent")

$$\mathbf{s}^k = -\nabla f(\mathbf{x}^k) \tag{6.8}$$

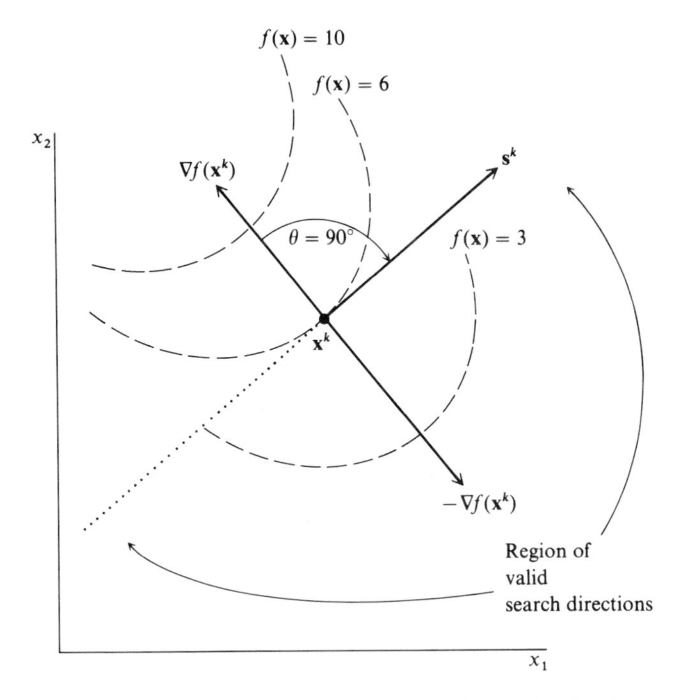

Figure 6.7 Identification of the region of possible search directions.

In steepest descent at the kth stage, the transition from point $\mathbf{x}^k$ to another point $\mathbf{x}^{k+1}$ can be viewed as given by the following expression:

$$\mathbf{x}^{k+1} = \mathbf{x}^k + \Delta\mathbf{x}^k = \mathbf{x}^k + \lambda^k\mathbf{s}^k = \mathbf{x}^k - \lambda^k \nabla f(\mathbf{x}^k) \tag{6.9}$$

where $\Delta\mathbf{x}^k$ = vector from $\mathbf{x}^k$ to $\mathbf{x}^{k+1}$

$\mathbf{s}^k$ = search direction, the direction of steepest descent

λ^k = scalar that determines the step length in direction $\mathbf{s}^k$

The negative of the gradient gives the direction for minimization but not the magnitude of the step to be taken, so that various steepest descent procedures are possible, depending upon the choice of λ^k. We assume that the value of $f(\mathbf{x})$ is continuously reduced. Because one step in the direction of steepest descent will not, in general, arrive at the minimum of $f(\mathbf{x})$, Eq. (6.9) must be applied repetitively until the minimum is reached. At the minimum, the value of the elements of the vector gradient will each be equal to zero.

Among the many methods of selecting λ^k that have appeared in the literature, we will mention two. One method employs a one-dimensional search along the negative gradient to select the step size. The second prespecifies the step size to be a constant value and uses this value for every iteration. Clearly, the main difficulty with the second approach is that often we do not know an appropriate step size a priori. How should it be changed as the direction of steepest-descent changes and how should it be reduced to permit convergence? Some illustrations will clarify the problem.

First, let us consider the well-scaled objective function, $f(\mathbf{x}) = x_1^2 + x_2^2$, whose contours are concentric circles as shown in Fig. 6.8. Suppose we calculate the gradient at the point $\mathbf{x}^T = [2 \quad 2]$

$$\nabla f(\mathbf{x}) = \begin{bmatrix} 2x_1 \\ 2x_2 \end{bmatrix} \qquad \nabla f(2, 2) = \begin{bmatrix} 4 \\ 4 \end{bmatrix} \qquad \mathbf{H}(\mathbf{x}) = \mathbf{H} = \begin{bmatrix} 2 & 0 \\ 0 & 2 \end{bmatrix}$$

The direction of steepest descent is $\mathbf{s} = -\begin{bmatrix} 4 \\ 4 \end{bmatrix}$. Observe that $\mathbf{s}$ is a vector pointing toward the optimum at (0, 0). In fact, the gradient at any point along a contour of this objective function will pass through the origin (the optimum).

On the other hand, for functions not so nicely scaled and which have nonzero off-diagonal terms in the Hessian matrix (corresponding to interaction terms such as x_1x_2), then the negative gradient direction is unlikely to pass through the optimum. Figure 6.9 illustrates the contours of a quadratic function of two variables which includes an interaction term. Observe that contours are tilted with respect to axes. Interaction terms plus poor scaling corresponding to narrow valleys, or ridges, cause the gradient method to exhibit poor performance and slow convergence.

If $f(\mathbf{x})$ is optimized exactly, as indicated by Eq. (6.6), successive steps of the steepest-descent method are orthogonal to each other illustrated in Fig. 6.9.

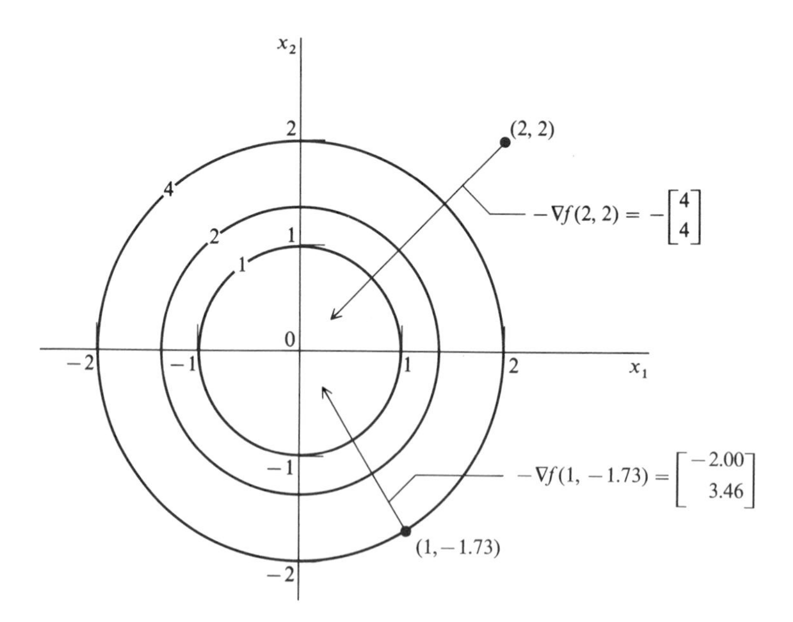

Figure 6.8 Gradient vector for $f(x) = x_1^2 + x_2^2$.

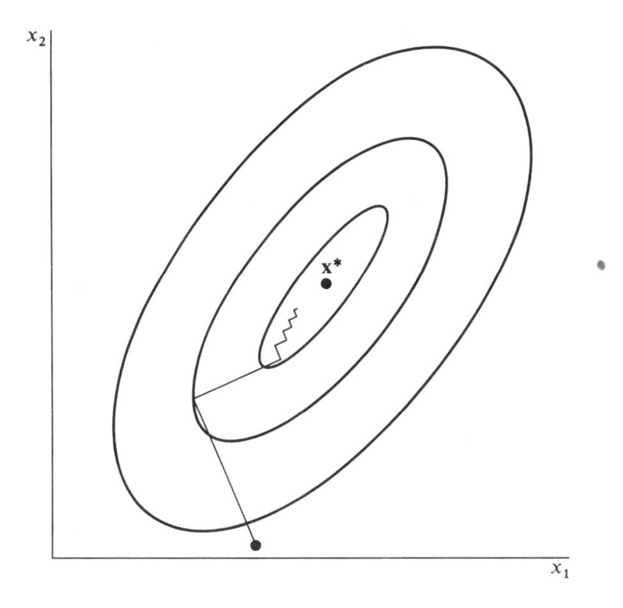

Figure 6.9 Steepest-descent method for a general quadratic function.

This seemingly peculiar result occurs for any $f(\mathbf{x})$ because the derivative of $f(\mathbf{x})$ along any line $\mathbf{s}(\lambda)$ by the chain rule of calculus is

$$\frac{df}{d\lambda} = \sum_i \frac{d}{d\lambda} x_i(\lambda) \frac{df}{dx_i} = \sum_i s_i \frac{\partial f}{\partial x_i} = \mathbf{s}^T \nabla f$$

At the end of the search we want $df/dx_i = 0$ so that $\mathbf{s}^T \nabla f(x^{k+1}) = 0$. In practice, it is not desirable to optimize too accurately. While the gradient method may yield satisfactory progress in reducing $f(\mathbf{x})$ in early iterations, it slows down significantly in later iterations because of the short step sizes taken.

The steepest-descent algorithm can be summarized in the following steps:

1. Choose an initial or starting point $\mathbf{x}^0$. Thereafter at the point $\mathbf{x}^k$:
2. Calculate (analytically or numerically) the partial derivatives

$$\frac{\partial f(\mathbf{x})}{\partial x_j} \qquad j = 1, \ldots, n$$

3. Calculate the search vector

$$\mathbf{s} = -\nabla f(\mathbf{x}^k)$$

4. Use the relation

$$\mathbf{x}^{k+1} = \mathbf{x}^k + \lambda^k \mathbf{s}^k$$

to obtain the value of $\mathbf{x}^{k+1}$. To get λ^k use Eq. (6.5) or minimize $f(\mathbf{x})$ numerically, as described in Chap. 5. Note that in the "fixed-step" gradient method, one sets $\lambda^k = \lambda$, a constant, for all iterations.

5. Compare $f(\mathbf{x}^{k+1})$ with $f(\mathbf{x}^k)$; if the change in $f(\mathbf{x})$ is smaller than some tolerance, stop. If not, return to step 2 and set $k = k + 1$. Termination can also be specified by setting a tolerance on the size of λ^k ($\lambda^k \leq \gamma$) or by specifying some tolerance on the norm of $\nabla f(\mathbf{x}^k)$.

A procedure of strictly steepest descent can terminate at any type of stationary point, i.e., at a point where the elements of the gradient of $f(\mathbf{x})$ are zero. Thus you must ascertain if the presumed minimum is indeed a local minimum (i.e., a solution) or a saddle point. If it is a saddle point, it is necessary to employ a nongradient method to move away from the point, after which the minimization may continue as before. The stationary point may be tested by examining the Hessian matrix of the objective function as described in Chap. 4. If the Hessian matrix is not positive definite, the stationary point is a saddle point. Perturbation from the stationary point followed by optimization should lead to $\mathbf{x}^*$ if $\mathbf{x}^*$ is a minimum (or a maximum).

The basic difficulty with steepest descent is that the method is too sensitive to the scaling of $f(\mathbf{x})$, so that convergence is very slow and what amounts to oscillation in the $\mathbf{x}$ space can easily occur. For these reasons steepest descent/ascent is not a very effective optimization technique, hence we turn to some other indirect methods.

6.2.2 Conjugate Gradient

This method was devised by Fletcher and Reeves (1964), and if $f(\mathbf{x})$ is quadratic and is minimized exactly in each search direction it has the desirable features of quadratic convergence [refer to Eq. (5.3) with $p = 2$] and of using conjugate directions (refer to Sec. 6.1.5). The method represents a major improvement over steepest descent with only a marginal increase in computational effort. The conjugate-gradient method essentially combines current information about the gradient vector with that of gradient vectors from previous iterations (a memory feature) to obtain the new search direction. What you do is to compute the search direction by a linear combination of the current gradient and the previous search direction. The main advantage of this method is that it requires only a small amount of information to be stored on each stage of calculation, and thus can be executed on small computers or even calculators. The computation steps are listed below.

Step 1. At $\mathbf{x}^0$ calculate $f(\mathbf{x}^0)$. Let

$$\mathbf{s}^0 = -\nabla f(\mathbf{x}^0)$$

Step 2. Save $\nabla f(\mathbf{x}^0)$ and compute

$$\mathbf{x}^1 = \mathbf{x}^0 + \lambda^0 \mathbf{s}^0$$

by minimizing $f(\mathbf{x})$ with respect to λ in the $\mathbf{s}^0$ direction (i.e., carry out a unidimensional search for λ^0).

Step 3. Calculate $f(\mathbf{x}^1)$, $\nabla f(\mathbf{x}^1)$. The new search direction is a linear combination of $\mathbf{s}^0$ and $\nabla f(\mathbf{x}^1)$:

$$\mathbf{s}^1 = -\nabla f(\mathbf{x}^1) + \mathbf{s}^0 \frac{\nabla^T f(\mathbf{x}^1) \nabla f(\mathbf{x}^1)}{\nabla^T f(\mathbf{x}^0) \nabla f(\mathbf{x}^0)}$$

For the kth iteration the relation is

$$\mathbf{s}^{k+1} = -\nabla f(\mathbf{x}^{k+1}) + \mathbf{s}^k \frac{\nabla^T f(\mathbf{x}^{k+1}) \nabla f(\mathbf{x}^{k+1})}{\nabla^T f(\mathbf{x}^k) \nabla f(\mathbf{x}^k)} \tag{6.10}$$

For a quadratic function it can be shown that these successive search directions are conjugate. After n iterations ($k = n$), the procedure cycles again with $\mathbf{x}^{n+1}$ becoming $\mathbf{x}^0$.

Step 4. Test for convergence to the minimum of $f(\mathbf{x})$. If convergence is not attained, return to step 3.

Step n. Terminate the algorithm when $\|(\mathbf{s}^k)\|$ is less than some prescribed tolerance.

Note that if the ratio of the inner products of the gradients from stage $k + 1$ relative to stage k is very small, the conjugate-gradient method will behave much like the steepest-descent method. Some of the same difficulties that occur with Powell's method arise also with conjugate gradients; the linear dependence of search directions can be resolved by periodically restarting the conjugate-gradient method with a steepest-descent search (step 1 above). The proof that Eq. (6.10) given above yields conjugate directions and quadratic convergence has been given by Fletcher and Reeves (1964).

EXAMPLE 6.3 APPLICATION OF THE FLETCHER-REEVES CONJUGATE-GRADIENT ALGORITHM

We will solve the same problem as solved in Example 6.2 (Rosenbrock's function)

$$\text{Minimize } f(\mathbf{x}) = 100(x_2 - x_1^2)^2 + (1 - x_1)^2$$

starting at $\mathbf{x}^{(0)} = [-1.2 \quad 1.0]^T$. The first few stages of the Fletcher–Reeves procedure are listed in Table E6.3 using the same line search as in Example 6.2. The

Table E6.3 Results for Example 6.3 using the Fletcher-Reeves method

Iteration	No. of function calls	$f(\mathbf{x})$	x_1	x_2	$\frac{\partial f(\mathbf{x})}{\partial x_1}$	$\frac{\partial f(\mathbf{x})}{\partial x_2}$
0	1	24.2	−1.2	1.0	−215.6	−88.00
1	4	4.377945	−1.050203	1.061141	−21.65	−8.357
2	7	4.127422	−1.029913	1.069013	−0.6437	1.658
3	9	3.947832	−0.963870	0.898872	−15.56	−6.034
5	14	3.165142	−0.777190	0.612232	−1.002	−1.6415
10	28	1.247687	−0.079213	−0.025322	−3.071	−5.761
15	41	0.556612	0.254058	0.063189	−1.354	−0.271
20	57	0.147607	0.647165	0.403619	3.230	−3.040
25	69	0.024667	0.843083	0.710119	−0.0881	−0.1339
30	80	0.0000628	0.995000	0.989410	0.2348	−0.1230
35	90	1.617×10^{-15}	1.000000	1.000000	-1.60×10^{-8}	-3.12×10^{-8}

trajectory is similar to that illustrated in Fig. E6.2 but uses fewer function evaluations.

For additional details concerning the application of conjugate-gradient methods, especially to large-scale and sparse problems, refer to Buckley (1978), Shanno (1978), Fletcher (1980), Hestenes (1980), Gill, Murray, and Wright (1981), Dembo, et al. (1982), and Griewank and Toint (1982).

6.3 INDIRECT METHODS—SECOND ORDER

From one viewpoint the search direction of steepest descent can be interpreted as being orthogonal to a linear approximation (tangent to) the objective function at point $\mathbf{x}^k$; examine Fig. 6.10*a*. Now suppose we make a quadratic approximation of $f(\mathbf{x})$ at $\mathbf{x}^k$

$$f(\mathbf{x}) \approx f(\mathbf{x}^k) + \nabla^T f(\mathbf{x}^k)\Delta\mathbf{x}^k + \tfrac{1}{2}(\Delta\mathbf{x}^k)^T\mathbf{H}(\mathbf{x}^k)\Delta\mathbf{x}^k \tag{6.11}$$

where $\mathbf{H}(\mathbf{x}^k)$ is the Hessian matrix of $f(\mathbf{x})$ defined in Chap. 4 (the matrix of second partial derivatives with respect to $\mathbf{x}$ evaluated at $\mathbf{x}^k$). Then it will be possible to take into account the curvature of $f(\mathbf{x})$ at $\mathbf{x}^k$ in determining a search direction as described next for Newton's method and its analogs.

6.3.1 Newton's Method

Newton's method makes use of the second-order (quadratic) approximation of $f(\mathbf{x})$ at $\mathbf{x}^k$, and thus employs second-order information about $f(\mathbf{x})$, that is, information obtained from the second partial derivatives of $f(\mathbf{x})$ with respect to the

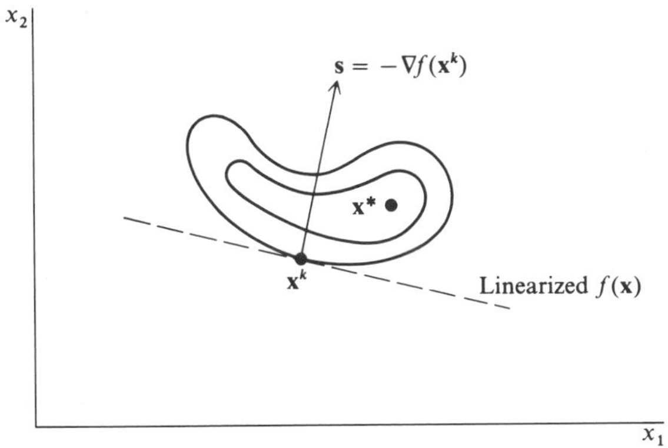

(*a*) Steepest descent: first-order approximation (linearization) of $f(\mathbf{x})$ at $\mathbf{x}^k$

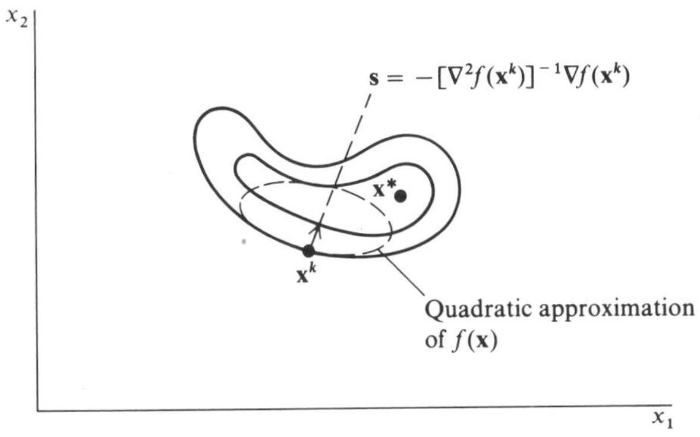

(*b*) Newton's method: second-order (quadratic) approximation of $f(\mathbf{x})$ at $\mathbf{x}^k$

Figure 6.10 Comparison of steepest-descent with Newton's method from the viewpoint of objective function approximation.

independent variables. Thus, it is possible to take into account the curvature of $f(\mathbf{x})$ at $\mathbf{x}^k$ and identify better search directions than can be obtained via the gradient method. Examine Fig. 6.10*b*.

The minimum of $f(\mathbf{x})$ in the direction of $\mathbf{x}^k$ is obtained by differentiating the quadratic approximation of $f(\mathbf{x})$ with respect to each of the components of $\mathbf{x}$ and equating the resulting expressions to zero to give

$$\nabla f(\mathbf{x}) = \nabla f(\mathbf{x}^k) + \mathbf{H}(\mathbf{x}^k)\Delta\mathbf{x}^k = \mathbf{0} \tag{6.12}$$

or

$$\mathbf{x}^{k+1} - \mathbf{x}^k = \Delta\mathbf{x}^k = -[\mathbf{H}(\mathbf{x}^k)]^{-1}\nabla f(\mathbf{x}^k) \tag{6.13}$$

where $[\mathbf{H}(\mathbf{x}^k)]^{-1}$ is the inverse of the Hessian matrix $\mathbf{H}(\mathbf{x}^k)$. Equation (6.13) reduces to Eq. (5.5) for a one-dimensional search.

Note that both the direction *and* step length are specified as a result of Eq. (6.12). If $f(\mathbf{x})$ is actually quadratic, only one step is required to reach the minimum of $f(\mathbf{x})$. However, for a general nonlinear objective function, the minimum of $f(\mathbf{x})$ will not be reached in one step, so that Eq. (6.13) can be modified to conform to Eq. (6.9) by introducing the parameter for the step length into (6.13).

$$\mathbf{x}^{k+1} - \mathbf{x}^k = -\lambda^k[\mathbf{H}(\mathbf{x}^k)]^{-1}\nabla f(\mathbf{x}^k) \tag{6.14}$$

Observe that the search direction $\mathbf{s}$ is now given (for minimization) by

$$\mathbf{s}^k = -[\mathbf{H}(\mathbf{x}^k)]^{-1}\nabla f(\mathbf{x}^k) \tag{6.15}$$

and that the step length is λ^k. The step length λ^k can be evaluated numerically as described in Chap. 5 or analytically via Eq. (6.5). Equation (6.14) is applied iteratively until some termination criteria are satisfied. For Newton's method, $\lambda = 1$ on each step.

Also note that to evaluate $\Delta\mathbf{x}$ in Eq. (6.13), a matrix inversion is not necessarily required. You can take its precursor, Eq. (6.12), and solve the following set of linear equations for $\Delta\mathbf{x}^k$

$$\mathbf{H}(\mathbf{x}^k)\Delta\mathbf{x}^k = -\nabla f(\mathbf{x}^k) \tag{6.16}$$

a procedure that often leads to less round-off error than calculating $\mathbf{s}$ via an inversion of a matrix.

EXAMPLE 6.4 APPLICATION OF NEWTON'S METHOD TO A CONVEX QUADRATIC FUNCTION

We will minimize the function

$$f(\mathbf{x}) = 4x_1^2 + x_2^2 - 2x_1x_2$$

starting at $\mathbf{x}^0 = [1 \quad 1]^T$

$$\nabla f(\mathbf{x}) = \begin{bmatrix} 8x_1 - 2x_2 \\ 2x_2 - 2x_1 \end{bmatrix}$$

$$\mathbf{H}(\mathbf{x}) = \begin{bmatrix} 8 & -2 \\ -2 & 2 \end{bmatrix} \qquad \mathbf{H}^{-1}(\mathbf{x}) = \begin{bmatrix} \frac{1}{6} & \frac{1}{6} \\ \frac{1}{6} & \frac{2}{3} \end{bmatrix}$$

with $\lambda = 1$,

$$\Delta\mathbf{x}^0 = -\mathbf{H}^{-1}\nabla f(\mathbf{x}^0) = -\begin{bmatrix} \frac{1}{6} & \frac{1}{6} \\ \frac{1}{6} & \frac{2}{3} \end{bmatrix}\begin{bmatrix} 6 \\ 0 \end{bmatrix} = \begin{bmatrix} -1 \\ -1 \end{bmatrix}$$

hence,

$$\mathbf{x}^1 = \mathbf{x}^* = \mathbf{x}^0 + \Delta\mathbf{x}^0 = \begin{bmatrix} 1 \\ 1 \end{bmatrix} + \begin{bmatrix} -1 \\ -1 \end{bmatrix} = \begin{bmatrix} 0 \\ 0 \end{bmatrix}$$

$$f(\mathbf{x}^*) = 0$$

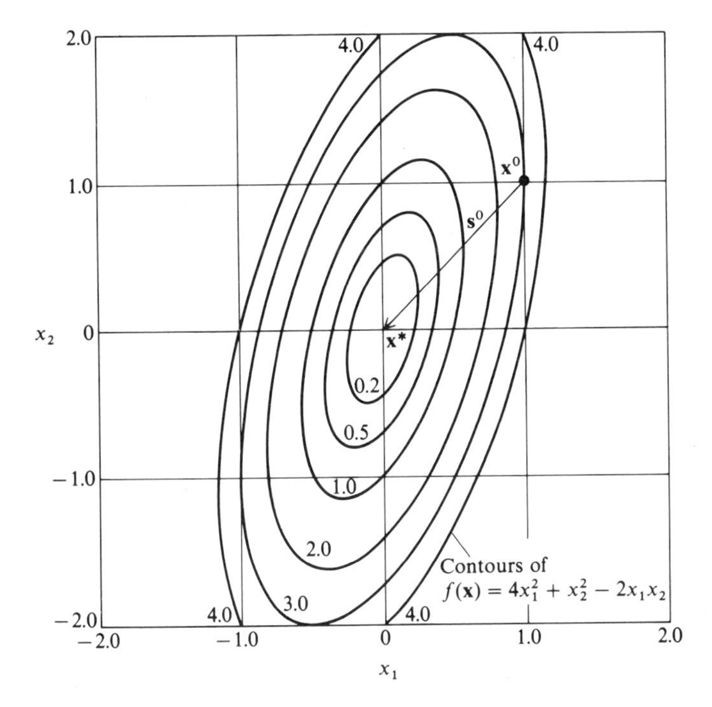

Figure E6.4

Instead of taking the inverse of **H**, we could solve Eq. (6.16)

$$\begin{bmatrix} 8 & -2 \\ -2 & 2 \end{bmatrix} \begin{bmatrix} \Delta x_1^0 \\ \Delta x_2^2 \end{bmatrix} = -\begin{bmatrix} 6 \\ 0 \end{bmatrix}$$

which gives

$$\Delta x_1^0 = -1$$

$$\Delta x_2^0 = -1$$

as before. The search direction $\mathbf{s}^0 = -\mathbf{H}^{-1}\,\nabla f(\mathbf{x}^0)$ is shown in Fig. E6.4.

EXAMPLE 6.5 APPLICATION OF NEWTON'S METHOD AND QUADRATIC CONVERGENCE

If we minimize the nonquadratic function

$$f(\mathbf{x}) = (x_1 - 2)^4 + (x_1 - 2)^2 x_2^2 + (x_2 + 1)^2$$

from the starting point of (1, 1), can you show that Newton's method exhibits quadratic convergence? *Hint:* Show that

$$\frac{\| \mathbf{x}^{k+1} - \mathbf{x}^* \|}{\| \mathbf{x}^k - \mathbf{x}^* \|^2} < c \qquad \text{(see Sec. 5.3)}$$

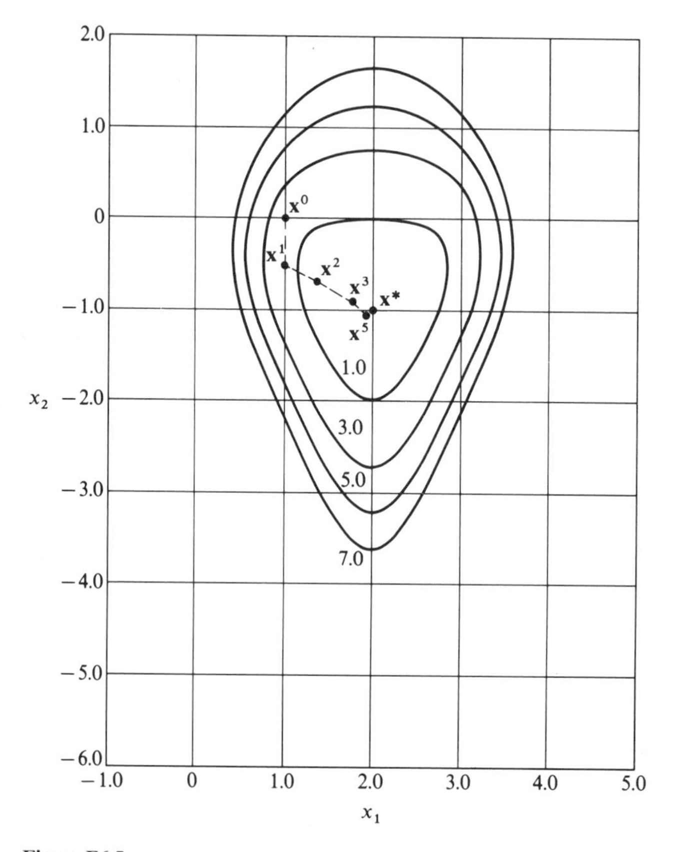

Figure E6.5

Solution. Newton's method produces the following sequences of values for x_1, x_2, and $[f(\mathbf{x}^{k+1}) - f(\mathbf{x}^k)]$ (you should try to verify the calculations shown below; the trajectory is traced in Fig. E6.5).

Iteration	x_1	x_2	$f(\mathbf{x}^{k+1}) - f(\mathbf{x}^k)$
0	1.000000	1.000000	6.000
1	1.000000	−0.500000	1.500
2	1.391304	−0.695652	4.09×10^{-1}
3	1.745944	−0.948798	6.49×10^{-2}
4	1.986278	−1.048208	2.53×10^{-3}
5	1.998734	−1.000170	1.63×10^{-6}
6	1.9999996	−1.000002	2.75×10^{-12}

You can calculate between iterations 2 and 3 that $c = 0.55$; and between 3 and 4 that $c \approx 0.74$. Hence, quadratic convergence can be demonstrated numerically.

Newton's method is the fastest method of minimization (or maximization) when it succeeds. But it has the disadvantages of:

1. The method does not necessarily find the global solution if multiple local solutions exist, but this is a characteristic of all the methods described in this chapter.
2. It requires matrix inversion or else the solution of a set of n linear equations.
3. It requires both analytical first and second partial derivatives which may not be practical to obtain.
4. The method may proceed to a saddle point if $\mathbf{H}(\mathbf{x})$ is not positive definite.

Difficulty 3 can be ameliorated by using (properly) finite difference approximation as substitutes for derivatives (see Sec. 6.5). Global convergence can be enhanced by

(*a*) Back-tracking if a full Newton step is unsatisfactory because it is too long and $f(\mathbf{x})$ does not decrease, and/or
(*b*) Use of trust regions, that is, estimating a region in which the local quadratic model can be trusted to be an adequate representation of the objective function, and taking a step to minimize the model in this region. See Sec. 6.3.2.

To avoid difficulty 4, you must be very careful to use a technique that continuously guarantees a positive definite $\mathbf{H}(\mathbf{x}^k)$ or inverse $\mathbf{H}^{-1}(\mathbf{x}^k)$. Some standard digital computer programs for matrix inversion or solving sets of linear equations may cause the condition number of $\mathbf{H}$ or $\mathbf{H}^{-1}$ to burgeon and hence may be unsatisfactory. The value of the objective function is guaranteed to be improved on each cycle only if the Hessian matrix of the objective function is positive definite. $\mathbf{H}(\mathbf{x})$ is positive definite for strictly convex functions, but for general functions $\mathbf{H}(\mathbf{x})$ changes from point to point and Newton's method may lead to search directions which have few or no components that point in a direction that reduces $f(\mathbf{x})$. Recall that a real symmetric matrix is positive definite if and only if its eigenvalues are positive. Consequently, in minimization, we observe from Figs. 4.12 to 4.16 that when the eigenvalues of $\mathbf{H}(\mathbf{x}^k)$ are all positive, the quadratic approximation corresponds locally to a circular or elliptical hollow having a minimum. On the other hand, if a pair of eigenvalues have opposite signs, we observe from Fig. 4.14 that a saddle exists that does not have a local minimum. In this latter case the direction given by Newton's method will indicate that the best search direction is toward the saddle point instead of away from it in the "downhill" direction.

Newton's method does have the distinct advantage of quadratic convergence (see Sec. 5.3) in the vicinity of the minimum of the objective function if $\mathbf{H}(\mathbf{x})$ is positive definite and if the objective function can be approximated reasonably well by a quadratic function, namely the region in which most descent methods perform the poorest. Far away from the minimum, other methods may temporarily converge faster than Newton's method.

We next consider what to do if the Hessian matrix of the quadratic approximating model is not locally positive definite (for a minimum).

6.3.2 Forcing the Hessian Matrix to be Positive Definite

Marquardt (1963), Levenberg (1944), and others have suggested that the Hessian matrix of $f(\mathbf{x})$ be modified on each stage of the search as needed to ensure that the modified $\mathbf{H}(\mathbf{x})$, $\tilde{\mathbf{H}}(\mathbf{x})$, is positive definite and well-conditioned. The procedure adds elements to the diagonal elements of $\mathbf{H}(\mathbf{x})$

$$\tilde{\mathbf{H}}(\mathbf{x}) = [\mathbf{H}(\mathbf{x}) + \beta\mathbf{I}] \tag{6.17}$$

where β is a positive constant large enough to make $\tilde{\mathbf{H}}(\mathbf{x})$ positive definite when $\mathbf{H}(\mathbf{x})$ is not. Also it is possible to use

$$[\tilde{\mathbf{H}}(\mathbf{x})]^{-1} = [\mathbf{H}^{-1}(\mathbf{x}) + \gamma\mathbf{I}] \tag{6.18}$$

where γ is scalar of sufficient value for the same purpose. To ascertain the value of β to use, you might estimate the smallest (most negative) eigenvalue of $\mathbf{H}(\mathbf{x})$

Table 6.1 A modified Marquardt method

Step 1

Pick $\mathbf{x}^0$ the starting point. Let ε = convergence criterion.

Step 2

Set $k = 0$. Let $\beta^0 = 10^3$.

Step 3

Calculate $\nabla f(\mathbf{x}^k)$.

Step 4

Is $\|\nabla f(\mathbf{x}^k)\| < \varepsilon$? If yes, terminate. If no, continue.

Step 5

Calculate $\mathbf{s}^k = -[\mathbf{H}^k + \beta^k\mathbf{I}]^{-1}\nabla f(\mathbf{x}^k)$.

Step 6

Calculate $\mathbf{x}^{k+1} = \mathbf{x}^k + \lambda^k\mathbf{s}^k$.

Step 7

Is $f(\mathbf{x}^{k+1}) < f(\mathbf{x}^k)$? If yes, go to step 8. If no, go to step 9.

Step 8

Set $\beta^{k+1} = \frac{1}{4}\beta^k$ and $k = k + 1$. Go to step 3.

Step 9

Set $\beta^k = 2\beta^k$. Go to step 5.

and make $\beta > -\min\{\alpha_1\}$, where α_1 is an eigenvalue of $\mathbf{H}(\mathbf{x})$. Note that with a β sufficiently large, $\beta\mathbf{I}$ can overwhelm $\mathbf{H}(\mathbf{x})$ and the minimization approaches a steepest-descent search.

A simpler procedure that may result in a suitable value of β is to apply the Gill and Murray modified Cholesky factorization as follows:

$$\mathbf{H}(\mathbf{x}^k) + \mathbf{D} = \mathbf{L}\mathbf{L}^T \tag{6.19}$$

where $\mathbf{D}$ is a diagonal matrix with nonnegative elements [$d_{ii} = 0$ if $\mathbf{H}(\mathbf{x}^k)$ is positive definite] and $\mathbf{L}$ is a lower triangular matrix. Upper bounds on the elements in $\mathbf{D}$ are calculated using the Gershgorin circle theorem [see Dennis and Schnabel (1983) for details].

A simple algorithm based on an arbitrary adjustment of β (a modified Marquardt's method) is listed in Table 6.1.

EXAMPLE 6.6 APPLICATION OF MARQUARDT'S METHOD

The algorithm listed in Table 6.1 is to be applied to Rosenbrock's function $f(\mathbf{x}) = 100(x_2 - x_1^2)^2 + (1 - x_1)^2$ starting at $\mathbf{x}^0 = [-1.2 \quad 1.0]^T$ with $\mathbf{H}^0 = \mathbf{H}(\mathbf{x}^0)$.

A quadratic interpolation subroutine was used to minimize in each search direction. Table E6.6 lists the values of $f(\mathbf{x})$, $\mathbf{x}$, $\nabla f(\mathbf{x})$, and the elements of $[\mathbf{H}(\mathbf{x}) + \beta\mathbf{I}]^{-1}$ for each stage of the minimization. A total of 96 function evaluations and 16 calls to the gradient evaluation subroutine were needed, and the CPU time on a Cyber 170 computer was 19 milli-seconds.

Table E6.6 Marquardt's method

					Elements of $[\mathbf{H}(\mathbf{x})^k) + \beta\mathbf{I}]^{-1}$			
$f(\mathbf{x})$	x_1	x_2	$\dfrac{\partial f(\mathbf{x})}{\partial x_1}$	$\dfrac{\partial f(\mathbf{x})}{\partial x_2}$	$\tilde{h}_{11}^{-1}$	$\tilde{h}_{12}^{-1}$	$\tilde{h}_{21}^{-1}$	$\tilde{h}_{22}^{-1}$
24.2000	−1.2000	1.0000	−215.6000	−88.0000	0.0005	−0.0002	−0.0002	0.0009
4.1498	−1.0315	1.0791	2.1844	3.0284	0.0005	−0.0002	−0.0002	0.0009
4.1173	−1.0289	1.0557	−5.2448	−0.5768	0.0014	−0.0013	−0.0013	0.0034
3.9642	−0.9412	0.9301	12.7861	8.8552	0.0037	−0.0059	−0.0059	0.0130
3.4776	−0.8542	0.7098	−10.5031	−3.9772	0.0195	−0.0341	−0.0341	0.0641
2.7527	−0.6028	0.3206	−13.5391	−8.5706	0.0399	−0.0669	−0.0669	0.1170
1.9132	−0.3167	0.0580	−7.9993	−8.4706	0.0464	−0.0557	−0.0557	0.0718
1.1890	−0.0313	−0.0344	−2.5059	−7.0832	0.0519	−0.0328	−0.0328	0.0258
0.6885	0.2278	0.0215	1.2242	−6.0759	0.0616	−0.0039	−0.0039	0.0052
0.3266	0.4570	0.2031	3.2402	−4.5160	0.0706	0.0322	0.0322	0.0196
0.1275	0.6846	0.4520	3.9595	3.3523	0.0906	0.0861	0.0861	0.0868
0.0237	0.8705	0.7495	2.6299	−1.6593	0.1148	0.1573	0.1573	0.2203
0.0006	0.9870	0.9721	0.7700	−0.4033	0.1880	0.3273	0.3273	0.5748
0.0000	0.9974	0.9949	−0.0589	0.0269	0.3563	0.7033	0.7033	1.3932
0.0000	0.9999	0.9999	−0.0004	0.0002	0.5138	1.0249	1.0249	2.0494
0.0000	1.0000	1.0000	0.0000	−0.0000	0.5001	1.0001	1.0001	2.0050
0.0000	1.0000	1.0000						

6.3.3 Movement in the Search Direction

Up to this point we have focused on calculating $\mathbf{H}$, $\mathbf{H}^{-1}$, $\tilde{\mathbf{H}}$, or $\tilde{\mathbf{H}}^{-1}$ from which the search direction $\mathbf{s}$ can be ascertained via Eq. (6.15) or $\Delta\mathbf{x}$ from Eq. (6.16) (for minimization). In this section we discuss briefly how far to proceed in the search direction, i.e., select a step length, for a general function $f(\mathbf{x})$. If $\Delta\mathbf{x}$ is calculated from Eq. (6.13) or (6.16), $\lambda = 1$ and the step is a Newton step. If $\lambda \neq 1$, then any procedure can be used to calculate λ as discussed in Chap. 5.

LINE SEARCH. The oldest and simplest method of calculating λ to obtain $\Delta\mathbf{x}$ is via a *unidimensional line search.* In a given direction that reduces $f(\mathbf{x})$, take a step, or sequence of steps yielding an overall step, that reduces $f(\mathbf{x})$ to some acceptable degree. This operation can be carried out by any of the one-dimensional search techniques described in Chap. 5. Early investigators always minimized $f(\mathbf{x})$ as accurately as possible in a search direction $\mathbf{s}$, but subsequent experience, and to some extent theoretical results, have indicated that such a concept is invalid. Good algorithms first calculate a full Newton step ($\lambda = 1$) to get $\mathbf{x}^{k+1}$, and if $f(\mathbf{x}^k)$ is not reduced, back-track in some systematic way toward $\mathbf{x}^k$. Failure to take the full Newton step in the first iteration leads to loss of the advantages of Newton's method near the minimum, where convergence is slow. To avoid very small decreases in $f(\mathbf{x})$ values relative to the length of the steps taken, most algorithms require that the average rate of descent from $\mathbf{x}^k$ to $\mathbf{x}^{k+1}$ be at least some prescribed fraction of the initial rate of descent in the search direction. Mathematically this means (Armijo, 1966)

$$f(\mathbf{x}^k + \lambda\mathbf{s}^k) \leq f(\mathbf{x}^k) + \alpha\lambda\nabla^T f(\mathbf{x}^k)\mathbf{s}^k \tag{6.20}$$

Examine Fig. 6.11. In practice α is often chosen to be very small, about 10^{-4}, so that just a small decrease in the function value occurs.

Back-tracking can be accomplished in any of the ways outlined in Chap. 5 but with the objective of locating an $\mathbf{x}^{k+1}$ for which $f(\mathbf{x}^{k+1}) < f(\mathbf{x}^k)$ but moving

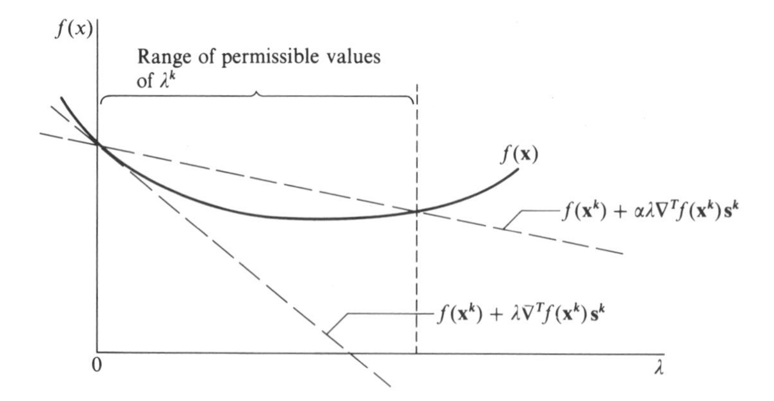

Figure 6.11 Range of acceptable values for choice of λ^k to meet criterion (6.20) with $\alpha = 0.02$.

as far as possible in the direction $\mathbf{s}^k$ from $\mathbf{x}^k$. The minimum of $f(\mathbf{x}^k + \lambda\mathbf{s}^k)$ does not have to be found exactly. As an example of one procedure, at $\mathbf{x}^k$ where $\lambda = 0$, you know two pieces of information about $f(\mathbf{x}^k + \lambda\mathbf{s}^k)$: the values of $f(\mathbf{x}^k)$ and $\nabla^T f(\mathbf{x}^k)\mathbf{s}^k$. After the Newton step ($\lambda = 1$) you know the value of $f(\mathbf{x}^k + \mathbf{s}^k)$. From these three pieces of information you can make a quadratic interpolation to get the value $\hat{\lambda}$ where the objective function $f(\lambda)$ has a minimum:

$$\hat{\lambda} = -\frac{\nabla^T f(\mathbf{x}^k)\mathbf{s}^k}{2[f(\mathbf{x}^k + \mathbf{s}^k) - f(\mathbf{x}^k) - \nabla^T f(\mathbf{x}^k)\mathbf{s}^k]} \tag{6.21}$$

After $\hat{\lambda}$ is obtained, if additional back-tracking is needed, cubic interpolation can be carried out. It is a good idea if $\hat{\lambda}$ is too small, say $\hat{\lambda} < 0.1$, try $\hat{\lambda} = 0.1$ instead.

TRUST REGIONS. The name *trust region* refers to the region in which the quadratic model can be "trusted" to represent $f(\mathbf{x})$ reasonably well. In the unidimensional line search, the search direction is retained but the step length is reduced if the Newton step proves to be unsatisfactory. In the trust region approach, a shorter step length is selected and *then* the search direction determined. We can only sketch the basic concept here without going into details because many heuristic factors are introduced into specific algorithms to make them effective. Refer to the following references for details: Powell (1970*a,b*), Fletcher (1980), Moré (1977), Gander (1981), Gay (1981), and Sorenson (1982).

The concept of the trust region approach is to estimate the length of a maximal successful step from $\mathbf{x}^k$. In other words, $\|\mathbf{x}\| < \rho$, the bound on the step. Figure 6.12 shows $f(\mathbf{x})$, the quadratic model of $f(\mathbf{x})$, and the desired trust region. First, an initial estimate of ρ or the step bound has to be determined. If knowledge about the problem does not help, Powell suggested using the distance to the minimizer of the quadratic model of $f(\mathbf{x})$ in the direction of steepest descent from $\mathbf{x}^k$, the so-called Cauchy point. Next, some curve or piecewise linear function is determined with an initial direction of steepest descent so that the tentative point $\mathbf{x}^{k+1}$ lies on the curve and is less than ρ. Figure 6.12 shows $\mathbf{s}$ as a straight line of one segment. The trust region is updated, and the sequence is continued. Heuistic parameters are usually required such as minimum and maximum step lengths, scaling $\mathbf{s}$, and so forth.

6.3.4 Termination

No single stopping criterion will suffice for Newton's method or any of the optimization methods described in this chapter. The following simultaneous criteria are recommended to avoid scaling problems:

$$\left|\frac{f(\mathbf{x}^{k+1}) - f(\mathbf{x}^k)}{f(\mathbf{x}^k)}\right| < \varepsilon_1 \tag{6.22a}$$

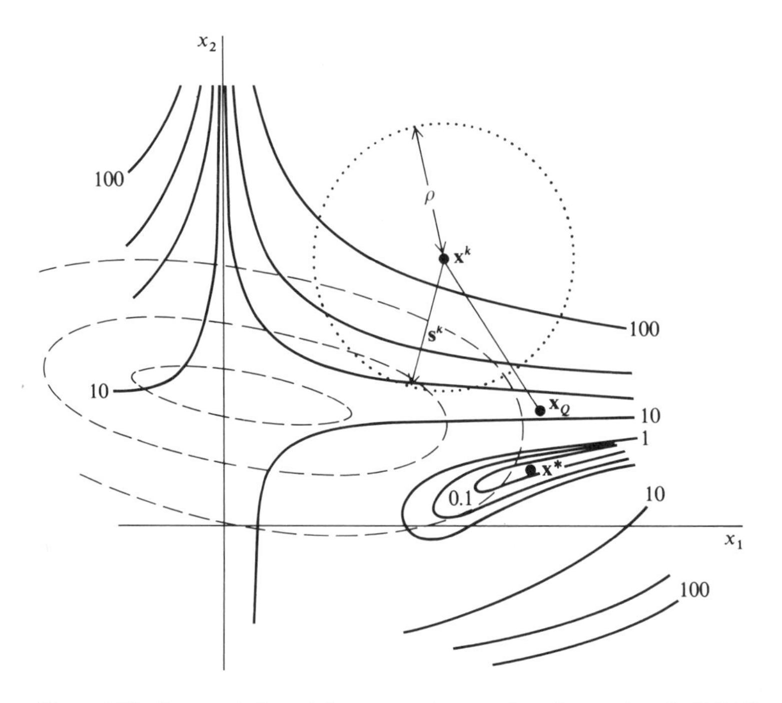

Figure 6.12 Representation of the trust region to select the step length. Solid lines are contours of $f(\mathbf{x})$. Dashed lines are contours of the convex quadratic approximation of $f(\mathbf{x})$ at $\mathbf{x}^k$. Dotted circle is the trust region boundary in which δ is the step length. $\mathbf{x}_0$ is the minimum of the quadratic model for which $\hat{\mathbf{H}}(\mathbf{x})$ is positive definite.

or as $f(\mathbf{x}^k) \to 0$,

$$|f(\mathbf{x}^{k+1}) - f(\mathbf{x}^k)| < \varepsilon_2 \tag{6.22b}$$

$$\left| \frac{x_i^{k+1} - x_i^k}{x_i^k} \right| < \varepsilon_3 \tag{6.23a}$$

or as $x_i^k \to 0$

$$|x_i^{k+1} - x_i^k| < \varepsilon_4 \tag{6.23b}$$

$$\| \mathbf{s}^{k+1} \| < \varepsilon_5 \tag{6.24a}$$

or

$$\| \nabla f(\mathbf{x}^k) \| < \varepsilon_6 \tag{6.24b}$$

6.3.5 Summary of Newton's Method

The basic Newton algorithm (or quasi-Newton if $\lambda \neq 1$ or if $\tilde{\mathbf{H}}$ is used) is as follows:

Step 0. Select $\mathbf{x}^0$.

Step k

(1) Calculate $f(\mathbf{x}^k)$, $\nabla f(\mathbf{x}^k)$, $\mathbf{H}(\mathbf{x}^k)$ [or $\tilde{\mathbf{H}}(\mathbf{x}^k)$] or $\mathbf{H}^{-1}(\mathbf{x}^k)$ [or $\tilde{\mathbf{H}}^{-1}(\mathbf{x}^k)$].

(2*a*) Calculate $\mathbf{s}^k = -\mathbf{H}^{-1}(\mathbf{x}^k)\,\nabla f(\mathbf{x}^k)$ and a tentative $\Delta\tilde{\mathbf{x}}^k = \lambda^k\mathbf{s}^k$, where $\lambda^k = 1$, or

(2*b*) Calculate $\Delta\tilde{\mathbf{x}}^k$ by solving the linear equations $\mathbf{H}(\mathbf{x}^k)\Delta\tilde{\mathbf{x}}^k = -\nabla f(\mathbf{x}^k)$.

(3) Calculate a tentative $\tilde{\mathbf{x}}^{k+1} = \mathbf{x}^k + \Delta\tilde{\mathbf{x}}^k$. If $f(\tilde{\mathbf{x}}^{k+1}) > f(\mathbf{x}^k)$ or λ^k is too large, reduce the step length as described in Sec. 6.3.3. until $\tilde{\mathbf{x}}^{k+1}$ is acceptable.

Step k + 1

(1) Calculate $f(\mathbf{x}^{k+1})$, $\nabla f(\mathbf{x}^{k+1})$.

(2) Test for convergence as indicated in Sec. 6.3.4. If convergence is not achieved, return to step k.

6.3.6 Relation Between Conjugate Gradient Methods and Quasi-Newton Methods

Toint (1984) has demonstrated that the connection between the conjugate gradient method and the quasi-Newton methods is as follows. Suppose that the $\nabla f(\mathbf{x}^k)$ is "preconditioned" by multiplying by $\mathbf{H}^k$ using an analog of Eq. (6.10)

$$\mathbf{s}^{k+1} = -[\mathbf{H}^k\nabla f(\mathbf{x}^k)] + \tau^k\mathbf{s}^k \tag{6.25}$$

where

$$\tau(k) = \frac{\nabla^T f(\mathbf{x}^k)\Delta\mathbf{g}^k}{(\mathbf{s}^k)^T\Delta\mathbf{g}^k}$$

and $\Delta\mathbf{g}^k = \nabla f(\mathbf{x}^{k+1}) - \nabla f(\mathbf{x}^k)$.

After some algebraic manipulations it can be shown for a quadratic function that

$$\mathbf{s}^{k+1} = -(\hat{\mathbf{H}}^k)^{-1}\nabla f(\mathbf{x}^k)$$

where $\hat{\mathbf{H}}$ is calculated by Eq. (6.28) (a formula to be discussed in the next section). Thus, a preconditioned conjugate gradient method gives the same search directions as the BFGS update for a quadratic function. For a nonlinear function this conclusion does not apply, of course, but preconditioning improves the performance of conjugate gradient methods.

6.4 SECANT METHODS

By analogy with the secant technique for functions of a single variable, the procedure described in this section minimizes $f(\mathbf{x})$ using only values of $f(\mathbf{x})$ and $\nabla f(\mathbf{x})$; the Hessian matrix of $f(\mathbf{x})$ is approximated from combinations of the two. Of course $\nabla f(\mathbf{x})$ may be evaluated by finite difference substitutes—hence only function values would be employed in minimization—but the secant itself is not used to evaluate $\nabla f(\mathbf{x})$. Secant methods can also be called *quasi-Newton methods* because $\mathbf{H}(\mathbf{x})$ is approximated by $\hat{\mathbf{H}}(\mathbf{x})$, or $\mathbf{H}^{-1}(\mathbf{x})$ is approximated by $\hat{\mathbf{H}}^{-1}(\mathbf{x})$. Most of the methods employ conjugate directions but not all. In principle, those methods employing conjugate directions should all be equivalent in effectiveness in minimizing $f(\mathbf{x})$, and they are if $f(\mathbf{x})$ is minimized *exactly* in a given search direction, and *no* numerical round-off errors are introduced into the sequence of computations. Of course, these two qualifications cannot be implemented in practice as they require too much computational effort so that each of the secant methods exhibits a different performance. Furthermore, the performance is affected by the characteristics of $f(\mathbf{x})$ and the location of the starting vector $\mathbf{x}^0$.

6.4.1 Determination of the Approximate Hessian Matrix

For the secant methods, we start with the same equation as used in Newton's method, Eq. (6.12)

$$\nabla f(\mathbf{x}) + \mathbf{H}(\mathbf{x}^k)\Delta\mathbf{x}^k = 0 \tag{6.12}$$

Note the analogy to Eq. (5.9). For objective functions other than quadratic functions, we want to approximate $\mathbf{H}$ by a matrix $\hat{\mathbf{H}}$, the Hessian matrix in the quadratic approximation of $f(\mathbf{x})$, but use only the first partial derivatives of $f(\mathbf{x})$ in evaluating $\hat{\mathbf{H}}$. If we want to use the approximate inverse matrix, the secant or quasi-Newton methods compute a new $\mathbf{x}$ vector from the one on the preceding stage by an equation analogous to Eq. (6.14):

$$\mathbf{x}^{k+1} - \mathbf{x}^k = -\lambda^k[\hat{\mathbf{H}}(\mathbf{x}^k)]^{-1}\nabla f(\mathbf{x}^k) \tag{6.26}$$

where $[\hat{\mathbf{H}}(\mathbf{x}^k)]^{-1}$ is sometimes called the *direction matrix* and represents an approximation of $\mathbf{H}^{-1}(\mathbf{x}^k)$. From this viewpoint, in steepest descent $\hat{\mathbf{H}}^{-1}(\mathbf{x}^k)$ is the identity matrix.

The various secant formulas can be evolved as follows. Suppose for the moment $f(\mathbf{x})$ is approximated by a quadratic function. Then we would pick two points, $\mathbf{x}^k$ and $\mathbf{x}^{k+1}$, to start the secant procedure by analogy with the discussion of the secant method in Chap. 5 ($\mathbf{x}^p$ is the reference point)

$$\nabla f(\mathbf{x}^{k+1}) = \nabla f(\mathbf{x}^p) + \hat{\mathbf{H}}(\mathbf{x}^{k+1} - \mathbf{x}^p)$$

$$\nabla f(\mathbf{x}^k) = \nabla f(\mathbf{x}^p) + \hat{\mathbf{H}}(\mathbf{x}^k - \mathbf{x}^p)$$

$$\nabla f(\mathbf{x}^{k+1}) - \nabla f(\mathbf{x}^k) = \hat{\mathbf{H}}(\mathbf{x}^{k+1} - \mathbf{x}^k) \tag{6.27}$$

For a nonquadratic function, the matrix $\hat{\mathbf{H}}$ in Eq. (6.27) would be obtained by solving the "secant equations" (6.27) using the points $\mathbf{x}^k$ and $\mathbf{x}^{k+1}$

$$\hat{\mathbf{H}}(\mathbf{x}^k)\Delta\mathbf{x}^k = \Delta\mathbf{g}^k \tag{6.28}$$

where

$$\Delta\mathbf{g}^k \equiv \nabla f(\mathbf{x}^{k+1}) - \nabla f(\mathbf{x}^k)$$

An equivalent relation to (6.28) is

$$\Delta\mathbf{x}^k = \hat{\mathbf{H}}^{-1}(\mathbf{x}^k)\Delta\mathbf{g}^k \tag{6.29}$$

To simplify the notation, at a number of places in the remainder of this section we will replace

$$\hat{\mathbf{H}}(\mathbf{x}^k) \quad \text{by } \hat{\mathbf{H}}^k$$

$$\hat{\mathbf{H}}^{-1}(\mathbf{x}^k) \quad \text{by } (\hat{\mathbf{H}}^k)^{-1}$$

Because for the n equations with n^2 unknown elements in $\hat{\mathbf{H}}^k$ [or $(\hat{\mathbf{H}}^k)^{-1}$] an infinite number of candidates exist for $\hat{\mathbf{H}}$ (or $\hat{\mathbf{H}}^{-1}$) when $n > 1$, we would like to choose $\hat{\mathbf{H}}$ (or $\hat{\mathbf{H}}^{-1}$) that is closest to $\mathbf{H}$ (or $\mathbf{H}^{-1}$) in some sense. Many suggestions have been proposed to recompute ("update") $\hat{\mathbf{H}}$ (or $\hat{\mathbf{H}}^{-1}$) from stage to stage for functions that are not quadratic. We can only describe in detail a few that have proved to be effective in practice.

Suppose you start at the point $\mathbf{x}^k$. There you know the values of $\mathbf{x}^k$, $f(\mathbf{x}^k)$, $\nabla f(\mathbf{x}^k)$, $\hat{\mathbf{H}}^k$, $(\hat{\mathbf{H}}^k)^{-1}$, and $\mathbf{s}^k$. Next you step to the point $\mathbf{x}^{k+1}$ at which you can compute $f(\mathbf{x}^{k+1})$ and $\nabla f(\mathbf{x}^{k+1})$. How can $\hat{\mathbf{H}}^{k+1}$ [or $(\hat{\mathbf{H}}^k)^{-1}$] be computed? Broyden (1967) suggested a "least change" secant update in the sense that the Frobenius norm[1]

$$\|\hat{\mathbf{H}}^{k+1} - \hat{\mathbf{H}}^k\|_F$$

is minimized subject to $\hat{\mathbf{H}}^{k+1}\Delta\mathbf{x}^k = \Delta\mathbf{g}^k$. The unique result is (Dennis and Schnabel, 1983)

$$\Delta\hat{\mathbf{H}}^k = \hat{\mathbf{H}}^{k+1} - \hat{\mathbf{H}}^k = \frac{[\Delta\mathbf{g}^k - \hat{\mathbf{H}}^k\Delta\mathbf{x}^k](\Delta\mathbf{x}^k)^T}{(\Delta\mathbf{x}^k)^T\Delta\mathbf{x}^k} \tag{6.30a}$$

Although (6.30a) represents an updating method that provides superlinear convergence (for a quadratic function), it has the major disadvantage that after the initial stage $k = 0$, $\hat{\mathbf{H}}^k$ stops being a symmetric matrix, and it is not feasible to represent $\mathbf{H}$ itself, a symmetric matrix, with a nonsymmetric matrix near $\mathbf{x}^*$.

Broyden (1967) described a rank one updating relation using $(\hat{\mathbf{H}}^k)^{-1}$ that remains positive definite if $(\hat{\mathbf{H}}^0)^{-1}$ was positive definite

$$\Delta(\hat{\mathbf{H}}^k)^{-1} = \frac{[\Delta\mathbf{x}^k - (\hat{\mathbf{H}}^k)^{-1}\Delta\mathbf{g}^k][\Delta\mathbf{x}^k - (\hat{\mathbf{H}}^k)^{-1}\Delta\mathbf{g}^k]^T}{[\Delta\mathbf{x}^k - (\hat{\mathbf{H}}^k)^{-1}\Delta\mathbf{g}^k]^T\Delta\mathbf{g}^k} \tag{6.31a}$$

Equation (6.31a) has proved quite effective. Powell (1970b) developed a symmetric secant update known as the Powell-Broyden update by projecting the Frobenius norm onto a subspace of symmetric matrices, but the updating formula shares the deficiency of (6.30a) in that it does not necessarily produce a positive definite $\hat{\mathbf{H}}^{k+1}$ even if $\hat{\mathbf{H}}^k$ is positive definite.

---

[1] The Frobenius norm of a square matrix $\mathbf{A}$ is $\|\mathbf{A}\|_F = (\sum_{i=1}^n \sum_{j=1}^n a_{ij}^2)^{1/2}$ where a_{ij} is an element of $\mathbf{A}$.

The best theoretical and practical results have been obtained by using symmetric positive definite secant updates for $\hat{\mathbf{H}}$. The theoretical convergence of these methods is no better than for (6.30*a*), but the approximations of $\mathbf{H}$ generated in practice are more accurate and hence performance is better. Also, the updates can be shown to be invariant under linear transformations of the variable space (Dennis and Schnabel, Chap. 9, 1983) which helps alleviate scaling difficulties. (This invariance property is shared by Newton's method but not by steepest descent.)

One of the earliest examples of a positive definite secant update was the relation of rank two now called the Davidon-Fletcher-Powell (DFP) update for $(\hat{\mathbf{H}}^k)^{-1}$:

$$\Delta(\hat{\mathbf{H}}^k)^{-1} = \frac{(\Delta\mathbf{x}^k)(\Delta\mathbf{x}^k)^T}{(\Delta\mathbf{x}^k)^T(\Delta\mathbf{g}^k)} - \frac{(\hat{\mathbf{H}}^k)^{-1}(\Delta\mathbf{g}^k)(\Delta\mathbf{g}^k)^T[(\hat{\mathbf{H}}^k)^{-1}]^T}{(\Delta\mathbf{g}^k)^T(\hat{\mathbf{H}}^k)^{-1}(\Delta\mathbf{g}^k)} \tag{6.31b}$$

and its analog

$$\Delta\hat{\mathbf{H}}^k = \frac{(\Delta\mathbf{g}^k - \hat{\mathbf{H}}^k\Delta\mathbf{x}^k)(\Delta\mathbf{g}^k)^T + \Delta\mathbf{g}^k(\Delta\mathbf{g}^k - \hat{\mathbf{H}}^k\Delta\mathbf{x}^k)^T}{(\Delta\mathbf{g}^k)^T(\Delta\mathbf{x}^k)} - \frac{(\Delta\mathbf{g}^k - \hat{\mathbf{H}}^k\Delta\mathbf{x}^k)^T\Delta\mathbf{x}^k\Delta\mathbf{g}^k(\Delta\mathbf{g}^k)^T}{[(\Delta\mathbf{g}^k)^T\Delta\mathbf{x}^k][(\Delta\mathbf{g}^k)^T\Delta\mathbf{x}^k]} \tag{6.30b}$$

Although widely known, these two relations are not as effective in practice as the BFGS relations that are described next.

All of the secant updating methods can be encompassed by the relations formulated in Table 6.2. Note the duality present in each set of relations in that substitution of

$$\Delta\mathbf{x}^k \Leftrightarrow \Delta\mathbf{g}^k$$

$$\hat{\mathbf{H}}^k \Leftrightarrow (\hat{\mathbf{H}}^k)^{-1}$$

$$\omega \Leftrightarrow \frac{1}{\omega}$$

in one of the resulting equations yields the other. Probably the most successful current updating relations as evidenced by comparative numerical tests were determined independently and essentially simultaneously by Broyden (1970), Fletcher (1970), Goldfarb (1970), and Shanno (1970), now called the BFGS update:

$$\Delta\hat{\mathbf{H}}^k = \frac{\Delta\mathbf{g}^k(\Delta\mathbf{g}^k)^T}{(\Delta\mathbf{g}^k)^T\Delta\mathbf{x}^k} - \frac{\hat{\mathbf{H}}^k\Delta\mathbf{x}^k(\Delta\mathbf{x}^k)^T\hat{\mathbf{H}}^k}{(\Delta\mathbf{x}^k)^T\hat{\mathbf{H}}^k\Delta\mathbf{x}^k} \tag{6.32}$$

$$\Delta(\hat{\mathbf{H}}^k)^{-1} = \frac{[\Delta\mathbf{x}^k - (\hat{\mathbf{H}}^k)^{-1}\Delta\mathbf{g}^k](\Delta\mathbf{x}^k)^T + \Delta\mathbf{x}^k[\Delta\mathbf{x}^k - (\hat{\mathbf{H}}^k)^{-1}\Delta\mathbf{g}^k]^T}{(\Delta\mathbf{g}^k)^T\Delta\mathbf{x}^k} - \frac{[\Delta\mathbf{x}^k - (\hat{\mathbf{H}}^k)^{-1}\Delta\mathbf{g}^k]^T\Delta\mathbf{g}^k\Delta\mathbf{x}^k(\Delta\mathbf{x}^k)^T}{[(\Delta\mathbf{g}^k)^T\Delta\mathbf{x}^k][(\Delta\mathbf{g}^k)^T\Delta\mathbf{x}^k]} \tag{6.33}$$

Table 6.2 Determination of $\hat{\mathbf{H}}^{k+1}$ and $(\hat{\mathbf{H}}^{k+1})^{-1}$ by secant updates

	For $\hat{\mathbf{H}}^{k+1}$	For $(\hat{\mathbf{H}}^{k+1})^{-1}$
(1) Require that this equation hold at $\mathbf{x}^{k+1}$(ω is a scalar usually equal to unity):	$\omega\hat{\mathbf{H}}^{k+1}\Delta\mathbf{x}^k = \Delta\mathbf{g}^k$ (6.28a)	$\frac{1}{\omega}\Delta\mathbf{x}^k = (\hat{\mathbf{H}}^{k+1})^{-1}\Delta\mathbf{g}^k$ (6.29a)
(2) Define:	$\Delta\hat{\mathbf{H}}^k = \hat{\mathbf{H}}^{k+1} - \hat{\mathbf{H}}^k$	$(\Delta\hat{\mathbf{H}}^k)^{-1} = (\hat{\mathbf{H}}^{k+1})^{-1} - (\hat{\mathbf{H}}^k)^{-1}$
(3) Then solve:	$(\Delta\hat{\mathbf{H}}^k + \hat{\mathbf{H}}^k)\Delta\mathbf{x}^k = \Delta\mathbf{g}^k\frac{1}{\omega}$ or $\Delta\hat{\mathbf{H}}^k\Delta\mathbf{x}^k = \frac{1}{\omega}\Delta\mathbf{g}^k - \hat{\mathbf{H}}^k\Delta\mathbf{x}^k$	$\frac{1}{\omega}\Delta\mathbf{x}^k = [(\Delta\hat{\mathbf{H}}^k)^{-1} + (\hat{\mathbf{H}}^k)^{-1}]\Delta\mathbf{g}^k$ or $(\Delta\hat{\mathbf{H}}^k)^{-1}\Delta\mathbf{g}^k = \frac{1}{\omega}\Delta\mathbf{x}^k - (\hat{\mathbf{H}}^k)^{-1}\Delta\mathbf{g}^k$
(4) The solution of (3) can be shown by direct substitution of (4) into (3) to be (where **y** and **z** are functions such as $\Delta\mathbf{x}^k$, $\Delta\mathbf{g}^k$, $\hat{\mathbf{H}}^k\Delta\mathbf{g}^k$, etc.):	$\Delta\hat{\mathbf{H}}^k = \frac{1}{\omega}\frac{\Delta\mathbf{g}^k\mathbf{y}^T}{\mathbf{y}^T\Delta\mathbf{x}^k} - \frac{\hat{\mathbf{H}}^k\Delta\mathbf{x}^k\mathbf{z}^T}{\mathbf{z}^T\Delta\mathbf{x}^k}$ (6.30)	$(\Delta\hat{\mathbf{H}}^k)^{-1} = \frac{1}{\omega}\frac{\Delta\mathbf{x}^k\mathbf{y}^T}{\mathbf{y}^T\Delta\mathbf{g}^k} - \frac{(\hat{\mathbf{H}}^k)^{-1}\Delta\mathbf{g}^k\mathbf{z}^T}{\mathbf{z}^T\mathbf{g}^k}$ (6.31)

These relations meet the two important criteria that

1. $\hat{\mathbf{H}}$ (or $\hat{\mathbf{H}}^{-1}$) remains positive definite if $\hat{\mathbf{H}}^0$[or $(\hat{\mathbf{H}}^0)^{-1}$] is positive definite
2. $\hat{\mathbf{H}}$ (or $\hat{\mathbf{H}}^{-1}$) is symmetric

Since both updates in Table 6.2 in effect rescale the variable space at every iteration, such methods are sometimes called *variable metric* methods. Broyden, Dennis, and Moré (1975) showed that the BFGS formulas exhibit superlinear convergence to $\mathbf{x}^*$. As mentioned in Sec. 6.3.5, the BFGS formulas give the same search directions for a quadratic function as does the preconditioned conjugate-gradient method.

Comparison of the updating relations developed so far reveal the specific quantities selected for ω, $\mathbf{y}$, and $\mathbf{z}$ in Table 6.2:

Name	Eq. No.	$\hat{\mathbf{H}}$ or $\hat{\mathbf{H}}^{-1}$	ω	$\mathbf{y}$	$\mathbf{z}$
Broyden	6.30*a*)	$\hat{\mathbf{H}}$	1	$\Delta\mathbf{x}$	$\Delta\mathbf{x}$
Broyden	6.31*a*)	$\hat{\mathbf{H}}^{-1}$	1	$(\Delta\mathbf{x} - \hat{\mathbf{H}}^{-1}\Delta\mathbf{g})$	$(\Delta\mathbf{x} - \hat{\mathbf{H}}^{-1}\Delta\mathbf{g})$
DFP	(6.30*b*)	$\hat{\mathbf{H}}$	1	$\Delta\mathbf{x}$	$\hat{\mathbf{H}}^{-1}\Delta\mathbf{g}$
DFP	(6.31*b*)	$\hat{\mathbf{H}}^{-1}$			
BFGS	(6.32)	$\hat{\mathbf{H}}$		$\Delta\mathbf{g}$	$\hat{\mathbf{H}}^T\Delta\mathbf{x}$ (note $\hat{\mathbf{H}}^T = \hat{\mathbf{H}}$)
BFGS	(6.33)	$\hat{\mathbf{H}}^{-1}$			

In computer codes, $\hat{\mathbf{H}}$ or $\hat{\mathbf{H}}^{-1}$ are often not updated directly by using Eqs. (6.32) or (6.33), but are instead computed from a Cholesky factorization of $\hat{\mathbf{H}}$ (or $\hat{\mathbf{H}}^{-1}$) in which the lower triangular matrix in $\hat{\mathbf{H}} = \mathbf{LL}^T$ (or $\hat{\mathbf{H}}^{-1} = \mathbf{LL}^T$) is updated, or $\mathbf{LDL}^T$ is updated where $\mathbf{D}$ is a diagonal matrix [see Goldfarb (1976) for the formulas].

Another secant-based algorithm that has performed well in practice is that of Davidon (1975). Davidon's method has three main characteristics that cause it to differ from most proposed secant methods. First, it eliminates all unidimensional searches, that is searches to determine λ^k. It instead uses projections of $\Delta\mathbf{x}$ and $\Delta\mathbf{g}$ to update $\hat{\mathbf{H}}^{-1}$. Second, the updated matrix $(\hat{\mathbf{H}}^{k+1})^{-1}$ is selected to have optimal conditioning (in the sense that the ratio of the largest to the smallest eigenvalue of $(\hat{\mathbf{H}}^{k+1})^{-1}$ is minimized. Third, instead of using $\hat{\mathbf{H}}^{-1}$ directly, $\hat{\mathbf{H}}^{-1}$ is factored as the product $\mathbf{JJ}^T$, and only $\mathbf{J}$ is stored and updated.

Let us now compare some of these updating relations.

EXAMPLE 6.7 COMPARISON OF FOUR UPDATING RELATIONS

The function in Examples 6.2 and 6.3, $f(\mathbf{x}) = 100(x^2 - x_1^2)^2 + (1 - x_1)^2$, is minimized starting at $\mathbf{x}^0 = [-1.2 \quad 1.0]^T$ using three algorithms (*a*) BFGS, (*b*) Broyden's symmetric update for $\hat{\mathbf{H}}^{-1}$, and (*c*) Davidon, each with the same unidimensional search (quadratic interpolation).

The values of $f(\mathbf{x})$, $\mathbf{x}$, $\nabla f(\mathbf{x})$, and $\hat{\mathbf{H}}^{-1}(\mathbf{x})$ are listed for comparison on each iteration in Tables E6.7*a*, E6.7*b*, and E6.7*c*, respectively. In addition in Table E6.7*d*,

Table E6.7*a* BFGS method

Stage	$f(\mathbf{x})$	x_1	x_2	$\frac{\partial f(\mathbf{x})}{\partial x_1}$	$\frac{\partial f(\mathbf{x})}{\partial x_2}$	$\hat{h}_{11}^{-1}$	$\hat{h}_{12}^{-1} = \hat{h}_{21}^{-1}$	$\hat{h}_{22}^{-1}$
0	24.2000	−1.2000	1.0000	−215.6000	−88.0000	0.0056	−0.0135	0.0374
1	4.7307	−1.1750	1.3814	−3.9886	0.1538	0.0056	−0.0132	0.0361
2	4.6246	−1.1505	1.3230	−4.5785	−0.1206	0.4787	−1.1185	2.6172
3	4.1254	−0.9014	0.7411	−29.5540	−14.2843	0.0427	−0.0928	0.2047
5	3.6478	−0.9036	0.8009	−9.4314	−3.1123	0.1321	−0.2492	0.4740
10	2.0812	−0.4203	0.1513	−7.0945	−5.0613	0.0204	−0.0160	0.0174
15	0.9571	0.0456	0.0194	−1.5162	−4.3015	0.0885	−0.0107	0.0121
20	0.3420	0.4559	0.1864	2.8204	−4.2870	0.0910	0.0759	0.0698
25	0.0693	0.8236	0.6588	6.0816	−3.9061	0.1197	0.1743	0.2566
30	0.0066	0.9226	0.8488	0.7290	−0.4790	0.3748	0.6537	1.1480
35	0.0000	1.0004	1.0007	0.0253	−0.0122	0.4970	0.9923	1.9860
36	0.0000	1.0000	1.0000	0.0002	−0.0001	0.4812	0.9614	1.9259
37	0.0000	1.0000	1.0000	0.0000	−0.0000	0.4998	0.9996	2.0043

45 function evaluations, 45 gradient evaluations.

Table E6.7*b* Broyden's symmetric matrix method

Stage	$f(\mathbf{x})$	x_1	x_2	$\frac{\partial f(\mathbf{x})}{\partial x_1}$	$\frac{\partial f(\mathbf{x})}{\partial x_2}$	$\hat{h}_{11}^{-1}$	$\hat{h}_{12}^{-1}$	$\hat{h}_{21}^{-1}$	$\hat{h}_{22}^{-1}$
0	24.2000	−1.2000	1.0000	−215.6000	−88.0000	0.0056	−0.0135	−0.0135	0.0374
1	0.1947	1.4408	2.0779	−0.2341	0.3872	0.0166	−0.0106	−0.0106	0.0381
2	0.1946	1.4410	2.0775	0.3402	0.1880	0.0024	0.0059	0.0059	0.0190
3	0.1318	1.3363	1.7720	7.9768	−2.7330	0.2137	0.5946	0.5946	1.6590
4	0.0885	1.2425	1.5266	9.0425	−3.4436	0.1171	0.3075	0.3075	0.8065
5	0.0579	1.2406	1.5393	0.3486	0.0534	0.1889	0.4691	0.4691	1.1698
6	0.0295	1.1514	1.3177	4.0428	−1.6240	0.1889	0.4692	0.4692	1.1653
7	0.0263	1.1233	1.2513	4.9898	−2.1112	0.1259	0.3024	0.3024	0.7240
8	0.0065	1.0719	1.1526	−1.4122	0.7258	0.3551	0.7832	0.7832	1.7326
9	0.0017	1.0361	1.0717	0.8485	−0.3729	0.2870	0.6222	0.6222	1.3519
10	0.0006	1.0092	1.0163	0.9251	−0.4492	0.1044	0.4624	0.4624	1.2120
11	0.0000	1.0068	1.0138	−0.0159	0.0147	0.3776	0.7608	0.7608	1.5380
12	0.0000	1.0004	1.0007	0.0281	−0.0137	0.5007	1.0048	1.0048	2.0216
13	0.0000	1.0000	1.0001	−0.0006	0.0003	0.5471	1.0969	1.0969	2.0245
14	0.0000	1.0000	1.0000	−0.0000	0.0000	0.4995	0.9991	0.9991	2.0035
15	0.0000	1.0000	1.0000						

2 function evaluations, 15 gradient evaluations.

Table E6.7c Davidon's method

Stage	$f(\mathbf{x})$	x_1	x_2	$\frac{\partial f(\mathbf{x})}{\partial x_1}$	$\frac{\partial f(\mathbf{x})}{\partial x_2}$	$\hat{h}_{11}^{-1}$	$\hat{h}_{12}^{-1}$	$\hat{h}_{21}^{-1}$	$\hat{h}_{22}^{-1}$
1	23.0638	−1.2375	1.1065	−214.8219	−84.9877	0.0055	−0.0140	−0.0132	0.0388
2	22.9760	−1.2412	1.1170	−214.8519	−84.7415	−0.0082	−0.0714	0.0254	0.2003
3	21.4568	−1.3609	1.4534	−221.6641	−79.7071	−0.0115	−0.0746	0.0349	0.2095
4	21.3979	−1.3751	1.4939	−223.0872	−79.3898	−0.0133	−0.0771	0.0415	0.2187
5	21.2802	−1.3931	1.5465	−224.5537	−78.8747	−0.0131	−0.0769	0.0445	0.2227
10	7.2634	−1.6482	2.6665	−38.2916	−10.0095	−0.0077	−0.0760	0.0958	0.2464
15	5.3139	−1.2575	1.5347	−27.9731	−9.3271	0.2504	0.1029	−0.6713	−0.1978
20	3.4419	−0.8061	0.6074	−17.2851	−8.4807	0.0037	0.2613	−0.0967	−0.4677
30	1.0944	−0.0429	−0.0064	−2.2272	−1.6487	−0.1376	−0.4169	0.0768	0.0613
40	0.2429	0.5641	0.2952	4.3162	−4.5983	0.2124	0.2466	0.1601	0.2667
50	0.0116	0.8992	0.8048	1.1440	−0.7483	−0.2840	0.0791	−0.4983	0.0837
55	0.0000	0.9949	0.9893	0.1831	−0.0972	−0.3573	−0.4016	−0.6545	−0.8340
56	0.0000	0.9972	0.9944	0.0452	−0.0254	−0.3812	−0.5492	−0.7035	−1.1359
57	0.0000	0.9997	0.9994	0.0018	−0.0012	−0.3937	−0.5957	−0.7288	−1.1975
58	0.0000	1.0000	1.0000	0.0002	−0.0001	−0.3919	−0.5882	−0.7250	−1.2155

69 function evaluations, 59 gradient evaluations.

Table E6.7*d* Results for Schnabel's code

Stage	$f(\mathbf{x})$	$\mathbf{x}$	$\nabla f(\mathbf{x})$	$\hat{\mathbf{H}}$	
0	24.2	−1.20	−215.6	1330	480
		1.00	−88.0	480	200
1	4.732	−1.175	4.637	1107	470
		−1.380	−0.122	470	200
2	4.206	−1.043	−11.63	879	417
		1.069	3.621	417	200
5	2.373	−0.540	−4.182	237	216
		0.286	−1.022	216	200
10	1.299	−0.0057	−2.133	23	2.28
		−0.0535	−10.72	2.28	200
15	0.256	0.497	0.164	202	−199
		0.241	−1.175	−199	200
20	0.00526	0.927	0.0180	690	−371
		0.826	−0.0878	−371	200
24	6.86×10^{-16}	1.000	-5.07×10^{-9}	802	−400
		1.000	-5.47×10^{-9}	−400	200

39 function evaluations, 73 gradient evaluations.

the same problem is solved using the code UNCMIN developed by Schnabel (refer to Table 6.3) based on the BFGS updating relation (6.30). Values of the elements of $\hat{\mathbf{H}}(x^k)$ are listed in Table E6.7*d*. Each of the searches was started with $\mathbf{H}^{-1}(\mathbf{x}^0)$ as the initial $\hat{\mathbf{H}}^{-1}$. Use of $\hat{\mathbf{H}}^{-1} = \hat{\mathbf{H}} = \mathbf{I}$ as the starting $\hat{\mathbf{H}}^0$ made little significant difference in the number of function and derivative evaluations.

For large dimensional problems with sparse structure, secant updates may not be computationally effective. See Dennis and Schnabel (1983), Chap. 11.

6.4.2 Movement in the Search Direction

All of the explanation about line searches and trust regions found in Sec. 6.3.3 applies equally well here. Another procedure that has been found effective with secant methods for partial minimization in the search direction is that described by Fletcher (1970). Fletcher's method does not minimize $f(\mathbf{x})$ to the very minimum. Instead a scalar $\hat{\lambda}$ is calculated from

$$\hat{\lambda} = \frac{2[f(\mathbf{x}^k) - f_{\text{low}}]}{\nabla^T f(\mathbf{x}^k)\mathbf{s}^k} \tag{6.34}$$

where f_{low} is the lowest estimated value of $f(\mathbf{x})$. [If $f(\mathbf{x})$ turns out to be lower than f_{low}, the calculations stop to permit the selection of a new f_{low}.] Then, a cubic interpolation is used between $\mathbf{x}^k$ and $\mathbf{x}^k + \hat{\lambda}\mathbf{s}^k$ to minimize $f(\mathbf{x})$ in the search direction. Because no bracket is required for a cubic interpolation and the result of the interpolation may be unsatisfactory if $\mathbf{x}^k$ is on a concave portion of the approximating polynomial, Fletcher limited the value of λ to be the smaller of (*a*) $\lambda = \lambda^s$ found by the cubic interpolation, or (*b*) $\lambda^s = 0.1\lambda^{k-1}$.

After calculating λ^s, the test

$$-\frac{f(\mathbf{x}^k) - f(\mathbf{x}^k + \lambda^s\mathbf{s}^k)}{\lambda^s \nabla^T f(\mathbf{x}^k)\mathbf{s}^k} \geq 10^{-4} \tag{6.35}$$

is carried out, and if satisfied, the kth stage determination of λ is completed with $\lambda^k = \lambda^s$. Otherwise, the cubic interpolation is repeated and the procedure to determine λ^k is repeated.

Fletcher's method of determining the value of λ has proved to be an excellent one over many years of experience. Apparently the cubic interpolation in a given search direction and the limited step length makes Fletcher's technique efficient.

6.4.3 Termination

Criteria for termination are the same as described in Sec. 6.3.4.

6.4.4 Summary of Secant Methods

The basic secant (quasi-Newton) algorithm (for minimization) is as follows:

Step 0. Scale the variables if feasible. Select $\mathbf{x}^0$, $\mathbf{s}^0$ [usually $\mathbf{s}^0 = -\nabla f(\mathbf{x}^0)$], and $\hat{\mathbf{H}}^0 \equiv \hat{\mathbf{H}}(\mathbf{x}^0)$ [often $\hat{\mathbf{H}}^0 = \mathbf{I}$ if $\mathbf{H}(\mathbf{x}^0)$ is not positive definite].

Step k

(1) Compute $\mathbf{x}^{k+1}$ by minimizing $f(\mathbf{x})$ to a certain extent in the search direction via (*a*) a line search or (*b*) use of a trust region.
(2) Compute $f(\mathbf{x}^{k+1})$ and $\nabla f(\mathbf{x}^{k+1})$.
(3) Compute $\Delta\mathbf{x}^k = \mathbf{x}^{k+1} - \mathbf{x}^k$ and $\Delta\mathbf{g}^k = \nabla f(\mathbf{x}^{k+1}) - \nabla f(\mathbf{x}^k)$.
(4) Compute $\hat{\mathbf{H}}^{k+1}$ using Eq. (6.30) or $(\hat{\mathbf{H}}^{k+1})^{-1}$ using Eq. (6.31).
(5) Compute $\mathbf{s}^{k+1}$ from
  (*a*) $\mathbf{s}^{k+1} = -(\hat{\mathbf{H}}^{k+1})^{-1}\nabla f(\mathbf{x}^{k+1})$ via matrix inversion, or
  (*b*) $\hat{\mathbf{H}}^{k+1}\mathbf{s}^{k+1} = -\nabla f(\mathbf{x}^k)$, by solving a set of linear equations.

Step k + 1

(1) Test for termination as explained in Sec. 6.3.4.
(2) If the convergence criteria are satisfied, stop. Otherwise return to step k.

Some of the specific numerical problems that arise in using secant (quasi-Newton) methods are:

1. $\hat{\mathbf{H}}^k$ or $(\hat{\mathbf{H}}^k)^{-1}$ may cease to be positive definite, in which case it is necessary to make sure that $\hat{\mathbf{H}}^{k+1}$ is made positive definite by one of the methods mentioned in Sec. 6.3.2, or by reinitialization of $\hat{\mathbf{H}}$ or $(\hat{\mathbf{H}}^k)^{-1}$ (which is generally not an effective procedure).
2. The correction $\hat{\mathbf{H}}^{k+1} - \hat{\mathbf{H}}^k$ may become unbounded (sometimes even for quadratic functions) because of round-off.
3. If $\Delta\mathbf{x}^k = -\lambda^k(\hat{\mathbf{H}}^k)^{-1}\nabla f(\mathbf{x}^k)$ by chance is in the direction of the preceding stage, $(\hat{\mathbf{H}}^{k+1})^{-1}$ becomes singular or undetermined. The same remark applies to the calculation involving $\hat{\mathbf{H}}^{k+1}$.

6.4.5 Summary of Indirect Methods

Indirect methods have been demonstrated to be more efficient than direct methods on all experimental testing of algorithms. By replacing analytical derivatives with their finite difference substitutes as discussed in the next section, you can avoid one of the supposed major deficiencies of indirect methods. Within the classification of indirect methods, procedures that use second-order information are far superior to those that use first-order information (gradients) alone, but keep in mind that usually the second-order information may be only approximate as it is based not on second derivatives themselves but approximates to second derivatives.

All of the methods described in this chapter are essentially executed in two identical phases: (1) determine a search direction, $\mathbf{s}$, and (2) move a certain distance in that direction, $\Delta\mathbf{x}$. All of the indirect methods to calculate $\mathbf{s}$ can be viewed as evolving from Newton's method:

$$\Delta\mathbf{x}^k = \lambda\mathbf{s}^k = -\lambda^k[\mathbf{H}(\mathbf{x}^k)]^{-1}\nabla f(\mathbf{x}^k)$$

or

$$\mathbf{H}(\mathbf{x}^k)\Delta\mathbf{x}^k = -\nabla f(\mathbf{x}^k)$$

In the gradient method $\mathbf{H} = \mathbf{H}^{-1} = \mathbf{I}$. In the secant methods $\mathbf{H}$ or $\mathbf{H}^{-1}$ is approximated by various combinations of first derivatives. Table 6.3 lists the formulas and/or equation numbers used in each algorithm to calculate $\mathbf{s}^k$ and includes some remarks as to how λ^k is determined.

Table 6.3 Summary of indirect optimization algorithms

Algorithm	Section	Formulas to calculate $\mathbf{H}$ or $\mathbf{H}^{-1}$ in the search direction	Method(s) of calculating the distance λ^k to move in the search direction
Gradient	6.2.1	$\mathbf{H} = \mathbf{H}^{-1} = \mathbf{I}$ (6.8)	*A*, *B*
Conjugate gradient	6.2.2	$\mathbf{H} = \hat{\mathbf{H}}$ (6.10); (6.25), (6.28)	*A*
Newton's method	6.3.1	$\mathbf{H}^{-1} \equiv \mathbf{H}^{-1}$ (6.14)	Fixed by (6.15), (6.16)
Modified Newton's method	6.3.1	$\mathbf{H}^{-1} \equiv \mathbf{H}^{-1}$ (6.14)	*A*, *C*, *D*
Marquardt	6.3.2	$\mathbf{H} = \mathbf{H} + \beta\mathbf{I}$ $\mathbf{H}^{-1} = \mathbf{H}^{-1} + \delta\mathbf{I}$ (6.17), (6.18)	*A*, *C*, *D*
Broyden	6.4.1	$\Delta\mathbf{H}^{-1} = \Delta\hat{\mathbf{H}}^{-1}$ (6.31*a*)	*C*, *D*
Davidon-Fletcher-Powell (DFP)	6.4.1	$\Delta\mathbf{H} = \Delta\hat{\mathbf{H}}$ $\Delta\mathbf{H}^{-1} = \Delta\hat{\mathbf{H}}^{-1}$ (6.30*b*), (6.31*b*)	*A*, *C*, *D*
BFGS	6.4.1	$\Delta\mathbf{H} = \Delta\hat{\mathbf{H}}$ $\Delta\mathbf{H}^{-1} = \Delta\hat{\mathbf{H}}^{-1}$ (6.32), (6.33)	*A*, *B*, *C*
Davidon	6.4.1	See reference to Davidon (1975)	Uses projections of $\Delta\mathbf{x}$ and $\Delta\mathbf{g}$

A: Minimize in search direction
B: Prespecified step size
C: Trust region
D: Line search other than using Eqs. (6.34) and (6.35)

6.5 FINITE DIFFERENCE APPROXIMATIONS AS SUBSTITUTES FOR DERIVATIVES

Why not replace derivatives with their finite difference approximates in the unconstrained optimization routines and avoid having to calculate derivatives analytically? One consequence of doing so is the cost of extra function evaluations vis-à-vis evaluating analytical derivatives. The main difficulty you experience is that truncation and round-off errors cause the optimization codes to terminate prematurely. Because errors in the approximation of derivatives by finite differences are unknown (if they were known the derivatives could be computed exactly!), in practice you must take care to select an appropriate perturbation step size and finite difference formula for the computer being used.

Truncation errors depend on higher derivatives of the function being differentiated, but bounds on the derivatives are not known a priori. In addition, rounding errors arise because numbers cannot in general be represented exactly in a computer using floating point arithmetic.

As an example of the difficulty, consider the central difference formula for a derivative of a function of one variable

$$\frac{df(x)}{dx} \approx \frac{1}{2h}[f(x_k + h) - f(x_k - h)]$$

The truncation error is $(h^2/6)\, f''(x)$ for $x_k - h < x < x_k + h$. If $f(x_k + h)$ and $f(x^k - h)$ can be represented with an error of $\pm\varepsilon(\varepsilon \approx 10^{-13}$ for a CDC Cyber computer) and h is represented exactly, then the rounding error for the central difference can be as large as ε/h. Hence, as h decreases, the truncation error for the derivative decreases but the rounding error increases.

Suppose it were possible to determine a value for h such that the rounding error and the truncation errors are roughly balanced, as recommended by several authors. Then the computer approximations to the derivatives will be optimal in the sense that the minimum total error is achieved. If the total error is 2 or 3 orders of magnitude less than the value of the various termination criteria employed (see Sec. 6.3.4), then finite difference substitutes will be successful, although yielding algorithms that are slower to converge than those employing analytical derivatives. If the truncation and round-off errors are larger than the convergence parameters, the introduction of substitutes will cause an algorithm to fail to converge or cause it to converge to the wrong solution.

Thus, it is the values of the derivatives, or their approximates, near the solution of the problem, that are of importance. (Keep in mind that the derivatives are approaching zero for unconstrained optimization.) Errors in the derivatives at the start of the minimization have little effect.

Himmelblau and Lindsay (1980) showed that simple finite difference formulas were as good as complex ones (and are more efficient in terms of function evaluations) if the step size is selected properly. What happens if some selection other than the optimal range is used for the step size, h? The more complex formulas are much more robust over a broader range of values of h than are the simpler difference relations. The latter have relatively narrow windows for values of h that yield satisfactory approximates for analytical derivatives. Thus, a user would be safer with respect to introducing undesirable numerical errors if he or she employed the more complex formulas at the expense of extra computations. However, if a single function evaluation takes considerable time, then substantial savings can be achieved by selecting a less complex difference formula with the proper step size, so as to balance truncation and round-off error. What this outcome means is that finite difference substitutes for derivatives can be employed in unconstrained optimization without significant loss of precision, but at the expense of additional function evaluations, and hence computation time. By proper selection of the step size, the penalty paid can be substantially reduced. Furthermore, the use of finite difference substitutes with an effective quasi-Newton algorithm is more efficient than employing a so-called derivative-free algorithm.

Some recommended finite-difference relations and step sizes are as follows (ε = round-off error, a function of the machine precision, which is related to the number of bits available to store a number; for a 64-bit word $\varepsilon \approx 10^{-13}$). The step size except for method 1 below is determined from the following relations where r is the power of h in the truncation error:

$$h > \varepsilon^{1/(r+1)}$$

$$h^r \sim 10^{-8} \text{ to } 10^{-7}$$

Method 1. Forward difference with variable step size

$$\frac{\partial f(\mathbf{x})}{\partial x_j} \approx \frac{f(\mathbf{x} + h\mathbf{e}_j) - f(\mathbf{x})}{h}$$

where $\mathbf{e}_j$ is the jth element in the vector $\mathbf{e}$ whose elements are all zero except for the jth element which is 1. This method was used by Lasdon et al. (1975) in their generalized reduced-gradient code with $h = 10^{-7}|x_j|$ for $|x_j| > 1$ and $h = 10^{-7}$ for $|x_j| < 1$. The forward-difference formula has a truncation error of $(h/2)f''(\mathbf{z})$ for those $\mathbf{x} - h\mathbf{e}_j < \mathbf{z} < \mathbf{x} + h\mathbf{e}_j$ while the rounding error can be as large as $2\varepsilon/h$.

Method 2. Forward difference with fixed step size. The same relation is used as in method 1 but with h fixed ($r = 1$).

Method 3. Central difference

$$\frac{\partial f(\mathbf{x})}{\partial x_j} \approx \frac{f(\mathbf{x} + h\mathbf{e}_j) - f(x - h\mathbf{e}_j)}{2h}$$

In this approximation, the truncation error is $(h^2/6)f^{(3)}(\mathbf{z})$ for those $\mathbf{x} - h\mathbf{e}_j < \mathbf{z} < \mathbf{x} + h\mathbf{e}_j$. Here $r = 2$.

Figure 6.13 illustrates the outcome of minimizing Powell's function (1964)

$$f(x) = (x_1 + 10x_2) + 5(x_3 - x_4)^2 + (x_2 - 2x_3)^4 + 10(x_1 - x_4)^4$$

on a CDC Cyber computer with a forward difference approximation of the analytical derivatives. The dashed line shows the analytical value of $\partial f(\mathbf{x})/\partial x_1$, and the dotted line is the value of the forward difference approximation for $|\partial f(\mathbf{x})/\partial x_1|$. Note how the value of the modulus of the derivative is much greater than the modulus of either the truncation error or the round-off error up to the vicinity of 190 function evaluations, at which point the truncation error has a value of about 10^{-7}. This value is about the same as the value of the derivative itself, hence no further precision can be obtained for the finite-difference substitute in numerical minimization.

What can be said about the merit of substituting finite-difference approximation relations for second partial derivatives? If the first-order partial derivative can be determined analytically, you can approximate second partial derivatives by

$$\frac{\nabla f(\mathbf{x}^k + h_j\mathbf{e}_j) - \nabla f(\mathbf{x}^k)}{h_j}$$

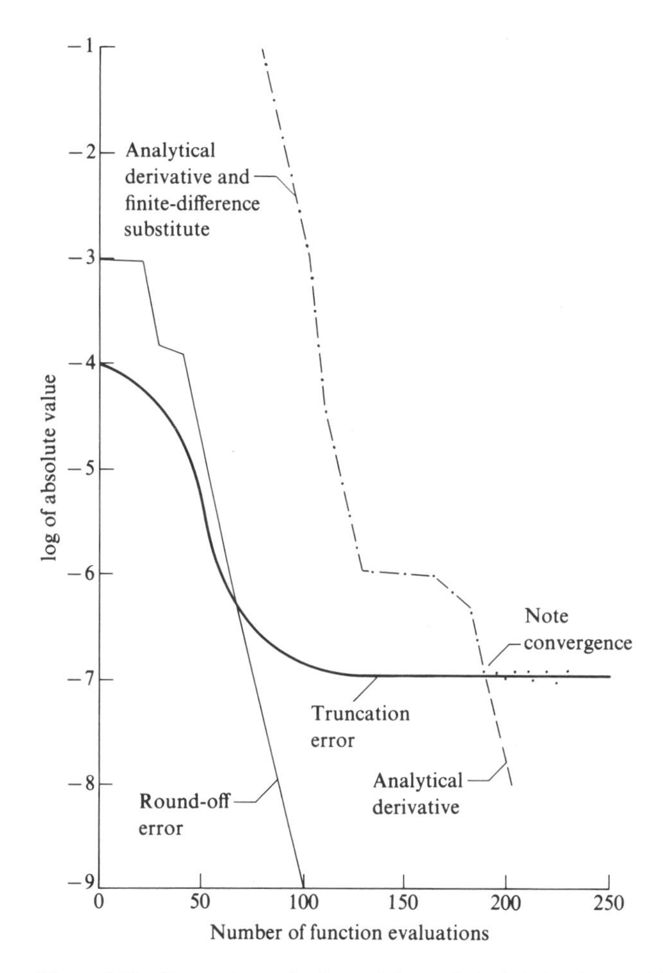

Figure 6.13 Comparison of values of the analytical derivatives and forward difference finite difference substitute for $\partial f(\mathbf{x})/\partial x_1$ for Powell's function, and the estimated truncation and round-off errors. Step length $h = 10^{-7}$.

Experience has shown that difference formulas of function values alone, such as

$$\frac{f(\mathbf{x}^k + h_i \mathbf{e}_i + h_j \mathbf{e}_j) - f(\mathbf{x}^k + h_i \mathbf{e}_i) - f(\mathbf{x}^k + h_j \mathbf{e}_j) + f(\mathbf{x}^k)}{h_i h_j}$$

may be unsatisfactory if $f(\mathbf{x})$ is noisy.

6.6 SOURCES OF COMPUTER CODES FOR UNCONSTRAINED OPTIMIZATION

Recent improvements in unconstrained nonlinear programming codes have made a number of robust codes available at quite reasonable cost. Table 6.4 lists some of the many sources of codes. Keep in mind that computer library sources

Table 6.4 Sources of main-frame computer codes for unconstrained optimization

Code names	Source	Method	Computers	Language	Price
M1FC1	INRIA, BP 105, 78153, Le Chesnay France	Many	IBM 3033, HB 68, CRAY 1, AMDAHL, CDC 6600	FORTRAN IV	Free for academic use; sale otherwise
OPTLIB	Mechanical & Aerospace Eng. Dept., University of Missouri, Columbia, MO 65211	About 10	DEC-10, VAX 11-780 CDC, DG, IBM 370	FORTRAN	Inquire
SYSOPT	Prof. M. Rijckaert, Dept. Chem. Eng., K.U. University of Leuven, B-3030 Heverlee, Belgium	Many	Apple II	BASIC	$200
UNCMIN	R. B. Schnabel, Dept. Computer Science, Univ. Colorado, Boulder CO 80309	Secant plus various combinations of linear search, trust region	CDC, VAX IBM	FORTRAN 66	$100
VA02 To VA14 (12)	Harwell Subroutine library Building 8.9, AERE Harwell, Didcot, Oxon, OX11 ORA, England	Several	CDC, IBM	FORTRAN	$50

Various (refer to ACM index)	ACM Distribution Service, c/o Intl. Math. Statistics Service, 7500 Bellaire Blvd. Houston, TX 77036	Several	CDC, IBM	FORTRAN	\$75
Various (refer to IMSL index)	IMSL, 7500 Bellaire Boulevard Houston, TX 77036. In IMSL Math. Library	Several	Numerous	FORTRAN	\$500
Various (refer to CPC index, Sec. 7.5)	CPC Program Library, Queen's University of Belfast Belfast BT7 1NN, Northern Ireland	Several	Several	FORTRAN	Inquire
Class E04	Numerical Algorithms Group (NAG) Mayfield House, 256 Banbury Rd. Oxford OX2 7DE, England (Similar to National Physics Laboratory)	Several	Numerous	FORTRAN, ALGOL	Inquire
Various (see catalog Sec. P)	National Energy Software Center Argonne National Laboratory 9700 S. Cass Avenue, Argonne, IL 60439	Several	IBM, CDC	FORTRAN	\$1500 for library
Various (see catalog)	Boeing Computer Services Co., Mail Stop 9C-01, P. O. Box 24346, Seattle, WA 98124	Several (based on Harwell Library)	CYBER, CRAY TSO	FORTRAN	Inquire

are more likely to be the most fruitful sources in the long run because of continued updating, but that because of technological lag, they may not contain the most recent algorithms. Berman (1985) lists 4000 articles in his bibliography that can be retrieved by key word indexes and many of these describe codes. Golden and Wasil (1986) describe and evaluate 11 computer codes for nonlinear programming that can be used on microcomputers.

6.7 EVALUATION OF CODES FOR UNCONSTRAINED OPTIMIZATION

Because of the proliferation of algorithms and codes for unconstrained nonlinear programming, a user is bound to be uncertain as to what code to use for a particular problem. Theoretical analysis is of very limited value. As for experimental testing, widespread disagreement exists regarding the performance criteria and the test problems to be used (real-life, artificial, randomly generated, size, degree of nonlinearity, etc.); the ranking and rating procedures to support the decision-making process; how to report and summarize the results; which coding of an algorithm to use; and so on. What makes the situation even worse is the implication underlying comparative studies, namely that the conclusions are valid for a wide range of computers and compilers and for other problems. The effect of the computational environment is mostly ignored. Moré, Garbow, and Hillstrom (1981) and Davidon and Nocedal (1985) have discussed evaluation of unconstrained optimization codes in some detail.

With the bewildering combination of methods, codes, and test problems, no clear-cut rating of computer codes can be provided. The general feeling among practitioners is that the BFGS algorithm, if properly coded, is probably the best one to use. Table 6.5 lists some results of analyses of tests on 15 functions (Himmelblau and Lindsay, 1980) extracted from the literature of test functions, all of small dimension. The plus sign following an algorithm with an additional name designates a particular line search method was used in place of quadratic interpolation.

Any experimental comparison of algorithms depends to a considerable extent on how the algorithms are programmed for the computer. Small details of the programming can exert a considerable influence on the effectiveness of an algorithm. In addition to slight changes in the termination criteria, the unidimensional search technique, tests of matrices for singularity, matrix inversion procedures, and the like make a big difference in the performance of an algorithm. Just a simple change in the initial step in the unidimensional search can be shown to exert quite an impact on the search trajectory.

With these qualifications in mind, the criteria used for Table 6.5 were

1. Robustness—success in obtaining an optimal solution (to within a certain precision)

Table 6.5 Relative rankings of eight algorithms on fifteen problems

Method‡‡	Rank based on time	Method	Rank based on function evaluations
Davidon†,‡	84.2	Davidon†,‡	91.3
BFGS†,‡	81.8	BFGS†,‡	77.6
Davidon†,§	67.4	Davidon†,§	66.8
BFGS + Cullum¶	51.0	Cullum‡	66.5
Cullum¶	48.3	HOMOG (Kowalik)†,‡	61.8
Cullum‡	46.2	BFGS§	54.7
BFGS§	44.5	May†,‡	53.0
Powell††	41.6	BFGS + Cullum¶	50.0
DFP + Stewart¶	40.7	Cullum¶	49.1
BFGS + Curtis Reid¶	30.9	BFGS + Stewart¶	42.0
BFGS + Stewart¶	29.2	Brent†,††	27.4
Brent†,††	26.3	DFP + Stewart¶	26.0
HOMOG (Kowalik)†,‡	22.7	BFGS + Curtis Reid¶	24.5
May†,††	20.9	Powell	21.0

† Solved all the problems.

‡ Using derivatives.

§ Using forward difference, variable h.

¶ Using finite differences.

†† Derivative free.

‡‡ Refer to reference list for citations.

2. Number of function or equivalent function evaluations
3. Central processing time to termination (to within the desired degree of precision)

In order to reduce the mass of data and reach some reasonably comprehensible conclusions, the algorithms were ranked according to their relative execution times, or number of function evaluations, and the rankings averaged over the 15 problems. The ranking scale ranged from 0 (worst) to 100 (best). The best performing algorithm for time and for function evaluations was used as the normalizing one; its time, or number of function evaluations, was divided by the respective values of the poorer performing algorithms, and the result multiplied by 100. If an algorithm failed on a problem, it was assigned an infinite time or number of function evaluations, hence its score on the problem would be zero.

Table 6.6 has been prepared from the number of equivalent function evaluations reported in the study of Le (1985) who tested 20 codes via 18 test functions from a number of starting points. In Table 6.6 the rank is based only on comparing the successful attempts at a solution (failures are not included).

Chapters 10–13 illustrate the application of various of the techniques described in this chapter to chemical engineering problems. Case studies involving heat exchange, curve fitting, and reactors can be found in Secs. 10.3, 11.2, and 13.4.

Table 6.6 Relative rankings of 20 codes based on 18 test problems

Name	Method	Rank based on equivalent function evaluations
L MINN	Conjugate gradient (Le)	100
QDFP	Quasi-Newton Davidon-Fletcher-Powell	80
QSBF	Quasi-Newton, BFGS method self-scaling	66
QOSS	Quasi-Newton-Oren-Spedicato self-scaling	55
QNAG	Quasi-Newton—NAG (1981) library	52
CGQN	Combined conjugate gradient quasi-Newton-Buckley	50
QIBF	Quasi-Newton-BFGS with initial scaling of $\hat{\mathbf{H}}^{(0)}$	46
QIDS	Quasi-Newton-Davidon switch with initial scaling of $\hat{\mathbf{H}}^{(0)}$	37
CPO	Conjugate gradient with Powell restart	35
QBBF	Quasi-Newton-Biggs	34
CFR	Conjugate gradient—Fletcher Reeves	33
BREN	Brent	32
QBFS	Quasi-Newton—BFGS method	31
CPEO	Conjugate gradient with restart-Perry	31
QFS	Quasi-Newton—Fletcher	30
QDS	Quasi-Newton-Davidon switch method	24
CNAG	Conjugate gradient—NAG (1981) library	23
CSBP	Conjugate gradient—Shanno with Beale restart	22
CPR	Conjugate gradient—Polak-Ribiere	16
CPE2	Conjugate gradient—Perry two-step method	15

E. M. L. Beale, "A derivation of conjugate gradients," in: *Numerical methods for nonlinear optimization* (ed. F. A. Lootsma), Academic Press, New York, 1972, pp. 39–43.

M. C. Biggs, "A note on minimization algorithms which make use of nonquadratic properties of the objective function," *J. Inst. Math. Its Appl.*, **12** (1973), pp. 337–338.

R. P. Brent, *Algorithms for minimization without derivatives*, Prentice-Hall, Englewood Cliffs, New Jersey, (1973).

A. G. Buckley, "A combined conjugate gradient quasi-Newton minimization algorithm," *Math. Program.* **15** (1978), pp. 200–210.

W. C. Davidon, "Optimally conditioned optimization algorithms without line searches," *Math. Program.*, **9** (1975), pp. 1–30.

R. Fletcher, "A new approach to variable metric algorithms," *Comput. J.*, **13** (1970), pp. 317–322.

R. Fletcher and M. J. D. Powell, "A rapidly convergent descent method for minimization," *Comput. J.*, **6** (1963), pp. 163–168.

R. Fletcher and C. M. Reeves, "Function minimization by conjugate gradients," *Comput. J.*, **7** (1964), pp. 149–154.

Le, D., "A Fast and Robust Unconstrained Optimization Method Requiring Minimum Storage," *Math. Prog.*, **32**: 41 (1985).

NAG, Numerical Algorithms Group, Inc., *NAG Fortran Library—Mark 8* (Downers Grove, Illinois, 1981).

S. S. Oren and E. Spedicato, "Optimal conditioning of self-scaling variable metric algorithms," *Math. Program.* **10** (1976), pp. 70–90.

A. Perry, "A modified conjugate gradient algorithm." *Oper. Res.*, **26** (1978), pp. 1073–1078.

M. J. D. Powell, "Restart procedures for the conjugate gradient method," *Math. Program.* **12** (1977), pp. 241–254.

G. Ribiere, "Sur la Methode de Davidon-Fletcher-Powell pour la minimisation des fonctions," *Manage. Sci.*, **16** (1970), p. 572.

D. F. Shanno, "Conjugate gradient methods with inexact searches," *Math. Oper. Res.*, **3** (1978), pp. 244–256.

6.8 DIAGNOSIS OF OPTIMIZATION CODE FAILURE TO SOLVE A PROBLEM

When an optimization code known to be reliable fails to solve your problem, you should carry out a series of tests to try to ascertain the cause(s) of and possible remedies for the failure. Table 8.10 in Sec. 8.11 provides a list of symptoms, causes, and remedies that can be of assistance in the diagnosis.

REFERENCES

Armijo, L., "Minimization of Functions Having Lipschitz Continuous First Partial Derivatives," *Pac. J. Math.*, **16**: 1 (1966).

Berman, G., *Nonlinear Optimization Bibliography with Two-level Key-word and Author Indexes*, Univ. of Waterloo Press, Waterloo, Ont. Canada, 1985.

Box, G. E. P., and J. S. Hunter, "A Useful Method for Model Building," *Technometrics* **4**: 301 (1962).

Brent, R. P., *Algorithms for Minimization Without Derivatives*, Prentice-Hall, Englewood Cliffs, New Jersey, 1973.

Broyden, C. G., "Quasi-Newton Methods and Their Application to Function Minimization," *Math. Comput.*, **21**: 368 (1967).

Broyden, C. G., "The Convergence of a Class of Double-Rank Minimization Algorithms," parts I and II, *J. Inst. Math. Appl.*, **6**: 76, 222 (1970).

Broyden, C. G., J. E. Dennis, and J. J. Moré, "On the Local and Superlinear Convergence of Quasi-Newton Methods," *J. Inst. Math. Appl.*, **12**: 223 (1975).

Buckley, A. G., "A Combined Conjugate Gradient Quasi-Newton Algorithm," *Math. Prog.*, **15**: 200 (1978).

Cullum, J., "Unconstrained Minimization of Functions without Explicit Use of Their Derivitives," Report, IBM, T. J. Watson Research Center, Yorktown Heights, New York, 1971.

Curtis, A. R., and J. K. Reid, "The Choice of Step Lengths When Using Differences to Approximate Jacobian Matrices," *J. Inst. Math. Appl.*, **13**: 121 (1974).

Davidon, W. C., "Optimality Conditioned Optimization Algorithms without Line Searches," *Math. Prog.*, **9**: 1 (1975).

Davidon, W. C., and J. Nocedal, "Evaluation of Step Directions in Optimization in Algorithms," *ACM Trans. Math. Software*, **11**: 12 (1985).

Dembo, R. S., S. C. Eisenstat, and T. Steihang, "Inexact Newton Methods," *S.I.A.M. J. Num. Anal.*, **19**: 400 (1982).

Dennis, J. E., and R. B. Schnabel, *Numerical Methods for Unconstrained Optimization and Nonlinear Equations*, Prentice-Hall, Englewood Cliffs, New Jersey (1983).

Dixon, L. C. W., and L. James, "On Stochastic Variable Metric Methods," in *Analysis and Optimization of Stochastic Systems*, (eds. Q. L. R. Jacobs et al.), Academic Press, London (1980).

Fletcher, R., "A New Approach to Variable Metric Algorithms," *Comput. J.*, **13**: 317 (1970).

Fletcher, R., and C. M. Reeves, "Function Minimization by Conjugate Gradients," *Comput. J.*, **7**: 149 (1964).

Fletcher, R., *Practical Methods of Optimization*, vol. 1, John Wiley, New York (1980).

Gander, W., "Least Squares with a Quadratic Constraint," *Num. Math.*, **36**: 291 (1981).

Gay, D. M., "Computing Optimal Locally Constrained Steps," *S.I.A.M. J. Sci. Stat. Comput.*, **2**: 186 (1981).

Gill, P. E., W. Murray, and M. H. Wright, *Practical Optimization*, Academic Press, New York (1981).

Golden, B. L., and E. A. Wasil, "Nonlinear Programming on a Microcomputer," *Computers and Operations Res.*, **13**, p. 149 (1986).

Goldfarb, D., "A Family of Variable Metric Methods Derived by Variational Means," *Math. Comput.*, **24**: 23 (1970).

Goldfarb, D., "Factorized Variable Metric Methods for Unconstrained Optimization," *Math. Comput.*, **30**: 796 (1976).

Griewank, A. O., and P. Toint, "Partitioned Variable Metric Updates for Large Sparse Optimization Problems," *Num. Math.*, **39**: 119 (1982).

Hestenes, M. R., *Conjugate-Direction Methods in Optimization*, Springer Verlag, New York (1980).

Himmelblau, D. M., *Applied Nonlinear Programming*, McGraw-Hill, New York (1972).

Himmelblau, D. M., and J. W. Lindsay, "An Evaluation of Substitute Methods for Derivatives in Unconstrained Optimization," *Oper. Res.*, **28**(II): 668 (1980).

Kowalik, J. S., personal communication, Washington State University, Pullman, Wash., 1976.

Lasdon, L. S., A. D. Warren, A. Jain, and M. W. Ratner, "Design and Testing of a Generalized Reduced Gradient Code for Nonlinear Optimization," Tech. Memo. 353, Dept. Operations Research, Case Western Reserve University, Cleveland, Ohio, March 1975.

Le, D., "A Fast and Robust Unconstrained Optimization Method Requiring Minimum Storage," *Math. Prog.*, **32**: 41 (1985).

Levenberg, K., "A Method for the Solution of Certain Problems in Least Squares," *Q. Appl. Math.*, **2**: 164 (1944).

Marquardt, D., "An Algorithm for Least-Squares Estimation of Nonlinear Parameters," *S.I.A.M. J. Appl. Math.*, **11**: 431 (1963).

May, J. H., "Solving Nonlinear Programs Without Using Analytical Derivatives, Part I," Graduate School of Business, University of Pittsburgh, Reprt WP153, April 1976.

Moré, J. J., B. S. Garbow, and K. E. Hillstrom, "Testing Unconstrained Optimization Software," *ACM Trans. Math. Software*, **7**: 17 (1981).

Moré. J. J., "The Levenburg-Marquardt Algorithm: Implementation and Theory," *Numerican Analysis*, (ed. G. A. Watson), Lecture Notes in Math. 630, Springer-Verlag, Berlin, p. 105 (1977).

Nelder, J. A., and R. Mead, "A Simplex Method for Function Minimization," *The Computer Journal*, **7**: 308 (1965).

Powell, M. J. D., "An Efficient Method for Finding the Minimum of a Function of Several Variables Without Calculating Derivatives," *Computer Journal*, **7**: 155 (1964); 303 (1965).

Powell, M. J. D., "A Hybrid Model for Nonlinear Equations," *Numerical Methods for Nonlinear Algebraic Equations*, (ed. P. Rabinowitz), Gordon & Breach, London, p. 87 (1970*a*).

Powell, M. J. D., "A New Algorithm for Unconstrained Optimization," in *Nonlinear Programming*, (eds. J. B. Rosen, O. L. Mangasarian, and K. Ritter), p. 31, Academic Press, New York (1970*b*).

Shanno, D. F., "Conjugate Gradient Methods with Inexact Searches," *Math. Oper. Res.*, **3**: 244 (1978).

Shanno, D. F., "Conditioning of Quasi-Newton Methods for Function Minimization," *Math. Comput.*, **24**: 647 (1970).

Sorensen, D. C., "Newton's Method with a Model Trust Region Modification," *S.I.A.M. J. Num. Anal.*, **19**: 409 (1982).

Spendley, W., G. R. Hext, and F. R. Himsworth, *Technometrics*, **4**: 44 (1962).

Stewart, G. W., "A Modification of Davidon's Minimization Method to Accept Difference Approximations of Derivatives," *J. Assoc. Comput. Mach.*, **14**: 72 (1967).

Toint, P., "Large-Scale Nonlinear Programming," paper presented at the 1984 NATO Conference on Computational Mathematical Programming, Bad Windsheim, 1984.

SUPPLEMENTARY REFERENCES

Beightler, C. S., D. T. Phillips, and D. G. Wilde, *Foundations of Optimization*, Prentice-Hall, Englewood Cliffs, New Jersey (1979).

Beveridge, G. S., and R. S. Schechter, *Optimization Theory and Practice*, McGraw-Hill, New York, 1970.

Boggs, P. T., and J. E. Dennis, "A Stability Analysis for Perturbed Nonlinear Iterative Methods," *Math. Comput.*, **30**: 1 (1976).
Boggs, P. T., R. H. Byrd, and R. B. Schnabel, *Numerical Optimization*, SIAM, Philadelphia, Philadelphia (1985).
De Haan, L., "Estimation of the Minimum of a Function Using Order Statistics," *J. Amer. Stat. Assoc., Theory and Methods Section*, **76**: 467 (1981).
Dennis, J. E., and R. B. Schnabel, *Numerical Methods for Unconstrained Optimization and Nonlinear Equations*, Prentice-Hall, Englewood Cliffs, New Jersey (1978).
Fletcher, R., *Practical Methods of Optimization*, vol. 1, John Wiley, New York (1980).
Fox, R. L. *Optimization Methods for Engineering Design*, Addison-Wesley, Reading, Massachusetts (1971).
Himmelblau, D. M., *Applied Nonlinear Programming*, Swift Publishing Co., Austin, Texas (1972).
Kan, A. H. G., and G. Th. Timer, "A Stochastic Approach to Global Optimization", in *Numerical Optim.*, 1984 (eds. P. T. Boggs, R. H. Byrd, and R. B. Schnabel), SIAM Philadelphia, 245 (1985).
Kan, A. H. G., C. G. E. Boender, and G. Th. Timer, "A Stochastic Approach to Global Optimization" (ed. K. Schittkowski), in *Comput. Math. Program.*, NATO ASI Series, vol. F15, Springer-Verlag, Berlin, p. 281 (1985).
Kantorovich, L. V., "Functional Analysis and Applied Mathematics," *Uspehi Math. Nauk*, **3**: 89 (1948), translated by C. Benster as N.B.S. Report 1509, Washington, D.C. (1952).
Lootsma, F. A., "Parallel Unconstrained Optimization Methods," REPT-84-30, Delft, The Netherlands (1984). (PB 85-235430/GAR from NTIS).
Lootsma, F. A., "Performance Evaluation of Nonlinear Optimization Methods Via Pairwise Comparison and Fuzzy Numbers," *Math. Prog.*, **33**: 93 (1985).
Murray, W. (ed.), *Numerical Methods for Unconstrained Optimization*, Academic Press, New York (1972).
Nazareth, J. L., "*Conjugate Gradient Methods Less Dependent on Conjugacy*," SIAM Rev., **28**, 501 (1986).
Palosaari, S., et al., *A Random Search Algorithm for Constrained Global Optimization*, Acta Polytech. Scand., Ser. 172, Helsinki (1986).
Powell, M. J. D., "Convergence Properties of Algorithms to Nonlinear Optimization," SIAM Rev., **28**, 487 (1986).
Reklaitis, G. V., A. Ravindran, and K. M. Ragsdel, *Engineering Optimization—Methods and Applications*, John Wiley, New York (1983).
Schittkowski, K., *Computational Mathematical Programming*, Springer-Verlag, Berlin (1985).
Schnabel, R. B., J. E. Koontz, and B. E. Weiss, "A Modular System of Algorithms for Unconstrained Minimization," *ACM Trans. Math. Software*, **11**: 419 (1985).
Shanno, D. F., "Large Scale Unconstrained Optimization," *Comput. Chem. Eng.*, **7**: 569 (1983).
Shanno, D. F., "Globally Convergent Conjugate Gradient Algorithms," *Math. Prog.*, 33: 61 (1985).
Solis, F. J., and R. J. B. Wets, "Minimization by random search techniques," *Math. Oper. Res.*, **6**: 19 (1981).
Van Laarhoven, P. J. M., "Parallel Variable Metric Algorithms for Unconstrained Optimization," *Math. Prog.*, **33**: 68 (1985).

PROBLEMS

6.1. If you want to locate the optimum of an objective function within a final interval of uncertainty of 1.0 unit and start with an initial interval of 10.0 units, the reduction is 1/10. Now suppose that you want to locate the optimum of an objective function of *two* independent variables. If the final interval of uncertainty is still 1.0 unit and the initial interval is 10.0 units for each variable, the required reduction is $(1/10)^2$ or 1/100 of the original region.

Develop a general formula for n variables each of which has the range $b_i - a_i$, $i = 1, 2, \ldots, n$, that gives the fractional reduction in interval if the final interval for each variable is δ_i.

6.2. If you carry out an exhaustive search (i.e., examine each grid point) for the optimum of a function of 5 variables, and each step is 1/20 of the interval for each variable, how many objective function calculations must be made?

6.3. For the objective function

$$f(\mathbf{x}) = (x_1 - 2)^2 + (x_2 - 5)^2$$

use the point (0, 0) as the first guess for $\mathbf{x}$, and show that a univariate search based on $\mathbf{s}_1 = [1 \quad 0]^T$ for the first search direction and $\mathbf{s}_2 = [0 \quad 1]^T$ for the second, will find the optimum of $f(\mathbf{x})$ in one cycle comprised of the two search directions.

6.4. Consider the following minimization problem:

$$\text{Minimize} \quad f(\mathbf{x}) = x_1^2 + x_1 x_2 + x_2^2 + 3x_1$$

(*a*) Find the minimum (or minima) analytically.

(*b*) Are they global or relative minima?

(*c*) Construct four contours of $f(\mathbf{x})$ [lines of constant value of $f(\mathbf{x})$].

(*d*) Would univariate search be a good numerical method for finding the optimum of $f(\mathbf{x})$? Why or why not?

(*e*) Suppose the search direction is given by $\mathbf{s} = [1 \quad 0]^T$. Start at (0, 0), find the optimum point P_1 in that search direction analytically, not numerically. Repeat the exercise for a starting point of (0, 4) to find P_2.

(*f*) Show graphically that a line connecting P_1 and P_2 passes through the optimum. To which algorithm is this procedure analogous?

6.5. Determine a regular simplex figure in a three-dimensional space such that the distance between vertices is 0.2 unit and one vertex is at the point (−1, 2, −2).

6.6. Carry out the four stages of the simplex method to minimize the function

$$f(\mathbf{x}) = x_1^2 + 3x_2^2$$

starting at $\mathbf{x} = [1 \quad 1.5]^T$. Use $\mathbf{x} = [1 \quad 2]^T$ for another corner. Show each stage on a graph.

6.7. A three-dimensional simplex optimal search for a minimum provides the following intermediate results:

x vector			Value of objective function
[0	0	$0]^T$	4
[−4/3	−1/3	$-1/3]^T$	7
[−1/3	−4/3	$-1/3]^T$	10
[−1/3	−1/3	$-4/3]^T$	5

What is the next point to be evaluated in the search? What point is dropped?

6.8. Find a direction orthogonal to the vector

$$\mathbf{s} = \left[\frac{1}{\sqrt{3}} \quad -\frac{1}{\sqrt{3}} \quad -\frac{1}{\sqrt{3}}\right]^T$$

at the point

$$\mathbf{x} = [0 \quad 0 \quad 0]^T$$

Find a direction conjugate to $\mathbf{s}$ with respect to the Hessian matrix of the objective function $f(\mathbf{x}) = x_1 + 2x_2^2 - x_1x_2$ at the same point.

6.9. Given the function $f(\mathbf{x}) = x_1^2 + x_2^2 + 2x_3^2 - x_1x_2$, generate a set of conjugate directions. Carry out two stages of the minimization in the conjugate directions minimizing $f(\mathbf{x})$ in each direction. Did you reach the minimum of $f(\mathbf{x})$? Start at (1, 1).

6.10. For what values of $\mathbf{x}$ are the following directions conjugate for the function $f(\mathbf{x}) = x_1^2 + x_1x_2 + 16x_2^2 + x_3^2 - x_1x_2x_3$?

$$\mathbf{s}^{(1)} = \begin{bmatrix} -\frac{1}{\sqrt{3}} \\ \frac{1}{\sqrt{3}} \\ -\frac{1}{\sqrt{3}} \end{bmatrix} \qquad \mathbf{s}^{(2)} = \begin{bmatrix} -\frac{1}{\sqrt{3}} \\ \frac{2}{\sqrt{3}} \\ 0 \end{bmatrix}$$

6.11. In the minimization of

$$f(\mathbf{x}) = 5x_1^2 + x_2^2 + 2x_1x_2 - 12x_1 - 4x_2 + 8$$

starting at (0, −2), find a search direction $\mathbf{s}$, conjugate to the x_1 axis. Find a second search vector $\mathbf{s}_2$ which is conjugate to $\mathbf{s}_1$.

6.12. List the search directions for Powell's method to minimize $f(\mathbf{x}) = 2x_1^2 + x_2^2 - x_1x_2$, starting from $\mathbf{x}^0 = [2 \quad 2]^T$.

6.13. Because Powell's method uses conjugate directions, can one guarantee to find the minimum of a quadratic objective function of n variables in n steps?

6.14. List the first four search directions for Powell's method to minimize $f(\mathbf{x}) = x_1^2 + \exp(x_1^2 + x_2^2)$, starting at the point $\mathbf{x}^0 = [2 \quad 2]^T$.

6.15. Explain how Powell's method obtains a set of conjugate search directions for a three-dimensional problem. After proceeding in two of the search directions, give the details of determining the third direction. After two stages, are all the search directions conjugate?

6.16. Will Powell's method give the same set of search directions as the Fletcher-Reeves method if you start at the same initial $\mathbf{x}$ vector?

6.17. Evaluate the gradient of the function

$$f(\mathbf{x}) = (x_1 + x_2)^3 x_3 + x_3^2 x_1^2 x_2^2$$

at the point $\mathbf{x} = [1 \quad 1 \quad 1]^T$.

6.18. You wish to minimize

$$f(\mathbf{x}) = 10x_1^2 + x_2^2$$

If you use steepest descent starting at (1, 1), will you reach the optimum in

(*a*) One iteration
(*b*) Two iterations
(*c*) More that two?

Explain.

6.19. Evaluate the gradient of the function

$$f(\mathbf{x}) = e^{x_1 x_2} - 2e^{x_1} + 2e^{x_2} + (x_1 x_2)^2$$

at the point (0, 0).

6.20. Consider minimizing the function $f(\mathbf{x}) = x_1^2 + x_2^2$. Use the formula $\mathbf{x}^{k+1} = \mathbf{x}^k - \lambda \nabla \mathbf{f}(\mathbf{x}^k)$, where λ is chosen to minimize $f(\mathbf{x})$. Show that $\mathbf{x}^{k+1}$ will be the optimum $\mathbf{x}$ after only one iteration. You should be able to optimize $f(\mathbf{x})$ with respect to λ analytically.

Start from $\mathbf{x}^{\text{old}} = \begin{bmatrix} 3 \\ 5 \end{bmatrix}$

6.21. Show that Eq. (6.10) is the correct weighting so that the conjugate gradient method uses conjugate directions.

6.22. Use the Fletcher-Reeves search to find the minimum of the objective function

$$(a) \quad f(\mathbf{x}) = 3x_1^2 + x_2^2$$

$$(b) \quad f(\mathbf{x}) = 4(x_1 - 5)^2 + (x_2 - 6)^2$$

starting at $\mathbf{x}^0 = [1 \quad 1]^T$.

6.23. Discuss the advantages and disadvantages of the following four search methods for the function shown in Fig. P6.23:

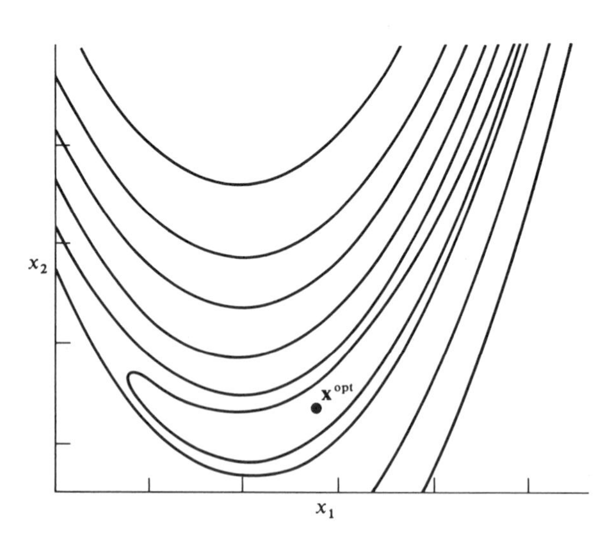

Figure P6.23

(*a*) Fixed-step gradient
(*b*) Steepest descent
(*c*) Conjugate gradient
(*d*) Powell's method

Discuss the basic idea behind each of the four methods (don't write out the individual steps, though).

6.24. Given the function $f(\mathbf{x}) = 3x_1^2 + 3x_2^2 + 3x_3^2$ to minimize, would you expect that steepest descent or Newton's method (in which adjustment of the step length is used for minimization in the search direction) would be faster in solving the problem from the same starting point $\mathbf{x} = [10 \quad 10 \quad 10]^T$. Explain the reasons for your answer.

6.25. Consider the following objective functions:

(*a*) $f(\mathbf{x}) = 1 + x_1 + x_2 + \dfrac{4}{x_1} + \dfrac{9}{x_2}$

(*b*) $f(\mathbf{x}) = (x_1 + 5)^2 + (x_2 + 8)^2 + (x_3 + 7)^2 + 2x_1^2x_2^2 + 4x_1^2x_3^2$

Will Newton's method converge for these functions?

6.26. Consider the minimization of the objective function

$$f(\mathbf{x}) = x_1^3 + x_1x_2 - x_2^2x_1^2$$

by Newton's method starting from the point $\mathbf{x}^0 = [1 \quad 1]^T$. A computer code carefully programmed to execute Newton's method has not been successful. Explain the probable reason(s) for the failure.

6.27. What is the initial direction of search determined by Newton's method for $f(\mathbf{x}) = x_1^2 + 2x_2^2$? What is the step length? How many steps are needed to minimize $f(\mathbf{x})$ analytically?

6.28. Will Newton's method minimize Rosenbrock's function $f(\mathbf{x}) = 100(x_2 - x_1^2)^2 + (1 - x_1)^2$ starting at $\mathbf{x}^0 = [-1.2 \quad 1.0]^T$ in one stage? How many stages will it take if you minimize $f(\mathbf{x})$ exactly on each stage? How many stages if you let the step length be unity on each stage?

6.29. Solve the following problems by Newton's method:

Minimize:

(*a*) $f(\mathbf{x}) = 1 + x_1 + x_2 + x_3 + x_4 + x_1x_2 + x_1x_3 + x_1x_4 + x_2x_3 + x_2x_4 + x_3x_4 + x_1^2 + x_2^2 + x_3^2 + x_4^2$

starting from $\mathbf{x}^0 = [-3 \quad -30 \quad -4 \quad -0.1]^T$ and also $\mathbf{x}^0 = [0.5 \quad 1.0 \quad 8.0 \quad -0.7]^T$.

(*b*) $f(\mathbf{x}) = x_1x_2^2x_3^3x_4^4[\exp - (x_1 + x_2 + x_3 + x_4)]$, starting from $\mathbf{x}^0 = [3 \quad 4 \quad 0.5 \quad 1]^T$.

6.30. How can the inverse of the Hessian matrix for the function $f(\mathbf{x}) = 2x_1^2 - 2x_2^2 - x_1x_2$ be approximated by a positive definite matrix by the method of Marquardt?

6.31. You are to minimize $f(\mathbf{x}) = 2x_1^2 - 4x_1x_2 + x_2^2$.
Is $\mathbf{H}(\mathbf{x})$ positive definite? If not, start at $\mathbf{x}^0 = [2 \quad 2]^T$ and develop an approximation of $\mathbf{H}(\mathbf{x})$ that is positive definite by Marquardt's method.

6.32. Answer whether the following statements are true or false and explain the reasons for your answer.

(*a*) All search methods that are based on conjugate directions (such as Powell's method) always use conjugate directions.

(*b*) The matrix, or its inverse, used in the BFGS method, is an approximation of the Hessian matrix, or its inverse, of the objective function $[\nabla^2 f(\mathbf{x})]$.

(*c*) The BFGS method has the advantage over Newton's method in that the latter requires second derivatives whereas the former requires only first derivatives to get the search direction.

6.33. For each of the secant methods discussed in Sec. 6.4 give the values of the elements of the approximate to the Hessian (inverse Hessian) matrix for the first two stages of search for the following problems:

(*a*) Maximize $f(\mathbf{x}) = -x_1^2 + x_1 - x_2^2 + x_2 + 4$

(*b*) Minimize

$$f(\mathbf{x}) = x_1^3 \exp\left[x_2 - x_1^2 - 10(x_1 - x_2)^2\right]$$

$$f(\mathbf{x}) = x_1^2 + x_2^2 + x_3^2 + x_4^2$$

starting from the point (1, 1) or (1, 1, 1, 1) as the case may be.

6.34. Given the initial matrix $\hat{\mathbf{H}}^0 = \mathbf{I}$, what is $(\hat{\mathbf{H}}^1)^{-1}$ by the Davidon-Fletcher-Powell algorithm if $f(\mathbf{x}) = x_1^2 + 2x_2^2$ and $\mathbf{x}^0 = [1 \quad 1]^T$?

6.35. What is the recursive equation to calculate the approximate Hessian (inverse Hessian) matrix by (*a*) the Davidon-Fletcher-Powell algorithm, (*b*) the BFGS algorithms, (*c*) Broyden's algorithm, if the objective function is to be maximized rather than minimized?

6.36. In using a Taylor series to calculate finite difference substitutes for partial derivatives, the following formulas can be obtained for $\partial f(x)/\partial x$ (single variable)

(*a*) Forward difference $\quad \dfrac{f(x+\delta) - f(x)}{\delta} + O(\delta)$

(*b*) Backward difference $\quad \dfrac{f(x) - f(x-\delta)}{\delta} + O(\delta)$

(*c*) Central difference $\quad \dfrac{f(x+\delta) - f(x-\delta)}{2\delta} + O(\delta)^2$

(*d*) Central difference $\quad \dfrac{f(x+\delta) - 2f(x) + f(x-\delta)}{\delta^2} + O(\delta^2)$

What should be the approximate step size used for each formula if the machine precision is (*a*) $\varepsilon = 10^{-13}$, (*b*) $\varepsilon = 10^{-8}$, and (*c*) $\varepsilon = 10^{-5}$?

6.37. Compare the forward difference formula (see Prob. 6.36) in computing the partial derivatives of $f(\mathbf{x}) = 100x_1^2 + x_2^2$ via (*a*) a hand-held calculator (such as HP or TI), (*b*) a personal computer mathematics code in single precision arithmetic, and (*c*) a mainframe computer. Use various step sizes and test at the points $\mathbf{x} = [1 \quad 1]^T$, $\mathbf{x} = [0.001 \quad 0.001]^T$ and $\mathbf{x} = [10^{-6} \quad 10^{-6}]^T$. What conclusions can you draw?

6.38. Estimate the values of the parameters k_1 and k_2 by minimizing the sum of the squares of the deviations

$$\phi = \sum_{i=1}^{n} (y_{\text{observed}} - y_{\text{predicted}})_i^2$$

where

$$y_{\text{predicted}} = \frac{k_1}{k_1 - k_2}(e^{-k_2 t} - e^{-k_1 t})$$

for the following data:

t	y_{observed}
0.5	0.263
1.0	0.455
1.5	0.548

Plot the sum-of-squares surface with the estimated coefficients.

6.39. Repeat Prob. 6.38 for the following model and data:

$$y = \frac{k_1 x_1}{1 + k_2 x_1 + k_3 x_2}$$

y_{observed}	x_1	x_2
0.126	1	1
0.219	2	1
0.076	1	2
0.126	2	2
0.186	0.1	0

6.40. Approximate the minimum value of the integral

$$\int_0^1 \left[\left(\frac{dy}{dx}\right)^2 - 2yx^2\right] dx$$

subject to the boundary conditions $dy/dx = 0$ at $x = 0$ and $y = 0$ at $x = 1$.

Hint: Assume a trial function $y(x) = a(1 - x)$ that satisfies the boundary conditions and find the value of a that minimizes the integral. Will a more complicated trial function that satisfies the boundary conditions improve the estimate of the minimum of the integral?

6.41. In a decision problem it is desired to minimize the expected risk defined as follows:

$$\mathscr{E}\{\text{risk}\} = (1 - P)c_1[1 - F(b)] + Pc_2\theta\left(\frac{b}{2} + \frac{2\pi}{4}\right)\mathrm{F}\left(\frac{b}{2} - \frac{\sqrt{2\pi}}{4}\right)$$

where $F(b) = \int_{-\infty}^{b} e^{-u^2/2\theta^2}\, du$ (normal probability function)

$$c_1 = 1.25 \times 10^5$$

$$c_2 = 15$$

$$\theta = 2000$$

$$P = 0.25$$

Find the minimum expected risk and b.

6.42. The function

$$f(\mathbf{x}) = (1 + 8x_1 - 7x_1^2 + \tfrac{7}{3}x_1^3 - \tfrac{1}{4}x_1^4)(x_2^2 e^{-x_2})F(\mathbf{x}_3)$$

has two maxima and a saddle point. For (*a*) $F(\mathbf{x}_3) = 1$ and (*b*) $F(\mathbf{x}_3) = x_3 e^{-(x_3+1)}$, locate the global optimum by a search technique.

Answer: (*a*) $\mathbf{x}^* = [4 \quad 2]^T$ and (*b*) $\mathbf{x}^* = [4 \quad 2 \quad 1]^T$.

6.43. By starting with (*a*) $\mathbf{x}^0 = [2 \quad 1]^T$ and (*b*) $\mathbf{x}^0 = [2 \quad 1 \quad 1]^T$, can you reach the solution for Prob. 6.42? Repeat for (*a*) $\mathbf{x}^0 = [2 \quad 2]^T$ and (*b*) $\mathbf{x}^0 = [2 \quad 2 \quad 1]^T$.

Hint: $[2 \quad 2 \quad 1]$ is a saddle point.

6.44. Estimate the coefficients in the correlation

$$y = ax_1^{b_1}x_2^{b_2}$$

from the following experimental data by minimizing the sum of the square of the deviations between the experimental and predicted values of y.

y_{exptl}	x_1	x_2
46.5	2.0	36.0
591	6.0	8.0
1285	9.0	3.0
36.8	2.5	6.25
241	4.5	7.84
1075	9.5	1.44
1024	8.0	4.0
151	4.0	7.0
80	3.0	9.0
485	7.0	2.0
632	6.5	5.0

6.45. The cost of refined oil when shipped via the Malacca Straits to Japan in dollars per kiloliter was given as the linear sum of the crude oil cost, the insurance, customs, freight cost for the oil, loading and unloading cost, sea berth cost, submarine pipe cost, storage cost, tank area cost, refining cost, and freight cost of products as[1]

$$c = c_c + c_i + c_x + \frac{2.09 \times 10^4 t^{-0.3017}}{360} + \frac{1.064 \times 10^6 a t^{0.4925}}{52.47q(360)}$$

$$+ \frac{4.242 \times 10^4 a t^{0.7952} + 1.813 ip(nt + 1.2q)^{0.861}}{52.47q(360)}$$

$$+ \frac{4.25 \times 10^3 a(nt + 1.2q)}{52.47q(360)} + \frac{5.042 \times 10^3 q^{-0.1899}}{360}$$

$$+ \frac{0.1049 q^{0.671}}{360}$$

where a = annual fixed charges, fraction (0.20)
c_c = crude oil price, \$/kL (12.50)
c_i = insurance cost, \$/kL (0.50)
c_x = customs cost, \$/kL (0.90)
i = interest rate (0.10)
n = number of ports (2)
p = land price, \$/m^2 (7000)
q = refinery capacity, bbl/day
t = tanker size, kL

Given the values indicated in parentheses, use a computer code to compute the minimum cost of oil and the optimum tanker size and refinery size by Newton's method and the secant methods (note that 1 kL = 6.29 bbl).

(The answers in the reference were

$$t = 427{,}000 \text{ dwt} \approx 485{,}000 \text{ kL}$$

$$q = 185{,}000 \text{ bbl/day})$$

[1] T. Uchiyama, *Hydrocarbon Process*, **47**(12): 85 (1968).

CHAPTER 7

LINEAR PROGRAMMING AND APPLICATIONS

Linear programming (LP) is one of the most widely used optimization techniques and one of the most effective. The term linear programming was coined by George Dantzig in 1947 to refer to the procedure of optimization in problems in which both the objective function and the constraints are linear (Dantzig, 1963). "Programming" does not specifically require computer coding, but you will find that the solution of almost all practical linear programming problems does involve the use of a computer code. Examples of LP problems which occur in plant management are:

1. Assign employees to schedules so that the work force is adequate each day of the week and worker satisfaction and productivity are as high as possible.
2. Select products to manufacture in the upcoming period, taking best advantage of existing resources and current prices to yield maximum profit.
3. Find a pattern of distribution from plants to warehouses that will minimize costs within the capacity limitations.
4. Submit bids on procurement contracts to take into account profit, competitors' bids, and operating constraints.

When stated mathematically, each of these problems potentially involves many variables, many equations, and inequalities. A solution must not only satisfy all of the equations, but also must achieve an extremum of the objective function, such as maximizing profit or minimizing cost. With the aid of computer codes you can solve LP problems with hundreds and even thousands of variables and constraints.

Linear programming (LP) problems are a type of convex programming problem, where the objective function is convex and the linear constraints form a convex set (see Chap. 4). This means that a local optimum will be a global optimum. Linear programming problems also exhibit the special characteristic that the optimal solution of the problem must lie on some constraint or at the intersection of several constraints, and not in the interior of the convex region where the inequality constraints can be satisfied. Examine Fig. 7.1*a* in which an unconstrained linear objective function $f(\mathbf{x}) = x_1 + x_2$ is plotted. The first derivatives $(\partial f/\partial x_1, \partial f/\partial x_2)$ for $f(\mathbf{x})$ can never be zero, so that for a linear objective function, no finite maximum can exist. However, suppose you add linear inequality constraints to the maximization problem, namely $x_1 \leq 2$ and $x_1 \leq 1$. Now by inspection we see that a finite maximum occurs at $x_1 = 2$ and $x_2 = 1$; note that this maximum lies at the intersection of the two constraints. If we remove either constraint, then f does not have a finite optimum. We conclude that to have a meaningful (well-posed) optimization problem, constraints must be introduced.

Suppose that an equality constraint is added to the problem stated above, namely,

$$2x_1 + x_2 = 2$$

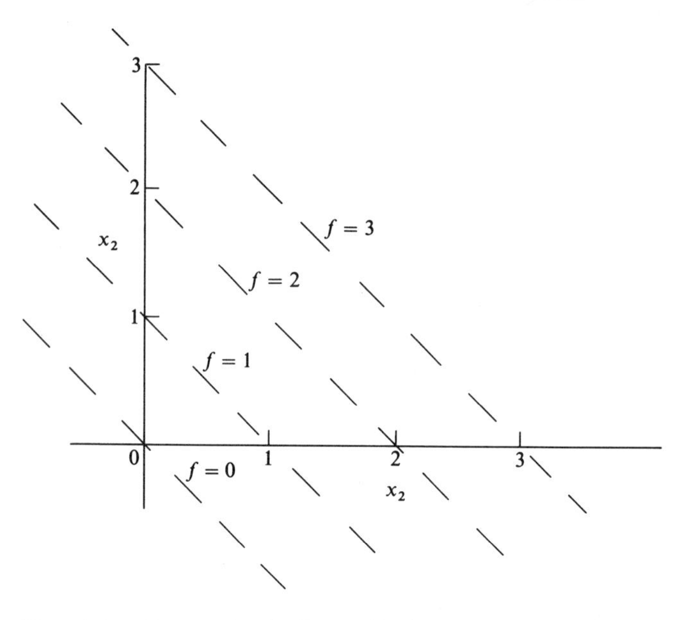

Figure 7.1a Contours for the linear objective function, $f = x_1 + x_2$.

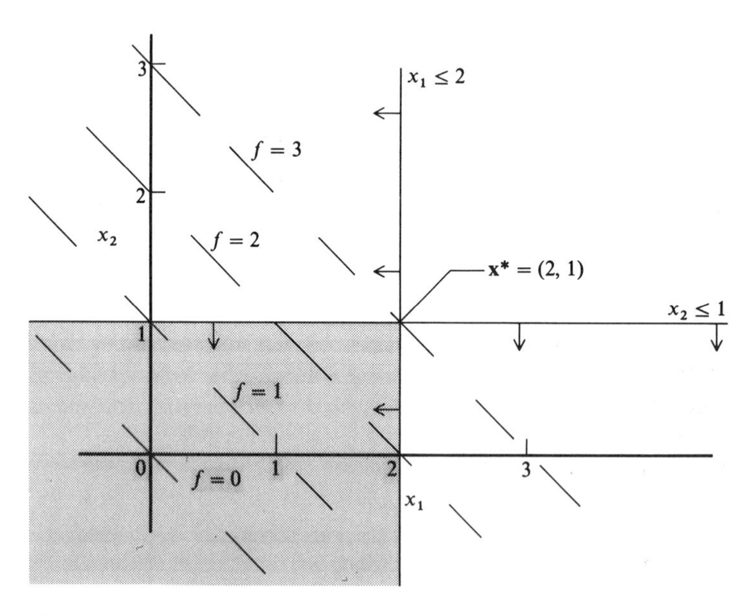

Figure 7.1b Addition of linear inequality constraints $x_1 \leq 2$, $x_2 \leq 1$ (arrows point to the feasible region which is shaded).

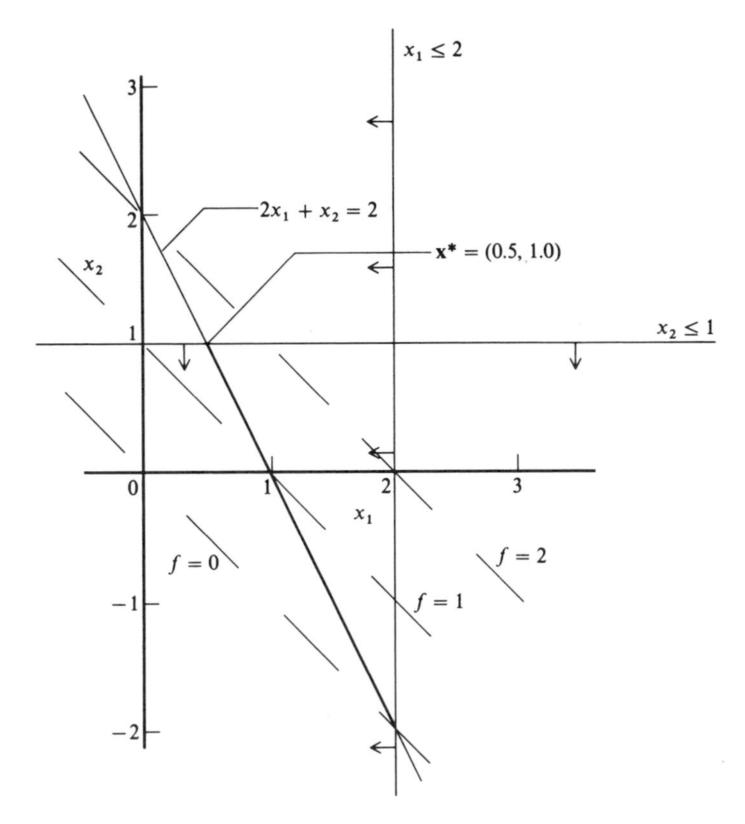

Figure 7.1*c* Addition of an inequality constraint $2x_1 + x_2 = 2$ (bold line indicates which points satisfy all constraints).

Figure 7.1*c* illustrates that the optimum value of f occurs at the intersection of the line $x_2 = 1$ and the equality constraint. At this point $x_1^* = 0.5$ and $f^* = 1.5$; note that the constraint $x_1 \leq 2$ is no longer binding as in the previous case. However, the optimum again appears at the intersection of two constraints.

The fact that the extremum in the solution of an LP problem lies on the intersection of constraints was recognized many years ago, and serves as the basis for a whole class of linear programming algorithms. A general linear programming problem can be stated as follows [multiply $f(\mathbf{x})$ by -1 for maximization]:

$$\text{Minimize} \quad f(\mathbf{x}) = \sum_{i=1}^{r} c_1 x_i = \mathbf{c}^T\mathbf{x} \tag{7.1}$$

$$\text{Subject to} \quad x_i \geq 0 \qquad i = 1, 2, \ldots, r \text{ (all } r \text{ of the variables are nonnegative)} \tag{7.2}$$

and

$$\sum_{i=1}^{r} a_{ji} x_i = b_j \qquad j = 1, 2, \ldots, m \tag{7.3}$$

or

$$\mathbf{A}_1\mathbf{x} = \mathbf{b}_1$$

and

$$\sum_{i=1}^{r} a_{ji}x_i \geq b_j \qquad j = m+1, \ldots, p \tag{7.4}$$

or

$$\mathbf{A}_2\mathbf{x} = \mathbf{b}_2$$

Hence there are r variables with r nonnegativity restrictions, $(p - m)$ inequality constraints, and m equality constraints.

We will use the observation that the optimum lies at the intersection of constraints; therefore we could move along constraints which improve the value of the objective function. This strategy is essentially the basis for the LP algorithm presented later in this chapter.

7.1 BASIC CONCEPTS IN LINEAR PROGRAMMING

Let us initiate the discussion of how to solve a linear programming problem by analyzing a detailed example. We will consider a simple version of a refinery blending and production problem, the solution for which can be illustrated via graphical techniques. Figure 7.2 is a schematic of feedstocks and products for the refinery (costs and selling prices are given in parentheses). Table 7.1 lists the information pertaining to the expected yields of the two types of crudes when processed by the refinery. Note that the product distribution from the refinery is quite different for the two crudes. Table 7.1 also lists the limitations on the established markets for the various products in terms of the allowed maximum daily production. In addition, processing costs are given.

To set up the complete linear programming problem, you must (1) formulate an objective function, and (2) formulate the constraints for the refinery operation. You can see from Fig. 7.2 that six variables are involved, namely the flow rates of each of the two raw materials and the four products.

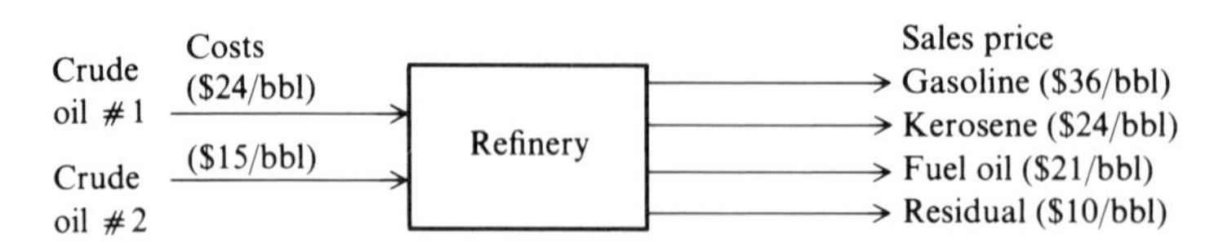

Figure 7.2 Refinery input and output schematic.

Table 7.1 Data for the refinery inputs and products

	Volume percent yield		Maximum allowable production bbl/day
	Crude #1	Crude #2	
Gasoline	80	44	24,000
Kerosene	5	10	2,000
Fuel oil	10	36	6,000
Residual	5	10	
Processing cost ($/bbl)	0.50	1.00	

Objective Function

Let the variables be:

$$x_1 = \text{bbl/day of crude \#1}$$
$$x_2 = \text{bbl/day of crude \#2}$$
$$x_3 = \text{bbl/day of gasoline}$$
$$x_4 = \text{bbl/day of kerosene}$$
$$x_5 = \text{bbl/day of fuel oil}$$
$$x_6 = \text{bbl/day of residual}$$

A typical linear objective function (to be maximized) is the difference between income and costs, namely profit:

$$\text{Maximize } f(\mathbf{x}) = \text{profit} = \text{income} - \text{raw material cost} - \text{processing cost} \tag{7.5}$$

$$\left.\begin{aligned} \text{Income} &= 36x_3 + 24x_4 + 21x_5 + 10x_6 \\ \text{Raw material cost} &= 24x_1 + 15x_2 \\ \text{Processing cost} &= 0.5x_1 + x_2 \end{aligned}\right\} \quad \text{(all in dollars per day)} \tag{7.6}$$

Constraints

Using the yield data, we can write four linear *equality constraints* (material balances) relating x_1 through x_6:

$$\text{Gasoline:} \quad 0.80x_1 + 0.44x_2 = x_3 \tag{7.7}$$
$$\text{Kerosene:} \quad 0.50x_1 + 0.10x_2 = x_4 \tag{7.8}$$
$$\text{Fuel oil:} \quad 0.10x_1 + 36x_2 = x_5 \tag{7.9}$$
$$\text{Residual;} \quad 0.05x_1 + 0.10x_2 = x_6 \tag{7.10}$$

The dimensionality of the problem can be reduced by eliminating the variables x_3, x_4, x_5, and x_6 via the equality constraints, leaving just two independent variables, x_1 and x_2. Introduction of (7.7) to (7.10) into the income function gives

$$\begin{aligned}\text{Total income} = &\quad (36)(0.80x_1 + 0.44x_2)\\ &+ (24)(0.50x_1 + 0.10x_2)\\ &+ (21)(0.10x_1 + 0.36x_2)\\ &+ (10)(0.05x_1 + 0.10x_2)\\ = &\quad \$(32.6x_1 + 26.8x_2)/\text{day}\end{aligned}$$

and the objective function (profit) to be maximized after subtracting raw material and processing costs is

$$f(\mathbf{x}) = 8.1x_1 + 10.8x_2 \tag{7.11}$$

Alternatively, you could express the objective function as minimize $(-8.1x_1 - 10.8x_2)$, but this step is not necessary in this illustration.

What other constraints exist or are implied in this problem? Table 7.1 lists certain restrictions on the x's in terms of production limits. You can formulate these as *inequality constraints*:

$$A.\ \text{Gasoline:}\quad x_3 \leq 24{,}000;\quad 0.80x_1 + 0.44x_2 \leq 24{,}000 \tag{7.12}$$

$$B.\ \text{Kerosene:}\quad x_4 \leq 2{,}000;\quad 0.05x_1 + 0.10x_2 \leq 2{,}000 \tag{7.13}$$

$$C.\ \text{Fuel oil:}\quad x_5 \leq 6{,}000;\quad 0.10x_1 + 0.36x_2 \leq 6{,}000 \tag{7.14}$$

One other set of constraints, although not explicitly stated in the formulation of the problem, is composed of the nonnegativity restrictions on x_1 and x_2, the independent variables:

$$\begin{aligned} x_1 &\geq 0\\ x_2 &\geq 0 \end{aligned} \tag{7.15}$$

The independent variables must be zero or positive because it is meaningless to have negative production rates. Many chemical and physical variables are by definition positive quantities, e.g., absolute pressure, concentration, absolute temperature. If for some reasons negative valued variables must be allowed, then new nonnegative variables which are linear transformations of the old variables must be used in the LP problem.

The formal statement of the linear programming problem is now complete, consisting of Eqs. (7.11) to (7.15). We now proceed to solve the LP problem. A graphical illustration of the solution procedure will be used, consisting of three steps:

1. Plot the constraints on the x_1-x_2 plane.
2. Determine the feasible region for x_1 and x_2 (those values of x_1 and x_2 that satisfy all of the constraints).

3. Find the point along the boundary of the feasible region that maximizes the objective function $f(\mathbf{x})$ by examining the constraint intersections; the slope of the objective function $f(\mathbf{x}) = e$, where e is a parameter indicating different values of f, can indicate which intersections are most favorable.

Figure 7.3 delineates the feasible region for the example problem by the bounds of constraints A, B, and C. The shaded area, the feasible region that satisfies all five constraints, is convex. Figure 7.4 is an enlargement of the feasible region on which contours of the profit function are superimposed. As x_1 and x_2 become large, the profit function of course increases; however, unlimited production levels are not feasible given the constraints of the problem. Therefore you must find the maximum profit level which also satisfies the constraints. Suppose you start at $x_1 = x_2 = 0$, moving along the constraint $x_2 = 0$ (along the x_1 axis). We can improve the objective function up to the intersection of constraint A with $x_2 = 0$. Next we move along constraint A, noting that the profit increases along this path also.

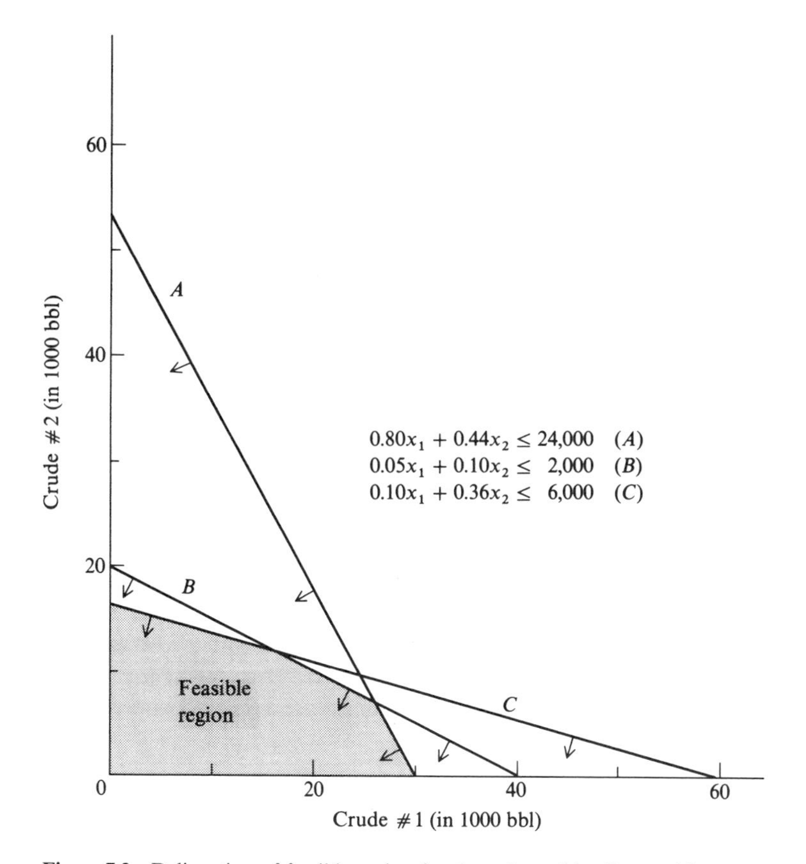

Figure 7.3 Delineation of feasible region for the refinery blending problem.

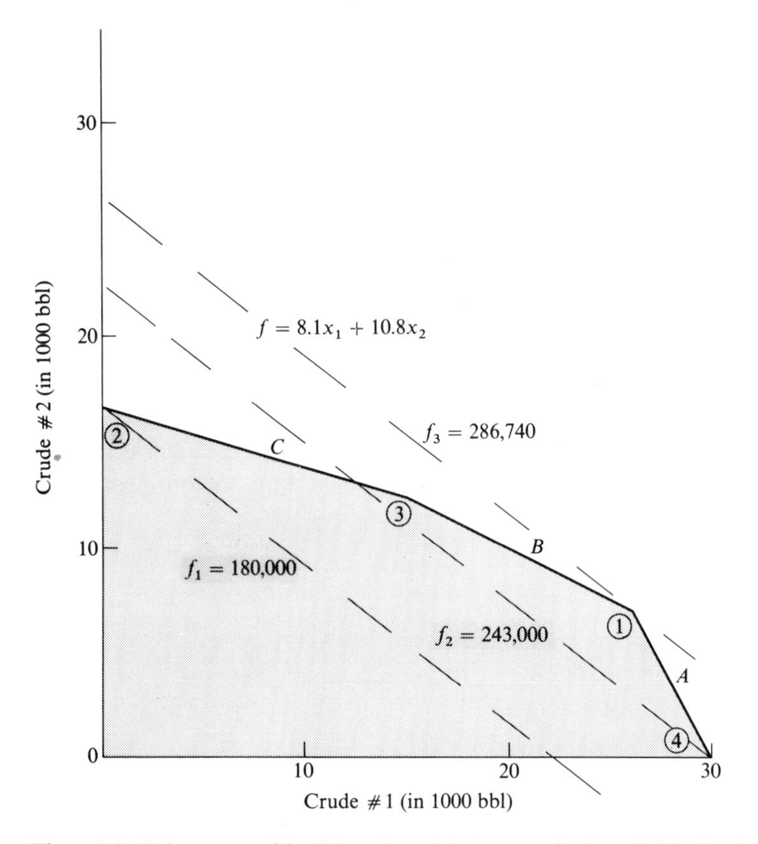

Figure 7.4 Enlargement of feasible region with parameterization of objective function.

It is helpful to examine the profit function to see which directions yield an improved value of f. Figure 7.4 identifies three lines corresponding to three different profit levels; in each case the equation for the profit contour is $f = 8.1x_1 + 10.8x_2$ and each line identifies possible values of x_1 and x_2 which give this profit. Note that it is easy to visualize from the contours which movements along the constraints yield improved values of f. In Fig. 7.4 the optimum occurs at a corner of the feasible region, with profit $= f_3$. With some experience in graphical LP solution, you can usually find the optimum corner by inspection after constructing the feasible region and drawing only one profit contour.

Regardless of how the feasible region is constructed and what slope the profit line has, the unique maximizing (or minimizing) state for $\mathbf{x}$ will always be at a corner of the feasible region for a well-posed problem. For a problem with two variables, the optimum will always occur at the intersection of two or more constraint bounds. These intersecting constraint bounds are said to be "active" if their value is 0. As discussed later, this idea can be generalized to n variables; in the n-dimensional case the optimum will lie at the intersection of the bounds of n or more different inequality constraints.

In the refinery blending problem, using Fig. 7.4, note the optimum $\mathbf{x}(\mathbf{x}^*)$ occurs roughly at

$$x_1^* \approx 26{,}000$$
$$x_2^* \approx 7{,}000$$

Because $\mathbf{x}^*$ is at the intersection of two constraints A and B, exact values of x_1 and x_2 can be found by solving inequalities (7.12) and (7.13) as equalities simultaneously

$$0.80x_1 + 0.44x_2 = 24{,}000$$
$$0.05x_1 + 0.10x_2 = 2{,}000$$

to get

(1) $x_1^* = 26{,}200 \qquad x_2^* = 6{,}900 \qquad f(\mathbf{x}^*) = 8.1x_1^* + 10.8x_2^*$

$$= \$286{,}700$$

Let us investigate the profit at other corners of the feasible region:

(2) $(0, 16{,}667)$ $\quad f(\mathbf{x}) = \$180{,}000$ $\quad$ Intersection C and $x_1 = 0$

(3) $(15{,}000, 12{,}500)$ $\quad f(\mathbf{x}) = \$256{,}500$ $\quad$ Intersection B and C

(4) $(30{,}000, 0)$ $\quad f(\mathbf{x}) = \$243{,}000$ $\quad$ Intersection A and $x_2 = 0$

Therefore corner 1 is indeed the maximizing point. Note that the intersection of A and B corresponds to maximum production of gasoline and kerosene, but below the maximum production of fuel oil.

7.2 DEGENERATE LP's—GRAPHICAL SOLUTION

If the linear programming problem is not properly posed, it may not have a unique or even finite solution. Such LP problems are termed degenerate.

(1) *A nonunique solution*
Figure 7.5 shows a feasible region with respect to four inequality constraints and contours of $f(\mathbf{x})$ for the following problem:

$$\text{Maximize} \quad f(\mathbf{x}) = 2x_1 + 0.5x_2$$
$$\text{Subject to} \quad 6x_1 + 5x_2 \leq 30 \quad (A)$$
$$4x_1 + x_2 \leq 12 \quad (B)$$
$$x_1, x_2 \geq 0$$

The objective function contours and constraint bound B are parallel, indicating that the objective function and the constraint B are not linearly independent (refer to Appendix B). Hence a unique solution cannot be found.

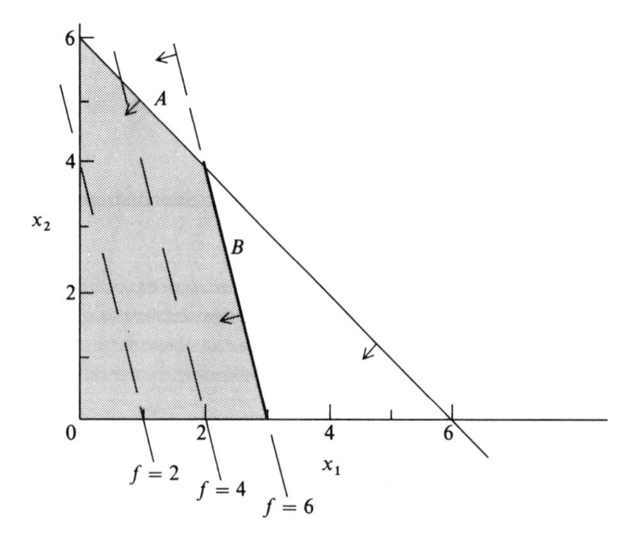

Figure 7.5 Occurrence of a nonunique optimum (shown by heavy line). A and B are inequality constraints.

Other degenerate cases include

(2) *An unbounded optimum* (*see Fig. 7.6*)

$$\begin{aligned} &\text{Minimize} \quad f(x) = -x_1 - x_2 \\ &\text{Subject to} \quad 3x_1 - x_2 \geq 0 \quad (A) \\ &\qquad\qquad\quad x_2 \leq 3 \qquad\quad (B) \\ &\qquad\qquad\quad x_1, x_2 \geq 0 \end{aligned}$$

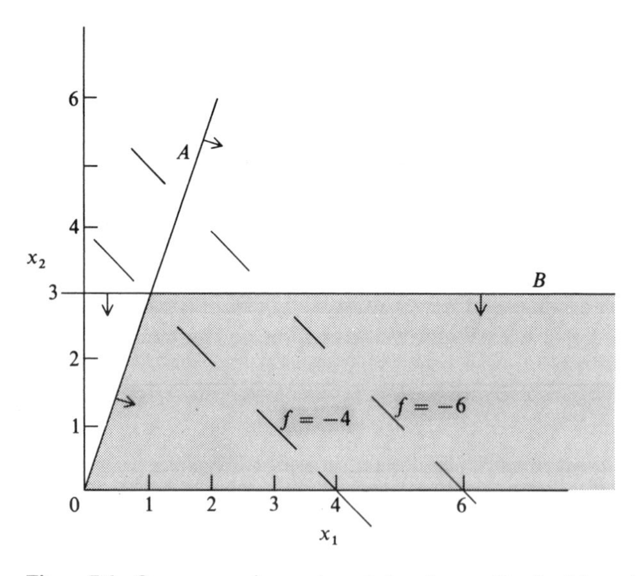

Figure 7.6 Occurrence of an unbounded optimum (the feasible region is partially bounded).

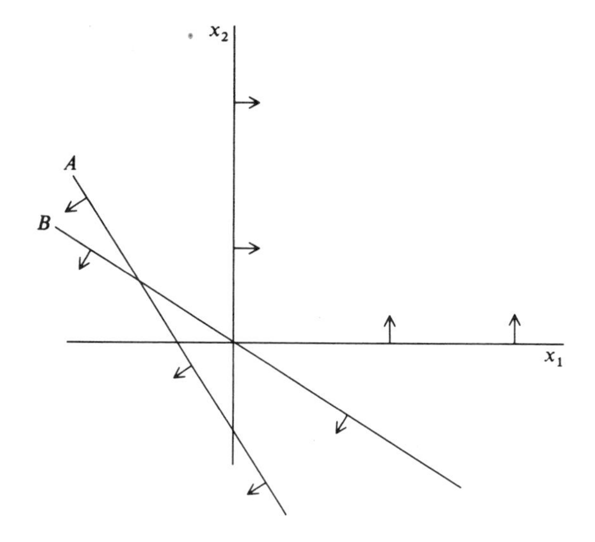

Figure 7.7 No feasible region exists. The arrows on the constraint show the direction of feasibility.

Note that x_1 is unbounded for positive values, hence $f(\mathbf{x})$ can decrease without bound.

(3) *No feasible region exists* (*see Fig. 7.7*)

$$\begin{aligned} \text{Minimize} \quad & f(\mathbf{x}) = -x_1 - x_2 \\ \text{Subject to} \quad & x_1 + x_2 \le -2 \quad (A) \\ & x_1 + 2x_2 \le 0 \quad (B) \\ & x_1, x_2 \ge 0 \end{aligned}$$

There is no solution to problem (3) because no feasible region exists for the constraints given.

7.3 NATURAL OCCURRENCE OF LINEAR CONSTRAINTS

Before taking up the strategy to solve linear programming problems numerically, several general comments need to be made about the occurrence of linear inequality constraints in industrial processes. For chemical plants several different kinds of restrictions may be present.

1. Production limitations arise because of equipment throughput restrictions, storage limitations, or market constraints (no additional product can be sold beyond some specific level); see Eqs. (7.12) to (7.14).
2. Raw material limitations occur because of limitations in feedstock supplies; these supplies often are determined by production levels of other plants within the same company. In the refinery example, you could have introduced a constraint on the feedstocks available.

3. Safety or operability restrictions exist because of limitations on allowable operating temperatures, pressures, and flowrates.

Linear equality constraints exist as well:

1. Material and energy balances exist because of the basic conservation laws. In many cases the material balance can be expressed as a yield matrix (see Table 7.1) which is linear in form.
2. Physical property specifications on products must be considered. In refineries we require that the vapor pressure or octane level of fuel products must satisfy some specification. For blends of various products, you usually assume that a composite property can be calculated through the averaging of pure component physical properties. For N components with physical property values V_i and volume fraction y_i, the average property $\bar{V}$ is

$$\bar{V} = \sum_{i=1}^{N} V_i y_i$$

Note that the above expression is linear.

EXAMPLE 7.1 FORMULATION OF A LINEAR INEQUALITY CONSTRAINT FOR BLENDING

Suppose three intermediates (light naphtha, heavy naphtha, and "catalytic" oil) made in a refinery are to be blended to produce an aviation fuel. The octane number of the fuel must be at least 95. The octane numbers for the three intermediates are as follows:

	Amount blended, bbl/day	Octane no.
Light naphtha	x_1	92
Heavy naphtha	x_2	86
Catalytic oil	x_3	97

Write an inequality constraint for the octane mumber of the aviation fuel assuming a linear mixing rule.

Solution. Assume the material balance can be based on conservation of volume (as well as mass). The production rate of aviation gas is $x_4 = x_1 + x_2 + x_3$. The volume-average octane number of the gasoline can be computed as

$$\frac{x_1}{x_1 + x_2 + x_3}(92) + \frac{x_2}{x_1 + x_2 + x_3}(86) + \frac{x_3}{x_1 + x_2 + x_3}(97) \geq 95 \qquad (a)$$

Multiplying Eq. (*a*) by $(x_1 + x_2 + x_3)$ and rearranging we get

$$-3x_1 - 9x_2 + 2x_3 \geq 0 \qquad (b)$$

This constraint ensures that the octane number specification is satisfied.

Because most LP problems will involve more than two variables, a method more versatile than graphical analysis must be employed to obtain the optimum. However, the understanding and analysis of an LP problem on a two-dimensional plot is very helpful in understanding how you ought to treat a higher dimensional problem. In formulating a numerical approach, we shall use the fact that the optimum for a linear programming problem always lies at a corner of the feasible region. An efficient search technique must be used to progress from one corner to the next; in so doing the objective function is continually improved.

7.4 THE SIMPLEX METHOD OF SOLVING LINEAR PROGRAMMING PROBLEMS

In 1947 George Dantzig first advanced a general analytical procedure for handling large-dimensional linear programming problems. The iterative procedure employed, called the Simplex algorithm (not to be confused with the simplex method explained in Sec. 6.1.4) seeks to improve the objective function by considering the value of $f(\mathbf{x})$ at one constraint intersection after another (Dantzig, 1963). The iterations are designed so that the value of the objective function is always being improved. We shall describe the steps for solving the general linear programming problem and then examine how the procedure works for a function of two variables, illustrating the operations via a graph.

The general procedure presented requires some understanding of matrix operations. If you are not familiar with these operations, refer to Appendix B for a general summary of the concepts involved. Keep in mind that a properly posed linear programming problem has a convex objective function and concave constraints (that form a convex region), hence it will have a unique solution.

The simplex algorithm entails the definition of additional variables that are introduced into the inequality constraints in the problem. These variables, called *surplus or slack variables* depending on the direction of the inequality, convert the inequalities to equality constraints. Given a general inequality constraint

$$\sum_{i=1}^{r} a_{ji}x_i \geq b_j \qquad (b_j \geq 0) \qquad (7.16)$$

it can be converted to an equality constraint by using a *surplus* variable $s_j \geq 0$ such that

$$\sum_{i=1}^{r} a_{ji}x_i - s_j = b_j \qquad (7.17)$$

If the surplus variable is zero, the jth constraint is at its bound (binding). If $s_j > 0$, the constraint is not binding (is inactive). Sometimes constraints are expressed as

$$\sum_{i=1}^{r} a_{ji}x_i \leq b_j \qquad (b_j \geq 0) \tag{7.18}$$

In this case a nonnegative *slack* variable is added, i.e.,

$$\sum_{i=1}^{r} a_{ji}x_i + \mathrm{s_j} = \mathrm{b_j} \tag{7.19}$$

The sample problem to be used for illustration of the Simplex algorithm is

$$\text{Minimize} \quad f = -x_1 + x_2 \tag{7.20a}$$

$$\text{Subject to} \quad 2x_1 - x_2 \geq -2 \quad (A) \tag{7.20b}$$

$$-x_1 + 3x_2 \geq -2 \quad (B) \tag{7.20c}$$

$$-x_1 - x_2 \geq -4 \quad (C) \tag{7.20d}$$

$$x_1 \geq 0, x_2 \geq 0$$

The LP problem is shown in Fig. 7.8. The optimum lies at the intersection of constraints B and C (corner 2 of the feasible region). In the steps below we show how the corners of the feasible region are identified and evaluated relative to minimization of the objective function.

Step 0 Convert all inequality constraints to the form in which the right-hand sides are positive. All three inequality constraints have negative values on the right-hand side so that each constraint must be multiplied by (-1):

$$-2x_1 + x_2 \leq 2 \tag{7.21a}$$

$$x_1 - 3x_2 \leq 2 \tag{7.21b}$$

$$x_1 + x_2 \leq 4 \tag{7.21c}$$

Step 1 Introduce slack/surplus variables and convert the inequality constraints to equality constraints. Let the slack variables be x_3, x_4, and x_5, the inequality constaints, (7.21) become respectively

$$-2x_1 + x_2 + x_3 = 2 \quad (A) \tag{7.22a}$$

$$x_1 - 3x_2 + x_4 = 2 \quad (B) \tag{7.22b}$$

$$x_1 + x_2 + x_5 = 4 \quad (C) \tag{7.22c}$$

Note that all variables (including slack variables) must be nonnegative. The optimization problem (augmented) now requires that you find the values of x_i $(i = 1, 5)$ that minimize $f(\mathbf{x}) = -x_1 + x_2$. This minimization must be performed

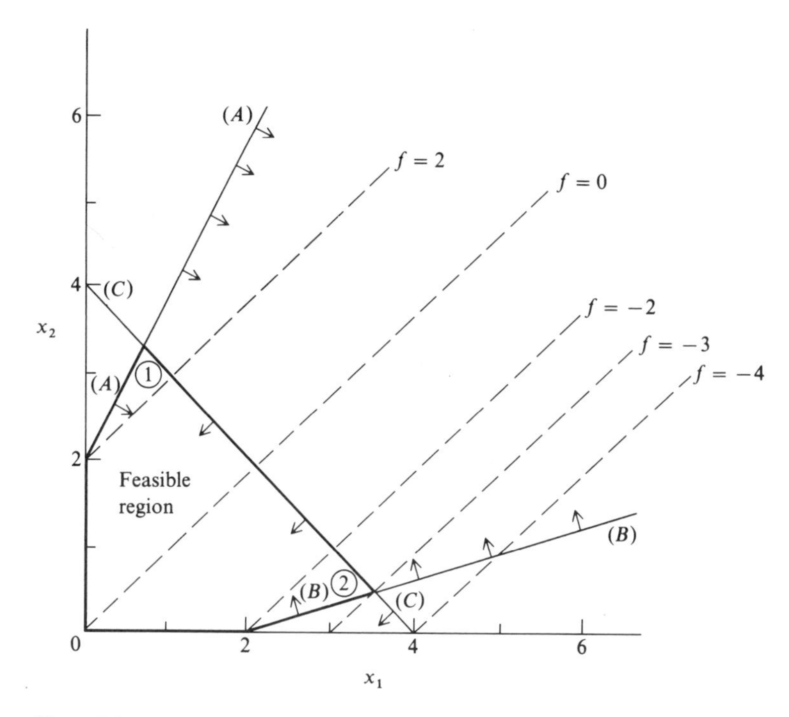

Figure 7.8 Representation of LP used to illustrate the Simplex method of solution (arrows point to the direction in which the inequality constraint is satisfied).

subject to five nonnegativity constraints and three equality constraints hence only two degrees of freedom exist.

Step 2 Define a basic solution corresponding to a vertex or corner of the feasible region. Let us define some new variables for the purposes of discussion, using Eqs. (7.3) and (7.4). Let n = total number of variables, including slack variables. Therefore

$$n = 2 + 3 = 5 \text{ for this example}$$

Let m = total number of equations. Then

$$m = 3.$$

The set of equations (7.22*a*) to (7.22*c*) involves three equations with five unknowns, hence it has no unique solution. If the values of two of the x_i variables are fixed (for example, set equal to zero), then you have a set of three equations which can be solved to obtain the values of the other three variables. The process of identifying and specifying the values of a subset of the five variables is called the formulation of the *basic solution.* A basic solution (also called the *basic feasible solution* when the m variables are nonnegative) corresponds to a solution for $\mathbf{x}$ obtained by solving for m variables in terms of the remaining $(n\text{-}m)$ variables and setting the $(n\text{-}m)$ variables equal to zero. The m nonzero variables are called the *basis* or *basic variables*. The $(n\text{-}m)$ variables are then called *nonbasic variables*.

Since there are two independent variables in this problem, it will be convenient to express the basic variables in terms of x_1 and x_2. Figure 7.8 shows the feasible region. The vertices or corners of the feasible region correspond to basic feasible solutions in which two of the three slack variables are zero (and the rest of the variables are nonzero). For example, the intersection of constraint bounds B and C corresponds to $x_4 = x_5 = 0$; however, the intersection of A and B $(x_3 = x_4 = 0)$ does not correspond to a feasible point (see Fig. 7.8). At the intersection of a constraint with either the x_1 or x_2 axes, one slack variable is zero and either x_1 or x_2 is zero. For this example problem we have assumed as a starting point that both x_1 and x_2 (the nonbasic variables) are zero; the vector of basic variables is (x_3, x_4, x_5). For $x_1 = x_2 = 0$, x_3, x_4, x_5 and $f(\mathbf{x})$ can be easily calculated by writing the constraints and objective function in the following form:

$$x_3 - 2x_1 + x_2 = 2 \tag{7.23a}$$

$$x_4 + x_1 - 3x_2 = 2 \tag{7.23b}$$

$$x_5 + x_1 + x_2 = 4 \tag{7.23c}$$

$$f + x_1 - x_2 = 0 \tag{7.23d}$$

By inspection $x_3 = 2$, $x_4 = 2$, $x_5 = 2$, and $f = 0$. Note that in all the constraints the right-hand sides are positive. This arrangement makes it easy to calculate the values of the basic variables and the objective function when x_1 and x_2 (the nonbasic variables) are assumed to be zero. In addition, the objective function is written only in terms of the nonbasic variables which are assumed to be zero initially.

Step 3 Selection of new basic and nonbasic variables. Normally the initial selection of the basic solution does not correspond to the optimum. Therefore it is necessary to change the basic solution in such a way that f is improved.

Examine the coefficients of the terms in f to determine which variable (x_1 or x_2) decreases the value of the objective function more when that variable is increased from zero. This variable will become a new basic variable and an old basic variable will be changed to the nonbasic category. In Eq. (7.23d) if x_1 is increased from zero, f decreases. If x_2 increases, f will increase, which is not desirable. As a general rule, it is most advantageous to select as a new basic variable that variable which has the largest positive coefficient in the objective function equation.

Once x_1 beomes nonzero, it is no longer a nonbasic variable and becomes a basic variable. Simultaneously, one of the existing basic variables needs to be converted into a nonbasic variable, that is, set equal to zero. Which basic variable would you select from (x_3, x_4, x_5) to be exchanged for x_1?

Figure 7.8, in which the feasible region is denoted by the heavy solid lines, shows the limitations placed on x_1 in changing its value. Note that the constraint bounds A, B, and C cross the x_1 axis at different points. Hence each constraint represents a different limit on the potential increase of x_1. Because the objective function equation (7.23*d*) is linear you would like to increase x_1 as much as possible (in fact you would like to make x_1 infinite if this were allowable). However, since there are three inequality constraints in the problem all of which must hold simultaneously, you must identify that constraint which is the limiting one, i.e., the most restrictive. You can see from the graph that the limiting constraint corresponds to $B(-x_1 + 3x_2 \geq 2)$.

Keep the nonbasic variable x_2 at 0 in Eqs. (7.23*a*) to (7.23*c*) and examine the effect of variations in x_1 relative to each constraint being satisfied:

(*a*) As x_1 is increased, x_3 must always be positive in order to satisfy Eq. (7.23*a*), which means the nonnegativity constraint is automatically satisfied. In Fig. 7.8, note that constraint bound A does not affect movement of x_1 along its axis.

(*b*) x_1 can be increased to 2.0 without causing x_4 to become negative in (7.23*b*). This increase corresponds to the intersection of B with the x_1 axis at $x_1 =$ 2.0.

(*c*) x_1 can be increased to 4.0 without forcing x_5 to become negative in (7.23*c*) (up to the intersection of C with the x_1 axis), but this would violate constraint B.

The limiting constraint is thus B; therefore we would choose x_4, the slack variable for constraint B, to be zero, corresponding to the intersection of that constraint with the x_1 axis ($x_2 = 0$). Constraint B is now active (binding).

The limiting value for x_1 can be calculated analytically for each equation in the set (7.23) by using the respective coefficient on the right-hand side divided by the corresponding coefficient of x_1. The limiting constraint is the one for which the ratio is positive and has the smallest value. In the example, from the set $2/-2$, $2/1$, and $4/1$, you can see that the second constraint B is thus the limiting constraint. The ratio test will henceforth be used to determined which variable to remove from the set of basic variables and make into a nonbasic variable.

Step 4 Transformation of the equations. Next we develop a set of equations in a format analogous to (7.23) with the nonbasic variables now $x_2 = x_4 = 0$ and the basic variable vector being (x_1, x_3, x_5). This transformation can be accomplished by substituting for x_1 as functions of x_2 and x_4 in all the equations (using the limiting constraint identified in step 3)

$$x_1 = 2 - x_4 + 3x_2 \tag{7.24a}$$

By substitution of x_1

$$x_3 = 2 + 2x_1 - x_2 = 6 - 2x_4 + 5x_2 \tag{7.24b}$$

$$x_5 = 4 - x_1 - x_2 = 2 + x_4 - 4x_2 \tag{7.24c}$$

$$f = -x_1 + x_2 = -2 + x_4 - 2x_2 \tag{7.24d}$$

Note that the value of the objective function has decreased from $f = 0$ to $f = -2$, which is desirable for a minimization problem. In the format used previously equations (7.24) are

$$x_3 + 2x_4 - 5x_2 = 6 \tag{7.25a}$$

$$x_1 + x_4 - 3x_2 = 2 \tag{7.25b}$$

$$x_5 - x_4 + 4x_2 = 2 \tag{7.25c}$$

$$f - x_4 + 2x_2 = -2 \tag{7.25d}$$

Rather than use algebraic substitution, Gauss-Jordan (or Gaussian) elimination can be used to systematically transform the equations from one basic solution to the next (refer to Appendix B). Let us write the original set of equations and objective function (7.23) in matrix notation $\mathbf{Ax} = \mathbf{b}$ (B designates basic variables and N nonbasic variables):

$$\begin{matrix} N & N & B & B & B & \\ \end{matrix}$$
$$\begin{bmatrix} -2 & 1 & 1 & 0 & 0 & 0 \\ 1 & -3 & 0 & 1 & 0 & 0 \\ 1 & 1 & 0 & 0 & 1 & 0 \\ 1 & -1 & 0 & 0 & 0 & 1 \end{bmatrix} \begin{bmatrix} x_1 \\ x_2 \\ x_3 \\ x_4 \\ x_5 \\ f \end{bmatrix} = \begin{bmatrix} 2 \\ 2 \\ 4 \\ 0 \end{bmatrix} \tag{7.26}$$

To use Gauss-Jordan elimination, the matrix $\mathbf{A}$ is augmented with the constant vector on the right-hand side of (7.26):

$$\begin{bmatrix} -2 & 1 & 1 & 0 & 0 & 0 & 2 \\ \textcircled{1} & -3 & 0 & 1 & 0 & 0 & 2 \\ 1 & 1 & 0 & 0 & 1 & 0 & 4 \\ 1 & -1 & 0 & 0 & 0 & 1 & 0 \end{bmatrix} \begin{matrix} \\ \leftarrow \text{pivot row} \\ \\ \\ \end{matrix} \tag{7.27}$$

↑ pivot column

In matrix (7.27) we have identified a *pivot column* and a *pivot row.* The pivot column is the column of coefficients corresponding to the new basic variable selected by analysis of the coefficients in the terms of f. The pivot row is the row of coefficients corresponding to the coefficients in the limiting constraint. The *pivot element* is the element at the intersection of the pivot row and the pivot column (circled in the matrix).

To exchange the basic variable x_4 for x_1, the first column is transformed by elementary row operations (refer to Appendix B) to $[0 \quad 1 \quad 0 \quad 0]^T$, and as a result of the operations the fourth column contains some additional nonzero elements. Elementary row operations are used to obtain zeros in elements a_{11}, a_{31}, and a_{41}, and $a_{21} = 1$ of the transformed matrix with the result

$$\mathbf{A}_1 = \begin{bmatrix} 0 & -5 & 1 & 2 & 0 & 0 & 6 \\ 1 & -3 & 0 & 1 & 0 & 0 & 2 \\ 0 & 4 & 0 & -1 & 1 & 0 & 2 \\ 0 & 2 & 0 & -1 & 0 & 1 & -2 \end{bmatrix} \tag{7.28}$$

Matrix multiplication yields the same set of equations (7.24) derived earlier by algebraic substitution.

To help interpret the matrix after it is transformed, a "tableau" can be employed;

(7.29)

	x_1	x_2	x_3	x_4	x_5	f	b
x_3	0	-5	1	2	0	0	6
x_1	1	-3	0	1	0	0	2
x_5	0	(4)	0	-1	1	0	2 ←
	0	2	0	-1	0	1	-2
		↑					

The variables listed on the left side identify the basic variables. We shall use this format in subsequent calculations; more elaborate tableaus can be constructed (Kim, 1971; Cooper and Steinberg, 1974; Reklaitis et al., 1983).

Step 5 Iterative improvement in the objective function. To continue with the Simplex procedure, we examine the objective function equation again. Note that the term $2x_2$ in (7.25*d*) indicates that an increase in x_2 will yield an improvement (a decrease) in the objective function. The largest coefficient in the bottom row of the tableau is $+2$ (the pivot column is indicated by an arrow). However, as before, you must determine the limiting constraint. The ratios of the b column to the x_2 column ($-6/5$, $-2/3$, $2/4$) indicate that the third constraint gives the smallest positive ratio. x_5, the slack variable for the third constraint, will now be set equal to zero. Eliminate x_2 from the nonbasic variables and replace it with x_5, to obtain the largest allowable increase in f.

Next, in tableau (7.29) the x_2 column (pivot column) must become a unit vector (x_2 is a new basic variable), and the column corresponding to x_5 will not be a unit vector. The pivot row is the one corresponding to x_5 in the basic variable column (marked with an arrow). The pivot element is circled in (7.29). Using Gauss-Jordan elimination, the new tableau is

(7.30)

	x_1	x_2	x_3	x_4	x_5	f	b
x_3	0	0	1	0.75	1.25	0	8.5
x_1	1	0	0	0.25	0.75	0	3.5
x_2	0	1	0	-0.25	0.25	0	0.5
	0	0	0	-0.5	-0.5	1	-3

The resulting set of equations is therefore

$$x_3 + 0.75x_4 + 1.25x_5 = 8.5 \tag{7.31a}$$

$$x_1 + 0.25x_4 + 0.75x_5 = 3.5 \tag{7.31b}$$

$$x_2 - 0.25x_4 + 0.25x_5 = 0.5 \tag{7.31c}$$

$$f - 0.5x_4 - 0.5x_5 = -3.0 \tag{7.31d}$$

The objective function has decreased to -3 from the value of -2 on the previous iteration.

At this stage we examine the equation for f. Note that increases in x_4 or x_5 do not have a beneficial effect on the objective function. Therefore the objective function cannot be increased any further, and the optimum has been found. Reading from (7.31), we have $x_3 = 8.5$, $x_1 = 3.5$, $x_2 = 0.5$ (the first three elements of the b column), $x_4 = 0$, $x_5 = 0$ (nonbasics), and $f = -3$ (the last element of the b column). In the tableau the optimum solution has been reached if all coefficients in the objective function row (except for f and b) are negative; then the procedure is terminated.

Other LP Formulations

A number of textbooks (e.g., Kim (1971) and Cooper and Steinberg (1974)) formulate the LP as a maximization problem (rather than minimization). You should be careful in using transformation rules formulated for a different objective function or constraint convention, since a number of variations exist in the literature.

7.5 STANDARD LP FORM

In the five steps above we have by way of motivation illustrated the procedure for a simple two-dimensional problem which could also be solved graphically. We now focus on the general LP format and procedure in which many variables and constraints occur in the problem statement, namely

$$\text{Minimize:} \quad f(\mathbf{x}) = c_1x_1 + c_2x_2 + \cdots + c_nx_n \tag{7.32}$$

$$\text{Subject to:} \quad a_{11}x_1 + a_{12}x_2 + \cdots + a_{1n}x_n = b_1 \tag{7.33}$$

$$a_{21}x_1 + a_{22}x_2 + \cdots + x_{2n}x_n = b_2$$

$$\vdots$$

$$a_{m1}x_1 + a_{m2}x_2 + \cdots + a_{mn}x_n = b_m$$

$$x_i \geq 0 \qquad i = 1, n$$

$$b_j \geq 0 \qquad j = 1, m$$

In more compact notation:

$$\begin{aligned} \min \quad & f = \mathbf{c}^T\mathbf{x} \\ \text{s.t.} \quad & \mathbf{A}\mathbf{x} = \mathbf{b} \\ & \mathbf{x} \geq \mathbf{0}, \mathbf{b} \geq \mathbf{0} \end{aligned} \tag{7.34}$$

Tableau (7.30) illustrated the *standard form* which is used for the constraints in the Simplex solution procedure for a linear programming problem with $m = 3$ and $n = 5$. In general, the canonical form for a system of n variables and m constraints is shown in Table 7.2; note that there are m basic variables and (n-m) nonbasic variables. Because the independent (nonbasic) variables are zero, the dependent (basic) variables can be calculated directly as $x_i = b_i$ ($i = 1, \ldots, m$). If any b_i is zero, this result indicates that some of the constraints are dependent.

You should recognize that in order to write an LP in the form of Table 7.2, you must express the objective function in terms of the nonbasic variables ($x_{m+1}, x_{m+2}, \ldots, x_n$). This functional dependence rarely occurs in the original problem statement, thus procedures must be developed to obtain the first basic feasible solution and to obtain the tableau in standard (or canonical) form. Section 7.6 discusses several methods for obtaining a basic solution.

With n variables and m constraints, there are a finite number of possible basic solutions that can be obtained for the LP problem, a number ψ which is given by

$$\psi = \binom{n}{m} = \frac{n!}{m!(n-m)!}$$

As long as n and m are relatively small, only a few possible solutions need be evaluated. The Simplex procedure attempts to make the search for the optimum an efficient one, but it does not necessarily take the shortest route to the optimum. Based on practical experience, the number of iterations to reach an optimal solution ranges generally between m and $3m$, depending more heavily on the number of constraints than on the number of variables (Reklaitis, et al., 1983).

Table 7.2 Canonical representation with basic variables $x_1, x_2, \ldots, x_m$

Dependent (basic) variables	Independent (nonbasic) variables	Constants
x_1	$+ a_{1,m+1}x_{m+1} + a_{1,m+2}x_{m+2} + \cdots + a_{1n}x_n =$	b_1
x_2	$+ a_{2,m+1}x_{m+1} + a_{2,m+2}x_{m+2} + \cdots + a_{2n}x_n =$	b_2
$\ddots$	$\vdots \quad \vdots$	$\vdots$
x_m	$+ a_{m,m+1}x_{m+1} + a_{m,m+2}x_{m+2} + \cdots + a_{mn}x_n =$	b_m

EXAMPLE 7.2 FORMULATION OF AN LP PROBLEM IN STANDARD FORM

Solve the refinery blending problem given in Eqs. (7.11) to (7.15) using the Simplex method. First formulate the problem as a minimization problem.

Solution. The blending problem stated in a minimization format is

$$\text{Minimize} \quad f = -8.1x_1 - 10.8x_2 \tag{7.35a}$$

$$\text{Subject to} \quad -0.80x_1 - 0.44x_2 \geq -24{,}000 \tag{7.35b}$$

$$-0.05x_1 - 0.10x_2 \geq -2000 \tag{7.35c}$$

$$-0.10x_1 - 0.36x_2 \geq -6000 \tag{7.35d}$$

Step 0. Multiply Eqs. (7.35*b, c, d*) by (-1).

Step 1. Introduce slack variables x_3, x_4, and x_5 into (7.35*b*) to (7.35*d*), respectively, and multiply each constraint equation by (-1) so that the constants on the right-hand side are positive:

$$0.80x_1 + 0.44x_2 + x_3 = 24{,}000 \tag{7.36a}$$

$$0.05x_1 + 0.10x_2 + x_4 = 2000 \tag{7.36b}$$

$$0.10x_1 + 0.36x_2 + x_5 = 6000 \tag{7.36c}$$

$$f + 8.1x_1 + 10.8x_2 \qquad = 0 \tag{7.36d}$$

Step 2. If x_1 and x_2 are selected as the nonbasic variables and (x_3, x_4, x_5) are the basic variables, then the LP is already in standard form and you can proceed with the tableau calculations. The initial tableau is

(7.37)

	x_1	x_2	x_3	x_4	x_5	f	b	
x_3	0.80	0.44	1	0	0	0	24,000	
x_4	0.05	0.10	0	1	0	0	2,000	
x_5	0.10	(0.36)	0	0	1	0	6,000	←
	8.10	10.80	0	0	0	1	0	
		↑						

Step 3. Note that increases in both x_1 and x_2 will improve the objective function. You should select the largest positive coefficient for the pivot column and use the ratio test to find the pivot row. Can you show that the arrows correctly mark the pivot row and pivot column in (7.37)?

Step 4/Step 5. Successive transformations are listed below:

(7.38)

	x_1	x_2	x_3	x_4	x_5	f	b	
x_3	0.68	0	1	0	−1.22	0	16,667	
x_4	(0.02)	0	0	1	−0.28	0	333.3	←
x_2	0.28	1	0	0	2.78	0	16,667	
	5.1	0	0	0	−30.00	1	−180,000	
	↑							

	x_1	x_2	x_3	x_4	x_5	f	b	
x_3	0	0	1	−30.5	(7.25)	0	6,500	←
x_1	1	0	0	45.0	−12.50	0	15,000	
x_2	0	1	0	−12.5	6.25	0	12,500	
	0	0	0	−229.5	33.75	1	−256,500	
					↑			

(7.39)

	x_1	x_2	x_3	x_4	x_5	f	b
x_5	0	0	0.14	−4.21	1	0	896.5
x_1	1	0	1.72	−7.59	0	0	26,207
x_2	0	1	−0.86	13.79	0	0	6,897
	0	0	−4.66	−87.52	0	1	−286,765

(7.40)

We terminate with (7.40) because no x_i coefficients in f are positive (in the bottom row). This example illustrates that the Simplex procedure does not necessarily take the shortest route to the optimum. If you had the foresight to select x_1 for a basis variable at the first iteration, you would reach the optimum after two cycles of the Simplex procedure. After the value obtained in the first iteration, f would increase to \$243,000, which is an improvement over the value obtained in the first iteration using the conventional rules. It is possible to put a provision in the transformation rules to perform more than one set of ratio tests (evaluate more than one column to find the limiting constraint which allows the largest increase in f). Use of such a strategy would save effort in transforming the arrays for this particular example, but it is difficult to predict whether this approach would save computation time on other problems.

7.6 OBTAINING A FIRST FEASIBLE SOLUTION

In many LP problems the choice of a first basic solution which satisfies all the constraints is not obvious. However, a basic solution can be obtained via a technique which utilizes a special type of variable called *artificial variable*. Thereafter, the Simplex procedure can be applied. Consider the problem:

$$\text{Minimize} \quad f(\mathbf{x}) = x_1 + 2x_2 \tag{7.41}$$

$$\text{Subject to} \quad 3x_1 + 4x_2 \geq 5$$

$$x_1 + x_2 \leq 4$$

Introduction of slack and surplus variables x_4 and x_3 changes the constraints to

$$3x_1 + 4x_2 - x_3 = 5 \tag{7.42a}$$

$$x_1 + x_2 + x_4 = 4 \tag{7.42b}$$

Suppose you tried $x_1 = x_2 = 0$ for a basic solution; this requires $x_3 = -5$, violating the nonnegativity restriction (however, $x_4 = 4$, which is satisfactory).

Therefore the origin is not in the feasible region and you must find a different starting point. Define an artificial slack variable, x_5, and introduce it into (7.42*a*) such that

$$3x_1 + 4x_2 - x_3 + x_5 = 5 \tag{7.43}$$

Now $x_1 = x_2 = 0$ is allowable, because positive values of x_3 and x_5 can satisfy Eq. (7.43). We next must drive the value of x_5 to zero, by adding a penalty function (see Sec. 8.5) to the original objective function

$$\bar{f} = f + Mx_5 \tag{7.44}$$

For M large, x_5 will hopefully approach zero in the minimization of $\bar{f}$ via the Simplex method. This technique, call the "Big M" method (Murty, 1983) allows you to initiate the minimization of f at a starting point outside the feasible region.

In practice, the "Big M" method suffers from several practical difficulties, especially for problems with many variables. Choice of the correct value of M is not at all clear. If the LP solution terminates with some artificial variables non-zero, is this because there is no feasible solution or because M is too small? If M is chosen too large, there may be problems with precision of computer arithmetic; round-off error may lead to a nonoptimal solution. Below we discuss an alternative approach, referred to as "Phase I-Phase II."

Murty (1983) and Schrage (1983) have discussed the solution of an LP problem starting from a nonfeasible starting point utilizing these two phases. To drive the artificial variables to zero in an LP problem, phase I is carried out in which the sum of the artificial variables is minimized to obtain a basic feasible solution, or demonstrate that no feasible solution exists. In phase II, the Simplex procedure is executed starting with feasible solution obtained in phase I.

The phase I procedure is summarized below:

1. Change the signs of any equations as needed (by multiplying by -1) so that the b's (the right-hand side coefficients) are all positive.
2. Convert the system of inequality constraints into a set of equalities as explained before using slack and/or surplus variables.
3. Augment the set of equations by one artificial variable for each equation to get a new standard form. After step 3, the set of constraints will appear as follows:

$$\begin{aligned}
a_{11}x_1 + a_{12}x_2 + \cdots + a_{1n}x_n + x_{n+1} \qquad\qquad\qquad &= b_1 \\
a_{21}x_1 + a_{22}x_2 + \cdots + a_{2n}x_n \qquad + x_{n+2} \qquad &= b_2 \\
\vdots \qquad\qquad \vdots \qquad\qquad \vdots \qquad\qquad\qquad\qquad\qquad & \ \ \vdots \\
a_{m1}x_1 + a_{m2}x_2 + \cdots + a_{mn}x_n \qquad\qquad\qquad + x_{n+m} &= b_m
\end{aligned} \tag{7.45}$$

A solution of (7.45) is:
(*a*) The basic variables are: $x_{n+i} = b_i, i = 1, \ldots, m$
(*b*) The nonbasic variables are: $x_j = 0, \; j = 1, \ldots, n$

4. Next proceed to drive all the artificial variables to zero by minimizing their sum, by using the following objective function.

$$w = x_{n+1} + x_{n+2} + \cdots + x_{n+m} \tag{7.46}$$

To minimize w using the Simplex method, you need to express w in terms of x_1 to x_n instead of x_{n+1} to x_{n+m}. To accomplish this step, subtract the first equation in the set (7.45) from the equation for w, Eq. (7.46), repeat for the second equation, so as to set up the following standard form:

$$\begin{aligned} a_{11}x_1 + a_{12}x_2 + \cdots + a_{1n}x_n + x_{n+1} \quad &= b_1 \\ \vdots \qquad \vdots \qquad\qquad \vdots \qquad\qquad &\ \ \vdots \\ a_{m1}x_1 + a_{m2}x_2 + \cdots + a_{mn}x_n \qquad + x_{n+m} &= b_m \end{aligned} \tag{7.47}$$

$$w + \underbrace{\left(\sum_{i=1}^{m} a_{i1}\right)}_{d_1}x_1 + \underbrace{\left(\sum_{i=1}^{m} a_{i2}\right)}_{d_2}x_2 + \cdots + \underbrace{\left(\sum_{i=1}^{m} a_i\right)}_{d_n}x_n = \left(\sum_{i=1}^{m} b_i\right) \tag{7.48}$$

Now use the Simplex procedure to minimize w.

5. Terminate phase I. If the minimum of w is nonzero, then there is no solution to the problem. An example of this outcome might be improper specification of the constraints so that the feasible region is not convex, or does not exist.

If the minimum w is reduced to zero, then all the artificial variables have been reduced to 0 because w is the sum of nonnegative variables only. You can now delete the artificial variables from further consideration because you have found a feasible solution for the original problem. The basis contains no artificial variable (except, perhaps, some with zero value only). If the artificial variables (at 0 value) are found to be basic variables at the end of the phase I calculations, it means that the original system of equations contained dependent equations (Schrage, 1983; Luenberger, 1973).

After completing the phase I procedure to generate an initial feasible solution, you proceed with phase II—the solution of the original LP problem using the original objective function. You

(*a*) Drop all the variables with $d_j \le 0$ from the tableau (generally this includes all of the artificial variables)

(*b*) Retain all the variables with $d_j \ge 0$

(*c*) Introduce the original objective function in standard form using the feasible solution from phase I.

For the problem considered earlier, Eq. (7.41), first add artificial variables to both constraints as follows:

$$\begin{aligned} 3x_1 + 4x_2 - x_3 + x_5 &= 5 \\ x_1 + x_2 + x_4 + x_6 &= 4 \end{aligned} \tag{7.49}$$

The objective function for phase I can be obtained from (7.48) by summing coefficients:

$$w + 4x_1 + 5x_2 - x_3 + x_4 = 9 \tag{7.50}$$

and the constraints are Eq. (7.49). Therefore the starting tableau for phase I is

	x_1	x_2	x_3	x_4	x_5	x_6	w	b	
x_5	3	(4)	−1	0	1	0	0	5	←
x_6	1	1	0	1	0	1	0	4	
	4	5	−1	1	0	0	1	9	
		↑							

Subsequent tableaus obtained are

	x_1	x_2	x_3	x_4	x_5	x_6	w	b	
x_2	0.75	1	−0.25	0	0.25	0	0	1.25	
x_6	0.25	0	(0.25)	1	−0.25	1	0	2.75	←
	0.25	0	0.25	1	−1.25	0	1	2.75	
			↑						

	x_1	x_2	x_3	x_4	x_5	x_6	w	b
x_2	1	1	0	1	0	1	0	4
x_3	1	0	1	4	−1	4	0	11
	0	0	0	0	−1	−1	1	0

Note that the final basic feasible solution is $x_1 = x_4 = x_5 = x_6 = 0$, $x_2 = 4$, $x_3 = 11$, and $w = 0$. Therefore phase I is completed and x_5 and x_6 are dropped because they have negative coefficients in the objective function (recall that x_5 and x_6 were the original artificial variables). Phase II would begin with $x_1 = x_4 = 0$ as the initial basic solution and the original objective function ($f = x_1 + 2x_2$).

Since x_1 and x_4 are now nonbasic, f must be expressed in terms of x_1 and x_4 using row 1 of the last tableau; therefore

$$\begin{aligned} f = x_1 + 2x_2 &= x_1 + 2[-x_1 - x_4 + 4] \\ &= -x_1 - 2x_4 + 8 \end{aligned} \tag{7.51}$$

The new tableau only differs from the last tableau by the third row (f) and the deletion of the x_5 and x_6 columns:

	x_1	x_2	x_3	x_4	f	b	
x_2	1	1	0	1	0	4	
x_3	1	0	1	(4)	0	11	←
	1	0	0	2	1	8	
				↑			

Subsequent transformations are

	x_1	x_2	x_3	x_4	f	b	
x_2	(0.75)	1	−0.25	0	0	1.25	←
x_4	0.25	0	0.25	1	0	2.75	
	0.50	0	−0.50	0	1	2.50	
	↑						

	x_1	x_2	x_3	x_4	f	b
x_1	1	1.33	−0.33	0	0	1.67
x_4	0	−0.33	0.58	1	0	2.33
	0	−0.67	−0.33	0	1	1.67

Based on the bottom row, the optimum is

$$x_1 = 1.67, x_2 = x_3 = 0, x_4 = 2.33, \text{ and } f = 1.67.$$

Unrestricted Variables

In some problems one or more variables which appear in the objective function can assume negative as well as positive values. In order to employ the Simplex procedure in these cases, you can define additional variables as shown in the example below.

EXAMPLE 7.3 USE OF THE SIMPLEX PROCEDURE WITH UNRESTRICTED VARIABLES

You are given the LP:

$$\text{Minimize} \quad f(x) = x_1 + 4x_2$$

$$\text{Subject to} \quad x_1 + x_2 \leq 3$$

$$-x_1 + x_2 \leq 1$$

$$x_1 \text{ unrestricted}, x_2 \geq 0$$

Find the optimal values of x_1 and x_2. The feasible region and objective function for $f = 4$ are shown in Fig. E7.3.

Solution. Introduce the slack variables x_3 and x_4 as follows:

$$x_1 + x_2 + x_3 = 3$$

$$-x_1 + x_2 + x_4 = 1$$

In addition, let

$$x_1 = x_5 - x_6$$

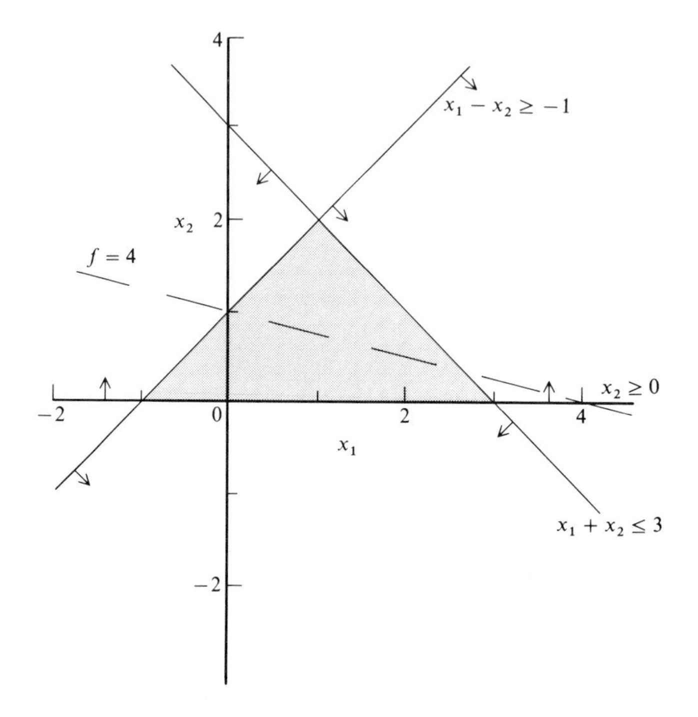

Figure E7.3 Feasible region and objective function contours for Example 7.3.

where $x_5 \geq 0$, $x_6 \geq 0$. Thus x_1 can assume either positive or negative values. Substitute for x_1 into the original LP and rearrange to get the standard form

$$x_2 + x_3 + x_5 - x_6 = 3$$

$$x_2 + x_4 - x_5 + x_6 = 1$$

$$f - 4x_2 - x_5 + x_6 = 0$$

The tableau with basic variables x_3 and x_4 and nonbasic variables $x_2 = x_5 = x_6 = 0$ is

	x_2	x_3	x_4	x_5	x_6	f	b	
x_3	1	1	0	1	−1	0	3	
x_4	1	0	1	−1	①	0	1	←
	−4	0	0	−1	1	1	0	
					↑			

The pivot element having been identified, exchange x_6 and x_4 (x_6 becomes basic) and transform:

	x_2	x_3	x_4	x_5	x_6	f	b
x_3	2	1	1	0	0	0	4
x_6	1	0	1	−1	1	0	1
	−5	0	−1	0	0	1	−1

Since all coefficients in the bottom row are negative, no further improvement in f can be obtained. Hence at the optimum,

$$x_2 = x_4 = x_5 = 0$$

$$x_3 = 4 \qquad x_6 = 1 \qquad x_1 = x_5 - x_6 = -1$$

$$f = -1$$

Some authors (e.g., Luenberger, 1973) have suggested eliminating unrestricted variables if possible by algebraic substitution, a step that reduces the number of equality constraints. In the above example, x_1 would be eliminated from f and the inequality constraints in terms of x_2, x_3, and x_4, leading to one constraint involving these three variables.

7.7 THE REVISED SIMPLEX METHOD

While the row operations illustrated in the previous sections are readily programmable on a digital computer, the actual procedure to execute the solution of an LP problem as implemented in computer codes is somewhat different (Murtagh, 1983). In general there is a great deal of inefficiency in the Simplex method; in a problem involving many variables, pivots may actually occur only in a small fraction of the columns so that calculations transforming all of the columns at each iteration are inefficient. The Revised Simplex Method (refer to Dantzig, 1963; Luenberger, 1973; Murtagh, 1983) was devised to avoid unnecessary calculations by ordering the calculations and utilizing a matrix-based technique. New tableaus are generated using the original set of equations rather than the previous tableau.

In order to illustrate the revised Simplex procedure, we partition the general LP (Eq. 7.34) as follows:

$\mathbf{x}^T = [\mathbf{x}_B \,\vdots\, \mathbf{x}_N]$ $\quad \mathbf{x}_B$ = vectors of basic variables ($m \times 1$),
$\mathbf{x}_N$ = vector of nonbasic variables, $(n - m) \times 1$

$\mathbf{A} = [\mathbf{B} \,\vdots\, \mathbf{N}]$ $\quad \mathbf{B}$ = basis matrix, corresponding to $\mathbf{x}_B$, comprised of the first m columns of $\mathbf{A}$, $m \times m$
$\mathbf{N}$ = remaining $(n - m)$ columns of $\mathbf{A}$

$\mathbf{c} = \begin{bmatrix} \mathbf{c}_B \\ \mathbf{c}_N \end{bmatrix}$ $\quad \mathbf{c}_B$ = vector of cost coefficients for basic variables
$\mathbf{c}_N$ = vector of cost coefficients for nonbasic variables

$\mathbf{t} = \mathbf{n} - \mathbf{m}$ $\quad \mathbf{t}$ = number of nonbasic variables

Now the standard LP becomes

$$\text{Minimize} \quad f = \mathbf{c}_B^T \mathbf{x}_B + \mathbf{c}_N^T \mathbf{x}_N \tag{7.52}$$

$$\text{Subject to} \quad \mathbf{B}\mathbf{x}_B + \mathbf{N}\mathbf{x}_N = \mathbf{b} \tag{7.53}$$

$$\mathbf{x}_B \geq \mathbf{0}, \mathbf{x}_N \geq \mathbf{0}$$

The dependent variables $\mathbf{x}_B$ can be evaluated by multiplying Eq. (7.53) from the left by $\mathbf{B}^{-1}$ to get

$$\mathbf{x}_B = \mathbf{B}^{-1}\mathbf{b} - \mathbf{B}^{-1}\mathbf{N}\mathbf{x}_N \tag{7.54}$$

If $\mathbf{x}_N = \mathbf{0}$ (as part of the basic feasible solution), then the corresponding values of the basic variables can be calculated from

$$\mathbf{x}_B = \mathbf{B}^{-1}\mathbf{b} \tag{7.55}$$

The cost function can also be expressed in terms of the nonbasic variables:

$$f = \mathbf{c}_B^T(\mathbf{B}^{-1}\mathbf{b} - \mathbf{B}^{-1}\mathbf{N}\mathbf{x}_N) + \mathbf{c}_N^T\mathbf{x}_N \tag{7.56}$$

$$= \mathbf{c}_B^T\mathbf{B}^{-1}\mathbf{b} + (\mathbf{c}_N^T - \mathbf{c}_B^T\mathbf{B}^{-1}\mathbf{N})\mathbf{x}_N \tag{7.57}$$

The first term in (7.57) on the right-hand side is the value of the objective function ($\mathbf{x}_N = 0$) while the second term involves the inner product of two vectors; the cost coefficients of the nonbasic variables are given by the vector ($\mathbf{c}_N^T - \mathbf{c}_B^T\mathbf{B}^{-1}\mathbf{N}$). These coefficients, called "relative" cost coefficients, determine which variable to bring into the set of basic variables.

The tableau corresponding to (7.54) to (7.57) would appear as follows:

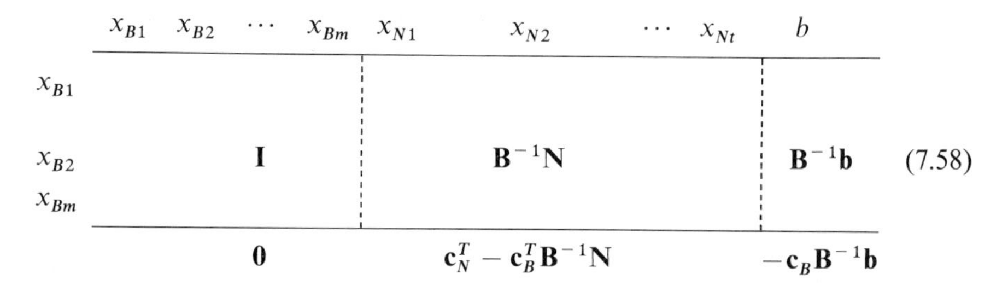

(7.58)

The revised Simplex procedure involves the following steps:

1. Select a basis $\mathbf{B}$ and calculate $\mathbf{B}^{-1}$.
2. Examine the relative cost coefficient vector ($\mathbf{c}_N^T - \mathbf{c}_B^T\mathbf{B}^{-1}\mathbf{N}$) in the bottom row and find the largest coefficient; it denotes the pivot column, which we call x_{Nj}. If no positive coefficients appear, then terminate the calculations.
3. Perform the ratio test using the pivot column of $\mathbf{B}^{-1}\mathbf{N}$ [see Tableau (7.58)] and the vector $\mathbf{B}^{-1}\mathbf{b}$, one row at a time. The row with the smallest positive ratio indicates which variable is to leave the basis.
4. With new basic variables redefine $\mathbf{B}$ and $\mathbf{N}$ and update $\mathbf{B}^{-1}$ using the original set of equations. Then recalculate $\mathbf{B}^{-1}\mathbf{N}$ and $\mathbf{B}^{-1}\mathbf{b}$. As shown by Cooper and Steinberg (1974), the matrix $\mathbf{B}^{-1}$ does not need to be completely recalculated, but rather only a subset of the columns needs to be modified. Return to step 2 to carry out further transformations.

Refer to the textbooks by Cooper and Steinberg (1974), Luenberger (1973), and Hadley (1962) for sample calculations.

One variant of the revised Simplex procedure is based on a product form of the inverse (Murty, 1983; Schrage, 1983). The principal advantage of this approach is that high-speed memory requirements can be reduced, which may be important for problems with over 100 variables. According to Schrage (1983), a typical real-world LP has about four nonzero elements per column, so **A** is a sparse matrix. The product form of the inverse exploits this sparsity. All commercial LP packages utilize some type of product form inverse; Table 7.3 lists a

Table 7.3 Sources of computer codes for linear programming

Code name	Source	Computers†	Language	Price in dollars. Academic rate in ( )
APEX-III	CDC	CDC 600/ 7000, Cyber 170	Fortran + control languages	Write
FMPS	Sperry Corp., P.O. Box 64942, St. Paul, MN 55164	Univac	Fortran	Write
LINDO	Prof. Linus Schrage, Graduate School Business, Univ. Chicago, Chicago, IL 60637	Numerous	Fortran	$750 and up
LP/PROTRAN	IMSL, 7500 Bellaire Blvd., Houston, TX 77036	Numerous + Data General	Fortran	$6000 (500)
MICROSOLVE	Holden-Day, Inc., 4432 Telegraph Ave., Oakland, CA, 94609	IBM PC	Basic	$49.95
MINI-LP	Scicon Ltd., Brick Close, Milton Keynes, MK11 3EF, UK	Numerous	Fortran	$5000
MINOS-HP	Tennessee Valley Authority, P.O. Drawer E, Norris, TN 37828	HP 1000	Fortran	Write
MPS III	IBM	IBM	Assembly + Fortran	$300/mo batch $500/mo interactive
MPSX	IBM	IBM	Assembly + Fortran	$1000–16,000
OMNI + OPTIMIZER	Haverly Systems, Inc. P.O. Box 919, Denville, NJ 07834	IBM 370, 303X, 43XX, CDC	Special	Write
SCICONIC	Scicon Ltd., Brick Close, Milton Keynes, MK11 3EF, UK	Numerous	Fortran	$15,000
WHIZARD	Kentron, 1700 N. Moore St., Arlington, VA 22209	Several	Assembly	$550/mo

† Numerous includes CDC 6000/7000, Cyber 70/170, DEC VAX-11, IBM 360/370/30XX/43XX, as well as personal computers.

number of such commercial software packages. Using mainframe codes, LP problems with thousands of variables and constraints can be solved efficiently (Shamir, 1987). However, special features such as matrix generators are necessary to formulate such large problems. See Barnes et al. (1975) for a review of these software packages, and Sharda (1984) for an analysis of 13 microcomputer codes. Various society newsletters and trade magazines should be read to stay current with the latest personal computer LP codes.

Recently new LP algorithms suggested by Khachian (1979) and Karmarkar (see Sec. 7.10) have for some types of problems offered improved performance over the Simplex method, but these approaches have not yet been demonstrated to be superior to established LP computer codes for all problems.

7.8 SENSITIVITY ANALYSIS

One of the main advantages of the Simplex algorithm is that sensitivity information is relatively easy to obtain from the tableau without recomputing the optimal solution (refer to Dantzig, 1963; Hughes and Grawoig, 1973). Calculation of the change in the value of the optimal solution in response to changes in coefficients in the objective function or constraints is known as postoptimality or sensitivity analysis. Such an analysis is important because the coefficients and/or limits in the constraints may be poorly known. Even when they are known fairly accurately, the effect of expanding capacity may be of interest, and the cost or profit can be evaluated directly. In addition, by postoptimality analyses you can examine how changes in costs of raw materials or the selling prices of products influence the optimum.

The refinery example presented earlier will serve as the vehicle for illustrating different facets of sensitivity analysis. The first component to be examined is the *shadow price*. A shadow price is the change in the optimum value of the objective function per unit change in a constraint limit (the right-hand side of an inequality constraint). In Example 7.2 the final tableau consisted of the following equations:

$$x_5 + 0.14x_3 - 4.21x_4 = 896.5 \tag{7.59}$$

$$x_1 + 1.72x_3 - 7.59x_4 = 26{,}207 \tag{7.60}$$

$$x_2 - 0.86x_3 + 13.79x_4 = 6897 \tag{7.61}$$

$$f = 4.66x_3 + 87.52x_4 - 286{,}765 \tag{7.62}$$

Recall that the slack variables x_3, x_4, and x_5 were introduced into the inequality constraints to represent the difference between the value of the constraint and the value at its bound with the units of barrels per day of gasoline (x_3), kerosene (x_4), and fuel oil (x_5), respectively. The coefficients in the objective function, Eq. (7.62), indicate in absolute units that f is most sensitive to a change in x_4

(kerosene capacity), while an incremental change in fuel oil production capacity (x_5) has *no* effect on f. Why? Because x_5 is a basic variable at the optimum and is not equal to zero, there is excess capacity in fuel oil production at the optimum; hence there is no effect in slightly changing b for this constraint (examine constraint C in Fig. 7.4).

Suppose the right-hand side of the gasoline constraint is increased by 1 bbl/day. Because this is tantamount to relaxing the constraint, we expect that the profit should increase. In terms of the slack variable x_3, relaxing the constraint by 1 bbl/day is equivalent to allowing x_3 to assume a negative value, in this case $x_3 = -1$. Referring to Eq. (7.62), if x_3 becomes -1, f decreases from $-286{,}770$ (\$/day), a change of $-\$4.66$. Note that $\Delta f = 4.66\Delta x_3$, where \$4.66 is the shadow price ($\Delta x_3 = -1.0$). Likewise, a 1 bbl/day increase in kerosene production ($x_4 = -1.0$) yields an incremental profit change of \$87.52/day. Therefore, the coefficients in the bottom row of the tableau indicate how the objective function changes in response to capacity changes.

While small relative changes in the coefficients can be analyzed easily from the tableau, larger changes must be examined with care. The evaluation of large-scale changes is referred to as "ranging" of the coefficients. As long as the coefficients do not change outside of a certain range, the same variables remain basic/nonbasic, although the value of the objective function and basic variables can change.

Suppose the crude oil prices change, causing the coefficients of the objective function to change also. Since x_1 and x_2 are the two basic variables in the final tableau, a \$1.00 increase in the cost coefficient of either x_1 and x_2 in the original problem statement can be found from the right-hand side of the tableau (the b column). This is because $x_1^{\text{opt}} = 26{,}207$ and $x_2^{\text{opt}} = 6897$ and the original objective function (profit) was $8.1x_1 + 10.8x_2$. Note that if the x_1 profit coefficient is changed by \$1.00 (to $9.1x_1$), this will increase f more than a \$1.00 change in the x_2 coefficient (to \$11.80). A change in the original objective function coefficient corresponds to a change in slope in the objective function (Hughes and Grawoig, 1973) and can be interpreted graphically for a two-variable problem.

If the original cost coefficient for x_2 in (7.35) is changed from 10.8 to 11.8, the slope of the objective function line changes from -0.75 to -0.69. This change does not influence the location of optimal corner of the feasible region but it does change the value of the objective function at that point. However, the location of the optimum can move from corner 2 to corner 3 (see Fig. 7.4) if the x_2 coefficient is changed enough.

Sometimes the sensitivity information can be plotted as a continuous function of one of the parameters. Beightler et al. (1979) called these plots "availability charts"; they discussed such charts in the context of a detailed refinery blending problem. Also shortcut methods exist for analyzing simultaneous changes in several parameters, e.g., Reklaitis et al. (1983).

The calculation of sensitivity information is a standard feature of LP software. Generally, data on shadow prices and ranging on the objective function and constraint capacity coefficients are given in the program output. You should keep in mind whether the LP solved is a maximization or minimization problem, since this aspect has a significant impact on the interpretation of the sensitivity data.

7.9 DUALITY IN LINEAR PROGRAMMING

In the previous section we discussed how the coefficients of the slack variables provided sensitivity information on changes in capacity. For the refinery example, gasoline capacity is worth \$4.66/bbl, kerosene capacity is valued at \$87.52/bbl, and fuel oil capacity does not contribute to the objective function (\$0/bbl). Let us multiply the capacity limit in the original constraints (Eqs. 7.35*b*, *c*, *d*) for each product times the shadow prices, which yields

$$4.66(24{,}000) + 87.52(2000) = \$286{,}880$$

This value is the same as the optimum profit calculated in the previous section (\$286,740), and differs only due to numerical round-off. This equivalence is an important property of the linear programming problem, known as *duality*. It implies that the original LP (called the "primal" problem) could be formulated in an alternative way, called the dual. The basis for the dual problem comes out of Lagrange multiplier theory, which is covered in Chap. 8. See Murty (1983), Hadley (1962), or Dantzig (1963) for the theoretical derivation of the primal and dual relationships in linear programming. For the general LP given in Eqs. (7.1 to 7.3), the dual problem is

$$\text{Maximize} \quad f^* = b_1 y_1 + b_2 y_2 + \cdots + b_q y_q \tag{7.63}$$

$$\begin{aligned} \text{Subject to} \quad & a_{11}y_1 + a_{21}y_2 + \cdots + a_{q1}y_q \le c_1 \\ & a_{12}y_1 + a_{22}y_2 + \cdots + a_{q2}y_q \le c_2 \\ & \vdots \qquad\quad \vdots \qquad\qquad\quad \vdots \qquad\quad \vdots \\ & a_{1r}y_1 + a_{2r}y_2 + \cdots + a_{qr}y_q \le c_r \\ & y_1, y_2, \ldots, y_q \ge 0 \end{aligned} \tag{7.64}$$

$y_1, y_2, \ldots, y_q$ are called the *dual variables* and $q = p - m$. Compare the primal and the dual versions of the LP. Note that the objective function coefficients of the primal problem have become the right-hand side constants of the dual. Similarly, the right-hand side constants of the primal have become the cost coefficients of the dual. In addition, the minimization problem is now one of maximization of f^*, and the sense of the inequalities is reversed. Instead of matrix $\mathbf{A}$ in the primal, $\mathbf{A}^T$ is used to form the dual. One dual variable exists for each primal constraint and one dual constraint exists for each primal variable.

The optimal solution of the decision variables (y_i) will correspond to the shadow prices obtained from solution of the primal problem. Note that the shadow prices are positive or zero in this formulation; however, in the primal tableau the coefficients which appear in the bottom row are either negative or zero, i.e., they must have signs opposite to those of the y_i's. The y's are also called Lagrange multipliers (as will be explained in Chap. 8). At the optimum, $f^{\text{opt}} = f^{*\text{opt}}$.

Consider the problem of

$$\begin{aligned} &\text{Minimize} \quad f = 3x_1 + 5x_2 \\ &\text{Subject to} \quad \begin{aligned} x_1 + 3x_2 &\geq 10 & \quad x_1 \geq 0 \\ 2x_1 - x_2 &> 4 & \quad x_2 \geq 0 \\ -x_1 + 4x_2 &\geq -2 \\ -x_1 - x_2 &\geq -20 \end{aligned} \end{aligned}$$

To convert this problem to the form given in (7.32) and (7.33), define slack and surplus variables x_3, x_4, x_5, and x_6

$$\begin{aligned} x_1 + 3x_2 - x_3 &= 10 \\ 2x_1 - x_2 - x_4 &= 4 \\ -x_1 + 4x_2 - x_5 &= -2 & \qquad x_1 - 4x_2 + x_5 &= 2 \\ & & \text{or} \qquad & \\ -x_1 - x_2 - x_6 &= -20 & \qquad x_1 + x_2 + x_6 &= 20 \end{aligned}$$

Note that $x_1 = x_2 = 0$ does not yield a basic feasible solution since x_3 and x_4 would be negative. A phase I procedure with artificial variables would need to be employed to obtain a first feasible solution.

On the other hand, the dual problem [see Eqs. (7.63) and (7.64)] is:

$$\begin{aligned} &\text{Maximize} \quad F = 10y_1 + 4y_2 - 2y_3 - 20y_4 \\ &\text{Subject to} \quad \begin{aligned} y_1 + 2y_2 - y_3 - y_4 &\leq 3 \\ 3y_1 - y_2 + 4y_3 - y_4 &\leq 5 \\ y_1 \geq 0 \qquad y_3 &\geq 0 \\ y_2 \geq 0 \qquad y_4 &\geq 0 \end{aligned} \end{aligned}$$

The slack variables y_5 and y_6 then would be added:

$$\begin{aligned} y_1 + 2y_2 - y_3 - y_4 + y_5 &= 3 \\ 3y_1 - y_2 + 4y_3 - y_4 + y_6 &= 5 \end{aligned}$$

Note that a basic feasible solution would be $y_1 = y_2 = y_3 = y_4 = 0$ and no phase I procedure is needed (Prob. 7.15 asks for the solution to this problem). Therefore we see that by solving the optimization problem in the dual formulation, a basic feasible solution can be quite straightforward to obtain. In some applications the number of constraints may be quite large in the primal problem; by solving the dual you can obtain some reduction in computational effort (Beveridge and Schechter, 1970).

If equality constraints are present in the linear programming problem statement, the primal-dual formulation is as follows. Suppose there are q inequality constraints and m equality constraints in the primal problem statement, with $\mathbf{r}_1$ variables required to be nonnegative and $\mathbf{r}_2$ variables unrestricted:

$$\text{Minimize} \quad f = \mathbf{c}_1^T\mathbf{x}_1 + \mathbf{c}_2^T\mathbf{x}_2 \tag{7.65}$$

$$\begin{aligned} \text{Subject to} \quad & \mathbf{A}_{11}^T\mathbf{x}_1 + \mathbf{A}_{12}^T\mathbf{x}_2 \geq \mathbf{b}_1 && (q) \\ & \mathbf{A}_{21}^T\mathbf{x}_1 + \mathbf{A}_{22}^T\mathbf{x}_2 = \mathbf{b}_2 && (m) \\ & \mathbf{x}_1 \geq \mathbf{0} && (r_1) \\ & \mathbf{x}_2 \text{ unrestricted} && (r_2) \end{aligned} \tag{7.66}$$

The dual formulation is

$$\text{Maximize} \quad F = \mathbf{b}_1^T\mathbf{y}_1 + \mathbf{b}_2^T\mathbf{y}_2 \tag{7.67}$$

$$\begin{aligned} \text{Subject to} \quad & \mathbf{A}_{11}\mathbf{y}_1 + \mathbf{A}_{12}\mathbf{y}_2 \leq \mathbf{c}_1 && (r_1) \\ & \mathbf{A}_{21}\mathbf{y}_2 + \mathbf{A}_{22}\mathbf{y}_2 = \mathbf{c}_2 && (r_2) \\ & \mathbf{y}_1 \geq 0 && (q) \\ & \mathbf{y}_2 \text{ unrestricted} && (m) \end{aligned} \tag{7.68}$$

The letters in parentheses indicate the respective number of equations or variables. Note that if $r_2 = 0$ and $m = 0$, the conditions reduce to the same ones in Eqs. (7.63) and (7.64). However, if $m = 0$ but $r_2 \neq 0$, the Lagrange multipliers $\mathbf{y}_1$ and $\mathbf{y}_2$ may still exist in the dual.

There is a rather large body of theory centered around the primal-dual relationship (Kim, 1971; Cooper and Steinberg, 1974) that cannot be covered here due to space limitations. For example, you can solve the primal and dual problems jointly to make sure that you converge from above and below to the optimal value of the objective function. From a computational point of view, there are other advantages to using solutions of the dual for a given problem, especially for problems with many variables where feasibility is difficult to attain. The dual Simplex algorithm (Murty, 1983; Hadley, 1962) does not require an initial basic feasible solution in $\mathbf{x}$, since the dual problem has different constraints than the primal problem. By repetitive solution of the dual, you satisfy the optimality criterion of an LP without being primal feasible (but $\mathbf{y}$ is feasible), thus avoiding the phase I procedure described in Sec. 7.6. The basis for $\mathbf{y}$ is changed until at termination you obtain a first feasible solution for $\mathbf{x}$ (the primal problem). In contrast, application of the Simplex method to the primal problem always maintains feasibility for $\mathbf{x}$ while gradually improving f (until all coefficients of the objective function are the proper sign) by changing the basis one vector at a time.

7.10 THE KARMARKAR ALGORITHM

Karmarkar proposed in 1984 an algorithm as an alternate to the Simplex method for solving large-scale problems. His method attempts to find improved search directions in the strict interior of the constrained region. This idea can be contrasted with the Simplex method which searches along the boundary from one feasible vertex (defined by the constraints) to an adjacent vertex until the optimal point is found. For large-scale problems, the number of vertices is quite large, hence the Simplex method would be prohibitively expensive in computer time if any substantial fraction of the vertices had to be evaluated. In practice, the Simplex method only requires the evaluation of a very small fraction of the total number of vertices. Also, only a small amount of computation is involved in each iteration so that the procedure is rapid, and also has strong properties of numerical stability.

In several tests of large size problems (Kozlov and Black, 1986), Karmarkar's algorithm showed a speed advantage on the average vs. the Simplex code in MINOS. Specific matrix manipulations and formulas can be found in Strang (1985) or in recent journal articles in which various implementations of the algorithm are described. In what follows we illustrate the general concept for a simple example

$$\begin{aligned} &\text{Minimize} \quad f(\mathbf{x}) = x_1 + 2x_2 + 3x_3 && f = \mathbf{c}^T\mathbf{x} \\ &\text{Subject to} \quad x_1 + x_2 + x_3 = 1 \qquad \text{or} && \mathbf{A}\mathbf{x} = \mathbf{b} \\ &\qquad x_i \geq 0 \qquad i = 1, 2, 3 && \mathbf{x} \geq \mathbf{0} \end{aligned}$$

By inspection, the optimum is at $x_1 = 1$, $x_2 = 0$, and $x_3 = 0$. Figure 7.9 illustrates the constraints and direction of steepest descent, $-\nabla f(\mathbf{x})$. Two main phases of the Karmarkar algorithm are carried out as follows:

Phase k

Start at the centroid of the simplex comprising the constraints $\mathbf{Ax} = \mathbf{b}$, a point that is feasible, and project $-\mathbf{c}$ onto the intersection (null space—see App. C) of

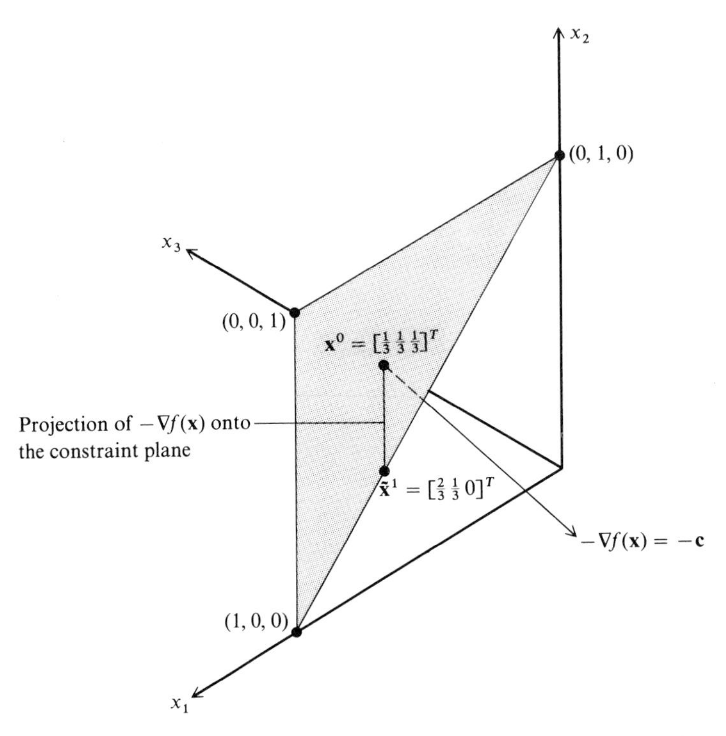

Figure 7.9 Illustration of Kalmarkar's algorithm. The shaded surface is the constraint $\mathbf{Ax} = \mathbf{b}$, and the coordinate planes are the other constraints. $\mathbf{c}^T\mathbf{x}$ is the objective function.

the equality constraints. In Fig. 7.9 the intersection is the shaded plane. From App. C, the projection matrix is $\mathbf{P} = \mathbf{I} - \mathbf{A}^T(\mathbf{AA}^T)^{-1}\mathbf{A}$ so that for this example

$$\mathbf{P} = \begin{bmatrix} 1 & 0 & 0 \\ 0 & 1 & 0 \\ 0 & 0 & 1 \end{bmatrix} - \begin{bmatrix} 1 \\ 1 \\ 1 \end{bmatrix} \left[[1 \quad 1 \quad 1] \begin{bmatrix} 1 \\ 1 \\ 1 \end{bmatrix} \right]^{-1} [1 \quad 1 \quad 1]$$

$$= \begin{bmatrix} 2/3 & -1/3 & -1/3 \\ -1/3 & 2/3 & -1/3 \\ -1/3 & -1/3 & 2/3 \end{bmatrix}$$

and the projected negative gradient is $\mathbf{P}(-\nabla f) = \mathbf{P}(-\mathbf{c}) = [1 \quad 0 \quad -1]^T$.

The search in the intersection of the constraints starts from $x(0) = [1/3 \quad 1/3 \quad 1/3]^T$, the centroid, and continues until a boundary is reached—almost. If the search did continue until it reached $x_3 = 0$, the point $\tilde{\mathbf{x}}^1$ would be found (see Fig. 7.9) determined by

$$1/3 + \lambda(1) = x_1^1$$

$$1/3 + \lambda(0) = x_2^1$$

$$1/3 + \lambda(-1) = x_3^1$$

With $x_3 = 0$ the last equation gives $\lambda = 1/3$, and $\tilde{\mathbf{x}}^1 = [2/3 \quad 1/3 \quad 0]^T$. However, you stop just before reaching the boundary at $\tilde{\mathbf{x}}^1$, say at $\lambda = 0.98(1/3)$, so that $\hat{\mathbf{x}}^1 = [1.98/3 \quad 1/3 \quad 0.02/3]^T$.

Phase $(k + 1)$

The basic idea of the second phase is to rescale the variables and transfer from the new $\hat{\mathbf{x}}^{k+1}$ back to the rescaled centroid. Rescaling is accomplished by dividing the elements of $\mathbf{x}^k$ by 1.98, 1, and 0.02, respectively, hence[1]

$$\mathbf{x}^{k+1} = \left[\frac{x_1^k}{1.98} \quad \frac{x_2^k}{1} \quad \frac{x_3^k}{0.02}\right]^T$$

Formally this step is

$$\mathbf{x}^{k+1} = \mathbf{D}^{-1}\mathbf{x}^k$$

where $\mathbf{D}$ is a diagonal matrix with elements here of (1.98, 1, 0.02).

In phase $(k + 1)$, the constraint $\mathbf{A}\mathbf{x}^k = \mathbf{b} = \mathbf{A}\mathbf{D}\mathbf{D}^{-1}\mathbf{x}^k = \mathbf{b}$ is changed to $\mathbf{A}\mathbf{D}\mathbf{x}^{k+1} = \mathbf{b}$, and the objective function $f(\mathbf{x}) = \mathbf{c}^T\mathbf{D}\mathbf{D}^{-1}\mathbf{x}^k$ becomes instead $f(\mathbf{x}) = \mathbf{c}^T\mathbf{D}\mathbf{x}^{k+1}$. In effect $\mathbf{AD}$ replaces $\mathbf{A}$ in the projection matrix and $\mathbf{Dc}$ replaces $\mathbf{c}$ in the objective function. After rescaling, the procedure on the kth stage is repeated.

Thus, for the next stage the search starts from $\mathbf{x}^1 = [0.168 \quad 0.333 \quad 16.67]^T$. The matrix $\mathbf{A}$ in the projection matrix is replaced by

$$\mathbf{AD} = [1 \quad 1 \quad 1]\begin{bmatrix} 1.98 & 0 & 0 \\ 0 & 1 & 0 \\ 0 & 0 & 0.02 \end{bmatrix} = [1.98 \quad 1 \quad 0.02]$$

and the objective function becomes

$$f(\mathbf{x}) = \mathbf{c}^T\mathbf{D}\mathbf{x} = [1 \quad 2 \quad 3]\begin{bmatrix} 1.98 & 0 & 0 \\ 0 & 1 & 0 \\ 0 & 0 & 0.02 \end{bmatrix}\begin{bmatrix} x_1 \\ x_2 \\ x_3 \end{bmatrix}$$
$$= 1.98x_1 + 2x_2 + 0.06x_3$$

The procedure then continues as explained in phase k.

Termination is tested by evaluating the vector of Lagrange multipliers $\mathbf{y}$ calculated from $\mathbf{A}\mathbf{A}^T\mathbf{y} = \mathbf{A}\mathbf{c}$. A positive y_i corresponds to an inactive constraint

---

[1] Karmarkar actually used a nonlinear transformation

$$\mathbf{x}^{k+1} = \frac{n\mathbf{D}^{-1}\mathbf{x}^k}{\mathbf{e}^T\mathbf{D}^{-1}\mathbf{x}^k}$$

where $\mathbf{e}$ is an n dimensional vector of 1's and $\mathbf{e}^T\mathbf{x}^0 = n$, or here $[1 \quad 1 \quad 1]\begin{bmatrix} x_1^0 \\ x_2^0 \\ x_3^0 \end{bmatrix}$.

and a zero y_i corresponds to an active constraint. $\mathbf{P}(-\nabla f)$ should be zero and $\mathbf{y} \geq \mathbf{0}$ at the solution.

7.11 LP APPLICATIONS

It has been estimated that a considerable fraction of the computer time expended at oil and chemical companies is devoted to solving LP's of various types. The kinds of problems solved and references include

1. Multiplant production/distribution (Hadley, 1962), including oil tanker routing and scheduling (Garvin, 1960).
2. Gasoline blending (Garvin, 1960; Johnson and Williamson, 1967)
3. Petroleum refinery operations (Beightler et al., 1979; Pike, 1986)
4. Power generation, steam systems (Bouillod, 1969)
5. Olefin manufacture (Sourander et al., 1984).

Case studies of the optimization of steam systems, a separation train, and olefin production by linear programming are described in Secs. 10.4, 11.5, and 13.4, respectively. General reviews of LP applications in a wide variety of fields have been given by Gass (1975) and Schrage (1983).

REFERENCES

Barnes, J. W., and R. M. Crisp, "Linear Programming. A Survey of General Purpose Algorithms," *AIIE Trans.*, 7(3) (1975).

Beightler, C. S., D. T. Phillips, and D. J. Wilde, *Foundations of Optimization*, Prentice-Hall, Englewood Cliffs, New Jersey (1979).

Beveridge, G. S., and R. S. Schechter, *Optimization: Theory and Practice*, McGraw-Hill, New York (1970).

Bouillod, P., "Compute Steam Balance by LP," *Hydroc. Proc.*, (August, 1969), p. 127.

Cooper, L., and D. Steinberg, *Methods and Applications of Linear Programming*, W. B. Saunders, Philadelphia (1974).

Dantzig, G. B., *Linear Programming and Extensions*, Princeton University Press, Princeton, New Jersey (1963).

Garvin, W. W., *Introduction to Linear Programming*, McGraw-Hill, New York (1960).

Gass, S. I., *Linear Programming*, 5th ed., McGraw-Hill, New York (1984).

Hadley, G., *Linear Programming*, Addison-Wesley, Reading, Massachusetts (1962).

Hughes, A. J., and D. E. Grawoig, *Linear Programming: An Emphasis on Decision Making*, Addison-Wesley, Reading, Massachusetts (1973).

IBM Corporation, "Mathematical Programming System—Extended (MPSX) and Generalized Upper Bounding (GUB) Program Description," SH20-0968-1, White Plains, New York.

Johnson, J. D., and C. Q. Williamson, "In-Line Gasoline Blending at Suntide Refinery," *IEEE Trans. Industry and General Applications*, March/April, 159–167 (1967).

Karmarkar, N., "A New Polynomial-Time Algorithm for Linear Programming," *Combinatoria*, 4, 373 (1984).

Khachian, L. G. "A Polynomial Algorithm in Linear Programming," *Sov. Math. Doklady*, *20*(1): 191 (1979).

Kim, C., *Introduction to Linear Programming*, Holt, Rinehart, and Winston, New York (1971).

Kozlov, A., and L. W. Black, "Berkeley Obtains New Results with the Karmarkar Algorithm," *SIAM News*, May, 1986, pp. 3, 20.

Luenberger, D. G., *Introduction to Linear and Nonlinear Programming*, Addison-Wesley, Reading, Massachusetts (1984).
Murtagh, B. A., *Advanced Linear Programming*, McGraw-Hill, New York (1983).
Murty, K. G., *Linear Programming*, John Wiley and Sons, New York (1983).
Pike, R. W., *Optimization for Engineering Systems*, Van Nostrand Reinhold, New York (1986).
Reklaitis, G. V., A. Ravindran, and K. M. Ragsdell, *Engineering Optimization*, Wiley-Interscience, New York (1984).
Shamir, R., "The Efficiency of the Simplex Method: A Survey," *Man. Sci.*, **33**: 301 (1987).
Schrage, L., *Linear, Integer, and Quadratic Programming with LINDO*, Scientific Press, Palo Alto, California (1983).
Sharda, R., "Linear Programming on Microcomputers: A Survey," *Interfaces*, **14**: 27 (1984).
Sourander, M. L., M. Kolari, J. C. Cugini, J. B. Poje, and D. C. White, "Control and Optimization of Olefin-Cracking Heaters," *Hydrocarbon Process.*, (June, 1984), p. 63.
Strang, G., *Introduction to Applied Mathematics*, Wellesley-Cambridge Press, Wellesley, Massachusetts, 1985.

SUPPEMENTARY REFERENCES

Barnes, E. R., "A Variation on Karmarkar's Algorithm for Solving Linear Programming Problems," *Math. Program.*, **36**, 174 (1986).
Best, M. J., and K. Ritter, *Linear Programming: Active Set Analysis and Computer Programs*, Prentice-hall, Englewood Cliffs, New Jersey, (1985).
Bozenhardt, H., "Multilevel Integrated Batch Control," *Chem. Eng. Prog.*, Dec., 35 (1985).
Cameron, N., *Introduction to Linear and Convex Programming*, Cambridge University Press, New York, (1985).
Fletcher, R. "Cancellation Errors in Quasi-Newton Methods," *SIAM J. Sci. Stat. Comput.*, **7**, 1387 (1986).
Gill, P. E., W. Murray, M. A. Saunders, J. A. Tomlin, and M. H. Wright, *Projected Newton Barrier Methods for Linear Programming and an Equivalence to Karmarkar's Projective Method*, Report SOL 85-11, Dept. OR., Standford Univ., July 1985.
Gill, P. E., W. Murray, M. A. Saunders, J. A. Tomlin, and M. H. Wright, "On Projected Newton Barrier Methods for Linear Programming and an Equivalence to Karmarkar's Projective Method," *Math Program.*, **36**, 183 (1986).
Hartley, R., *Linear and Nonlinear Programming*, John Wiley, New York, 1985.
Hooker, J. N., "Karmarkar's Linear Programming Algorithm," *Interfaces*, **16**, 75 (1986).
Jeter, M. W., *Math. Program.*, Marcel Dekker, New York (1986).
National Technical Information Service, *Linear Programming Algorithms for Optimization. 1974-November 1985 (Citations from the INSPEC: Information Services for the Physics and Engineering Communities Data Base)*. Rept. for 1974—November 85. PB 86-85 1664 GAR.
Ostrovsky, G. M., Y. M. Mikhailova, and T. A. Berezhinsky, "Large Scale System Optimization," *Comput. Chem. Eng.*, **10**, 123 (1986).
Perry, C. and R. Crellin, "The Precise Meaning of a Shadow Price," *Interfaces* **12**: 61 (1982).
Schmeck, J. "The Karmarkar Algorithm—A Fundamental Breakthrough in Linear Programming," 1986 Annual Conf. Proceed. ASEE, p. 1616 (1986).
Schrijver, A., *Theory of Linear and Integer Programming*, John Wiley, New York (1986).
Snee, R. D., "Developing Blending Models for Gasoline and Other Mixtures," *Technometrics*, **23**: 119 (1981).
Walsh, G. R., *An Introduction to Linear Programming*, 2d ed., John Wiley, New York, 1985.

PROBLEMS

7.1. A refinery has available two crude oils that have the yields shown in the following table. Because of equipment and storage limitations, production of gasoline, kerosene, and fuel oil must be limited as also shown in this table. There are no plant limitations on the production of other products such as gas oils.

The profit on processing crude No. 1 is \$1.00/bbl and on crude No. 2 it is \$0.70/bbl. Find the approximate optimum daily feed rates of the two crudes to this plant via a graphical method.

	Volume percent yields		Maximum allowable product rate bbl/day
	Crude 1	Crude 2	
Gasoline	70	31	6,000
Kerosene	6	9	2,400
Fuel oil	24	60	12,000

7.2. A confectioner manufactures two kinds of candy bars: Ergies (packed with energy for the kiddies) and Nergies (the "lo-cal" nugget for weight watchers without will power). Ergies sell at a profit of 50¢ per box while Nergies have a profit of 60¢ per box. The candy is processed in three main operations, blending, cooking, and packaging. The following table records the average time in minutes required by each box of candy, for each of the three activities.

	Blending	Cooking	Packing
Ergies	1	5	3
Nergies	2	4	1

During each production run, the blending equipment is available for a maximum of 14 machine hours, the cooking equipment for at most 40 machine hours, and the packaging equipment for at most 15 machine hours. If each machine can be allocated to the making of either type of candy at all times that it is available for production, determine how many boxes of each kind of candy the confectioner should make in order to realize the maximum profit. Use a graphical technique.

7.3. A problem has been converted to standard canonical form by the addition of slack variables, and has a basic feasible solution (with $x_1 = x_2 = 0$) as shown in the following tableau:

	x_1	x_2	x_3	x_4	x_5	f	b
Constraints x_3	-2	2	1	0	0	0	3
x_4	5	2	0	1	0	0	11
x_5	1	-1	0	0	1	0	4
Obj. function	4	2	0	0	0	1	0

Answer the following questions:

(*a*) Which variable should be increased first?

(*b*) Which row and which column designate the pivot point?

(*c*) What will be the limiting value of the variable you designated in a?

7.4. For the problem given in 7.3, find the next basis. Show the steps you take to calculate the new tableau, and indicate what the basic variables and nonbasic variables are in the new tableau. (Just a single step from one vertex to the next is asked for in this problem.)

7.5. You are given the following LP tableaus:

	x_1	x_2	x_3	x_4	f	b
x_3	3	−1	1	0	0	−6
x_4	4	−3	0	1	0	−4
	1	3	0	0	1	0

Why is the above formulation incorrect?

	x_1	x_2	x_3	x_4	f	b
x_3	1	−2	1	0	0	7
x_4	1	−3	0	1	0	4
	1	3	0	0	1	0

Is the problem that leads to the above formulation solvable? How do you interpret this problem geometrically?

	x_1	x_2	x_3	x_4	f	b
x_3	4	2	1	0	0	6
x_4	6	3	0	1	0	9
	1	3	0	0	1	0

Apply the Simplex rules to minimize f for the above formulation. Is the solution unique?

	x_1	x_2	x_3	x_4	f	b
x_3	4	2	1	0	0	7
x_4	6	3	0	1	0	5
	−1	0	0	0	1	0

Can you find the minimum of f? Why or why not?

7.6. A company has two alkylate plants, A_1 and A_2, from which a given product is distributed to customers C_1, C_2, and C_3. The transportation costs are given as follows:

Refinery	A_1	A_1	A_1	A_2	A_2	A_2
Customer	C_1	C_2	C_3	C_1	C_2	C_3
Cost, $/ton	25	60	75	20	50	85

The maximum refinery production rates and minimum customer demand rates are fixed and known to be as follows:

Customer or refinery	A_1	A_2	C_1	C_2	C_3
Rate, tons/day	1.6	0.8	0.9	0.7	0.3

The cost of production for A_1 is $30/ton for production levels less than 0.5 ton/day; for production levels greater than 0.5 ton/day, the production cost is $40/ton. A_2's production cost is uniform at $35/ton.

Find the optimum distribution policy to minimize the company's total costs.

7.7. A chemical plant makes three products and utilizes three raw materials in limited supply as shown in Fig. P7.7. Each of the three products is produced in a separate process (1, 2, 3) according to the schematic shown in the figure.

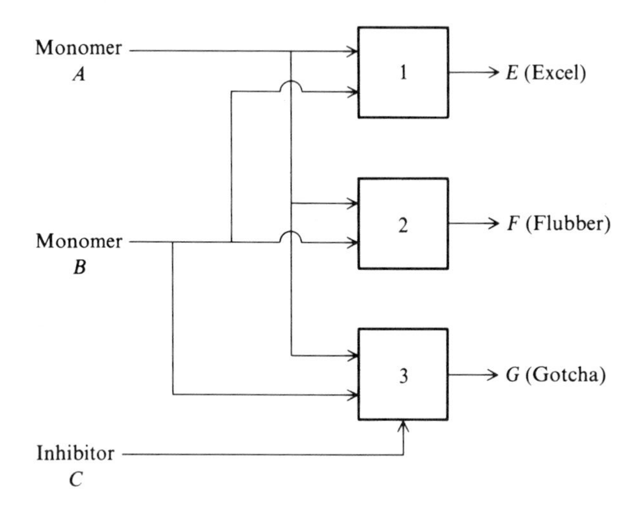

Figure P7.7

The available *A*, *B*, and *C* do not have to be totally consumed.

Process data:

Raw material	Maximum available, lb/day	Cost $/100 lb
A	4000	1.50
B	3000	2.00
C	2500	2.50

Process	Product	Reactants needed (lb) per lb product	Operating cost	Selling price of product
1	*E*	2/3 *A*, 1/3 *B*	$1.00/100 lb *A* (consumed in 1)	$4.00/100 lb *E*
2	*F*	2/3 *A*, 1/3 *B*	$0.50/100 lb *A* (consumed in 2)	$3.30/100 lb *F*
3	*G*	1/2 *A*, 1/6 *B*, 1/3 *C*	$1.00/100 lb *G* (produced in 3)	$3.80/100 lb *G*

Set up the linear profit function and linear constraints to find the optimum product distribution and apply the Simplex technique to obtain numerical answers.

7.8. A gentleman farmer with 100 acres wishes to optimize his crop of tomatoes, cabbage, and lettuce. He can sell tomatoes at 20 ¢ a pound, lettuce at 18 ¢ a head, and cabbage at 25 ¢ a head. The average yield per acre is 1500 pounds of tomatoes, 2500 heads of lettuce, and 1500 heads of cabbage. Labor required per acre during the growing season is five man-days for tomatoes and six man-days for both lettuce and cabbage. Within the cooperative, 400 man-days of labor are available and each laborer who is a member of the cooperative is paid $30 per man-day. Fertilizer costs 15 ¢ per pound; fertilizer requirements are: 125 pounds per acre of tomatoes and 50 pounds per acre of lettuce and cabbage. Based on this information, we desire to find the best way to use land and manpower to maximize profits.

(*a*) Set up this problem as a linear programming problem.
(*b*) Solve using the Simplex algorithm.
(*c*) Which constraint is not active; why is not all of the acreage being used?
(*d*) Would higher prices for cabbage or lettuce
 (1) change the optimum?
 (2) leave it unchanged?
 (3) do either, depending upon the effect of the price changes?

Discuss your conclusions regarding the effect of these price changes.

7.9. A fertilizer company mixes three different types of chemicals in their Premium blend which sells for $1.00 per pound. Their less expensive blend (Regular) sells for $0.80 per pound and contains only two types of chemicals. The mixing formula for both blends is shown below.

	Chemicals		
Blends:	**Type *A***	**Type *B***	**Type *C***
Premium	10%	5%	5%
Regular	0%	15%	5%
Amount available	400 lb	300 lb	500 lb

The remainder of each blend is inert filler material which comes from a nearby pit mine at a cost of $0.10/lb. An unlimited supply of filler is available.

The cost per pound of type *A*, type *B* and type *C* chemicals is $0.30, $0.25, and $0.21, respectively. Determine the optimal amount of each blend that should be prepared; leftover chemicals can be used again. Solve using liner programming.

7.10. Maximize $y = x_1 + 3x_2 - x_3$

Subject to $x_1 + 2x_2 + x_3 = 4$

$$2x_1 + x_2 \leq 5$$

$$x_1, x_2, x_3 \geq 0$$

using the Simplex method. Find the optimum values of x_1, x_2, x_3, and y.

Solve the problem by

(1) guessing to obtain an initial basis
(2) using a phase I/phase II method.

7.11. Examine the following problem

$$\text{Minimize} \quad y = 3x_1 + x_2 + x_3$$

$$\text{Subject to} \quad x_1 - 2x_2 + x_3 \le 11$$

$$-4x + x_2 + 2x_3 \ge 3$$

$$2x_1 - x_3 = -1$$

$$x_1, x_2, x_3 \ge 0$$

Is there a basic feasible solution to the problem? Answer yes or no and explain.

7.12. You are asked to solve the following problem:

$$\text{Maximize} \quad f = 5x_1 + 2x_2 + 3x_3$$

$$\text{Subject to} \quad x_1 + 2x_2 + 2x_3 + x_4 = 8$$

$$3x_1 + 4x_2 + x_3 - x_5 = 7$$

$$x_1, \ldots, x_5 \ge 0$$

Explain in detail what you would do to obtain the first feasible solution to this problem. Show all equations. You do not have to calculate the feasible solution —just explain in detail how you would calculate it.

7.13. Solve the following LP:

$$\text{Minimize} \quad f = x_1 + x_2$$

$$x_1 + 3x_2 \le 12$$

$$x_1 - x_2 \le 1$$

$$2x_1 - x_2 \le 4$$

$$2x_1 + x_2 \le 8$$

$$x_i \ge 0$$

Does the solution via the Simplex method exhibit cycling (a degenerate solution)?

7.14. Alkylate, cat cracked gasoline, and straight run gasoline are blended to make aviation gasolines *A* and *B* and two grades of motor gasoline. The specifications on motor gasoline are not as rigid as for aviation gas. Physical property and production data for the inlet streams are as follows:

Stream	RVP	ON(0)	ON(4)	Available bbl/day
Alkylate	5	94	108	4000
Cat cracked gasoline	8	84	94	2500
S.r. gasoline	4	74	86	4000

RVP = Reid vapor pressure (measure of volatility)
ON = octane number; in parentheses, number of mL/gal of tetraethyl lead (TEL)
S.r. = straight run

For the blended products:

Product	RVP	TEL level	ON	Profit, $/bbl
Aviation gasoline *A*	≤7	0	≥80	5.00
Aviation gasoline *B*	≤7	4	≥91	5.50
Leaded motor gasoline	....	4	≥87	4.50
Unleaded motor gasoline	....	0	≥91	4.50

Set up this problem as an LP and solve using a standard LP computer code.

7.15. Minimize $f = 3x_1 + 5x_2$

Subject to

$$
\begin{aligned}
x_1 + 3x_2 &\geq 10 \\
2x_1 - x_2 &\geq 4 \\
-x_1 + 4x_2 &\geq -2 \qquad x_1 \geq 0 \\
-x_1 - x_2 &\geq -20 \qquad x_2 \geq 0
\end{aligned}
$$

(*a*) Use artificial variables and a phase I/phase II procedure to obtain $\mathbf{x}^*$.

(*b*) Formulate the dual problem and solve it. Compare your answers with *a*.

(*c*) Identify where the shadow prices appear in parts *a* and *b*.

7.16. Resolve Prob. 7.9 using the dual formulation.

7.17. In Prob. 7.1 what are the shadow prices for incremental production of gasoline, kerosene, and fuel oil? Suppose the profit coefficient for crude #1 is increased by 10 percent and crude #2 by 5 percent. Which change has a larger influence on the objective function?

7.18. The problem of optimizing production from several plants with different cost structures and distributing the products to several distribution centers is a common one in the chemical industry. Newer plants often yield lower cost products because we learn from the mistakes made in designing the original plant. Due to plant expansions, rather unusual cost curves can result. The key cost factor is the incremental variable cost, which gives the cost per pound of an additional pound of production. Ordinarily, this variable cost is a function of production level.

Let us consider three different plants producing a product called DAB. The Frag plant located in Europe has an original design capacity of 100×10^6 lb/year but has been expanded to produce as high as 170×10^6 lb/year. The incremental variable cost for this plant decreases slightly up to 120×10^6 lb/year, but for higher production rates severe reaction conditions cause the yields to deteriorate, and cause a gradual increase in the variable cost, as shown by the equation below. There are no significant by-products sold from this plant. Using VC = variable cost in \$/100 lb and x = production level $\times 10^{-6}$ lb/year

$$VC = 4.5 - (x - 100)(0.005) \qquad 100 \leq x \leq 120$$

$$VC = 4.4 + (x - 120)(0.02) \qquad 120 \leq x \leq 170$$

The Swung-Lo plant, located in the Far East, is a relatively new plant which has an improved reactor/recycle design. This plant can be operated between 80×10^6 and 120×10^6 lb/year and has a constant variable cost of \$5.00/100 lb.

The Hogshooter plant, located in the United States, has a range of operation from 120×10^6 to 200×10^6 lb/year. The variable cost structure is rather complicated due to the effects of extreme reaction conditions, separation tower limitation, and several by-products, which are affected by environmental considerations. These considerations cause a discontinuity in the incremental variable cost curve at 140×10^6 lb-year as given by the equations below:

$$VC = 3.9 + (x - 120)(0.005) \qquad 120 \le x \le 140$$

$$VC = 4.6 + (x - 140)(0.01) \qquad 140 \le x \le 200$$

There are three main customers for the DAB, located in the Europe ($C1$), the Far East ($C2$), and the United States ($C3$), respectively. The matrix below shows the transportation costs of (¢/lb) and total demand to the customers ($C1$, $C2$, $C3$) with plant locations denoted as $A1$ (Frag), $A2$ (Swung-Lo), and $A3$ (Hogshooter). The closest pairing geographically is $A1$-$C1$; $A2$-$C2$; and $A3$-$C3$.

	A1	**A2**	**A3**	**Total demand**
$C1$	0.2	0.7	0.6	140
$C2$	0.7	0.3	0.8	100
$C3$	0.6	0.8	0.2	170

Use an iterative method based on successive linearization of the objective function to determine the optimum distribution plan for the product, DAB. Use an LP code to minimize total cost at each iteration.

CHAPTER 8

NONLINEAR PROGRAMMING WITH CONSTRAINTS

In Chap. 1, you saw some examples of the constraints that occur in optimization problems. In any problem, constraints should be identified either as being inequality constraints or equality constraints, and linear or nonlinear. Chapter 7 described the Simplex method for solving problems with linear objective functions subject to linear constraints. This chapter treats more difficult problems involving minimization (or maximization) of a nonlinear objective function subject to linear and/or nonlinear constraints:

$$\begin{aligned} &\text{Minimize} \quad f(\mathbf{x}) && \mathbf{x} = [x_1 \quad x_2 \quad \cdots \quad x_n]^T \\ &\text{Subject to} \quad h_j(\mathbf{x}) = 0 && j = 1, 2, \ldots, m \\ &\qquad\qquad\quad g_j(\mathbf{x}) \geq 0 && j = m+1, \ldots, p \end{aligned} \tag{8.1}$$

The inequality constraints in problem (8.1) are frequently transformed into equality constraints as explained in the introduction to Sec. 8.4 so that we focus first on problems involving only equality constraints.

One method of handling just one or two linear or nonlinear equality constraints is to solve explicitly for one variable and eliminate that variable from the problem formulation by substitution for it in all the functions and equations in the problem. In many problems elimination of a single equality constraint will often be a superior method to an approach in which the constraint is retained and some constrained optimization procedure is executed. For example, suppose you want to minimize the following objective function which is subject to a single equality constraint

$$\text{Minimize} \quad f(\mathbf{x}) = 4x_1^2 + 5x_2^2 \tag{8.2a}$$

$$\text{Subject to} \quad 2x_1 + 3x_2 = 6 \tag{8.2b}$$

Either x_1 or x_2 can be eliminated without difficulty. Let us solve for x_1

$$x_1 = \frac{6 - 3x_2}{2} \tag{8.3}$$

and then eliminate it in Eq. (8.2*a*). If the expression for x_1 is substituted into the objective function, the new equivalent objective function in terms of a single variable x_2 is

$$f(x_2) = 14x_2^2 - 36x_2 + 36 \tag{8.4}$$

The constraint in the original problem has been eliminated, and $f(x_2)$ is an unconstrained function with one degree of freedom (one independent variable).

We can now minimize the objective function (8.4), say by setting the first derivative equal to zero, and solving for the optimum value of x_2

$$\frac{\partial f(x_2)}{\partial x_2} = 28x_2 - 36 = 0 \qquad x_2^* = 1.286$$

Once x_2^* is obtained, then, x_1^* can be directly obtained via the constraint (8.2*b*):

$$x_1^* = \frac{6 - 3x_2^*}{2} = 1.071$$

The geometric interpretation of the above problem requires visualizing the objective function as the surface of a paraboloid in three-dimensional space. See Fig. 8.1. The projection of the intersection of the paraboloid and the plane representing the constraint onto the $f(x_2) - x_2$ plane is a parabola. We then find the minimum of the resulting parabola. The elimination procedure described above is tantamount to projecting the intersection locus onto the x_2 axis. The intersection locus could also be projected onto the x_1 axis (by elimination of x_2). Would you obtain the same result for $\mathbf{x}^*$ as before?

In problems in which there are n variables and m equality constraints, we could attempt to eliminate m variables by direct substitution. If all the equality constraints can be removed, and there are no inequality constraints, the objective function can then be differentiated with respect to each of the remaining $(n-m)$ variables and the derivatives set equal to zero. Or, a computer code for unconstrained optimization can be employed to obtain $\mathbf{x}^*$. If the objective function is convex (as in the example above) and the constraints form a convex region, then there is only one stationary point, which is the global minimum.

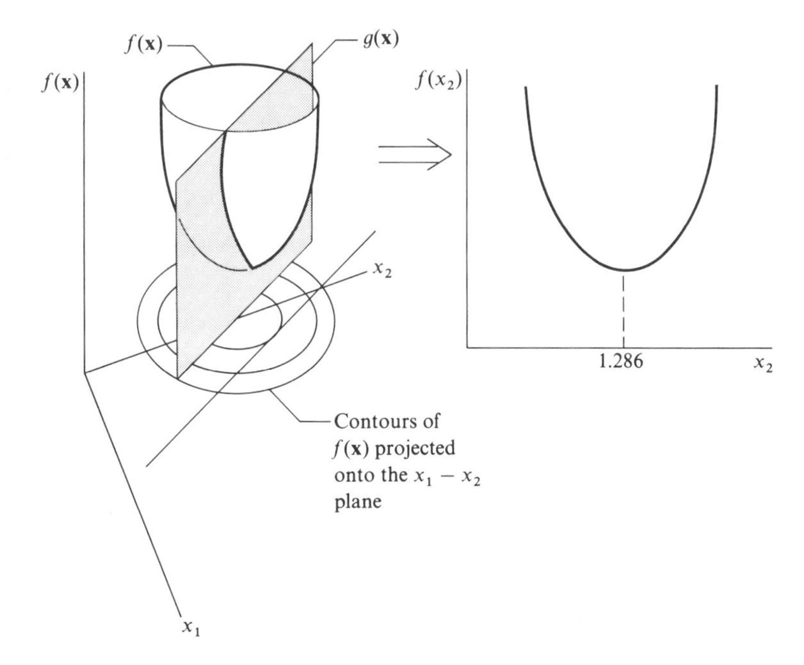

Figure 8.1 Graphical representation of a function of two variables reduced to a function of one variable by substitution for one variable in the constraint.

Unfortunately, very few problems in practice assume this simple form or even permit the elimination of all equality constraints.

Consequently, in this chapter we will discuss the four major approaches for solving nonlinear programming problems with constraints:

1. Lagrange multiplier methods
2. Iterative linearization methods
3. Iterative quadratic programming methods
4. Penalty function methods

As we will show in this chapter, the last three methods are better for general problems than the classical method based on Lagrange multipliers.

8.1 THE LAGRANGE MULTIPLIER METHOD

Let us now consider an optimization problem involving two variables and a single equality constraint. The statement of the problem is as follows:

$$\text{Minimize} \quad f(x_1, x_2) \tag{8.5}$$

$$\text{Subject to} \quad \bar{h}(x_1, x_2) = e \qquad (e \text{ is a constant}) \tag{8.5a}$$

i.e.,

$$\text{Subject to} \quad h(x_1, x_2) = \bar{h} - e = 0 \tag{8.5b}$$

At the constrained optimum the following two necessary conditions must hold simultaneously

$$df = 0 = \frac{\partial f}{\partial x_1} dx_1 + \frac{\partial f}{\partial x_2} dx_2 \tag{8.6}$$

$$dh = 0 = \frac{\partial h}{\partial x_1} dx_1 + \frac{\partial h}{\partial x_2} dx_2 \tag{8.7}$$

If $f(x_1, x_2)$ were an unconstrained function, both partial derivatives of $f(\mathbf{x})$ would have to be zero (that is, $\partial f/\partial x_1 = 0$ and $\partial f/\partial x_2 = 0$ for all dx_1 and dx_2) at the optimum. However, the variables x_1 and x_2 are constrained (dx_1 and dx_2 are not independent), so we cannot arbitrarily set both partial derivatives of $f(\mathbf{x})$ equal to zero. However, $f(\mathbf{x})$ must be an extremum at the optimum, hence $df(\mathbf{x}) = 0$. The second condition, namely $dh(\mathbf{x}) = 0$, exists because the constraint $\bar{h}(x_1, x_2)$ was assumed to be equal to a constant.

In Eqs. (8.6) and (8.7), we can envision that dx_1 and dx_2 are differential perturbations or variations from the optimum point x_1^* and x_2^*, and rewrite the equations in the form of two equations in two unknown variables:

$$a_{11}\, dx_1 + a_{12}\, dx_2 = 0 \tag{8.8a}$$

$$a_{21}\, dx_1 + a_{22}\, dx_2 = 0 \tag{8.8b}$$

Because the right-hand sides of both equations are 0, either a trivial solution ($dx_1 = dx_2 = 0$) or a nontrivial (nonunique) solution can be chosen (refer to Appendix B). In order for a nontrivial solution to exist, the determinant of the coefficient matrix (a_{ij}) must be 0 in (8.8), or $a_{11}a_{22} - a_{21}a_{12} = 0$. Expressing this condition in another way:

$$\frac{a_{11}}{a_{21}} = \frac{a_{12}}{a_{22}} \qquad \text{or} \qquad \frac{a_{11}}{a_{12}} = \frac{a_{21}}{a_{22}} \tag{8.9}$$

Therefore, in general

$$a_{11} = -\omega a_{21} \qquad \text{and} \qquad a_{12} = -\omega a_{22} \tag{8.10}$$

where ω is a parameter which relates the partial derivative of $f(\mathbf{x})$ and $h(\mathbf{x})$; ω is commonly referred to as a *Lagrange multiplier*. If an augmented objective function, called the *Lagrangian*, is defined as

$$L(\mathbf{x}, \omega) = f(\mathbf{x}) + \omega h(\mathbf{x}) \qquad [\text{note } h(\mathbf{x}) = 0]$$

then Eqs. (8.10) can be written in the following way (for two variables x_1 and x_2)

$$\frac{\partial f}{\partial x_1} + \omega \frac{\partial h}{\partial x_1} = \frac{\partial L}{\partial x_1} = 0 \tag{8.11a}$$

$$\frac{\partial f}{\partial x_2} + \omega \frac{\partial h}{\partial x_2} = \frac{\partial L}{\partial x_2} = 0 \tag{8.11b}$$

In the conditions (8.11) describing the optimum, we have not explicitly included the effect of the constraint. Therefore, in addition to the necessary conditions for a constrained optimum the following equation must hold:

$$\frac{\partial L(\mathbf{x}, \omega)}{\partial \omega} = h(\mathbf{x}) = 0 \tag{8.12}$$

In summary, the procedure of using Lagrange multipliers to locate an optimum point requires that an additional variable ω be defined as well as a Lagrange function. One of the necessary conditions for an extremum of $f(\mathbf{x})$ is that the partial derivatives of $L(\mathbf{x})$ with respect to x_1, x_2, and ω must vanish (rather than those solely with respect to the original two variables). Therefore, we have increased the dimensionality of the optimization problem by one corresponding to one equality constraint, but we have eliminated the constraint explicitly. In addition, if the constraint $h(\mathbf{x}) = 0$ is satisfied, then the values of $f(\mathbf{x})$ and $L(\mathbf{x}, \omega)$ at the constrained optimum are the same.

EXAMPLE 8.1 USE OF LAGRANGE MULTIPLIERS

We want to

$$\text{Minimize} \quad f(\mathbf{x}) = 4x_1^2 + 5x_2^2 \tag{a}$$

$$\text{Subject to} \quad h(\mathbf{x}) = 0 = 2x_1 + 3x_2 - 6 \tag{b}$$

Solution. Let

$$L(x, \omega) = 4x_1^2 + 5x_2^2 + \omega(2x_1 + 3x_2 - 6) \tag{c}$$

Apply the necessary conditions (8.11) and (8.12)

$$\frac{\partial L(\mathbf{x}, \omega)}{\partial x_1} = 8x_1 + 2\omega = 0 \tag{d}$$

$$\frac{\partial L(\mathbf{x}, \omega)}{\partial x_2} = 10x_2 + 3\omega = 0 \tag{e}$$

$$\frac{\partial L(\mathbf{x}, \omega)}{\partial \omega} = 2x_1 + 3x_2 - 6 = 0 \tag{f}$$

By substitution, $x_1 = -\omega/4$ and $x_2 = -3\omega/10$, and therefore Eq. (f) becomes

$$2(-\omega/4) + 3(-3\omega/10) - 6 = 0$$

$$\omega = -30/7$$

$$x_1^* = 1.071$$

$$x_2^* = 1.286$$

We must check to see $\mathbf{x}^*$ is a minimum and not a stationary point (or maximum) because the Lagrangian function itself exhibits a saddle point with respect to $\mathbf{x}$ and $\boldsymbol{\omega}$ at the optimum (often referred to as a "minimax" problem; see Bazaraa and Shetty, 1979, sec. 6.2). The sufficient conditions to check for a minimum are fairly involved and are listed in Sec. 8.2. Because in Example 8.1 the three equations arising from the necessary conditions were simple linear equations, we did not have to resort to solving a system of three nonlinear equations. However, if the character of the objective function and constraints was such that (d), (e), and (f) in Example 8.1 were nonlinear equations, then the method of Lagrange multipliers is not particularly useful because the solution of a set of nonlinear equations is as difficult, or more so, than the direct solution of the original optimization problem by one of the methods described in Secs. 8.4 and 8.6 below.

The theory of Lagrange multipliers can readily be extended to constrained optimization problems involving more than two variables. In general, if the objective function contains n decision variables, and there are m equality constraints involving the decision variables (the m constraints must be independent and m must be less than n), then the Lagrange multiplier formulation of an optimization problem requires the use of m Lagrange multipliers. The final unconstrained augmented objective function, the Lagrangian, will be a function of $n + m$ variables, and $n + m$ necessary conditions corresponding to each of these $n + m$ variables must be derived and solved in order to obtain a solution at the constrained optimum.

The general nonlinear programming problem can be converted to a problem that contains only equality constraints as follows:

$$\begin{aligned} &\text{Minimize} && f(\mathbf{x}) && \\ &\text{Subject to} && h_j(\mathbf{x}) = 0 && j = 1, \ldots, m \\ & && g_j(\mathbf{x}) - \sigma_j^2 = 0 && j = m+1, \ldots, p \end{aligned} \tag{8.13}$$

By substracting the square of a slack variable, σ_j^2, from $g_j(\mathbf{x})$, $j = m + 1, \ldots, p$, we can guarantee that the inequality constraints are satisfied but convert them to equality constraints. Note that by use of a slack variable, σ_j^2, a positive value for any σ_j, we avoid adding any inequality constraints $\sigma_j \geq 0$ to the nonlinear programming problem as is done in linear programming. Then, we can define the Lagrange function

$$L(\mathbf{x}, \omega) = f(\mathbf{x}) + \sum_{j=1}^{m} \omega_j h_j(\mathbf{x}) + \sum_{j=m+1}^{p} \omega_j [g_j(\mathbf{x}) - \sigma_j^2] \tag{8.14}$$

where the ω_j, $i = 1, \ldots, p$ are weighting factors independent of $\mathbf{x}$ identifiable as Lagrange multipliers. The necessary and sufficient conditions (Dennis, 1959) for x^* to be a solution of the nonlinear programming problem (8.13) are that $f(\mathbf{x}^*)$ be convex, that the constraint set be convex in the vicinity of $\mathbf{x}^*$, and that the following set of equations yielding a stationary solution of (8.14) be satisfied at $\mathbf{x}^*$

$$\begin{aligned} &\frac{\partial L(\mathbf{x}^*)}{\partial x_i} = 0 && \text{for } i = 1, \ldots, n \\ &\frac{\partial L(\mathbf{x}^*)}{\partial \omega_j} = 0 && \text{for } j = 1, \ldots, p \\ &\frac{\partial L(\mathbf{x}^*)}{\partial \sigma_j} = 2\omega_j \sigma_j = 0 && \text{for } j = m+1, \ldots, p \\ &\omega_j \leq 0 && j = 1, \ldots, p \end{aligned} \tag{8.15}$$

For a maximum of $f(\mathbf{x})$, replace $\omega_j \leq 0$ with $\omega_j \geq 0$. If the set of equations (8.15) is a nonlinear set, it is preferable to select one of the other optimization methods discussed in subsequent sections.

EXAMPLE 8.2 APPLICATION OF THE LAGRANGE MULTIPLIER METHOD WITH NONLINEAR INEQUALITY CONSTRAINTS

Solve the problem

$$\text{Minimize} \quad f(\mathbf{x}) = x_1 x_2$$

$$\text{Subject to} \quad g(\mathbf{x}): 25 - x_1^2 - x_2^2 \geq 0 \tag{a}$$

by the Lagrange multiplier method.

Solution. The Lagrange function is

$$L(\mathbf{x}, \omega) = x_1 x_2 + \omega(25 - x_1^2 - x_2^2 - \sigma_1^2) \tag{b}$$

The necessary conditions for a stationary point are

$$\begin{aligned} \frac{\partial L}{\partial x_1} &= x_2 - 2\omega x_1 = 0 \\ \frac{\partial L}{\partial x_2} &= x_1 - 2\omega x_2 = 0 \\ \frac{\partial L}{\partial \omega} &= 25 - x_1^2 - x_2^2 - \sigma_1^2 = 0 \\ \frac{\partial L}{\partial \sigma_1} &= 2\omega\sigma_1 = 0 \end{aligned} \tag{c}$$

The simultaneous solutions of Eqs. (*c*) for $\omega = 0$ and for $\omega \neq 0$ are listed in Table E8.2. How would you calculate these values?

Columns two and three of Table E8.2 list the components of $\mathbf{x}^*$ that are the stationary solutions of problem (*a*). Note that the solutions for $\omega_1 < 0$ are minima, those for $\omega_1 > 0$ are maxima, and $\omega_1 = 0$ is a saddle point of problem (*a*). Figure E8.2 illustrates the functions in problem (*a*). The contours of the objective function (hyperbolas) are represented by broken lines, and the feasible region is bounded by the shaded area enclosed by the circle $g(\mathbf{x}) = 0$. Points *B* and *C* correspond to the two minima, *D* and *E* to the two maxima, and *A* to the saddle point of $f(\mathbf{x})$.

Table E8.2 Solutions of problem (*a*) by the Lagrange multiplier method

ω	x_1	x_2	**Point**	σ_1	$f(\mathbf{x})$	**Remarks**
0	0	0	*A*	5	0	Saddle
−0.5	+3.54	−3.54	*B*	0	−12.5	Minimum
	−3.54	+3.54	*C*	0	−12.5	Minimum
0.5	+3.54	+3.54	*D*	0	+12.5	Maximum
	−3.54	−3.54	*E*	0	+12.5	Maximum

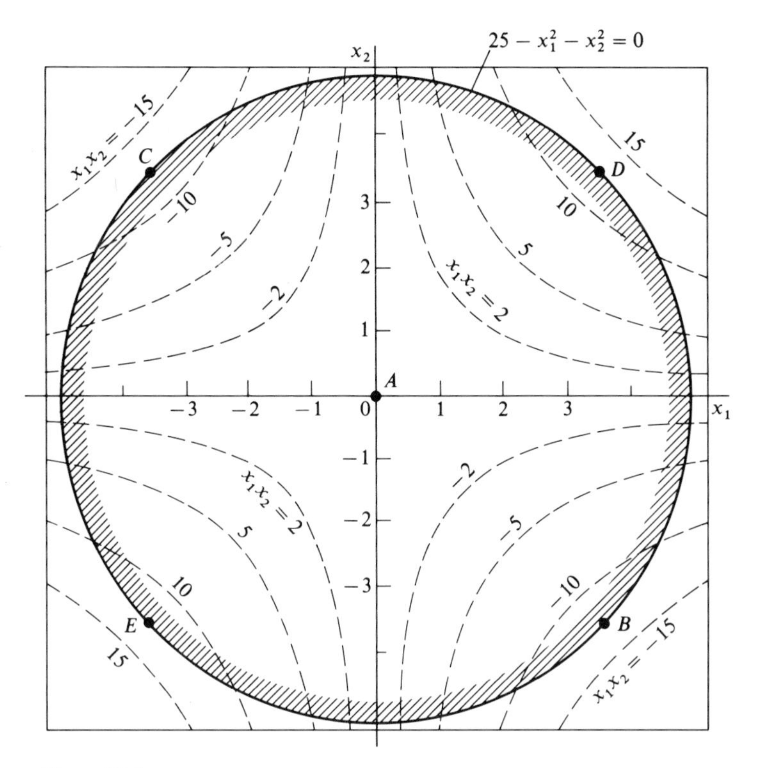

Figure E8.2

Lagrange multipliers have an important special interpretation in optimization as discussed in Sec. 7.8. The Lagrange multiplier for a given constraint indicates how much the objective function will change for a differential change in the constraint constant [the parameter e in Eq. (8.5*a*)]. Look again at Eq. (8.5*a*). We would like to determine how much the value of the objective function at the optimum will change as e varies. Form the partial derivative of the Lagrangian with respect to the parameter e

$$\frac{\partial L}{\partial e} = \frac{\partial}{\partial e}[f(\mathbf{x}) + \omega(\bar{h}(\mathbf{x}) - e)] = -\omega \tag{8.16}$$

Equation (8.16) can be approximated using difference notation

$$\frac{\Delta L}{\Delta e} = -\omega \qquad \text{or} \qquad \Delta L = -\omega \Delta e$$

Therefore, a small change in the constant e results in a change in L that is proportional to the Lagrange multiplier. Since, as previously stated, the Lagrangian L is equal to f when the constraint is satisfied ($h = 0$), then the Lagrange multiplier can be interpreted as giving the change in f (as well as L) for a change in the constant e at the optimal solution.

If we designate by ω^* the value of the Lagrange multiplier at $\mathbf{x}^*$, the optimal solution, where $\bar{h}(\mathbf{x}^*) = e$ and $f(\mathbf{x}^*) = L(\mathbf{x}^*, \omega^*)$, you can see that the values of $\mathbf{x}^*$ and ω^* are explicit functions of e, which is interpreted as the quantity of a limiting resource. In economics and operations research, Lagrange multipliers are termed *shadow prices* of the constraints because the change in the optimal value of the objective function per unit increase in the right-hand side of the equality constraint is given by ω^*. The value of $f(\mathbf{x})$ increases or decreases from $f(\mathbf{x}^*)$ with an increase or decrease in e depending on the sign of ω^*.

EXAMPLE 8.3 SENSITIVITY ANALYSIS USING LAGRANGE MULTIPLIERS

Consider the following general quadratic objective function with a linear constraint:

$$\text{Minimize} \quad f(\mathbf{x}) = c_1 x_1^2 + c_2 x_2^2 \tag{a}$$

$$\text{Subject to} \quad a_1 x_1 + a_2 x_2 = e \tag{b}$$

What is the change in $f(\mathbf{x})$ with respect to e at the stationary point?

Solution. The Lagrangian is

$$L(\mathbf{x}, \omega) = c_1 x_1^2 + c_2 x_2^2 + \omega(a_1 x_1 + a_2 x_2 - e) \tag{c}$$

Some of the necessary conditions are

$$\frac{\partial L}{\partial x_1} = 0 = 2c_1 x_1 + a_1 \omega \qquad x_1^* = \frac{-a_1 \omega^*}{2c_1}$$

$$\frac{\partial L}{\partial x_2} = 0 = 2c_2 x_2 + a_2 \omega \qquad x_2^* = \frac{-a_2 \omega^*}{2c_2}$$

$$\frac{\partial L}{\partial \omega} = 0 = a_1 x_1 + a_2 x_2 - e$$

These expressions lead to the following relations

$$0 = \frac{-a_1^2 \omega^*}{2c_1} - \frac{a_2^2 \omega^*}{2c_2} - e$$

$$\omega^* = \frac{e}{(-a_1^2/2c_1) - (a_2^2/2c_2)}$$

$$x_1^* = \frac{a_1 e}{2c_1[(a_1^2/2c_1) + (a_2^2/2c_2)]}$$

$$x_2^* = \frac{a_2 e}{2c_2[(a_1^2/2c_1) + (a_2^2/2c_2)]}$$

At $\mathbf{x}^*$ with $a_1x_1^* + a_2x_2^* - e = 0$

$$f(\mathbf{x}^*) = L(\mathbf{x}^*, \omega^*) = \frac{[(a_1^2e^2/4c_1) + (a_2^2e^2/4c_2)]}{[(a_1^2/2c_1) + (a_2^2/2c_2)]^2} \tag{d}$$

The sensitivity of $f(\mathbf{x}^*)$ with respect to e can be found by differentiation:

$$\frac{\partial f(x^*)}{\partial e} = \frac{e[(a_1^2/2c_1) + (a_2^2/2c_2)]}{[(a_1^2/2c_1) + (a_2^2/2c_2)]^2} = \frac{e}{(a_1^2/2c_1) + (a_2^2/2c_2)} = -\omega^* \tag{e}$$

the expected result.

The Lagrange method is quite helpful in analyzing parameter sensitivities in problems with multiple constraints. In a typical refinery, a number of different products are manufactured; these products usually must meet (or exceed) certain specifications in terms of purity as required by the customers. Suppose we carry out a constrained optimization for an objective function that includes the many variables which occur in the refinery, including variables in the fluid catalytic cracker, in the distillation column, and so on, and arrive at some economic optimum subject to the constraints on product purity. With the optimum values of the variables plus Lagrange multipliers corresponding to the product purity, we could then pose the question: How will the profits change if the product specification is either relaxed or made more stringent? To answer this question simply requires examining the Lagrange multiplier for each constraint. As an example, consider the case in which there are three major products (A, B, and C) and the Lagrange multipliers corresponding to each of the three demand constraints were calculated to be:

$$\omega_A = -0.001$$

$$\omega_B = -1.0$$

$$\omega_C = -0.007$$

These values for ω_i show (ignoring scaling) that profit of the refinery is much more sensitive to the production of product B than for the other two products.

8.2 NECESSARY AND SUFFICIENT CONDITIONS FOR A LOCAL MINIMUM

In this section we list the necessary and sufficient conditions for $\mathbf{x}^*$ to be a local minimum of the general constrained nonlinear programming problem

$$\begin{aligned} &\text{Minimize} && f(\mathbf{x}) && \mathbf{x} = [x_1 \quad x_2 \quad \cdots \quad x_n]^T \\ &\text{Subject to} && h_j(\mathbf{x}) = 0 && j = 1, \ldots, m \\ & && g_j(\mathbf{x}) \geq 0 && j = m+1, \ldots, p \end{aligned} \tag{8.17}$$

The evolution of the concepts underlying these conditions (stated below) and the proofs related to the conditions are fairly involved and can be found in Gill, Murray, and Wright (1981), Sec. 3.4, and Ben-Israel, Ben-Tal, and Zlobec (1981), Chap. 5.

Recall from Sec. 8.1 that a Lagrange function can be formed including the objective function and equality constraints. Here we include the inequality constraints with additional multipliers

$$L(\mathbf{x}, \boldsymbol{\omega}, \mathbf{u}) = f(\mathbf{x}) + \sum_{j=1}^{m} \omega_j h_j(\mathbf{x}) - \sum_{j=m+1}^{p} u_j g_j(\mathbf{x}) \tag{8.18}$$

(If the inequality constraints had been expressed as $g_j(\mathbf{x}) \le 0$ in Eq. (8.17), then we would add terms involving the product of the Lagrange multipliers u_j and $g_j(\mathbf{x})$ in the right-hand sum instead of subtract.) The ω_j can be positive or negative but the u_j must be ≥ 0. For proof, see Bazaraa and Shetty (1979 Chap. 4). Table 8.1 lists one form of the necessary and sufficient conditions. You can change the original problem (8.1) or (8.17) into a maximization problem by letting the new objective function be $-f(\mathbf{x})$.

The first-order necessary conditions, conditions (*a*), (*c*), (*d*), (*e*), (*f*), and (*g*) in Table 8.1, are the well-known *Kuhn-Tucker conditions* for optimality. These conditions serve as the basis for the design of some algorithms and as termination criteria for others. Recall from Sec. 6.2 that a valid search direction must satisfy the condition $\nabla^T f(\mathbf{x})\mathbf{s} < 0$. Any small movement along a search direction $\mathbf{s}$ for an NLP problem with constraints must also satisfy the constraints. Examine Fig. 8.2, which illustrates the concepts involved in two dimensions. At $\mathbf{x}^*$, the true minimum, $-\nabla f(\mathbf{x}^*)$ can be expressed as a non-negative (that is, $u_j^* \ge 0$) linear combination of the gradients of the active constraints at $\mathbf{x}^*$, or

$$-\nabla f(\mathbf{x}^*) = u_1^*[-\nabla g_1(\mathbf{x}^*)] + u_2^*[-\nabla g_2(\mathbf{x}^*)]$$

All feasible directions of search lie in the cone comprised of the negative gradients of the active constraints at $\mathbf{x}^*$.

Why this statement is so is illustrated by Figs. 8.2*a* and 8.2*b*. A small move in a direction $\mathbf{s}$ making an angle less than 90° with $-\nabla f(\mathbf{x}^*)$ will decrease $f(\mathbf{x})$ so that at a valid $\mathbf{x}^*$ no feasible search direction can have an angle of less than 90° from $-\nabla f(\mathbf{x}^*)$. Suppose that $-\nabla f(\mathbf{x}^*)$ lay above $-\nabla g_2(\mathbf{x}^*)$ as in Fig. 8.2*b*. Then $-\nabla f(\mathbf{x}^*)$ will make an angle of less than 90° with respect to $g_2(\mathbf{x})$ and hence allow a feasible move. Similarly, if $-\nabla f(\mathbf{x}^*)$ lies below $-\nabla g_1(\mathbf{x}^*)$, a feasible move (at least differentially) is possible with respect to $g_1(\mathbf{x})$. Because neither outcome should occur at an optimal point, a necessary condition is that $-\nabla f(\mathbf{x}^*)$ is restricted to fall within the cone generated by the negative gradients of the active constraints.

Table 8.1 The necessary and sufficient conditions for $\mathbf{x}^*$ to be a local minimum of (8.17)

The *necessary conditions* for $\mathbf{x}^*$ to be a local minimum of $f(\mathbf{x})$ are:

(*a*) $f(\mathbf{x})$, $h_j(\mathbf{x})$, $g_i(\mathbf{x})$ are all twice differentiable at $\mathbf{x}^*$

(*b*) The so-called "second-order constraint qualification" holds (it contains information about the curvature of the constraints that is taken into account at $\mathbf{x}^*$ as explained in Example 8.4 below); the sufficient conditions for this requirement are that the gradients of the binding constraints ($g_j(\mathbf{x}^*) = 0$), $\nabla g_j(\mathbf{x}^*)$, and the equality constraints, $\nabla h_j(\mathbf{x}^*)$, because $h_j(\mathbf{x}^*) = 0$, are linearly independent (refer to Fig. E8.5 for an illustration).

(*c*) The Lagrange multipliers exist; they do if (*b*) holds

(*d*) The constraints are satisfied at $\mathbf{x}^*$

(1) $h_j(\mathbf{x}^*) = 0$

(2) $g_i(\mathbf{x}^*) \geq 0$

(*e*) The Lagrange multipliers u_j^* (at $\mathbf{x}^*$) for the inequality constraints are not negative (ω_j can be positive or negative)

$u_j^* \geq 0$

(*f*) The binding (active) inequality constraints are zero; the inactive inequality constraints are >0, and the associated u_j's are 0 at $\mathbf{x}^*$

$u_j^* g_j(\mathbf{x}^*) = 0$

(*g*) The Lagrangian function is at a stationary point

$\nabla L_{\mathbf{x}}(\mathbf{x}^*, \boldsymbol{\omega}^*, \mathbf{u}^*) = \mathbf{0}$

(*h*) The Hessian matrix of L is *positive semidefinite* for those $\mathbf{v}$'s for which $\mathbf{v}^T \nabla g_j(\mathbf{x}^*) = 0$, and $\mathbf{v}^T \nabla h_j(\mathbf{x}^*) = 0$, that is, for all the active constraints

$\mathbf{v}^T \nabla^2 [L(\mathbf{x}^*, \boldsymbol{\omega}^*, \mathbf{u}^*)] \mathbf{v} \geq 0$

The *sufficient conditions* for $\mathbf{x}^*$ to be a local minimum are:

(*a*) The necessary conditions (*a*), (*b*) by implication, (*c*), (*d*), (*e*), (*f*), and (*g*)

(*b*) Plus a modification of necessary condition (*h*):

The Hessian matrix of L is *positive definite* for these vectors $\mathbf{v}$ such that

$$\left.\begin{aligned} \mathbf{v}^T \nabla g_j(\mathbf{x}^*) &= 0 \\ \mathbf{v}^T \nabla h_j(\mathbf{x}^*) &= 0 \end{aligned}\right\} \quad \text{for the binding constraints}$$

$$\mathbf{v}^T \nabla g_j(\mathbf{x}^*) \geq 0 \quad \text{for the inactive constraints}$$

$$\mathbf{v}^T \nabla^2 [L(\mathbf{x}^*, \boldsymbol{\omega}^*, \mathbf{u}^*)] \mathbf{v} > 0$$

When stated in algebraic terms, the above concepts are:

$$-\nabla f(\mathbf{x}^*) = \sum_j u_j^* [-\nabla g_j(\mathbf{x}^*)]$$

$$u_j^* \geq 0 \qquad \text{for the set of binding constraints}$$

To include all the inequality constraints, u_j^* is made equal to zero if $g_j(\mathbf{x}^*) > 0$, hence *complementarity condition* (*f*) arises, namely

$$u_j^* \geq 0 \qquad \text{if} \qquad g_j(\mathbf{x}^*) = 0$$

$$u_j^* = 0 \qquad \text{if} \qquad g_j(\mathbf{x}^*) > 0$$

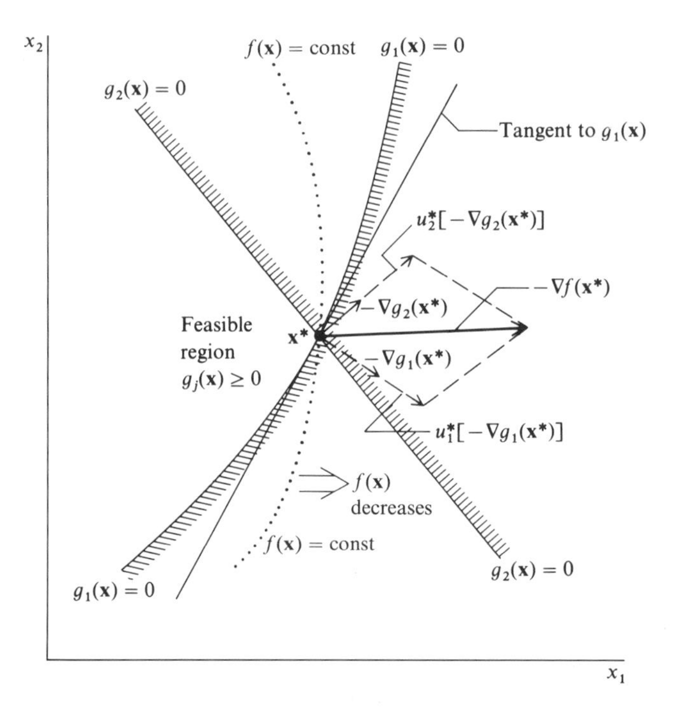

Figure 8.2*a* Interpretation of the Lagrange multipliers as weighting factors of the gradients of the active inequality constraints $g_j(\mathbf{x}) = 0$. The cone generated by $-\nabla g_j(\mathbf{x}^*)$ contains $-\nabla f(\mathbf{x}^*)$.

and the product $u_j^* g_j^*(\mathbf{x}^*) = 0$ for all the inequality constraints. Condition (*g*) arises because

$$\nabla L(\mathbf{x}^*, \mathbf{u}^*) = \nabla f(\mathbf{x}^*) - \sum u_j^* \nabla g_j(\mathbf{x}^*)$$

and the terms on the right-hand side are equal to each other at the optimum.

Condition (*b*) must be satisfied because of certain rare cases such as instances in which the gradients of the constraints fall along the same line in the same direction or in the opposite direction. To properly characterize the topography of a local minimum, it is necessary to take into account the curvature of the constraints at the presumed $\mathbf{x}^*$. If no reduction in $f(\mathbf{x})$ can be obtained along a smooth arc extending from $\mathbf{x}^*$ into the feasible region (i.e., a point on the arc satisfies all constraints), then $\mathbf{x}^*$ is a local minimum of $f(\mathbf{x})$. To demonstrate that feasible perturbations from $\mathbf{x}^*$ do not reduce $f(\mathbf{x})$, the constraint functions must meet certain theoretical conditions called constraint qualifications. Several ways exist in which these constraint qualifications can be specified, but the most practical is condition (*b*) in Table 8.1. In effect (*b*) means that the matrix whose rows are the gradients of the active constraints, namely the Jacobian matrix of the linearized constraints at $\mathbf{x}^*$, is of full rank.

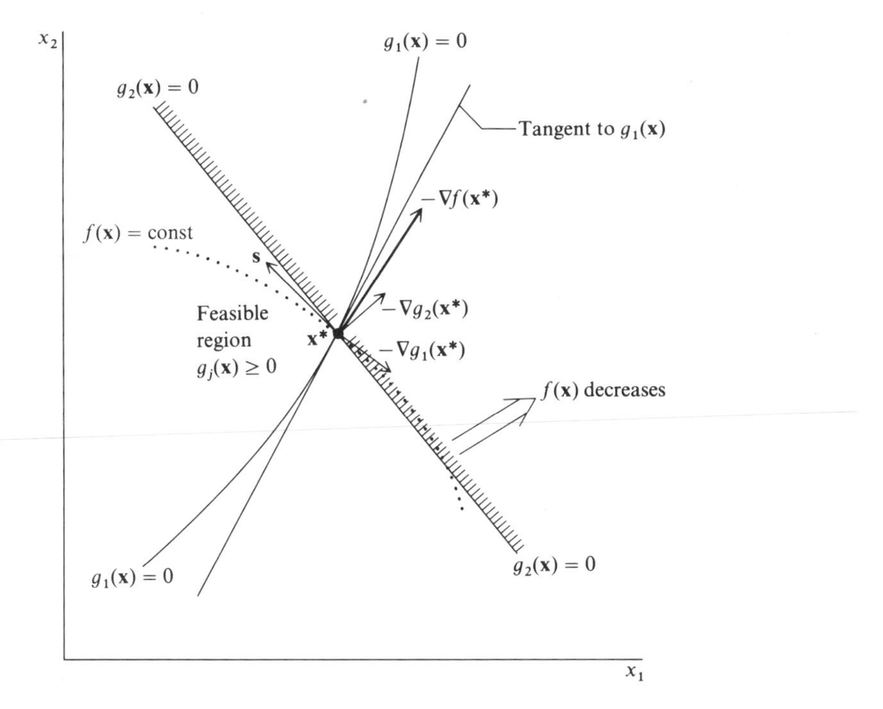

Figure 8.2*b* Interpretation of the Lagrange multipliers as weighting factors of the gradients of the active inequality constraints $g_j(\mathbf{x}) = 0$. Case when $-\nabla f(\mathbf{x})$ is not orthogonal to **s**, and lies above $-\nabla g_2(\mathbf{x})$.

Condition (*h*) means that the Lagrange function is convex with respect to those tangents to the active constraints at $\mathbf{x}^*$. Example 8.4 illustrates the significance of condition (*h*).

If you can demonstrate that the sufficient conditions are satisfied, then $\mathbf{x}^*$ is guaranteed to be a local minimum. Of course, a local minimum can exist even if the sufficient conditions are *not* met so that the direct use of the sufficient conditions as the strategy to search for a minimum rarely proves effective.

A special case of a nonlinear programming problem known as a *convex programming* problem exists as follows:

Minimize	$f(\mathbf{x})$	$f(\mathbf{x})$ is convex	
Subject to	$h_j(\mathbf{x}) = 0$	$j = 1, \ldots, m$	$h_j(x)$ are linear
	$g_j(\mathbf{x}) \geq 0$	$j = m + 1, \ldots, p$	$g_j(\mathbf{x})$ are concave
	$x_i \geq 0$		

For these conditions the constraints form a convex set and it can be shown that the *local minimum is also the global minimum* (Avriel, 1976, Sec. 4.5). Analogously a local maximum is also the global maximum if the objective function is concave and the constraints form a convex set.

EXAMPLE 8.4 TEST FOR NECESSARY AND SUFFICIENT CONDITIONS

Fiacco and McCormick (1968) used the following problem to illustrate the necessity of considering the curvature of the constraints in determining if $\mathbf{x}^*$ is a local minimum. Examine Fig. E8.4. The problem is to

$$\text{Minimize} \quad f(\mathbf{x}) = (x_1 - 1)^2 + x_2^2$$

$$\text{Subject to} \quad \begin{cases} \text{either} & (a)\ g_1(x) = -x_1 + \dfrac{x_2^2}{4} \geq 0 \\ \text{or} & (b)\ g_2(x) = -x_1 + x_2^2 \geq 0 \end{cases}$$

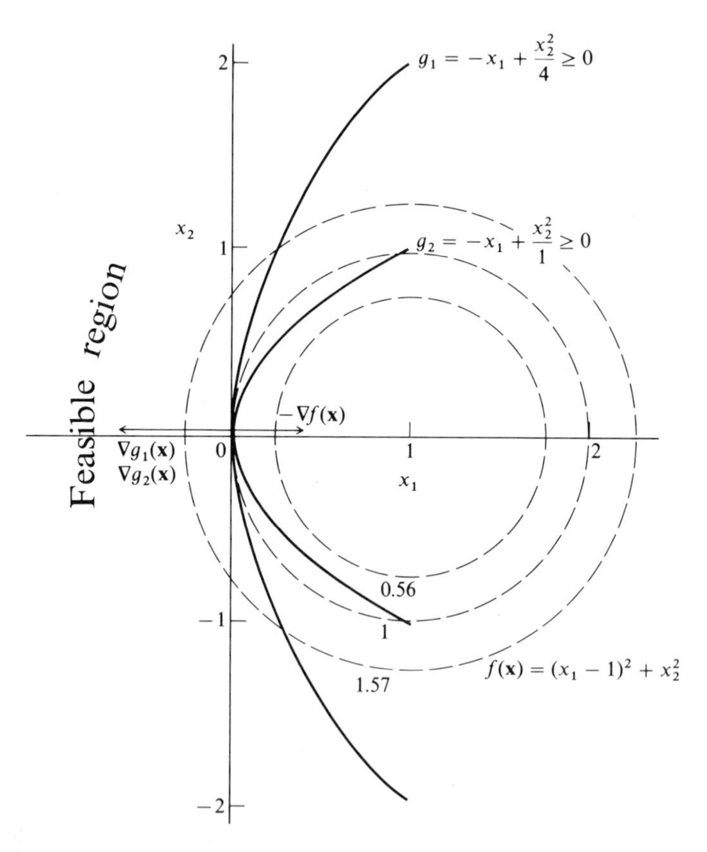

Figure E8.4

To illustrate how the necessary conditions might be used to find a possible solution to the above problem, we will find a solution when constraint (*a*) applies. To determine $\mathbf{x}^*$ we set $\nabla_{\mathbf{x}} L = 0$

$$\nabla_{\mathbf{x}} L = \begin{bmatrix} 2(x_1 - 1) \\ 2x_2 \end{bmatrix} - u \begin{bmatrix} -1 \\ \dfrac{x_2}{2} \end{bmatrix} = \begin{bmatrix} 0 \\ 0 \end{bmatrix} \tag{a}$$

$$2x_1 - 2 + u = 0 \tag{b}$$

$$2x_2 - \tfrac{1}{2}ux_2 = 0 \tag{c}$$

Is $u = 0$ or is $u > 0$? If $u = 0$, $x_1 = 1$ and $x_2 = 0$ (this point is the unconstrained optimum). However $g_1(\mathbf{x})$ is not satisfied because

$$-1 + 0 \ngeq 0$$

hence $u > 0$ and $g_1(\mathbf{x}) = 0$. Thus $x_1 = (x_2^2/4)$ and $u = 2 - 2x_1$ from Eq. (*b*). Introduction of these two relations into Eq. (*c*) yields $x_2 = 0$, $x_1 = 0$, $u = 2$, or $x_2 = \pm 2$, $x_1 = 1$, $u = 0$, but the latter triple is invalid (u must be >0), so that $\mathbf{x}^* = [0 \quad 0]^T$ and $u^* = 2$. Inspection of the contours of $f(\mathbf{x})$ indicates $\mathbf{x}^* = [0 \quad 0]^T$ is a local minimum of $f(\mathbf{x})$ subject to $g_1(\mathbf{x})$. Check to see if $\nabla^2 L$ is positive definite.

However, if $g_2(\mathbf{x})$ is substituted for $g_1(\mathbf{x})$, the situation changes drastically. A smooth arc exists from (0, 0) into the feasible region and $f(\mathbf{x})$ can be reduced, yet (0, 0) meets conditions (*a*), (*b*), (*c*), (*d*), (*e*), (*f*), and (*g*) in Table 8.1 as can be seen as follows:

(*a*) $f(\mathbf{x})$ and $g(\mathbf{x})$ are twice differentiable at (0, 0)
(*b*) Only one constraint exists so (*b*) is satisfied
(*c*) Hence u exists
(*d*) $g_2(0, 0)$ is satisfied
(*g*) $L(\mathbf{x}, u) = (x_1 - 1)^2 + x_2^2 - u(-x_1 + x_2^2)$

$$\nabla L(\mathbf{x}, u) = \begin{bmatrix} 2(x_1 - 1) + u \\ 2x_2 - 2ux_2 \end{bmatrix} = \begin{bmatrix} 0 \\ 0 \end{bmatrix}$$

hence at (0, 0), $2(0 - 1) + u = 0$, or $\hat{u} = 2$ ($\hat{u}$ is a presumed optimal multiplier)
(*e*) $\hat{u} = 2 \geq 0$
(*f*) $\hat{u}g_2(0, 0) = 0$ because $g_2(0, 0) = 0$

However, $\nabla^2 L(0, 0)$ is not positive semidefinite

$$\nabla^2 L(\mathbf{x}, u) = \begin{bmatrix} 2 & 0 \\ 0 & 2(1 - u) \end{bmatrix}$$

$$\nabla^2 L(0, 0, 2) = \begin{bmatrix} 2 & 0 \\ 0 & -2 \end{bmatrix}$$

The vector $\mathbf{v}$ that satisfies $\mathbf{v}^T \nabla g(0, 0) = 0$ is

$$[v_1 \quad v_2] \begin{bmatrix} -1 \\ 0 \end{bmatrix} = 0 \qquad \text{or} \qquad -v_1 + (0)v_2 = 0$$

hence v_2 can be any value and $v_1 = 0$. Thus $\mathbf{v}$ is a vector at $x_1 = 0$ along the x_2 axis, and

$$[0 \quad v_2]\begin{bmatrix} 2 & 0 \\ 0 & -2 \end{bmatrix}\begin{bmatrix} 0 \\ v_2 \end{bmatrix} = -2v_2^2 \not> 0$$

Let us now again consider minimizing $f(\mathbf{x})$ subject to constraint $g_2(\mathbf{x})$ but with $\mathbf{x}^*$ is located at $(\frac{1}{2}, \pm\sqrt{\frac{1}{2}})$. If we set $\nabla L(\frac{1}{2}, \pm\sqrt{\frac{1}{2}}) = 0$, we find

$$2(\tfrac{1}{2} - 1) + u = 0 \qquad \text{or} \qquad u^* = 1$$

Thus

$$\nabla^2 L(\tfrac{1}{2}, \pm\sqrt{\tfrac{1}{2}}, 1) = \begin{bmatrix} 2 & 0 \\ 0 & 0 \end{bmatrix}$$

which also is not positive definite. However, when we calculate the vector $\mathbf{v}^T \nabla g(\mathbf{x}^*) = 0$

$$[v_1 \quad v_2]\begin{bmatrix} -1 \\ \pm\sqrt{2} \end{bmatrix} = 0 \qquad \text{or} \qquad -v_1 \pm \sqrt{2}v_2 = 0$$

and we conclude $v_1 = \sqrt{2}v_2$ so that $\mathbf{v}^T = [\sqrt{2}v_2 \quad v_2]$.
Then

$$[\sqrt{2}v_2 \quad v_2]\begin{bmatrix} 2 & 0 \\ 0 & 0 \end{bmatrix}\begin{bmatrix} \sqrt{2}v_2 \\ v_2 \end{bmatrix} = 4v_2^2 > 0$$

so that condition (*h*) is satisfied and the sufficient condition (*b*′) is also satisfied.

EXAMPLE 8.5 A TEST FOR A MINIMUM

Determine if the potential minimum $\mathbf{x}^* = [1.000 \quad 4.900]^T$ is indeed a local minimum of the problem

$$\begin{aligned}
\text{Minimize} \quad & f(\mathbf{x}) = 4x_1 - x_2^2 - 12 \\
\text{Subject to} \quad & h_1(\mathbf{x}) = 25 - x_1^2 - x_2^2 = 0 \\
& g_1(\mathbf{x}) = 10x_1 - x_1^2 + 10x_2 - x_2^2 - 34 \geq 4 \\
& g_2(\mathbf{x}) = (x_1 - 3)^2 + (x_2 - 1)^2 \geq 0 \\
& g_3(\mathbf{x}) = x_1 \geq 2 \\
& g_4(\mathbf{x}) = x_2 \geq 0
\end{aligned}$$

Figure E8.5 shows the contours of these functions on the $x_1 - x_2$ plane.

Solution. Test each of the necessary and sufficient conditions:

1. That the functions are twice differentiable at $\mathbf{x}^*$ can be seen by inspection.

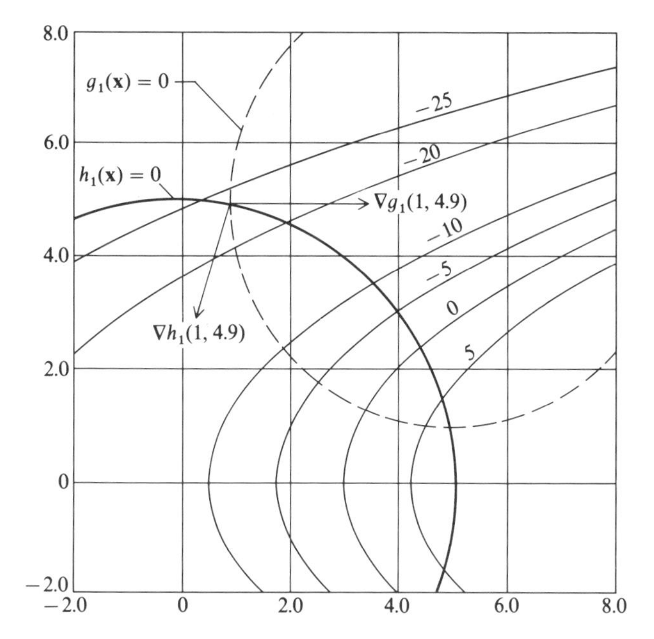

Figure E8.5 Contours of $f(\mathbf{x})$, $h_1(\mathbf{x}) = 0$, and $g_1(\mathbf{x}) = 0$.

2. Are the constraints satisfied?

$$h_1: \quad 25 - 1 - (4.9)^2 = 0$$

$$25 - 1 - 24.0 = 0$$

$$g_1: \quad 10 - 1 + 49 - 24.0 - 34 \geq 0 \qquad \text{yes, } g_1 = 0 \text{ and is binding}$$

$$g_2: \quad (1 - 3)^2 + (4.9 - 1)^2 \geq 0 \qquad \text{yes, } g_2 > 0 \text{ and is not active}$$

$$g_3 > 0, g_4 > 0$$

Hence the constraints are satisfied within reasonable precision.

3. Does the second-order constraint qualification hold? Check for linear independence of ∇h_1 and ∇g_1:

$$c_1 \begin{bmatrix} (10 - 2x_1^*) \\ (10 - 2x_2^*) \end{bmatrix} + c_2 \begin{bmatrix} -2x_1^* \\ -2x_2^* \end{bmatrix} = 0$$

$$c_1 \begin{bmatrix} 8 \\ 0.2 \end{bmatrix} + c_2 \begin{bmatrix} -2 \\ -9.8 \end{bmatrix} = 0$$

A solution to the set of 2 homogeneous equations exists only if the determinant of the coefficients is zero. Does the $\det \begin{bmatrix} 8 & -2 \\ 0.2 & -9.8 \end{bmatrix} = 0$? No, hence $c_1 = c_2 = 0$ is the only solution of the homogeneous equations. Consequently the gradients of the active constraints are linearly independent and the first- and second-order constraint qualifications are satisfied at $\mathbf{x}^*$.

4. Are all the $u_j^* g_j(\mathbf{x}^*) = 0$, $j = 1, 2, 3$, and 4?

$$u_1^* \text{ can be any value because } g_1(\mathbf{x}^*) = 0$$
$$u_2^* \text{ must be zero because } g_2(\mathbf{x}^*) > 0$$
$$u_3^* \text{ must be zero because } g_3(\mathbf{x}^*) > 0$$
$$u_4^* \text{ must be zero because } g_4(\mathbf{x}^*) > 0$$

5. Is $\nabla L(\mathbf{x}^*, \boldsymbol{\omega}^*, \mathbf{u}^*) = 0$?

$$\begin{bmatrix} 4 \\ -2x_2^* \end{bmatrix} + \omega_1^* \begin{bmatrix} -2x_1^* \\ -2x_2^* \end{bmatrix} - u_1^* \begin{bmatrix} 10 - 2x_1^* \\ 10 - 2x_2^* \end{bmatrix} - 0 \overset{?}{=} 0$$

$$4 - 2\omega_1^* - 8u_1^* = 0$$

$$-9.8 - 9.8\omega_1^* - 0.2u_1^* = 0$$

We use Cramer's rule to solve for ω_1^* and u_1^*

$$\omega_1^* = \left| \frac{\begin{matrix} 4 & 8 \\ -9.8 & 0.2 \end{matrix}}{\begin{matrix} 2 & 8 \\ 9.8 & 0.2 \end{matrix}} \right| \qquad u_1^* = \left| \frac{\begin{matrix} 2 & 4 \\ 9.8 & -9.8 \end{matrix}}{\begin{matrix} 2 & 8 \\ 9.8 & 0.2 \end{matrix}} \right|$$

$$= \frac{+77.2}{-76.0} = -1.016 \qquad = \frac{-58.8}{-76.0} = 0.754$$

∇L is 0 for $\omega_1^* = -1.016$ and $u_1^* = 0.754$. Note $u_1 \geq 0$ is satisfied; ω_1 can be positive or negative.

6. Is the Hessian matrix of $L(\mathbf{x}^*, \boldsymbol{\omega}^*, \mathbf{u}^*)$ positive definite for those $\mathbf{v}$ satisfying $\mathbf{v}^T \nabla g(\mathbf{x}^*) = 0$ and $\mathbf{v}^T \nabla h(\mathbf{x}^*) = 0$?

$$\frac{\partial^2 L}{\partial x_1^2} = -2\omega_1 + 2u_1$$

$$\frac{\partial^2 L}{\partial x_2^2} = -2 - 2\omega_1 + 2u_1$$

The cross partial derivatives vanish.

$$\nabla^2 L(\mathbf{x}^*, \boldsymbol{\omega}^*, \mathbf{u}^*) = \begin{bmatrix} [(-2)(-1.016) + 2(0.754)] & 0 \\ 0 & [-2 - (2)(-1.016) + 2(0.754)] \end{bmatrix}$$

$$= \begin{bmatrix} 3.54 & 0 \\ 0 & 1.54 \end{bmatrix}$$

so that $\nabla^2 L$ is positive definite at $\mathbf{x}^*$.

However, if you try to find the vector(s) $\mathbf{v}$ of

$$[v_1 \quad v_2] \begin{bmatrix} 10 - 2(1) \\ 10 - 2(4.900) \end{bmatrix} = 0$$

$$[v_1 \quad v_2] \begin{bmatrix} -2(1) \\ -2(4.900) \end{bmatrix} = 0$$

or

$$8v_1 + 0.2v_2 = 0$$

$$-2v_1 - 9.8v_2 = 0$$

the only vector satisfying the two equations is $\mathbf{v} = [0 \quad 0]^T$.

Consequently, we can show that

$$\mathbf{v}^T\nabla^2 L(\mathbf{x}^*, \boldsymbol{\omega}^*, \mathbf{u}^*)\mathbf{v} \geq 0 \qquad \text{(necessary condition)}$$

but not

$$\mathbf{v}^T\nabla^2 L(\mathbf{x}^*, \boldsymbol{\omega}^*, \mathbf{u}^*)\mathbf{v} > 0 \qquad \text{(sufficient condition)}$$

The difficulty here is that the original problem has two variables and two binding independent constraints at $\mathbf{x}^*$ so that in truth the problem is one of solving two simultaneous equations; zero degrees of freedom exist for such an optimization problem.

You can see that the computations needed to evaluate the sufficient conditions might be quite extensive for large-scale problems. Furthermore, because the results may be inconclusive, you may not be able to show that the sufficient conditions are satisfied, but on the other hand you cannot show that they are not satisfied. Most computer codes do not carry out a test for sufficiency once $\mathbf{x}^*$ has been computed.

8.3 QUADRATIC PROGRAMMING

Quadratic programming (QP) is the name given to the procedure that minimizes a quadratic function of n variables subject to m linear inequality or equality, or both types, of constraints. A quadratic programming problem is the simplest form of a nonlinear programming problem with inequality constraints. In this text we discuss QP as a subproblem to solve general nonlinear programming problems. In addition, a number of practical optimization problems are naturally posed as QP problems, such as constrained least squares, optimal control of linear systems with quadratic cost functions, and the solution of algebraic and differential equations. The techniques proposed for the solution of a quadratic programming problem bear many similarities to those used in solving the linear programming problems discussed in Chap. 7. Specifically, each inequality constraint must either be satisfied as an equality (i.e., is binding) or it is not involved in the solution of the QP problem, so that once the binding constraints are identified, the QP technique can reduce to a vertex-searching procedure examining the intersection of linear equations as in LP. This idea is the fundamental link between the two methods.

In compact notation, the quadratic programming problem is

$$\begin{aligned} \text{Minimize} \quad & f(\mathbf{x}) = \mathbf{c}^T\mathbf{x} + \tfrac{1}{2}\mathbf{x}^T\mathbf{Q}\mathbf{x} \\ \text{Subject to} \quad & \mathbf{A}\mathbf{x} \geq \mathbf{b} \\ & \mathbf{x} \geq \mathbf{0} \end{aligned} \qquad (8.19)$$

where $\mathbf{c}$ is a vector of constant coefficients, $\mathbf{A}$ is an $(m \times n)$ matrix, and it is generally assumed that $\mathbf{Q}$ is a symmetric matrix.

Because the constraints (8.19) are linear and presumably independent, the constraint qualification condition (Sect. 8.2) is always satisfied, hence the Kuhn-Tucker conditions are the necessary conditions to obtain an optimal solution of the QP problem. In addition, if $\mathbf{Q}$ is positive definite, the Kuhn-Tucker conditions are also the sufficient conditions for an extremum, and a solution meeting these conditions will yield the global optimum. If $\mathbf{Q}$ is not positive definite, the problem may have unbounded solutions and local minima. For (8.19) the Kuhn-Tucker conditions can be written down directly from conditions (*a*), (*c*), (*e*), (*f*), and (*g*) in Table 8.1.

We start with the Lagrange function

$$L = \mathbf{x}^T\mathbf{c} + \tfrac{1}{2}\mathbf{x}^T\mathbf{Q}\mathbf{x} - \mathbf{v}^T(\mathbf{A}\mathbf{x} - \mathbf{b}) - \mathbf{u}^T\mathbf{x}$$

and equate the gradient of L (with respect to $\mathbf{x}^T$) to zero (note $\mathbf{v}^T(\mathbf{A}\mathbf{x} - b) = (\mathbf{A}\mathbf{x} - \mathbf{b})^T\mathbf{v} = (\mathbf{x}^T\mathbf{A}^T - \mathbf{b}^T)\mathbf{v}$ and $\mathbf{u}^T\mathbf{x} = \mathbf{x}^T\mathbf{u}$)

$$\nabla_{\mathbf{x}^T} L = \mathbf{c} + \mathbf{Q}\mathbf{x} - \mathbf{A}^T\mathbf{v} - \mathbf{u} = \mathbf{0}$$

The nonnegative slack variables $\mathbf{y}$ can be inserted into the constraints $\mathbf{A}\mathbf{x} - \mathbf{b} \geq 0$ so that $\mathbf{A}\mathbf{x} - \mathbf{b} - \mathbf{y} = 0$ or $\mathbf{y} = \mathbf{A}\mathbf{x} - \mathbf{b}$ as explained in Sec. 7.4. Then the Kuhn-Tucker conditions reduce to the following set of linear equations:

$$\mathbf{c} + \mathbf{Q}\mathbf{x} - \mathbf{A}^T\mathbf{v} - \mathbf{u} = \mathbf{0} \tag{8.20}$$

$$\mathbf{A}\mathbf{x} - \mathbf{b} - \mathbf{y} = \mathbf{0} \tag{8.21}$$

$$\mathbf{x} \geq \mathbf{0} \qquad \mathbf{u} \geq \mathbf{0} \qquad \mathbf{v} \geq \mathbf{0} \qquad \mathbf{y} \geq \mathbf{0} \tag{8.22}$$

$$\mathbf{u}^T\mathbf{x} = 0 \qquad \mathbf{v}^T\mathbf{y} = 0 \tag{8.23}$$

$$(\text{or } \mathbf{u}^T\mathbf{x} + \mathbf{v}^T\mathbf{y} = 0) \tag{8.24}$$

where the u_i and v_j are the Lagrange multipliers and the y_j are the slack variables. The set of variables $(\mathbf{x}^*, \mathbf{u}^*, \mathbf{v}^*, \mathbf{y}^*)$ that satisfies (8.20) to (8.24) is the optimal solution to (8.19).

It can be shown in a straightforward way (Cottle, 1968) that if $(\mathbf{x}, \mathbf{u}, \mathbf{v}, \mathbf{y})$ is the solution of the QP problem (8.19), then $f(\mathbf{x})$ in (8.19) is equivalent to

$$f = \tfrac{1}{2}(\mathbf{c}^T\mathbf{x} + \mathbf{b}^T\mathbf{y}) \tag{8.25}$$

Minimization of function (8.25), subject to (8.20) to (8.24), form what is called the *linear complementarity problem* (Eaves, 1971), the solution of which is the solution of the original QP problem (8.19). The term *complementarity* arises from the idea that in (8.23) if one of the "complementary pair" (u_i, x_i) is zero, then the "complementary condition" (8.23) is satisfied. Note that with the exception of Eq. (8.23), the Kuhn-Tucker conditions are nothing more than constraints for a linear programming problem involving $2(n + m)$ variables. All the complementary restriction says is that it is not permissible for both x_i and u_i to be basic variables in a basic feasible solution.

A problem in which the linear constraints $\mathbf{Ax} > \mathbf{b}$ are replaced with $\mathbf{Ax} = \mathbf{b}$ in (8.19) is also often called a quadratic programming problem. The Kuhn-Tucker conditions are the same as (8.20) to (8.24) except $\mathbf{y} \equiv 0$. For a mixture of equality and inequality constraints, an element of $\mathbf{y}$ would have to be retained for each inequality. Also, by way of interest, if $\mathbf{Q}$ is at least positive semidefinite, a dual problem of (8.19) can be formulated.

$$\begin{aligned} \text{Maximize} \quad & F = -\tfrac{1}{2}\mathbf{z}^T\mathbf{Qz} + \mathbf{b}^T\mathbf{w} \\ \text{Subject to} \quad & \mathbf{Qz} - \mathbf{A}^T\mathbf{w} + \mathbf{p} \geq \mathbf{0} \\ & \mathbf{w} \geq \mathbf{0} \\ & \mathbf{z} \text{ unrestricted} \end{aligned} \tag{8.19a}$$

and the max $F = \min f$ at $\mathbf{x}^*$. For further details refer to Cottle (1963).

Pang (1983) reviewed various methods that have been used to solve quadratic programming problems, and identified four families of methods:

1. LP Simplex-related
2. Active set strategy
3. Internal representation of the feasible set
4. Shrinking ellipsoid

The first two classifications cover most of the existing methods, but as Pang has pointed out, many methods are essentially the same and therefore it makes little difference as to which method is used as long as the computer code is properly prepared. Some practitioners prefer to use the dual formulation of the QP problem because the origin is a feasible point, and a phase I procedure (finding an initial feasible point) can be avoided. Table 8.2 lists some quadratic

Table 8.2 Quadratic programming codes

Code Name	Source	Computers	Language	Price
CONQUA	Econometric Institute Erasmus University, P.O. Box 1738, 3000 DR Rotterdam, The Netherlands	DEC 2060	Fortran	10,000 Dutch guilders
DUAL QP	OCI, P.O. Box 144, Park Ridge, NJ 07656	IBM 370, 4300, 3000	Fortran	Inquire
QPSOL	Office of Technology Licensing, Stanford University, Stanford, CA 94305	Numerous	Fortran 66	$1000 (academic discount exists)
ZQPCVX	MJD Powell, DAMTP, Silver St., Cambridge, CB3 9EW England	IBM 3081	Fortran	None

programming codes. In practice, you can use a general nonlinear programming code to solve quadratic programming problems reasonably efficiently without requiring a separate quadratic programming code.

We turn now to techniques that have been effective in solving the general constrained nonlinear programming problem (8.1).

8.4 THE GENERALIZED REDUCED GRADIENT METHOD

Certainly, the most direct approach to solving the general nonlinear programming problem would be to linearize the problem and successively apply linear programming techniques by such means as:

1. Formulate a model with a nominal operating point and linearize all the constraints and objective function about the operating point so that they fit the linear programming format. Use linear programming once to solve the linearized problem.
2. Successively linearize the constraints and objective function of a nonlinear problem as successive improved feasible solutions are reached. Once the solution of the nominal LP is obtained and the nominal optimal point proves to be nonfeasible, locate the nearest (or some) feasible point, and again linearize the constraints and objective function about this new point.
3. Linearize the functions in a piecewise fashion so that a series of straight-line segments are used to approximate the objective function and constraints.

There is no guarantee of convergence for any of these methods. Refer to Sec. 8.9 for more information on this approach.

In practice the best current general algorithm (best on the basis of many tests—see Sec. 8.7) using iterative linearization is the Generalized Reduced Gradient algorithm (GRG). The GRG algorithm (Abadie and Carpentier, 1969; Faure and Huard, 1965; Abadie and Guigou, 1969; Liebman, et al., 1986) is an extension of the Wolfe algorithm (Wolfe, 1962) for linear constraints modified to accommodate both a nonlinear objective function and nonlinear constraints. In essence the method employs linear or linearized constraints, defines new variables that are normal to the constraints, and expresses the gradient (or other search direction) in terms of this normal basis. (Wolfe has shown that the original reduced gradient method is related to the Simplex method of linear programming.) Although the problem solved by the GRG method is

$$\begin{array}{lll} \text{Minimize} & f(\mathbf{x}) & \mathbf{x} = [x_1 \quad x_2 \quad \cdots \quad x_n]^T \\ \text{Subject to} & h_j(\mathbf{x}) = 0 & j = 1, \ldots, m \\ & L_i \le x_i \le U_i & i = 1, \ldots, n \end{array} \qquad (8.1a)$$

where L_i and U_i are the lower and upper bounds on x_i, respectively, the upper and lower bounds are treated as separate vectors rather than being classified as inequality constraints because they are treated differently in determining the step length in a search direction.

Nonlinear inequality constraints can be accommodated by subtracting (or adding, as the case may be) the square of slack variables from the inequality constraints thus:

$$h_j(\mathbf{x}) = g_j(\mathbf{x}) - \sigma_j^2 = 0$$

and permitting the bounds on the σ_j's to be $-\infty \leq \sigma_j \leq \infty$. (The σ_j's are added to the set of n variables.) An alternate method is to subtract (or add) the slack variable itself and make the slack variable nonnegative by adding bounds to the problem

$$h_j(\mathbf{x}) = g_j(\mathbf{x}) - \sigma_j = 0$$

$$\text{Subject to:} \quad \sigma_j \geq 0$$

Finally, what is called the "active constraint set" strategy might be used, that is, at each stage of the search the active inequality constraints are added and the inactive ones removed from the equality constraint set, the selection based on Lagrange multiplier estimates. Recall that in problem (8.1*a*), in general you cannot directly reduce the dimensionality of the problem by substitution because the nonlinear equality constraints implicitly connect the variables. Hence the equations cannot be solved for a set of dependent variables that can be substituted into the objective function, leaving only independent variables. However, a method of local linearization permits a reduction of dimensionality to take place.

Reduced-gradient algorithms seek to maintain feasibility on a set of the nonlinear constraints while reducing the value of the objective function. Thus, the dimensionality of the search is reduced to the total number of variables less the number of independent constraints. As you might expect, one difficulty with this procedure is to return to a feasible point if the solution of a subproblem with linearized active constraints yields a nonfeasible point with respect to the original constraints. Differences among the generalized reduced-gradient algorithms result from differences in viewpoint as to how to carry out the search, how to reduce the objective function, and how to return to a feasible point.

8.4.1 Concept of the Reduced Gradient

Suppose that the m constraints in problem (8.1*a*) are linear or are linearized approximates to $h_j(\mathbf{x}) = 0$. Imposition of these constraints reduces the number of degrees of freedom associated with $\mathbf{x}$ from n to $(n\text{-}m)$. In carrying out the search for the solution of problem (8.1*a*), only $(n\text{-}m)$ variables will be independent variables; m variables will be dependent ones. This reduction in dimensionality is described more formally in Appendix C in terms of two subspaces, but here we just use a simple example involving two variables to explain the concept of the reduced gradient.

Consider the following problem:

$$\begin{aligned} \text{Minimize} \quad & f(x_1, x_2) \\ \text{Subject to} \quad & h(x_1, x_2) = 0 \end{aligned} \tag{8.26}$$

The total derivatives of each function are

$$df(\mathbf{x}) = \frac{\partial f(\mathbf{x})}{\partial x_1} dx_1 + \frac{\partial f(\mathbf{x})}{\partial x_2} dx_2 \tag{8.27}$$

$$dh(\mathbf{x}) = \frac{\partial h(\mathbf{x})}{\partial x_1} dx_1 + \frac{\partial h(\mathbf{x})}{\partial x_2} dx_2 = 0 \tag{8.28}$$

Note that the components of $\nabla f(\mathbf{x})$ and $\nabla h(\mathbf{x})$ are the coefficients in the respective total derivatives.

Suppose that x_1 is designated to be the dependent variable, and x_2 to be the independent variable. Then we can eliminate dx_1 from Eq. (8.27) by use of Eq. (8.28) because

$$dx_1 = -\frac{[\partial h(\mathbf{x})/\partial x_2]}{[\partial h(\mathbf{x})/\partial x_1]} dx_2 \tag{8.29}$$

so that

$$df(\mathbf{x}) = \left\{ -\left[\frac{\partial f(\mathbf{x})}{\partial x_1}\right]\left[\frac{\partial h(\mathbf{x})}{\partial x_1}\right]^{-1}\left[\frac{\partial h(\mathbf{x})}{\partial x_2}\right] + \left[\frac{\partial f(\mathbf{x})}{\partial x_2}\right]\right\} dx_2 \tag{8.30}$$

The expression in the braces is called the *reduced gradient*. Of course, in this problem the reduced gradient contains only one element because there is only one independent variable, hence here the reduced gradient g_R is a scalar. But if $f(\mathbf{x})$ were a function of five variables and was constrained by two constraints, then the reduced gradient would be composed of three elements.

As shown in Appendix C, a search direction confined to the intersection of the constraints can be selected, and is thus a feasible direction. Of course, if the

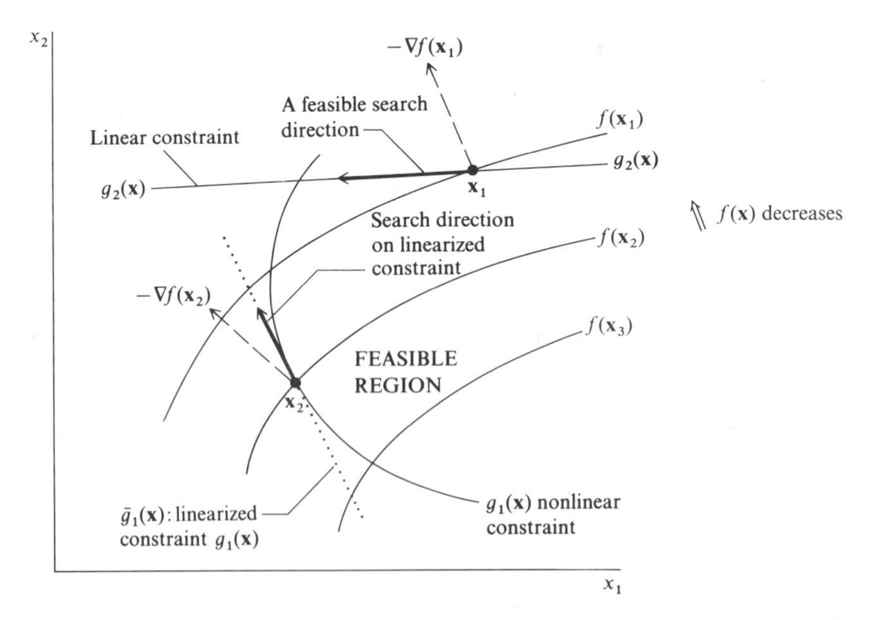

Figure 8.3 Search directions satisfying the constraints are valid for some distance for linear constraints but only locally along a linearized constraint.

constraints in (8.26) are linearized constraints, then the search direction chosen is only locally feasible as indicated in Fig. 8.3, and a search may end at a non-feasible point (with respect to the original constraints). If the constraints in (8.26) are linear, then the reduced gradient can be evaluated at each stage of the search, and when the elements of the gradient reach zero, the point so located is a local minimum as illustrated in Example 8.6.

Table 8.3 lists the same relations as (8.27) to (8.30) for the general matrix **A** but with the $(n - m)$ independent variables identified by the subscript I and the m dependent variables identified by the subscript D for clarity.

Table 8.3 Reduced gradient for n variables and m constraints

Notation: $\mathbf{x}_I$ is the vector of independent variables; $\mathbf{x}_D$ is the vector of dependent variables

$$\mathbf{h} \equiv \begin{bmatrix} h_1(\mathbf{x}) \\ \vdots \\ h_m(\mathbf{x}) \end{bmatrix} \quad \text{an } m \times 1 \text{ vector } \mathbf{x}^T = [\mathbf{x}_I \;\vdots\; \mathbf{x}_D]$$

$$\frac{\partial \mathbf{h}}{\partial \mathbf{x}_D} \equiv \begin{bmatrix} \dfrac{\partial h_1(\mathbf{x})}{\partial x_1} \cdots \dfrac{\partial h_1(\mathbf{x})}{\partial x_m} \\ \cdots\cdots\cdots\cdots \\ \dfrac{\partial h_m(\mathbf{x})}{\partial x_1} \cdots \dfrac{\partial h_m(\mathbf{x})}{\partial x_m} \end{bmatrix} \quad \text{an } m \times m \text{ matrix (basis matrix)}$$

$$\left[\frac{\partial f(\mathbf{x})}{\partial \mathbf{x}_I}\right]^T \equiv [\nabla^T_{\mathbf{x}_I} f] = \left[\frac{\partial f(\mathbf{x})}{\partial x_{m+1}} \quad \cdots \quad \frac{\partial f(\mathbf{x})}{\partial x_n}\right] \quad \text{a } 1 \times (n\text{-}m) \text{ vector}$$

$$\left[\frac{\partial f(\mathbf{x})}{\partial \mathbf{x}_D}\right]^T \equiv [\nabla^T_{\mathbf{x}_D} f] = \left[\frac{\partial f(\mathbf{x})}{\partial x_1} \quad \cdots \quad \frac{\partial f(\mathbf{x})}{\partial x_m}\right] \quad \text{a } 1 \times m \text{ vector}$$

$$\frac{d\mathbf{x}_D}{d\mathbf{x}_I} \equiv \begin{bmatrix} \dfrac{dx_1}{dx_{m+1}} \cdots \dfrac{dx_1}{dx_n} \\ \cdots\cdots\cdots\cdots \\ \dfrac{dx_m}{dx_{m+1}} \cdots \dfrac{dx_m}{dx_n} \end{bmatrix} \quad \text{an } m \times (n\text{-}m) \text{ matrix}$$

$$\frac{d\mathbf{h}}{d\mathbf{x}_I} \equiv \begin{bmatrix} \dfrac{dh_1(\mathbf{x})}{dx_{m+1}} \cdots \dfrac{dh_1(\mathbf{x})}{dx_n} \\ \cdots\cdots\cdots\cdots \\ \dfrac{dh_m(\mathbf{x})}{dx_{m+1}} \cdots \dfrac{dh_m(\mathbf{x})}{dx_n} \end{bmatrix} \quad \text{an } m \times (n\text{-}m) \text{ matrix}$$

The components of the reduced gradient are

$$\mathbf{g}_R^T = \text{reduced gradient} = \left[\frac{\partial f(\mathbf{x})}{\partial \mathbf{x}_I}\right]^T - \left[\frac{\partial f(\mathbf{x})}{\partial \mathbf{x}_D}\right]^T \left[\frac{\partial \mathbf{h}}{\partial \mathbf{x}_D}\right]^{-1} \left[\frac{\partial \mathbf{h}}{\partial \mathbf{x}_I}\right]$$

or

$$\mathbf{g}_R = \left[\frac{\partial f(\mathbf{x})}{\partial \mathbf{x}_I}\right] - \left[\frac{\partial \mathbf{h}(\mathbf{x})}{\partial \mathbf{x}_I}\right]^T \left[\left[\frac{\partial \mathbf{h}(\mathbf{x})}{\partial \mathbf{x}_D}\right]^{-1}\right]^T \left[\frac{\partial f(\mathbf{x})}{\partial \mathbf{x}_D}\right]$$

EXAMPLE 8.6 CALCULATION OF THE GENERALIZED REDUCED GRADIENT

Let us examine a very simple problem involving two variables and an objective function $f(x_1, x_2) = 2x_1^2 + 2x_2^2$ subject to one constraint. We will linearize the constraint about the last known $\mathbf{x}$ so that it becomes locally

$$a_{11}x_1 + a_{12}x_2 - b = 0 \qquad \text{or} \qquad \mathbf{Ax} = \mathbf{b}$$

or, specifically

$$x_1 + 2x_2 - 2 = 0$$

Which variable is independent and which is dependent at the start is arbitrary. Let $x_1 = x_D$ be the dependent variable and $x_2 = x_I$ be the independent variable. Then

$$\frac{\partial h(\mathbf{x})}{\partial x_1} = 1 \qquad \frac{\partial h(\mathbf{x})}{\partial x_2} = 2$$

$$\frac{\partial f(\mathbf{x})}{\partial x_1} = 4x_1 \qquad \frac{\partial f(\mathbf{x})}{\partial x_2} = 4x_2$$

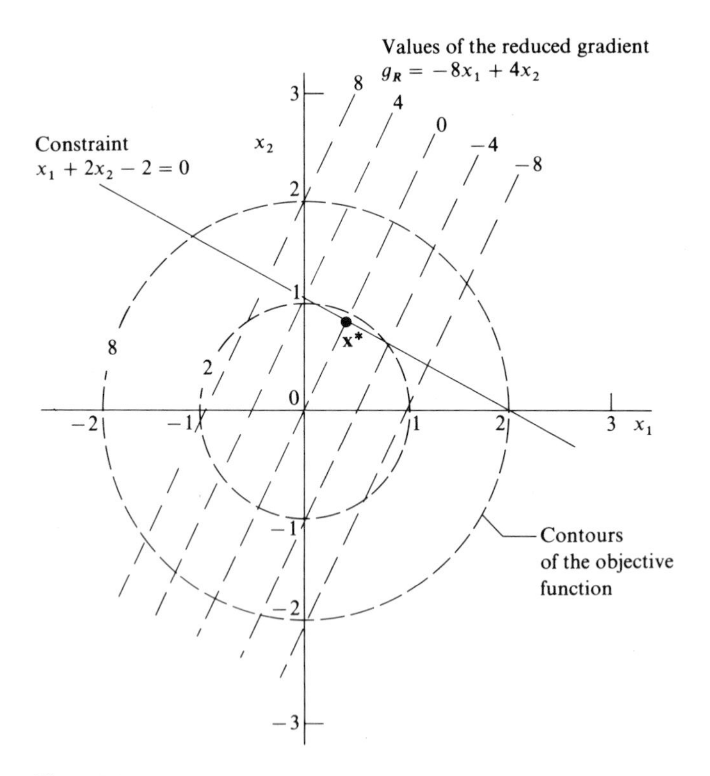

Figure E8.6

Consequently, the reduced gradient (element) is given by Eq. (8.30)

$$g_R = -[4x_1][1]^{-1}[2] + 4x_2 = -8x_1 + 4x_2 \tag{a}$$

If the constraint was actually linear, then you would pick a feasible search direction, i.e., search along the constraint. From any starting point, g_R can be evaluated as the search proceeds until $g_R = 0$, where $\mathbf{x}^* = [\frac{2}{5} \;\; \frac{4}{5}]^T$ occurs (see Fig. E8.6). If the constraint is nonlinear, $x^T = [\frac{2}{5} \;\; \frac{4}{5}]$ will not be a feasible point and you have to return to a feasible point to start the linearization and search over again as described below.

The relation of Lagrange multipliers [refer to Eq. (8.18)] and the reduced gradient is as follows. If

$$L(\mathbf{x}, \mathbf{u}) = f(\mathbf{x}) - \mathbf{u}^T\mathbf{h}(\mathbf{x}) \tag{8.31}$$

then

$$\frac{\partial L}{\partial \mathbf{x}_I} = \frac{\partial f(\mathbf{x})}{\partial \mathbf{x}_I} - \mathbf{u}^T \frac{\partial \mathbf{h}(\mathbf{x})}{\partial \mathbf{x}_I} \tag{8.32}$$

$$\frac{\partial L}{\partial \mathbf{x}_D} = \frac{\partial f(\mathbf{x})}{\partial \mathbf{x}_D} - \mathbf{u}^T \frac{\partial \mathbf{h}(\mathbf{x})}{\partial \mathbf{x}_D} = 0 \tag{8.33}$$

The Lagrangian in Eq. (8.33) is equated to zero because Eqs. (8.6) and (8.7) hold at a constrained optimal point. If Eq. (8.33) is solved for $\mathbf{u}$, we find

$$\mathbf{u}^T = \left[\frac{\partial f(\mathbf{x})}{\partial \mathbf{x}_D}\right]\left[\frac{\partial \mathbf{h}(\mathbf{x})}{\partial \mathbf{x}_D}\right]^{-1}$$

which when substituted into (8.32) yields the reduced gradient

$$\frac{\partial L}{\partial \mathbf{x}_I} = \left[\frac{\partial f(\mathbf{x})}{\partial \mathbf{x}_I}\right] - \left[\frac{\partial f(\mathbf{x})}{\partial \mathbf{x}_D}\right]\left[\frac{\partial \mathbf{h}(\mathbf{x})}{\partial \mathbf{x}_D}\right]^{-1}\left[\frac{\partial \mathbf{h}(\mathbf{x})}{\partial \mathbf{x}_I}\right]$$

8.4.2 The Generalized Reduced-Gradient Algorithm

We outline here some of the essential phases of a generalized reduced-gradient algorithm. Any one computer code will contain variations in the listed procedure because of special matrix manipulation procedures, selection of search direction elements, and heuristic decisions based on experience. For any iteration it is assumed that $[\partial\mathbf{h}(\mathbf{x})/\partial\mathbf{x}_D]$ is nonsingular.

For convenience in explanation the algorithm can be divided into five phases. To simplify the notation, we will let $f(\mathbf{x}^k) = f^k$ and $\mathbf{h}(\mathbf{x}^k) = \mathbf{h}^k$. Note that the components of the vectors $\mathbf{x}_I$ and $\mathbf{x}_D$ may well change from stage to stage, hence these vectors bear superscripts k.

Phase 1 Determine the search components for the independent variables. At stage k linearize the constraints at the feasible point $\mathbf{x}^k$ and compute the reduced gradient

$$(\mathbf{g}_R^k)^T = \left[\frac{\partial f^k}{\partial \mathbf{x}_I^k}\right]^T - \left[\frac{\partial f^k}{\partial \mathbf{x}_D^k}\right]^T \left[\frac{\partial \mathbf{h}^k}{\partial \mathbf{x}_D^k}\right]^{-1} \left[\frac{\partial \mathbf{h}^k}{\partial \mathbf{x}_I^k}\right]$$

Note: At $\mathbf{x}^0$, the designations of independent and dependent variables must be assigned. Usually the independent variables are selected from among the controllable variables in a process.

The search direction components *in the space of the independent variables* are established from the elements of the reduced gradient g_{R_i} as follows

(*a*) If x_i is at one of its bounds, the search direction component is $\Delta_i^k = 0$ if the step would exceed the bound, that is,

$$\Delta_i^k = 0 \qquad \text{if} \quad x_i^k = U_i \qquad g_{R_i} < 0$$
$$= L_i \qquad g_{R_i} > 0$$

(*b*) If $L_i < x_i < U_i$, the search direction is the negative of the corresponding element of the reduced gradient, $\Delta_i^k = -g_{R_i}$

Note: In the Abadie algorithm only the initial search direction is $-g_{R_i}$. Subsequent search directions on iteration use the Fletcher–Reeves conjugate gradient formula as follows.

$$\boldsymbol{\Delta}_I^{k+1} = -\mathbf{g}_R^{k+1} + \boldsymbol{\Delta}_I^k \frac{(\mathbf{g}_R^{k+1})^T(\mathbf{g}_R^{k+1})}{(\mathbf{g}_R^k)^T(\mathbf{g}_R^k)}$$

It can be demonstrated that these search directions are constrained to the hyperplanes of the locally linearized active constraints and thus are equivalent to those that would be obtained by using the Δ's determined from $\mathbf{g}_R$.

Phase 2 Determine the search components for the dependent variables. To maintain feasibility with respect to the linearized constraints, calculate the search direction components in the *space of the dependent variables* as follows

$$\boldsymbol{\Delta}_D^k = -\left[\frac{\partial \mathbf{h}^k}{\partial \mathbf{x}_D}\right]^{-1}\left[\frac{\partial \mathbf{h}^k}{\partial \mathbf{x}_I}\right]\boldsymbol{\Delta}_I^k$$

Note: The relation for $\boldsymbol{\Delta}_D^k$ comes from a generalization of Eq. (8.29).

$$d\mathbf{h} = \frac{\partial \mathbf{h}}{\partial \mathbf{x}_I} d\mathbf{x}_I + \frac{\partial \mathbf{h}}{\partial \mathbf{x}_D} d\mathbf{x}_D = 0$$

$$\frac{d\mathbf{x}_D}{d\mathbf{x}_I} = -\left[\frac{\partial \mathbf{h}}{\partial \mathbf{x}_D}\right]^{-1}\left[\frac{\partial \mathbf{h}}{\partial \mathbf{x}_I}\right]$$

so that $d\mathbf{x}_D$ can be calculated from $d\mathbf{x}_I$.

Note: If the equality constraints are linear (only), the net result of carrying out a search in a search direction given by components $\mathbf{\Delta}_I$ and $\mathbf{\Delta}_D$ is to start at a feasible point and maintain feasibility in the same way as a projection of the gradient would force the search to be carried out in the intersection of the constraints.

Phase 3 Improving the value of the objective function. Minimize $f(\mathbf{x}_I^k + \lambda \mathbf{\Delta}_I^k, \mathbf{x}_D^k + \lambda \mathbf{\Delta}_D^k)$ with respect to $\lambda (\lambda > 0)$ by a one-dimensional search as discussed in Chap. 5 (maintaining feasibility with respect to the trivial bounds on x_i).

Note: Various techniques have been proposed to limit the magnitude of λ^k because the search in general for the case of linear or nonlinear equality constraints and upper and lower bounds on x_i can lead to a nonfeasible point.

Phase 3 yields

$$\mathbf{x}_I^{k+1} = \mathbf{x}_I^k + \lambda^k \mathbf{\Delta}_I^k$$

$$\tilde{\mathbf{x}}_D^{k+1} = \mathbf{x}_D^k + \lambda^k \mathbf{\Delta}_D^k \qquad \tilde{\mathbf{x}}_D \text{ is a tentative point}$$

Phase 4 Use Newton's method to regain feasibility of dependent variables. Because some of the components of $\tilde{\mathbf{x}}_D^{k+1}$ will not be feasible, hence $\mathbf{h}(\mathbf{x}_I^{k+1}, \tilde{\mathbf{x}}_D^{k+1}) \neq \mathbf{0}$, $\tilde{\mathbf{x}}_D^{k+1}$ is modified by Newton's method to return from a point away from a set of constraints to a point satisfying the constraints.

Note: Newton's method finds the desired root $\mathbf{x}_D^{k+1}$ starting with $\tilde{\mathbf{x}}_D^{k+1}$. If you expand $\mathbf{h}$ and equate to zero

$$\mathbf{h}(\mathbf{x}_I^{k+1}, \mathbf{x}_D^{k+1}) \approx \mathbf{h}(\mathbf{x}_I^{k+1}, \tilde{\mathbf{x}}_D^{k+1}) + \frac{\partial \mathbf{h}(\mathbf{x}_I^{k+1}, \tilde{\mathbf{x}}_D^{k+1})}{\partial \mathbf{x}_D^k} (\mathbf{x}_D^{k+1} - \tilde{\mathbf{x}}_D^{k+1}) = \mathbf{0}$$

Newton's method is

$$\mathbf{x}_D^{k+1} = \tilde{\mathbf{x}}_D^{k+1} - \left[\frac{\partial \mathbf{h}(\mathbf{x}_I^{k+1}, \tilde{\mathbf{x}}_D^{k+1})}{\partial \mathbf{x}_D^k} \right]^{-1} \mathbf{h}(\mathbf{x}_I^{k+1}, \tilde{\mathbf{x}}_D^{k+1})$$

Note: Several iterations of Newton's method may be required to satisfy the constraints adequately. Figure 8.4 illustrates the move from $\mathbf{x}^k$ to $(\mathbf{x}_I^{k+1}, \tilde{\mathbf{x}}_D^{k+1})$ and thence by Newton's method to $(\mathbf{x}_I^{k+1}, \mathbf{x}_D^{k+1})$.

Phase 5 Procedure on lack of convergence of Newton's method. Various possible outcomes of Newton's method are

(*a*) If $\mathbf{x}^{k+1}$ is a feasible point and $f(\mathbf{x}_I^{k+1}, \mathbf{x}_D^{k+1}) < f(\mathbf{x}_I^k, \mathbf{x}_D^k)$, adopt $\mathbf{x}^{k+1}$, and start with phase 1 again.

(*b*) If $\mathbf{x}^{k+1}$ is a feasible point but if $f(\mathbf{x}_I^{k+1}, \mathbf{x}_D^{k+1}) > f(\mathbf{x}_I^k, \mathbf{x}_D^k)$, reduce λ by a factor, say 1/10, and start phase 4 over again.

(*c*) If Newton's method fails after, say, 20 iterations to yield a feasible $\mathbf{x}_D^{k+1}$, reduce λ by a factor, say 1/2 or 1/10, and start phase 4 over again.

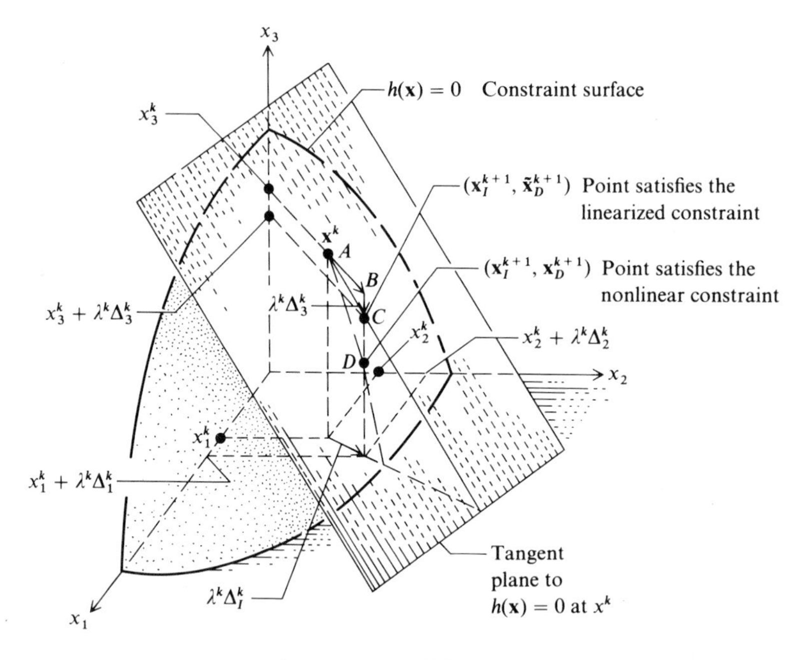

Figure 8.4 One step from $\mathbf{x}^k$ (point A) to $\mathbf{x}^{k+1}$ (point D) with x_1 and x_2 the independent variables and x_3 the dependent variable. Also shown are the intermediate steps (a) the step of the independent variables only (to point B) and (b) the step in x_3 to return (i) to the tangent hyperplane (at point C) and (ii) to the constraint surface of the nonlinear constraint (point D).

(d) If neither (a), (b) nor (c) occur, or (b) and (c) are unsuccessful after two or three passes, change the basis by exchanging a dependent variable with a former independent variable. Examine Fig. 8.5 in which x_2 becomes the new independent variable replacing x_3. This process is termed a *change of basis*. Various rules exist to decide for any change in basis what variable should be removed from the basis and what independent variable added, but these are too complicated to list here and can be found in the references cited at the beginning of Sec. 8.4.

Hitting a bound. The variable x_i from the set $\mathbf{x}_I$ that hits a bound is removed from the basis, takes its boundary value, and is replaced by another variable x_i from the set $\mathbf{x}_D$. Newton's iteration is then carried out for the new set of dependent variables.

Termination. Any of the usual criteria for termination described in Chaps. 4, 5, and 6 can be applied, but a simple one is to check after phase 3 whether

$$|\Delta_i^k| < \varepsilon_i \qquad i = 1, \ldots, n$$

where ε_i is a small number whose magnitude depends on the scaling of x_i.

An example will now illustrate some phases of the algorithm.

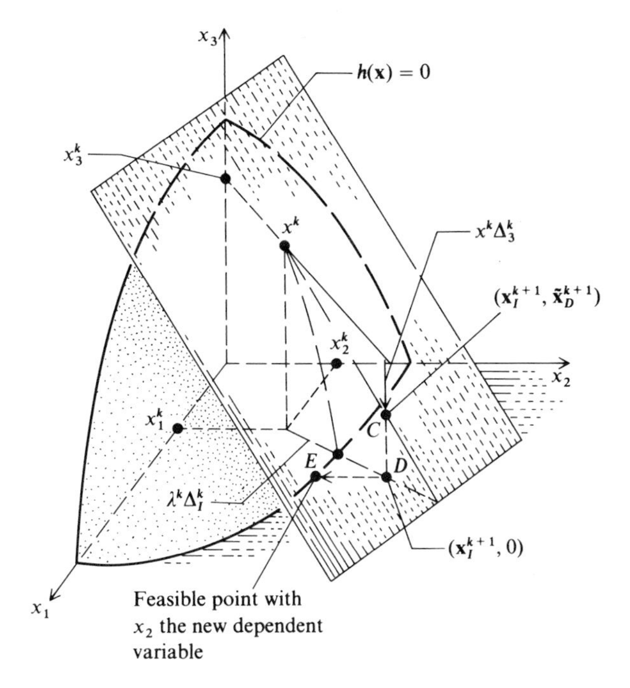

Figure 8.5 This figure illustrates the need for a change in basis (dependent variable). From point C satisfying the linearized constraint, successive applications of Newton's method do not return the search trajectory to the nonlinear constraint $h = 0$. A change in basis occurs at point D by which x_2 becomes the dependent variable, and x_3 the independent variable, leading to a feasible point for $h = 0$, point E.

EXAMPLE 8.7 GENERALIZED REDUCED GRADIENT ALGORITHM

The problem is

$$\text{Maximize} \quad f(\mathbf{x}) = (2x_1 - 0.5x_1^2) + (3x_2 - 0.5x_2^2)$$

$$\text{Subject to} \quad h(\mathbf{x}) = x_1^2 + x_2^2 + x_3 - 1 = 0$$

$$-2 \le x_j \le 2 \qquad j = 1, 2, 3$$

We minimize $-f(\mathbf{x})$.

Phase 1. Let $\mathbf{x}_I^T = [x_1 \quad x_3]$ and $\mathbf{x}_D = [x_2]$. Start at $(\mathbf{x}^0)^T = [0.5 \quad 0.5 \quad 0.5]$ so that $(\mathbf{x}_I^0)^T = [0.5 \quad 0.5]$ and $\mathbf{x}_D^0 = 0.5$. The reduced gradient from Table 8.3 is

$$\mathbf{g}_R = \left[\frac{\partial f(\mathbf{x})}{\partial \mathbf{x}_I}\right] - \left[\frac{\partial \mathbf{h}(\mathbf{x})}{\partial \mathbf{x}_I}\right]^T \left[\left[\frac{\partial \mathbf{h}(\mathbf{x})}{\partial \mathbf{x}_D}\right]^{-1}\right]^T \left[\frac{\partial f(\mathbf{x})}{\partial \mathbf{x}_D}\right]$$

where

$$\frac{\partial f(\mathbf{x}^0)}{\partial \mathbf{x}_I^0} = -\begin{bmatrix} 2 - x_1 \\ 0 \end{bmatrix} \qquad \frac{\partial f(\mathbf{x})}{\partial \mathbf{x}_D^0} = -(3 - x_2)$$

$$\frac{\partial h(\mathbf{x}^0)}{\partial \mathbf{x}_I^0} = [2x_1 \quad 1]^T \qquad \frac{\partial h(\mathbf{x})}{\partial \mathbf{x}_D^0} = 2x_2$$

Thus

$$\mathbf{g}_R^0 = -\begin{bmatrix} 2 - x_1 \\ 0 \end{bmatrix} - \begin{bmatrix} 2x_1 \\ 1 \end{bmatrix}\begin{bmatrix} \dfrac{1}{2x_2} \end{bmatrix}[-(3 - x_2)]$$

$$= -\begin{bmatrix} \dfrac{x_2(2 - x_1) - x_1(3 - x_2)}{x_2} \\ \\ \dfrac{x_2 - 3}{2x_2} \end{bmatrix} = \begin{bmatrix} 1 \\ 2.5 \end{bmatrix}$$

Consequently, the search direction for the independent variables is

$$\boldsymbol{\Delta}_I^0 = \begin{bmatrix} \Delta_1^0 \\ \Delta_3^0 \end{bmatrix} = \begin{bmatrix} -1 \\ -2.5 \end{bmatrix}$$

Phase 2. The corresponding change for the dependent variable is

$$\boldsymbol{\Delta}_D^k = -\left[\frac{\partial \mathbf{h}^k}{\partial \mathbf{x}_D}\right]^{-1}\left[\frac{\partial \mathbf{h}^k}{\partial \mathbf{x}_I}\right]\boldsymbol{\Delta}_I^0$$

$$\Delta_D^0 = \Delta_2^0 = -\begin{bmatrix} \dfrac{1}{2x_2} \end{bmatrix}[2x_1 \quad 1]\begin{bmatrix} \Delta_1^0 \\ \Delta_3^0 \end{bmatrix}$$

$$= -[1][2(0.5)(-1) + 1(-2.5)] = 3.5$$

Phase 3. We minimize $-f(\mathbf{x}^0 + \lambda\boldsymbol{\Delta}^0)$ with respect to λ (here analytically to save space)

$$-f(\mathbf{x}^0 + \lambda\boldsymbol{\Delta}^0) = -\{[2(0.5) + \lambda(-1)] - 0.5[0.5 + \lambda(-1)]^2 + [3(0.5 + \lambda(3.5)] - 0.5[0.5 + \lambda(3.5)]\}^2$$

Set $(\partial f/\partial \lambda) = 0$ and solve the resulting linear equation to get

$$\lambda^0 = 0.547$$

Calculate

$$\mathbf{x}_I^1 = \begin{bmatrix} 0.5 \\ 0.5 \end{bmatrix} + 0.547\begin{bmatrix} -1.0 \\ -2.5 \end{bmatrix} = \begin{bmatrix} -0.047 \\ -0.867 \end{bmatrix}$$

$$\tilde{\mathbf{x}}_D^1 = [0.5] + 0.547[3.5] = 2.415$$

Check to see if these components satisfy the constraint h_1:

$$h_1 = (-0.047)^2 + (2.415)^2 + (-0.867) - 1 \stackrel{?}{=} 0$$

The sum is 3.967 hence $\tilde{x}_2^1$ must be modified.

Phase 4. Apply Newton's method to modify $\tilde{x}_2^1$:

$$\frac{\partial h_1(-0.047, 2.415, -0.867)}{\partial x_2} = 2(2.415) = 4.829$$

$$h_1(-0.047, 2.415, -0.867) = 3.967$$

$$\hat{x}_2^1 = 2.415 - \left(\frac{1}{4.829}\right)(3.967) = 1.594$$

where the ^ superscript designates the approximate desired x_2^1. Successive application of Newton's method starting each time with the previously determined $\hat{x}_2^1$ leads to the desired result (a result can be obtained analytically here by solving $(-0.047)^2 + x_2^2 + (-0.867) - 1 = 0$ for x_2^1:

$$x_2^1 = 1.366$$

We now have a feasible point $\mathbf{x}^1 = [-0.047 \quad 1.366 \quad -0.867]^T$.

Phase 5. Is $f(\mathbf{x}^1) > f(\mathbf{x}^0)$?

$$[2(-0.047) - 0.5(-0.047)^2] + [3(1.366) - 0.5(1.366)^2] \overset{?}{>} [2(0.5) - 0.5(0.5)^2] + [3(0.5) - 0.5(0.5)^2]$$

The answer is yes because $3.070 > 2.25$.

At this stage of the solution of the problem, phase 1 starts over with $\mathbf{x}^1$ being the reference point.

Phase 1.

$$\frac{\partial f(\mathbf{x}^1)}{\partial \mathbf{x}_I^1} = -\begin{bmatrix} 2.047 \\ 0 \end{bmatrix} \qquad \frac{\partial f(\mathbf{x}^1)}{\partial \mathbf{x}_D^1} = -1.634$$

$$\frac{\partial h(\mathbf{x}^1)}{\partial \mathbf{x}_I^1} = [-0.154 \quad 1] \qquad \frac{\partial h(\mathbf{x}^1)}{\partial \mathbf{x}_D^1} = 2.732$$

$$\mathbf{g}_R^1 = -\begin{bmatrix} 2.047 \\ 0 \end{bmatrix} - \begin{bmatrix} -0.154 \\ 1 \end{bmatrix}\begin{bmatrix} \dfrac{1}{2.732} \end{bmatrix}[-1.634]$$

$$= \begin{bmatrix} -2.139 \\ 0.598 \end{bmatrix}$$

$$\mathbf{\Delta}_I^1 = \begin{bmatrix} 2.139 \\ -0.598 \end{bmatrix} = \begin{bmatrix} \Delta_1^1 \\ \Delta_3^1 \end{bmatrix}$$

Phase 2.

$$\mathbf{\Delta}_D^1 \equiv \Delta_2^1 = -\begin{bmatrix} \dfrac{1}{2.732} \end{bmatrix}[-0.154 \quad 1]\begin{bmatrix} 2.139 \\ -0.598 \end{bmatrix} = 0.339$$

Phase 3.

$$\lambda^1 = 0.976$$

$$\mathbf{x}_I^2 = \begin{bmatrix} -0.047 \\ -0.867 \end{bmatrix} + 0.976\begin{bmatrix} 2.139 \\ -0.598 \end{bmatrix} = \begin{bmatrix} 2.041 \\ -1.451 \end{bmatrix}$$

$$\mathbf{x}_D^2 = 2.145 + 0.976[0.339] = 2.476$$

Check to see $h = 0$:

$$(2.041)^2 + (2.476)^2 + (-1.451) - 1 = 10.75 \neq 0$$

Phase 4. Application of Newton's method does not lead to a feasible $\mathbf{x}_D^2$. You can see from the analytical solution of

$$x_2 = \sqrt{1 - x_1^2 - x_3} = \sqrt{1 - (2.04)^2 - (-1.451)} = \sqrt{-1.711}$$

that no such root exists.

Phase 5. At this stage λ is reduced by 1/10 to $\lambda^1 = 0.0976$.

$$\mathbf{x}_I^2 = \begin{bmatrix} -0.047 \\ -0.867 \end{bmatrix} + 0.0976\begin{bmatrix} 2.139 \\ -0.598 \end{bmatrix} = \begin{bmatrix} 0.162 \\ -0.925 \end{bmatrix}$$

$$\mathbf{x}_D^2 = 2.145 + 0.0976[0.339] = 2.178$$

From this nonfeasible point Newton's method leads to $x_2^2 = 1.378$, and a new iteration commences.

If on reduction of λ, a feasible x_2 still cannot be obtained, then the basis must be changed, and a new dependent variable used, either x_1 or x_2, as dictated by certain tests that can be found in the basic references to the GRG method.

8.4.3 Sources of Computer Codes

Table 8.4 lists some information as of 1985 concerning generalized reduced-gradient codes available from various sources. In most of the comparative studies of computer codes implementing NLP algorithms, (Schittkowski, 1980; Sandgren, 1977), GRG methods are comparable in performance to successive quadratic programming (discussed in Sec. 8.6), and these two classes of methods are superior to the others found in the literature.

8.5 PENALTY FUNCTION AND AUGMENTED LAGRANGIAN METHODS

The second major technique to solve the general nonlinear programming problem is to employ penalty functions or their analogs. The essential idea of a penalty function method is to transform problem (8.1) into a problem in which a single unconstrained function is minimized

$$\left.\begin{array}{ll} \text{Minimize} & f(\mathbf{x}) \\ \text{Subject to} & \mathbf{g}(\mathbf{x}) \geq \mathbf{0} \\ & \mathbf{h}(\mathbf{x}) = \mathbf{0} \end{array}\right\} \Rightarrow \text{minimize } P(f, \mathbf{g}, \mathbf{h})$$

where $P(f, \mathbf{g}, \mathbf{h})$ is a "penalty function" or an "augmented function." After the penalty function is formulated, it is minimized by stages for a series of values of parameters associated with the penalty, values that are altered on each successive unconstrained minimization so as to force the sequence of $\mathbf{x}$ vectors to approach the constrained optimum of problem (8.1). Thus, the strategy of penalty-function methods converts problem (8.1) from a constrained problem to a systematic series of minimizations of unconstrained functions of the same form but with different parameters.

Table 8.4 Sources of computer codes using the generalized reduced-gradient method

Code name	Source	Computers	Language	Price
GRG2	Prof. Leon Lasdon, Dept. General Business, Univ. of Texas, Austin, TX 78712	IBM, CDC, Univac DEC, Prime IBM-PC	Fortran $1250 PC; $350 academic use	$15,000 mainframe, $1250 PC; $350 academic use
GRG	Dr. Jean Abadie, 29 Blvd. Edgar Quinet 75014 Paris, France	IBM, CDC	Fortran	Write
GINO	Scientific Press, 540 University Ave., Palo Alto CA 94301	Several	LINDO (for commands)	$700; $400 academic use
SIMUSOLV	Mitchell/Gauthier Assoc. 73 Junction Square Dr. Concord, MA 01742	IBM, Amdahl, NAS, DEC (includes process simulators)	Fortran	$18,000 to $35,000 (there is an educational discount

As an example, consider the problem

$$\text{Minimize} \quad f(\mathbf{x}) = (x_1 - 1)^2 + (x_2 - 2)^2$$
$$\text{Subject to} \quad h(\mathbf{x}) = x_1 + x_2 - 4 = 0$$

We might formulate a new unconstrained objective function

$$P(\mathbf{x}, r) = (x_1 - 1)^2 + (x_2 - 2)^2 + r(x_1 + x_2 - 4)^2$$

where r is a scalar weight, by adding the square of the constraint as a "penalty." As long as the value of the penalty is zero at the solution $\mathbf{x}^*$, the solution is $P(\mathbf{x}^*) = f(\mathbf{x}^*)$ as desired. The advantage of minimizing an unconstrained function instead of a constrained function should be clear. Figure 8.6 illustrates the above problem and the corresponding transformation for various values of r.

The idea of approximating a constrained optimization problem by an unconstrained problem through the use of a penalty function is a fairly old one. It has been systematically employed in numerical optimization for many years and was popularized mainly through the work of Fiacco and McCormick (1966, 1968). Despite criticism directed at their slow convergence properties, unboundedness from below of P (i.e., in some formulations the value of P can approach $-\infty$), ill-conditioning, and numerical instabilities, penalty functions have been employed primarily because they are general and simple relative to other methods for constrained minimization. Also, penalty functions are related by theory to successive quadratic programming, a topic discussed in the next section.

A classic penalty function is the quadratic loss function

$$P_1(\mathbf{x}, r) = f(\mathbf{x}) + \frac{r}{2} \sum_{j=1}^{m} h_j^2(\mathbf{x}) + \frac{r}{2} \sum_{j=m+1}^{p} [\min \{0, g_j(\mathbf{x})\}]^2$$

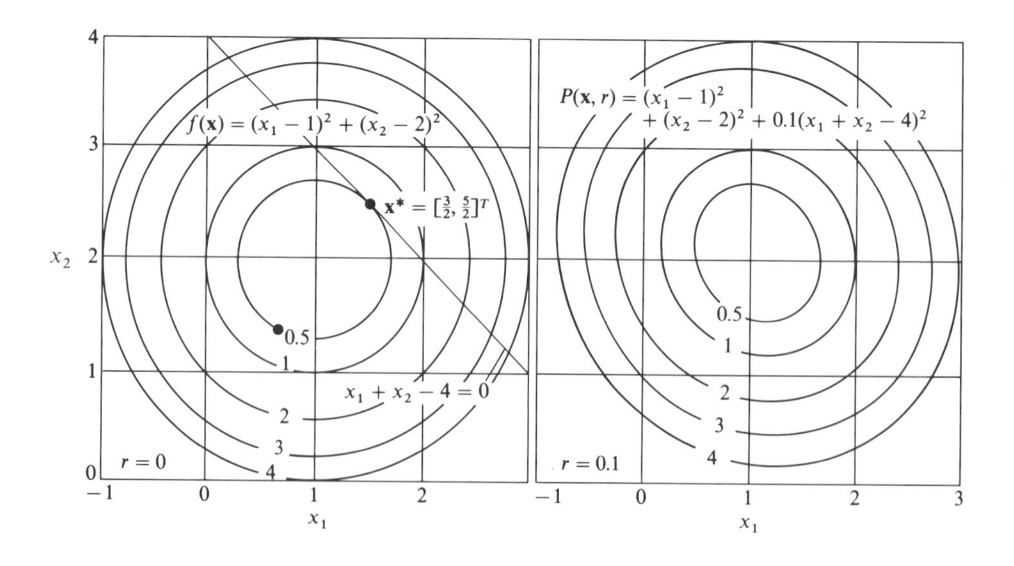

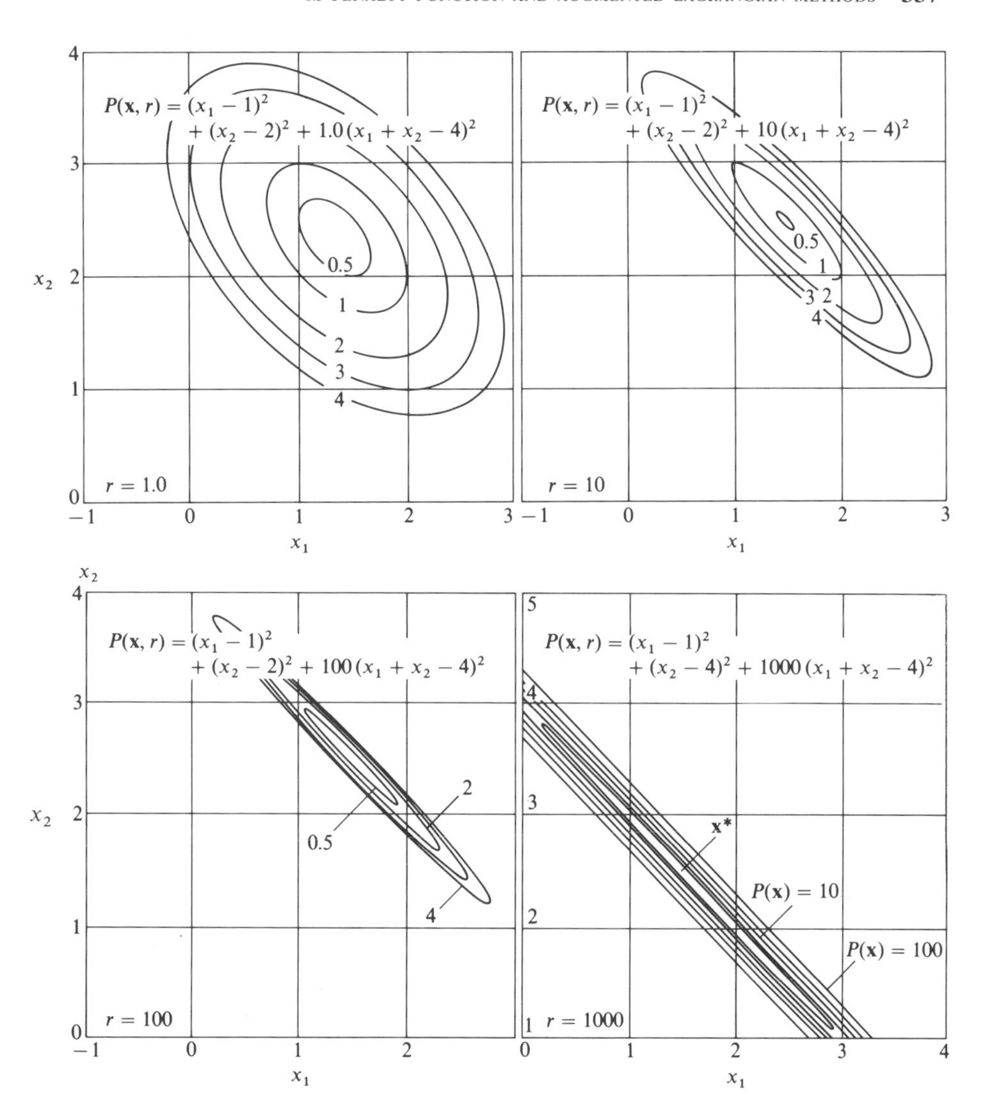

Figure 8.6 Transformation of a constrained problem to an unconstrained equivalent problem.

Another function, a nonsmooth one, is the exact l_1 penalty function

$$P_2(\mathbf{x}, r) = f(\mathbf{x}) + r \sum_{j=1}^{m} |h_j(\mathbf{x})| + r \sum_{j=m+1}^{p} \max\{0, g_j(\mathbf{x})\}$$

P_2 is called exact because under mild conditions there is a finite value of r, $\bar{r}$, such that $\mathbf{x}^*$ is a local minimum of P_2 as long as $r > \bar{r}$. (For an example, examine Fig. 8.7 on page 340.) Other related functions are discussed in the review by Fletcher (1981).

Although the quadratic (squared) loss type of penalty shown in Fig. 8.5 is simple in concept, it is not easy to determine what values of the weights r to use in minimizing $P(\mathbf{x}, r)$. Also, a sequence of increasingly larger weights is required to drive the penalty term to a very small value relative to the value of $f(\mathbf{x})$. It can be demonstrated that as the constrained minimum of problem (8.17) is approached, the Hessian matrix of a penalty function becomes increasingly ill-conditioned, thus causing the successive unconstrained minimizations to become increasingly sensitive to numerical truncation and round-off errors that often cause failure of the method to solve an NLP problem. Ill-conditioning of a square symmetric matrix can be measured by a "condition number," defined as the ratio of the absolute value of the largest to the smallest eigenvalue of the matrix. As the condition number increases, the ill-conditioning of the matrix increases. Sometimes the penalty function is unbounded below, hence convergence will not occur. Consequently, we will not discuss simple penalty function methods here as they are not recommended. For information about these methods and various types of penalty functions, refer to the supplementary references at the end of this chapter.

Several methods have been suggested to decrease the detrimental effects mentioned above. One class, generally known as acceleration techniques, uses extrapolation. At the end of each stage of minimization of $P(\mathbf{x}, r)$, the newest minimum point and two or three minimum points from prior stages of minimization with the precursor values of r are fit by a quadratic or cubic function and extrapolated to yield a new starting point for the next stage of minimization.

A more satisfactory procedure is usually called the *augmented Lagrangian method* or sometimes termed the *method of multipliers* (because of the involvement of Lagrange multipliers). Powell (1969) and Hestenes (1969) independently suggested the new type of penalty as did Haarhoff and Buys (1970), in which a quadratic penalty function "loss" term was added to the usual Lagrangian function to obtain the augmented Lagrange function

$$P(\mathbf{x}, r_A, \boldsymbol{\omega}) = f(\mathbf{x}) + \sum_{j=1}^{m} \omega_j h_j(\mathbf{x}) + \frac{r_A}{2} \sum_{j=1}^{m} h_j^2(\mathbf{x}) \tag{8.34}$$

where ω is a Lagrange multiplier. Inequality constraints would be treated in one of the ways described in connection with the generalized reduced-gradient method; refer to the comments in connection with Eq. (8.1a), Sec. 8.4. From another viewpoint, Eq. (8.34) represents a linear term added to quadratic-loss penalty function. Hestenes suggested that solving a sequence of unconstrained minimizations of (8.34), each followed by updating the values of ω_j by a simple formula that in effect maximized a dual problem, would solve problem (8.1).

Function (8.34) has some desirable properties, at least compared with simple penalty functions. Both the quadratic penalty term and its gradient vanish at the solution $\mathbf{x}^*$. The linear term and its gradient vanish at $\mathbf{x}^*$. Consequently, if $\omega_j = \omega_j^*$, $\mathbf{x}^*$ represents a stationary point with respect to $\mathbf{x}$ of $P(\mathbf{x}^*, r_A, \boldsymbol{\omega}^*)$.

Furthermore, the Hessian matrix of the first two terms on the right-hand side of (8.34) is

$$\mathbf{W}(\mathbf{x}, \boldsymbol{\omega}) \equiv \nabla^2 f(\mathbf{x}) + \sum_{j=1}^{m} \omega_j \nabla^2 h_j(\mathbf{x})$$

and although $\mathbf{W}(\mathbf{x}^*, \boldsymbol{\omega}^*)$ may not be positive definite so that a minimum of $P(\mathbf{x}, r_A, \boldsymbol{\omega})$ exists, the projected Hessian of $\mathbf{W}$ is positive definite

$$\mathbf{Z}^T(\mathbf{x}^*)\mathbf{W}(\mathbf{x}^*, \boldsymbol{\omega}^*)\mathbf{Z}(\mathbf{x}^*)$$

where $\mathbf{Z}$ is the matrix whose columns form a basis for the set of vectors in the null space orthogonal to the rows of $\mathbf{A}$ in the linear expansion of each constraint; the vector for the ith row in $\mathbf{A}$ is $\mathbf{a}_j \equiv \nabla h_j^T$

$$h_j(\mathbf{x} + \boldsymbol{\delta}) = h_j(\mathbf{x}) + \mathbf{a}_j(\mathbf{x})\boldsymbol{\delta} + \tfrac{1}{2}\boldsymbol{\delta}^T \nabla^2 h_j(\mathbf{x})\boldsymbol{\delta}$$

Consequently, $\mathbf{x}^*$ is a local minimum of the Lagrange function within the space of the vectors orthogonal to the gradients of the active constraints. The Hessian matrix of the quadratic penalty term in Eq. (8.34) is

$$\frac{r_A}{2}\left[\sum_{j=1}^{m} h_j(\mathbf{x})\nabla^2 h_j(\mathbf{x}) + \mathbf{A}^T(\mathbf{x})\mathbf{A}(\mathbf{x})\right]$$

Because $h_j(\mathbf{x}^*) = 0$, the Hessian of the quadric penalty term reduces to $\mathbf{A}^T(\mathbf{x})\mathbf{A}(\mathbf{x})$, which is a positive semidefinite matrix with strictly positive eigenvalues corresponding to eigenvectors in the range space of $\mathbf{A}^T(\mathbf{x}^*)$. In effect, adding a scalar $r_A/2$ times the penalty term increases the eigenvalues (which might be negative) of $\mathbf{W}$ in the range space of $\mathbf{A}^T(x^*)$, but leaves the other eigenvalues unchanged. Thus $\mathbf{x}^*$ can be shown to be an unconstrained minimum of the augmented Lagrange function $P(\mathbf{x}, \boldsymbol{\omega}^*, r_A)$ if r_A exceeds some value.

Some of these concepts are illustrated in Fig. 8.7 (adapted from Gill, Murray, and Wright, 1981, p. 227) for a function of one variable

$$\text{Minimize} \quad x^3$$

$$\text{Subject to} \quad x + 1 = 0$$

Clearly the solution is $x^* = -1$. We can calculate the optimal Lagrange multiplier from

$$\nabla L(x^*, \omega) = 3x^2 + \omega = 0$$

so that $\omega^* = -3$ at $x^* = -1$. Other functions derived from or components of $P(x, \omega, r_A) = x^3 + \omega(x + 1) + (r_A/2)(x + 1)^2$ are

Function	**Value at** $\omega^* = -3$, $x^* = -1$
$\nabla_x L(x, \omega) = 3x^2 + \omega$	0
$\nabla_x[(r_A/2)(x + 1)^2 = 2(r_A/2)(x + 1)$	0
$\nabla_x P(x, \omega, r_A) = 3x^2 + \omega + r_A(x + 1)$	0
$\nabla_x^2 L(x, \omega) = 6x$	-6 (not positive definite)
$\nabla_x^2 P(x, \omega, r_A) = 6x + r_A$	Positive definite only for $r_A > 6$
$(r_A/2)(x + 1)^2$	0

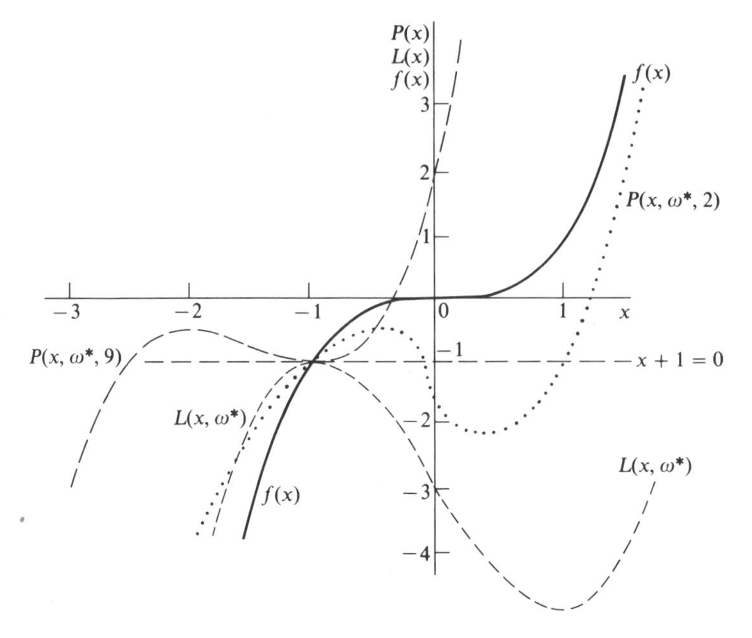

Figure 8.7 Trajectories of $f(x)$, $L(x, -3)$, and $P(x, -3, 9)$. Note that $f(x)$ intersects the constraint at a point $x^* = -1$ where $L(x, -3)$ is a maximum! $P(x^*, \omega^*, r_A) \equiv P(-1, 3, 9)$ exhibits only a local minimum in the vicinity of $x^* = -1$; it has a global minimum at $-\infty$. A change in r_A so it is less than 6, say $r_A = 2$, shifts the local minimum of $P(x, -3, r_A)$ to a nonfeasible point.

Although the concept is clear—minimize P as an unconstrained function—the difficulty is that the desirable properties of P mentioned above exist only if $\boldsymbol{\omega}^*$ is known, and the vector of Lagrange multipliers at $\mathbf{x}^*$ cannot be determined until $\mathbf{x}^*$ is calculated. Thus, a number of different algorithms exist to solve the NLP problem (8.1) in which different methods are used to calculate $\boldsymbol{\omega}$, r_A, and carry out the minimization. Some codes to implement these algorithms are listed in Table 8.5. The crucial element for success is that $\boldsymbol{\omega} \to \boldsymbol{\omega}^*$ as $\mathbf{x} \to \mathbf{x}^*$.

Newell (1975) proposed the following augmented Lagrange function

$$P(\mathbf{x}, \theta^k, r_A^k) = f(\mathbf{x}) + r_A^k \sum_{j=1}^{m} (h_j^2(\mathbf{x}) - 2h_j(\mathbf{x})\theta_j^k) \tag{8.35}$$

where θ_j^k is a parameter associated with the ith equality constraint, and r_A^k is a scalar positive weighting factor. Rather than use slack variables, inequality constraints are included in the penalty function by transforming them into a loss function that maintains continuity of the gradient of P with respect to $\mathbf{x}$

$$h_j(\mathbf{x}, \theta_j) = \min \{g_j(\mathbf{x}) - \theta_j, 0\} + \theta_j \tag{8.36}$$

Function (8.35) is equivalent to Powell's augmented penalty function (1969)

$$P(\mathbf{x}, \boldsymbol{\theta}^k, \mathbf{r}^k) = f(\mathbf{x}) + \sum_{j=1}^{m} r_j(h_j(\mathbf{x}) - \theta_j)^2 \tag{8.37}$$

Table 8.5 Sources of computer codes for augmented Lagrangian methods

Code name	Source	Computers	Language	Price
GAPF	D. M. Himmelblau, Department of Chemical Engineering, University of Texas, Austin, TX 78712	CDC	Fortran 66	$300
TWOFAS	Econometric Institute, Erasmus University, P.O. Box 1738, 3000 DE Rotterdam, The Netherlands	DEC, VAX	Fortran	Write

insofar as the minimization of $P(\mathbf{x}, \boldsymbol{\theta}, \mathbf{r})$ with respect to $\mathbf{x}$ is concerned, but differs because (*a*) a single scalar value of r is used, $r_A = r_1 = r_2 = \cdots = r_m$, and (*b*) the term $r_A \sum_{j=1}^{m} \theta_j^2$ is suppressed. Function (8.35) is the same as (8.34) if the nonoptimal Lagrange multipliers are $\omega_j^k = 2r_A^k \theta_j^k$.

If between unconstrained minimizations of (8.35) with respect to $\mathbf{x}$, r_A remains unaltered and $\boldsymbol{\theta}$ is adjusted in manner such that $2r_A\boldsymbol{\theta}$ converges to $\boldsymbol{\omega}^*$, the source of the ill-conditioning of the Hessian matrix of the penalty function can be alleviated.

Based on the properties of $P(\mathbf{x}, \boldsymbol{\theta}, r_A)$, Newell has shown that the solution of

$$\max_{\boldsymbol{\theta}} \left[\min_{\mathbf{x}} P(\mathbf{x}, \boldsymbol{\theta}, r_A)\right] = \max_{\boldsymbol{\theta}} P(\hat{\mathbf{x}}, \boldsymbol{\theta}, r_A) \tag{8.38}$$

where $\hat{\mathbf{x}}$ is a local minimum of P for a given $\boldsymbol{\theta}$, satisfies the necessary conditions for a constrained minimum of problem (8.1). In other words, the Lagrangian multipliers (here $\boldsymbol{\theta}$) are changed between steps executing the exact unconstrained minimizations of the augmented Lagrange function with respect to $\mathbf{x}$, and the changes should occur in a step in which the function (8.35) is maximized with respect to $\boldsymbol{\theta}$ in $\boldsymbol{\theta}$ space. It is quite effective to compute the Lagrangian multiplier changes by any efficient unconstrained maximization technique.

The algorithm comprises six steps (r_A is suppressed in the arguments of the functions to save space).

Step 1. Let $k = 0$ initially. Given the starting vector $\mathbf{x}^0$, which may be a nonfeasible vector, select $\boldsymbol{\theta}^0$, r_A^0, and $\hat{\mathbf{H}}^0$, where $\hat{\mathbf{H}}^0$ is the direction matrix. Typical examples are: $r_A^0 = 1$, $\boldsymbol{\theta}^0 = \mathbf{0}$, and $\hat{\mathbf{H}}^0 = -\mathbf{I}$. Set in Eq. (8.35) $P(\hat{\mathbf{x}}^{-1}) = -10^{10}$ and $\|\mathbf{h}(\hat{\mathbf{x}}^{-1})\| = 10^{10}$.

Step 2. After the initial stage, let $k = k + 1$. Minimize $P(\mathbf{x}, \boldsymbol{\theta}^0)$ with respect to $\mathbf{x}$ in search direction $\mathbf{s}^k$ using any unconstrained minimization method to yield $(\hat{\mathbf{x}}, \boldsymbol{\theta}^0)$, an intermediate minimum. In Newell's code for the algorithm, the BFGS unconstrained method (see Sec 6.3) proved to be the best minimization technique. A new $\mathbf{x}^{k+1}$ is computed from $\mathbf{x}^k$ as follows:

$$\mathbf{x}^{k+1} = \mathbf{x}^k - \lambda^k(\hat{\mathbf{H}}^k)^{-1}\nabla P(\mathbf{x}^k, \boldsymbol{\theta}^k)$$

where λ is a scalar step length.

Step 3. Terminate the method if $|h_i(\mathbf{x}^k)| < \varepsilon_t$, $i = 1, \ldots, m$, where $0 < \varepsilon_t \ll 1$, that is, ε_t is some preselected convergence tolerance such as 10^{-5}; otherwise continue to step 4.

Step 4. If $P(\hat{\mathbf{x}}^k) > P(\hat{\mathbf{x}}^{k+1})$ and $\|\mathbf{h}(\hat{\mathbf{x}}^k)\| \le \|\mathbf{h}(\mathbf{x}^k)\|$ continue to step 5. Otherwise let $r_A^k = 10r_A^k$ and $\boldsymbol{\theta}^k = 0.1\boldsymbol{\theta}^k$, and return to step 2 with $\hat{\mathbf{x}}^k$ being the starting vector.

Step 5. Let $\boldsymbol{\theta}^{k+1} = \boldsymbol{\theta}^k + \lambda^k\mathbf{s}^k$, where $\mathbf{s}^k$ is a vector in the direction of increasing values of $P(\hat{\mathbf{x}}^k, \boldsymbol{\theta})$ with respect to $\boldsymbol{\theta}$, and λ^k is a positive scalar chosen to maximize (or increase) $P(\hat{\mathbf{x}}^k, \boldsymbol{\theta})$ in the direction $\mathbf{s}^k$. Take one step to maximize $P(\hat{\mathbf{x}}^k, \boldsymbol{\theta})$ with respect to $\boldsymbol{\theta}$ yielding $(\hat{\mathbf{x}}, \boldsymbol{\theta}^{k+1})$. As a specific example, one might use Broyden's method (see Sec. 6.3) to maximize $P(\hat{\mathbf{x}}^k, \boldsymbol{\theta})$. Evaluate $(\hat{\mathbf{H}}^k)^{-1}$. Calculate $\boldsymbol{\theta}^{k+1} = \boldsymbol{\theta}^k + (\hat{\mathbf{H}}^k)^{-1}\nabla\mathbf{h}(\mathbf{x}^k, \boldsymbol{\theta})$.

Step 6. Return to step 2 with the starting $\mathbf{x}$ vector for the next stage $(k + 1)$ being $\hat{\mathbf{x}}^k$ and the starting values for $\boldsymbol{\theta}$ and r_A being the values from stage 5.

Although the BFGS and Broyden's methods were recommended by Newell for steps 2 and 5, respectively, any unconstrained minimization technique could be used for steps 2 and 5 instead.

So far the role of the positive weighting factor r_A has not been discussed. As r_A increases, the penalty function as a function of $\mathbf{x}$ becomes increasingly poorly scaled for the minimization phase of the algorithm. On the other hand, as r_A increases, the penalty function, as a function of $\boldsymbol{\theta}$, becomes increasingly better-scaled for the maximization phase. It is recommended that a moderate initial choice of r_A^0 be made (say $r_A^0 = 1$), and that r_A be increased as the algorithm proceeds only if an unconstrained minimization fails to converge or if the maximization phase is not monotonically convergent with respect to $\boldsymbol{\theta}$.

Inasmuch as augmented Lagrange methods have not proved as satisfactory as generalized reduced-gradient methods and successive quadratic-programming methods, a detailed numerical example is omitted.

8.6 SUCCESSIVE (SEQUENTIAL, RECURSIVE) QUADRATIC PROGRAMMING

The last method of solving nonlinear programming problems to be described in this chapter is the most recent, and perhaps the best, a method that uses quadratic programming sequentially. Wilson (1963) long ago described a program called SOLVER that approximated the objective function locally by a quadratic function, and the constraints by linear functions, so that quadratic programming could be used recursively. Quadratic programming was discussed in Sec. 8.3. What has evolved over many years of revision and research is the following strategy for problem (8.1). At the start of an iteration $\mathbf{x}$, an estimate of $\mathbf{x}^*$, is known. A direction of search $\mathbf{s}$, to be used to get a better estimate of $\mathbf{x}^*$, is obtained by solving a problem of this general form:

$$\begin{aligned} &\text{Minimize} \quad \text{a quadratic merit function in terms of } \mathbf{s} \\ &\text{Subject to} \quad \left.\begin{cases}\text{linearized equality constraints} \\ \text{linearized inequality constraints}\end{cases}\right\} \text{functions of } \mathbf{s} \end{aligned} \tag{8.39}$$

Then a search is carried out in the direction $\mathbf{s}$ to reduce $f(\mathbf{x})$.

8.6.1 Form of the Quadratic-Programming Subproblem

An optimal solution to problem (8.1) must satisfy the Kuhn-Tucker conditions listed in Table 8.1 [conditions (*a*), (*c*), (*d*), (*e*), (*f*), and (*g*)]. If you apply Newton's method (for solving equations) to the Kuhn-Tucker necessary conditions for an NLP problem containing only equality constraints, the Lagrange function is

$$L(\mathbf{x}, \boldsymbol{\omega}) = f(\mathbf{x}) + \sum_j \omega_j h_j(\mathbf{x})$$

and the Kuhn-Tucker necessary conditions are

$$\nabla L = \nabla f(\mathbf{x}) + \sum \omega_j \nabla h_j(\mathbf{x}) = 0$$

$$h_j(\mathbf{x}) = 0$$

Newton's method applied to the above two equations yields

$$\begin{bmatrix} \nabla^2 L & \mathbf{J} \\ -\mathbf{J}^T & 0 \end{bmatrix} \begin{bmatrix} \Delta \mathbf{x} \\ \Delta \boldsymbol{\omega} \end{bmatrix} = \begin{bmatrix} -\nabla L \\ \mathbf{h} \end{bmatrix}$$

where $\mathbf{J}$ stands for the Jacobian matrix of the equality constraints. This system of linear equations is solved for $\Delta\mathbf{x}$ and $\Delta\boldsymbol{\omega}$. It can be shown that if $\Delta\mathbf{x}$ and $\Delta\boldsymbol{\omega}$ satisfy the two linear equations, then they satisfy the necessary conditions for optimality of the following quadratic-programming problem (Powell, 1978*a*) to determine $\mathbf{s}$

$$\begin{aligned} &\text{Minimize} \quad F(\mathbf{s}) = \mathbf{s}^T \nabla f(\mathbf{x}) + \tfrac{1}{2}\mathbf{s}^T \mathbf{B}\mathbf{s} \\ &\text{Subject to} \quad g_j(\mathbf{x}) + \mathbf{s}^T \nabla g_j(\mathbf{x}) \geq 0 \qquad j = m+1, \ldots, p \\ &\qquad\qquad\quad h_j(\mathbf{x}) + \mathbf{s}^T \nabla h_j(\mathbf{x}) = 0 \qquad j = 1, \ldots, m \end{aligned} \tag{8.40}$$

where $\mathbf{s}$ is the search direction and $\mathbf{B}$ is not the Hessian matrix of the Lagrange function $L(\mathbf{x})$ but a positive definite approximation of the Hessian. The constraints are linearized constraints and the objective function is quadratic. If $\mathbf{B}$ is used in (8.40) instead of $\nabla^2 f(\mathbf{x})$, one can show superlinear convergence takes place (Stoer, 1985).

Inequality constraints can be treated by including them in the Lagrange function. Because the Hessian matrix of the Lagrange function may not be positive definite [so that (8.40) might not have a solution], in (8.40) **B** is used instead via a suitable updating formula such as the BFGS formula discussed in Sec. 6.4, a formula that requires only first derivatives so that second derivatives do not have to be computed.

Two extremes exist with respect to the quadratic programming subproblem to get **s**:

1. The *inequality constrained version* (*IQP*), problem (8.40) above, in which the inequality constraints are retained as such (refer to Fletcher, 1975; Han, 1977; Powell, 1978*b*; Tapia, 1977; Fletcher, 1982; Gill et al., 1984).
2. The *equality constrained version* (*EQP*), problem (8.41) below, in which the inequality constraints that are members of the *active set* (i.e., are close to being equality constraints—see phase 2 below) become equality constraints, and the "subject to" constraints are solely equality constraints (refer to Biggs, 1975; Murray and Wright, 1978):

$$\begin{aligned} &\text{Minimize} && \mathbf{s}^T\nabla f(\mathbf{x}) + \tfrac{1}{2}\mathbf{s}^T\mathbf{B}\mathbf{s} \\ &\text{Subject to} && g_j(\mathbf{x}) + \mathbf{s}^T\nabla g_j(\mathbf{x}) = 0 \\ & && h_j(\mathbf{x}) + \mathbf{s}^T\nabla h_j(\mathbf{x}) = 0 \qquad j = 1, \ldots, q \end{aligned} \tag{8.41}$$

where q is the number of active inequality plus equality constraints (linearized at $\mathbf{x}^{(k)}$). In between these extremes are methods that use features from both (refer to Schittkowski, 1981; Gill, 1984).

The IQP formulation for the search direction has been the subject of the majority of the literature on sequential quadratic programming. The main difference in the two procedures is that in the EQP solution the set of active constraints is unknown so that an *active set strategy* must be determined by an additional procedure. Usually the active set of the IQP predicts correctly the active set of the NLP problem itself in the neighborhood of $\mathbf{x}^*$ for any bounded positive definite **B** (Robinson, 1974). Also, the Lagrange multipliers in problem (8.40) approach the multipliers of the NLP problem as $\mathbf{x} \to \mathbf{x}^*$. To sum up, the respective characteristics of problems (8.40) and (8.41) are:

	IQP (problem 8.40)	**EQP (problem 8.41)**
Objective function	Same	Same
Lagrange multipliers from	QP subproblem	Least squares
Matrix **B**	Full approximate	Projected approximate

A relationship exists between the EQP subproblem to get $\mathbf{s}$ and the penalty function discussed in Sec. 8.5

$$P(\mathbf{x}, r) = f(\mathbf{x}) + \frac{1}{r}\hat{\mathbf{h}}^T(\mathbf{x})\hat{\mathbf{h}}(\mathbf{x}) \tag{8.42}$$

where $\hat{\mathbf{h}}(\mathbf{x})$ designates the vector of all the equality constraints plus the active inequality constraints (see Van der Hoek, 1980). As a result of the relationship, the right-hand side of the set of constraints in (8.41), the 0's, might be replaced by $(-\frac{1}{2}ru_j)$, where the u_j are approximations of the Lagrange multipliers; analogously the right-hand side of the set of constraints in (8.40) might be replaced by $(-\frac{1}{2}ru_j)$ as indicated by Bartholomew-Biggs (1982). Tests on problem sets indicate that this replacement improves the performance of EQP algorithms. Thus, the equality quadratic-programming subproblem for one stage k (the superscript (k) is suppressed here to save space) becomes

$$\begin{aligned} &\text{Minimize} \quad \mathbf{s}^T\nabla f(\mathbf{x}) + \tfrac{1}{2}\mathbf{s}^T\mathbf{B}\mathbf{s} \\ &\text{Subject to} \quad \mathbf{A}\mathbf{s} = -\frac{r}{2}\mathbf{u} - \hat{\mathbf{h}}(\mathbf{x}) \end{aligned} \tag{8.41a}$$

where $\mathbf{u} = [u \quad \cdots \quad u_m]^T$, $\mathbf{x} = [x_1, \ldots, x_n]^T$, $\mathbf{s} = [s_1, \ldots, s_n]^T$, and $\mathbf{A}$ is the Jacobian matrix of the active constraints (the row $\mathbf{a}_i = \nabla^T g_j$ or $\nabla^T h_j$). The vector of estimates of the Lagrange multipliers is given by the solution of m equations (equal to the number of active constraints)

$$\left(\frac{r}{2}\mathbf{I} + \mathbf{A}\mathbf{B}^{-1}\mathbf{A}^T\right)\mathbf{u} = \mathbf{A}\mathbf{B}^{-1}\nabla f(\mathbf{x}) - \hat{\mathbf{h}}(\mathbf{x}) \tag{8.43}$$

and the solution of (8.41a) can be written explicitly (Bartholomew-Biggs, 1982) as

$$\mathbf{s} = \mathbf{B}^{-1}[\mathbf{A}^T\mathbf{u} - \nabla f(\mathbf{x})] \tag{8.44}$$

An especially useful feature of formulation (8.41a) is that for $r > 0$, the Lagrange multipliers $\mathbf{u}$ and the search direction $\mathbf{s}$ are defined even if linear dependence exists among the rows of $\mathbf{A}$. The REQP (recursive EQP) algorithm of Bartholomew-Biggs just requires for satisfactory search that r at each stage k satisfy

$$-\frac{r}{2}\hat{\mathbf{h}}^T(\mathbf{x})\mathbf{u} - \hat{\mathbf{h}}^T(\mathbf{x})\hat{\mathbf{h}}(\mathbf{x}) \le 0 \tag{8.45}$$

This inequality means that $\hat{\mathbf{h}}(\mathbf{x})\mathbf{A}\mathbf{s} \le 0$ because of the constraints in (8.41a), and consequently $\mathbf{s}$ is in a direction along which the active constraint violation is nonincreasing.

Some debate exists as to whether the IQP or the EQP formulation is the most effective. Certainly (8.41) is a smaller size problem than (8.40), and it is desirable that an analytical solution can be written down for (8.41) or (8.41a) whereas the solution to (8.40) must be obtained by successive iteration. However, in the EQP formulation, at each stage of search the active constraints have to be identified. Of course, in the IQP formulation one must essentially get the same information via Lagrange multiplier estimates. For large problems, there is some indication that the number of function evaluations for the EQP formulation is greater than for the IQP method.

Another consideration is that in linearization of the constraints in (8.40) a feasible point of the quadratic programming problem may not exist and/or the gradients of the active constraints may not be linearly independent. For example, consider the following modification of the problem in Example 8.2 according to formulation (8.40).

EXAMPLE 8.8 COMPARISON OF AN EQP WITH AN IQP SUBPROBLEM

$$\text{Minimize} \quad x_1 x_2$$

$$\text{Subject to} \quad h(\mathbf{x}) = 25 - x_1^2 - x_2^2 = 0$$

$$g(\mathbf{x}) = x_1 + x_2 \geq 0$$

as illustrated in Fig. E8.8*a*. The solution is at point *C*, where $x_1 = -3.536$ and $x_2 = +3.536$. Here the gradients of the constraints (both active) are linearly independent

$$\nabla g = \begin{bmatrix} 1 \\ 1 \end{bmatrix} \qquad \nabla h = \begin{bmatrix} -7.071 \\ 7.071 \end{bmatrix}$$

Now, suppose that a search direction is to be obtained at point *E* where $\mathbf{x} = [-3.536 \quad -3.536]^T$, a nonfeasible point with respect to $g(\mathbf{x})$, by solving a

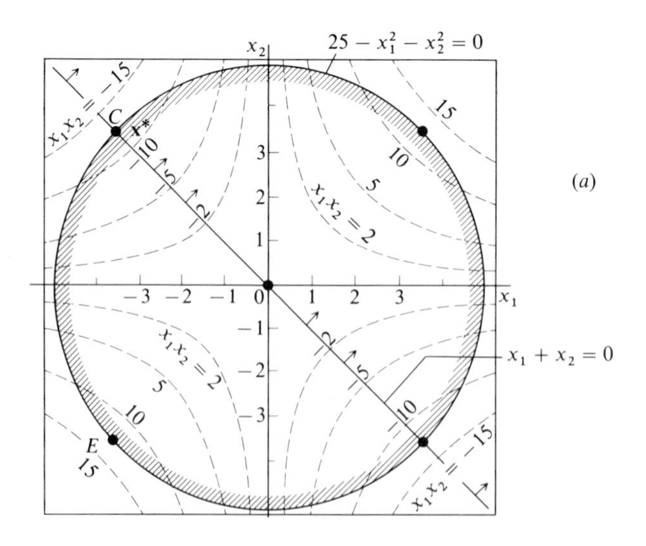

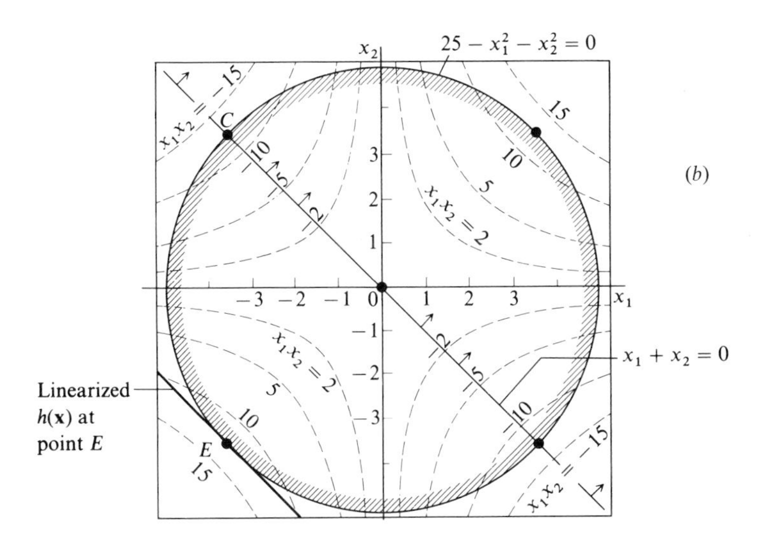

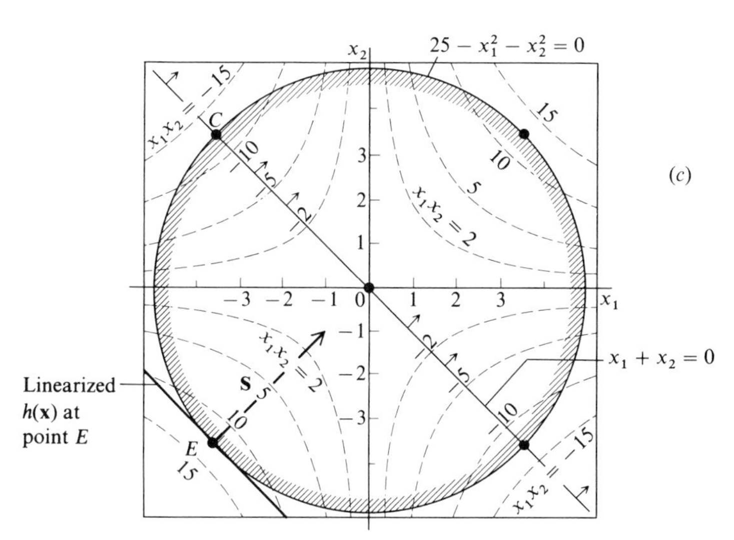

Figure E8.8 Quadratic programming subproblems: arrows on $g(\mathbf{x}) \geq 0$ point into the feasible region.

quadratic-programming problem in the formulation (8.40) in which $h(\mathbf{x})$ is linearized at $x_1 = -3.536$ and $x_2 = -3.536$. We will let $\mathbf{B} = \mathbf{I}$ for convenience here. The QP problem is

$$\text{Minimize} \quad \mathbf{s}^T \nabla f(\mathbf{x}) + \tfrac{1}{2}\mathbf{s}^T\mathbf{s}$$

$$\text{Subject to} \quad h(\mathbf{x}) = 50 + 7.071x_1 + 7.071x_2 \begin{Bmatrix} \leq 0 \\ \geq 0 \end{Bmatrix}$$

$$g(\mathbf{x}) = x_1 + x_2 \geq 0$$

Can you show that the gradients of the constraints are not linearly independent? From Fig. E8.8*b* you can see that the constraint set is not compatible so that no quadratic-programming solution can exist, hence a straightforward IQP algorithm will fail.

Now consider instead the same problem but put in the EQP formulation (8.41*a*) for the same nonfeasible starting point $\mathbf{x} = [-3.536 \quad -3.536]^T$ with $r = 40$ and the initial estimate of $\mathbf{B} = \mathbf{I}$

$$\text{Minimize} \quad \mathbf{s}^T\nabla f(\mathbf{x}) + \tfrac{1}{2}\mathbf{s}^T\mathbf{s}$$

$$\text{Subject to} \quad \begin{bmatrix} -2x_1 & -2x_2 \\ 1 & 1 \end{bmatrix}\begin{bmatrix} s_1 \\ s_2 \end{bmatrix} = -\frac{40}{2}\begin{bmatrix} u_1 \\ u_2 \end{bmatrix} - \begin{bmatrix} 0 \\ -7.071 \end{bmatrix}$$

The matrix $\mathbf{A}$ is

$$\mathbf{A} = \begin{bmatrix} 7.071 & 7.071 \\ 1 & 1 \end{bmatrix}$$

and the $\nabla f(\mathbf{x})$ is

$$\nabla f(-3.536, -3.536) = \begin{bmatrix} -3.536 \\ -3.536 \end{bmatrix}$$

The Lagrange multiplier estimates are obtained from the solution of Eq. (8.43)

$$\left\{20\mathbf{I} + \begin{bmatrix} 7.071 & 7.071 \\ 1 & 1 \end{bmatrix}\mathbf{I}\begin{bmatrix} 7.071 & 1 \\ 7.071 & 1 \end{bmatrix}\right\}\begin{bmatrix} u_1 \\ u_2 \end{bmatrix}$$

$$= \begin{bmatrix} 7.071 & 7.071 \\ 1 & 1 \end{bmatrix}\mathbf{I}\begin{bmatrix} -3.356 \\ -3.536 \end{bmatrix} - \begin{bmatrix} 0 \\ -7.071 \end{bmatrix}$$

$$\mathbf{u} = [-0.451 \quad 0.290]^T$$

Consequently the search direction exists and is (from Eq. 8.44)

$$\mathbf{s} = [0.637 \quad 0.637]^T$$

Examine Fig. E8.8*c*. You can see that this direction at least represents a direction in which a feasible value of $\mathbf{x}$ can be obtained.

8.6.2 Successive Quadratic-Programming Algorithm

Once the search direction has been established by solving the quadratic-programming subproblem, a minimization algorithm must be used to calculate a step size in the search direction. Various kinds of line searches have been employed in which a quadratic-loss penalty function, Eq. (8.42), is the objective function, others use the "exact" penalty function (see Sec. 8.5), while others use the Lagrange function or an approximation thereof. Refer to Chaps. 5 and 6 for specific methods that might be employed for line searches embedded in a multidimensional unconstrained NLP problem.

Another important phase of the solution strategy is the updating of the matrix **B** at each iteration. No general agreement exists as to what is the best updating formula to use to build up a good estimate of $\nabla^2 L$ and avoid **B** becoming singular or unbounded.

An essential aspect of implementing a sequential QP algorithm is coding so as to make the calculation of **s** efficient. You should take into account the fact that $\mathbf{B}^{k+1}$ differs only by a low rank correction from $\mathbf{B}^k$. Also, near the solution, you expect to have available a good estimate of the active constraints.

A typical EQP algorithm is represented approximately by the following phases. The variables and constraints should be prescaled insofar as possible.

Phase 1 Initialization. Let $k = 0$. Pick r^0, $\mathbf{x}^0$, $\mathbf{B}^0$ (usually $\mathbf{B}^0 = \mathbf{I}$). Continue.

Phase 2 Calculate members of active set of constraints. Calculate $\hat{\mathbf{h}}(\mathbf{x}^k)$ comprised of all the equality constraints plus the active (binding and violated) inequality constraints at $\mathbf{x}^k$ or in the neighborhood of $\mathbf{x}^k$.

If $k = 0$, go to phase 5. Otherwise continue.

Note (a): Bounds on the variables can be included in the QP subproblem or treated as general inequality constraints (the latter is probably a more inefficient procedure).

Note (b): A suitably sized active set must be used as a working set. Too few constraints, such only as those violated at $\mathbf{x}^k$, will cause too many additions and deletions, but too many constraints (i.e., not deleting a constraint) may waste time minimizing a distorted penalty function. A good compromise would be to include all the equality constraints, all the inequality constraints that are violated at $\mathbf{x}^{(k)}$ by some tolerance, and all the inequality constraints with a negative estimated Lagrange multiplier. Omit those with positive multipliers as they satisfy the Kuhn–Tucker conditions.

Phase 3 Termination Typical termination criteria are

$$\frac{\|\mathbf{x}^k - \mathbf{x}^{k-1}\|}{\|\mathbf{x}^k\| + 1} < \varepsilon_1 \qquad \text{for some prespecified tolerance } \varepsilon_1 > 0$$

and

$$|\hat{h}_j(\mathbf{x}^k)| < \varepsilon_j \qquad \text{for some prespecified tolerance } \varepsilon_j > 0$$

Stop if both criteria are satisfied; otherwise continue.

Phase 4 Update B. Use a secant method such as BFGS or Broyden's method as described in Sec. 6.4 to update the approximate Hessian matrix. But instead of $\Delta \mathbf{g}^k = \nabla f(\mathbf{x}^{k+1}) - \nabla f(\mathbf{x}^k)$, use $\Delta \mathbf{g}^k = \nabla_{\mathbf{x}} L(\mathbf{x}^{k+1}, \mathbf{u}^{k+1}) - \nabla_{\mathbf{x}} L(\mathbf{x}^k, \mathbf{u}^{k+1})$ and make sure that $\mathbf{s}^k \Delta \mathbf{g} > 0$.

Continue.

Note: The objective is to maintain a positive definite approximation to $\nabla^2 L$.

Phase 5 Estimate the Lagrange multipliers. Use Eq. (8.43) to estimate the vector of Lagrange multipliers $\mathbf{u}^k$. Continue.

Phase 6 Calculate a new search direction. Calculate a new search direction $\mathbf{s}^k$ at $\mathbf{x}^k$ from Eq. (8.44).

If a satisfactory search direction cannot be found from the solution of the QP program, calculate a new r^k that satisfies inequality (8.45) and repeat solution of the QP program, phases 5 and 6.

Continue.

Phase 7 Move a step in the $\mathbf{s}^k$ direction. Calculate the step length λ^k along $\mathbf{s}^k$ by minimizing $P(\mathbf{x}^k + \lambda \mathbf{s}^k, r^k, \mathbf{u}^k)$ where P is given by Eq. (8.42) or one of the penalty functions mentioned in Sec. 8.5.

Let $\mathbf{x}^{k+1} = \mathbf{x}^k + \lambda^k \mathbf{s}^k$.

Set $k = k + 1$ and go to phase 2.

Note (*a*): Maratos (1978) demonstrated that $\lambda^k = 1$ (a full step size) may not be possible when $\mathbf{x}^k$ is close to $\mathbf{x}^*$, hence Q-superlinear convergence will be inhibited. Alternate selections of λ^k can be made by (1) using a second-order correction step to reduce the degree of constraint violations, (2) replacing one penalty function with another such as replacing the "exact" penalty function with a differentiable augmented Lagrangian function, (3) adding a second line search function, the so-called "watch-dog" technique of Chamberlain et al. (1982).

Note (*b*): Experience has shown that a full step size may be far too big on the first few iterations of the quadratic subproblem, and these large steps cannot be reduced because r is too small. It may be advantageous to impose a bound on λ^k for the first few iterations, say $k \le 4$. Another possibility is to introduce the use of a trust region (see Chap. 6). The step $\Delta \mathbf{x}^k$ would be obtained by solving the quadratic programming subproblem

$$\begin{aligned}
\text{Minimize} \quad & f(\mathbf{x}) + \mathbf{s}^T \nabla f(\mathbf{x}) + \tfrac{1}{2} \mathbf{s}^T \mathbf{B} \mathbf{s} + k\rho^2 \\
\text{Subject to} \quad & h_j(\mathbf{x}) + \mathbf{s}^T \nabla h_j(\mathbf{x}) = \rho h_j(\mathbf{x}) \\
& g_j(x) + \mathbf{s}^T \nabla g_j(\mathbf{x}) \ge \rho g_j(\mathbf{x}) \\
& 0 \le \rho \le 1 \\
& \|\mathbf{s}\| \le D, \text{ an arbitrary constant}
\end{aligned}$$

where ρ is the dimension of the trust region.

The major difference between an IQP and an EQP algorithm is that phase 2 is different (or omitted) for the IQP method. In an IQP, one solves an inequality constrained QP problem in which all of the constraints are included as inequality constraints (including the equality constraints as $h_i \geq 0$ and $h_i \leq 0$ simultaneously). The solution of the IQP must be iterative rather than come from Eqs. (8.43) and (8.44). The estimates of the Lagrange multipliers will form the solution of the IQP problem. Because the sequence of quadratic programs may not change very much near the solution, it may be possible to save and use the same active set of constraints and parts of **B** if factored appropriately.

From a practical viewpoint, the fact that both the EQP algorithm outlined above and the IQP analog generate an intermediate sequence of nonfeasible points may cause a user difficulties if premature termination occurs with a given code.

EXAMPLE 8.9 COMPARISON OF IQP AND EQP TREATMENTS ON THE SAME PROBLEM

An interesting comparison between an IQP treatment and an EQP treatment (for linear constraints) was reported by Gill (1984). He solved the assignment problem of Bracken and McCormick (1968) using both an EQP-based code based on Eq. (8.41) named LCSOL and an IQP code based on Eq. (8.40) named NPSOL. The problem is to minimize

$$f(\mathbf{x}) = \sum_{j=1}^{20} c_j\left(\prod_{i=1}^{5} a_{ij}^{x_{ij}} - 1\right) \tag{a}$$

subject to 12 linear inequality constraints of the type ($g_j(\mathbf{x}) \leq 0$), and seven of the type $g_j(\mathbf{x}) \geq 0$. A table of all the values of c_j and a_{ij} can be found in the reference (Bracken and McCormick, p. 26) and will not be listed here. One hundred variables are involved.

Table E8.9*a* lists the results for the EQP code and Table E8.9*b* for the IQP. The abbreviated column headings are defined in footnotes to the respective parts of Table E8.9.

The major noticeable differences in the values listed in the two tables are that the IQP code starts with 72 degrees of freedom whereas the EQP code begins with 0 degrees of freedom. At termination, both have 18 degrees of freedom (100 variables—82 binding constraints, including bounds). The condition number of the projected Hessian of the Lagrange function for the IQP code is not significantly worse than that for the EQP code at any iteration. In some nonlinear problems the condition number of the projected Hessian in the range space, $\mathbf{Z}^T\mathbf{B}\mathbf{Z}$, is several orders of magnitude smaller than the condition number of **B** itself, but that was not the case in this example.

Example 13.6 also illustrates the solution of a problem with a nonlinear objective function and nonlinear constraints by NPSOL.

Table E8.9*a*   Results obtained via the EQP code LCSOL

ITN	JDEL	JADD	STEP	NUMF	OBJECTIVE	BND	LC	NZ	NORM GZ	MIN LM	COND HZ	COND T
0	0	0	0.0D − 01	1	−1.9539D 02	91	9	0	0.00D − 01	....	1.D 00	4.D 00
1	72L	0	1.0D 00	2	−3.1565D 02	90	9	1	5.28D 00	−2.7D 01	1.D 00	4.D 00
2	101U	0	1.0D 00	3	−3.6372D 02	90	8	2	4.44D 00	−2.4D 01	1.D 00	1.D 00
3	67L	0	1.0D 00	4	−5.2490D 02	89	8	3	5.47D 00	−2.2D 01	2.D 00	1.D 00
4	97L	0	1.0D 00	5	−6.8450D 02	88	8	4	4.60D 00	−1.9D 01	2.D 00	2.D 00
5	87L	0	1.0D 00	6	−8.1237D 02	87	8	5	3.43D 00	−1.3D 01	3.D 00	2.D 00
6	77L	96L	2.1D − 01	7	−8.4275D 02	87	8	5	7.90D 00	−1.3D 01	5.D 00	2.D 00
7	64L	46L	1.4D − 01	8	−8.7937D 02	87	8	5	1.21D 01	−1.2D 01	3.D 00	2.D 00
8	0	0	1.0D 00	9	−9.9438D 02	87	8	5	5.28D 00	....	3.D 00	2.D 00
9	82L	0	1.0D 00	10	−1.1043D 03	86	8	6	3.20D 00	−1.3D 01	3.D 00	3.D 00
10	92L	0	1.0D 00	11	−1.2014D 03	85	8	7	2.53D 00	−1.3D 01	3.D 00	3.D 00
20	22L	0	1.0D 00	21	−1.5083D 03	78	8	14	1.42D 00	−3.5D 00	6.D 00	3.D 00
30	0	0	1.0D 00	31	−1.5892D 03	75	8	17	1.51D 00	....	3.D 01	3.D 00
40	0	87L	5.5D − 01	41	−1.6614D 03	78	7	15	1.11D 00	....	1.D 01	3.D 00
50	0	0	1.0D 00	51	−1.6935D 03	77	6	17	1.08D 00	....	1.D 01	3.D 00
60	0	0	1.0D 00	62	−1.7100D 03	79	7	14	4.36D − 01	....	4.D 01	2.D 00
70	0	0	1.0D 00	72	−1.7152D 03	78	8	14	3.36D − 01	....	3.D 01	2.D 00
80	0	0	1.0D 00	82	−1.7190D 03	78	8	14	2.49D − 01	....	8.D 01	2.D 00
90	0	0	1.0D 00	92	−1.7242D 03	78	8	14	2.05D − 01	....	3.D 02	2.D 00
100	0	0	1.0D 00	102	−1.7267D 03	77	8	15	2.74D − 01	....	2.D 01	2.D 00
110	0	0	1.0D 00	112	−1.7289D 03	77	6	17	2.18D − 01	....	3.D 01	2.D 00
120	0	0	1.0D 00	122	−1.7309D 03	78	6	16	1.09D − 01	....	4.D 01	2.D 00

130	0	0	1.0D 00	132	−1.7315D 03	77	6	17	1.09D − 01	....	3.D 01	2.D 00
140	0	0	1.0D 00	142	−1.7322D 03	79	6	15	6.96D − 02	....	1.D 02	2.D 00
150	0	0	1.0D 00	152	−1.7326D 03	78	6	16	5.98D − 02	....	2.D 02	2.D 00
160	0	0	1.0D 00	162	−1.7329D 03	78	6	16	3.99D − 02	....	7.D 01	2.D 00
170	0	0	1.0D 00	172	−1.7331D 03	77	6	17	3.22D − 02	....	8.D 01	2.D 00
180	0	0	1.0D 00	182	−1.7333D 03	76	6	18	6.32D − 02	....	9.D 01	2.D 00
190	73U	0	1.0D 00	192	−1.7335D 03	75	6	19	2.27D − 02	−2.0D − 02	2.D 02	2.D 00
200	0	0	1.0D 00	202	−1.7337D 03	75	6	19	4.69D − 02	....	2.D 02	2.D 00
210	0	0	1.0D 00	212	−1.7344D 03	75	7	18	5.69D − 02	....	2.D 01	2.D 00
220	0	0	1.0D 00	222	−1.7347D 03	75	7	18	2.36D − 02	....	3.D 01	2.D 00
230	0	0	1.0D 00	232	−1.7348D 03	74	7	19	6.15D − 03	....	3.D 01	2.D 00
240	0	0	1.0D 00	242	−1.7350D 03	75	7	18	1.76D − 04	....	4.D 01	2.D 00
241	0	0	1.0D 00	243	−1.7350D 03	75	7	18	5.82D − 05	....	4.D 01	2.D 00
242	0	0	1.0D 00	244	−1.7350D 03	75	7	18	1.45D − 05	....	4.D 01	2.D 00
243	0	0	1.0D 00	245	−1.7350D 03	75	7	18	1.93D − 06	....	4.D 01	2.D 00

EXIT LC PHASE. INFORM = 0 ITER = 243 NFEVAL = 245

ITN — The major iteration count, k.
JDEL — Index of constraint deleted from working set (0 indicates no deletion).
JADD — Index of constraint added to working set (0 indicates no addition).
STEP — The step $\lambda^{(k)}$ taken along the computed search direction.
NUMF — The total number of evaluations of the problem functions.
OBJECTIVE — The value of the objective function, $f(\mathbf{x}^{(k)})$.
BND — The number of bounds in the predicted active set.
LC — The number of linear constraints in the predicted active set.
NZ — N minus the number of constraints in the predicted active set, i.e., the number of degrees of freedom.
NORM GZ — The Euclidean norm of the projected gradient.
MIN LM — Lagrange multiplier at minimum for the iteration.
COND HZ — A lower bound on the condition number of the projected Hessian of the Lagrange function.
COND T — A lower bound on the condition number of the matrix of predicted active constraints.

Table E8.9*b*   Results obtained via the IQP code NPSOL

ITN	ITQP	STEP	NUMF	OBJECTIVE	BND	LC	NZ	NORM GF	NORM GZ	COND HZ	COND H	COND T
0	58	0.0	1	−2.2839D + 02	25	3	72	5.9D + 01	5.41D + 01	2.D + 01	1.D + 00	1.D + 00
1	10	1.0D + 00	2	−1.1092D + 03	16	3	81	1.9D + 01	1.84D + 01	3.D + 01	2.D + 00	1.D + 00
2	3	1.0D + 00	3	−1.4382D + 03	16	3	81	9.2D + 00	9.16D + 00	2.D + 01	2.D + 00	1.D + 00
3	15	1.0D + 00	4	−1.5673D + 03	29	4	67	5.7D + 00	5.56D + 00	2.D + 01	2.D + 00	1.D + 00
4	3	1.0D + 00	5	−1.6401D + 03	27	4	69	3.7D + 00	3.43D + 00	4.D + 01	3.D + 00	1.D + 00
5	4	1.0D + 00	6	−1.6631D + 03	29	5	66	2.6D + 00	2.36D + 00	3.D + 01	4.D + 00	5.D + 00
6	2	1.0D + 00	7	−1.6796D + 03	28	5	67	2.0D + 00	1.77D + 00	1.D + 02	4.D + 00	1.D + 00
7	8	1.0D + 00	8	−1.6886D + 03	34	6	60	1.5D + 00	1.35D + 00	2.D + 02	4.D + 00	1.D + 00
8	2	1.0D + 00	9	−1.6915D + 03	35	6	59	1.4D + 00	1.22D + 00	2.D + 02	4.D + 00	1.D + 00
9	2	1.0D + 00	10	−1.6931D + 03	36	6	58	1.3D + 00	1.12D + 00	2.D + 00	4.D + 00	1.D + 00
10	1	1.0D + 00	11	−1.6945D + 03	36	6	58	1.3D + 00	1.06D + 00	2.D + 02	4.D + 00	1.D + 00
11	2	2.7D + 00	13	−1.6978D + 03	37	6	57	1.1D + 00	9.54D − 01	2.D + 02	4.D + 00	1.D + 00
12	2	1.0D + 00	15	−1.6989D + 03	38	6	56	1.1D + 00	9.21D − 01	2.D + 02	4.D + 00	1.D + 00
13		1.2D + 00	17	−1.7001D + 03	38	6	56	1.1D + 00	8.92D − 01	2.D + 02	4.D + 00	1.D + 00
14	2	1.7D + 00	19	−1.7016D + 03	39	6	55	1.1D + 00	8.43D − 01	2.D + 02	4.D + 00	1.D + 00
15	2	4.1D + 00	21	−1.7049D + 03	40	6	54	8.9D − 01	6.61D − 01	7.D + 01	4.D + 00	1.D + 00
16	2	3.6D + 00	23	−1.7071D + 03	41	6	53	8.5D − 01	6.26D − 01	9.D + 01	4.D + 00	1.D + 00
17	2	3.6D + 00	25	−1.7088D + 03	42	6	52	8.4D − 01	6.25D − 01	9.D + 01	4.D + 00	1.D + 00
18	2	6.8D + 00	28	−1.7113D + 03	43	6	51	7.8D − 01	5.77D − 01	9.D + 01	4.D + 00	1.D + 00
19	2	6.9D + 00	31	−1.7133D + 03	44	6	50	7.4D − 01	5.49D − 01	9.D + 01	4.D + 00	1.D + 00
20	2	1.9D + 00	33	−1.7137D + 03	45	6	49	7.3D − 01	5.37D − 01	9.D + 01	4.D + 00	1.D + 00
30	1	1.0D + 00	53	−1.7281D + 03	56	6	38	4.0D − 01	2.42D − 01	6.D + 01	3.D + 01	1.D + 00
40	3	6.0D + 00	73	−1.7323D + 03	62	7	31	3.4D − 01	1.37D − 01	2.D + 02	8.D + 01	2.D + 00
50	2	1.0D + 00	84	−1.7331D + 03	63	7	30	3.4D − 01	1.34D − 01	1.D + 03	1.D + 02	2.D + 00
60	1	1.0D + 00	94	−1.7337D + 03	64	7	29	2.8D − 01	7.20D − 02	5.D + 02	1.D + 02	1.D + 00

70	5	1.0D + 00	104	−1.7340D + 03	65	7	28	2.9D − 01	9.86D − 02	6.D + 02	1.D + 02	1.D + 00
80	1	1.0D + 00	114	−1.7345D + 03	69	7	24	2.7D − 01	6.03D − 02	1.D + 02	9.D + 01	1.D + 00
90	1	1.0D + 00	124	−1.7346D + 03	69	7	24	2.6D − 01	3.48D − 02	6.D + 02	1.D + 02	1.D + 00
100	1	1.0D + 00	134	−1.7346D + 03	70	7	23	2.6D − 01	2.03D − 02	3.D + 02	1.D + 02	1.D + 00
110	1	1.0D + 00	144	−1.7346D + 03	71	7	22	2.6D − 01	1.58D − 01	2.D + 02	1.D + 02	1.D + 00
120	1	1.0D + 00	154	−1.7347D + 03	72	7	21	2.5D − 01	2.48D − 02	5.D + 02	1.D + 02	2.D + 00
130	1	1.0D + 00	164	−1.7347D + 03	72	7	21	2.5D − 01	2.09D − 02	9.D + 02	1.D + 02	2.D + 00
140	1	1.0D + 00	174	−1.7347D + 03	72	7	21	2.7D − 01	2.87D − 02	8.D + 03	2.D + 02	1.D + 00
150	1	1.0D + 00	184	−1.7349D + 03	73	7	20	2.8D − 01	3.39D − 02	2.D + 03	2.D + 03	1.D + 00
160		1.0D + 00	194	−1.7349D + 03	73	7	20	2.8D − 01	3.59D − 02	2.D + 03	3.D + 03	1.D + 00
170	2	1.0D + 00	204	−1.7350D + 03	75	7	18	2.5D − 01	2.10D − 02	8.D + 02	5.D + 03	2.D + 00
180	1	1.0D + 00	214	−1.7350D + 03	75	7	18	2.5D − 01	9.33D − 03	8.D + 02	5.D + 03	2.D + 00
190	1	1.0D + 00	224	−1.7350D + 03	75	7	18	2.5D − 01	3.87D − 03	6.D + 02	5.D + 03	2.D + 00
200	1	1.0D + 00	234	−1.7350D + 03	75	7	18	2.5D − 01	1.02D − 03	5.D + 02	5.D + 03	2.D + 00
210	1	1.0D + 00	244	−1.7350D + 03	75	7	18	2.5D − 01	2.47D − 04	5.D + 02	5.D + 03	2.D + 00
211	1	1.0D + 00	245	−1.7350D + 03	75	7	18	2.5D − 01	1.31D − 04	5.D + 02	5.D + 03	2.D + 00
212	1	1.0D + 00	246	−1.7350D + 03	75	7	18	2.5D − 01	6.18D − 05	5.D + 02	5.D + 03	2.D + 00
213	1	1.0D + 00	247	−1.7350D + 03	75	7	18	2.5D − 01	4.90D − 05	5.D + 02	5.D + 03	2.D + 00
214	1	1.0D + 00	248	−1.7350D + 03	75	7	18	2.5D − 01	4.78D − 05	5.D + 02	5.D + 03	2.D + 00

EXIT NP PHASE. INFORM = 0 MAJITS = 214 NFEVAL = 248

ITN	The major iteration count, k.
ITQP	The number of minor iterations needed to solve the QP subproblem.
STEP	The step $\lambda^{(k)}$ taken along the computed search direction.
NUMF	The total number of evaluations of the problem functions.
OBJECTIVE	The value of the objective function, $f(\mathbf{x}^{(k)})$.
BND	The number of bounds in the predicted active set.
LC	The number of linear constraints in the predicted active set.
NZ	N minus the number of constraints in the predicted active set, i.e., the number of degrees of freedom.
NORM GF	The norm of the gradient of the objective function with respect to the free variables.
NORM GZ	The Euclidean norm of the projected gradient.
COND HZ	A lower bound on the condition number of the projected Hessian of the Lagrange function.
COND H	A lower bound on the condition number of the Hessian approximation of the Lagrange function.
COND T	A lower bound on the condition number of the matrix of predicted active constraints.

EXAMPLE 8.10 RECURSIVE QUADRATIC PROGRAMMING USING REQP

REQP is an EQP algorithm based on Eqs. (8.41*a*), (8.43), and (8.44) developed by Bartholomew-Biggs (1980). When applied to the following problem (Bartholomew-Biggs, 1984)

$$\text{Minimize} \quad f(\mathbf{x}) = \tfrac{1}{2}(x_1^2 + x_2^2 + x_3^2)$$

$$\text{Subject to} \quad g_1(\mathbf{x}) = x_1 + x_2 - \tfrac{3}{2} \geq 0$$

$$g_2(\mathbf{x}) = -x_1 + x_3 + 1 \geq 0$$

$$g_3(\mathbf{x}) = -x_2 - x_3 + \tfrac{2}{5} \geq 0$$

the following sequence of results occur starting from $\mathbf{x}^0 = [0.81 \quad 0.61 \quad -0.21]^T$.

$$\mathbf{A} = \begin{bmatrix} 1 & 1 & 0 \\ -1 & 0 & 1 \\ 0 & -1 & -1 \end{bmatrix} \qquad \mathbf{g}(\mathbf{x}) = \begin{bmatrix} -0.8 \\ -0.02 \\ 0.02 \end{bmatrix}$$

$$\nabla f(\mathbf{x}^0) = [0.81 \quad 0.61 \quad -0.21]^T$$

Let $\mathbf{B} = \mathbf{I}$ and $r = 0.02$. Solve Eq. (8.43) for $\mathbf{u}$

$$\begin{bmatrix} 2.01 & -1 & -1 \\ -1 & 2.01 & -1 \\ -1 & -1 & 2.01 \end{bmatrix} \qquad \mathbf{u} = \begin{bmatrix} 1.5 \\ -1.0 \\ -0.4 \end{bmatrix}$$

to get $\mathbf{u}^0 = [3.821 \quad 2.990 \quad 3.189]^T$ and a new $\mathbf{x}^1$ of

$$\mathbf{x}^1 = [0.8306 \quad 0.6312 \quad -0.1993]^T$$

Let $r = 0.01$ on the next iteration leading to

$$\mathbf{u}^1 = [7.155 \quad 6.323 \quad 6.522]^T$$

$$\mathbf{x}^2 = [0.8319 \quad 0.6323 \quad -0.1997]^T$$

Additional stages give the results listed in Table E8.10. With $r = 6 \times 10^{-5}$, $\mathbf{x} = [0.833 \quad 0.633 \quad -0.200]^T$, the constraints are $\mathbf{g} = [-0.0333 \quad -0.0333 \quad -0.0333]^T$.

Detailed inspection of the problem shows that it has no feasible solution, but the solution obtained is the "best" solution in the sense of a least-squares solution to the constaints if transformed to equality constraints with a residue of 0.0333 for each equation.

Table E8.10

Stage	r	$\mathbf{u}$			$\left(\frac{r}{2}\mathbf{u}\right)$		
0	0.02	3.82	2.99	3.18	0.0382	0.0299	0.0319
1	0.01	7.15	6.32	6.52	0.0358	0.0316	0.0326
2	0.045	15.30	14.47	14.67	0.0344	0.0326	0.0330
3	0.018	37.07	36.24	36.44	0.0338	0.0330	0.0332
4	0.00066	100.8	100.0	100.2	0.0335	0.0332	0.0333

Table 8.6 Sources of computer codes for successive quadratic-programming methods

Code name	Source	Computers	Language	Price
ALRQP	Dr. M. C. Bartholomew-Biggs, School of Information Sciences, Hatfield Polytechnic, Hatfield, Herts, England	DEC 10	Fortran 77	F.O.C.
IDESIGN	Dr. J. S. Arora, Dept. Civil & Environmental Engr., University of Iowa, Iowa City, IA 52242	Prime 750	Fortran 77	$2000
GINO	The Scientific Press, 540 University Ave., Palo Alto, CA 94301	CDC, IBM, VAX, PC	Fortran + input language	$300
MINOS/AUGMENTED	Office Technology Licensing, Stanford Univ., Stanford, CA 94305	Numerous	Fortran 66;77	Nonprofit $300 Industrial $5000
NLPQL	Dr. K. Schittkowski, Institut für Informatik, Universität Stuttgart, Azenbergstr. 12, 7000 Stuttgart 1, W.G.	Several	Fortran	$100
NPSOL	Office Technology Licensing, Stanford Univ., Stanford, CA 94305	Numerous	Fortran 66	Nonprofit $300 Industrial $2000
TWOFAS	Dr. G. van der Hoek, Economic Inst., Erasmus Univ. Rotterdam, P.O. Box 1738, 3000 DR Rotterdam, The Netherlands	DEC 20, VAX 11/750	Fortran	Tape charge
VF02AD	Librarian, Harwell, Oxford, Oxon, England	IBM	Fortran 66	$70
VMCON	National Energy Center, Argonne NL, Argonne IL(NTIS Doc. DE83-048922) ANL-80-64	IBM-370	Fortran	$2000
WIEM	Dr. G. van der Hoek, Economic Inst., Erasmus Univ. Rotterdam, P.O. Box 1738, 3000 DR Rotterdam, The Netherlands	DEC 20	Fortran	Tape charge

A class of methods termed Projected Lagrange Algorithms which conceptually might be viewed as a modification of problem (8.39) is based on the technique proposed by Robinson (1972), namely to use a general objective function and linear (or linearized) constraints (Murtagh and Saunders, 1983; Best, Braüningern, Ritter, and Robinson, 1981; Rosen, 1978). In the method of Robinson the objective function is replaced by a shifted Lagrange function so that the optimization problem is posed as [the diacritic $\sim$ (tilde overlay) denotes linearized function]; $\boldsymbol{\delta}$ is the perturbation in $\mathbf{x}$

$$\begin{aligned} &\text{Minimize} \quad f(\mathbf{x}^k + \boldsymbol{\delta}) - (\boldsymbol{\omega}^k)^T[\mathbf{h}(\mathbf{x}^k + \boldsymbol{\delta}) - \tilde{\mathbf{h}}^k(\boldsymbol{\delta})] \\ &\text{Subject to} \quad \tilde{h}_j^k(\boldsymbol{\delta}) = 0 \qquad j = 1, \ldots, m \\ &\qquad\qquad\quad \tilde{g}_j^k(\boldsymbol{\delta}) \geq 0 \qquad j = m + 1, \ldots, p \end{aligned} \tag{8.46}$$

On one stage, $\mathbf{x}^{k+1} = \mathbf{x}^k + \boldsymbol{\delta}^k$, and $\boldsymbol{\omega}^{k+1}$, are obtained by solving (8.46).

In the projected Lagrangian method, the constraint linearization is fixed during a major iteration while the approximation to the Lagrangian function is updated. In contrast, in sequential quadratic programming the quadratic function and linearized constraints are fixed. For additional details refer to Gill, Murray, and Wright (1981, Sec. 6.5).

An excellent code that makes use of the Simplex method and a projected Lagrangian algorithm is MINOS 5.0 by Murtagh and Saunders (1983) [available from Office of Technology Licensing, Stanford University, Stanford, CA 94305, Fortran 77, $5000 (with a large educational discount)].

8.6.3 Successive Quadratic-Programming Codes

One of the first generally available successive quadratic programming codes was that of Powell (1978b) who modified the algorithm developed by Han (1977). Pshenichny (1970) developed a different algorithm which has been coded by Belegundu and Arora (1984). Although quite a few algorithms have been proposed, only a few of those that have been coded are generally available for distribution. Many of the best codes are listed in Table 8.6 on page 357.

8.7 RANDOM SEARCH METHODS

Random search methods date back to the earliest methods proposed for optimization. From an engineering viewpoint, case studies are a simple form of heuristic random search [Gaines and Gaddy (1976); Heuckroth, et al. (1976)]. A number of surveys exist, including Bekey and Maeri (1983), Dobbie (1968), McMurtry (1970), Mihail and Maria (1986), and Pronzato et al. (1984), that explain the specific details of proposed optimization methods.

One of the earliest computer codes was the completely random search of Kelley and Wheeling (1962) at Mobil Oil. Given a starting point $\mathbf{x}^0$, the code generates a random path by constructing a sequence of steps, a step being taken from the $\mathbf{x}$ vector yielding the lowest value of $f(\mathbf{x})$ discovered in the random search to another yielding a lower value. Each step comprised the weighting of a "history vector" which indicated the average direction of search in prior steps and a vector of random normal deviates, the sum of which was multiplied by λ, the step size, calculated as in steepest descent. The vector $\mathbf{x}^{k+1}$ was accepted or rejected according to whether or not $f(\mathbf{x}^{k+1}) < f(\mathbf{x}^k)$. After $\mathbf{x}^{k+1}$ was accepted (or rejected), λ^k was increased (or decreased) by a factor roughly depending upon whether the search had been easy or difficult.

Another procedure is to choose random points of hyperspheres of fixed (or random) radius from $\mathbf{x}^k$. A vector from $\mathbf{x}^k$ through the point on the hypersphere that has the lowest value of $f(\mathbf{x})$—for minimization—is selected as the search direction. The step length can be calculated by any of the methods described in Chap. 5, or be random.

As the circle gets larger, but not too large of course, the best $\mathbf{x}$ on the hypersphere will tend to be closer to $\mathbf{x}^*$. Any point on the circle segment between the broken lines in Fig. 8.8 will yield a more favorable $\mathbf{x}$ than that determined by a steepest-descent minimization, but not as favorable an $\mathbf{x}$ as obtained by a second-order information method. To have satisfactory convergence, the

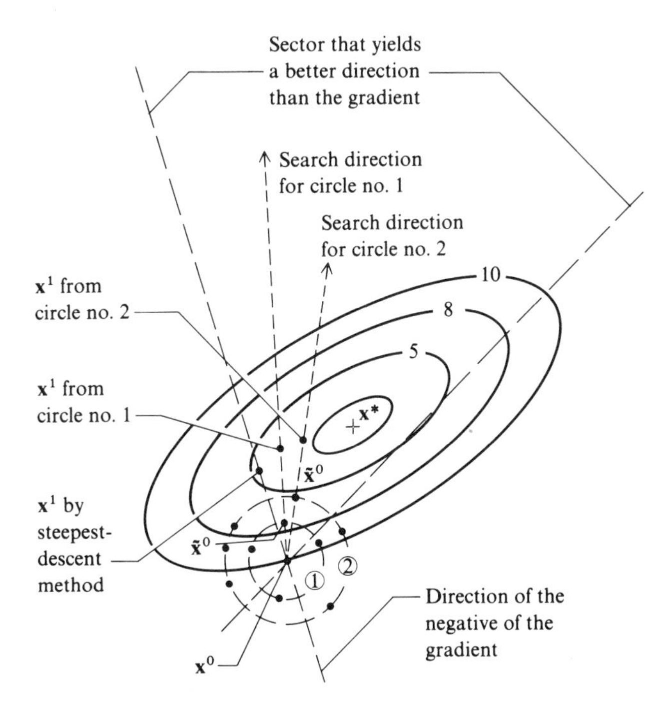

Figure 8.8 Random search illustrating the effect of changing the radius of the hypersphere used to select $\mathbf{x}^{k+1}$.

sequence of radii of the hyperspheres is reduced periodically. Also, the search can be accelerated if the points on the hypersphere are chosen at random but constrained to lie at least at a minimum angle from each other and from the previous direction of search.

A third procedure that has been employed in some applications in chemical engineering is that of Box (1965). Box's method evolved from the Simplex method of Spendley et al. and of Nelder and Mead discussed in Sec. 6.1.4. The difficulty with the methods of Spendley and Nelder and Mead when they encounter a constraint is that it is necessary to withdraw the nonfeasible vertex until it becomes feasible. After many such withdrawals the polyhedron will collapse into $(n - 1)$ or fewer dimensions, and the search will be quite slow. Furthermore, if the constraint ceases to be active, the collapsed polyhedron cannot readily expand back into the full n-dimensional space again. To avoid these difficulties, Box selected a polyhedron with more than $(n + 1)$ vertices, which he termed a *complex*.

The complex method employes $(n + 1)$ or more vertices p (each of which must satisfy the constraints at all k stages). An initial $\mathbf{x}_1^0$ is chosen, and $(p - 1)$ additional vertices are selected one at a time through the use of pseudorandom numbers and the following relation:

$$\mathbf{x}_i^0 = \mathbf{L}_i + \mathbf{r}_i^0(\mathbf{U}_i - \mathbf{L}_i) \qquad i = 2, \ldots, p$$

where $\mathbf{L}_i$ and $\mathbf{U}_i$ are, respectively, the lower and upper bounds on $\mathbf{x}_i$, and $\mathbf{r}_i^0$ is a diagonal matrix of pseudo-random numbers uniformly distributed over the interval (0, 1). If no bounds are known, the initial polyhedron should be chosen to cover the search region.

Next, the objective function is evaluated at each vertex. The vertex having the worst value of $f(\mathbf{x})$ is replaced by a new vertex located along the line joining the rejected point and the centroid of the remaining points at a distance equal to or greater than the distance from the rejected point to the centroid. If a new vertex turns out to give the worst value of $f(\mathbf{x})$ of all the vertices in the new polyhedron, it is replaced by another vertex located half the distance from the new vertex to the centroid. (If a constraint is violated, the new vertex is also moved halfway in toward the centroid.) Box recommended from empirical studies that the overexpansion be a factor of 1.3 and that $p = 2n$ vertices be used in the search. The overexpansion and the use of more than $(n + 1)$ vertices are features of the procedure designed to prevent the polyhedron from flattening near constraints. The search continues until the polyhedron is reduced essentially to the centroid to within a given precision.

All numerical tests of random methods have indicated that they are less efficient than the deterministic methods explained in Chaps. 5, 6, 7, 8, and 9 in this book. Random methods do have the possibility of locating a global minimum rather than just a local minimum, and a number of investigations of random type algorithms to minimize unconstrained multiextremal functions indicate high probabilities of success (Kan and Timmer, 1985).

8.8 COMPARATIVE EVALUATION OF GENERAL NONLINEAR PROGRAMMING CODES

Many studies have been carried out to evaluate the performance of nonlinear programming codes on sets of test problems. In 1980 Schittkowski published the results of a very extensive survey of 26 codes developed in the 1970s. What criteria to use in evaluation are somewhat subjective, but Table 8.7 lists Schittkowski's criteria together with three sets of weighting factors to permit relative emphasis of various criteria. Table 8.8 lists the best in Schittkowski's study of types of codes discussed in this chapter. Only a few codes based on sequential QP were tested so that no clear-cut resolution of the IQP vs. EQP issue occurred. MINOS was not included in the tests but has been reported to be effective particularly if the problem contains many linear constraints and is of large scale.

Table 8.7 Weighting factors for evaluation

	Weights		
Performance criteria	**I**	**II**	**III**
Efficiency (weighted measure of mean time of execution divided by accuracy, function calls per unit accuracy, gradient calls per unit accuracy)	0.32	0.18	0.14
Reliability (weighted combination of fraction of wrong solutions, objective-function values, constraint violations, problem failures—because of excessive calculation time, undetermined arguments, overflow, etc.)	0.23	0.18	0.36
Global convergence (weighted combination of fraction of cases reaching global solution, mean of objective function values, mean of constraint violations)	0.08	0.08	0.20
Performance in solving degenerate problems (rated as efficiency)	0.05	0.03	0.03
Performance in solving ill-conditioned problems (rated as efficiency)	0.05	0.06	0.03
Performance in solving indefinite problems (rated as efficiency)	0.03	0.03	0.03
Sensitivity to variations in the problem (fraction of deviations)	0.03	0.03	0.06
Sensitivity to location of the starting point (ratio of efficiency for close and far points)	0.07	0.06	0.06
Ease of use (good documentation, provision of problem data and functions, program organization, sensitivity to input parameters)	0.14	0.35	0.09
	1.00	1.00	1.00

Table 8.8 The Best nonlinear programming codes (Schittkowski)

Authors	Type	Name	Date
	Linearization methods		
Abadie & Guigou	Generalized reduced gradient	GRGA	1970–75
Lasdon, Waren, Ratner	Generalized reduced gradient	GRG2	1980
	Quadratic approximation methods		
Powell	Recursive quadratic programming	VF02AD	1978
Bartholomew-Biggs	Exterior penalty with quadratic subproblem	OPRQP	1978
	Penalty function methods		
Best & Bowler	Method of multipliers	ACDPAC	1978
Newell	Method of multipliers	QAPQF	1975
Ragsdell & Root	Minimize augmented lagrangian	BIAS(1)	1978
	Other		
Best	Method of Robinson	FCDPAK	1975
Kreuser & Rosen	Gradient projection plus penalty	GPM/GPMNLC	1971

It proved to be quite evident from the study, and has been also demonstrated by more recent but less exhaustive studies, that recursive quadratic-programming codes and generalized reduced-gradient codes surpassed all the others quite clearly. Table 8.9 lists the relative performance of the codes listed in Table 8.8 evaluated according to the criteria in Table 8.7. Codes developed more recently should perform similarly. Applications of the techniques discussed in this chapter can be found in case studies involving liquid extraction, distillation, gas transmission networks, reactors, and refrigeration in Secs. 11.1, 11.2, 13.4, and Examples 12.4, 14.1, 14.2, 14.3, and 14.4.

Table 8.9 Overall performance of the best codes
(100 = best, 0 = worst in the overall classification)

	Weights		
Code	I	II	III
VF02AD	100	100	100
GRGA	87	91	99
GRG2	84	92	86
FCDPAK	82	82	78
OPRQP	76	73	71
ACDPAC	68	76	77
BIAS(1)	65	70	71
QAPQF	50	52	49

8.9 SUCCESSIVE LINEAR PROGRAMMING

As mentioned in Sec. 8.4, one of the more obvious methods to solve a nonlinear programming problem is to use repeated linearization of both the objective function and constraints at some estimate of the solution. This procedure is the basis for a method called *successive linear programming* (SLP), in which at each stage a linear program is solved using the Simplex method (discussed in Chap. 7).

Linear programming has been used in the oil and chemical industry for over 20 years, so it is not surprising that one of the earliest reported SLP algorithms was employed in refinery optimization, the so-called Method of Approximate Programming (MAP, Griffith and Stewart, 1961). Suppose that $\tilde{\mathbf{x}}$ is a feasible solution to the nonlinear programming problem. Linearization of each nonlinear function around $\tilde{\mathbf{x}}$ gives

$$\begin{aligned} &\text{Minimize} && f(\mathbf{x}) \approx f(\tilde{\mathbf{x}}) + (\mathbf{x}\text{-}\tilde{\mathbf{x}})^T\nabla f(\tilde{\mathbf{x}}) \\ &\text{Subject to} && h_j(\tilde{\mathbf{x}}) \approx \mathbf{h}_j(\tilde{\mathbf{x}}) + (\mathbf{x}\text{-}\tilde{\mathbf{x}})^T\nabla h_j(\tilde{\mathbf{x}}) = 0 \\ & && g_j(\mathbf{x}) \approx g_j(\tilde{\mathbf{x}}) + (\mathbf{x}\text{-}\tilde{\mathbf{x}})^T\nabla g_j(\tilde{\mathbf{x}}) \geq 0 \end{aligned} \tag{8.47}$$

Each component of the vector $(\mathbf{x}\text{-}\tilde{\mathbf{x}})$ is restricted by arbitrary upper and lower bounds to remain in a region that is not too far from $\tilde{\mathbf{x}}$.

The procedure is to start at $\tilde{\mathbf{x}}^k$, and if the optimal solution of (8.47) via linear programming is feasible for the original nonlinear problem (8.1), then the new $\tilde{\mathbf{x}}^{k+1}$ is used as the reference point for the next linearization. However, if the new $\tilde{\mathbf{x}}^{k+1}$ is infeasible [as is most likely the case if equality constraints exist in (8.1)], the new $\tilde{\mathbf{x}}^{k+1}$ is not accepted and the linear program is repeated for a narrower range of upper and lower bounds on $\mathbf{x}$. Various heuristics are employed to determine if $\tilde{\mathbf{x}}^{k+1}$ is acceptable and how the step bounds can be increased. Reklaitis et al. (1983) have presented several simple examples where this procedure is used to solve an NLP.

A more sophisticated algorithm was proposed by Beale (1974) using a combination of SLP and reduced-gradient concepts. Beale's method facilitates satisfying the constraints on $\mathbf{x}$ and yields a search direction using the nonbasic variables, thus restoring feasibility. This version of SLP has been used in a number of large-scale applications, such as oilfield development (Cheifett, 1974), refinery applications (Bodington, 1979), and olefin plants (Buzby, 1974). In some of the successful applications for refinery optimization, the product properties such as octane number and vapor pressure were nonlinear functions of the mole fractions of each component in the blend, while the remainder of the problem assumed the conventional LP form. Also, SLP can be used in conjunction with a large process simulator (Colville, 1963); see Chap. 14 for more details. Lasdon and Waren (1980) have described some of these applications in more detail.

A recent variation of SLP, called PSLP (penalty SLP), attempts to deal with restoring feasibility through use of penalty functions of the constraint violations (Lasdon and Waren, 1983; Palacios-Gomez et al., 1982; Lasdon et al.,

1983). The algorithm also utilizes trust regions (see Sec. 6.4.2) which are varied for each iteration.

LP-based algorithms share the same advantages, namely that they are easy to implement, can handle very large numbers of variables and constraints, and will achieve rapid convergence when the optimum lies at the vertex of the constraints. SLP has been demonstrated to be successful on problems with a "moderate" degree of nonlinearity. For large-scale problems with few nonlinear constraints the technique holds promise. Typical examples are quarterly production planning and distribution in which only small shifts in operations take place from time period to time period, the assumption being that during the current time period operations are at an optimal point. On the other hand, disadvantages of SLP methods which may prevent their successful use are as follows:

1. SLP may converge *very slowly* on problems where the optimum lies in the interior of the feasible region and where there are a large number of nonlinear variables.
2. SLP will usually violate nonlinear constraints until convergence, often by large amounts, hence exceedingly small steps may be required, resulting in high computational costs. We have not included SLP in the comparisons of various NLP methods discussed in the previous section because to date no SLP algorithm has been involved in testing on a wide class of nonlinear programming problems.

8.10 OPTIMIZATION OF DYNAMIC PROCESSES

In previous sections of this chapter we have considered problems in which the constraints were algebraic equations and/or inequalities. Another important class of optimization problems exists for which the equality constraints are ordinary differential equations. This type of problem is generally referred to as an *optimal-control* problem and is an element of the field of control theory which has experienced significant development since the 1950s. Because the solution of optimal-control problems involves so many theoretical and practical features, here we will only cover ways nonlinear programming can be employed to solve such problems.

Consider the minimization of the following objective functions with respect to the control variable $\mathbf{u}(t)$ over the time interval $[t_0, t_f]$:

$$f(\mathbf{u}) = \Theta[\mathbf{x}(t_f)] + \int_{t_0}^{t_f} \Phi(\mathbf{x}(t), \mathbf{u}(t), t)\, dt \tag{8.48}$$

where $\mathbf{x}(t)$ and $\mathbf{u}(t)$ are related via the ordinary differential equations

$$\frac{d\mathbf{x}}{dt} = \boldsymbol{\phi}(\mathbf{x}, \mathbf{u}, t) \tag{8.49}$$

The initial conditions $\mathbf{x}(t_0)$ are given, and Θ, Φ, and $\boldsymbol{\phi}$ are assumed to be continuous with continuous first partial derivatives (with respect to $\mathbf{x}(t)$, $\mathbf{u}(t)$, and t). The independent variable is t (usually time), although a spatial distance variable can also be employed.

Selection of various functional forms for f is governed by criteria such as:

1. *Minimum yield of a contaminant.* If x_1 is a concentration or yield of an undesirable compound in a batch reaction system, then $x_1(t_f)$ is the concentration at the end of a reaction time t_f. In this case $\Phi = 0$ and letting $\Theta = x_1(t_f)$ is tantamount to minimizing yield.

2. *Maximum production of a key component.* In a continuously operating process such as distillation, we might seek to maximize the production of a valuable component x_2 over a period of time t_f. This can be done by setting $\Phi = -Dx_2(t)$, where D is the distillate flow-rate. In order to include a cost of control effort, such as reflux flow (say u_1), then f could be adjusted by setting $\Phi = -C_2Dx_2(t) + C_1u_1(t)$. C_2 is the selling price of x_2 and C_1 is the utility cost for reflux in this case. Both quantities are integrated over time via Eq. (8.48).

3. *Start-up or shut-down.* In the above two cases, the terminal time t_f is assumed fixed although there are some other important cases where it is not fixed. In a start-up or shut-down problem the objective function could be the time to carry out the change in operating conditions, hence the name of the "minimum time" problem. In (8.48), set $\Phi = 1$ and $\Theta = 0$.

Some books refer to the optimal control problem as an infinite dimensional problem because $\mathbf{u}$ is a function of time (rather than a specific value) at the optimum. Necessary conditions for a minimum can be developed using Lagrange multiplier theory, which leads to the so-called minimum principle for dynamic systems (Pontryagin et al., 1962; Lapidus and Luus, 1967; Bryson and Ho, 1969; Sage, 1968). These books contain many details on theoretical and computational aspects of the minimum principle.

In some cases, it is efficient to employ nonlinear programming rather than the minimum principle to compute the optimal control. Consider a problem with a single control variable $u(t)$. Conceptually you can think of the true optimal control being approximately represented by

1. A set of piecewise constant (discrete) values of u: $u(0), u(1), \ldots, u(n-1)$. In this case, an optimization problem with n variables results with the computed optimum solution approximating the optimal control according to some criterion (see Fig. 8.9).

2. Some functional form,

$$\hat{u}(t) = \sum_{i=0}^{p} a_i \Psi_i(t) \tag{8.50}$$

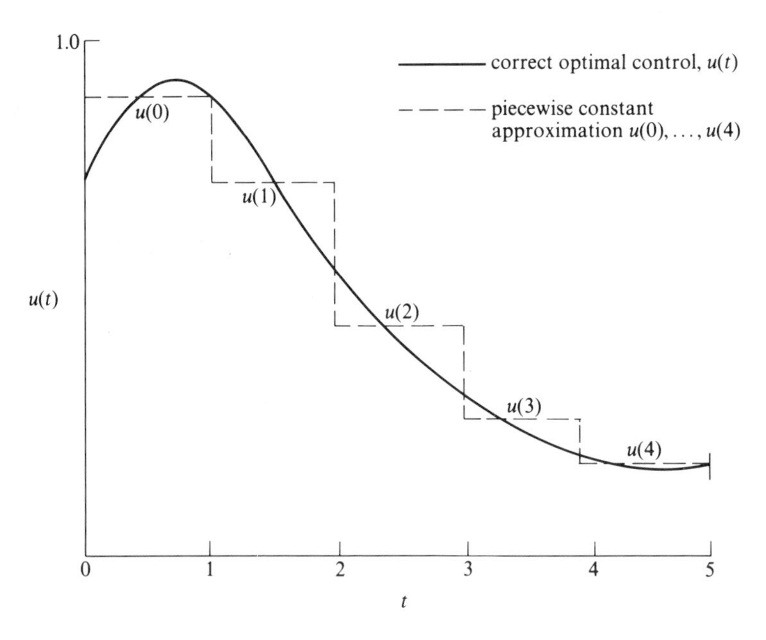

Figure 8.9 Piecewise constant approximation of an optimal control.

In (8.50), $\hat{u}(t)$ is the sum of a convergent series of linearly independent functions of time. In this approach, using judgment or experience, you specify the form of $\Psi_i(t)$ and then evaluate the coefficients $a_0, a_1, \ldots, a_p$ to fit the desired $u(t)$ according to some criterion. Consequently, $(p + 1)$ variables can be adjusted. We discuss each of the two methods below, followed by an example.

Discretization of the Control Function

Suppose there is a single control variable $u(t)$. If $u(t)$ is approximated by piecewise constant values, then it will be necessary to compute search directions and step sizes based on the value of $f(u)$. This procedure involves repeated integration of the set of differential equations (8.49) and then evaluating the objective function (8.48), a rather time-consuming operation. It is tempting to approximate the left-hand side of the differential equations (8.49) by the first-order Eulerian difference expression

$$\frac{d\mathbf{x}}{dt} \approx \frac{\mathbf{x}(t_i) - \mathbf{x}(t_{i-1})}{\Delta t} \tag{8.51}$$

where $\Delta t = t_i - t_{i-1} = (t_f - t_0)/n$, the grid interval for t. If the right-hand side of (8.49) is evaluated at $\mathbf{x}(t_{i-1})$, the explicit discrete form of (8.49) results:

$$\mathbf{x}(t_i) = \mathbf{x}(t_{i-1}) + \Delta t \boldsymbol{\phi}(\mathbf{x}(t_{i-1}), u(t_{i-1})) \tag{8.52}$$

Note that (8.52) is now a set of algebraic equality constraints (rather than a differential equation), and the optimal control problem is now expressed in the standard form of a constrained nonlinear programming algorithm.

However, the weakness of the above approach lies in the Eulerian finite-difference approximation of $d\mathbf{x}/dt$. In general, the integration of ordinary differential equations using this method will only be accurate as long as Δt is made very small (or n very large). This requirement in turn leads to an optimization problem of large scale (say 1000 variables), hence it is not a very practical approach. Lower levels of discretization for most engineering problems do not allow accurate integration.

An alternate to (8.52) is to use a more accurate integration scheme by combining a high-quality computer library integrator with an unconstrained optimization code. Such a combination is also attractive because the process model (8.49) does not have to be modified. In order to employ an unconstrained optimization method, it is necessary to integrate the set of differential equations (8.49) and evaluate f in Eq. (8.48) at each iteration k. The search method then updates the values of $u^k(0), u^k(1), \ldots, u^k(n-1)$. Figure 8.10 is a flowsheet for the necessary calculations.

Several practical issues exist in using a discretized control. First, you will need to try several levels of discretization to ensure that the solution for $u(t)$ is

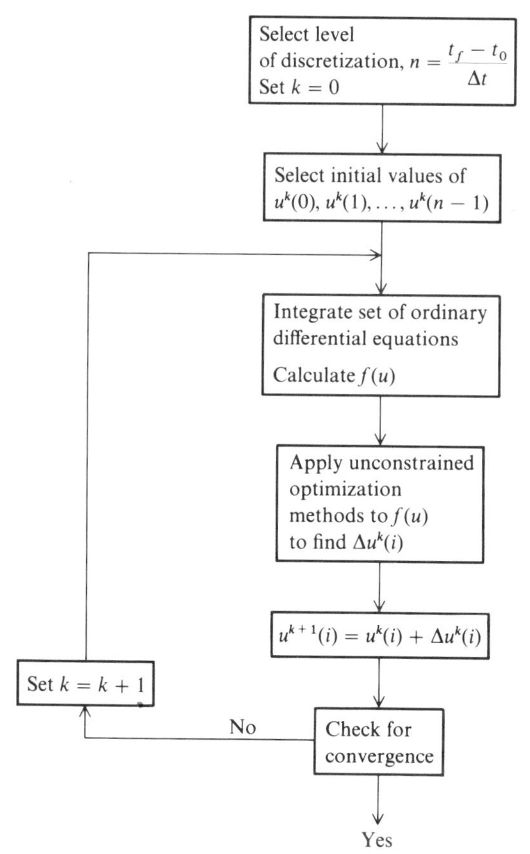

Figure 8.10 Calculation of optimal control via unconstrained optimization combined with iterative solution of the differential equations comprising the model.

close enough to the optimal function and provides the level of detail desired. In that regard, Fig. 8.9 shows a satisfactory approximation. If the level of discretization (n) is doubled and there are inconsequential changes in the computed discrete control, then you have a close approximation to the desired optimal control. On the other hand, you should try several initial guesses to make sure they converge to the same answer (although this does not guarantee that you have found the global optimum).

If algebraic constraints on **u** and **x** coexist with (8.49), then the procedure that must be employed is more complicated; refer to Sargent and Sullivan (1978).

Parameterization of the Control Function

Parameterization of the optimal control function is related to well-known mathematical methods involving Fourier series or Laguerre functions, although the functions $\Psi_i(t)$ in Eq. (8.50) are finite in number and not necessarily orthogonal. A simple polynomial form may suffice such as

$$\hat{u}(t) = a_0 + a_1 t + a_2 t^2 + \cdots + a_p t^p \tag{8.53}$$

In some problems, however, it may be advantageous to use orthogonal functions in order to improve the rate of convergence to the optimum. As suggested by Ray (1982) and Sargent and Sullivan (1978), the time horizon $[t_0, t_f]$ can be divided into several different subintervals to allow more flexibility in the selection of $\Psi_i(t)$. For example, if you know the optimal control is varying rapidly in a subinterval, you can allow for this by proposing a more complicated set of functions.

The selection of the form $\Psi_i(t)$ requires more physical insight than does discretization of the control function discussed above. If you have no experience with the problem, then a general form such as a polynomial may be used. Prior to selecting the functional form, it may be beneficial to solve the problem first using a coarse level of discretization (in fact, the discretization approach can be considered as a subset of parameterization of the control function). However, even if you use several trial functions $\Psi_i(t)$, there is no guarantee that the computed control law will be close to the optimum.

The computation of the optimum $\hat{u}(t)$ proceeds analogously to that used in the discretization procedure. The parameters $a_0, a_1, \ldots, a_p$ are updated based on accurate integration of the differential equations followed by computation of f. An unconstrained optimization code can be employed as illustrated in Fig. 8.10.

A large amount of literature exists on the computational use of the minimum principle to find an optimal control. Certainly anyone who must solve dynamic optimization problems should become familiar with this material. Due to the limitations of space we have only presented here two simple approaches to calculating $\hat{u}(t)$. For further information consult the supplemental references given at the end of this chapter.

Below we illustrate the use of both discretization and parameterization to obtain the control law function for a reactor optimization.

EXAMPLE 8.11 OPTIMAL CONTROL VIA NONLINEAR PROGRAMMING

Figure E8.11*a* is a diagram of a tubular reactor in which component A decomposes into B which in turn decomposes into C:

$$A \overset{k_1}{\rightarrow} B \overset{k_2}{\rightarrow} C$$

Thus, the feed and product stream compositions differ. The objective is to maximize the amount of B present in the effluent of the reactor by adjusting the temperature profile in the reactor.

For the purposes of illustration we assume the reactions are first-order and irreversible, the total number of moles does not change, and that the rate constants are given by

$$k_1 = k_{10} e^{-E_1/RT} \qquad k_2 = k_{20} e^{-E_2/RT}$$

where T is the absolute temperature, E the activation energy, k_{10} and k_{20} the preexponential constants, and R the ideal gas constant.

The two differential equations that model the system are

$$v \frac{dx_1}{dz} = -k_1 x_1 \tag{a}$$

$$v \frac{dx_2}{dz} = k_1 x_1 - k_2 x_2 \tag{b}$$

where x_1 = concentration of A, x_2 = concentration of B, and v = plug-flow fluid velocity. The amount of component C can be obtained by a material balance. We shift to residence time as the independent variable by letting $t = z/v$ and noting

$$v \frac{dx}{dz} = v \frac{dx}{dt} \frac{dt}{dz} = \frac{dx}{dt}$$

so that the reactor equations become

$$\begin{aligned} \frac{dx_1}{dt} &= -k_{10} x_1 e^{-E_1/RT} \\ \frac{dx_2}{dt} &= -k_{10} x_1 e^{-E_1/RT} - k_{20} x_2 e^{-E_2/RT} \end{aligned} \tag{c}$$

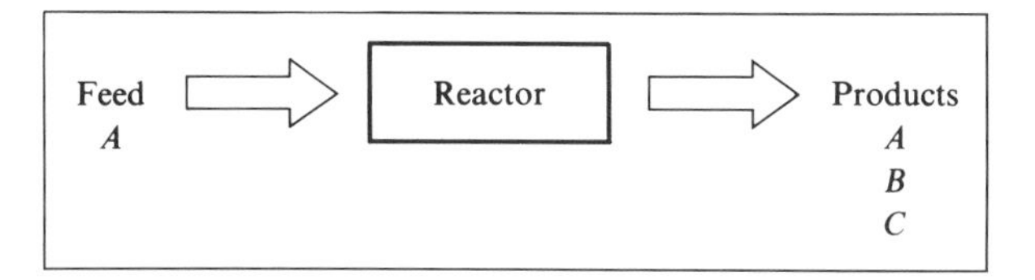

Figure E8.11*a* Schematic of a plug-flow reactor.

In Eqs. (*c*), x_1 and x_2 represent the state variables and T the (single) control variable. The initial concentrations are

$$x_1(0) = x_{10} \qquad x_2(0) = x_{20} \tag{d}$$

For this example, we want to determine the optimal temperature profile so as to maximize the concentration of component B in the product subject to matching the terminal residence time (i.e., the reactor length) with the calculated t_f. Therefore the objective function to be minimized is

$$f = -x_2(t_f) \tag{e}$$

The final time is $t_f = L/v = 8$ min (the residence time), based on $L = 2$ m and $v = 0.25$ m/min. The terminal conditions on the other components are not fixed. If this problem is to have a solution, we must ensure that $E_2 > E_1$. Common sense indicates that the reactor should be at a low temperature near the outlet where the most B is present so that only a little C forms by the time stream exits, but be at a high temperature near the inlet to increase the concentration of B. Therefore the temperature should decrease with residence time (length). Use the following parameters in finding the temperature profile:

$$E_1 = 75{,}000 \text{ J/gmol}$$

$$E_2 = 125{,}000 \text{ J/gmol}$$

$$R = 8.31 \text{ J/(gmol)(K)}$$

$$x_{10} = 0.7 \text{ mol/L}$$

$$x_{20} = 0.0$$

Table E8.11 Optimal piece-wise constant control for tubular reactor example for two levels of discretization

	Temperature	
z, m	$n = 5$	$n = 10$
0.0	365.4	369.1
0.2	365.4	361.8
0.4	358.4	359.0
0.6	358.4	357.3
0.8	355.9	356.1
1.0	355.9	355.2
1.2	354.3	354.5
1.4	354.3	353.9
1.6	353.3	353.4
1.8	353.3	352.9
$x_2(t_f)$	0.3687	0.3696
maximum gradient (absolute value)	1.528×10^{-8}	1.821×10^{-9}
Computation time (s)	8.26	39.77

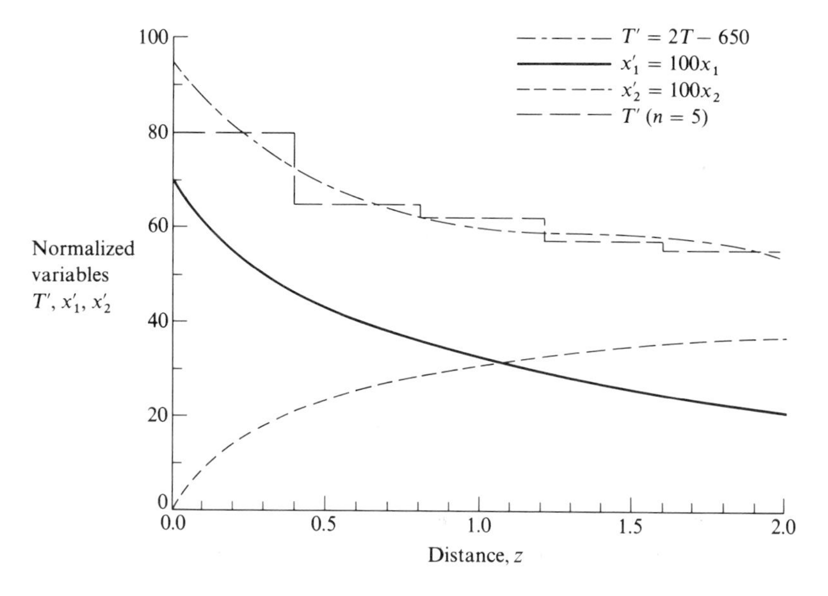

Figure E8.11*b* Optimal trajectories for tubular reactor using cubic polynomial for T.

Solution. First we illustrate the discretization method. If we let the number of subintervals be five, we have five degrees of freedom in the search. The quasi-Newton (secant) method was used to obtain the optimal solution. Table E8.11 shows the calculated values, the maximum gradient, and computation time for $n = 5$ and $n = 10$; note that there is not much change in the temperature profile resulting from the finer level of discretization, although there was a slight improvement in final conversion.

Next the optimal temperature was obtained via a parameterized temperature profile:

$$T(t) = a_0 + a_1 t + a_2 t^2 + a_3 t^3$$

Using the same method, the optimal values of the parameters and the gradients computed (shown in parentheses) were

$$a_0 = 371.9 \qquad (-1.667 \times 10^{-6})$$

$$a_1 = -37.0 \qquad (-2.206 \times 10^{-6})$$

$$a_2 = 26.7 \qquad (-6.300 \times 10^{-6})$$

$$a_3 = -6.59 \qquad (-1.567 \times 10^{-6})$$

The maximum value of $x_2(t_f)$ is 0.3699 ($x_1(t_f) = 0.2071$). The computation time required was 3.24 seconds. Figure E8.11b shows the temperature profile plotted as a function of z ($0 \leq z \leq 2.0$) as well as the concentrations of A and B for the parameterization approach. Also shown is the discretized control for $n = 5$.

8.11 DIAGNOSIS OF THE FAILURE OF OPTIMIZATION CODES TO SOLVE PROBLEMS

Table 8.10 lists some common symptoms experienced by users of optimization codes and possible causes and remedies. A one-to-one identification between symptom and cause of a malfunction is rarely possible by inspection of the code and its output. You must carry out a series of tests and look at intermediate values of matrices, vectors, parameters, and so on in one or more subroutines so as to eliminate potential causes and focus on possible causes of failure. In the list in Table 8.10 we presume that the algorithm and associated heuristic factors you are using are coded correctly, at least insofar as the code has been demonstrated to solve a substantial set of test problems correctly. Table 8.10 applies to both unconstrained and constrained problems.

Table 8.10 Symptoms, causes, and remedies for optimization code failures

Symptoms
Premature termination at infeasible point (constraints not satisfied)
Termination at wrong local minimum
Abortion at intermediate point in execution
No progress from initial conditions
Objective function diverges from proper trend
Solution oscillates
Indefinite argument of a function

Possible causes	**Possible remedies**
Problem function(s) not entered correctly	Review entries
Problem data not entered correctly; improper initialization	Review entries
Program parameters not entered correctly	Review entries
Tolerance criterion too small in line search/ overall search	Use double precision or increase criterion value
Indefinite Jacobian in LP, QP subroutine	Use different starting point; change code
Hessian matrix of f, P, or L becomes ill-conditioned (ratio of maximum to minimum eigenvalue becomes too big)	Change parameter values; properly scale variables, functions, change function form; weight functions differently; use Cholesky decomposition
Current Hessian/inverse Hessian is not a good approximation of true Hessian particularly near solution	Reset Hessian to identity matrix or diagonal matrix
Cycling in LP, QP, main code; degeneracy exists	Active set strategy needs revision or flexibility; change parameters; remove degeneracy cause

Possible causes	Possible remedies
Wrong active set retained	Change active set strategy or parameters; check binding constraints for overdefinition
Oscillation of solution	Check active set strategy; check for noncontinuous functions, derivatives, discontinuities; change search direction algorithm; possible multiple optima
Wrong starting point	Replace infeasible point with feasible one; select point nearer solution
Newton step fails to improve f	Choose another line search algorithm; scale properly
Number of function evaluations, CP time or line searches greater than that specified	Relax specification or tolerances; check other causes
Line search fails because of nonconvexity $(\mathbf{s}^T)[f(\mathbf{x}^{k+1}) - f(\mathbf{x}^k)] > 0$	Change starting point; change algorithm for line search
Successive search directions become linearly dependent	Raise stopping criterion; check functions; rearrange order of functions in user-supplied routines
Hessian matrix/inverse Hessian stop being positive definite	See remarks under ill-conditioning
Linearization of functions does not yield acceptable approximation	Add bounds on variables at each iteration
Precision of solution inadequate	Use double precision; revise tolerances
Unable to obtain feasible point in a subroutine	Use different starting point; check condition of matrices; use double precision; change code
Functions too sensitive/too insensitive to change in variables	Rescale; reformulate functions; check Lagrange multilpliers
Objective function continues to increase/decrease (unbounded)	Add bounds

REFERENCES

Abadie, J., and J. Carpentier, "Generalization de la methode du gradient réduit de Wolfe au cas de contraintes nonlineaires," Proc. IFORS Conf., rev., and in English in chap. 4, *Optimization*, (ed. R. Fletcher), Academic Press Inc., London (1969).

Abadie, J., and J. Guigou, Gradient réduit generalise, Électricité de France Note HI 069/02, Apr. 15 (1969).

Avriel, M., *Nonlinear Programming*, Prentice-Hall, Englewood, New Jersey, 1976.

Bartholomew-Biggs, M. C., "Recursive Quadratic Programming Based on Penalty Functions for Constrained Minimization," in Nonlinear Optimization Theory & Algorithms (eds. L. C. W. Dixon, E. Spedicato, and G. P. Szego), Birkhauser Books (1980).

Bartholomew-Biggs, M. C., Recursive Quadratic-Programming Methods for Nonlinear Constraints," in *Nonlinear Optimization 1981* (ed. M. J. D. Powell), Academic Press, New York, 1982, chap. 4.2.

Bartholomew-Biggs, M. C., Numerical Optimization Centre Tech. Report 140, Hatfield Polytechnic, Hatfield, England (1984).

Bazaraa, M. S., and C. M. Shetty, *Nonlinear Programming*, John Wiley, New York (1979).

Beale, E. M. L., "A Conjugate Gradient Method of Approximation Programming," in *Optim. Meth.* (eds. R. W. Cottle and J. Krarup), English Universities Press, London (1974).

Bekey, G. A., and S. F. Maeri, "Random Search Techniques for Optimization of Nonlinear Systems with Many Parameters," *Math. Comput. Simul.*, **25**: 210 (1983).

Belegundu, A. D., and J. S. Arora, "A Recursive Quadratic-Programming Method with Active Set Strategy for Optimal Design," *Int. J. Numer. Meth. in Eng.*, **20**: 803 (1984).

Ben-Israel, A., A. Ben-Tal, and S. Zlobec, *Optimality in Nonlinear Programming: A Feasible Directions Approach*, John Wiley, New York (1981).

Best, M. J., J. Braüninger, K. Ritter, and S. M. Robinson, "A Globally and Quadratically Convergent Algorithm for General Nonlinear Programming Problems," *Comput.*, **26**: 141 (1981).

Biggs, M. C., "Constrained Minimization Using Recursive Quadratic Programming," in *Toward Global Optimization* (eds. L. C. W. Dixon and G. P. Szego), North Holland, Amsterdam (1975).

Bodington, C. E., "Nonlinear Program for Product Blending," Joint National TIMS/ORSA Meeting, New Orleans, (April/May, 1979).

Box, M. J., "A New Method of Constrained Optimization and a Comparison with Other Methods," *Comput. J.*, **8**: 42 (1965).

Bracken, J., and G. P. McCormick, *Selected Applications of Nonlinear Programming*, John Wiley, New York (1968).

Bryson, A. E., and Y. C. Ho, *Applied Optimal Control*, 2d ed., Halsted Press, New York (1969).

Buzby, B. R., "Techniques and Experience Solving Really Big Optimization Programs," in *Optim. Meth.* (eds. R. W. Cottle and J. Krarup), English Universities Press, London (1974).

Chamberlain, R. M., C. Lemarechal, H. C. Pedersen, and M. J. Powell, *Math Program. Study 16*, **1** (1982).

Cheifett, S. A. "An Oilfield Development Model Using Approximation Programming," 4th ORSA/TIMS National Conference, Boston, Massachusetts (April, 1974).

Colville, A. R., "Process Optimization Program—POP," IBM Program Library, 7040 H 9 IBM 007 (December, 1963).

Cottle, R. W., "Symmetric Dual Quadratic Programs," *Q. Appl. Math.*, **21**: 237 (1963).

Cottle, R. W., "The Principal Pivoting Method of QP," in *Math. Decision Sciences, Part I*, (eds. G. B. Dantzig and A. F. Veinott Jr.), Amer. Math. Soc., Providence, Rhode Island (1968), p. 144.

Dennis, J. B., *Mathematical Programs and Electron Networks*, John Wiley & Sons, New York (1959).

Dobbie, J. M., "A Survey of Search Theory," *Oper. Res.*, **16**: 525 (1968).

Eaves, B. C., "On the Basic Theorem of Complementarity," *Math. Program.*, **1**: 68 (1971).

Faure, P., and P. Huard, *Rev. Franc Recherche Operationelle*, **9**: 167 (1965).

Fiacco, A. V., and G. P. McCormick, "Extension of SUMT for Nonlinear Programming: Equality Constraints and Extrapolation," *Manage. Sci.*, **12**: 816 (1966).

Fiacco, A. V., and G. P. McCormick, *Nonlinear Programming: Sequential Unconstrained minimization Techniques*, Wiley, New York (1968).

Fletcher, R., "An Ideal Penalty Function for Constrained Optimization," in *Nonlinear Programming*, vol. 2 (eds. O. L. Mangasarian, R. R. Myers, and S. M. Robinson), Academic Press, New York (1975).

Fletcher, R., "Methods for Nonlinear Constraints," in *Nonlinear Optimization 1981* (ed. M. J. D. Powell), Academic Press, New York (1982).

Gaines, L. D., and J. L. Gaddy, "Process Optimization by Flow Sheet Simulation," *IEC Proc. Des. Dev.*, **15** (1976).

Gill, P. E., W. Murray, and M. H. Wright, *Practical Optimization*, Academic Press, New York (1981).

Gill, P. E., Talk presented at the NATO Advanced Study Institute on Computational Mathematical Programming, Bad Windsheim, Germany, July, 1984.

Gill, P. E., W. Murray, M. A. Saunders, and M. H. Wright, "Software and Its Relationship to Methods," Report SOL 84-10, Dept. OR, Stanford University, Stanford, California (1984).

Griffith, R. E., and R. A. Stewart, "A Nonlinear Programming Technique for the Optimization of Continuous Processing Systems," *Manage. Sci.*, **7**: 379 (1961).

Haarhoff, P. C., and J. D. Buys, "A New Method for the Optimization of a Nonlinear Function Subject to Nonlinear Constraints," *Comput. J.*, **13**: 178 (1970).

Han, S. P., "A Globally Convergent Method for Nonlinear Programming," *J. Opt. Theory Appl.*, **22**: 279 (1977).

Hestenes, M. R., "Multiplier and Gradient Methods," *J. Opt. Theory Appl.*, **4:** 303 (1969).

Heuckroth, M. W., M. W. Gaddy, and J. L. Gaines, *AIChE J.*, **22**: 744 (1976).

Kan, A. H. G. R., and G. T. Timmer, "A Stochastic Approach to Global Optimization," in *Numerical Optimization 1984* (ed., P. T. Boggs, R. H. Byrd and R. B. Schnabel), SIAM, Philadelphia (1985), p. 245.

Kelley, R. J., and R. F. Wheeling, "A Digital Computer Program for Optimizing Nonlinear Functions," Mobil Oil Corp., Research Dept., Central Research Div., Princeton, New Jersey, July (1962).

Lapidus, L., and R. Luus, *Optimal Control of Engineering Processes*, Ginn/Blaisdell, Waltham, Massachusetts (1967).

Lasdon, L. S., and A. D. Waren, "Large-Scale Nonlinear Programming," *Comp. Chem. Eng.*, **7**(5): 595 (1983).

Lasdon, L. S., and A. D. Waren, "A Survey of Nonlinear Programming Applications," *Oper. Res.*, **28**(5): 34 (1980).

Lasdon, L. S., N. Kim, and J. Zhang, "An Improved Successive Linear Programming Algorithm," Department of General Business, University of Texas, Austin, Texas (1983).

Liebman, J., L. Lasdon, L. Schrage, and A. Waren, *GINO*, The Scientific Press, Palo Alto, California, 1986.

McMurtry, G. J., "Adaptive Optimization Procedures," in Adaptive, Learning and Pattern Recognition Systems, (eds. J. M. Mendel and K. S. Fu), Academic Press, New York (1970).

Maratos, N., Ph.D. Thesis, Imperial College (1978).

Mihail, R., and G. Maria, "A Modified Matyas Algorithm (MMA) for Random Process Optimization" *Compt. Chem. Eng.*, **10**, 539 (1986).

Murray, W., and M. H. Wright, "Projected Lagrangian Methods Based on The Trajectories of Penalty and Barrier Functions," Report SOL 78-23, Dept. Operations Research, Stanford Univ., Stanford, California (1978).

Murtagh, B. A., and M. A. Saunders, *MINOS 5.0 Users Guide*, Report SOL 83-20, Dept. of Operations Research, Stanford University, Stanford, California (1983).

Newell, J. S., and Himmelblau, D. M., "A New Method for Nonlinearly Constrained Optimization," *AIChE J.*, **21**: 479 (1975).

Palacios-Gomez, F., L. Lasdon, and M. Engquist, "Nonlinear Optimization by Successive Linear Programming," *Manage. Sci.*, **28**: 1106 (1982).

Pang, J-S., "Methods for Quadratic Programming: A Survey," *Comp. Chem. Eng.*, **7**(5): 583 (1983).

Pontryagin, L. D., et al., *The Mathematical Theory of Optimal Processes*, Wiley Interscience, New York (1962).

Powell, M. J. D., "A Method for Nonlinear Constraints in Minimization Problems," in *Optimization* (ed. R. Fletcher), Academic Press, New York (1969), p. 283.

Powell, M. J. D., "Algorithms for Nonlinear Constraints That Use Lagrange Functions," *Math. Program.*, **14**: 224 (1978*a*).

Powell, M. J. D., "A Fast Algorithm for Nonlinearly Constrained Optimization Calculations," in *Numerical Analysis, Dundee, 1977*, Lecture Notes in Math. 630 (ed. G. A. Watson), Springer-Verlag (1978*b*).

Pronzato, L., E. Walter, A. Venot, and J. F. Lebruche, "A General Purpose Global Optimizer: Implementations and Applications," *Math. Comput. Simul.*, **25**: 412 (1984).

Pshenichny, B. N., "Algorithms for the General Problem of Mathematical Programming, *Kibernetica* (*No.* 5), **120** (1970).

Reklaitis, G. V., A. Ravindran, and K. M. Ragsdell, *Engineering Optimization-Methods and Applications*, John Wiley, New York (1983).

Robinson, S. M., "A Quadratically Convergent Algorithm for Generaal NLP Problems", *Math. Prog.*, **3**, 145 (1972).

Robinson, S. M., "Perturbed Khun–Tucker Points and Rates of Convergence for a Class of NLP Algorithms", *Math. Prog.*, **3**, 1 (1974).

Rosen, J. B., "Two-phase Algorithm for Nonlinear Constraint Problems," in *Nonlinear Programming 3* (eds. O. L. Mangasarian, R. R. Meyer, and S. M. Robinson), Academic Press, New York (1978).

Ray, W. H., *Advanced Process Control*, McGraw-Hill, New York (1982).

Sage, A. P., *Optimum Systems Control*; Prentice-Hall, Englewood Cliffs, New Jersey (1968).

Sandgren, E., "The Utility of Nonlinear Programming Algorithms," Ph.D. Dissertation, Purdue University, W. Lafayette, Indiana (Dec. 1977).

Sargent, R. W. H. and G. R. Sullivan, "The Development of an Efficient Optimal Control Package," in *Proceed. 8th IFIP Conf. on Optimization Techniques*, J. Stoer, (ed.) p. 158, Springer-Verlag, Berlin (1978).

Schittkowski, K., *Nonlinear Programming Codes: Information, Tests, Performance*, Lecture Notes in Economics and Mathematical Systems, vol. 183, Springer-Verlag, New York (1980).

Schittkowski, K., "The Nonlinear Programming Method of Wilson, Han, and Powell with an Augmented Lagrangian Type Line Search Function," *Numerische Math.*, **38**: 83 (1981).

Stoer, J., "Foundations of Recursive Quadratic Programming Method for Solving Nonlinear Programs" in *Computational Mathematical Programming* (ed. K. Schittowski), Springer-Verlag (1985).

Tapia, R. A., "Diagonalized Multiplier Methods and Quasi-Newton Methods for Constrained Optimization," *J. Opt. Theory & Appl.*, **22**: 135 (1977).

Van der Hoek, G., *Reduction Methods in Nonlinear Programming*, Math. Centrum., Amsterdam, (1980), chap. 3.

Wilson, R. B., "A Simplicial Algorithm for Concave Programming," Ph.D. Thesis, Harvard University (1963).

Wolfe, P., Notices Am. Math. Soc., **9**: 308 (1962); Methods of Nonlinear Programming, pp. 76–77 in *Recent Advances in Mathematical Programming* (eds.) R. L. Graves and P. Wolfe, McGraw-Hill, New York (1963).

SUPPLEMENTARY REFERENCES

General

Bartels, R. H., and N. Mahdavi–Amiri, "On Generating Test Problems for Nonlinear Programming Algorithms," *SIAM J. Sci. Stat. Comput.*, **7**: 769 (1986).

Berman, G., *Nonlinear Optimization Bibliography with Two-level Key-Word and Author Indexes*, University of Waterloo Press, 1985. (ISBN: 0-88898-046-9) (bibliography of 4000 articles).

Conn, A. R., "Nonlinear Programming, Exact Penalty Functions and Projection Techniques for Non-Smooth Functions," Report CS-84-26, July, 1984.

Ecker, J. G., and M. Kupferschmid, "A Computational Comparison of the Ellipsoid Algorithm with Several Nonlinear Programming Algorithms," *SIAM J. Control and Opt.*, **23**: 657 (1985).

Goodman, J., "Newton's Method for Constrained Optimization," *Math. Prog.*, **35**: 162 (1985).

Hartley, *Linear and Nonlinear Programming*; *An Introduction to Linear Methods in Mathematical Programming*, John Wiley and Sons, New York, (1985).

Lasdon, L. S., and A. D. Waren, "Large-Scale Nonlinear Programming," *Comput. Chem. Eng.*, **7**: 595 (1983).

Luenberger, D. G., *Linear and Nonlinear Programming*, 2d. ed., Addison Wesley, Amsterdam, 1984.

Powell, M. J. D., and Y. Yuan, "A Trust Region Algorithm for Equality Constrained Optimization," DAMTP 1986/NAZ, Dept. of Appl. Math. & Theor. Phys., University of Cambridge, England, 1986.

Rangaiah, G. P., "Studies in Constrained Optimization of Chemical Process Problems," *Comput. Chem. Eng.*, **9**: 395 (1985).

Simon, J. D., and H. M. Azma, "Exxon Experience with Large-Scale Linear and Nonlinear Programming Applications", *Comput. Chem. Eng.*, **7**: 605 (1983).

Vardi, A., "A Trust Region Algorithm for Equality Constrained Minimization," *SIAM J. Num. Anal.*, **22**: 575 (1984).

Generalized networks

Ahlfeld, D. P., R. S. Dembo, J. M. Mulvey, and J. A. Zenios, "Nonlinear Programming on Generalized Networks," Report EES-85-7, Dept. Civil Engr., Princeton Univ., Princeton, New Jersey (1985).

Elam, J., F. Glover, and D. Klingman, "Strongly Convergent Primal Simplex Algorithm for Generalized Networks," *Math. of O.R.*, **4**: 39–59 (1979).

Grigoriadis, M., "Numerical Methods for Basic Solutions of Generalized Flow Networks," Rutger University Laboratory for Computer Science Research Report LCSR-TR-43.

McBride, R. D., and A. Gapta, "Solving Embedded Generalized Network Problems," Working Paper, University of Southern California (1982).

Generalized reduced-gradient method

Abadie, J., "The GRG Method for Nonlinear Programming," in *Design and Implementation of Optimization Software* (ed. H. J. Greenberg), Sijthoff and Noordhoff, Netherlands, pp. 335–362 (1978).

Lasdon, L. S., A. D. Waren, A. Jain, and M. Ratner, "Design and Testing of a GRG Code for Nonlinear Optimization," *ACM Trans. Math. Software*, **4**: 34 (1978).

Pibouleau, L., P. Floquet, and S. Domenech, "Optimization de Procédés Chemiques par une Méthod de Gradient Réduit," *Rech. opérationnelle*, **19**: 247 (1985).

Sargent, R. W. H., "Reduced-Gradient and Projection Methods for Nonlinear Programming," in *Numerical Methods for Constrained Optimization* (eds. P. E. Gill and W. Murray), Academic Press, New York, pp. 149–174 (1974).

Schittkowski, K., "On the Convergence of a Generalized Reduced Gradient Algorithm," *Optimization*, **17**: 731 (1986).

Nonlinear networks

Beck, P., L. Lasdon, and M. Engquist, "A Reduced-Gradient Algorithm for Nonlinear Network Problems," *ACM Trans. on Math. Software*, **9**: 57–70 (1983).

Dembo, R., and J. Klincewicz, "A Scaled Reduced-Gradient Algorithm for Network Flow Problems with Convex Separable Costs," *Math. Prog. Study* **15**: 125–147 (1981).

Feijoo, B., and R. R. Meyer, "Piecewise-Linear Approximation Methods for Nonseparable Convex Optimization," University of Wisconsin—Madison Computer Sciences Department Technical Report (1984).

Meyer, R. R., "Two-Segment Separable Programming," *Man. Sci.*, **25**: 385–395 (1979).

Optimal control

Athans, M., and P. L. Falb, *Optimal Control*, McGraw-Hill, New York (1966).

Bellman, R., *Dynamic Programming*, Princeton University Press, Princeton, New Jersey (1957).

Denn, M. M., *Optimization by Variational Methods*, McGraw-Hill, New York (1969).

Kalman, R. E., "The theory of optimal control and the calculus of variations," in *Mathematical Optimization Techniques* (ed. R. Bellman), University of California Press, Berkeley (1963) chap. 16.

Kelley, H. J., "Method of gradients," in *Optimization Techniques* (ed. G. Leitman), Academic Press, New York (1962) chap. 7.

Kopp, R. E., "Pontryagin maximum principle," in *Optimization Techniques* (ed. G. Leitmann), Academic Press, New York (1962) chap. 7.

Kwakernaak, H., and R. Sivan, *Linear Optimal Control Systems*, Wiley-Interscience, New York (1972).

Ray, W. H., and J. Szekely, *Process Optimization*, Wiley-Interscience, New York (1973).

Rozonoer, L. I., "L. S. Pontryagin's maximum principle in optimal control," *Automation and Remote Control*, **20**(10), 1320; (11), 1441; (12), 1561 (1959).

Penalty-function methods

Coleman, T. F., and A. R. Conn, Nonlinear Programming via an Exact Penalty Function Method, Reports CS-80-30 and CS-80-31, University of Waterloo, Ontario, Canada (1980).

Conn, A. R., "Penalty-Function Methods," in *Nonlinear Optimization 1981* (ed. M. J. D. Powell), Academic Press, New York (1982), chap. 4.4.

Lootsma, F. A., "A Survey of Methods for Solving Constrained Optimization Problems via Unconstrained Minimization," in *Numerical Methods for Nonlinear Optimizations*, (ed. F. A. Lootsma), pp. 313–347, Academic Press, New York (1972).

Murray, W., "Analytical Expressions for the Eigenvalues and Eigenvectors of the Hessian Matrices of Barrier and Penalty Functions," *J. Opt. Theory Appl.*, **7**: 189 (1971).

Pietrzykowski, T., "An Exact Potential Method for Constrained Maxima," *SIAM J. Numer. Anal.*, **6**: 299 (1969).

Pillo, G. di, and L. Grippo, "A Continuously Differentiable Exact-Penalty Function for Nonlinear Programming Problems with Inequality Constraints," *SIAM J. Control Opt.*, **23**: 72 (1985).

Swann, W. H., "Constrained Optimization by Direct Search," in *Numerical Methods for Constrained Optimization*, (eds. P. E. Gill and W. Murray), pp. 191–217, Academic Press, New York (1974).

Quadratic programming

Beale, E. M. L., and R. Beneviste, "Quadratic Programming" in *Design and Implementation of Optimization Software* (ed. H. J. Greenberg), Sijthoff & Noordhoff, The Netherlands (1978).

Fletcher, R., "A General Quadratic Programming Algorithm," *J. Inst. Math. Appl.*, **7**: 76 (1971).

Goldfarb, D., and A. Idani, "Dual and Primal-Dual Methods for Solving Strictly Convex Quadratic Programs," in Proceed. IIMAS Workshop on Numerical Analysis, Mexico, Lecture Notes in Math; Springer-Verlag (1981).

Random-search methods

Ariyahansa, K. A., and J. G. C. Templeton, "On Statistical Control of Optimization," *Math. Oper. Statis., ser. opt.*, **14:** 393 (1983).

Baba, N., "Convergence of a Random Optimization Method for Constrained Optimization Problems," *J. Optim. Theory Appl.* **33**: 451 (1981).

Brooks, S. H., "A Discussion of Random Methods for Seeking Maxims," *J. Oper. Res.*, **6**: 244 (1958).

Brooks, S. H., "A Comparison of Maximum Seeking Methods," *J. Oper. Res. Soc. Am.*, **7** (1959).

Devroye, L. P., "Progressive Global Random Search of Continuous Functions," *Math. Prog.*, **15**: 330 (1978).

Dorea, C. C. Y., "Expected Number of Steps of a Random Optimization Method," *J. Opt. Theory Appl.*, **39**: 165 (1983).

Lapidus, L., E. Shapiro, S. Shapiro, and R. E. Stillman, "Optimization of Process Performance," *AIChE J.*, **7**: 288 (1961).

McArthur, D. S., "Strategy in Research: Alternative Methods for the Design of Experiments," *IRE Trans.*, EM-8:34 (1961).
Rubinstein, Y. R., and R. M. Samorodnitsky, "Efficiency of the Random Search Method," *Math. Comput. Simul.*, **23**: 257 (1982).
Shimuzu, T., "A Stochastic Approximation Method for Optimization Problems," *J. Assoc. Computer Mach.*, **16**, 511 (1969).
Tsypkin, Y. Z., A. S. Poznyak, and A. M. Pesin, "Search Algorithms for Criterial Optimization Under Conditions of Uncertainty," *Sov. Phys. Dokl.*, **28**(5): 367 (1983) (Translation).
Zellnik, H. E., N. E. Sondak, and R. S. Davis, "Gradient Search Optimization," *Chem. Eng. Prog.*, **58**(8): 35 (1962).

Successive quadratic programming

Biegler, L. T., and R. R. Hughes, "Feasible Path Optimization With Sequential Modular Simulators," *Comput. Chem. Eng.*, **9**: 379 (1985).
Bartholomew-Biggs, M. C., "ALRQP for Inequality Constraints," Numerical Optimization Centre, Hatfield Polytechnic, England, 1984.
Biggs, M. C., "Constrained minimization using recursive equality quadratic programming," in *Numerical Methods for Non-Linear Optimization* (ed. F. A. Lootsma), pp. 411–428, Academic Press, London and New York (1972).
Fletcher, R., *Practical Methods of Optimization, Volume 2, Constrained Optimization*, John Wiley & Sons, New York and Toronto (1981).
Fukushima, M., "A successive quadratic programming algorithm with global and superlinear convergence properties," *Math. Programming*, **35**, 253 (1986).
Gill, P. E., W. Murray, M. A. Saunders, and M. H. Wright, Procedures for optimization problems with a mixture of bounds and general linear constraints, Report SOL 82-6, Dept. of Operations Research, Stanford University, California (1982).
Gill, P. E., W. Murray, and M. H. Wright, *Practical Optimization*, Academic Press, London and New York (1981).
Hoek, W., and K. Schittkowski, *Test Examples for Nonlinear Programming Codes*, Lecture Notes in Economics and Mathematical Systems 187, Springer-Verlag, Berlin and New York (1981).
Powell, M. J. D. (ed.), *Nonlinear Optimization 1981*, Academic Press, New York (1982).
Powell, M. J. D., and Y. Yuan, "A recursive quadratic programming algorithm that uses differentiable exact penalty functions," *Math. Programming*, **35**, 265 (1986).
Schittkowski, K., "The nonlinear programming method of Wilson, Han and Powell with an augmented Lagrangian type line search function. Part 2: An efficient implementation with linear least-squares subproblems," *Numer. Math.* **38**: 115–127 (1986).

PROBLEMS

8.1. Does the following solution $\mathbf{x}^* = [\frac{1}{3} \quad \frac{5}{3}]^T$ meet the sufficient conditions for a minimum of the following problem:

$$\text{Minimize} \quad f(\mathbf{x}) = -\ln(1 + x_1) - \ln(1 + x_2)^2$$

$$\text{Subject to} \quad g_1(\mathbf{x}) = x_1 + x_2 - 2 \le 0$$

$$g_2(\mathbf{x}) = x_1 \ge 0$$

$$g_3(\mathbf{x}) = x_2 \ge 0$$

8.2. Determine whether or not the point $x = [0 \quad 0 \quad 0]^T$ is a local minimum of the problem:

$$\text{Minimize} \quad f(\mathbf{x}) = \tfrac{4}{3}(x_1^2 - x_1x_2 + x_2^2)^{3/4} + x_3$$

$$\text{Subject to} \quad x_1^2 + x_2^2 + x_3^2 = 0$$

$$x_1 \geq 0$$

$$x_2 \geq 0$$

$$x_3 \geq 0$$

Show all computations.

8.3. Test whether the solution $\mathbf{x}^* = [2 \quad 2]^T$ meets the sufficient conditions for a minimum of the following problem:

$$\text{Minimize} \quad f(\mathbf{x}) = -x_1^2x_2$$

$$\text{Subject to} \quad h_1(x) = x_1x_2 + (x_1^2/2) = 6$$

$$g_1(x) = x_1 + x_2 \geq 0$$

8.4. Do (*a*) the necessary and (*b*) the sufficient conditions hold at the point $\mathbf{x}^* = [0.82 \quad 0.91]^T$ for the following problem?

$$\text{Minimize} \quad f(\mathbf{x}) = (x_1 - 2)^2 + (x_2 - 1)^2$$

$$\text{Subject to} \quad g_1(\mathbf{x}) = -\frac{x_1^2}{4} - x_2^2 + 1 \geq 0$$

$$h_2(\mathbf{x}) = x_1 - 2x_2 + 1 = 0$$

(where $\mathbf{x}^*$ is the minimum $\mathbf{x}$).

Note: $\mathbf{x}^* = [(-1 + \sqrt{7})/2 \quad (1 + \sqrt{7})/4]^T$ exactly.

8.5. Is the problem

$$\text{Minimize} \quad f(\mathbf{x}) = x_1^3 + 4x_2^2 - 4x_1$$

$$\text{Subject to} \quad 2x_2 - x_1 \geq 12$$

a convex programming problem?

8.6. Solve the following problem via the Lagrange multiplier method:

Find the maximum and minimum distances from the origin to the curve $5x_1^2 + 6x_1x_2 + 5x_2^2 = 8$

Hint: The distance $\sqrt{x_1^2 + x_2^2}$ is the objective function.

8.7. Show that Lagrange multipliers do not exist for the following problem:

$$\text{Minimize} \quad f(\mathbf{x}) = x_1^2 + x_2^2$$

$$\text{Subject to} \quad (x_1 - 1)^2 - x_2^2 = 0$$

8.8. Examine the reactor in Fig. P8.8. The objective function, $f(c, T) = (c - c_r)^2 + T^2$ is subject to the constraint $c = c_0 + e^T$ and also $c_0 < K$, where c_r is the set point for the outlet concentration, a constant, and K is a constant.

Find the minimum value of the objective function using Lagrange multipliers for the case in which $K = c_r - 2$.

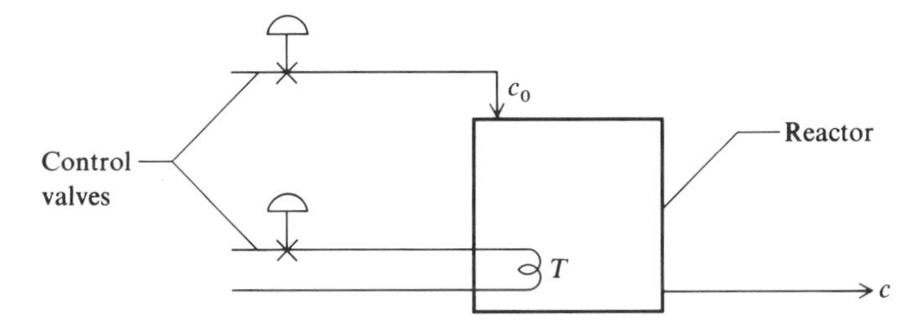

Figure P8.8

8.9. Maximize $f = x_1^2 + x_2^2 + 4x_1x_2$ subject to $x_1 + x_2 = 8$.

(*a*) Form the Lagrangian, L. Set up the necessary conditions for a maximum and solve for the optimum.

(*b*) If the constraint is changed to $x_1 + x_2 = 8.01$, compute f and L without resolving as in part *a*.

8.10. Minimize

(*a*) $f = x_1^2 + x_2^2 + 10x_1 + 20x_2 + 25$ subject to the constraint $x_1 + x_2 = 0$ using the Lagrange multiplier technique. Calculate the optimum values of x_1, x_2, ω, and f.

(*b*) Using sensitivity analysis, determine the increase in f^{opt} when the constraint is changed to $x_1 + x_2 = 0.01$.

(*c*) Let the constraint be added to f by a penalty function:

$$P = f + r(x_1 + x_2)^2$$

Find the optimum of P with respect to x_1 and x_2 (an unconstrained problem), noting that x_1^{opt} and x_2^{opt} will be functions of r.

(*d*) Is there a relationship between r, x_1^{opt}, x_2^{opt}, and ω^{opt}?

(*e*) Perform the second derivative test on P; is it convex for $P \gg 1$?

8.11. At stage $k = 2$, the generalized reduced gradient method is to be applied to the following problem at the point $\mathbf{x}^2 = [0 \quad 1 \quad 1]^T$.

$$\text{Minimize} \quad f(\mathbf{x}) = 2x_1^2 + 2x_2^2 + x_3^2 - 2x_1x_2 - 4x_1 - 6x_2$$

$$\text{Subject to} \quad x_1 + x_2 + x_3 = 2$$

$$x_1^2 + 5x_2 = 5$$

$$x_i \geq 0, i = 1, 2, 3$$

1. Compute the component step direction (+ or −) and magnitude for each of the independent variables.

2. Compute the component step direction (+ or −) and magnitude for each of the independent variables.

3. Reduce $f(\mathbf{x})$ in the search direction.

Explain (only) in detail how you would reach the feasible point to start the next stage ($k = 3$) of optimization.

8.12. Solve the following problems by the generalized reduced-gradient method. Also, count the number of function evaluations, gradient evaluations, constraint evaluations, and evaluations of the gradient of the constraints.

(*a*)

$$\text{Minimize} \quad f(\mathbf{x}) = -(x_1^2 + x_2^2 + x_3^2)$$

$$\text{Subject to} \quad x_1 + 2x_2 + 3x_3 - 1 = 0$$

$$x_1^2 + \frac{x_2^2}{2} + \frac{x_3^2}{4} - 4 = 0$$

Use various starting points.
Start at $\mathbf{x}^0 = [n \quad n \quad n]$ where $n = 2, 4, 6, 8, 10, -2, -4, -6, -8, -10$.

(*b*)

$$\text{Minimize} \quad f(\mathbf{x}) = (x_1 - 1)^2 + (x_1 - x_2)^2 + (x_2 - x_3)^2 + (x_3 - x_4)^4 + (x_4 - x_5)^4$$

$$\text{Subject to} \quad x_1 + x_2^2 + x_3^3 - 2 - 3\sqrt{2} = 0$$

$$x_2 - x_3^2 + x_4 + 2 - 2\sqrt{2} = 0$$

$$(x_1)(x_5) - 2 = 0$$

Start at $\mathbf{x}^0 = [n \quad n \quad n \quad n \quad n]^T$ where $n = 2, 4, 6, 8, 10, -2, -4, -6, -8, -10$.

8.13. Find the stationary point of the function $f(\mathbf{x}) = x_1^2 + x_2^2 + 4x_1x_2$ subject to the constraint, $x_1 + x_2 = 8$. Use direct substitution. What kind of stationary point is it?

For the same objective function and constraint, form a new function

$$P = x_1^2 + x_2^2 + 4x_1x_2 + r(x_1 + x_2 - 8)^2$$

This is called a penalty function, where r is a large number. We then optimize P.

(*a*) Find the stationary point of P with respect to x_1 and x_2, solving for x_1^{opt} and x_2^{opt} in terms of r.

(*b*) Find x_1^{opt}, x_2^{opt} as $r \to \infty$

(*c*) Does $P^{\text{opt}} \to f^{\text{opt}}$ for $r \to \infty$?

8.14. Minimize $x_2^2 - x_1^2$ subject to $x_1^2 + x_2^2 = 4$.

(*a*) Use Lagrange multipliers

(*b*) Use a penalty function.

8.15. The problem is to

$$\text{Minimize} \quad f(\mathbf{x}) = x_1^2 + 6x_1 + x_2^2 + 9$$

$$\text{Subject to} \quad g_i(\mathbf{x}): x_i \geq 0 \qquad \text{for } i = 1, 2$$

From the starting vector $\mathbf{x}^0 = [1 \quad 0.5]^T$.

(*a*) Formulate a penalty function suitable to use for an unconstrained optimization algorithm.

(*b*) What should be the value of r^0?

(*c*) Is the penalty function convex?

8.16. Formulate the following problems as

(*a*) Penalty-function problems

(*b*) Augmented Lagrange problems

(1)

$$\text{Minimize} \quad f(\mathbf{x}) = 2x_1^2 - 2x_1x_2 + 2x_2^2 - 6x_1 + 6$$
$$\text{Subject to} \quad h(\mathbf{x}) = x_1 + x_2 - 2 = 0$$

(2)

$$\text{Minimize} \quad f(\mathbf{x}) = x_1^3 - 3x_1x_2 + 4$$
$$\text{Subject to} \quad g(\mathbf{x}) = 5x_1 + 2x_z \geq 18$$
$$h(\mathbf{x}) = -2x_1 + x_2^2 - 5 = 0$$

8.17. Comment on the following proposed penalty functions suggested for use with the problem:

$$\text{Minimize} \quad f(\mathbf{x})$$
$$\text{Subject to} \quad g_i(\mathbf{x}) \geq 0 \qquad i = 1, 2, \ldots, m$$

starting from a feasible point. The P functions are

(*a*)

$$P(\mathbf{x}, \mathbf{x}^k) = \frac{1}{f(\mathbf{x}^k) - f(\mathbf{x})} + \sum_{j=1}^{m} \frac{1}{g_j(\mathbf{x})}$$

(*b*)

$$P(\mathbf{x}, r) = f(\mathbf{x}) - r \sum_{j=1}^{m} \ln g_j(\mathbf{x})$$

(*c*)

$$P(\mathbf{x}, r^k) = f(\mathbf{x}) + r^k \sum_{j=1}^{m} \frac{1}{g_j(\mathbf{x})}$$

What advantages might they have in comparison with each other? What disadvantages?

8.18. Solve the following over-constrained problem:

$$\text{Minimize} \quad f(\mathbf{x}) = x_1^2 + x_2^2 + x_3^2$$

$$\begin{aligned}
\text{Subject to} \quad g_1(\mathbf{x}) &= -2x_1 - x_2 \geq -5 \\
g_2(\mathbf{x}) &= -x_1 - x_3 \geq -2 \\
g_3(\mathbf{x}) &= -x_1 - 2x_2 - x_3 \geq -10 \\
h_1(\mathbf{x}) &= 2x_1 - 2x_2 + x_3 = -2 \\
h_2(\mathbf{x}) &= 10x_1 + 8x_2 - 14x_3 = 26 \\
h_3(\mathbf{x}) &= -4x_1 + 5x_2 - 6x_3 = 6
\end{aligned}$$

$$x_1 \geq 1 \qquad x_2 \geq 2 \qquad x_3 \geq 0$$

Starting point: $\mathbf{x}^0 = [1 \quad 1 \quad 1]^T$.

8.19. Solve the following problems using

(*a*) A generalized reduced-gradient code

(*b*) A sequential quadratic-programming code and compare your results.

(*c*) Objective function:

$$\text{Minimize} \quad f(\mathbf{x}) = \sum_{k=1}^{10} \left(\frac{1}{k} x_k^2 + kx_k + k^2\right)^2$$

$$\begin{aligned}
\text{Constraints} \quad h_1(\mathbf{x}) &= x_1 + x_3 + x_5 + x_7 + x_9 = 0 \\
h_2(\mathbf{x}) &= x_2 + 2x_4 + 3x_6 + 4x_8 + 5x_{10} = 0 \\
h_3(\mathbf{x}) &= 2x_2 - 5x_5 + 8x_8 = 0 \\
g_1(\mathbf{x}) &= -x_1 + 3x_4 - 5x_7 + x_{10} \geq 0 \\
g_2(\mathbf{x}) &= -x_1 - 2x_2 - 4x_4 - 8x_8 \geq -100 \\
g_3(\mathbf{x}) &= -x_1 - 3x_3 - 6x_6 + 9x_9 \geq -50
\end{aligned}$$

$$-10^3 \leq x_i \leq 10^3 \qquad i = 1, 2, \ldots, 10$$

Starting point (feasible):

$$x_i^0 = 0 \qquad i = 1, 2, \ldots, 10$$

$$f(\mathbf{x}^0) \quad = \quad 25{,}333.0$$

(*b*)

$$\text{Minimize} \quad f(\mathbf{x}) = \sum_{i=1}^{11} x_i + \sum_{i=1}^{10} (x_i + x_{i+1})$$

$$\begin{aligned}
\text{Subject to} \quad & x_i \geq 0, \qquad i = 1, \ldots, 11 \\
h_1(x) &= 0.1x_1 + 0.2x_7 + 0.3x_8 + 0.2x_9 + 0.2x_{11} = 1.0 \\
h_2(x) &= 0.1x_2 + 0.2x_8 + 0.3x_9 + 0.4x_{10} + 1.0x_{11} = 2.0 \\
h_3(x) &= 0.1x_3 + 0.2x_8 + 0.3x_9 + 0.4x_{10} + 2.0x_{11} = 3.0
\end{aligned}$$

$$g_4(x) = x_4 + x_8 + 0.5x_9 + 0.5x_{10} + 1.0x_{11} \geq 1.0$$

$$g_5(x) = 2.0x_5 + x_6 + 0.5x_7 + 0.5x_8 + 0.25x_9 + 0.25x_{10} + 0.5x_{11} \geq 1.0$$

$$g_6(x) = x_4 + x_6 + x_8 + x_9 + x_{10} + x_{11} \geq 1.0$$

$$g_7(x) = 0.1x_1 + 1.2x_7 + 1.2x_8 + 1.4x_9 + 1.1x_{10} + 2.0x_{11} \geq 1.0$$

Starting point (feasible):

$$x_i = 1.0 \qquad i = 1, 2, \ldots, 11$$

(*c*)

$$\text{Maximize} \quad f(\mathbf{x}) = 3x_1 e^{-0.1x_1x_6} + 4x_2 + x_3^2 + 7x_4 + \frac{10}{x_5} + x_6$$

$$\text{Subject to} \quad -x_4 + x_5 - x_6 = 0.1$$

$$x_1 + x_2 + x_3 + x_4 + x_5 + x_6 = 10$$

$$2x_1 + x_2 + x_3 + 3x_4 \geq 2$$

$$-8x_1 - 3x_2 - 4x_3 - x_4 - x_5 \geq -10$$

$$-2x_1 - 6x_2 - x_3 - 3x_4 - x_6 \geq -13$$

$$-x_1 - 4x_2 - 5x_3 - 2x_4 \geq -18$$

$$-20 \leq x_i \leq +20 \qquad i = 1, \ldots, 6$$

The nonfeasible starting point is:

$$x_i = 1.0 \qquad i = 1, \ldots, 6$$

(*d*)

$$\text{Minimize} \quad f(\mathbf{x}) = x_1^2 + 2x_2^2 + 3x_3^2 + 4x_4^2 + 5x_5^2$$

$$\text{Subject to} \quad h_1(\mathbf{x}) = 2x_1 + x_2 - 4x_3 + x_4 - x_5 = 0$$

$$h_2(\mathbf{x}) = 5x_1 - 2x_3 + x_4 - x_5 = 0$$

$$g_1(\mathbf{x}) = x_1 + 2x_2 + x_3 \geq 6$$

$$g_2(\mathbf{x}) = 4x_3 + x_4 - 2x_5 \leq 0$$

Starting point (nonfeasible):

$$x_i = 1 \qquad i = 1, \ldots, 5$$

(*e*)

$$\text{Minimize} \quad f(\mathbf{x}) = (x_1 - x_2)^2 + (x_2 - x_3)^2 + (x_3 - x_4)^4 + (x_4 - x_5)^4$$

$$\text{Subject to} \quad x_1 + 2x_2 + 3x_3 - 6 = 0$$

$$x_2 + 2x_3 + 3x_4 - 6 = 0$$

$$x_3 + 2x_4 + 3x_5 - 6 = 0$$

Starting point (feasible):

$$\mathbf{x}^0 = [35 \quad -31 \quad 11 \quad 5 \quad -5]^T$$

(*f*)

$$\text{Minimize} \quad f(\mathbf{x}) = (x_1 - 1)^2 + (x_1 - x_2)^2 + (x_3 - 1)^2 + (x_3 - 1)^2 + (x_4 - 1)^4 + (x_5 - 1)^6$$

$$\text{Subject to} \quad x_1^2 x_4 + \sin(x_4 - x_5) - 2\sqrt{2} = 0$$

$$x_2 + x_3^4 x_4^2 - 8 - \sqrt{2} = 0$$

Starting point

$$\mathbf{x}^0 = [2 \quad 2 \quad 2 \quad 2 \quad 2]^T$$

(*g*)

$$\text{Minimize} \quad f(\mathbf{x}) = (x_1 - 1)^2 + (x_1 - x_2)^2 + (x_2 - x_3)^2 + (x_3 - x_4)^4 + (x_4 - x_5)^4$$

$$\text{Subject to} \quad x_1 + x_2^2 + x_3^3 - 2 - 3\sqrt{2} = 0$$

$$x_2 - x_3^2 + x_4 + 2 - 2\sqrt{2} = 0$$

$$x_1 x_5 - 2 = 0$$

Starting points:

$$x_1 = \pm 10, \; x_2 = \pm 8, \; x_3 = \pm 6, \; x_4 = \pm 4, \; x_5 = \pm 2$$

8.20. The cost of constructing a distillation column can be written

$$C = C_p AN + C_s HAN + C_f + C_d + C_b + C_L + C_x \tag{a}$$

where C = Total cost, \$
C_p = cost per square foot of plate area, \$/ft^2
A = column cross-sectional area, ft^2
N = number of plates
$N_{\min}$ = minimum number of plates
C_s = cost of shell, \$/ft^3
H = distance between plates, ft
C_f = cost of feed pump, \$
C_d = cost of distillate pump, \$
C_b = cost of bottoms pump, \$
C_L = cost of reflux pump, \$
C_x = other fixed costs, \$

The problem is to minimize the total cost, once produce specifications and the throughput are fixed and the product and feed pumping costs are fixed; that is, C_f, C_d, C_L, and C_b are fixed. After selection of the material of construction, the costs are determined; that is, C_p, C_s, C_x are also fixed.

The process variables can be related through two empirical equations:

$$\frac{L}{D} = \left[\frac{1}{1 - (N_{\min}/N)}\right]^{\alpha} \left(\frac{L}{D}\right)_{\min} \tag{b}$$

$$A = K(L + D)^{\beta} \tag{c}$$

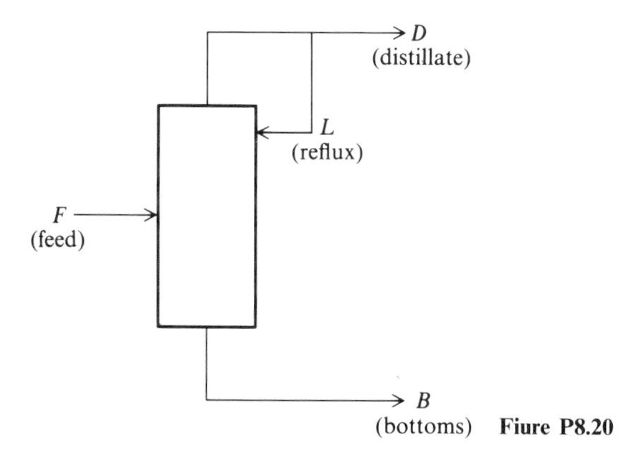

Fiure P8.20

For simplicity choose $\alpha = \beta = 1$; then

$$\frac{L}{D} = \left[\frac{1}{1 - (N_{\min}/N)}\right]\left(\frac{L}{D}\right)_{\min} \qquad (b')$$

$$A = K(L + D) \qquad (c')$$

For a certain separation and distillation column the following parameters are known to apply:

$$C_p = 30 \qquad C_x = 8000$$

$$C_s = 10 \qquad F = 1500$$

$$H = 2 \qquad D = 1000$$

$$C_f = 4000 \qquad N_{\min} = 5$$

$$C_d = 3000 \qquad \left(\frac{L}{D}\right)_{\min} = 1$$

$$C_b = 2000 \qquad K = \frac{1}{100}\,\frac{(\mathrm{h})(\mathrm{ft}^2)}{\mathrm{lb}}$$

The pump cost for the reflux stream can be expressed as

$$C_L = 5000 + 0.7L \qquad (d)$$

(*a*) Determine the process decision or independent variables. Which variables are dependent?

(*b*) Find the minimum total cost and corresponding values of the variables.

8.21. A chemical manufacturing company sells three products and has found that its revenue function is $f = 10x + 4.4y^2 + 2z$, where x, y, and z are the monthly production rates of each chemical. It is found from breakeven charts that it is necessary to impose the following limits on the production rates:

$$x \geq 2$$

$$\tfrac{1}{2}z^2 + y^2 \geq 3$$

In addition, only a limited amount of raw material is available; hence the following restrictions must be imposed upon the production schedule:

$$x + 4y + 5z \leq 32$$

$$x + 3y + 2z \leq 29$$

Determine the best production schedule for this company and find the best value of the revenue function.

8.22. A problem in chemical equilibrium is to minimize

$$f(\mathbf{x}) = \sum_{i=1}^{n} x_i \left(w_i + \ln P + \ln \frac{x_i}{\sum_{i=1}^{n} x_i} \right)$$

subject to the material balances

$$x_1 + 2x_2 + 2x_3 + x_6 + x_{10} = 2$$

$$x_4 + 2x_5 + x_6 + x_7 = 1$$

$$x_3 + x_7 + x_8 + 2x_9 + x_{10} = 1$$

Given $P = 750$ and w_i,

i	w_i	i	w_i
1	−10.021	6	−18.918
2	−21.096	7	−28.032
3	−37.986	8	−14.640
4	−9.846	9	−30.594
5	−28.653	10	−26.111

what is $\mathbf{x}^*$ and $f(\mathbf{x}^*)$?

8.23. The objective is to fit a fifth-order polynomial to the curve $y = x^{1/3}$. To avoid fluctuations from the desired curve, you are to divide the curve up into 10 points

$$\mathbf{x}_i(i = 1, \ldots, 10) = (0.5,\ 1,\ 4.5,\ 8,\ 17.5,\ 27,\ 45.5,\ 64,\ 94.5,\ 125)$$

and fit the polynomial (find the values of a_i)

$$P(\mathbf{a}, \mathbf{x}) = a_1 x + a_2 x^2 + a_3 x^3 + a_4 x^4 + a_5 x^5$$

by solving the following problem

$$\text{Minimize} \quad f(\mathbf{x}) = \sum_{i=1}^{10} (P(a, x_i) - x_i^{1/3})^2$$

$$\text{Subject to} \quad 0 \leq P(a, j) \leq 5 \qquad j = 1, 8, 27, 64$$

$$P(a, 125) = 5$$

8.24. The Williams-Otto process as posed in this problem involves 10 variables and 7 constraints leaving 3 degrees of freedom. Three starting points are shown below. Find the maximum Q and the values of the 10 variables from one of the starting points (S.P.). The minimum is very flat.

Figure P8.24 shows a simplified block diagram of the process. The plant consists of a perfectly stirred reactor, a decanter, and a distillation column in series. There is recycle from the column reboiler to the reactor.

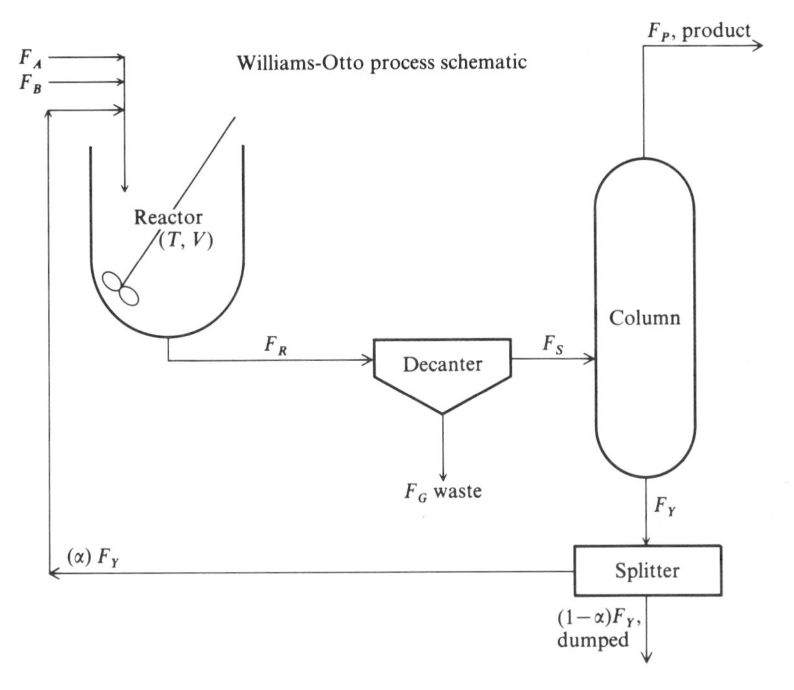

Figure P8.24

The mathematical descriptions of each plant unit are summarized in Table P8.24.3. The return function for this process as proposed by Williams and Otto and slightly modified by DiBella and Stevens to a variable reactor volume problem is

Table P8.24.1 Starting points

Variable	x	x^1	x^2	x^3
F_{RA}	1	18187	6000	22381
F_{RB}	2	60815	35000	72297
F_{RC}	3	3331	10000	4391
F_{RE}	4	60542	15000	78144
F_G	5	3609	5000	3140
F_{RP}	6	10817	6000	12557
α	7	0.761	0.789	0.81
F_A	8	13546	10000	12876
F_B	9	31523	40000	29416
T	10	656	610	648

$$\text{Maximize: } Q = \frac{100}{600V\rho}[8400(0.3F_P + 0.0068F_D - 0.02F_A - 0.03F_B - 0.01F_G)$$
$$-2.22F_R - (0.124)(8400)(0.3F_P + 0.0068F_D) - 60V\rho]$$
$$= \text{Return } (\%)$$

Table P8.24.2

g_i = residual of mass balance on component i, $\quad i = A, B, C, E, G$

Constraints

$$g_1 = F_A + F_{TA} - F_{RA} - P_1 = 0$$
$$g_2 = F_B + F_{TB} - F_{RB} - P_1 - P_2 = 0$$
$$g_3 = F_{TC} + 2P_1 - F_{RC} - 2P_2 - P_3 = 0$$
$$g_4 = F_{TE} + 2P_2 - F_{RE} = 0$$
$$g_5 = F_{TG} + P_2 - F_{RG} - 0.5P_3 = 0$$
$$g_6 = \text{overall mass balance} = F_A + F_B - F_D - F_G = 0$$
$$g_7 = \text{production requirement} = F_P - 4763 = 0$$
$$0 \le \alpha \le 1 \qquad 500 \le T \le 1000$$

Table P8.24.3 Williams Otto unit mathematical models

Decanter			
$F_{Si} = F_{Ri}$		$i = A, B, C, E, P$	
$F_{SC} = 0$			
$F_{C_b} = F_{RC}$			
Distillation column			
$F_{Yi} = F_{Si}$		$i = A, B, C, E, G$	
$F_P = F_{SP} - 0.1F_{SE}$			
$F_{YP} = F_{SP} - F_P$			
Splitter			
$F_{Ti} = \alpha F_{Yi}$		$i = A, B, C, E, G, P$	
$F_{Di} = (1 - \alpha)F_{Yi}$			
Reactor ($V = 0.0002964F_R$)	i	a_i	b_i
$K_i = a_i \exp(-b_i/T)V/F_R^2$			
$P_1 = K_1 F_{RA} F_{RB}$	1	5.9755×10^9	12,000
$P_2 = K_2 F_{RB} F_{RC}$	2	2.5962×10^{12}	15,000
$P_3 = K_3 F_{RC} F_{RG}$	3	9.6283×10^{15}	20,000

Table P8.24.4 Williams Otto process nomenclature

F_A, F_B	Fresh feeds of components A, B (lb/h)
F_R	Total reactor output flow rate
F_{Ri}	Reactor output flow rate of component i
F_{Si}	Decanter output flow rate of component i
F_G	Decanter bottoms flow rate of component G
F_P	Column overhead flow rate of component P
F_{Yi}	Column bottoms flow rate of component i
F_D	Total column bottoms takeoff flow rate
F_{Di}	Column bottoms takeoff flow rate of component i
F_T	Total column bottoms recycle flow rate
F_{Ti}	Column bottoms recycle flow rate of component i
α	Fraction of column bottoms recycled to reactor
V	Reactor volume (ft^3)
ρ	Density of reaction mixture (assumed constant, 50 lb/ft^3)

8.25. Klein and Klimpel [M. Klein and R. R. Klimpel, *J. Indus. Eng.*, **18**:90 (1967).] described a nonlinear programming problem involving the optimal selection of plant sites and plant sizes over a period of time. The functions representing fixed and working capital were of the form

$$\text{Fixed capital:} \quad \text{Cost} = a_0 + a_1 S^{a_2}$$

$$\text{Working capital:} \quad \text{Cost} = b_0 + b_1 P + b_2 S^{a_2}$$

where S = plant size
P = annual production
a's, b's = known constants obtained empirically

Variable annual costs were expressed in the form of

$$\text{Cost} = P(c_1 + c_2 S + c_3 S^{c_4})$$

Transportation costs were assumed proportional to the size of the shipments for a given source and destination.

The objective function is the net present value, NPV (sum of the discounted cash flows), using a discount rate of 10 percent. All flows except capital were assumed to be uniformly distributed over the year; working capital was added or subtracted instantaneously at the beginning of each year, and fixed capital was added only in the zero year.

The continuous discounting factors were:

1. For instantaneous funds,

$$F_i = e^{-ry} \qquad (r = \text{interest rate, } y = \text{years hence})$$

2. For uniformly flowing funds,

$$F_u = \frac{e^r - 1}{r} e^{-ry}$$

Table A Net percent value

1. Contribution of fixed capital (plant 1)

Year	Fixed capital	Discount factor	Discounted cash flow
0(1967)	$0.7\bar{S}_1 + 1.5S_1^6$	1.0517	$-0.7362\bar{S}_1 - 1.5775S_1^6$

2. Contribution of working capital (plant 1)

Year end	Working capital	Discount factor	Discounted cash flow at 10% discount rate
0	$0.4\bar{S}_1 + 0.2P_{101} + 0.05S_1^6$	1.000	$-0.4\bar{S}_1 - 0.2P_{101} - 0.05S_1^6$
1	$0.2(P_{102} - P_{101})$	0.9048	$-0.1810P_{102} + 0.1810P_{101}$
2	$0.2(P_{103} - P_{102})$	0.8187	$-0.1637P_{103} + 0.1637P_{102}$
3	$-0.4\bar{S}_1 - 0.2P_{103} - 0.05S_1^6$	0.7408	$+0.2963\bar{S}_1 + 0.1482P_{103} + 0.0370S_1^6$

3. Contribution of operational cost (plant 1)
a. Cost tabulation (excluding shipping)

Year	Amount	Depreciation†	Other costs
1	P_{101}	$0.4667\bar{S}_1 + 1.0S_1^6$	$0.03\bar{S}_1 - 0.01S_1 + 0.05S_1^{0.45} + 0.07S_1^{0.6} + 0.1P_{101}$ $-0.05P_{101}S_1 + 0.4P_{101}S_1^{-0.55}$
2	P_{102}	$0.1167\bar{S}_1 + 0.25S_1^6$	$0.03S_1 - 0.01S_1 + 0.05S_1^{0.45} + 0.07S_1^{0.6} + 0.095P_{102}$ $-0.0048P_{102}S_1 + 0.38P_{102}S_1^{-0.55}$
3	P_{103}	$0.1166\bar{S}_1 + 0.25S_1^6$	$0.03S_1 - 0.01S_1 + 0.05S_1^{0.45} + 0.07S_1^{0.6} + 0.0903P_{103}$ $-0.0045P_{103}S_1 + 0.361P_{103}S_1$

b. Discounted cash flow of costs (plant 1)*

Year	Discount factor	Discounted cost flow at 10% discount rate
1	0.9516	$0.1983S_1 + 0.0049\bar{S}_1 - 0.0247S_1^{0.45} + 0.4221S_1^{0.6} - 0.0495P_{101}$ $+0.0025P_{101}S_1 - 0.1979P_{101}S_1^{-0.55}$
2	0.8611	$0.0348S_1 + 0.0045\bar{S}_1 - 0.0224S_1^{0.45} + 0.0720S_1^{0.6} - 0.0425P_{102}$ $+0.0020P_{102}S_1 - 0.1702P_{102}S_1^{-0.55}$
3	0.7791	$0.0315S_1 + 0.0041\bar{S}_1 - 0.0203S_1^{0.45} + 0.0651S_1^{0.6} - 0.0366P_{103}$ $+0.0017P_{103}S_1 - 0.1463P_{103}S_1^{0.55}$

4. Contribution of shipping costs (from plant 1)

Year	Discount factor	Shipping cost	Discounted cash flow at 10% discount rate
1	0.9516	$0.8P_{121} + 0.5P_{121}$	$-0.396P_{121} - 0.247P_{131}$
2	0.8611	$0.7P_{122} + 0.45P_{132}$	$-0.313P_{122} - 0.201P_{132}$
3	0.7791	$0.6P_{123} + 0.4P_{133}$	$-0.243P_{123} - 0.162P_{133}$

† Method of double rate-declining balance and straight-line crossover was used.

The variable y may be positive (after year zero) or negative (before year zero) or zero (for year ending with point zero in time).

As prices and revenue were not considered, maximization of net present value was equivalent to minimization of net cost.

Let P_{ijk} be the amount of product shipped from location $i(i = 1, 2, 3, 4)$ to market $j(j = 1, 2, 3)$ in year $k(k = 0, 1, 2, 3)$. Let S_i and $\bar{S}_1$ be, respectively, the size of plant in location i, and a variable restricted to 0 or 1, depending upon whether or not S_i is 0. Furthermore, let M_{ojk} be the market demand at center j in year k. Finally, for the sake of convenience, let P_{iok} denote the total production in plant i during year k.

The nonlinear programming problem is: Find S_i and P_{ijk} that will

$$\text{Maximize} \quad \sum_i \text{NPV (including shipping)}$$

$$\text{Subject to} \quad \sum_i P_{ijk} = M_{ojk}$$

$$S_i \geq 0, \; P_{ijk} \geq 0$$

Table A indicates how the net present value was determined for location 1; NPV relations for the other locations were similarly formed. Table B lists the overall objective function, and Table C lists (1) the 22 constraints, (2) one equation constraining the total plant capacity to be 10 million pounds per year, (3) nine equations requiring satisfaction of the three markets every year, and (4) 12 inequalities calling for plant production not to exceed plant capacity. In addition, there are the nonnegativity constraints applicable to all 40 variables. Thus the problem has 10 linear equality constraints and 52 inequality constraints.

Table B The objective function

$$
\begin{aligned}
Z_{\max} = & -0.5753\bar{S}_1 - 1.0313S_1^{0.6} - 0.0685P_{101} - 0.0597P_{102} - 0.0522P_{103} + 0.0135S_1 \\
& - 0.0674S_1^{0.45} + 0.0025P_{101}S_1 + 0.0020P_{102}S_1 + 0.0017P_{103}S_1 \\
& - 0.1979P_{101}S_1^{-0.55} - 0.1702P_{102}S_1^{-0.55} - 0.1463P_{103}S_1^{-0.55} - 0.396P_{121} \\
& - 0.247P_{131} - 0.313P_{122} - 0.202P_{132} - 0.243P_{123} - 0.162P_{133} - 0.3428\bar{S}_2 \\
& - 0.8920S_2^{0.6} - 0.0685P_{201} - 0.0597P_{202} - 0.0522P_{203} + 0.0135S_2 - 0.0809S_2^{0.45} \\
& + 0.0025P_{201}S_2 + 0.0020P_{202}S_2 + 0.0017P_{203}S_2 - 0.02227P_{201}S_2^{-0.55} \\
& - 0.1914P_{202}S_2^{-0.55} - 0.1645P_{203}S_2^{-0.55} - 0.396P_{211} - 0.495P_{231} - 0.313P_{212} \\
& - 0.448P_{232} - 0.243P_{213} - 0.405P_{233} - 0.3164\bar{S}_3 - 1.2987S_3^{0.6} - 0.0942P_{301} \\
& - 0.0819P_{302} - 0.0712P_{303} - 0.0539S_3^{0.45} + 0.0030P_{301}S_3 + 0.0026P_{302}S_3 \\
& + 0.0022P_{303}S_3 - 0.2227P_{301}S_3^{-0.55} - 0.1914P_{302}S_3^{-0.55} - 0.1645P_{303}S_3^{-0.55} \\
& - 0.247P_{311} - 0.495P_{321} - 0.202P_{312} - 0.448P_{322} - 0.162P_{313} - 0.405P_{323} \\
& - 0.2441\bar{S}_4 - 1.3707S_4^{0.6} - 0.0577P_{401} - 0.0504P_{402} - 0.0440P_{403} \\
& + 0.0020P_{401}S_4 + 0.0017P_{402}S_4 + 0.0015P_{403}S_4 - 0.1484P_{401}S_4^{-0.55} \\
& - 0.1276P_{402}S_4^{-0.55} - 0.1097P_{403}S_4^{-0.55} - 0.495P_{411} - 0.099P_{421} - 0.040P_{431} \\
& - 0.448P_{412} - 0.090P_{422} - 0.040P_{432} - 0.405P_{413} - 0.088P_{423} - 0.041P_{433}
\end{aligned}
$$

Table C The constants

(1) $S_1 + S_2 + S_3 + S_4 = 10$	(2) $P_{111} + P_{211} + P_{311} + P_{411} = 1$	
(3) $P_{112} + P_{212} + P_{312} + P_{412} = 4$	(4) $P_{113} + P_{213} + P_{313} + P_{413} = 5$	
(5) $P_{121} + P_{221} + P_{321} + P_{421} = 2$	(6) $P_{122} + P_{222} + P_{322} + P_{422} = 3$	
(7) $P_{123} + P_{223} + P_{323} + P_{423} = 2$	(8) $P_{131} + P_{231} + P_{331} + P_{431} = 4$	
(9) $P_{132} + P_{232} + P_{332} + P_{432} = 3$	(10) $P_{133} + P_{233} + P_{323} + P_{433} = 2$	
(11) $P_{101} - S_2 \leq 0$	(12) $P_{102} - S_1 \leq 0$	(13) $P_{103} - S_1 \leq 0$
(14) $P_{201} - S_2 \leq 0$	(15) $P_{202} - S_2 \leq 0$	(16) $P_{203} - S_2 \leq 0$
(17) $P_{301} - S_3 \leq 0$	(18) $P_{302} - S_3 \leq 0$	(19) $P_{303} - S_2 \leq 0$
(20) $P_{401} - S_4 \leq 0$	(21) $P_{402} - S_4 \leq 0$	(22) $P_{403} - S_4 \leq 0$

8.26. Consider the problem of minimizing the purchase of fuel oil when it is desired to produce an output of 50 MW from a two-boiler turbine-generator combination which can use fuel oil or blast furnace gas (BFG) or any combination of these. The maximum BFG that is available is specified.

By applying nonlinear curve-fitting we obtained the fuel requirements for the two generators explicitly in terms of MW produced. For generator 1 we have the fuel requirements for fuel oil in tons per hour (x_{11})

$$f_1 = 1.4609 + 0.15186x_{11} + 0.00145x_{11}^2$$

and for BFG in fuel units per hour (x_{12})

$$f_2 = 1.5742 + 0.1631x_{12} + 0.001358x_{12}^2$$

where $(x_{11} + x_{12})$ is the output in MW of generator 1. The range of operation of the generator is

$$18 \leq (x_{11} + x_{12}) \leq 30$$

Similarly for generator 2 the requirement for fuel oil is

$$g_1 = 0.8008 + 0.2031x_{21} + 0.000916x_{21}^2$$

and for BFG,

$$g_2 = 0.7266 + 0.2256x_{22} + 0.000778x_{22}^2$$

where $(x_{21} + x_{22})$ is the output in MW of generator 2. The range of operation of the second generator is

$$14 \leq (x_{21} + x_{22}) \leq 25$$

It is assumed that only 10.0 fuel units of BFG are available each hour and that each generator may use any combination of fuel oil or BFG. It is further assumed that when a combination of fuel oil and BFG is used the effects are additive.

The problem is to produce 50 MW from the two generators in such a way that the amount of fuel oil consumed is minimum. Use successive linear programming.

8.27. Solve Prob. 7.18 in Chap. 7 by the SLP method.

CHAPTER 9

OPTIMIZATION OF STAGED AND DISCRETE PROCESSES

Many process plants and pieces of equipment are represented as multistage processes even if the apparatus itself is not comprised of separate units. Each stage of the configuration is characterized by five parameters:

1. The value of the objective function or return, R_n
2. The stage number and/or type, n
3. The value(s) of the input variable(s), x_n
4. The value(s) of the output (dependent) variable(s), x_{n-1}
5. The value(s) of the decision (independent) variable(s), d_n

Both x_n and d_n can be vectors. Figure 9.1 illustrates a typical simple multistage process with serial flow of material. Typical criteria are to maximize the sum of the return revenues from the stages, minimize the total costs, and so on. The stages may be physically identifiable stages, imaginary ones, or time periods.

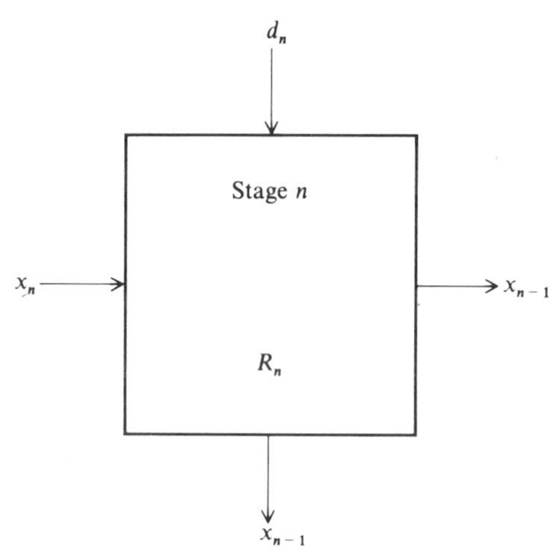

One stage of the process

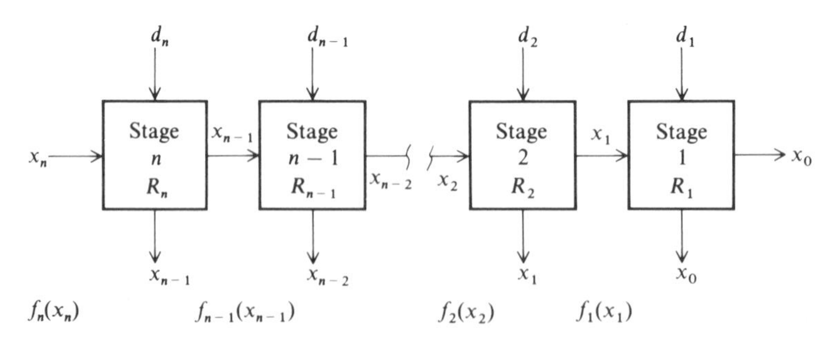

Figure 9.1 A multistage process; $f(x_k)$ is the overall objective function for the stages up to and including the kth stage; d_k is the decision (independent) variable vector; R_k is the return from stage k.

Table 9.1 Examples of multistage decision process-parameters

Parameter	Cross-current Extraction	Reactor	Distillation Column
1. Objective function	Maximize total solute extracted	Maximize yield over time	Maximize separation per unit cost of feed
2. Stage type	Extractor	Time period	Tray (stage, ideal plate)
3. Inputs	Solvent stream rate	Composition of feed	Composition of feed
4. Outputs	Solute composition in wash water	Composition of products	Composition of products
5. Decision variables	Wash water rates	Temperature, pressure, catalyst	Stage efficiency, temperature

Table 9.1 lists some examples of multi-stage decision processes in the chemical industry and their parameters. Determination of the optimal operating conditions or design for such processes is complicated because the stages themselves are integer quantities and the conditions to be set for each stage are interrelated. Section 9.1 below describes one way to optimize relatively simple configurations of units interconnected by flows of mass, energy, and information, namely by a method termed *dynamic programming*.

A second type of optimization problem, analyzed in Sec. 9.2, treats problems involving integer rather than continuous variables. Problems involving the design of plant apparatus in which an integer number of units must be specified, or problems in which specific pipe sizes must be determined, are examples of optimization problems involving discrete variables. Many scheduling, allocation, expansion, and construction problems are posed naturally in terms of discrete (or integer) variables. Such problems exhibit combinatorial properties in that the number of potential feasible solutions may approximate the number of possible combinations of the discrete variables.

9.1 DYNAMIC PROGRAMMING

Dynamic programming is used occasionally in chemical engineering design and operations. The essence of the procedure is to decompose a multivariate interconnected optimization problem into a sequence of subproblems that can be solved serially. Each of the subproblems contains one, or a few, decision variables. Therefore, the procedure can be conceptually applied to any process that can be broken down into stages of time, process components, or cost accounting units. Each individual stage must be optimized by one of the techniques discussed in Chaps. 6, 7, or 8. The development of dynamic programming has been surveyed by Bellman and Lee (1984).

Dynamic programming is based upon what has become known as Bellman's principle of optimality (Bellman, 1957; Bellman and Dreyfus, 1962). The concept is based on the *flow of information* in a process, and can be paraphrased as:

> If a decision forms an optimal solution at one stage in a process, then any remaining decisions must be optimal with respect to the outcome of the given decision. This concept can be best implemented if the decision-making from stage to stage takes place *in the direction of the information flow* in the stages which may be in the *opposite* direction of the materials flow.

Figure 9.2 illustrates the principle graphically. The information flow is in the opposite direction to the materials flow if the final state 5 is specified and stage 0 is to be determined. Suppose the decision made at point A results in the selection of the optimal path 1 with cost C_1. At B a choice must be made between paths 3 or 4 with costs C_3 and C_4 respectively. If a choice is made at B between 3 and 4, the choice between 1 and 2 is not affected.

To execute dynamic programming, because every stage in a serial structure influences another stage, the last stage is optimized first (Bellman and Dreyfus, 1962) for each possible state of the input(s) it receives. Once the last stage is optimized, the last two stages are grouped together and optimized for each possible input to the next to last stage. This procedure continues until the entire structure is included in the optimization. An example will make the procedure clear.

Figure 9.3 shows the routing for gasoline via a pipeline network. The gasoline can only flow toward the terminal H, and at each pumping station A through G only one pipeline branch can be selected for transport. Suppose you are the decision-maker located at C. Instead of trying all the possible allowable paths from each pumping station (an exhaustive search), you apply the principle of optimality. In this example the state is the pumping station and the decision is the choice of pipeline segment in order to minimize total cost.

At C you can select line CD or CF. Let C_{CD} and C_{CF} denote the cost of transport from C to D and C to F, respectively. Assume you know or can calculate the minimum costs C^*_{DH} and C^*_{FH} of transport from D and F to H,

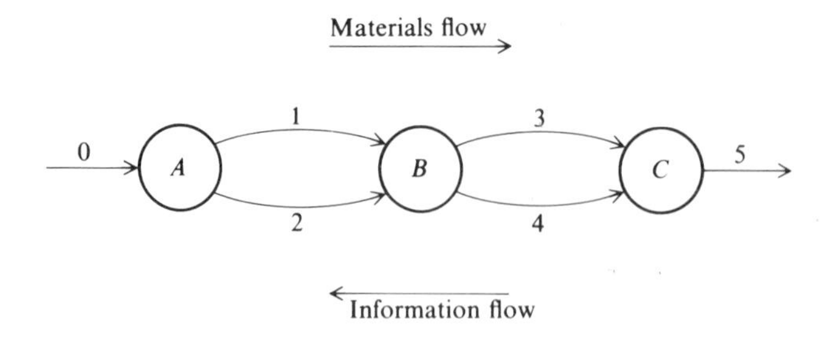

Figure 9.2 Materials and information flow between stages.

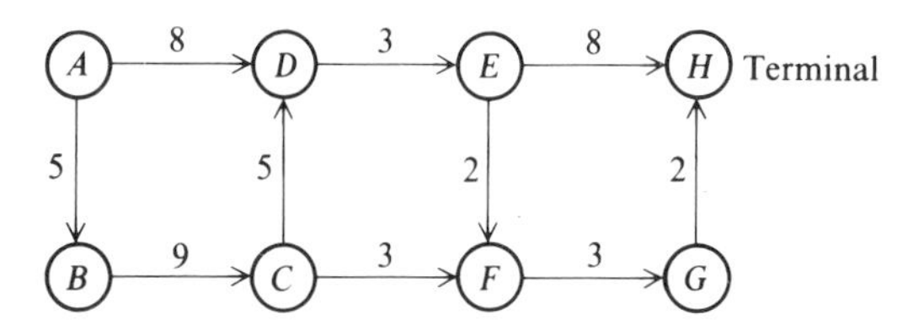

Figure 9.3 A pipeline network with pumping stations A to G and costs of transport.

respectively. (Here $C^*_{DH} = 10$ and $C^*_{FH} = 5$ as we shall see). Then the minimum cost C^*_{CH} to reach the terminal H is the smaller of

$$\begin{aligned} C_{CDH} &= C_{CD} + C^*_{DH} \\ &= 5 + 10 \end{aligned} \qquad \text{and} \qquad \begin{aligned} C_{CFH} &= C_{CF} + C^*_{FH} \\ &= 3 + 5 \end{aligned}$$

or

$$C^*_{CH} = \min\{C_{CDH}, C_{CFH}\} = \min\{15, 8\} = 8$$

and the optimal routing at C is to go to F.

Now how did we obtain values for C^*_{DH} and C^*_{FH}? We calculated these values by working backward from the terminal H. By inspection of Fig. 9.3 you can see $C^*_{GH} = 2$ (only one route exists) and $C^*_{FH} = 3 + 2 = 5$. Similarly

$$C^*_{EH} = \min\{C_{EH}, (C_{EF} + C^*_{FH})\} = 7$$

and so on. Table 9.2 lists all of the minimal costs from each pumping station via allowable flows to H.

Table 9.2 Optimal routes from each pumping station to the terminal

Pumping station ⓚ	Path	Next pumping station ⓙ	Minimum cost from ⓚ to H via ⓙ	Minimum cost to reach H from ⓚ	Optimal path at ⓚ
G	GH	H	$2 + 0 = 2$	2	GH
F	FG	G	$3 + 2 = 5$	5	FG
E	EH	H	$8 + 0 = 8$		
	EF	F	$2 + 5 = 7$	7	EF
D	DE	E	$3 + 7 = 10$	10	DE
C	CD	D	$5 + 10 = 15$		
	CF	F	$3 + 5 = 8$	8	CF
B	BC	C	$9 + 8 = 17$	17	BC
A	AD	D	$8 + 10 = 18$	18	AD
	AB	B	$5 + 17 = 22$		

Note how the optimal routes are built up backward toward the upstream pumping stations. The table can be used to reallocate the routing if for some reason a pipeline segment becomes unavailable. Note also that at A, selection of the initial low-cost route AB is misleading in spite of the local optimal result because the overall cost of transport is higher even if the optimal routing from B is followed thereafter.

Other applications (see Aris, 1961; Mitten and Nemhauser, 1963; Roberts, 1964) of dynamic programming have been to (*a*) stirred tank reactors in series with bypass in feed and multiple reactions, (*b*) multibed adiabatic reactors, (*c*) crosscurrent liquid extractors, (*d*) multistage gas compressors, (*e*) small heat exchanger networks, (*f*) inventory and production control, (*g*) catalyst replacement and regeneration, (*g*) equipment replacement, and (*h*) optimal control calculations.

The characteristics of nonlinear optimization problems to which the dynamic programming approach can be applied are:

1. The problem must be decomposable into stages with one or two variables at each stage for which decisions have to be made (decision variables, independent variables). The stages may represent sequential points in space as in Fig. 9.1 or sequences in time as in construction scheduling.
2. For each stage, the rest of the variables (a finite number, the fewer the better) are known as nondecision variables (stage variables, dependent variables) for the stage.
3. The state and decision variables may be continuous, discrete, or continuous in reality, but divided into arbitrary levels for optimization. By fixing the values of one or more decision variables in a stage of the process you can solve the model equations for that stage of the process and use the values of the entering states to calculate the values of the exit states of the stage. However, the exit states are the entering stages of the adjoining stage(s). The equation(s) used to relate the various states and decision variables in a stage is called the *transition function* (usually it is the process model for the state). Thus, a decision as to how much to produce of component A in stage k will determine the input into an adjacent stage.
4. Each state has associated with its variables a benefit, cost, or return function that serves as a measure of the effectiveness of the decisions made for the stage.
5. Bellman's principle of optimality can be applied; that is, for a given state of a stage, the optimal sequence of decisions for the remaining stages is independent of the decisions made in the previous stages. It is not until the optimal return has been calculated for the entire system that the optimum values for the intermediate stages can be determined.

A formal statement of Bellman's principle of optimality (refer to Fig. 9.1) for maximization is:

$$f_n(x_n) = \max_{d_n} \left[R_n(x_n, d_n) + f_{n-1}(x_{n-1})\right] \tag{9.1}$$

where n = number of stages remaining in the process
x_n = input to the nth stage
d_n = decision at the nth stage
$f_n^*(x_n)$ = maximum return from an n stage process with input x_n to the nth stage
$R_n(x_n, d_n)$ = return from stage n with input x_n and decision d_n
x_{n-1} = output from stage n and input to stage $n-1$
$f_{n-1}^*(x_{n-1})$ = maximum return from stages 1 through n

The variables x_n and d_n may each be vectors.

In summary, dynamic programming is not a prescription for solving a single-stage optimization problem—it is a procedure for decomposition of an optimization problem into a sequence of single-stage optimization problems. Thus dynamic programming has both advantages and disadvantages. Some advantages are:

1. It can treat nonconvex, nonlinear problems involving discontinuous objective functions and constraints meeting the requirements cited above.
2. It can treat problems involving integer-valued variables and mixed integer and continuous variables.
3. Unimodality of the cost surface is not essential and constraints are incorporated in the model for the elements forming the stages.
4. It requires reduced computational effort relative to complete enumeration. For example, for a process of N serial stages with n independent variables to be determined at each stage, each of which has r_j levels of distinction, the number of objective function evaluations would be r_j^{Nn} for complete enumeration and only Nr_j^n for dynamic programming.
5. It nearly always reaches a global extremum.

However, dynamic programming has some associated disadvantages that inhibit its widespread use:

1. The number of state variables on each stage is restricted to a very small number, say 3 or 4 or less. For example, at stage k let there be m state variables (dependent variables) and let each state variable have r_j feasible levels, $j = 1, 2, \ldots, m$. Then the number of possible combinations of state variables for which an objective function must be computed is $\prod_{j=1}^{m} r_j$, or if we let $r_j = r$, the number is r^m. If there are N stages and n decision variables at each stage, then the total storage requirements in the computer are $N(n + 2)r^m$. For $N = 6$ stages with $n = 1$ decision variable and $r = 10$ levels for each of $m = 4$ state variables, we would have to calculate and store

$$6(3)10^4 = 1.8 \times 10^5 \quad \text{computation results}$$

2. The optimization problem must be decomposable to a sequence of calculations. Processes having branches or recycle present severe problems and involve iterative calculations. Dynamic programming requires that output from

a stage be dependent only on the input and decision variables for that stage. Therefore, if recycle exists, some technique (refer to the supplementary references, e.g., Wilde (1965) must be applied to eliminate the feedback (much as in the solution of sets of flowsheet equations).

We now illustrate the application of dynamic programming to a simple process. Example 12.2 is another illustration for a serial process.

EXAMPLE 9.1 APPLICATION OF DYNAMIC PROGRAMMING TO A WATER DISTRIBUTION SYSTEM

Examine the water distribution system in Fig. E9.1*a*. Assume that the maximum value S_3 can attain is 3000 m^3 (per unit time). S_3 is to be divided into three distribution points in the delivered quantities D_1, D_2, and D_3. Assume that the value of the water at each outlet depends only on the amount of water supplied at that outlet. The problem is to determine the allocation of water that yields the maximum return from the system.

The first step is to convert the problem into a stagewise diagram (see Fig. E9.1*b*) in which the stages correspond to splitting points in the water distribution system. The state variables are S_3, S_2, and S_1 in cubic meters, and the decision variables are D_3, D_2, and D_1 in cubic meters.

Each stage has a transformation function (or model) of the same form, a material balance (each balance is an equality constraint)

$$S_i = S_{i+1} - D_{i+1} \qquad i = 1, 2 \tag{a}$$

$$S_1 = D_1$$

An additional inequality constraint on the whole process is

$$0 \le \sum D_i \le 3000 \tag{b}$$

Note that the constraint $D_1 + D_2 + D_3 = S_3$ is redundant with the stage material balances.

Our objective is to find the set of decision variables D_1, D_2, and D_3 (sometimes called a "policy") that maximizes the total return from the three stages

$$\text{Maximize} \quad f_N(S_N) = \sum_{i=1}^{N} f_i(S_i, D_i) \tag{c}$$

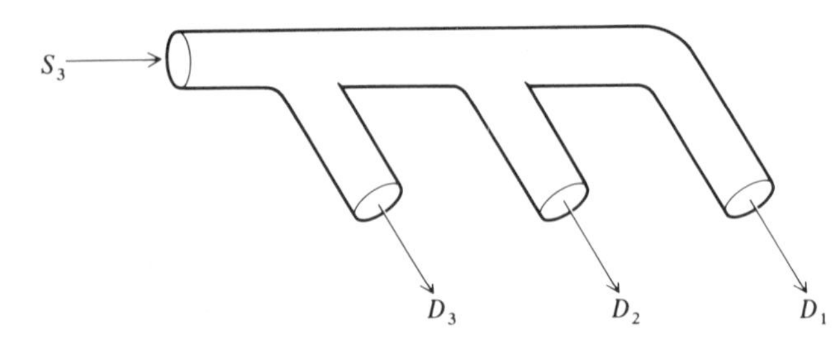

Figure E9.1*a* Water distribution system with one supply and three delivery locations.

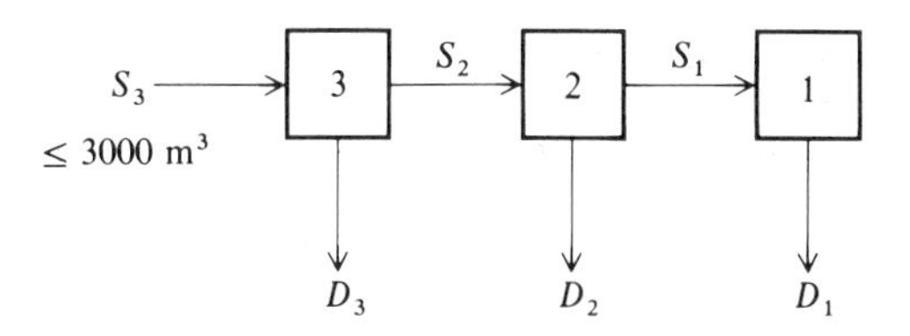

Figure E9.1*b* Functional diagram for the water distribution system of Fig. E9.1*a*.

subject to the constraints listed above. The revenues (f_i) for each stage are listed in Table E9.1*a* for three arbitrary levels of D_i; more levels could be chosen, of course, but a very coarse discretization of D_i will serve just as well to illustrate the method and keep the number of computations reasonable for this example.

To apply dynamic programming, we start with stage 1 and work backward against the water flow, carrying out the computations stage by stage. For stage 1 the revenues are

$$f_1^*(S_1) = \max_{D_1} f(S_1, D_1) \tag{d}$$

Subject to $0 \le S_1 \le 3$

$0 \le D_1 \le S_1$ (for symmetry in the resulting analysis; actually $D_1 = S_1$)

Table E9.1*a* lists revenues for various values of D_1 in the second column.

Next, we turn to stage 2. The total input S_2 must be shared between D_1 and D_2. How should S_2 be allocated? We shall examine all feasible distributions in which S_2 is divided between D_1 and D_2 to ascertain the maximum revenues from both stage 1 and stage 2. Our objective function for stage 2 is

$$f_2^*(S_2) = \max_{D_2} \{f_2(S_2, D_2) + f_1^*(S_1)\} \tag{e}$$

and (*e*) is subject to

$$0 \le S_2 \le 3$$

$$0 \le D_2 \le S_2$$

To compute values for $f_2(S_2)$ we need to fix S_2, say, at three levels, and at each level include as sublevels the levels of S_1. Examine Table E9.1*b*. The sum of the revenues from the two stages is listed in column (*f*) of Table E9.1*b*.

Table E9.1*a* Revenues for water

D_i, $m^3 \times 10^{-3}$	$f_1(D_1)$, $\$ \times 10^{-3}$	$f_2(D_2)$, $\$ \times 10^{-3}$	$f_3(D_3)$, $\$ \times 10^{-3}$
1	2	1	4
2	5	4	5
3	6	7	6

Table E9.1*b* Computations for stage 2

(*a*) S_2 $m^3 \times 10^{-3}$	(*b*) D_2 $m^3 \times 10^{-3}$	(*c*) $D_1 = S_1$ $m^3 \times 10^{-3}$	(*d*) $f_2(S_2, D_2)$ $\$ \times 10^{-3}$	(*e*) $f_1(S_1)$ $\$ \times 10^{-3}$	(*f*) $f_2(S_2) = (d) + (e)$ $\$ \times 10^{-3}$
1	1	0	1	0	1
	0	1	0	2	2†
2	2	0	4	0	4
	1	1	1	2	3
	0	2	0	5	5†
3	3	0	7	0	7†
	2	1	4	2	6
	1	2	1	5	6
	0	3	0	6	6

† Maximum value for levels of S_2.

Lastly we turn to stage 3. Again S_3 must be divided between D_1, D_2, and D_3 so the revenues are a maximum. Thus, the objective function for stage 3 is

$$f_3^*(S_3) = \max_{D_2} \{f_3(S_3, D_3) + f_2^*(S_2)\} \tag{f}$$

and (f) is subject to

$$0 \leq S_3 \leq 3$$

$$0 \leq D_3 \leq S_3$$

Table E9.1*c* is analogous to Table E9.1*b*. Note that the values for column (*e*) in the table are taken from the maximum values denoted by asterisks in Table E9.1*b*—you can omit function evaluations for state variables leading to lesser values of $f_2(S_2)$.

Table E9.1*c* Computations for stage 3

(*a*) S_3 $m^3 \times 10^{-3}$	(*b*) D_3 $m^3 \times 10^{-3}$	(*c*) S_2 $m^3 \times 10^{-3}$	(*d*) $f_3(S_3, D_3)$ $\$ \times 10^{-3}$	(*e*) $f_2(S_2)$ $\$ \times 10^{-3}$	(*f*) $f_2(S_3) = (d) + (e)$ $\$ \times 10^{-3}$
1	1	0	4	0	4†
	0	1	0	2	2
2	2	0	5	0	5
	1	1	4	2	6†
	0	2	0	5	5
3	3	0	6	0	6
	2	1	5	2	7
	1	2	4	5	9†
	0	3	0	7	7

† Maximum value for levels of S_3.

We have now completed the allocation process and found the optimal policy, namely

$$D_1^* = 2000 \qquad D_2^* = 0 \qquad D_3^* = 1000 \qquad f_3^* = \$9000$$

Furthermore, if we change the input S_3 to a maximum of $2000\ \mathrm{m}^3$ the optimal policy can be observed to be

$$D_1^* = 1000 \qquad D_2^* = 0 \qquad D_3^* = 1000 \qquad f_3^* = \$6000$$

If we need more precision in D_i and S_i, we have to increase the scope of the computations by adding intermediate levels of D_i and S_i. One can start with a coarse grid and then, using the optimal solution as a guide, reduce the feasible region and use a smaller grid mesh in the reduced region. If all the variables are continuous it may be possible to minimize $f(S_i, D_i)$ analytically or numerically at each stage.

At any stage N we can express the objective function as

$$f_N^*(S_N) = \max_{D_N} \{f_N(S_N, D_N) + f_{n-1}^*(S_{n-1})\} \qquad (g)$$

and (g) is subject to

(a) the transition function (process model) $T_N(S_N, D_N, S_{N-1}) = 0$

(b) $D_{N\min} \le D_N \le D_{N\max}$

(c) $S_{N\min} \le S_N \le S_{N\max}$

To reiterate, Eq. (g) states that the total maximum revenue (or minimum cost) for the system in state S_N at stage N is obtained by adding the return of stage N produced by decision N to the optimal return from all the upstream stages in the information flow.

The problem sketched in Fig. E9.1b that we have used as an example is known as an *initial state problem*. An initial state (value) problem occurs when the initial (input) stage of stage N is prespecified and the final (output) state from stage 1 is free (is determined by the solution of the problem). If the initial state at stage N *and* the final state at stage 1 are both fixed for a sequence of stages, we have a *two-point boundary value problem*. The optimization problem is to find the optimal return from the N stages as a function of the given input state of stage N and the given output from stage 1, $f_N(S_N, S_0)$, where S_0 would be the output variable of stage 1. At stage 1 you would

$$\underset{D_1}{\text{Maximize}} \quad f_1(S_1, D_1, S_1)$$

$$\text{Subject to} \quad T_1(S_1, D_1, S_0) = 0$$

Presumably T_1 can be solved explicitly for D_1

$$D_1 = T_1(S_0, S_1)$$

so that D_1 can be eliminated from f_1, and f_1 changed into a function of S_1 alone (with S_0 as a parameter). The only difference in procedure is that S_0 is carried along as an additional parameter, and the objective function at stage N is

$$f_N^*(S_N, S_0) = \max_{D_N} \{f_N(S_N, D_N) + f_{N-1}^*(S_{N-1}, S_0)\} \tag{9.2}$$

A third type of dynamic programming problem occurs when the final state of stage 1 is prespecified and the initial state of stage N is free to be determined in the problem. One way to solve the problem is to treat it as a two-point boundary value problem. However, if the state transition function can be solved ("inverted") so that

$$S_i = \hat{T}_i(S_{i-1}, D_i) \qquad i = 1, \ldots, N \tag{9.3}$$

then you can work backward from stage N to stage 1 in solving the problem because the information flow is reversed relative to the initial value problem. Consult the supplementary references for techniques to decompose these three types of problems.

Numerous methods have been proposed to reduce the dimensionality involved in dynamic programming. Refer to Morin (1978) who summarizes the work of about 150 investigators into nine different categories.

To summarize, no standard mathematical formulation exists for dynamic programming problems. The particular equations used must be developed to fit each individual problem. By understanding the general structure of dynamic programming, you can recognize when a problem can be solved by dynamic programming and how the solution procedure should be formulated. Dynamic programming is an iterative procedure but requires only a reasonable number of computer instructions to implement. Constraints imposed on the process can be used to reduce the number of feasible policies and therefore the time required to complete the optimization.

9.2 INTEGER AND MIXED INTEGER PROGRAMMING

Many problems in plant operation, design, location, and scheduling involve variables that are not continuous but instead have integer values. Decision variables for which the levels are a dichotomy—to install or not install a new piece of equipment, for example—are termed "0–1" variables. Other integer variables might be real numbers 0, 1, 2, 3, etc. Sometimes we can treat integer variables as if they were continuous, especially when they assume large values, and round the optimal solution to the nearest integer value. This step leads to a suboptimal solution, yet one that is quite acceptable from a practical viewpoint. However, when the answer is a number such as 1.3, we have less confidence in rounding. In this section we will illustrate some examples of problem formulation and subsequent solution in which one or more variables are treated as integer variables. Most of the integer programming problems are inherently combinatorial in nature, that is, the integer variables in the objective function correspond to combinations of n elements taken m at a time so that the number of possible combinations to consider is $\binom{n}{m}$, or correspond to permutations of n elements so that the number of possibilities is $n!$.

First let us classify the types of problems that are encountered in optimization with discrete variables. The most general case is a *mixed-integer programming* (MIP) problem in which the objective function depends on two sets of variables, $\mathbf{x}$ and $\mathbf{y}$; $\mathbf{x}$ is the vector of integer variables and $\mathbf{y}$ denotes the continuous variables. If only integer variables are involved ($y_j = 0$), we have an *integer programming* (IP) problem. Finally, a special case of IP is *binary integer programming* (BIP), where all variables x_i are either 0 or 1. Many MIP problems are linear in the objective function and constraints, hence are subject to solution by linear programming. These problems are called MILP problems.

Some formulations of typical integer programming problems are as follows (the popular names cited are those usually attached to the problem):

1. *The knapsack problem.* We have n articles each designated by x_i; the weight of the ith article is w_i and its value is s_i. The problem is to select the articles such that their total weight does not exceed W and their total value is a maximum.

$$\text{Maximize} \quad f(\mathbf{x}) = \sum_{i=1}^{n} s_i x_i$$

$$\text{Subject to} \quad \sum_{i=1}^{n} w_i x_i \le W \qquad x_i = 0, 1 \qquad i = 1, 2, \ldots, n$$

   In the last constraint, the binary variable indicates whether an article x_i is selected ($x_i = 1$) or it is not selected ($x_i = 0$). The knapsack problem has other interpretations: (1) if a given space is to be packed with items of different value and volume, the objective is to choose the most valuable packing; or (2) if a given item is to be divided into portions of different value, the objective is to find the most valuable division of the item.

2. *The traveling salesman problem.* The problem is to assign values of 0 or 1 to variables x_{ij}, where x_{ij} is 1 if the salesman travels from city i to city j and 0 otherwise (Little, 1963). The constraints in the problem are that the salesman must start at a particular city, visit each of the other cities only once, and return to the original city. Some cost (here distance) c_{ij} is associated with traveling from city i to city j, and the objective function is to minimize the total cost of the trips to each city, i.e.,

$$f(\mathbf{x}) = \sum_{i=1}^{n} \sum_{j=1}^{n} c_{ij} x_{ij} = \sum_{i=1}^{n} \mathbf{c}_i^T \mathbf{x}_i$$

subject to the $2n$ constraints:

$$\sum_{i=1}^{n} x_{ij} = 1 \qquad \sum_{j=1}^{n} x_{ij} = 1 \qquad \begin{array}{ll} x_{ij} = 0, 1 & i, j = 1, \ldots, n \\ x_{ij} = 0 & i = j \end{array}$$

where

$$\mathbf{c}_i^T = [c_{i1}, \ldots, c_{in}] \qquad \mathbf{x}_i^T = [x_{i1}, \ldots, x_{in}]$$

The two types of equality constraints ensure that each city is only visited once in any direction. We define $x_{ij}(i = j) = 0$ since no trip is involved. If you write $\mathbf{c}$ as a matrix, it is symmetric with zero diagonal terms. The equality constraints (the summations) indicate that the sums of all columns of the matrix $\mathbf{x}$ (x_{ij}) must be unity; likewise, the sums of all rows must be unity.

The problem can also be interpreted as that of processing n items in batches on a piece of equipment in which the equipment is reset between processing the ith and jth batches.

3. *The scheduling problem.* The problem is to assign integer values to variables x_{ij}, where x_{ij} is the starting time of job i on machine j, $j = 1, \ldots, n$. Given the time t_{ij} that it takes to complete the work of job i on machine j, the problem is to schedule the jobs on each machine so that the total time for the completion of all the jobs is a minimum. Because $(x_{in} + t_{in})$ is the time at which job i is completed on machine n, the maximum of these numbers is the time at which the latest job is completed. It is that time which is to be minimized. The criterion is

$$\text{Minimize} \quad f(\mathbf{x}) = \max_i \, (t_{in} + x_{in})$$

The constraints in the problem are that a job cannot be processed on machine n before it has been completed on machine (n-1), and it cannot be processed on machine (n-1) before it has been completed on machine (n-2), and so forth. These are called procedure constraints. Other constraints can include time limits on groups of machines and interchangeability of certain machines (see Plane and Macmillan, 1971).

4. *Location of oil wells (plant location problem).* It is assumed that a specific production-demand versus time relation exists for a reservoir. Several sites for new wells have been designated. The probem is how to select from among the well sites the number of wells to be drilled, their locations, and the production rates from the wells so that the difference between the production-demand curve and flow curve actually obtained is minimized. Refer to Rosenwald and Green (1974) and Murray and Edgar (1978) for a mathematical formulation of the problem. The integer variables are the well identities (0 = not drilled, 1 = drilled). This problem is related to the plant location problem (Salkin, 1975) and also the fixed charge problem (Hillier and Lieberman, 1986).

5. *Blending problem.* You are given the possible ingredients from a list containing the weight and analysis of each ingredient. The objective is to select from

the list of items and their weights so as to have a satisfactory total weight and analysis at minimum cost for a blend. Let x_j be the quantity of ingredient available in continuous amounts and y_k represent ingredients available in discrete quantities v_k ($y_k = 1$ if used and $y_k = 0$ if not used). Let c_j and d_k be the respective costs of the ingredients, and a_{ij} and a_{ik} be the fraction of component i. The problem is

$$\text{Minimize} \quad \sum_j c_j x_j + \sum_k d_k y_k$$

$$\text{Subject to} \quad \sum_j x_j + \sum_k v_k y_k \geq W^l$$

$$\sum_j x_j + \sum_k v_k y_k \leq W^u$$

$$\sum_j a_{ij} x_j + \sum_k a_{ik} v_k y_k \geq A_i^l$$

$$\sum_j a_{ij} x_j + \sum_k a_{ik} v_k y_k \leq A_i^u$$

$$0 \leq x_j \leq u_j \text{ for all } j$$

$$y_k = (0, 1) \text{ for all } k$$

where W^l and W^u are the lower and upper bounds on the weights, respectively, and A_i^l and A_i^u are the lower and upper bounds on the analysis for component i, respectively.

Innumerable other problems can be formulated as integer programming problems; refer to Schrage (1984), Plane and MacMillan (1971) and the supplementary references for additional examples.

To illustrate the characteristics inherent in integer programming problems, let us look at a very simple assignment problem, similar to the traveling salesman problem posed above. We want to assign streams to heat exchangers and the cost (in some measure) of doing so is listed in the matrix

$$\begin{array}{cc} & \text{Exchanger number} \\ & \begin{array}{cccc} 1 & 2 & 3 & 4 \end{array} \\ \text{Stream} \begin{array}{c} A \\ B \\ C \\ D \end{array} & \begin{bmatrix} 94 & 1 & 54 & 68 \\ 74 & 10 & 88 & 82 \\ 73 & 88 & 8 & 76 \\ 11 & 74 & 81 & 21 \end{bmatrix} \end{array} \tag{9.4}$$

Each element in the matrix represents the cost of transferring stream i to exchanger j. How can the cost be minimized if each stream goes to only one exchanger?

First let us write the problem statement. The toal number of streams n is 4. Let c_{ij} be an element of the cost matrix, which is the cost of assigning stream i to exchanger j. Then we have the following integer programming problem:

$$\text{Minimize} \quad f(\mathbf{x}) = \sum_{i=1}^{n} \sum_{j=1}^{n} c_{ij} x_{ij}$$

$$\sum_{i=1}^{n} x_{ij} = 1 \qquad j = 1, \ldots, n$$

$$\sum_{i=1}^{n} x_{ij} = 1 \qquad i = 1, \ldots, n \qquad \text{(Problem 1)}$$

$$x_{ij} = 0, 1 \qquad i, j = 1, \ldots, n$$

For the example matrix, you may detect upper and lower bounds on the cost. By computing the sum of the maximum and minimum costs in each column we find

$$\text{Upper bound} \quad f(x) = 94 + 88 + 88 + 82 = 352$$

$$\text{Lower bound} \quad f(x) = 11 + 1 + 8 + 21 = 41$$

but neither of these assignments is feasible because for the upper bound stream B is assigned to two exchangers and for the lower bound stream D is assigned to two exchangers. We will solve this example problem without complete enumeration shortly. How to reduce the scale of the required computations so that all the possible optima do not have to be examined is described below.

What makes integer programming problems hard to solve is their computational complexity. Search directions are meaningless in this type of problem. Some problems that on the surface seem to be easy may take hours of computing time without ever reaching optimality. For this example $4! = 24$ possible assignments exist. The number of integer vectors that might yield a feasible solution to a bounded integer programming problem in n-dimensional space grows exponentially in n (at least for linear integer programming problems), and the computational effort has been shown by experiments to rise as the square of the number of integer variables for many types of problems. Thus complete enumeration is not a practical tool for even modestly sized problems.

It may be attractive to use algorithms for continuous variables (such as LP) and then round up or down to obtain the integer values. However, this may be totally unsatisfactory and often misleading. Suppose the following two inequality constraints were binding in solving a continuous problem by linear programming:

$$-x_1 + x_2 \leq 3.5$$

$$x_1 + x_2 \leq 10.5$$

The solution ($x_1 = 3.5$, $x_2 = 7$) has the property that when x_1 is rounded up or down, one of the constraints is violated. There are procedures to accommodate integer and continuous variables that we discuss later (branch and bound); however, one must be careful about how such algorithms are used for mixed-integer programming problems.

9.2.1 Implicit Enumeration

An implicit enumeration algorithm comprises a coordinated, heuristically structured search of the space of all feasible solutions. The strategy involved is simply to search over various possible combinations of the discrete variables in such a manner as to eliminate certain sets of solutions to the discrete problem from consideration. Thus, an optimal solution to an integer optimization problem can be obtained with less than a complete enumeration of all the possible solutions. Plane and Macmillan (1971) provide a brief, simple summary of the strategy of the technique. Fundamentally, it is an *enumerative procedure*.

First let us solve the heat exchanger assignment problem described in connection with matrix (9.4) by implicit enumeration to illustrate the basic ideas. This problem is fairly simple in that it does not involve inequality constraints. Later we will solve two problems that do involve inequality constraints. The assignment problem given in matrix (9.4) is to assign n streams to n heat exchangers, with each heat exchanger to be assigned a different stream.

The general strategy is as follows. Four initial branches are made by assigning different streams to exchanger 1. The objective function in subproblem 1 is evaluated, and the value of $f(\mathbf{x})$ in the first feasible solution found is used as an upper reference bound. Additional assignments yield other values of $f(\mathbf{x})$, and the reference upper bound is reduced as better values are discovered from solutions of new subproblems. Lower bounds on $f(\mathbf{x})$ are evaluated for subproblems by choosing lowest cost assignments of streams, and these lower bounds, even if for infeasible solutions, can be used to discard some subproblems from further consideration if the lowest bounds exceed the reference upper bound for $f(\mathbf{x})$. The procedure continues until no feasible solution is found that has a value for the objective function that is less than the most recent reference upper bound.

Figure 9.4 outlines some of the subproblems to be solved in the form of a tree with branches. With four free variables, the first level of decisions involves four possible branches to four nodes. Once one variable is assigned, three free variables exist so that on the next lower level, three possible branches occur, and so on. Not all of the branches yield feasible solutions, of course. At each node the number designates the exchanger and the letter the stream, with an underline designating a fixed assignment. For example, $\underline{1\text{-}A}$ means exchanger 1 is assigned stream A; the other exchangers can be assigned the various streams which remain.

We start (arbitrarily) with the assignment of exchanger 1 to stream A. For the lowest cost assignment, we could assign exchanger 2 to stream A. However, stream A has already been allocated, so that we pick the next best cost, and assign stream B to exchanger 2. Exchanger 3 gets stream C and exchanger 4 gets

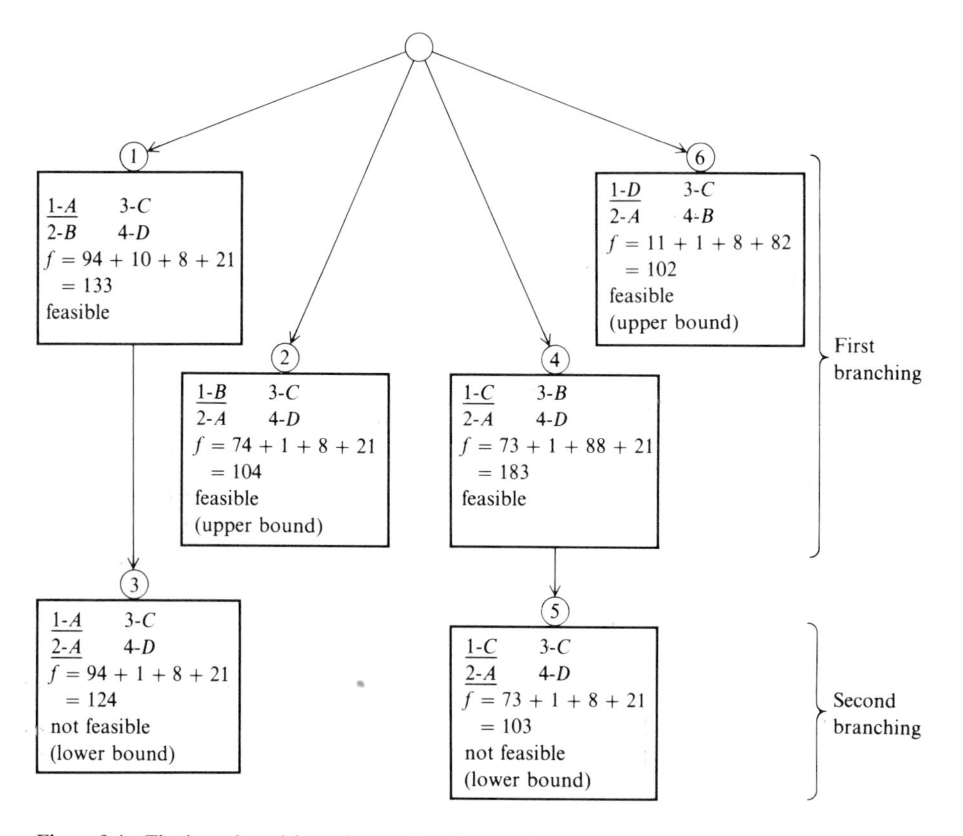

Figure 9.4 The branch and bound procedure for matrix (9.4).

stream D, both being the lowest costs in the respective columns of matrix (9.4). Thus, the solution to the subproblem in the first node is a feasible one with $f(\mathbf{x}) = 133$ which becomes the reference upper bound.

Next we (arbitrarily) start with 1-B and assign lowest costs giving 2-A, 3-C, and 4-D to get a feasible solution with $f = 104$ which becomes the new upper bound (node 2). Suppose we now go back and look at the assignment starting with 1-A and assign each stream to an exchanger that gives the lowest cost even if the solution is not feasible, namely 1-A, 2-A, 3-C, 4-D for which $f = 124$ (node 3). This infeasible solution has a lower bound greater than $f = 104$, hence *no* combination, feasible or not, starting with 1-A will yield a value of $f(\mathbf{x}) \leq 104$. Consequently all possible combinations of streams to exchangers starting with 1-A can be omitted from further consideration since they will be inferior to node 2.

We next start with 1-C, and assign the streams using the lowest cost when permissible, yielding a feasible solution with $f = 183$. This node is inferior to node 2. We can create a new node with 1-C but assign the least cost exchanger regardless of feasibility (node 5). The cost ($f = 103$) is not feasible but represents a new lower bound on f. Finally, we assign 1-D and allocate the rest of the exchangers to streams, in each case using the lowest cost, to get 1-D, 2-A, 3-C, and 4-B with $f = 102$ (node 6). This assignment is feasible, is a lower bound on f, and results in a new upper bound as well. Consequently, we do not have to consider any branches of the tree in Fig. 9.4 starting with 1-D or 1-B as the lowest costs of each assignment have been used. Also we can now omit from consideration all subproblem assignments branching from an initial assignment of 1-C since node 5 is infeasible but has a higher cost than node 6.

Figure 9.4 shows that by using the bounds as criteria, it is possible to exclude sections of the tree from the search for an optimum. In the limit, partial enumeration could involve complete enumeration, but in practice usually about two-thirds of the subproblems can be eliminated from consideration. In this example, there are 24 possible assignments but only six cases needed to be evaluated explicitly.

In the above example, we say that a node has been *fathomed* when further branching from that node has been ruled out (e.g., node 3 in Fig. 9.4). Therefore further subbranches (and nodes) have been *implicitly enumerated.* After finding that a partial solution is fathomed due to infeasibility, we then *backtrack* to other nodes and their branches. This general procedure will be useful for mixed-integer programming problems with constraints, as we will see later.

9.2.2 Branch and Bound Technique

The branch and bound technique can be applied to mixed-integer as well as general integer programming problems. It uses a partitioning method for the independent variables to divide up all the possible feasible solutions into subsets and serially searches for the optimum solution in the most promising subsets. The basic concept in employing partitioning is that most of the feasible solutions can be enumerated implicitly and only a few need be enumerated explicitly. Branch and bound analysis is a general procedure, and implicit enumeration discussed in the previous section is a special case of the branch and bound procedure.

The term "branch" derives from the concept of successively adding constraints to the independent integer variables in the problem. Partitioning an integer variable x_a into two branches involves selecting one branch to be greater than or equal to some integer value, while the other branch considers x_a to be less than that integer value. Bounding occurs because each subproblem solution forms a bound on the value of the objective function. With the addition of tighter constraints in the integer variable (branching), you know that the objective function value for the subproblem cannot improve, hence certain branches can be eliminated from consideration (or implicitly enumerated) in the search for the optimum.

The branch and bound procedure is comprised of the following steps:

1. Obtain the continuous (noninteger) solution of the problem; designate this problem "C."

2. Does the solution to C satisfy all integer requirements? If yes, then you are finished. If not, proceed to step 3.
3. From a review of the solution to problem C select one integer variable which has a noninteger value (say α). Consider the value of x_a, α, to be a starting node. Construct two nodes which branch from α by adding constraints:
   (*a*) $x_a \geq$ smallest integer greater than α
   (*b*) $x_a \leq$ largest integer less than α
   Call problem C with the added constraints problems C^1 and C^2. Next solve the continuous problem for C^1 or C^2, say C^1. We return to the other branch later.
4. Review the solution to C^1, and branch on a new variable x_b using two partitioning constraints as before. Select the branch which initially appears to be more promising for the next subproblem to solve. It may be necessary to branch on x_a again if later solutions require it. Continue branching on integer variables until one of three things happens:
   (*a*) You reach a solution which satisfies all the integer constraints (this may require fixing all of the variables at integer values)
   (*b*) The objective function for a solution of a subproblem becomes inferior to any previously obtained solution which satisfies the integer constraints
   (*c*) No feasible solution is found to exist
   In any case, this terminal node is deemed fathomed.
5. Next you backtrack to a previous node on the same branch, use the alternative constraint, and solve the optimization subproblem. Evaluate subproblem solutions as additional nodes as in step 4. Once a branch is fathomed, backtract again until you finally reach the original solution (node 1). Record the best value for the branch just analyzed for comparison with other branches.

You continue in this manner until all possible combinations of integer constraints have either been actually imposed upon the problem or fathomed. How to decide on which bound to change at a node, and thus determine what sequence of nodes to follow in the optimization, is problem-dependent. Common sense and in some cases luck dictate efficient branching. Refer to Salkin (1975) for more information on branch and bound analysis.

For the case of linear problems only (mixed-integer linear programming problems (MILP) and integer programming (IP) problems) in the first step you solve the MILP via linear programming assuming all the variables are continuous (Hillier and Lieberman, 1986). Usually the computed optimum contains some noninteger values for variables that should be integer values. Thus, the LP solution is nonoptimal but provides a lower bound for the objective function in a a minimization problem (upper bound in maximization). By using branch and bound analysis, the global optimum for an MILP can be found. Table 9.3 lists sources of computer codes based on branch and bound analysis which can be used to solve MILP's and IP's. Several examples of branch and bound analysis in IP and MILP follow.

Table 9.3 Sources of computer codes for mixed integer and integer programming (linear problems)

Code name	Type	Source	Computers	Language	Price
Algorithm 449	0-1	*Commu. ACM*, Vol. *16*, No. 7, (1973)	IBM 370	Fortran	Write
APEX III	MI	Control Data Corp., Publ. No. 76070000G (1979)	CDC, Cyber	Fortran + Assembly	Ask
ILLIP-2	0-1	Prof. S, Muroga, Dept. Computer Sci., Univ. of Illinois, Urbana, IL 61801	IBM 360/75	Fortran IV, PL 1	Handling only
LINDO	MI	Scientific Press, 540 University Ave., Palo Alto, CA 94301	Various IBM, IBM-PC	Compiled code only	Ask
MIP III	MI	Kentron, 1700 N. Moore St., Arlington, VA 22209	Several	Dataform	\$550/mo
MPSX	MI	IBM Scientific Marketing; MSPS/370 Program Ref. Manual, SH19-1094	IBM 370	Extended control language (PL/1)	Ask
ZOOM/XMP	MI, 0-1	Prof. Roy Marsten, Dept. Management Info. Systems, Univ. of Arizona, Tucson, AZ 85721	Several	Fortran	Ask

0-1: 0-1 variables in LP; MI: mixed integer.

EXAMPLE 9.2 BRANCH AND BOUND ANALYSIS OF IP

$$\text{Maximize} \quad f = 86x_1 + 4x_2 + 40x_3$$

$$\text{Subject to} \quad 774x_1 + 76x_2 + 42x_3 \leq 875$$

$$67x_1 + 27x_2 + 53x_3 \leq 875$$

$$x_1, x_2, x_3 = 0, 1$$

Figure E9.2 shows the sequence of decisions for this problem. A description of the branching analysis for each node is as follows:

Node 1. Solution to LP with all variables ≤ 1, yielding a fractional value of x_2.

Node 2. Set $x_2 = 1.0$ since it has a fractional value in node 1, and solve the LP again (we will examine the alternative branch $x_2 = 0.0$ later).

Node 3. Since x_1 has a noninteger value set it to 1.0 also; resolve the LP.

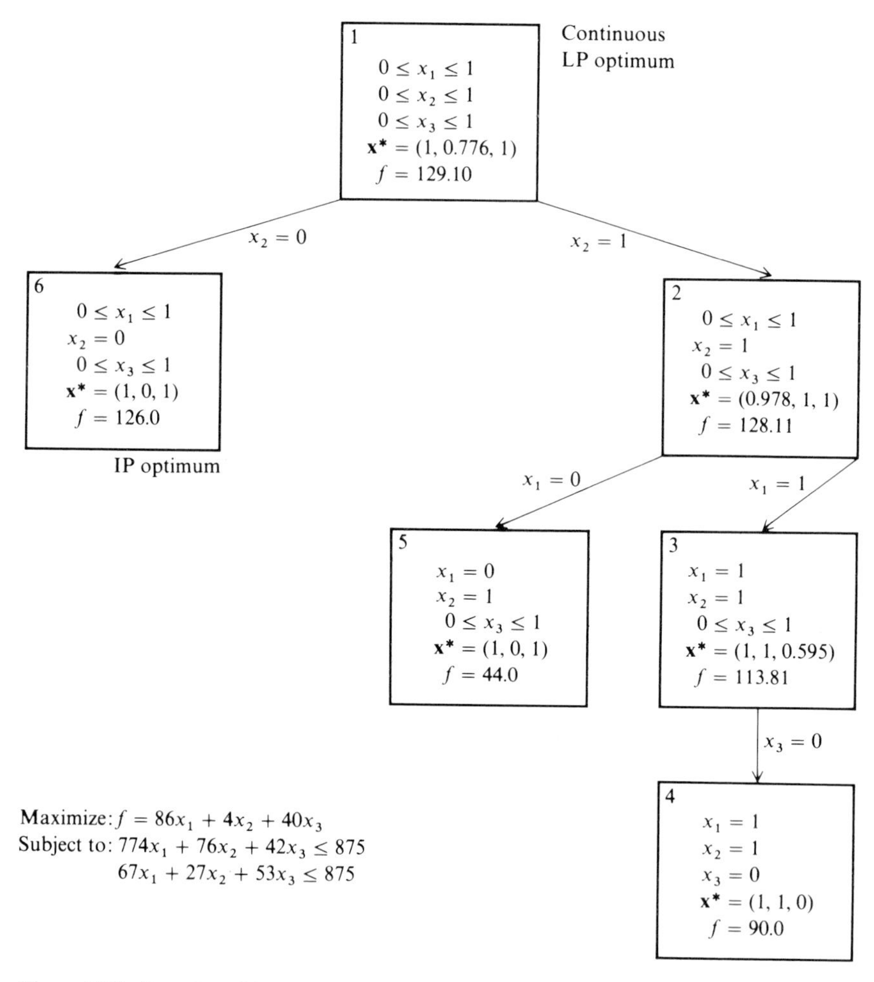

Figure E9.2 Branch-and-bound analysis with six nodes, integer linear program.

Node 4. Since x_3 is the remaining integer variable to be examined, set it to zero since $x_3 = 1$ is infeasible (violates the first constraint). This is the first feasible integer solution, and the value of f is 90.0. However, we do not know that it is optimal.

Node 5. Now backtrack from node 4 and change x_1 to zero, which is the alternative value to node 3. Since f $(= 44.0)$ is lower than f in node 4, we do not need to consider other branches from node 5 (it is fathomed).

Node 6. We backtrack to the original branch from node 1 and set $x_2 = 0$; the LP solution gives a higher value of f than node 4 and in fact is an integer solution. Hence, further combinations of x_1 and x_3 with $x_2 = 0$ will not provide an improved solution. Therefore, the optimum has been found.

Note in Fig. E9.2 that if we would have branched to node 6 first (by luck), fewer total nodes would have to be examined.

EXAMPLE 9.3 APPLICATION OF THE BRANCH AND BOUND PROCEDURE

In this example we have two units in a plant, No. 1 and No. 2, making products 1 and 2, respectively, from feedstocks as shown in Fig. E9.3*a*. Unit 1 has a maximum capacity of 8000 lb/day and unit 2 of 10,000 lb/day. To make 1 lb of product 1 requires 0.4 lb of A and 0.6 lb of B; to make 1 lb of product 2 requires 0.3 lb of B and 0.7 lb of C. A maximum of 6000 lb/day of B is available. Assume the net revenue after expenses from the production of product 1 is \$0.16/lb and of product 2 is \$0.20/lb. How much of product 1 should be produced per day, and how much of 2, assuming that each must be made in batches of 2000 lb?

We can formulate the optimization problem as follows:

$$\text{Maximize} \quad f(\mathbf{x}) = 0.16x_1 + 0.20x_2 \tag{a}$$

$$\text{Subject to} \quad x_1 \leq 8000 \tag{b}$$

$$x_2 \leq 10{,}000 \tag{c}$$

$$0.6x_1 + 0.3x_2 \leq 6000 \tag{d}$$

$$x_1, x_2 \geq 0 \tag{e}$$

$$x_1 = 0, 2000, 4000, 6000, 8000 \tag{f}$$

$$x_2 = 0, 2000, 4000, 6000, 8000, 10{,}000 \tag{g}$$

Figure E9.3*b* shows the feasible region (the grid intersection nodes represent the feasible region for the integers x_1 and x_2) and some of the contours of the objective function.

We want to avoid searching each node comprising the feasible region. Where should we start? We can relax the assumption that x_1 and x_2 are specific integers, and solve the resulting linear programming problem to reach the vicinity of the optimum. The subsequent search starts from the nearest node. (This choice of starting point is not guaranteed to be the best but the LP solution does place an upper bound on the value of $f(\mathbf{x})$ for maximization). The LP solution of the relaxed problem is point A in Fig. 9.3*b*, where $x_1 = 5000$ lb/day, $x_2 = 10{,}000$ lb/day, and

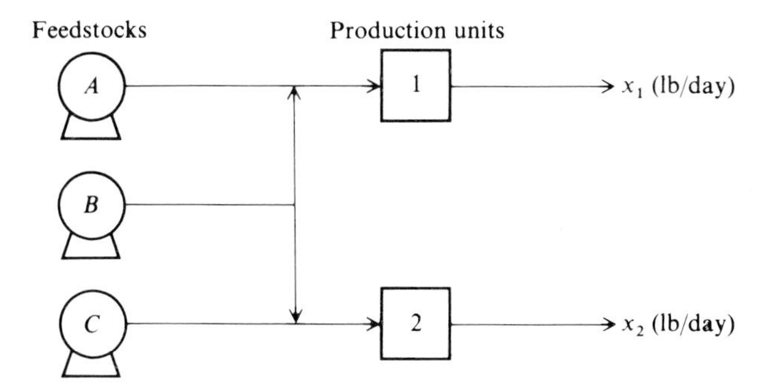

Figure E9.3*a* Flowsheet of batch plant.

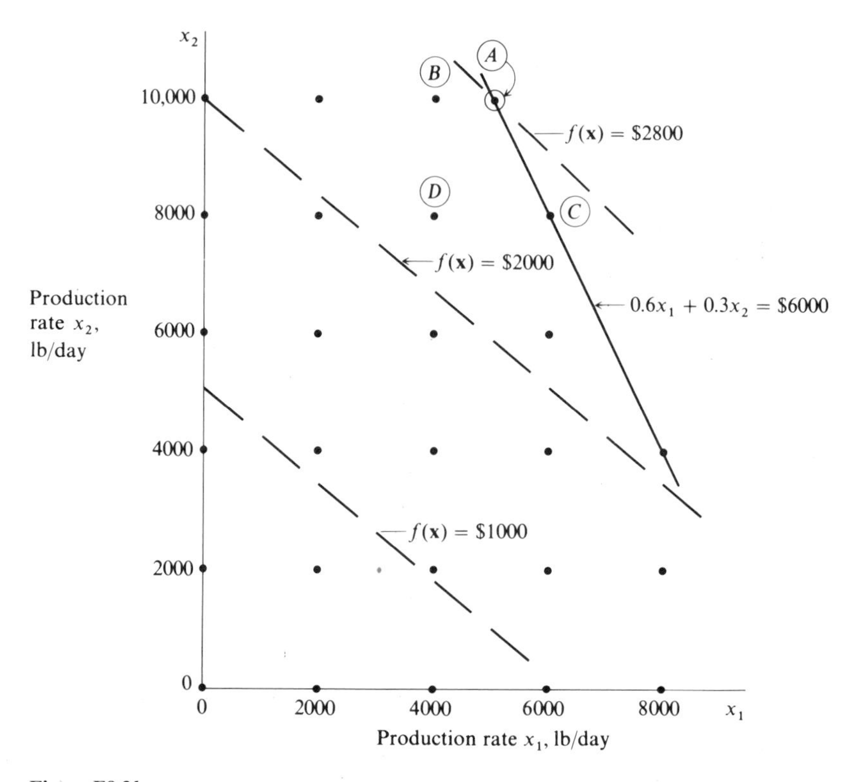

Figure E9.3*b*

$f(\mathbf{x}) = \$2800$. The most likely feasible points for the solution of the integer problem will be B or C; one of these appears to give the maximum for $f(\mathbf{x})$. After computation we find

$$f(\mathbf{x}_B) = \$2640$$

$$f(\mathbf{x}_C) = \$2560$$

hence $\mathbf{x}_B$ is the solution of the integer problem.

Let us now proceed to the same solution via a branch and bound analysis.

Problem 1 is comprised of Eqs. (*a*) through (*e*). We add the following constraints on the integer variables x_1 and x_2:

$$0 \leq x_1 < 8000 \qquad (h)$$

$$0 \leq x_2 \leq 10{,}000 \qquad (i)$$

and start to solve a new problem comprised of (*h*) and (*i*) added to (*a*) through (*e*). To save computations for this example we will use some common sense and initially tighten the lower bounds to

$$4000 \leq x_1 \leq 8000 \qquad (h')$$

$$6000 \leq x_2 \leq 10{,}000 \qquad (i')$$

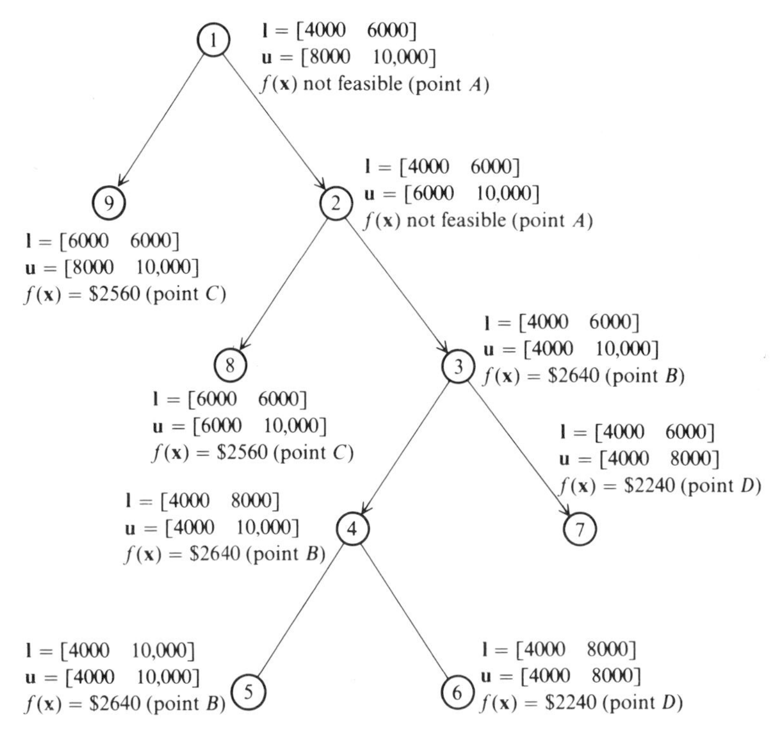

Figure E9.3*c* Branch and bound analysis, Example 9.3. (l = lower-bound vector; u = upper bound vector.)

The solution to problem 1 by LP is $\mathbf{x} = [5000 \quad 10{,}000]^T$, a nonfeasible point for the integer problem. Figure E9.3*c* shows the tree for the branch and bound procedure. With two free variables, two branches exist for each node, but no branch occurs if the bounds are tight.

The tree consists of vertices, denoted by circles, and edges which connect the vertices. The numbers in the circles show the eventual sequence of search. The lower left-hand branches show increases in lower bounds on the variables while the lower right-hand branches show decreases in the upper bounds on the variables. Only one change in one bound on one variable is made at a time. For example, from node 1 the original problem is modified (arbitrarily) to node 2 by lowering the upper bounds on x_1. Then a decision has to be made as to whether to proceed down the left-hand branch at node 1 or continue to branch from node 2. For the purpose of illustration, suppose we lower the upper bound on x_1 once more to 4000 (node 3), then raise the lower bound of x_2 to 8000 (node 4)—the bound on x_1 cannot be raised—and then to 10,000, and so on as shown in Fig. E9.3*c*. We evaluate $f(\mathbf{x})$ at each node (subproblem) as we go along using an LP code.

Nodes 1 and 2 do not yield feasible solutions so that node 3 represents the first feasible solution with $f(\mathbf{x}) = \$2650$, a value that becomes the reference lower bound (recall we are maximizing). Note we stop branching at nodes 5 and 6 because the upper and lower bounds are identical. We stop branching at node 7 because $f(\mathbf{x})$ at node 7 is less than our best bound on $f(\mathbf{x})$ in the tree, the bound calculated at node 3 of \$2640, and any further tightening of the constraints on x_2 will not increase $f(\mathbf{x})$. To this point the best $f(\mathbf{x}) = \$2640$.

In backtracking up the tree from node 7 to node 2 and thence to node 8, we find $f(\mathbf{x})$ at node 8 is less than \$2640 and so do not continue branching from node 8. Once more we backtrack up the tree to node 1 and thence to node 9 for which $f(\mathbf{x}) < \$2640$. Thus, the solution to the integer problem is $\mathbf{x} = [4000 \quad 10{,}000]^T$ and $f(\mathbf{x}) = \$2640$ at point B.

EXAMPLE 9.4 PROJECT SELECTION IN MANUFACTURING

A microelectronics manufacturing facility is considering six projects to improve operations as well as profitability. However, not all of these projects can be implemented due to expenditure limitations as well as engineering manpower constraints. The table below gives projected cost, manpower, and profitability data for each project.

Project	Description	First year expenditure	Second year expenditure	Engineering hours	Net present value
1	Modify existing production line with new etchers	\$300,000	0	4000	\$100,000
2	Build new production line	\$100,000	\$300,000	7000	\$150,000
3	Automate new production line	0	\$200,000	2000	\$ 35,000
4	Install plating line	\$ 50,000	\$100,000	6000	\$ 75,000
5	Build waste recovery plant	\$ 50,000	\$300,000	3000	\$125,000
6	Subcontract waste disposal	\$100,000	\$200,000	600	\$ 60,000

The resource limitations are:

First year expenditure: \$450,000
Second year expenditure: \$400,000
Engineering hours: 10,000

A new or modernized production line must be provided (project 1 or 2). Automation is feasible only for the new line. Either project 5 or project 6 can be selected, but not both. Determine which projects maximize the net present value subject to the various constraints.

Solution. Define the variables as x_j, $j = 1, 2, \ldots, 6$, where $x_j = 1$ means the jth project is selected, and $x_j = 0$ means the jth project is omitted. The objective function to be maximized is

$$f = 100{,}000x_1 + 150{,}000x_2 + 35{,}000x_3 + 75{,}000x_4 + 125{,}000x_5 + 60{,}000x_6 \quad (a)$$

subject to the following constraints:

First year expenditure:

$$g_1 = 300{,}000x_1 + 100{,}000x_2 + 0x_3 + 50{,}000x_4 + 50{,}000x_5 + 100{,}000x_6 \leq 450{,}000 \tag{b}$$

Second year expenditure:

$$g_2 = 0x_1 + 300{,}000x_2 + 200{,}000x_3 + 100{,}000x_4 + 300{,}000x_5 + 200{,}000x_6 \leq 400{,}000 \tag{c}$$

Engineering hours:

$$g_3 = 4000x_1 + 7000x_2 + 2000x_3 + 6000x_4 + 3000x_5 + 600x_6 \leq 10{,}000 \tag{d}$$

Production line is required:

$$g_4 = x_1 + x_2 \leq 1 \tag{e}$$

Automation is available only with new line:

$$g_5 = x_2 - x_3 \geq 0 \tag{f}$$

Waste recovery option:

$$g_6 = x_5 + x_6 \leq 1 \tag{g}$$

The branch and bound analysis begins by solving the LP problem with no integer restrictions on the variables, with the following result:

$$x_1 = 0.88$$
$$x_2 = 0.12$$
$$x_3 = 0.12$$
$$x_4 = 0.40$$
$$x_5 = 1.00$$
$$x_6 = 0.00$$
$$f = \$265{,}200$$

Note that several variables (x_1, x_2, x_3, x_4) in this solution are not integers. Based on this result, we set $x_1 = 1$ and branch using the x_4 variable. The branch and bound analysis was carried out using LINDO (Schrage, 1984). The final (optimal) integer solution is:

$$x_1 = 1$$
$$x_2 = 0$$
$$x_3 = 0$$
$$x_4 = 0$$
$$x_5 = 1$$
$$x_6 = 0$$
$$f = \$225{,}000$$

This indicates that the project 2 with the highest net present value is *not* selected because of the constraints in the problem. Note that the first noninteger solution achieves a larger value of f than the integer solution, as is expected.

9.2.3 Nonlinear Mixed-Integer Programming Algorithms

Most of the techniques that have been proposed for nonlinear discrete optimization are centered around one or more of the following five basic concepts: (1) rounding-off the continuous optimum, (2) adaptation of nonlinear optimization techniques, (3) linear approximation, (4) binary representation of variables, (5) direct search. The relative effectiveness of any technique is quite problem-dependent and no single procedure can claim a uniform advantage over all others for all problems, or even claim to be generally effective.

The most common approach to solving nonlinear discrete-value problems in practice has been to treat the variables as continuous ones. Once the continuous optimum has been determined by one of the methods outlined in earlier chapters, you select a feasible set of values of the discrete variables near the optimal point for the continuous variables. The selected point may not represent the discrete optimum and be misleading (Glover and Sommer, 1975) but frequently represents a quite suitable solution for practical purposes.

In the second approach, nonlinear discrete-valued problems are treated as nonlinear programming problems in which the discreteness of the variables is one of the restrictions. In the third approach, linear programming is used in conjunction with integer programming with any nonlinear functions being linearized. The fourth approach involves the transformation of a nonlinear function into a polynomial function of binary variables and then the transformation of the polynomial function into a linear function of binary variables such as proposed by Watter (1967), Glover and Woolsey (1973, 1974) and Garfinkel and Nemhauser (1972).

As to the last approach, discrete search techniques have not been demonstrated to be particularly reliable. The method most commonly referred to as sectioning search (Beveridge and Schechter, 1970) is known to be inefficient and may not find even a local optimum if it encounters a sharp ridge or valley in the objective function surface. Table 9.4 lists a few techniques that have been proposed in the literature for nonlinear problems. Of the various methods, only branch and bound has, over the past 20 years, been incorporated in commercial codes, but even it has limitations. In tests of linear integer programming codes (Lin and Rardin, 1980; Land and Powell, 1973), branch and bound and cutting plane methods (adding narrower constraints at each stage to reduce the feasible region) seem to be best. Crowder et al. (1983) reported that large-scale 0-1 MIP problems can be solved by a combination of problem preprocessing, clever branch and bound techniques, and cutting planes.

Table 9.4 Algorithms for Nonlinear Integer Programming

Classification	Reference	Procedure
0-1 variables only	Lawler and Bell (1966)	Partial enumeration; branch and bound
	Ross, G. T. (1975)	Branch and bound
Integer variables only		
Separable objective function and constraints	Morin (1978)	Survey of dynamic programming
	Morin and Marsten (1976)	Branch and bound
	Yormark (1979)	Recursive enumeration
	Cooper (1973)	Dynamic programming
	Aust (1976)	Branch and bound plus dynamic programming
Other problems	Faaland (1974)	Partial enumeration plus cutting planes
	Kunzi and Oettli (1973)	Approximate method via cutting planes
	Reiter and Sherman (1975)	Random search plus directed search
	Gisvold and Moe (1972)	Discretized penalty function with weighting factors
Mixed integer variables		
Quadratic objective function, linear constraints	Balas (1969)	Based on duality theory
Linear fractional (ratios)	Chandra and Chandramohan (1979)	Branch and bound
General	Gorry et al. (1972)	Relaxation (subdivision of regions)
	Kamaluddin and Bugess (1967)	Modified Lagrange multiplier technique
	Weinstein and Yu (1972)	Lagrange multiplier plus dynamic programming
	Geoffrion (1972)	Generalized Bender's decomposition
	Duran and Grossman (1986)	Outer-approximation algorithm

In conclusion, a number of types of integer programming problems can be solved by specialized computer codes. Small dimensional problems can be solved by enumerative techniques such as branch and bound. But often problems that are superficially amenable to integer programming cannot be solved in a reasonable amount of time. And, perhaps more importantly, one mathematical formulation may lead to very quick solutions while another formulation of the same physical problem may result in very slow execution, or be even virtually unsolvable. To know when and how to avoid incorrect problem formulations that are intractable requires considerable integer programming skill.

REFERENCES

Aust, R. J., "A Dynamic Programming Branch and Bound Algorithm for Pure Integer Programming," *Comput. Oper. Res.*, **3**: 17 (1976).

Aris, R. J., *The Optimal Design of Chemical Reactors*, Academic Press, New York (1961).

Balas, E., "Duality in Discrete Programming: The Quadratic Case," *Manage. Sci.*, **16**: 14 (1969).

Bellman, R. E., *Dynamic Programming*, Princeton University Press, Princeton, New Jersey (1957).

Bellman, R. E., and S. E. Dreyfus, *Applied Dynamic Programming*, Princeton University Press, Princeton, New Jersey (1962).

Bellman, R. E., and E. S. Lee, "History and Development of Dynamic Programming," *Control Systems Magazine*, p. 24, (November 1984).

Beveridge, G. S., and R. S. Schechter, *Optimization: Theory and Practice*, McGraw-Hill (1970).

Chandra, S., and M. Chandramohan, "An improved Branch and Bound Method for Mixed Integer Fractional Programming, *ZAMM*, **59**: 575 (1979).

Cooper, M. W., "An Improved Algorithm for Non-Linear Integer Programming," in *Computers and Mathematics with Applications*, p. 128 (1973).

Crowder, H., E. L. Johnson, and M. W. Padberg, "Solving Large Scale Zero-One Linear Programming Problems," *Oper. Res.*, **31**: 803 (1983).

Duran, M. A., and I. Grossmann, "A Mixed-Integer Nonlinear Programming Algorithm for Process Systems Synthesis," *AIChE J.*, **32**(4): 592 (1986).

Faaland, B., "An Integer Programming Algorithm for Portfolio Selection," *Manage. Sci.*, **20**: 1376 (1974).

Garfinkel, R. S., and G. L. Nemhauser, *Integer Programming*, John Wiley and Sons (1972).

Geoffrion, A. M., "Generalized Benders Decomposition," *JOTA*, **10**(4): 237 (1972).

Gisvold, K. M., and J. Moe, "A Method for Nonlinear Mixed-Integer Programming and Its Application to Design Problems," *ASME Transactions on Journal of Engineering for Industry*, **94**(2): 353 (1972).

Glover, F., and D. Sommer, "Pitfalls of Rounding in Discrete Management Decision Problems," *Decision Sci.*, **22**: 42 (1975).

Glover, F., and R. E. Woolsey, "Further Reduction of Zero-One Polynomial Programming Problems to Zero-One Linear Programming Problems," *Oper. Res.*, **21**: 156 (1973).

Glover, F., and R. E. Woolsey, "Converting the Zero-One Polynomial Programming Problem to a Zero-One Linear Program," *Oper. Res.*, **22**: 180 (1974).

Gorry, G. A., J. F. Shapiro, and L. A. Wolsey, "Relaxation Methods for Pure and Mixed-Integer Programming Problems," *Manage. Sci.*, **18**(5) 413, (1972).

Hillier, F. S., and G. J. Lieberman, *Introduction to Operations Research*, 4th ed., Holden-Day, San Francisco, California, p. 582 (1986).

Kamaluddin, B. A., and A. R. Bugess, "Economic Design of Discrete Physical Systems Subject to Fluctuating Environment," *J. Ind. Eng.*, **18**: 58 (1967).

Kunzi, H. P., and W. Oettli, "Integer Quadratic Programming," in *Recent Advances in Mathematical Programming*, (eds. R. Graves and P. Wolfe), McGraw-Hill, New York (1973).

Land, A. H., and S. Powell, *Fortran Codes for Mathematical Programming: Linear, Quadratic and Discrete*, Wiley, New York (1973).

Lawler, E. L., and M. D. Bell, "A Method for Solving Discrete Optimization Problems," *Oper. Res.*, **14**: 1098 (1966).

Lin, B. W., and R. L. Rardin, "Controlled Experimental Design for Statistical Comparison of Integer Programming Algorithms," *Manage. Sci.*, **25**: 1258 (1980).

Little, J. D. C., K. G. Murty, D. W. Sweeney, and C. Karel, "An algorithm for the Traveling Salesman Problem," *Oper. Res.*, **11**: 972 (1963).

Mitten, L. G., and G. L. Nemhauser, "Multistage Optimization," *Chem. Eng. Prog.*, **59**(1): 53 (1963).

Morin, T. L., "Computational Advances in Dynamic Programming," *Dynamic Programming and Its Applications* (ed. M. L. Puterman), Academic Press, New York, 53 (1978).

Morin T. L., and R. E. Marsten, "Branch and Bound Strategies for Dynamic Programming," *Oper. Res.*, **24**: 611 (1976).

Murray, J. E., and T. F. Edgar, "Optimal Scheduling of Production and Compression in Gas Fields," *J. Petrol. Technol.*, **3**: 109 (January, 1978).
Plane, D. R., and C. MacMillan, *Discrete Optimization*, Prentice-Hall, Englewood Cliffs, New Jersey (1971).
Reiter, S., and G. Sherman, "Discrete Optimizing," *J. Soc. Ind. Appl. Math.*, **13**: 214 (1975).
Roberts, S. M., *Dynamic Programming in Chemical Engineering and Process Control*, Academic Press, New York (1964).
Rosenwald, G. W., and D. W. Green, "A Method for Determining the Optimal Location of Wells," *Soc. Petrol. Eng. J.*, 44 (February 1974).
Ross, G. T., "A Branch and Bound Algorithm for the Generalized Assignment Problem," *Math. Program.*, **8**: 91 (1975).
Salkin, H. M., *Integer Programming*, Addison-Wesley, Reading, Massachusetts (1975).
Schrage, L., *Linear, Integer, and Quadratic Programming with LINDO*, The Scientific Press, Palo Alto, California, (1984).
von Randow, R., *Integer Programming and Related Areas: A Classified Bibliography 1981–1987*, Springer-Verlag, Berlin (1985).
Watter, L. J., "Reduction of Integer Polynomial Programming Problems to Zero-One Linear Programming Problems," *Oper. Res.*, **15**: 1171 (1967).
Weinstein, I. J., and S. O. Yu, "Comment on an Integer Maximization Problem," *Oper. Res.*, **21**: 648 (1972).
Wilde, D. J., "Strategies for Optimizing Macrosystems," *Chem. Eng. Prog.*, **65**(3), 86 (1965).
Yormark, J. S. Efficient Frontiers for Nonlinear Multidimensional Knapsack Problems, Graduate School of Business Administration, University of Southern California, (March, 1979).

SUPPLEMENTARY REFERENCES

Dynamic programming

Aris, R., G. L. Nemhauser, and D. J. Wilde, "Optimization of Multistage Cyclic and Branching Systems by Serial Procedures," *AIChE J.*, **10**: 913 (1964).
Beightler, C. S., and W. L. Meier, "Design of an Optimum Branched Allocation System," *Ind. Eng. Chem.*, **60**(2): 44 (1968).
Bertele, U., and F. Brioschi, *Nonserial Dynamic Programming*, Academic Press, New York (1972).
Cooper, L., and M. W. Cooper, *Introduction to Dynamic Programming*, Pergamon Press Inc., New York (1981).
Dolcetta, I. C., W. H. Fleming, and T. Zolezzi (eds.), *Recent Mathematical Methods in Dynamic Programming*, Wiley, New York (1985).
Dranoff, J. S., L. G. Mitten, W. F. Stevens, and L. A. Wanniger, "Application of Dynamic Programming to Countercurrent Flow Processes," *Oper. Res.*, **9**: 388 (1961).
Dreyfus, S. E., and A. M. Law, *The Art and Theory of Dynamic Programming*, Academic Press, New York (1977).
Larson, R. E., "A Survey of Dynamic Programming Computational Procedures," *IEEE Trans.*, **AC-12**: 767 (1967).
Lee, W., "Dynamic Optimization of Catalyst Make-Up Rate for Catalytic Systems," *Ind. Eng. Chem. Process Des. Dev.*, **9**: 154 (1970).
Morin, T. L., "Computational Advances in Dynamic Programming," in *Dynamic Programming and Its Applications*" (ed. M. L. Puterman), Academic Press, New York (1978).
Serji, I., and L. I. Stiel, "Optimal Design of Multiple-Effect Evaporators with Vapor Bleed Streams," *Ind. Eng. Chem. Process Des. Dev.*, **7**: 6 (1968).
Thomas, M. E., "A Survey of Dynamic Programming," *AIIE Trans.*, **8**(1): 59 (1976).
Van Cauenberghe, A. R., "Optimization of Multistage Recycle System," *Chem. Eng. Sci.* **22**: 193 (1967).

Wang, C. L., "The Principle and Models of Dynamic Programming," *J. Math. Anal. Appl.*, **118**: 287 (1986).

Wong, F. C., and M. Chandrashekar, "Application of Dynamic Programming to Systems with Structures," in *Large Engineering Systems* (ed. A. Wexler), Pergamon Press, Oxford, (1977).

Wong, P. J., and R. E. Larson, "Optimization of Natural-Gas Pipeline Systems Via Dynamic Programming," *IEEE Trans.*, **AC-13**: 475 (1968).

Mixed integer and integer programming

Beale, E. M. L., "Integer Programming" in *The Stage of the Art in Numerical Analysis* (ed. D. Jacobs), Academic Press, New York (1977).

Benichou, M., J. Gauthier, P. Girodet, G. Hentges, G. Ribiere, and O. Vicent, "Experiments in Mixed-Integer Linear Programming," *Math. Prog.*, **1**: 76 (1971).

Brown, G. C., and G. W. Graves, "Real-Time Dispatch of Petroleum Tank Trucks," *Man. Sci.*, **27**: 19 (1981).

Burdick, G. R., and D. M. Rasmuson, "Optimization of a Nonlinear System Consisting of Weakly Connected Subsystems," *Ind. Eng. Chem. Process Des. Dev.*, **19**: 195 (1980).

Dakin, R. J., "A Tree-Search Algorithm for Mixed-Integer Programming Problems," *Comput. J.*, **8**(3): 250 (1965).

Drezner, Z., and G. Gavish, "ε-Approximations for Multidimensional Weighted Location Problems," *Oper. Res.* **33**: 772 (1985).

Federgruen, A., and H. Groenevelt, "Optimal Flows in Networks with Multiple Sources and Sinks, with Applications to Oil and Gas Lease Investment Programs", *Oper. Res.*, **34**, 218 (1986).

Fisher, M. L., "The Lagrangian Relaxation Method for Solving Integer Programming Problems," *Manage. Sci.*, **27**: 581 (1981).

Glover, F., "Improved Linear Integer Programming Formulations of Nonlinear Integer Problems," *Manage. Sci.*, **22**: 455 (1975); Han, S. P., *Math. Program.*, **20**: 1 (1981).

Greenberg, H., *Integer Programming*, Academic Press, New York (1971).

Grossmann, I. E., "Mixed-Integer Programming Approach for the Synthesis of Integrated process Flowsheets," *Comput. Chem. Eng.*, **9**: 463 (1985).

Gupta, O. K., and A. Ravindran, "Branch and Bound Experiments in Convex Nonlinear Programming," *Manage. Sci.*, **31**: 1533 (1985).

Hansen, E., and G. W. Walster, "Global Optimization in Nonlinear Mixed-Integer Problems," *Proceed. 10th IMACS World Congress, Montreal*, August, 1982, 379 (1982).

Lazimy, R., "Mixed-Integer Quadratic Programming," *Math. Program.*, **22**: 332 (1982).

Rosenwald, G. W., and D. W. Green, "A Method of Determining Optimum Location of Wells in a Reservoir Using Mixed-Integer Programming," *Soc. Petrol. Eng. J.*, **253**: 44 (1974).

Shimizu, Y., and T. Takamatsu, "Application of Mixed-Integer Linear Programming in Multiterm Expansion Planning under Multiobjectives," *Comput. Chem. Eng.* **9**: 367 (1985).

Schrivjer, A., *Theory of Linear and Integer Programming* John Wiley, New York (1986).

Taha, H. A., *Integer Programming*, Academic Press, New York (1975).

Thomas, G. S., J. C. Jennings, and P. Abbott, "A Blending Problem Using Integer Programming On-Line," *Math. Program. Study*, **9**: 30 (1978).

Watson-Gandy, C. D. T., "The Solution of Distance Constrained Mini-Sum Location Problems," *Oper. Res.* **33**: 784 (1985).

Williams, H. P., "Experiments in the Formulation of Integer Programming Problems," *Math. Program. Study*, **2**: 180 (1974).

PROBLEMS

Solve the following problems via Dynamic Programming.

9.1. For the purposes of planning you are asked to determine the optimal heat exchanger areas for the sequence of three exchangers as shown in Fig. P9.1.

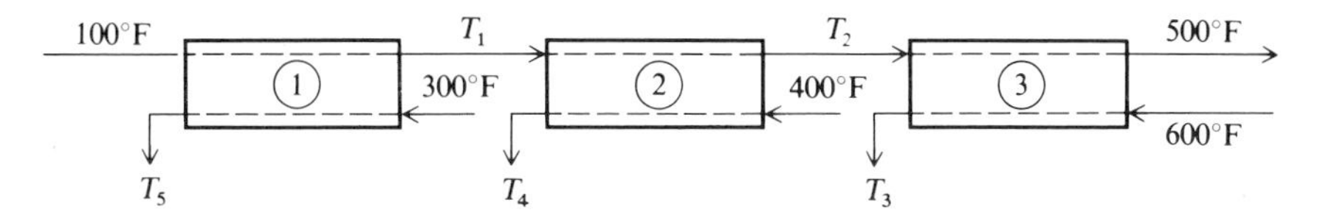

Figure P9.1

Data:

Exchanger	Overall heat transfer coefficient, U Btu/(h)(ft²)(°F)	Area required, ft²	Duty, Btu/h
1	$U_1 = 120$	A_1	Q_1
2	$U_2 = 80$	A_2	Q_2
3	$U_3 = 40$	A_3	Q_3

$wCp = 10^5$ Btu/(h)(°F)

Hint: Find the temperatures T_1, T_2, T_3 such that $\sum A_i$ is a minimum.

9.2. Cooling water is to be allocated to three distillation columns. Up to 8 million gallons per day are available, and any amount up to this limit may be used. The costs (the dollar amounts are negative) of supplying water to each column are

$$\text{Col. 1:} \quad f_1 = |1 - D_1| - 1 \qquad 0 \le D_1 \le 2$$
$$= 0 \qquad \text{otherwise}$$
$$\text{Col. 2:} \quad f_2 = -e^{-1/2(D_2-5)^2} \qquad 0 \le D_2 \le \infty$$
$$\text{Col. 3:} \quad f_3 = D_3^2 - 6D_3 + 8 \qquad 2 \le D_3 \le 4$$

Minimize $\sum f_i$ to find D_1, D_2, and D_3.
Use dynamic programming.

9.3. A proposed new method for desalting brackish water is as follows. The salt is to be adsorbed on to a patented solid adsorbent in a three-stage process shown in Fig. P9.3.

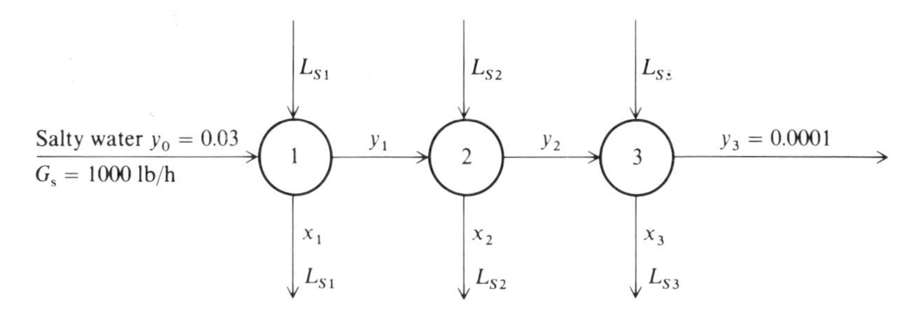

Figure P9.3

Determine the distribution of the weight in pounds per hour of pure adsorbent, L_{Si}, and y_1 and y_2 so as to minimize the total number of pounds of adsorbent used per hour.

Data:

y_i = lb salt/lb H_2O in the water stream

x_i = lb salt/lb adsorbent

L_{Si} = lb of pure adsorbent/h

Equilibrium follows the Freundlich equation: $y = X^2/100$

9.4. A mixture of solids of varying densities is to be separated in a series of 4 flotation cells. At the first stage, stage $n = 5$ in Fig. P9.4, it is assumed that the flotation agent is chosen so that as far as possible the material of greatest density will settle out and be removed as product while all lighter material will be retained on the surface and be sent to the next stage for further processing. In reality, because of factors such as nonuniform particles, an ideal separation will not be achieved even if equilibrium is achieved between the cell outputs.

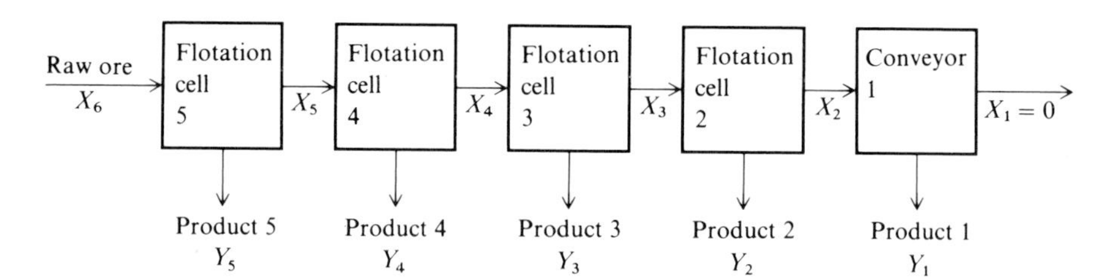

Figure P9.4

To represent the actual process, the operating characteristics of the stage will be represented by

$$\alpha_n p_{n,j} x_{n+1,j} = x_{n,j} \tag{a}$$

In each cell some of the ore is settled out and removed for sale; the remaining material is sent to the next stage for further processing. It is assumed that the operation of the system can be described by the model given in Eq. (*a*). Find the value of the operating variable α_n for each cell which will maximize the total return from the process. Also list the weight in pounds of each grade of material in the production from each of the five stages.

The composition of the raw ore is:

Grade	1	2	3	4	5
lb	2000	3000	5000	4000	2500

The product specifications and the market value of the products are:

Product Y_n removed at stage	Specifications	Market value ($/100 lb total material)
1	At least 70% grade 1, no more than 20% grade 2, 4% grade 3, no grade 4 or 5	80
2	At least 50% grade 2, no more than 20% grade 3, 2% grade 4, no grade 5	30
3	At least 50% grade 3, no more than 10% grade 4 and traces of grade 5	10
4	At least 40% grade 4, no more than 5% grade 5	8
5	At least 40% grade 5	5

The flotation cells have been designed to meet the specifications given above, provided that the raw ore is processed through every cell in the order shown in the flow-sheet. The constants $p_{n,j}$ for the respective cells and grades are:

	Grade				
Cell	1	2	3	4	5
1	0	0	0	0	0
2	0.9	0.2	0.1	0	0
3	0.95	0.8	0.2	0.1	0
4	1	0.9	0.8	0.2	0.05
5	1	1	0.9	0.8	0.1

The operating costs for the cells are functions of the cell inputs and α's:

Cell	Operating costs
1	0 (conveyor)
2	$\left(\dfrac{2\alpha_2}{1-\alpha_2}\right)\dfrac{X_3}{100}$
3	$\left(\dfrac{\alpha_3}{1-\alpha_3}\right)\dfrac{X_4}{100}$
4	$-0.2\,\dfrac{X_5}{100}\ln(1-\alpha_4)$
5	$-0.1\,\dfrac{X_6}{100}\ln(1-\alpha_5)$

α_n is the operating variable for the nth cell, the efficiency $0 \le \alpha_n \le 1$. $X_n = \sum_{j=1}^{5} x_{n,j}$ is the total amount of feed leaving the nth cell. Similarly, $Y_n = \sum_{j=1}^{5} y_{n,j}$ is the total amount of product removed from the nth cell.

Other notation:

j = index for components, i.e., grade $j = 1, 2, 3, 4, 5$
n = index for stage number, $n = 1, 2, 3, 4, 5$
$p_{n,j}$ = operating constant for cell, a given constant
$\mathbf{X}_i$ = vector, as for example

$$\mathbf{X}_1 = \begin{bmatrix} x_{1,1} \\ x_{1,2} \\ x_{1,3} \\ x_{1,4} \\ x_{1,5} \end{bmatrix} \quad \text{the vector of components or grades for the first-stage output}$$

9.5. This problem is taken from Chap. 5 of the Ph.D. Dissertation of G. L. Nemhauser, Northwestern University, August, 1961.

A chemical company is faced with making several design and operational decisions in regard to the production of its new product. The flowsheet given in Fig. P9.5 represents the basic method of production. Two reactors in series are shown on the flow-sheet. However, the second one may not be needed, depending upon the degree of conversion achieved in the first. The raw materials are mixed, heated to reaction temperature, converted to the desired product, and the product is separated from the remaining raw materials and the waste material produced in the reactors.

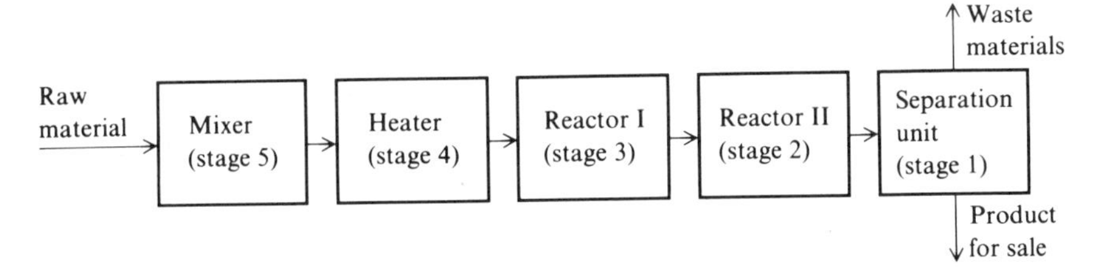

Figure P9.5

The measure of effectiveness upon which the best alternative will be chosen is maximum net profit over a five-year-period of operation. We define net profit to be:

Net profit = (profit from sale of products) − (operating costs + cost of equipment)

Net profit = (gross profit) − (total costs)

The design is based on 50,000 lb of raw material processed per year.

The expected selling price of the product will depend upon how much is produced. The demand is elastic, and greater production will cause a lowering of prices. Table P9.5*a*, the results of a market analysis, gives the expected selling price of the product for given production rates. The production rates are all based on 50,000 lb of raw material processed per year. They depend upon the degree of conversion achieved in the reactors.

Table P9.5*a* Expected selling prices of the product

1000 lb/year produced	47.5	45.0	42.5	40.0	37.5	30.0	25.0	22.5	20.0	15.0
Expected selling price ($/lb)	3.2	3.3	3.4	3.6	3.8	4.6	5.0	5.2	5.3	5.5

Three types of mixers are being considered. Mixers *A*, *B*, and *C* cost 10, 15, and 25 thousand dollars, respectively. Table P9.5*b* gives the operating costs for the mixing operation as a function of the mixing efficiency. In the following tables all costs are given in $1000 units, and operating costs are on a yearly basis.

Table P9.5*b* Annual operating costs for the mixer (in 1000's of $)

	(Output) mixing efficiency				
Mixer	1.0	0.8	0.6	0.5	Fixed cost of mixer
A	12.0	6.0	3.0	2.0	10
B	8.0	4.0	2.5	1.5	15
C	5.0	3.0	2.0	1.0	25

The costs involved in heating the feed to a given reaction temperature depend upon the mixing efficiency. Temperatures of above 700 degrees are very difficult to achieve. A specially designed heater which costs $20,000 must be used for these temperatures. A standard heater for temperatures of 700 degrees or less costs $5000. Table P9.5*c* gives the operating costs to attain the feasible reaction temperatures as a function of the mixing efficiency.

Table P9.5*c* Annual heater operating costs

Mixing efficiency (input)	Temperature (output)			
	650°	700°	750°	800°
1.0	0.5	1.0	6.0	10.0
0.8	1.0	1.5	8.0	12.0
0.6	1.5	2.5	10.0	16.0
0.5	2.0	3.0	12.0	20.0

Reactor I converts a portion of the raw materials to product and waste material. Stoichiometrically, one pound of raw material can be converted into one pound of product. Table P9.5*d* gives the percent conversion achieved in reactor I as

Table P9.5*d* Percent conversion in reactor I

Reactor	Temperature 650°		700°		750°		800°		5-year reactor and catalyst costs
	Catalyst 1	2	1	2	1	2	1	2	
I_A	30	25	40	30	50	45	60	50	I_A ① 110, ② 80
I_B	25	20	30	25	45	40	50	45	I_B ① 80, ② 50
I_C	20	15	25	20	40	30	45	40	I_C ① 60, ② 30

a function of temperature, catalyst, and the type of reactor used. Percent conversion refers to the percentage of raw material converted to product on a weight basis (e.g., 50% conversion means for every pound of raw material we get $\frac{1}{2}$ lb of product).

Reactor I_A costs \$40,000 with \$4000/year operating expense. Reactor I_B costs \$20,000 with \$2000/year operating expense. Catalyst 1 costs \$10,000/year and catalyst 2 costs \$4000/year.

Reactor II is used to increase the percent conversion. Since the market analysis indicated that a minimum of 15,000 lb of product should be produced yearly, the minimum acceptable conversion from the two reactors is 30%. Thus, if 30% conversion has been achieved in reactor I, the second reactor may be eliminated. Table P9.5*e* gives the percent conversion from the two reactors as a function of the conversion achieved in the first reactor and the possible designs for the second reactor.

Table P9.5*e* Total conversion from the two reactors

Total percent conversion	Conversion in reactor I: 15	20	25	30	40	45	50	60	5-year reactor cost
II_A	30	40	50	60	80	85	90	95	110
II_B	45	60	75	85	90	95	95	95	180

Reactor II_A costs \$60,000 with \$10,000/year operating expense. Reactor II_B costs \$80,000 with \$20,000/year operating expense.

Finally, the product is separated from the waste and raw materials. Completely pure product is removed from the separation unit. Design alternatives for the

Table P9.5f Costs for the separation unit

	One large separator			Two small separators — Per separator		
Percent conversion	Fixed cost	Operating cost	5-year cost	Fixed cost	Operating cost	5-year cost
30	12	2.5	24.5	7.5	1.5	30.0
40	12	3.0	27.0	7.5	1.5	30.0
45	12	4.0	32.0	7.5	1.5	30.0
50	15	4.0	35.0	9.0	1.5	33.0
60	15	5.0	40.0	9.0	2.0	38.0
75	20	6.0	50.0	12.0	2.0	44.0
80	20	6.5	52.5	12.0	2.0	44.0
85	20	7.0	55.0	12.0	2.0	44.0
90	20	7.5	57.5	12.0	2.5	49.0
95	20	8.0	60.0	12.0	3.0	54.0

separation unit are either one large separator or two smaller ones. The design of the separators and hence their costs depend upon the percent conversion achieved in the reactors. Table P9.5f gives the fixed and operating costs for the separators as a function of the percent conversion in the reactors.

You are asked to determine the best design and the optimal operating conditions of all the units in the design using dynamic programming.

9.6. A pipeline is to be built between an oil field and a refinery. The distance to be traversed is great enough to require that intermediate pumping stations be built. A number of feasible sites have been located for these stations (see Fig. P9.6). While the sites are distributed over an area and are not on a straight-line path, they are desirable because of ready access to electric power, low land cost, etc. To standardize the pipeline and lower costs through large purchases, only five pump sizes may

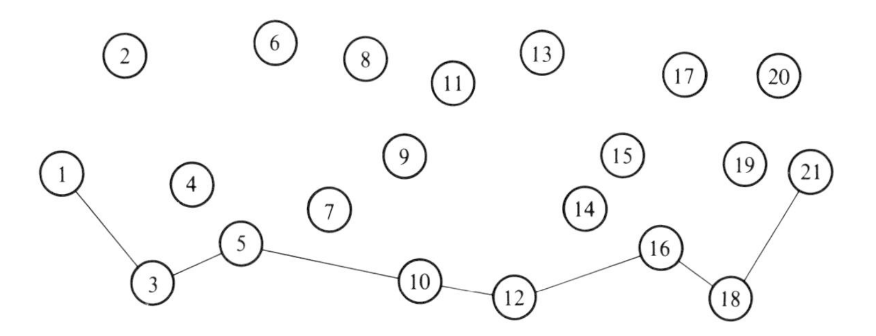

Figure P9.6 Site locations used as data with optimum route indicated (not to scale).

Table P9.6*a*   Distances† between pairs of sites

To \ From	1	2	3	4	5	6	7	8	9	10	11	12	13	14	15	16	17	18	19	20
2	35																			
3	62	60																		
4	75	51	40																	
5	95	82	40	37																
6	91	57	73	33																
7	122	107	73	53	33	64														
8	125	93	85	50	58	41	30													
9	149	121	103	75	68	72	30	30												
10	160	142	102	92	65	103	42	70												
11	161	129	125	91	92	72	60	40	34											
12	....	169	140	121	101	124	71	82	54	40	40									
13	....	162	149	126	112	109	75	70	47	71	70									
14	....	....	161	140	123	133	87	92	66	69	105	27	33							
15	....	....	175	150	140	140	104	100	78	91	95	54	30	28						
16	....	....	....	....	155	170	121	129	101	98	98	56	67	36	42					
17	....	....	....	....	168	168	131	130	105	116	139	75	60	48	28	37				
18	....	....	....	....	....	....	155	162	136	127	137	89	95	69	68	32	50			
19	....	....	....	....	....	....	163	168	140	139	154	99	99	76	68	42	43	17		
20	....	....	....	....	....	....	....	....	163	168	178	127	119	102	87	72	59	48	31	
21	....	....	....	....	....	....	....	....	181	180	172	140	140	119	109	86	82	53	43	30

† All distances are in miles.

be used and only five diameters of pipe are available, though each diameter comes in five thicknesses.

Determination of optimum pipeline route

The accounting department has broken down the costs of investment in land and equipment to yearly rates, while the surveying department has measured distances between sites. Distances are not shown where it would be infeasible to join the two sites. Tables P9.6(*a*) through P9.6(*e*) contain:

(*a*) Distances between pairs of sites
(*b*) Sizes of available pipes, sizes of available pumps
(*c*) The annual cost of pumps, the annual costs of the various pipes, the annual cost of land used for the piping
(*d*) Cost adjustment factors for various pairs of sites (which take into account the varieties of terrain between sites)
(*e*) Costs of piping sites on an annual basis

Develop a method to select the route so as to minimize the yearly cost. The unadjusted cost of land used for piping between sites: $10,000/year/mile.

Table P9.6*b* Size data

Number of sets: 20
Number of pipe diameters: 5
Number of pipe thicknesses: 5
Number of pump sizes: 5
Pipe constants, C (lb-ft^5/in^2-mile): 2.60, 4.10, 5.65, 7.25, 8.27
Pipe diameters, D(ft): 0.4, 0.6, 0.8, 1.0, 1.2
Pipe thicknesses, t(ft): 0.060, 0.065, 0.070, 0.075, 0.080
Pipe thickness constants, C_t (lb-ft/in^3): 6500, 6350, 6250, 6100, 6000
Pump pressure, Δp (psia): 100, 200, 300, 400, 500
Pump cost, C_{p_u} ($/year): 10,000, 23,000, 37,000, 53,000, 70,000.

Table P9.6*c* Cost of pipe (dollars per mile per year)

Diameter \ Thickness	0.060	0.065	0.070	0.075	0.080
0.4	6,700	7,100	7,600	8,000	8,400
0.6	10,300	10,900	11,600	12,300	12,900
0.8	13,800	14,700	15,900	16,700	17,600
1.0	17,400	18,800	20,200	21,300	22,500
1.2	21,800	23,300	24,600	26,000	27,400

Table P9.6*d* Cost adjustment factor for land use between sites

To \ From	1	2	3	4	5	6	7	8	9	10	11	12	13	14	15	16	17	18	19	20
2	1.12																			
3	1.02	1.25																		
4	1.31	1.25	1.25																	
5	1.04	1.22	1.04	1.25																
6	1.15	1.09	1.20	1.25																
7	1.30	1.13	1.07	1.30	1.07	1.21														
8	1.29	1.15	1.21	1.30	1.20	1.09	1.35													
9	1.40	1.25	1.21	1.35	1.21	1.30	1.35	1.35												
10	1.20	1.25	1.09	1.30	1.03	1.29	1.12	1.30	1.10											
11	1.20	1.13	1.21	1.30	1.21	1.04	1.15	1.20	1.30	1.07										
12	....	1.25	1.11	1.25	1.03	1.32	1.15	1.25	1.25	1.20										
13	....	1.13	1.19	1.30	1.20	1.06	1.20	1.25	1.30	1.09	1.05									
14	....	....	1.21	1.30	1.04	1.30	1.35	1.30	1.35	1.20	1.15	1.07	1.20							
15	....	....	1.25	1.30	1.20	1.06	1.35	1.30	1.35	1.08	1.04	1.15	1.09	1.20						
16	....	....	....	....	1.07	1.32	1.30	1.30	1.35	1.20	1.20	1.05	1.20	1.10	1.20					
17	....	....	....	....	1.20	1.06	1.35	1.30	1.35	1.07	1.04	1.15	1.09	1.19	1.15	1.15				
18	....	....	....	....	....	....	1.35	1.30	1.35	1.09	1.25	1.04	1.20	1.10	1.15	1.15	1.20			
19	....	....	....	....	....	....	1.35	1.30	1.35	1.15	1.25	1.12	1.20	1.12	1.15	1.15	1.20	1.25		
20	....	....	....	....	....	....	....	....	1.35	1.14	1.09	1.15	1.10	1.15	1.15	1.15	1.10	1.25	1.20	
21	....	....	....	....	....	....	....	....	1.35	1.14	1.15	1.09	1.15	1.12	1.15	1.15	1.19	1.20	1.35	1.20

Table P9.6e Annual costs of sites

Site	Cost, $	Site	Cost, $
1	43,000	11	50,000
2	40,000	12	27,000
3	37,000	13	59,000
4	75,000	14	62,000
5	36,000	15	62,000
6	35,000	16	47,000
7	55,000	17	37,000
8	55,000	18	38,000
9	65,000	19	75,000
10	40,000	20	42,000

Solve the following problems via the branch and bound method.

9.7. Maximize $f(\mathbf{x}) = 75x_1 + 6x_2 + 3x_3 + 33x_4$

Subject to $774x_1 + 76x_2 + 22x_3 + 42x_4 \leq 875$

$67x_1 + 27x_2 + 794x_3 + 53x_4 \leq 875$

x_1, x_2, x_3, x_4 either 0 or 1

9.8. Maximize $f(\mathbf{x}) = 2x_1 + x_2$

Subject to $x_1 + x_2 \leq 5$

$x_1 - x_2 \geq 0$

$6x_1 + 2x_2 \leq 21$

$x_1, x_2 \geq 0$ and integer

9.9. Minimize $f(\mathbf{x}) = x_1 + 4x_2 + 2x_3 + 3x_4$

Subject to $-x_1 + 3x_2 - x_3 + 2x_4 \geq 2$

$x_1 + 3x_2 + x_3 + x_4 \geq 3$

$x_1, x_2 \geq 0$ and integer

$x_3, x_4 \geq 0$

9.10. Determine the minimum sum of transportation costs and fixed costs associated with two plants and two customers based on the following data

Plant	Annual capacity (in thousands)	Annual fixed charges (in 10^4)
1	2	1
2	1	1

Customer (j)	Demand (j)
1	1
2	1

Plant (i)	Customer (j)	
	1	2
1	3	1
2	0	1

Hint: The mathematical statement is

$$\text{Minimize} \quad f(\mathbf{x}) = \sum_i \sum_j C_{ij}^T x_{ij} + \sum_i C_i^F y_i$$

$$\text{Subject to} \quad \sum_i x_{ij} = D_j \qquad j = 1, \ldots, n$$

$$\sum_j x_{ij} \le A_i \qquad i = 1, \ldots, m$$

$$x_{ij} - \min\{D_j, A_j\} \le 0 \qquad i = 1, \ldots, m \qquad j = 1, \ldots, n$$

$$y_i = 0, 1 \quad \text{(integers)} \qquad i = 1, \ldots, m$$

where C_{ij}^T = unit transportation cost from plant i to customer j
C_i^F = fixed cost associated with plant i
x_{ij} = quantity supplied to customer j from plant i
y_i = 1 (plant operates); = 0 (plant is closed)
A_i = capacity of plant i
D_j = demand of customer j

9.11. Four streams are to be allocated to four extractors. The costs of each stream are

	Extractor			
Stream	**1**	**2**	**3**	**4**
1	45	*P*	5	56
2	27	2	82	74
3	19	55	3	*P*
4	3	10	4	84

The symbol P means the transfer is prohibited. Minimize the total costs.

9.12. Minimize $f(\mathbf{x}) = 10x_1 + 11x_2$

Subject to $9x_1 + 11x_2 \ge 29$

$\mathbf{x} \ge 0$ and integer

PART III

APPLICATIONS OF OPTIMIZATION

This section of the book is devoted to representative applications of the optimization techniques presented in Chaps. 4 through 9. Chapters 10 through 14 include five major application areas:

1. Heat transfer and energy conservation (Chap. 10)
2. Separations (Chap. 11)
3. Fluid flow (Chap. 12)
4. Reactors (Chap. 13)
5. Process plants (Chap. 14)

In each chapter we review examples of optimization that can be found in the literature for the specific applications area, as well as present several detailed studies illustrating various techniques. The matrix on the following page shows the classification of the examples with respect to specific techniques. Truly optimal design of process plants cannot be performed by considering each unit operation separately. Hence, in Chap. 14 we discuss the optimization of large-scale plants including those represented by flowsheet simulators.

We have not included any homework problems in Chaps. 10 through 14. As a general suggestion for classroom use, parameters or assumptions in each example can be changed to develop a modified problem. Students can also change the numerical method employed, particularly for one-dimensional search, unconstrained optimization, and nonlinear programming, to achieve a variety of problem variations.

Classification of optimization applications (Section or Example number is in parentheses)

Method	Chapter 10	11	12	13	14
Analytical solution	Waste heat recovery (10.1)		Pipe diameter (Example 12.1)	Residence time, batch reactor (13.2; Example 13.1)	
One-dimensional search	Multistage evaporator (10.4)	Reflux ratio of distillation column (11.2; Example 11.4)	Fixed-bed filter (Example 12.3)	Bioreactor (13.2, Example 13.2)	
Unconstrained optimization	Heat exchanger network (10.3)	Nonlinear regression of VLE data (11.2; Example 11.3)		Reactor holding time (13.4, Example 13.4)	

Linear programming	Boiler/ turbo generator system (10.5)	Separation train (11.2, Example 11.5)		Thermal cracker (13.3, Example 13.3)	
Nonlinear programming	Heat exchanger (10.2)	Liquid extraction column (11.1, Example 11.1) Staged distillation column (11.2; Example 11.2)	Gas transmission network (Example 12.4)	Reactor yield (13.4; Example 13.4) Ammonia reactor (13.4; Example 13.5) Alkylation reactor (13.4; Example 13.6)	Refrigeration process (14.2; Example 14.1) Distillation column (14.2; Example 14.2) Propylene chlorination (14.3; Examples 14.3, 14.4)
Dynamic programming			Gas compressor (Example 12.2)		
Discrete variable	Heat exchanger (10.2)		Gas transmission network (Example 12.4)		

CHAPTER 10

HEAT TRANSFER AND ENERGY CONSERVATION

A variety of energy conservation measures are available that can be adopted to optimize energy usage throughout a chemical plant or refinery. The following is a representative list of design or operating factors related to heat transfer and energy utilization that can involve optimization:

1. Fired heater combustion controls
2. Heat recovery from stack gases
3. Fired heater convection section cleaning
4. Heat exchanger network configuration
5. Extended surface heat exchanger tubing to improve heat transfer
6. Scheduling of heat exchanger cleaning
7. Air cooler performance
8. Fractionating towers—optimal reflux ratio, heat exchange, etc.
9. Instrumentation for monitoring of energy usage
10. Reduced leakage in vacuum systems and pressure lines, condensers
11. Cooling water savings
12. Efficient water treating for steam raising plants
13. Useful work from steam pressure reduction
14. Steam traps, tracing, and condensate recovery
15. CO boilers on catalytic cracking units
16. Electrical load leveling
17. Power factor improvement
18. Power recovery from gases or liquids
19. Loss control in refineries
20. Catalyst improvements

Many of the conservation measures require detailed process analysis plus optimization. For example, the efficient firing of fuel (category 1) is extremely important in all applications. For any rate of fuel combustion, a theoretical quantity of air (for complete combustion to carbon dioxide and water vapor) exists under which the most efficient combustion will occur. Reduction of the amount of air available leads to incomplete combustion and a rapid decrease in efficiency. In addition, carbon particles may be formed which can lead to accelerated fouling of heater tube surfaces. To allow for small variations in fuel composition and flow rate and in air flow rates that are inevitable in industrial practice, it is usually desirable to aim for operation with a small amount of excess air, say 5 to 10 percent, above the theoretical amount for complete combustion. However, too much excess air leads to increased sensible heat losses through the stack gas.

In practice, the efficiency of a fired heater is controlled by monitoring the oxygen concentration in the combustion products in addition to the stack gas temperature. Dampers are used to manipulate the air supply. By tying the measuring instruments into a feedback loop with the mechanical equipment, optimization of operations can take place in real-time to account for variations in the fuel flow rate or heating value.

As a second example (category 4), a typical plant will contain large numbers of heat exchangers used to transfer heat from one process stream to another. It is important to continue to utilize the heat in the streams efficiently throughout the process. Figure 10.1 shows a simplified flow diagram of a crude distillation unit. The incoming crude oil is heated against various product and reflux streams before entering the fired heater in order to be brought to the desired fractionating column flash zone temperature. Figure 10.1 indicates one particular flow scheme; a large number of other combinations of flow paths are obviously possible. In designing a plant unit two major considerations exist:

1. What should be the configuration of flows (the order of heat exchange for the crude oil)?
2. How much heat exchange surface should be supplied within the chosen configuration?

Additional heat exchange surface area leads to improved heat recovery in the crude oil unit but increases capital costs so that increasing heat transfer surface area soon reaches diminishing returns. The optimal configuration and areas selected, of course, are strongly dependent on fuel costs. As fuel costs rise, existing plants can usually profit from the installation of additional heat exchanger surface in circumstances previously considered only marginally economic.

As a final example (category 6), although heat exchangers may be very effective when first installed, many such systems become dirty in use and heat transfer rates deteriorate significantly. It is therefore often useful to establish optimal heat exchanger cleaning schedules. Although the schedules can be based on observations of the actual deterioration of the overall heat transfer of the exchanger in question, it is also possible to optimize the details of the cleaning schedules depending on an economic assessment of each exchanger.

In this chapter we illustrate the application of various optimization techniques in heat-transfer-system design. First we shall show how simple rules of thumb on boiler temperature differences can be derived (Sec. 10.1). Then a more complicated design of a heat exchanger is examined (Sec. 10.2) leading to a constrained optimization problem involving some discrete-valued variables. Next the optimization of a simple heat-exchanger network is treated (Sec. 10.3). Section 10.4 discusses the use of optimization in the design and operation of evaporators, and we conclude this chapter by demonstrating how linear programming can be employed to optimize a steam/power system (Sec. 10.5).

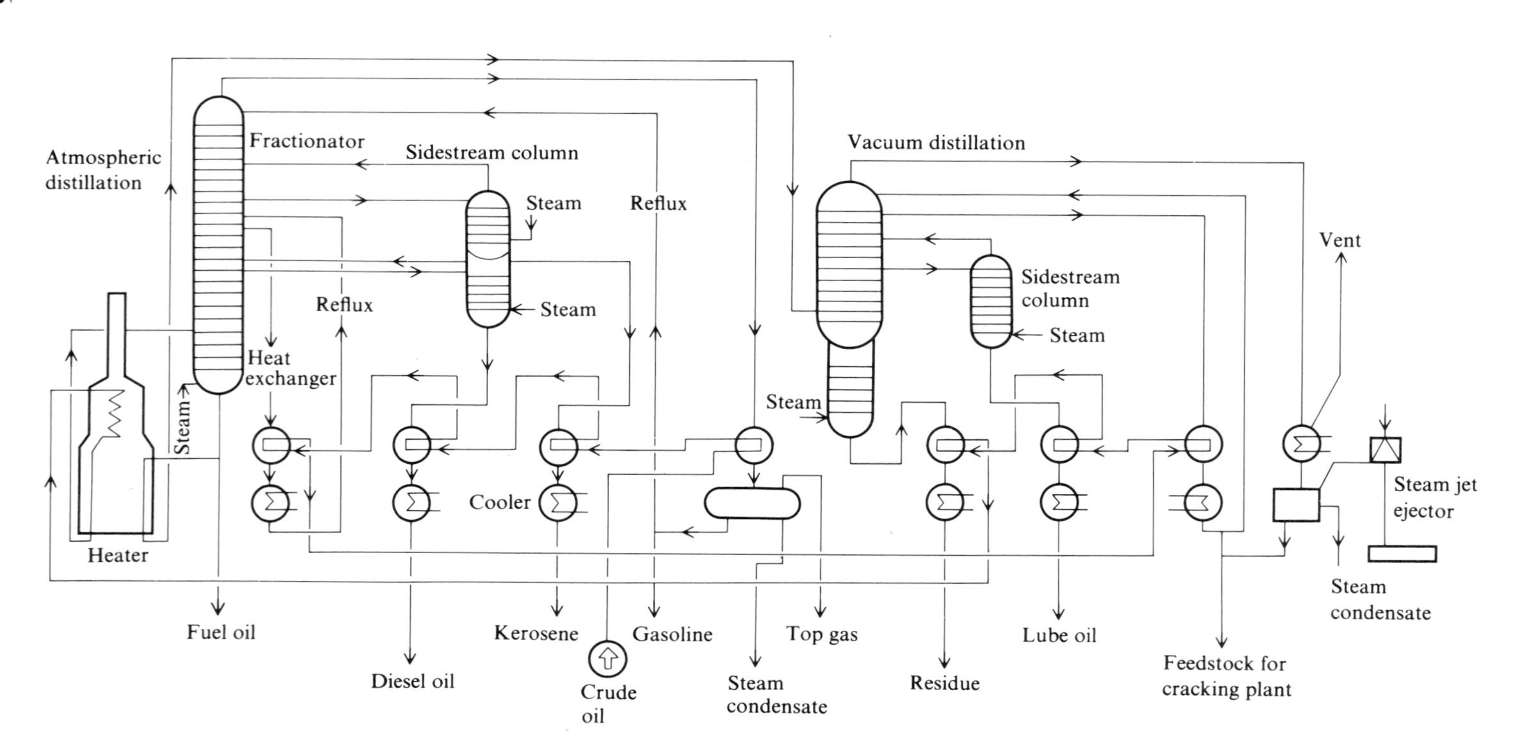

Figure 10.1 Typical crude oil distillation unit.

10.1 OPTIMIZING RECOVERY OF WASTE HEAT

A variety of sources of heat at elevated temperatures exist in a typical chemical plant that may be economically recoverable for production of power using steam or other working fluids such as freon or light hydrocarbons. Figure 10.2 shows a schematic of such a system. The system power output can be increased by using larger heat-exchanger surface areas for both the boiler and the condenser. However, there will be a trade-off between power recovery and capital cost of the exchangers. Sama (1983), Swearingen and Ferguson (1984), and Steinmeyer (1984) have proposed some simple rules of thumb based on analytical optimization of the boiler ΔT. You should recall that a simple optimal relationship for steam recovery was derived in Example 3.7.

In a power system the availability expended by any exchanger is equal to the net work that could have been accomplished by having each stream exchange heat with the surroundings through a reversible heat engine or heat pump. In the boiler in Fig. 10.2, heat is transferred at a rate Q (the boiler load) from the average hot fluid temperature T_s to the working fluid at T_H. The working fluid then exchanges heat with the condenser at temperature T_2. If we ignore mechanical friction and heat leaks, the reversible work available from Q at temperature T_s with the condensing (cold-side) temperature at T_2 is

$$W_1 = Q\left(\frac{T_s - T_2}{T_s}\right) \tag{10.1}$$

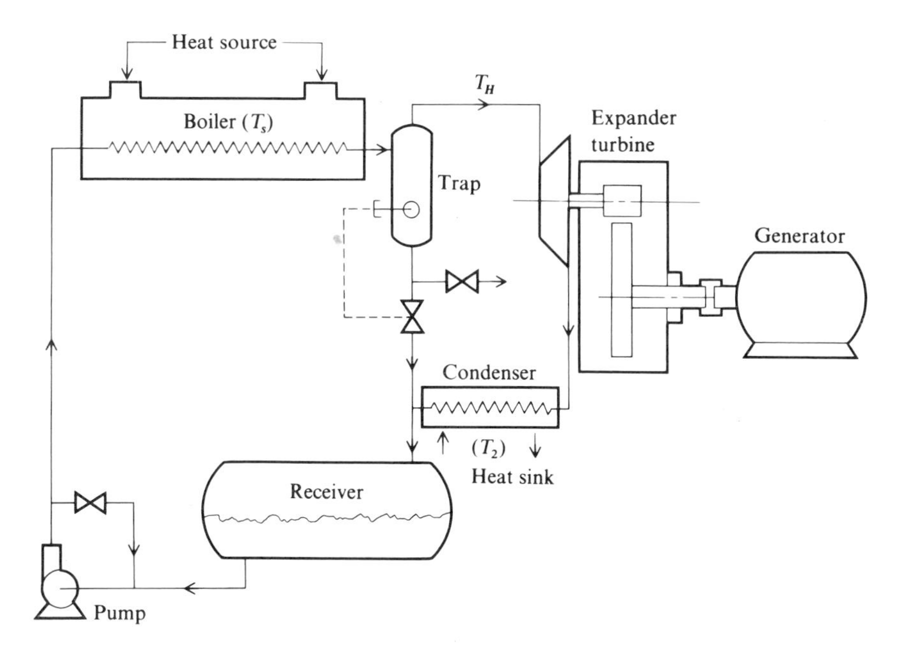

Figure 10.2 Schematic of power system.

The reversible work available from the condenser using the working fluid temperature T_H (average value) and the heat sink temperature T_2 is

$$W_2 = Q\left(\frac{T_H - T_2}{T_H}\right) \tag{10.2}$$

Hence the ideal power available from the boiler can be found by subtracting W_2 from W_1

$$W_2 - W_1 = \Delta W = Q\left(\frac{T_2}{T_H} - \frac{T_2}{T_s}\right) \tag{10.3}$$

In this expression T_s and T_2 are normally specified, and T_H is the variable to be adjusted. If Q is expressed in Btu/h, and the operating cost is C_{op}, then the value of the available power is

$$C_{op} = C_H \eta y Q\left(\frac{T_2}{T_H} - \frac{T_2}{T_s}\right) \tag{10.4}$$

where η = overall system efficiency (0.7 is typical), y = number of hours per year of operation, and C_H amalgamates the value of the power in \$/kWh and the necessary conversion factors to have a consistent set of units.

You can see, using Eq. (10.4) only, that C_{op} is minimized by setting $T_H = T_s$ (infinitesimal boiler ΔT). However, this outcome would increase the required boiler heat transfer area to an infinite area as can be noted from the calculation for the area

$$A = \frac{Q}{U(T_s - T_H)} \tag{10.5}$$

(In Eq. (10.5) an average value for the heat transfer coefficient U is assumed, ignoring the effect of pressure drop. U will depend on the working fluid and the operating temperature.) Let the cost per unit area of the exchanger be C_A and the annualization factor for capital investment be denoted by r. Then the annualized capital cost for the boiler is

$$C_c = \frac{C_A Q r}{U(T_s - T_H)} \tag{10.6}$$

Finally, the objective function to be minimized with respect to T_H, the working fluid temperature, is the sum of the operating cost and surface area costs:

$$f = C_H \eta y Q\left(\frac{T_2}{T_H} - \frac{T_2}{T_s}\right) + \frac{C_A Q r}{U(T_s - T_H)} \tag{10.7}$$

To get an expression for the minimum of f we differentiate Eq. (10.7) with respect to T_H and equate the derivative to zero to obtain

$$C_H \eta y Q\left(-\frac{T_2}{T_H^2}\right) + \frac{C_A Q r}{U(T_s - T_H)^2} = 0 \tag{10.8}$$

To solve the quadratic equation for T_H, let

$$\alpha_1 = C_H \eta y T_2 U$$

$$\alpha_2 = C_A r$$

Q cancels in both terms. On rearrangement, the resulting quadratic equation is

$$(\alpha_1 - \alpha_2)T_H^2 - 2\alpha_1 T_s T_H + \alpha_1 T_s^2 = 0 \tag{10.9}$$

The solution to (10.9) for $T_H < T_s$ is

$$T_H = T_s\left(\frac{\alpha_1 - \sqrt{\alpha_1 \alpha_2}}{\alpha_1 - \alpha_2}\right) \tag{10.10}$$

For a system with $C_A = \$25/\text{ft}^2$, a power cost of \$0.06/kWh ($C_H = 1.76 \times 10^{-5}$). $U = 95$ Btu/(h)(°F)(ft^2), $y = 8760$ h/year, $r = 0.365$, $\eta = 0.7$, $T_2 = 600$°R, and $T_s = 790$°R, the optimal value T_H is 760.7°R, giving a ΔT of 29.3°R. Swearingen and Ferguson (1984) showed that Eq. (10.8) can be expressed implicitly as

$$\Delta T = T_s - T_H = T_H\left(\frac{\alpha_1}{\alpha_2}\right)^{1/2} \tag{10.11}$$

In this form, it appears that the allowable ΔT increases as the working fluid temperature increases. This suggests that the optimum ΔT for a heat source at 400°F is lower than that for a heat source at 600°F. In fact, Eq. (10.10) indicates that the optimum ΔT is directly proportional to T_s. Sama (1983) argues that this is somewhat counterintuitive since the Carnot "value" of a high-temperature source would imply using a smaller ΔT to reduce lost work.

The working fluid must be selected based on the heat source temperature, as discussed by Swearingen and Ferguson (1984). See Sama (1983) for a discussion of optimal temperature differences for refrigeration systems; use of Eq. (10.11) leads to ΔT's ranging from 8 to 10°R.

10.2 OPTIMUM SHELL-AND-TUBE HEAT-EXCHANGER DESIGN

In this section we examine a procedure for optimizing the process design of a baffled shell-and-tube single-pass counterflow heat exchanger (see Fig. 10.3), in which the tube fluid is in turbulent flow but there is no change of phase of fluids in the shell or tubes. Usually the following variables are specified a priori by the designer:

1. Process fluid rate (the hot fluid passes through the tubes), W_i
2. Process fluid temperature change, $T_2 - T_1$
3. Coolant inlet temperature (the coolant flows through the shell), t_1
4. Tube spacing and tube inside and outside diameters (D_i, D_o).

Conditions 1 *and* 2 *imply the heat duty* Q of the exchanger is known.

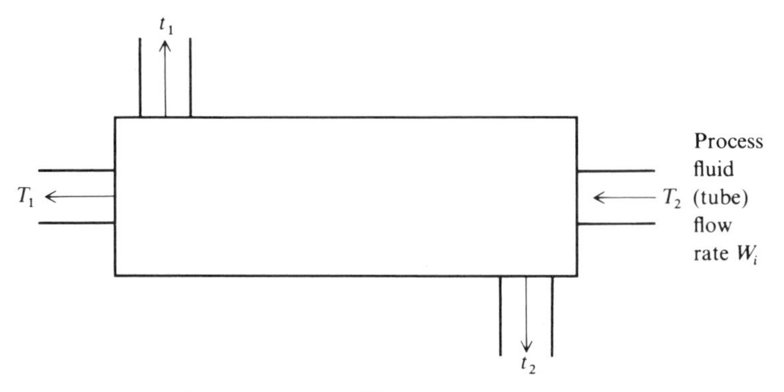

$\Delta t_1 = T_1 - t_1$ cold-end temperature difference
$\Delta t_2 = T_2 - t_2$ warm-end temperature difference

Figure 10.3 Process diagram of shell-and-tube counterflow heat exchanger.

The variables to be calculated via optimization include:

1. Total heat transfer area, A_o
2. Warm end temperature approach, Δt_2
3. Number and length of tubes, N_t and L
4. Number of baffle spacings, n_b
5. Tube-side and shell-side pressure drop
6. Coolant flow, W_c

Not all of these variables are independent, as discussed below.

In contrast to the analysis outlined in Sec. 10.1, the objective function does not make use of reversible work. Rather, a cost is assigned to the usage of coolant as well as to power losses because of the pressure drops of each fluid. In addition, annualized capital cost terms are included. The objective function in dollars per year is formulated using the notation in Table 10.1

$$C = C_c W_c y + C_A A_o + C_i E_i A_o + C_o E_o A_o \qquad (10.12)$$

Cichelli and Brinn (1956) suggested minimizing the objective function using the following set of four variables, a set slightly different from the list above:

1. Δt_2: warm end temperature difference
2. A_o: tube outside area
3. h_i: tube inside heat transfer coefficient
4. h_o: tube outside heat transfer coefficient

Only three of the four variables are independent. If A_o, h_i, and h_o are known, then Δt_2 can be found from the heat duty of the exchanger, Q:

$$Q = F_t U_o A_o \frac{\Delta t_2 - \Delta t_1}{\ln(\Delta t_2/\Delta t_1)} \tag{10.13}$$

F_t is unity for a single-pass exchanger. U_o is given by the values of h_o, h_i and the fouling coefficient h_t as follows:

$$\frac{1}{U_o} = \frac{1}{f_A h_i} + \frac{1}{h_o} + \frac{1}{h_t} \tag{10.14}$$

Cichelli and Brinn also showed that the annual pumping loss terms in Eq. (10.12) could be related to h_i and h_o by using friction-factor and *j*-factor relationships for tube flow and shell flow:

$$E_i = \phi_i h_i^{3.5} \tag{10.15}$$

$$E_o = \phi_o h_o^{4.75} \tag{10.16}$$

The coefficients ϕ_i and ϕ_o depend on fluid specific heat (c), thermal conductivity (k), density (ρ), and viscosity (μ) as well as the tube diameters. ϕ_o is based on either in-line or staggered tube arrangements.

If we solve for W_c from the energy balance

$$W_c = \frac{Q}{c(\Delta t_1 - \Delta t_2 + T_2 - T_1)} \tag{10.17}$$

and substitute for E_i, E_o, and W_c in Eq. (10.12), the resulting objective function is

$$f = \frac{C_c y Q}{c(\Delta t_1 - \Delta t_2 + T_2 - T_1)} + C_A A_o + C_i \phi_i h_i^{3.5} A_o + C_o \phi_o h_o^{4.75} A_o \tag{10.18}$$

To accommodate the constraint (10.13), a Lagrangian function (L) is formed by augmenting f with Eq. (10.13), using a Lagrange multiplier ω

$$L = f + \omega \left[\frac{F_t(\Delta t_2 - \Delta t_1)}{Q \ln(\Delta t_2/\Delta t_1)} - \frac{1}{U_o A_o} \right] \tag{10.19}$$

Equation (10.19) can be differentiated with respect to four variables (h_i, h_o, Δt_2, and A_o) as explained in Sec. 8.1. After some rearrangement, you can obtain a relationship between the optimum h_o and h_i, namely

$$h_o = \left(\frac{0.74 C_i \phi_i f_A}{C_o \phi_o} \right)^{0.17} h_i^{0.78} \tag{10.20}$$

This is the same result as derived by McAdams (1942), having the interpretation that the friction losses in the shell and tube sides and the heat transfer resistances must be balanced economically. The value of h_i can be obtained by solving

$$C_A - 2.5 C_i \phi_i h_i^{3.5} - 2.91 (C_o \phi_o)^{0.17} (C_i \phi_i f_A)^{0.83} h_i^{3.72} - \frac{3.5 C_i \phi_i f_A h_i^{4.5}}{h_t} = 0 \tag{10.21}$$

Table 10.1 Nomenclature for heat exchanger optimization

A_{lm}	Log mean of inside and outside tube surface areas
A_i	Inside tube surface area, ft^2
A_o	Outside tube surface area, ft^2
C	Total annual cost, \$/year
C_A	Annual cost of heat exchanger per unit outside tube surface area, \$/(ft²)(year)
C_c	Cost of coolant, \$/lb mass
C_i	Annual cost of supplying 1(ft)(lb$_f$)/h to pump fluid flowing inside tubes, (\$)(h)/(ft)(lb$_f$)(year)
C_o	Annual cost of supplying 1(ft)(lb$_f$)/h to pump shell-side fluid, (\$)(h)/(ft)(lb$_f$)(year)
c	Specific heat at constant pressure, Btu/(lb$_m$)(°F)
D_i	Tube inside diameter, ft
D_o	Tube outside diameter, ft
E_i	Power loss inside tubes per unit outside tube area, (ft)(lb$_f$)/(ft²)(h)
E_o	Power loss outside tubes per unit outside tube area, (ft)(lb$_f$)/(ft²)(h)
f	Friction factor, dimensionless
f_A	A_i/A_o
F_t	Multipass exchanger factor
g_c	Conversion factor, (ft)(lb$_m$)/(lb$_f$)(h²) = 4.18×10^8
h_f	Fouling coefficient
h_i	Coefficient of heat transfer inside tubes, Btu/(h)(ft²)(°F)
h_o	Coefficient of heat transfer outside tubes, Btu/(h)(ft²)(°F)
h_t	Combined coefficient for tube wall and dirt films, based on tube outside area Btu/(h)(ft²)(°F) $\frac{1}{h_t} = \frac{L'A_o}{k_w A_{lm}} + \frac{1}{h_{f_i}}\frac{A_o}{A_i} + \frac{1}{h_{f_o}}$
k	Thermal conductivity, Btu/(h)(ft)(°F)
L	Lagrangian function
L_t	Length of tubes, ft
L'	Thickness of tube wall, ft
n_b	Number of baffle spacing on shell side = number of baffles plus 1
N_c	Number of clearances for flow between tubes across shell axis

(Continued)

The simultaneous solution of Eqs. (10.17), (10.20), and (10.21) yields another expression:

$$\frac{C_c y U_o}{c(C_A + C_i E_i + C_o E_o)} = \left(1 + \frac{T_2 - T_1}{\Delta t_2 - \Delta t_1}\right)^2 \left(\ln\left(\frac{\Delta t_2}{\Delta t_1}\right) - 1 + \frac{\Delta t_2}{\Delta t_1}\right) \quad (10.22)$$

The following algorithm can be used to obtain the optimal values of h_i, h_o, A_o, and Δt_2 without the explicit calculation of ω:

1. Solve for h_i from Eq. (10.21)
2. Obtain h_o from Eq. (10.20)
3. Calculate U_o from Eq. (10.14)
4. Determine E_i and E_o from h_i and h_o using (10.15) and (10.16) and obtain Δt_2 by solving Eq. (10.22)
5. Calculate A_o from Eq. (10.13)
6. Find W_c from Eq. (10.17)

Table 10.1 Nomenclature for heat exchanger optimization (*continued*)

N_t	Number of tubes in exchanger
Δp_i	Pressure drop for flow through tube side, lb_f/ft^2
Δp_o	Pressure drop for flow through tube side, lb_f/ft^2
Q	Heat-transfer rate in heat exchanger, Btu/h
S_0	Minimum cross-sectional area for flow across tubes, ft^2
T_1	Outlet temperature of process fluid, °F
T_2	Inlet temperature of process fluid, °F
t_1	Inlet temperature of coolant, °F
t_2	Outlet temperature of coolant, °F
Δt_1	$T_1 - t_1$, = cold-end temperature difference
Δt_2	$T_2 - t_2$, = warm-end temperature difference
U_o	Overall coefficient of heat transfer, based on outside tube area, Btu/(h)(ft^2)(°F)
v_i	Average velocity of fluid inside tubes, ft/h
v_o	Average velocity of fluid outside tubes, ft/h at shell axis
W_c	Coolant rate, lb/h
W_i	Flow rate of fluid inside tubes, lb_m/h
W_o	Flow rate of fluid outside tubes, lb_m/h
y	Operating hours per year
ρ_i	Density of fluid inside tubes, lb_m/ft^3
ρ_o	Density of fluid outside tubes, lb_m/ft^3
μ	Viscosity of fluid, lb_m/(h)(ft)
ϕ_i	Factor relating friction loss to h_i
ϕ_o	Factor relating friction loss to h_o
ω	Lagrange multiplier
Subscripts	
c	Coolant
f	Film temperature, midway between bulk fluid and wall temperature
i	Inside the tubes
o	Outside the tubes
w	Wall

Note that steps 1 to 6 require that several nonlinear equations be solved one at a time.

Once these variables are known, the physical dimensions of the heat exchanger can be determined.

7. Determine the optimal v_i and v_o from h_i and h_o using the appropriate heat transfer correlations (see McAdams (1942)); recall that the inside and outside tube diameters are specified a priori.
8. The number of tubes N_t can be found from a mass balance:

$$v_i N_t \frac{\pi D_i^2}{4} = W_i \tag{10.23}$$

9. The length of the tubes L_t can be found from

$$A_o = N_t \pi D_o L_t \tag{10.24}$$

10. The number of clearances N_c can be found from N_t, based on either square pitch or equilateral pitch. The flow area S_o is obtained from v_o (flow normal to a tube bundle). Finally baffle spacing (or the number of baffles) is computed from S_o, A_o, N_t, and N_c.

Having presented the pertinent equations and the procedure for computing the optimum, let us check the approach by computing the degrees of freedom in the design problem.

Design variables	*Status (number of variables)*
W_i, T_1, T_2, t_1, tube spacing, D_i, D_o, Q	Given (8)
Δt_2, W_c, A_o, N_t, L_t, U_o, n_b Δp_t, Δp_s, v_i, v_o, h_i, h_0	Unspecified (13)

Total number of variables = 8 + 13 = 21

Design relationships	*Number of equations*
(*a*) Equations (10.13), (10.14), (10.15), (10.16) (10.17), (10.23), (10.24)	7
(*b*) Heat transfer correlations for h_i and h_o (step 7)	2
(*c*) $W_c = \rho_o v_o s_o$ (step 10)	1
Total number of relationships	10

Degrees of freedom for optimization = total number of variables − number of given variables − number of equations = 21 − 8 − 10 = 3

Note this agrees with Eq. (10.19) in that four variables are included in the Lagrangian, but with one constraint corresponding to three degrees of freedom.

Several simplified cases may be encountered in heat-exchanger design.

Case 1. U_o is specified and pressure drop costs are neglected in the objective function. In this case C_i and C_o can be set equal to zero and Eq. (10.22) can be solved for Δt_2 (see Peters and Timmerhaus (1980) for a similar equation for a condensing vapor). Figure 10.4 shows a solution to Eq. (10.22) (Cichelli and Brinn).

Case 2. Coolant flow rate is fixed. Here Δt_2 is known, so the tube side and shell-side coefficients and area are optimized. Use Eqs. (10.20) and (10.21) to find h_o and h_i. A_o is then found from Eq. (10.13).

Cichelli and Brinn discussed other special cases. See Jenssen (1969) for a discussion of the sensitivity of optimal designs for pressure drop and heat transfer in a heat exchanger.

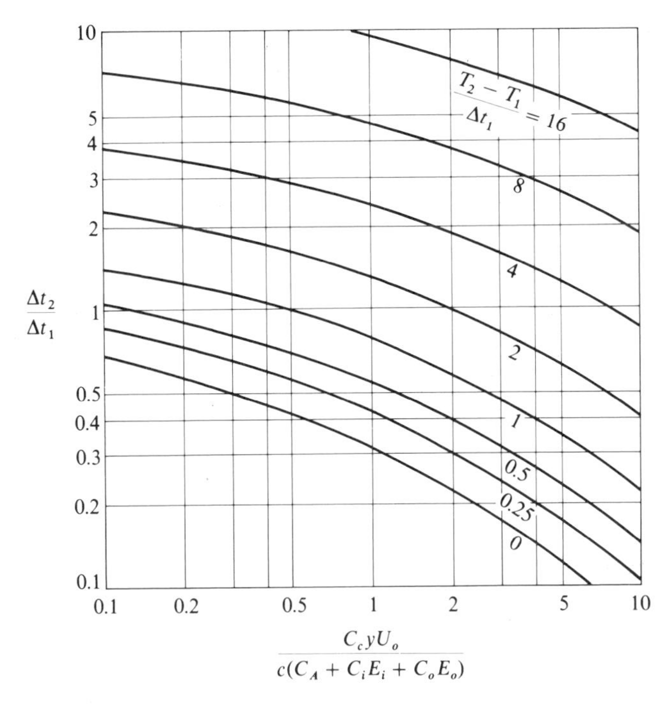

Figure 10.4 Solution to Eq. (10.22) for the case in which U_0 is specified and pressure drop costs are neglected.

In the analysis above no inequality constraints were introduced. As a practical matter the following inequality constraints may apply:

1. Maximum velocity on shell or tube side
2. Longest practical tube length
3. Closest practical baffle spacing
4. Maximum allowable pressure drops (shell or tube side)

The velocity on the tube side can be modified by changing the single-pass design to a multiple-pass configuration. In this case $F_t \neq 1$ in Eq. (10.13). From formulas in McCabe and Smith (1976), F_t depends on t_2 (or Δt_2), hence the necessary conditions derived previously would have to be changed. The fluids could be switched (shell vs. tube side) if constraints are violated, but there may well be practical limitations such as one fluid being quite dirty or corrosive so that the fluid must flow in the tube side (to facilitate cleaning or to reduce alloy costs).

Other practical features that must be taken into account are the fixed and integer lengths of tubes (8, 12, 16 and 20 feet), and the maximum pressure drops allowed. While a 20 psi drop may be typical for liquids such as water, higher values are employed for more viscous fluids. Exchanging shell/tube sides may mitigate pressure drop restrictions. The tube outside diameter is specified a priori in the optimization procedure described earlier; usually $\frac{3}{4}$- or 1-inch outside diameter (o.d.) tubes are used because of their greater availability and ease of cleaning.

Tarrer et al. (1971) described a computer program that combined the optimization procedure of Cichelli and Brinn with some of the practical considerations described above. Limits on operating variables such as maximum exit temperature of the coolant, maximum and minimum velocities for both streams, and maximum allowable shell area were included in the code. The number of tube passes must be specified, but the program handles discrete values for the number of tubes and the length of the exchanger in the vicinity of the optimum. Table 10.2 lists the specifications for a sample exchanger, and Table 10.3 gives the results of optimization for the two discrete variables in the first two rows for

Table 10.2 Design specifications for one case of heat exchanger optimization

Variables	
Process fluid	Gas
Inlet temperature of process fluid, °F	150
Outlet temperature of process fluid, °F	100
Process fluid flow-rate, lb/h	20,000
Maximum process fluid velocity, ft/s	160
Minimum process fluid velocity, ft/s	0.001
Utility fluid	Water
Inlet utility fluid temperature, °F	70
Maximum allowable utility fluid temperature, °F	140
Maximum utility fluid velocity, ft/s	8
Minimum utility fluid velocity, ft/s	0.5
Shellside fouling factor	2000
Tubeside fouling factor	1500
Cost of pumping process fluid, $\$/(ft)(lb_f)$	0.7533×10^{-8}
Cost of pumping utility fluid, $\$/(ft)(lb_f)$	0.7533×10^{-8}
Cost of utility fluid, $\$/lb_m$	0.5000×10^{-5}
Factor for pressure	1.45
Cost index	1.22
Fractional annual fixed charges	0.20
Fractional cost of installation	0.15
Tube material	Steel
Type of tube layout	Triangular
Construction type	Fixed tube sheet
Maximum allowable shell diameter, in	40
Bypassing safety factor	1.3
Constant for evaluating outside film coat	0.33
Hours operation per year	7000
Thermal conductivity of metal Btu/(h)(ft^2)(°F)	26
Number of tube passes	1

Source: Tarrer et al. (1971).

Table 10.3 Optimal solution for the heat exchanger and results of a search involving discrete variables

Variables	Optimal Design	Standard size			
		1	2	3	4
Tube length, ft	10.5	8	8	12	12
Number of tubes	66	110	85	64	42
Total area, ft^2	193.3	230	178	201	132
Total cost, $/year	734	908	923	738	784
Heat transfer coefficients, Btu/(h)(ft^2)(°F)					
Outside	554	561	649	512	617
Inside	56.2	37.1	45.9	57.4	80.5
Overall	41.0	28.4	34.5	41.5	56.2
Outlet utility fluid temperature (°F)	117.1	102.1	96.5	120.1	112.4
Utility fluid flow-rate, lb_m/h	5306	7790	9422	4993	5897
Inside pressure drop, psi	0.279	0.086	0.138	0.318	0.701
Outside pressure drop, psi	6.45	5.24	7.91	4.98	9.13
Number of baffle spaces	119	85	79	121	119
Shell diameter, in	12	16	14	12	10

Tube layout: 1.00-inch outside diameter
0.834-inch inside diameter
0.25-inch clearance
0.083-inch wall thickness
1.25-inch pitch

Source: Tarrer et al. (1971).

two standard lengths, 8 and 12 ft. The minimum cost occurs for a 12-ft tube length with 64 tubes (case 3).

For complicated exchanger design (multiple passes, condensing fluids, etc.), we recommend that commercial software be used, especially if mechanical specifications are to be calculated. One such program, B-JAC (1985), performs such calculations and includes some optimization capability via user interaction.

10.3 OPTIMIZATION OF HEAT-EXCHANGER NETWORKS

Heat-exchanger networks represent a significant factor in determining the energy efficiency of a process. How best to match process streams to be heated with process streams to be cooled, and how many exchangers are to be employed and in what configuration is a major synthesis problem. Because of the significance of this problem to industry, a substantial number of techniques to resolve it have been published in the literature (Lee, et al., 1970; Linnhoff and Flower, 1978; Grossmann and Sargent, 1978; Ponton and Donaldson 1974; Rathore and Powers, 1975; Umeda, et al., 1978; Umeda, et al., 1979). Refer to the review by Nishida et al. (1981) or articles by Umeda and Ichikawa (1971) or Shah and

Westerberg (1977). Many different solutions have been proposed—some are based on extensions of familiar optimization techniques, and others on completely new concepts of synthesis.

One goal is to automatically generate via a computer code a network configuration that minimizes utility and capital costs. Floudas, Ciric, and Grossmann (1986) proposed a method called MAGNETS of optimizing a network superstructure based on nonlinear programming that seemed effective when tested in several examples. They also review procedures published in the literature involving

1. Acyclic nets without stream splitting
2. Cyclic nets without stream splitting
3. Networks with stream splitting.

To overcome the combinatorial nature of the network synthesis problem Hohmann (1971) and Linnhoff and Flower (1978) employed a key factor, namely the concept of a minimum utility target. Papoulias and Grossmann (1983) used this idea and linear programming and mixed integer programming to make matches of streams, predict consumption of utilities, and locate pinch points. Many current design methods are based on a blend of physical insight and interaction by the engineer (Linnhoff, 1979; Linnhoff, 1980).

Once the network of exchangers is assembled, changes may occur in feeds, utility costs, interacting streams, etc. so that the optimal operating conditions for the given network must be redetermined. Optimization of the operating conditions and areas of a given configuration of exchangers is a direct extension of the optimization of a single exchanger. In the next example we show an optimization of a simple heat-exchanger network by differential calculus.

EXAMPLE 10.1 OPTIMAL ALLOCATION OF TEMPERATURES IN A SEQUENCE OF HEAT EXCHANGERS

Figure E10.1a shows a sequence of heat exchangers for which the total area is to be minimized by establishing the optimal values of the interstage temperatures given the values of N, the heat-transfer characteristics (U_i) of each exchanger, the entering flow rate W, the entering and exit temperatures T_0 and T_N, and each of the inlet temperatures T'_{i1} of the utility streams (designated W_{ui}).

The two relations to be used to calculate the area of one exchanger i are:

Energy balance:

$$WCp_i(T_i - T_{i-1}) = W_{ui}Cp_{ui}(T'_{i1} - T'_{i2}) = Q_i \qquad (a)$$

definition of Q:

$$Q_i = U_i A_i \, \Delta T_{\mathrm{lm},i} \qquad (b)$$

where

$$\Delta T_{\mathrm{lm},i} = \frac{(T'_{i1} - T_i) - (T'_{i2} - T_{i-1})}{\ln\,[(T'_{i1} - T_i)/(T'_{i2} - T_{i-1})]}$$

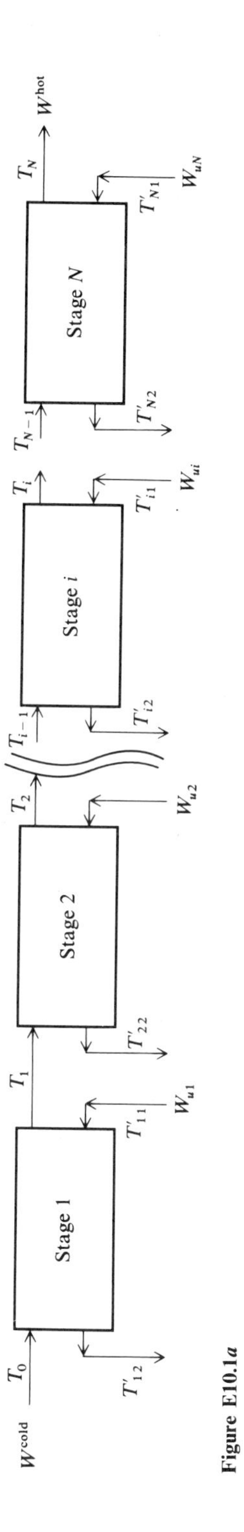

Figure E10.1*a*

We can solve Eq. (*b*) for A_i and use one of Eqs. (*a*) to eliminate Q_i:

$$A_i = \frac{(WCp)_i(T_i - T_{i-1})}{U_i \Delta T_{\text{lm},i}} \tag{c}$$

and the total area of all the exchangers (the objective function) is

$$A_T = \sum_{i=1}^{N} A_i = \sum_{i=1}^{N} \frac{(WCp)_i(T_i - T_{i-1})}{U_i \Delta T_{\text{lm},i}} \tag{d}$$

Ait-Ali and Wilde (1980) expressed A_T as a function of two parameters

$$\delta_i = T_i - T_{i-1} \tag{e}$$

$$\Delta_i = T'_{i1} - T'_{i2} \tag{f}$$

By algebra it can be shown that

$$\begin{aligned} (T'_{i1} - T_i) &= \Delta_i - \delta_i \\ (T'_{i2} - T_{i-1}) &= \Delta_i - R_i\delta_i \end{aligned} \qquad \text{where} \qquad R_i = \frac{(WCp)_i}{(WCp)_{ui}}$$

$$\frac{T_i - T_{i-1}}{\Delta T_{\text{lm},i}} = \frac{\delta_i}{\Delta T_{\text{lm},i}} = \frac{\delta_i \ln[(\Delta_i - \delta_i)/(\Delta_i - R_i\delta_i)]}{[(\Delta_i - \delta_i) - (\Delta_i - R_i\delta_i)]}$$

$$= \frac{\ln[(\Delta_i - \delta_i)/(\Delta_i - R_i\delta_i)]}{[R_i - 1]}$$

Thus

$$A_T = \sum_{i=1}^{N} \left(\frac{WCp}{U}\right)_i \frac{\ln[(\Delta_i - \delta_i)/(\Delta_i - R_i\delta_i)]}{(R_i - 1)} \tag{g}$$

We can minimize A_T with respect to the interstage temperature T_i by using the necessary condition

$$\frac{\partial A_T}{\partial T_i} = 0$$

(The second derivative of A_T with respect to T_i evaluated at T_i is positive.) For this example, only the two stages that involve T_i have to be considered, the *i*th and the $(i+1)$ stages (examine Fig. E10.1*b*):

$$\frac{\partial A_T}{\partial T_i} = \left(\frac{WCp}{U}\right)_i \frac{\Delta_i}{(\Delta_i - R_i\delta_i)(\Delta_i - \delta_i)} - \left(\frac{WCp}{U}\right)_{i+1} \frac{1}{(\Delta_{i+1} - R_{i+1}\delta_{i+1})} = 0 \tag{h}$$

From Eq. (*g*) we find

$$\frac{(WCp/U)_i}{(WCp/U)_{i+1}} = \left[\frac{(\Delta_i - \delta_i)}{(\Delta_i)}\right]\left[\frac{\Delta_i - R_i\delta_i}{(\Delta_{i+1} - R_{i+1}\delta_{i+1})}\right] \tag{i}$$

$$= \left[\frac{(T'_{i1} - T_i)}{(T_i - T_{i-1})}\right]\left[\frac{(T'_{i2} - T_{i-1})}{(T'_{i+1,2} - T_i)}\right] \tag{j}$$

$$= \frac{\Delta T_{\text{exit},i}}{\Delta T_{\text{stage},i}} \frac{\Delta T_{\text{enter},i}}{\Delta T_{\text{enter},i+1}} \tag{k}$$

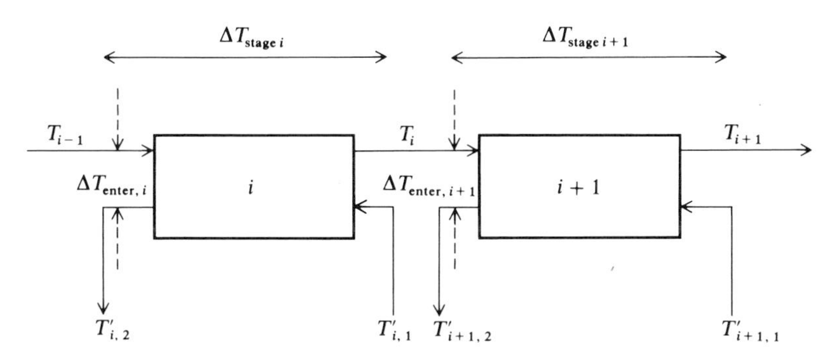

Figure E10.1*b*

To solve for the interstage temperatures you could solve the set of $(N-1)$ coupled quadratic equations given by Eq. (i) after substituting Eqs. (e) and (f) for δ_i and Δ_i, respectively. Successive substitution can be used to reduce the work involved. Heat-transfer areas for individual exchangers can then be obtained from Eq. (c). Or you can carry out an unconstrained optimization of Eq. (d) directly in terms of $T_1, T_2, \ldots, T_{N-1}$ by one of the algorithms described in Chap. 6.

As a simple example, suppose we want to minimize the heat-transfer area of a two-stage gas regeneration system which requires heating a process stream from 100 to 500°F. The values of the parameters are listed in Table E10.1. This problem has been previously solved by Boas (1968) using dynamic programming.

The optimal value of T_1^* is calculated from Eq. (j)

$$T_1^* = T'_{11} - \left[(T'_{11} - T_0)(T'_{21} - T_2)\frac{(WCp/U)_1}{(WCp/U)_2}\right]^{1/2} \qquad (l)$$

Substitution of the known parameters from Table E10.1 gives $T_1^* = 158.6°F$, $A_1^* = 518\ \text{ft}^2$, $A_2^* = 4268\ \text{ft}^2$, and $A_T^* = 4786\ \text{ft}^2$. For a single stage for the same temperature increase (100 to 500°F), $A_1 = 5000\ \text{ft}^2$. For three stages, with the same data except $T'_{11} = 300°\text{F}$, $T'_{21} = 400°\text{F}$, and $T'_{31} = 600°\text{F}$; $T_1^* = 116.5°\text{F}$, $T_2^* = 213.6°\text{F}$, and $A_1^* = 112\ \text{ft}^2$, $A_2^* = 864\ \text{ft}^2$, and $A_3^* = 3355\ \text{ft}^2$.

Table E10.1 Parameters for Example 10.1

Stage i	T'_{i1} °F	$R_i\left(\frac{WCp}{U_2}\right)_i$, ft²	$\left(\frac{WCp}{WCp}\right)_u$ dimensionless	WCp, $\frac{\text{Btu}}{(\text{h})(°\text{F})}$	U_i, $\frac{\text{Btu}}{(\text{h})(°\text{F})(\text{ft}^2)}$
1	300	1250	1	10^5	80
2	600	1250	1	10^5	80

10.4 OPTIMIZATION OF EVAPORATOR DESIGN

When a process requires an evaporation step, the problem of evaporator design needs serious examination. Although the subject of evaporation and the equipment to carry out evaporation have been studied and analyzed for many years, each application has to receive individual attention. No evaporation configuration and its equipment can be picked from a stock list and be expected to produce trouble-free operation.

An engineer working on the selection of optimal evaporation equipment must list what is "known," "unknown," and "to be determined." Such analysis should at least include the following:

Known

- Production rate and analysis of product
- Feed flow rate, feed analysis, feed temperature
- Available utilities (steam, water, gas, etc.)
- Disposition of condensate (location) and its purity
- Probable materials of construction

Unknown

- Pressures, temperatures, solids, compositions, capacities, and concentrations
- Number of evaporator effects
- Amount of vapor leaving the last effect
- Heat transfer surface

Features to be determined

- Best type of evaporator body and heater arrangement
- Filtering characteristics of any solids or crystals
- Equipment dimensions, arrangement
- Separator elements for purity of overhead vapors
- Materials, fabrication details, instrumentation

Utility Consumption

- Steam
- Electric power
- Water
- Air

In multiple effect evaporation, as shown in Fig. 10.5, the total capacity of the system of evaporation (of the same size) is no greater than that of a single-

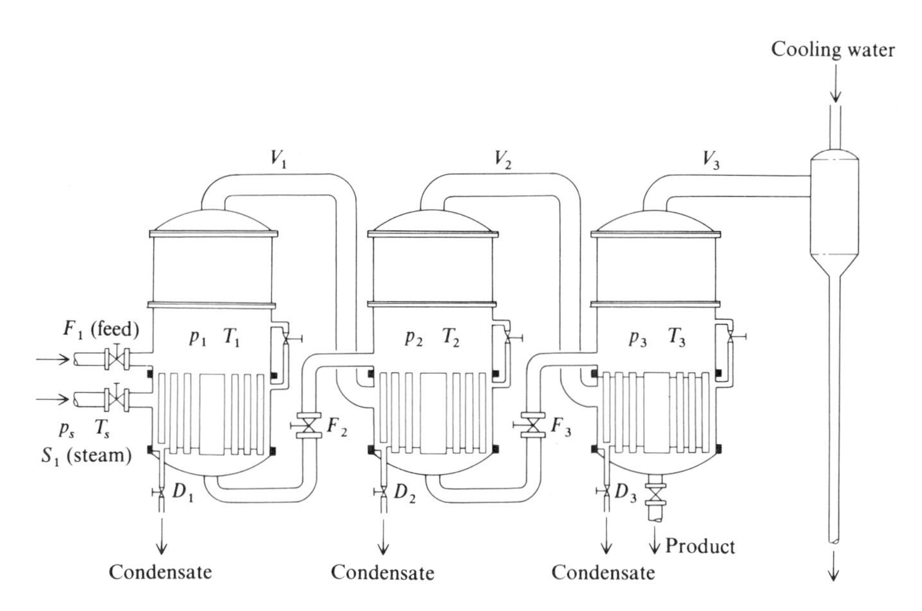

Figure 10.5 Multiple-effect evaporators with forward feed.

effect evaporator having a heating surface equal to one effect and operating under the same terminal conditions. The amount of water vaporized per unit surface area in n effects is roughly $1/n$ that of a single effect. Furthermore, the boiling point elevation causes a loss of available temperature drop in every effect thus reducing capacity. Why, then, are multiple effects often economic? It is because the cost of an evaporator per square foot of surface area decreases with total area (and asymptotically becomes a constant value) so that to achieve a given production, the cost of heat-exchange surface can be balanced with the steam costs.

Steady state mathematical models of single- and multiple-effect evaporators involving material and energy balances can be found in McCabe et al. (1985) and Esplugas and Mata (1983). The classical simplified optimization problem for evaporators (Schweyer (1955)) is to determine the most suitable number of effects given (*a*) an analytical expression for the fixed costs in terms of the number of effects n, and (*b*) the steam (variable) costs also in terms of n. Analytic differentiation yields the optimal n^*, as shown in Example 10.2 below.

EXAMPLE 10.2 OPTIMIZATION OF A MULTISTAGE EVAPORATOR

Assume we are concentrating an inorganic salt in the range of 0.1 to 1.0 weight percent using a plant capacity of 0.1 to 10 million gallons/day. Initially we will treat the number of stages n as a continuous variable. Figure E10.2*a* shows a single effect in the process.

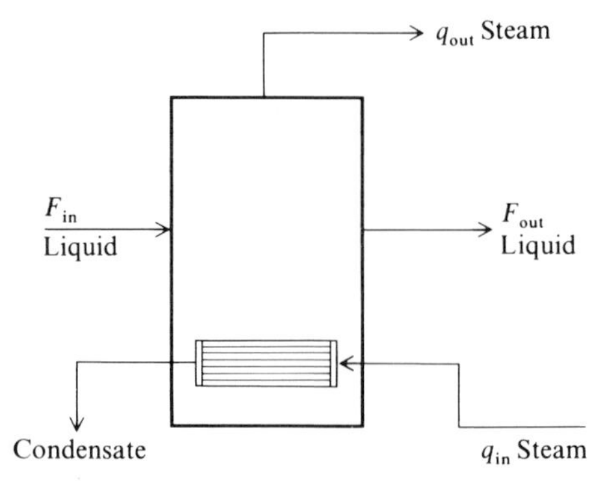

Figure E10.2*a*

Prior to discussions of the capital and operating costs, we need to define the temperature driving force for heat transfer. Examine the notation in Fig. E10.2*b*; by definition the log mean temperature difference ΔT_{lm} is

$$\Delta T_{lm} = \frac{T_i - T_d}{\ln (T_i/T_d)} \tag{a}$$

Let T_i be equal to constant K for a constant performance ratio P. Because $T_d = T_i - \Delta T_f/n$

$$\Delta T_{lm} = \frac{\Delta T_f/n}{\ln [K/(K - (T_f/n))]} \tag{b}$$

Optimum number of stages n^*

Let

A = condenser heat transfer areas, ft^2
c_p = liquid heat capacity, 1.05 Btu/(lb_m)(°F)
C_C = cost per unit area of condenser, \$6.25/ft^2
C_E = cost per evaporator (including partitions), \$7000/stage
C_S = cost of steam, \$/lb at the brine heater (first stage)
F_{out} = liquid flow out of evaporator, lb/h
K = T_i, a constant ($T_i = \Delta T - T_b$ at inlet)
n = number of stages
P = performance ratio, lb of H_2O evaporated/Btu supplied to brine heater
Q = heat duty, 9.5×10^8 Btu/h (a constant)
q_e = total lb H_2O evaporated/h
q_r = total lb steam used/h
r = capital recovery factor
S = lb steam supplied/h
T_b = boiling point rise, 4.3°F
ΔT_f = flash down range, 250°F
U = overall heat transfer coefficient (assumed to be constant), 625 Btu/(ft^2)(h)(°F)
ΔH_{vap} = heat of vaporization of water, about 1000 Btu/lb

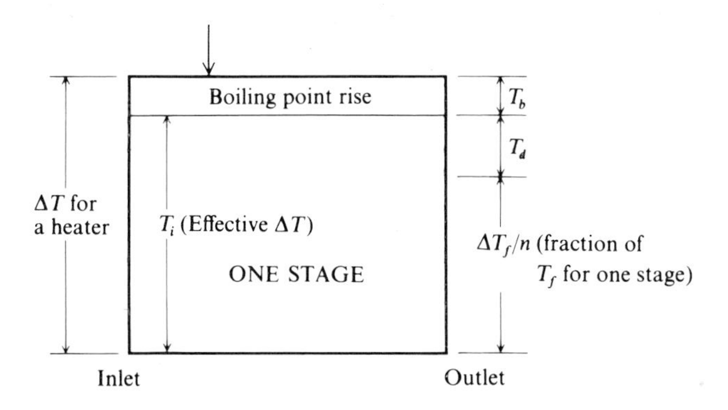

Figure E10.2*b*

For a *constant performance ratio* the total cost of the evaporator is

$$f_1 = C_E n + C_C A \tag{c}$$

For A we introduce

$$A = \frac{Q}{U(\Delta T_{\mathrm{lm}})}$$

Then we differentiate f_1 in Eq. (c) with respect to n and set the resulting expression equal to zero (Q and U are constant):

$$C_E + C_C \frac{Q}{U}\left[\frac{\partial(1/\Delta T_{\mathrm{lm}})}{\partial n}\right]_P = 0 \tag{d}$$

With the use of Eq. (b)

$$\left[\frac{\partial(1/\Delta T_{\mathrm{lm}})}{\partial n}\right]_P = -\frac{1}{nK(1 - \Delta T_f/nK)} - \frac{\ln(1 - \Delta T_f)}{\Delta T_f} \tag{e}$$

Substitution of (e) into (d) plus introduction of the values of Q, U, ΔT_f, C_E, and C_C we get

$$7000 - \left[\frac{(6.25)(9.5 \times 10^8)}{625}\right]\left[\frac{1}{nK(1 - \Delta T_f/nK)} + \frac{\ln(1 - \Delta T_f/nK)}{\Delta T_f}\right] = 0$$

Rearranging,

$$\frac{(625)(7000)(250)}{(6.25)(9.5 \times 10^8)} = 0.184 = \frac{250}{nK - 250} + \ln\left(1 - \frac{250}{nK}\right) \tag{f}$$

In practice, as the evaporation plant size changes (for constant Q), the ratio of the stage condenser area cost to the unit evaporator cost remains essentially constant so that the number 0.184 is treated as a constant for all practical purposes. Equation (f) can be solved for nK for constant P

$$nK = 590 \tag{g}$$

Next, we eliminate K from Eq. (g) by replacing K with a function of P so that n becomes a function of P. The performance ratio (with constant liquid heat capacity at 347°F) is defined as

$$P = \frac{(\Delta H_{\text{vap}})(q_e)}{(F_{\text{out}} c_{pF} \Delta T_{\text{heater}})_{\text{first stage}}} = \frac{1000}{1.05(4.3 + K)} \frac{q_e}{F_{\text{out}}} \tag{h}$$

The ratio q_e/F can be calculated from

$$\frac{q_e}{F_{\text{out}}} = 1 - \left(\frac{1194 - 322}{1194 - 70}\right)^{1.49} = 0.31$$

where ΔH_{vap} (355°F, 143 psia) = 1194 Btu/lb, $\Delta H_{\text{liq}\,H_2O}$ (350°F) = 322 Btu/lb, and $\Delta H_{\text{liq}\,H_2O}$ (100°F) = 70 Btu/lb. Equations (g) and (h) can be solved together to eliminate K and obtain the desired relation

$$\frac{300}{P} - 4.3 = \frac{590}{n^*} \tag{i}$$

Equation (i) shows how the boiling point rise ($T_b = 4.3$°F) and the number of stages affects the performance ratio.

Optimal Performance Ratio

The optimal plant operation can be determined by minimizing the total cost function, including steam costs, with respect to P (liquid pumping costs are negligible)

$$f_2 = [C_C A + C_E n]r + C_s S \tag{j}$$

$$rC_C \frac{\partial A}{\partial P} + rC_E \frac{\partial n}{\partial P} + C_s \frac{\partial S}{\partial P} = 0 \tag{k}$$

The quantity for $\partial A/\partial P$ can be calculated by using the equations already developed and could be expressed in terms of a ratio of polynomials in P such as

$$\frac{a(1 + 1/P)}{(1 - bP)^2}$$

where a and b are determined by fitting experimental data. The relation for $\partial n/\partial P$ can be determined from Eq. (i). The relation for $\partial S/\partial P$ can be obtained from equation (l) below

$$P = \frac{q_e}{Q} = \frac{q_e}{(\Delta H_{\text{vap}})S} = \frac{q_e}{1000S}$$

or

$$S\left(\frac{\text{lb}}{\text{h}}\right) = \frac{q_e}{1000P}$$

or

$$S(\text{lb}) = \frac{\alpha(8760)q_e}{1000P} \qquad (l)$$

where α is the fraction of hours per year (8760) during which the system operates.

Equation (k), given the costs, cannot be explicitly solved for P^*, but P^* can be obtained by any effective root-finding technique.

If a more complex mathematical model is employed to represent the evaporation process, you must shift from analytic to numerical methods. The material and enthalpy balances become complicated functions of temperature (and pressure). Usually all of the system parameters are specified except for the heat transfer areas in each effect (n unknown variables) and the vapor temperatures in each effect excluding the last one ($n - 1$ unknown variables). The model introduces n independent equations that serve as constraints, many of which are nonlinear plus nonlinear relations among the temperatures, concentrations, and physical properties such as the enthalpy and the heat transfer coefficient. King (1980) showed how to use Lagrange multipliers to solve for the n unknown areas. King (1963) also gave the results of optimizing a seawater conversion plant in which the influence of changes in the heat transfer coefficient, the boiling point elevation, and the need for purging overhead vapor were taken into account.

Because the number of evaporators represents an integer-valued variable and many engineers use tables and graphs as well as equations for evaporator calculations, dynamic programming is a suitable tool for optimization. Itahara and Stiel (1968) solved the same problem as King (1963) by this method.

10.5 BOILER/TURBO GENERATOR SYSTEM OPTIMIZATION

Linear programming is often used in the design and operation of steam systems in the chemical industry (Putman, 1980). Figure 10.6 shows a steam and power system for a small power house fired by wood pulp. To produce electric power, this system contains two turbo-generators whose characteristics are listed in Table 10.4. Turbine 1 is a double-extraction turbine with two intermediate streams leaving at 195 and 62 psig; the final stage produces condensate which is used as boiler feed water. Turbine 2 is a single-extraction turbine with one intermediate stream at 195 psig, and an exit stream leaving at 62 psig with no condensate being formed. The first turbine is more efficient due to the energy released from the condensation of steam, but it cannot produce as much power as the second turbine. Excess steam may bypass the turbines to the two levels of steam through pressure-reducing valves.

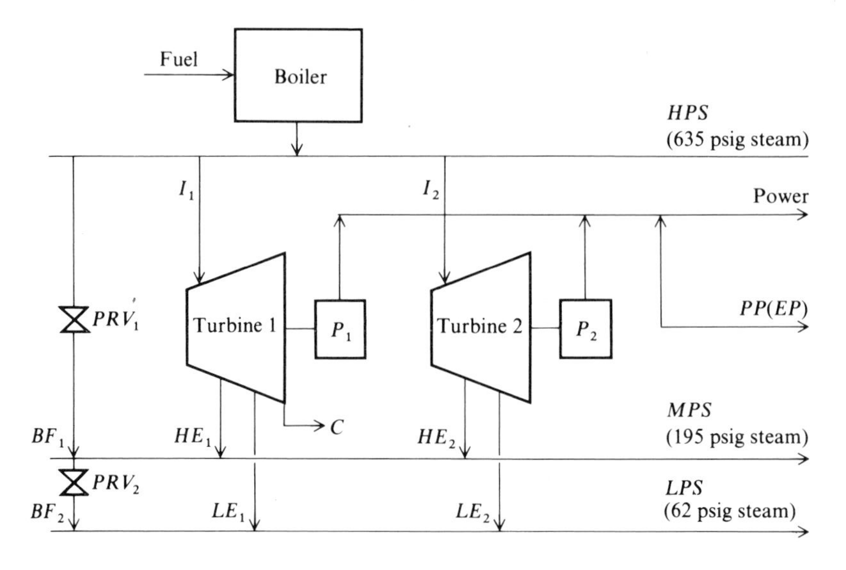

Notation:
I_i = inlet flowrate for turbine i [lb_m/h]
HE_i = exit flowrate from turbine i to 195 psig header [lb_m/h]
LE_i = exit flowrate from turbine i to 62 psig header [lb_m/h]
C = condensate flowrate from turbine 1 [lb_m/h]
P_i = power generated by turbine i [kW]
BF_1 = bypass flowrate from 635 psig to 195 psig header [lb_m/h]
BF_2 = bypass flowrate from 195 psig to 62 psig header [lb_m/h]
HPS = flowrate through 635 psig header [lb_m/h]
MPS = flowrate through 195 psig header [lb_m/h]
LPS = flowrate through 62 psig header [lb_m/h]
PP = purchased power [kW]
EP = excess power [kW] (difference of purchased power from base power)
PRV = pressure-reducing valve

Figure 10.6 Boiler/turbogenerator system.

Table 10.5 lists information about the different levels of steam and Table 10.6 gives the demands on the system. To meet the electric power demand, electric power may be purchased from another producer with a minimum base of 12,000 kW. If the electric power required to meet the system demand is less than this base, the power that is not used will be charged at a penalty cost. Table 10.7 gives the costs of fuel for the boiler and additional electric power to operate the utility system.

The system shown in Figure 10.6 may be modeled as linear constraints and combined with a linear objective function. The objective is to minimize the operating cost of the system by choice of steam flow-rates and power generated or purchased, subject to the demands and restrictions on the system. The objective function (10.25) shown below is the cost to operate the system per hour,

Table 10.4 Turbine data

Turbine 1	
Maximum generative capacity	6250 kW
Minimum load	2500 kW
Maximum inlet flow	192,000 lb_m/h
Maximum condensate flow	62,000 lb_m/h
Maximum internal flow	132,000 lb_m/h
High pressure extraction at	195 psig
Low pressure extraction at	62 psig

Turbine 2	
Maximum generative capacity	9000 kW
Minimum load	3000 kW
Maximum inlet flow	244,000 lb_m/h
Maximum 62 psi exhaust	142,000 lb_m/h
High pressure extraction at	195 psig
Low pressure extraction at	62 psig

Table 10.5 Steam header data

Header	Pressure	Temperature	Enthalpy
HP steam	635 psig	720°F	1359.8 Btu/lb_m
MP steam	195 psig	130°F superheat	1267.8 Btu/lb_m
LP steam	62 psig	130°F superheat	1251.4 Btu/lb_m
Feedwater (condensate)	....	....	193.0 Btu/lb_m

Table 10.6 Demands on the system

Resource	Demand
MP steam (195 psig)	271,536 lb_m/h
LP steam (62 psig)	100,623 lb_m/h
Electric power	24,550 kW

Table 10.7 Energy data

Fuel cost	\$1.68/10^6Btu
Boiler efficiency	0.75
Steam cost (635 psig)	\$2.24/10^6Btu
	= \$2.24 (1359.8 − 193)/10^6
	= \$0.002614/lb_m
Purchased electric power	\$0.0239/kWh average
Demand penalty	\$0.009825/kWh
Base purchased power	12,000 kW

namely the sum of steam produced (HPS), purchased power required (PP), and excess power (EP):

$$\text{Minimize:} \quad f = 0.00261HPS + 0.0239PP + 0.00983EP \tag{10.25}$$

The constraints are gathered into specific subsets below

Turbine 1

$$\begin{aligned} P_1 &\leq 6250 \\ P_1 &\geq 2500 \\ HE_1 &\leq 192{,}000 \\ C &\leq 62000 \\ I_1 - HE_1 &\leq 132{,}000 \end{aligned} \tag{10.26}$$

Turbine 2

$$\begin{aligned} P_1 &\leq 9000 \\ P_2 &\geq 3000 \\ I_2 &\leq 244{,}000 \\ LE_2 &\leq 142{,}000 \end{aligned} \tag{10.27}$$

Material balances

$$\begin{aligned} HPS - I_1 - I_2 - BF_1 &= 0 \\ I_1 + I_2 + BF_1 - C - MPS - LPS &= 0 \\ I_1 - HE_1 - LE_1 - C &= 0 \\ I_2 - HE_2 - LE_2 &= 0 \\ HE_1 + HE_2 + BF_1 - BF_2 - MPS &= 0 \\ LE_1 + LE_2 + BF_2 - LPS &= 0 \end{aligned} \tag{10.28}$$

Power purchased

$$EP + PP \geq 12{,}000 \tag{10.29}$$

Demands

$$\begin{aligned} MPS &\geq 271{,}536 \\ LPS &\geq 100{,}623 \\ P_1 + P_2 + PP &\geq 24{,}550 \end{aligned} \tag{10.30}$$

Energy balances

$$\begin{aligned} 1359.8I_1 - 1267.8HE_1 - 1251.4LE_1 - 192C - 3413P_1 &= 0 \\ 1359.8I_2 - 1267.8HE_2 - 1251.4LE_2 - 3413P_2 &= 0 \end{aligned} \tag{10.31}$$

Table 10.8 Optimal solution to steam system LP

Variable	Name	Value	Status
1	I1	134,915.64	BASIC
2	I2	244,000	BOUND
3	HE1	27,536	BASIC
4	HE2	244,000	BASIC
5	LE1	100,623	BASIC
6	LE2	0	ZERO
7	C	6,756.64	BASIC
8	BF1	0	ZERO
9	BF2	0	ZERO
10	HPS	378,915.64	BASIC
11	MPS	271,536	BASIC
12	LPS	100,623	BASIC
13	P1	6,250	BOUND
14	P2	6,577.20	BASIC
15	PP	11,722.79	BASIC
16	EP	277.20	BASIC

Value of objective function = 1271.87 \$/h
BASIC = basic variable
ZERO = 0
BOUND = variable at its upper bound

Table 10.8 lists the optimal solution to the linear program posed by Eqs. (10.25) to (10.31). Basic and nonbasic (zero) variables are identified in the table; the minimum cost is \$1271.87/h. Note that $EP + PP$ must sum to 12,000 kWh; in this case excess power is reduced to 277 kWh.

REFERENCES

Ait-Ali, M. A., and D. J. Wilde, "Optimal Area Allocation in Multistage Heat Exchanger Systems," J. Heat Transfer, **102**: 199–201 (1980).

B-JAC Computer Services, "An Integrated Set of Heat Exchanger Design Programs," Midlothian, Virginia, 1985.

Boas, A. H., "Optimization via Linear and Dynamic Programming," *Chem. Eng.*, **70**, 85 (April, 1968).

Cichelli, M. T., and M. S. Brinn, "How to Design the Optimum Heat Exchanger," *Chem. Eng.*, 196 (May 1956).

Esplugas, S., and J. Mata, "Calculator Design of Multistage Evaporators," *Chem. Eng.*, 59 (Feb. 7, 1983).

Floudas, C. A., A. R. Ciric, and I. E. Grossman, "Automatic Synthesis of Optimum Heat Exchanger Network Configurations", *AIChEJ.*, **32**: 276 (1986).

Grossman, I. E., and R. W. H. Sargent, "Optimum Design of Heat Exchanger Networks," *Comput. Chem. Eng.*, **2**: 1 (1978).

Hohmann, E. C., "Optimum Networks for Heat Exchange", Ph.D. Thesis, University of Southern California (1971).

Itahara, S., and L. I. Stiel, "Optimal Design of Multiple Effect Evaporators with Vapor Bleed Streams" *Ind. Eng. Chem. Proc. Des. Devel.*, **7**: 6 (1968).
Jenssen, S. K., "Heat Exchanger Optimization," *Chem. Eng. Prog.*, **65**(7): 59 (1969).
King, C. J., *Separation Processes*, McGraw-Hill, New York, (1980), Appendix B.
King, C. J., "Fresh Water from Sea Water," *A Case in Process Design* (ed. T. K. Sherwood), MIT Press, Cambridge, Massachusetts (1963).
Lee, K. F., A. H. Masso, and D. F. Rudd, "Branch and Bound Synthesis of Integrated Designs," *Ind. Eng. Chem. Fund.*, **9**: 48 (1970).
Linnhoff, B., "Entropy in Practical Process Design" in Proceed. of FOCAPD, (R. S. A. Mah and W. D. Seider, eds.), 537, Henniker, New Hampshire, (June 1980).
Linnhoff, B., D. R. Mason, and I. Wardle, CACE '79 Proceed. (214th event of the European Federation of Chemical Engineers), Montreux, Switzerland, April 8–11, 1979, 537.
Linnhoff, B., and J. R. Flower, "Synthesis of Heat Exchanger Networks," *AIChE J.*, **24**: 633 (1978).
McAdams, W. H., *Heat Transmission*, McGraw-Hill, New York (1942).
McCabe, W. L., J. Smith, and P. Harriott, *Unit Operations in Chemical Engineering*, 4th ed., McGraw-Hill, New York (1985).
Nishida, N., G. Stephanopoulos, and A. W. Westerberg, "Journal Review: Process Synthesis", *AIChE J.*, **27**, 321 (1981).
Papoulias, S. A., and I. E. Grossmann, "A Structural Optimization Approach in Process Synthesis II: Heat Recovery Networks", *Comput. Chem. Eng.* **7**: 707 (1983).
Peters, M., and K. Timmerhaus, *Plant Design and Economics for Chemical Engineers*, 3d ed., McGraw-Hill, New York (1980).
Ponton, J. W., and R. A. B. Donaldson, "A Fast Method for the Synthesis of Optimal Heat Exchanger Networks," *Chem. Eng. Sci.*, **29**: 2375 (1974).
Putman, R. E. J., "Boiler/Turbo-Generator Return Analysis," Proprietary Report, Westinghouse Electric Corporation, Pittsburgh, Pennsylvania, (1980).
Rathore, R. N. S., and G. J. Powers, "A Forward Branching Scheme for the Synthesis of Energy Recovery Systems," *Ind. Eng. Chem. Proc. Des. Dev.*, **14**: 175 (1975).
Sama, D. A., "Economic Optimum LMTD at Heat Exchangers," AIChE National Meeting, Houston, Texas (March 1983).
Schweyer, H. E., *Process Engineering Economics*, McGraw-Hill, New York, (1955), p. 214.
Shah, J. V., and A. W. Westerberg, "Process Synthesis Using Structural Parameters: A Problem with Inequality Constraints," *AIChE J.*, **23**: 378 (1977).
Steinmeyer, D. E., "Process Energy Conservation," Kirk-Othmer Encyclopedia Supplemental Volume, 3d ed., Wiley, New York, (1984).
Swearingen, J. S., and J. E. Ferguson, "Optimized Power Recovery from Waste Heat," *Chem. Eng. Prog.* (1984).
Tarrer, A. R., H. C. Lim, and L. B. Koppel, "Finding the Economically Optimum Heat Exchanger," *Chem. Eng.*, 79 (Oct. 4, 1971).
Umeda, T., and A. Ichikawa, *Ind. Eng. Proc. Des. Dev.* **10**: 229 (1971).
Umeda, T., and A. Ichikawa, T. Harada, and K. Shiroko, "A Thermodynamic Approach to the Synthesis of Heat Inte-gration Systems in a Chemical Process," CACE '79 Proceed. (214th event of the European Federation of Chemical Engineers), Montreux, Switzerland (April 8–11, 1979), pp. 487–499.
Umeda, T., and A. Ichikawa, J. Itoh, and K. Shiroko, "Heat Exchanger System Synthesis," *Chem. Eng. Prog.*, 70 (July, 1978).

SUPPLEMENTARY REFERENCES

Colmenares, T. R., and W. D. Seider, "Heat and Power Integration of Chemical Processes," *AIChE J*, **33**: 898 (1987).
Ganapathy, V., "Evaluating Waste Heat Recovery Projects," *Hydrocarbon Process.* (*Aug.*), 101 (1982).

Kröger, D. G., "Design Optimization of an Air-Oil Heat Exchanger," *Chem. Eng. Sci.*, **38**: 329 (1983).

Kürby, H. J., H. H. Erdmann, and K. H. Simmrock, "Reducing Energy Consumption of Multi-Effect Evaporators by Optimal Design," in *Proceed. Chem. Comp.* 1982 (ed. G. F. Froment), Konin Vlaamse Ing. V. Z. W. Genoot. Chem. Tech., Antwerp, 1982.

Le Goff, P., and M. Giulietti, "Comparison of Economic and Energy Optimizations for a Heat Exchanger," *Int. Chem. Eng.*, **22**: 252 (1982).

Nath, R., D. J. Libby, and H. J. Duhan, "Joint Optimization of Process Units and Utility Systems," *Chem. Eng. Prog.* (*May*), 31 (1986).

Newell, R. B., "A Comparative Study of Model and Goal Coordination in the Multi-Level Optimization of a Double-Effect Evaporator," *Can. J. Chem. Eng.*, **58**: 275 (1980).

Poje, J. B., and A. M. Smart, "On-Line Energy Optimization in a Chemical Complex," *Chem. Eng. Prog.* (*May*), 39 (1986).

Uraniec, K., *Optimal Design of Process Equipment*, John Wiley, New York (1986).

CHAPTER 11

SEPARATION PROCESSES

Separations are an important phase of almost all chemical engineering processes. Separations are needed because the chemical species from a single source stream must be sent to multiple destinations with specified concentrations. The sources usually are raw material inputs and reactor effluents; the destinations are reactor inputs and product and waste streams. To achieve a desired species allocation you have to determine the best types and sequence of separators to be used, evaluate the physical and/or chemical property differences to be exploited at each separator, fix the phases at each separator, and prescribe operating conditions for the entire process. Optimization is involved both in the design of the equipment and in the determination of the optimal operating conditions for the equipment.

This chapter contains examples of optimization techniques applied to the design and operation of two staged and continuous processes, namely extraction and distillation. We also illustrate the use of parameter estimation for fitting a function to thermodynamic data.

11.1 OPTIMIZATION OF LIQUID-LIQUID EXTRACTION PROCESSES

Liquid-liquid extraction is carried out either (*a*) in a series of well-mixed vessels, such as well-mixed tanks or plate columns, so that the process can be treated as a staged process akin to distillation (see Sec. 11.2), or (*b*) in continuous processes, such as spray columns, packed columns, and rotating disk columns. Staged processes, as mentioned in several earlier chapters, are more difficult to optimize if the number of stages is treated as an integer variable. For either cocurrent or countercurrent flow in continuous process equipment, integer variables are avoided and optimization of any reasonable objective function can be carried out by several of the techniques presented in previous chapters. Example 11.1 below illustrates optimization applied to a plug flow model; Table 11.1 summarizes a number of examples that have appeared in the literature for staged and continuous models of extraction.

EXAMPLE 11.1 OPTIMIZATION OF LIQUID EXTRACTION COLUMN FLOWRATES

Steady state continuous countercurrent liquid extraction can be modeled in a variety of ways, the most common of which are (*a*) a plug flow model and (*b*) an axial dispersion model. Figure E11.1a illustrates a typical column. Jackson and Agnew (1980) examined the accuracy of several models for a continuous pilot-scale extraction column in which water was used to extract acetic acid from amyl alcohol, and concluded that a plug flow model was sufficiently accurate to represent the data collected. Once a model is specified, it can be used to determine the maximum extraction rate. Below we describe the process model, the constraints, and the objective function.

Table 11.1 Techniques of optimization applied to steady state staged liquid extraction

Authors	Ref.	Independent variables	Process model staged (S) or continuous (C) (plug flow)	Objective function	Optimization technique
Salem & Jeffreys	[A]	Temperature in stages	S	Value of extracted solute less equipment and operating costs	Discrete maximum principle
Chen	[B]	Number of stages Ratio of flow rate of entering solvent to retention rate of liquid or solids	S	Value of recovered solute less costs of equipment, raw materials, and solvent	Analytical differentiation and a graphical method
Olson	[C]	Number of stages Extraction factor = (m) × (feed rate of phase Y divided by feed rate of phase X)	S	Recovery factor	Analytical differentiation and equating derivatives to zero
Jensen & Jeffreys	[D]	Number of stages Solvent flow rate	S	Value of product less cost of feed, solvent and fixed charges	Analytical differentiation and equating derivatives to zero

Treybal	[E]	Scale up factor Ratio of extract to raffinate flow rate	S	Capital costs, value of unextracted solvent, operating costs, cost of lost solvent, and labor costs	Analytical differentiation and equating derivatives to zero
Jackson & Agnew	[F]	Velocity of the two continuous phases Stirrer speed	C	Total extraction rate and other more complex functions of independent and dependent variables	Modified gradient projection using the Broyden-Fletcher-Shanno formula
Hartland	[G]	Extraction factor in one stage of column	C & S	Overall separation factor	Analytical differentiation and equating derivatives to zero
Ellingsen	[H]	Feed rates Reflux rates Purge rates	S	Cost of materials	Adaptive random search

[A] Salem, A. B., and G. V. Jeffreys, "Find Optimum Temperature for Countercurrent Extraction," *Hydrocarbon Process.* (October 1982) p. 93.
[B] Chen, N. H., "Optimal Theoretical Stages in Countercurrent Leaching," *Chem. Eng.* (Aug. 24, 1970) p. 71.
[C] Olson, R. S., "Optimum Economic Extraction," *Chem. Eng.* (Oct. 6, 1958) p. 143.
[D] Jensen, V. G., and G. V. Jeffreys, "The Economic Analysis of Liquid-Liquid Extraction Processes," *British Chem. Eng.*, **6**: 676 (1961).
[E] Treybal, R. E., "The Economic Design of Mixer-Settler Extractors," *AIChE J.*, **5**: 474 (1959).
[F] Jackson, P. J., and J. B. Agnew, "A Model Based Scheme for the On-Line Optimization of a Liquid-Liquid Extraction Process," *Comput. Chem. Eng.*, **4**: 241 (1983).
[G] Hartland, S., "The Optimum Operation of an Existing Forward and Backward Extractor," *Chem. Eng. Sci.*, **24**: 1075 (1969).
[H] Ellingsen, W. R., "Operating Cost Optimization Using a Dynamic Process Model," *Chem. Eng. Prog.* (January 1983) p. 43.

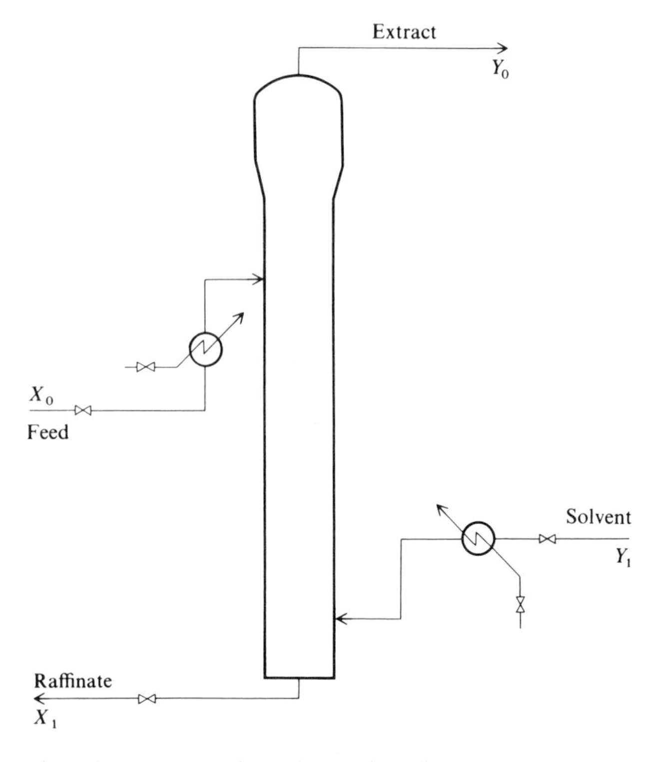

Figure E11.1*a* Extraction column schematic, Example 11.1.

1 The process model. Under certain conditions, the plug flow model for an extraction process has an analytical solution. Under other conditions, numerical solutions of the equations must be used. As a practical matter, specifying the model so that an analytical solution exists means assuming that the concentrations are expressed on a solute free mole basis, that the equilibrium relation between Y and X is a straight line $Y^* = mX + B$ (i.e., not necessarily through the origin), and that the operating line is straight, i.e., the phases are insoluble. Then the model is

$$\frac{dX}{dZ} - N_{OX}(X - Y) = 0 \qquad (a)$$

$$\frac{dY}{dZ} - FN_{OX}(X - Y) = 0 \qquad (b)$$

where F = extraction factor (mv_X/v_Y)
m = distribution coefficient
N_{OX} = number of transfer units
v_X, v_Y = superficial velocity in raffinate, extract phase
X = dimensionless raffinate phase concentration
Y = dimensionless extract phase concentration
Z = dimensionless contactor length

Figure E11.1a shows the boundary conditions X_0 and Y_1. Given values for m, N_{OX}, and the length of the column, a solution for Y_0 in terms of v_X and v_Y can be obtained; X_1 is related to Y_0 and F via a material balance: $X_1 = 1 - (Y_0/F)$. Hartland and Mecklenburgh (1975) list the solutions for the plug flow model (and also the axial dispersion model) for a linear equilibrium relationship, in terms of F:

$$Y_0 = \frac{F[1 - \exp\{N_{OX}(1 - F)\}]}{1 - F \exp[N_{OX}(1 - F)]} \tag{c}$$

In practice, N_{OX} would be calculated from experimental data by least squares or from an explicit relation for the plug flow model (Treybal, 1963)

$$N_{OX} = \left(\frac{1 - X_1}{X_1 + Y_0 - 1}\right) \ln\left(\frac{X_1}{1 - Y_0}\right) \tag{d}$$

Jackson and Agnew (1980) summarized a number of correlations for N_{OX} such as

$$N_{OX} = 4.81\left(\frac{v_X}{v_Y}\right)^{0.24} \tag{e}$$

2 Inequality constraints. Implicit constraints exist because of the use of dimensionless variables

$$\begin{aligned} X_0 &\leq X \leq X_1 \\ Y_1 &\leq Y \leq Y_0 \end{aligned} \tag{f}$$

Constraints on v_X and v_Y would be upper and lower bounds such as

$$\begin{aligned} 0.05 &< v_X < 0.25 \\ 0.05 &< v_Y < 0.30 \end{aligned} \tag{g}$$

and the flooding constraint

$$v_X + v_Y \leq 0.20 \tag{h}$$

3 Objective function. We will use for the objective function the same one proposed by Jackson and Agnew, namely to maximize the total extraction rate for constant disk rotation speed subject to the inequality and equality constraints:

$$\text{Maximize:} \quad f = v_Y Y_0 \tag{i}$$

The value of $m = 1.5$.

Results of the optimization. Figure E11.1*b* illustrates contours of the objective function for the plug flow model; the objective function (*i*) was optimized by the reduced gradient constrained optimization method. For small values of v_x (<0.01) the contours drop off quite rapidly. The starting point (point 1) is infeasible and is

$$v_x = 0.03$$

$$v_y = 0.10$$

which is infeasible. Points 2, 3, and 4 in Fig. E11.1*b* show the progress of the optimization method as it moves toward the optimum. Point 2 indicates the first feasible values of v_x and v_y (0.08, 0.10), point 3 indicates where the flooding

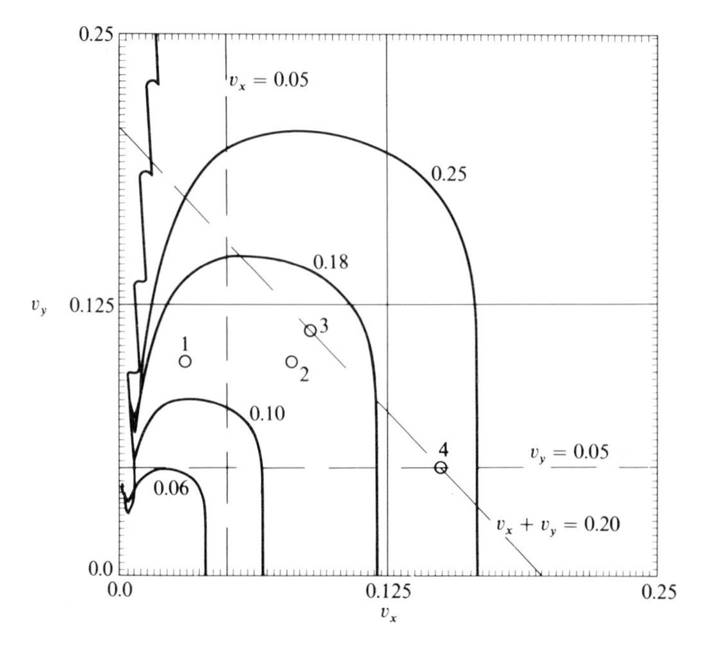

Figure E11.1*b* Contours for the objective function of extraction process. Points 1, 2, 3, and 4 indicate the progress of reduced-gradient method toward the optimum (point 4).

constraint (*h*) is active, and point 4 is the constrained optimum (0.15, 0.05). The objective function at point 4 is 0.2250.

11.2 OPTIMAL DESIGN AND OPERATION OF STAGED DISTILLATION COLUMNS

Distillation is probably the most widely used separation process in industry. Although the examples in this section focus on optimization procedures for distillation, absorption and extraction usually can be treated analogously. Table 11.2 summarizes a number of examples that in themselves contain sufficient detail to be useful if one or more optimization techniques are to be applied to steady state stagewise distillation.

In general we might classify optimization problems for steady state distillation into categories of increasing order of complexity:

1. Determine the optimal operating conditions for an existing column to achieve specific performance at minimum cost (or minimum energy usage) given the feed(s). Usually the manipulated (independent) variables are indirect heat inputs, cooling stream inputs, and product flow rates. The number of degrees of freedom is most likely equal to the number of product streams. Specific performance means specified component concentrations or fractional recoveries

Table 11.2 Techniques used to optimize conventional steady state staged distillation columns

Authors	Ref.	Independent column variables	Objective function	Optimization technique
Al-Haj-Ali & Holland	[A]	Reflux ratio Distillate (product) rate Ratio of molar flow rate of (1) light key and (2) heavy key in bottom product to distillate	Various	Box complex method with nested solution of material and energy balances and equilibrium relations
Sargent & Gaminibandara	[B]	One heat input Product flow rate, or reflux ratio	Reboiler heat input	Variable metric projection with nested solution of material and energy balances and equilibrium relations
Happel & Jordan	[C]	Reflux ratio	Capital costs of column, condenser, reboiler plus operating costs of steam	Analytical use of necessary conditions
Jafarey, Douglas, & McAvoy	[D]	Reflux ratio	Capital costs of column, condenser, reboiler plus operating costs of steam	Analytical use of necessary conditions
Malone & Douglas	[E]	Vapor rate (in terms of feed and distillate light key compositions, feed rate and separation factor Column configuration (for multiple columns)	Capital costs of column, condenser, reboiler plus operating costs of steam	Analytical use of necessary conditions
Cerda & Westerberg	[F]	Absorption factor ($L_k/K_{ik}V_k$) for heavy key component Stripping factor ($K_{ik}V_k/L_k$) for light key component Equilibrium constant, bottom tray, for heavy key Feed tray component location	Total number of theoretical stages	Analytical use of necessary conditions

(*Continued*)

Table 11.2 Techniques used to optimize conventional steady state staged distillation columns *(continued)*

Authors	Ref.	Independent column variables	Objective function	Optimization technique
Van Winkle & Todd	[G]	Minimum number of plates (via Fenske method, or minimum reflux ratio (via Underwood's method) Compositions of light and heavy key in distillate product and bottoms Fenske separation factor	Capital and operating costs	Statistical correlation
Prokopakis & Seider	[H]	Boil-up rate Fractional recovery of one component in the bottoms Composition of other two components in the bottoms	Minimize sum of squares of composition deviations from product specifications	Powell's algorithm together with material and energy balances and equilibrium relations
Krivosheev & Popkov	[I]	Flow rates of distillate, reflux, and two phases	Profit Maximize separation Maximize productivity of top, bottoms Minimize energy consumption	Two-stage decomposition with higher level relaxation controlling lower level of optimization
Tedder & Rudd	[J], [K]	Feed composition Pressure Feed enthalpy Minimum vapor flow ratio	Cost	Analytical and heuristic methods for synthesis; dynamic programming with variable metric or conjugate gradient search
Sophos, Stephanopoulos, & Morari	[L]	Various	Capital costs and costs of hot and cold streams	Decomposition and various topological rules
Fauth & Shinskey	[M]	Distillate flow Heat input	Incremental profit from product yield Product loss plus heating costs	Case study Unidimensional search Analytical differentiation

Martin, Latour, & Richard	[N]	Reflux rate	Product recovery value minus energy costs	Analytical differentiation and equating derivative to zero
Srygley & Holland	[O]	Number of plates below and above feed (and side streams)	Sum of squares of differences between calculated and specified values of selected design variables	Hooke & Jeeves method
Quadri	[P]	Column top pressure Internal reflux Compressor and distillate temperatures Composition of light key in bottoms	Annual cost = capital costs plus cooling water, electrical energy, lost value of light key in bottoms	Various
Zahradnik, Archer, & Rothfus	[Q]	Steam rate Distillate rate	Value of distillate and bottoms streams less feed and steam	Analytical differentiation and equating derivatives to zero
Melli & Spekuljak	[R]	Percent flooding	Ratio of concentration of heavy component in bottoms to other component	Analytical differentiation and equating derivatives to zero
Mazna et al.	[S]	Vapor flow rate from reboiler Liquid flow rate	Minimum (squared) deviation of the vapor temperature above a critical plate from a specified value	Analytically via Lagrange multipliers
Petlyuk, Platonov, & Slavinskii	[T]	Concentration at intermediate point of energy input	Minimum ideal work of separation	Analytical differentiation and equating result to zero

(Continued)

Table 11.2 Techniques used to optimize conventional steady state staged distillation columns *(continued)*

Authors	Ref.	Independent column variables	Objective function	Optimization technique
Waller & Gustafsson	[U]	Ratio distillate rate/feed rate Ratio vapor flow rate in stripping section/feed rate Total number of plates	Economic loss of difference between light composition in top product and bottoms plus similar loss for heavy component leaving in top product vs. bottoms plus cost of vapor generated	Unidimensional search

[A] Al-Hag-Ali, N. S., and C. D. Holland, "Way to Find Distillation Optimum," *Hydrocarbon Process.* (July, 1979) p. 165.
[B] Sargent, R. W. H., and K. Gaminibandara, "Optimum Design of Plate Distillation Columns," in *Optimization in Action* (ed. L. D. W. Dixon), Academic Press, New York (1976) p. 266.
[C] Happel, J., and D. G. Jordan, *Chemical Process Economics*, 2d ed., Marcel Dekker, New York 1975, p. 385
[D] Jafarey, A., J. M. Douglas, and T. J. McAvoy, "Short-Cut Techniques for Distillation Column Design and Control. 1. Column Design," *Ind. Eng. Chem. Process Des. Dev.*, **18**: 197 (1979).
[E] Malone, M. F., and J. M. Douglas, "Simple, Analytical Criteria for Sequencing of Distillation Columns for Multicomponent Separation," Personal communication, 1980.
[F] Cerda, J., and A. W. Westerberg, "Shortcut Method for Complex Distillation Column," *Ind. Eng. Chem. Process Des. Dev.*, **20**: 546 (1981).
[G] Van Winkle, M., and W. G. Todd, "Optimum Fractionation Design by Simple Graphical Methods," *Chem. Eng.* (Sept. 20, 1971) p. 136.
[H] Prokopakis, G. J., and W. D. Seader, "Feasible Specifications in Azeotropic Distillation," *AIChE J.*, **29**: 49 (1983).
[I] Krivosheev, V. P., and V. F. Popkov, "Optimization of Static Regimes of Complex Fractionation Units Based on the Decomposition Approach," *Teor. Osnoy Khim. Teckh.*, **16**: 381 (1982) (English translation).
[J] Tedder, W. D., and D. F. Rudd, "Parametric Studies in Industrial Distillation," *AIChE J.*, **24**: 303 (1978).
[K] Tedder, W. D., and D. F. Rudd, "Computer Aided Design of an Optimal Deethanizer Sequence," paper presented at the 1981 SCSC, Washington, D.C. (1981).
[L] Sophos, A., G. Stephanopoulos, and M. Morari, "Synthesis of Optimum Distillation Sequences with Heat Integration," Paper presented at the National AIChE Meeting, Miami, Florida, November, 1978.
[M] Fauth, G. F., and F. G. Shinskey, "Advanced Control of Distillation Columns," *Chem. Eng. Prog.*, **71**(6): 49 (June, 1975).
[N] Martin, G. D., P. R. Latour, and L. A. Richard, "Closed Loop Optimization of Distillation Energy," *Chem. Eng. Prog.*, **77** (9): 33 (1981).
[O] Srygley, J. M., and C. D. Holland, "Optimal Design of Conventional and Complex Distillation Columns," *AIChE J.* **11**: 695 (1965).
[P] Quadri, G. P., "Use Heat Pump for P-P Splitter," *Hydrocarbon Process.* (February 1981) p. 119.
[Q] Zahradnik, R. L., D. H. Archer, and R. R. Rothfus, "Dynamic Optimization of a Distillation Column," *Chem. Eng. Prog. Symp. Ser. 46*, **59**: 132 (1960).
[R] Melli, T. R., and Z. Spekuljak, "Optimal Design of Packed Distillation Towers," *Ind. Eng. Chem. Process Des. Devel.*, **22**: 220 (1983).
[S] Mazina, S. G. et al., "Static Optimization of Rectification from the Temperature of a Characteristic Plate," *J. Appl. Chem.*, **52**: 2022 (1979) (Translation).
[T] Petlyuk, F. B., V. M. Platanov, and D. M. Slavinskii, "Thermodynamically Optimal Method for Separating Multicomponent Mixtures," *Int. Chem. Eng.*, **5**: 555 (1965).
[U] Waller, K. V., and T. K. Gustafsson, "On Optimal Steady-State Operation in Distillation," *Ind. Eng. Chem. Process Des. Dev.*, **17**: 313 (1978).

from the feed (specifications leading to equality constraints), or minimum (or maximum) concentrations and recoveries (specifications leading to inequality constraints). In principle any of the specified quantities as well as costs can be calculated from the values of the manipulated variables given the mathematical model (or computer code) for the column. When posed as described above, the optimization problem is a nonlinear programming problem often with implicit nested loops for calculation of physical properties. If the number of degrees of freedom is reduced to zero by specifications placed on the controlled variables, the optimization problem reduces to the classic problem of distillation design that requires just the solution of a set of nonlinear equations.

2. A more complex problem is to not only determine the values of the operating conditions as outlined in 1 above, but to also determine the (minimum) number of stages required for the separation. Because the stages are discrete (although in certain examples in this book we have treated them as continuous variables), the problem outlined in 1 becomes a nonlinear mixed integer programming problem (see Chap. 9). In this form of the design problem, the costs include both capital costs and operating costs. Capital costs increase with the number of stages and internal column flow rates whereas operating costs decrease up to a certain point.
3. An even more difficult problem is to determine the number of stages and the optimal locations for the feed(s) and side stream(s) withdrawal. Fortunately, the range of candidates for stage locations for feed and withdrawals is usually small, and from a practical viewpoint the objective function is usually not particularly sensitive to a specific location within the appropriate range.

Keep in mind that an optimal design for one set of presumed input conditions may not be optimal for another. Suppose, for example, that the feed composition changes from an earlier composition. Then the operating conditions will have to be adjusted to meet the process constraints, but the new optimal set of conditions may result in higher cost, more energy consumption, and so on, than before the change. In addition to low cost, one would like to see a relatively flat objection function with respect to changes (or uncertainty) in process inputs because a flat response means the design is more resilient and flexible than a sensitive objective function. Figure 11.1 illustrates the trade-offs among several possible design configurations *A*, *B*, and *C*.

A factor to keep in mind in carrying out the optimization of a distillation column is that predicting cost or energy savings is not simple. In a processs that has a high throughput of valuable products worth, say, $500,000 per day, it is not necessarily true that a 1 percent improvement in yield will result in $5000 per day in benefits. An absolute minimum heat duty is required to operate a column with zero reflux, and if an existing column is operating at a low reflux ratio (say ~ 1) then only one-half of the reboiler duty is available for reflux variations. A 10 percent reduction in reboiler (and condenser) duty means a 20 percent reduction in reflux which may not provide a feasible solution.

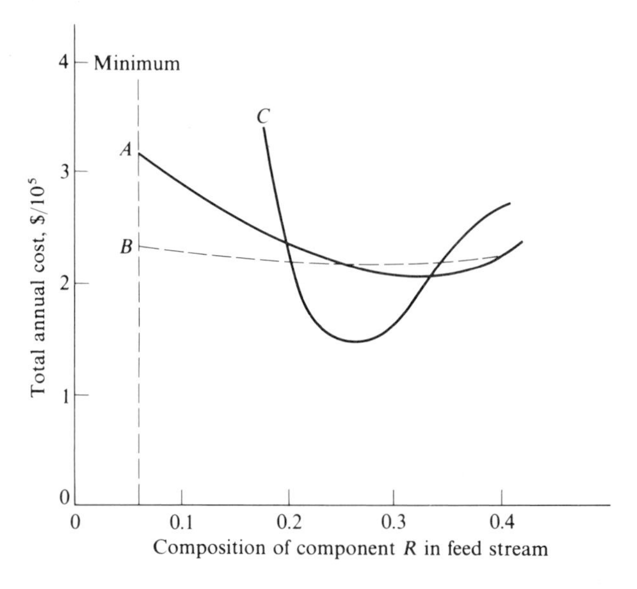

Figure 11.1 Minimum cost surfaces for three different distillation column designs.

EXAMPLE 11.2 OPTIMAL DESIGN AND OPERATION OF CONVENTIONAL STAGED DISTILLATION COLUMNS

In this example we optimize the operating conditions for a staged distillation column. Figure E11.2 shows a typical column and some of the notation we will use, while Table E11.2*a* lists the other variables and parameters. Feed will be denoted by superscript *F*. Withdrawals take the subscripts of the withdrawal stage. Superscripts *V* for vapor and *L* for liquid are used as needed to distinguish between phases. If we number the stages from the bottom of the column (the reboiler) upward with $k = 1$, then $V_0 = L_1 = 0$, and at the top of the column, or the condenser, $V_n = L_{n+1} = 0$. We first formulate the equality constraints, then the inequality constraints, and lastly the objective function.

1 The equality constraints. The process model comprises the equality constraints. For a conventional distillation column we would have the following typical relations:

(*a*) *Total material balances* (*one for each stage k*)

$$F_k^L + F_k^V + V_{k-1} + L_{k+1} = V_k + L_k + W_k^V + W_k^L \qquad (a)$$

(F_k and W_k would ordinarily not be involved in most of the stages)

(*b*) *Component material balances* (*one for each component i for each stage k*)

$$x_{i,k}^F F_k^L + y_{i,k}^F F_k^V + y_{i,k-1} V_{k-1} + x_{i,k+1} L_{k+1}$$

$$= y_{i,k} V_k + x_{i,k} L_k + y_{i,k} W_k^V + x_{i,k} W_k^L \qquad (b)$$

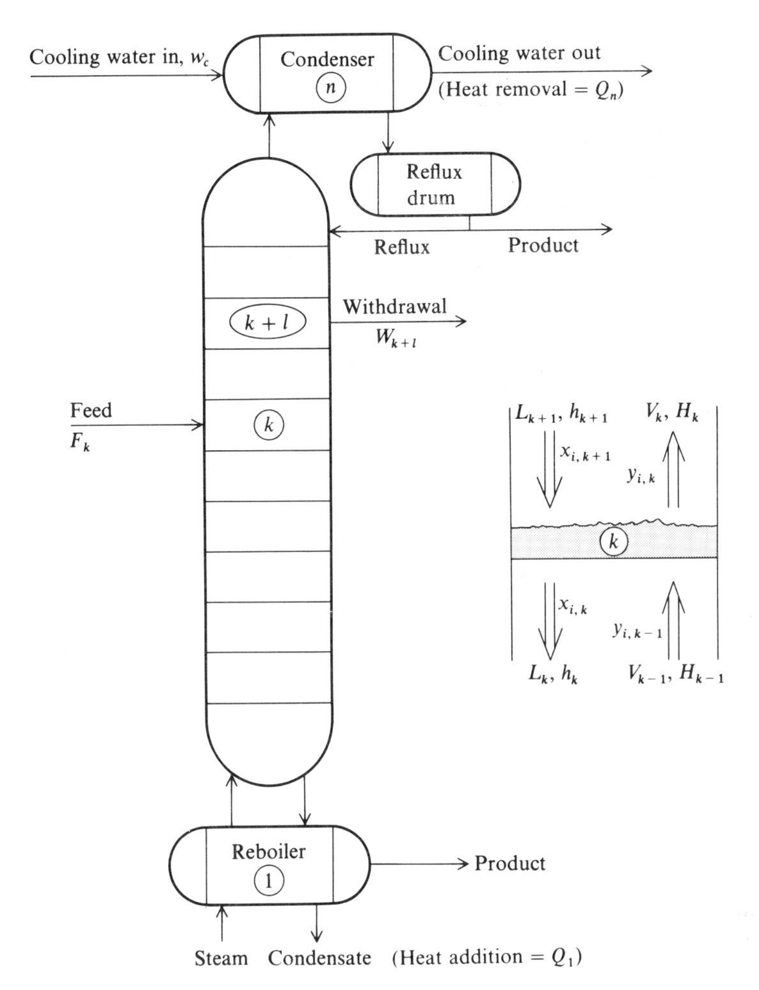

Figure E11.2 Schematic of staged distillation column.

(c) *Energy balance (one for each stage)*

$$Q_k + h_k^F F_k + H_{k-1}V_{k-1} + h_{k+1}L_{k+1} = H_k V_k + h_k L_k + H_k W_k^V + h_k W_k^L \tag{c}$$

(d) *Equilibrium relations for liquid and vapor at each stage (one for each stage)*

$$y_{i,k} = K_{i,k} x_{i,k} \tag{d}$$

(e) *Relation between equilibrium constant and p, T, x, y (one for each stage)*

$$K_{i,k} = K_i(p_k, T_k, \mathbf{x}_k, \mathbf{y}_k) \tag{e}$$

(f) *Relation between enthalpies and p, T, x, y (one for each stage)*

$$h_k = h(p_k, T_k, \mathbf{x}_k) \tag{f}$$

$$H_k = H(p_k, T_k, \mathbf{y}_k) \tag{g}$$

Table E11.2a Notation for distillation example

F_k	flow of feed into stage k, moles
h_k	liquid enthalpy (a function of p_k, T_k, and $\mathbf{x}_k$) on stage k
H_k	vapor enthalpy (a function of p_k, T_k, and $\mathbf{y}_k$) on stage k
k	stage index number, $k = 1, \ldots, n$.
$K_{i,k}$	equilibrium constant for component i for the mixture on stage k (a function of p_k, T_k, $\mathbf{x}_k$, $\mathbf{y}_k$)
L_k	flow of liquid from stage k, moles
m	number of components, $i = 1, \ldots, m$
p_k	pressure on stage k
Q_k	heat transfer flow to stage k (positive when into stage)
T_k	temperature on stage k
V_k	flow of vapor from stage k, moles
W_k	withdrawal stream from stage k, moles
$x_{i,k}$	mole fraction of component i on stage k in the liquid phase
$y_{i,k}$	mole fraction of component i on stage k in the vapor phase

The above classic set of algebraic equations form a well-defined sparse structure that has been analyzed extensively. Innumerable techniques of solution have been proposed for problems with zero degrees of freedom, i.e., the column operating or design variables are completely specified (see Holland (1975) or Wang and Wang (1981)).

Our interest here in posing an optimization problem is to have one or more degrees of freedom left after prespecifying the values of most of the independent variables. Frequently values are given for the following parameters:

1. Number of stages
2. Flow rate, composition, and enthalpy of the feed(s)
3. Location of the feed(s) and side stream withdrawal(s)
4. Flow rate of the side stream(s)
5. Heat input rate to each stage except one
6. Stage pressures (based on column detailed design specifications)

If the controllable parameters remaining to be specified, namely (1) one heat input, and (2) the flow rate of the product (or the reflux ratio), are determined via optimization, all of the values of V_k, L_k, T_k, $x_{i,k}$, and $y_{i,k}$, and the enthalpies can be calculated. More than two degrees of freedom might be introduced by eliminating some of the prespecified parameter values.

(*g*) *Certain implicit equality constraints exist*

Because of the way the model is specified, you must take into account the following additional equations in the solution of the column model:

$$\sum_{i=1}^{m} x_{i,k} = 1 \qquad (h)$$

$$\sum_{i=1}^{m} y_{i,k} = 1 \qquad (i)$$

2 The inequality constraints. Various kinds of inequality constraints exist such as requiring that all the $x_{i,k}$, $y_{i,k}$, Q_k, F_k, W_k, etc., be positive, upper and lower bounds on some of the product stream concentrations, and minimum recovery factors. A recovery factor for stage k is the ratio

$$\frac{x_{i,k} W_k^L + y_{i,k} W_k^V}{\sum_i (x_{i,k} F_k^L + y_{i,k} F_k^V)}$$

3 The objective function. The main costs of operation are the heating and cooling costs, Q_1, and Q_n, respectively. We will assume all the other values of Q_k are zero. Q_n is determined from the energy balance, so Q_1 is the independent variable. The cost of operation per annum will be assumed to be directly proportional to Q_1 since the maintenance and cooling costs are relatively small and the capital costs per annum are already fixed. Consequently, the objective function is relatively simple:

$$\text{Minimize:} \quad Q_1 \qquad (j)$$

As posed above, the problem is a nonlinear programming problem involving nested loops of calculations, the outer loop of which is Eq. (j) subject to Eqs. (a) through (i) and subject to the inequality constraints.

4 Results for a specific problem with five degrees of freedom. We use for illustration the data of Sargent and Gaminibandara (1976) for the objective function (j). The problem is to determine the location and individual amounts of the feeds given the following information.

A column of four stages exists analogous to that shown in Fig. E11.2 except that more than one feed can exist (the reboiler is stage 1 and the condenser is stage 4). Feed and product specifications are:

$$\text{Total feed} = 100 \text{ lb mol/h liquid}$$

$$h_F = 4000 \text{ Btu/lb mol}$$

$$x_1 = 0.05\ (C_3H_8)$$

$$x_2 = 0.15\ (i\text{-}C_4H_{10})$$

$$x_3 = 0.25\ (n\text{-}C_4H_{10})$$

$$x_4 = 0.20\ (i\text{-}C_5H_{12})$$

$$x_5 = 0.35\ (n\text{-}C_5H_{12})$$

$$\text{Top product} = 10 \text{ lb mol/h liquid}$$

$$x_5 \leq 0.07$$

The equality constraints are Eqs. (a) through (i) plus

$$\sum_{k=1}^{4} F_k = 100 \qquad (k)$$

The inequality constraints are ($k = 1, \ldots, 4$)

$$Q_1 \geq 0 \tag{l}$$

$$Q_4 \leq 0 \tag{m}$$

$$x_{i,k} \geq 0 \tag{n}$$

$$y_{i,k} \geq 0 \tag{o}$$

$$F_k \geq 0 \tag{p}$$

$$x_{5,4} \leq 0.07 \tag{q}$$

This problem has five degrees of freedom, representing the five variables Q_1, F_1, F_2, F_3, and F_4.

Various rules of thumb and empirical correlations exist to assist in making initial guesses for the independent variables. All the values of the feeds here can be assumed to be equal initially. If the reflux ratio is selected as an independent variable, a value of 1 to 1.5 times the minimum reflux ratio is generally appropriate (Holland, 1963; Peters and Timmerhaus, 1980).

To solve the problem a successive quadratic programming code was used in the outer loop of calculations. Inner loops were used to evaluate the physical properties. The results shown in Table E11.2*b* were essentially the same as those obtained by Sargent and Gaminibandara. Forward finite differences with a step size of $h = 10^{-7}$ were used on a CDC computer as substitute for derivatives. Equilibrium data were taken from Holland (1963).

We can conclude that it is possible to use some of the cold feed as reflux in the top stage without voiding the product composition specification. This outcome would not be an obvious choice for the problem specifications.

If capital costs are to be included in the objective function, the problem becomes much more complex. Al-Hag-Ali and Holland (1979) and Van Winkle and Todd (1971) have discussed the procedures for formulating such a cost function and minimization algorithms which can be used to solve distillation design problems.

Table E11.2*b* Results of optimization

Variable	Initial guess for the variable	Optimal values for the variable
F_1	25	23.7
F_2	25	0
F_3	25	0
F_4	25	76.3
Q_1	5.0×10^6	3.38×10^5
$x_{5,4}$	....	0.07

In the previous example there was a brief mention of the need to calculate physical property values. Valid physical property relationships thus form an important feature of the process model. To validate a model, representative data must be fit by some type of correlation using an optimization technique. Nonlinear regression instead of linear regression may be involved in the fitting. We illustrate the procedure in the next example.

EXAMPLE 11.3 NONLINEAR REGRESSION TO FIT VAPOR-LIQUID EQUILIBRIUM DATA

Separation systems include in their mathematical models various vapor-liquid equilibrium (VLE) correlations that are specific to the binary or multicomponent system of interest. Such correlations are usually obtained by fitting VLE data by least squares. The nature of the data can depend on the level of sophistication of the experimental work. In some cases it is only feasible to measure the total pressure of a system as a function of the liquid phase mole fraction (no vapor phase mole fraction data are available). Various approaches to fit such data by nonlinear regression (e.g., least squares, maximum likelihood) have been discussed by Prausnitz et al. (1980). In most cases an unconstrained optimization algorithm (see Chap. 6) can be employed.

Vapor-liquid equilibria data are often correlated using two adjustable parameters per binary mixture. In many cases, multicomponent vapor-liquid equilibria can be predicted using only binary parameters. For low pressures, the equilibrium constraint is

$$x_i \gamma_i p_i^{\text{sat}} = y_i p \qquad (i = 1, 2) \tag{a}$$

where p is the total pressure, p_i^{sat} is the saturation pressure of component i, x_i is the liquid phase mole fraction of component i, γ_i is the activity coefficient, and y_i is the vapor phase mole fraction. The van Laar model for a binary mixture is

$$\ln \gamma_1 = A_{12}\left[\frac{A_{21}x_2}{A_{12}x_1 + A_{21}x_2}\right]^2 \tag{b}$$

and

$$\ln \gamma_2 = A_{21}\left[\frac{A_{12}x_1}{A_{12}x_1 + A_{21}x_2}\right]^2 \tag{c}$$

where A_{12} and A_{21} are binary constants which are adjusted by optimization to fit the calculated data for x_i. To use total pressure measurements we write

$$p = y_1 p + y_2 p \tag{d}$$

or, using Eqs. (a) to (c)

$$p = x_1 \exp\left[A_{12}\left(\frac{A_{21}x_2}{A_{12}x_1 + A_{21}x_2}\right)^2\right]p_i^{\text{sat}} + x_2 \exp\left[A_{21}\left(\frac{A_{12}x_1}{A_{12}x_1 + A_{21}x_2}\right)^2\right]p_2^{\text{sat}} \tag{e}$$

The saturation pressures can be predicted at a given temperature using the Antoine equation. For a given temperature and a binary system ($x_2 = 1 - x_1$)

$$p = p(x_1, A_{12}, A_{21}) \tag{f}$$

so that the two binary coefficients may be determined from experimental values of p vs. x_1 by nonlinear least squares estimation (regression), i.e., by minimizing the objective function

$$f = \sum_{j=1}^{n} (p_j^{\text{calc}} - p_j^{\text{expt}})^2 \qquad (g)$$

where n is the number of data points.

In the book, *Vapor-Liquid Equilibrium Data Collection*, DECHEMA (1981), nonlinear regression has been applied to develop several different vapor-liquid equilibria relations suitable for correlating numerous data systems. As an example, p vs. x_1 data for the system water (1), 1,4 dioxane (2) at 20.00°C are listed in Table E11.3. The Antoine equation coefficients for each component are also shown in Table E11.3. A_{12} and A_{21} were calculated by Gmehling et al. (1981) using the Nelder-Mead Simplex method (see Sec. 6.1.4) to be 2.0656 and 1.6993, respectively. The vapor phase mole fractions, total pressure, and the deviation between predicted and experimental values of the total p

$$\Delta p_j = p_j^{\text{calc}} - p_j^{\text{expt}}$$

are listed in Table E11.3 for increments of $x_1 = 0.10$. The mean Δp is 0.09 mmHg for pressures ranging from 17.5 to 28.10 mmHg. Figure E11.3 shows the predicted y_1 vs. x_1 data; note that the model predicts an azeotrope at $x_1 = y_1 = 0.35$.

Table E11.3 Experimental VLE data for the system
(1) Water. (2) 1,4 dioxane at 20°C.

Experimental data		Predicted values		
x_1	p^{expt}(mmHg)	p^{calc}	Δp	y^{calc}
0.00	28.10	28.10	0.00	0.0
0.10	34.40	34.20	−0.20	0.2508
0.20	36.70	36.95	0.25	0.3245
0.30	36.90	36.97	0.07	0.3493
0.40	36.80	36.75	−0.05	0.3576
0.50	36.70	36.64	−0.06	0.3625
0.60	36.50	36.56	0.06	0.3725
0.70	35.40	35.36	−0.04	0.3965
0.80	32.90	32.84	−0.06	0.4503
0.90	27.70	27.72	0.02	0.5781
1.00	17.50	17.50	0.00	1.0

Antoine constants: $\log p^{\text{sat}} = a_1 - \dfrac{a_2}{T + a_3}$ p^{sat}: mmHg, T: °C

	a_1	a_2	a_3	Range
(1) Water	8.07131	1730.630	233.426	(1–100°C)
(2) 1,4 dioxane	7.43155	1554.679	240.337	(20–105°C)

Note: Data reported by Hororka et al. (1936).

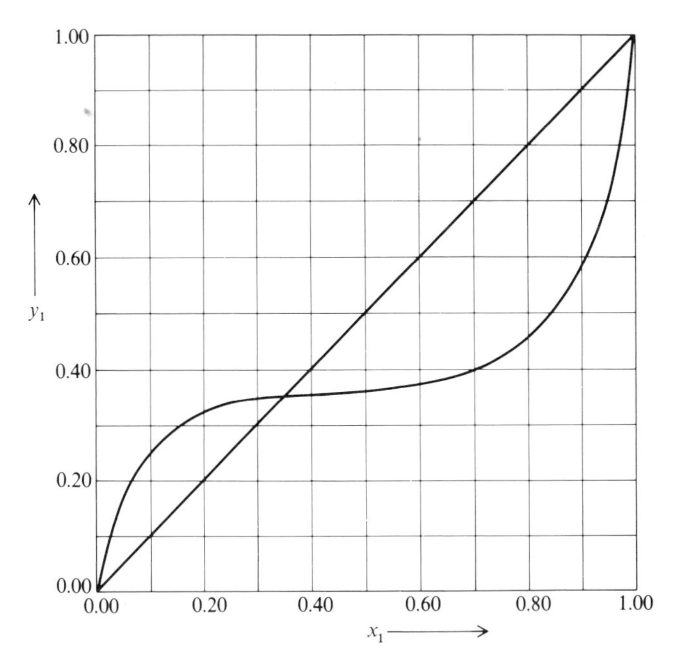

Figure E11.3 Experimental vapor-liquid equilibrium data, Example 11.3. [*Source: Gmehling et al.* (*1981*).]

Example 11.2 illustrated optimization involving five independent variables. Optimization of some measure of column performance in terms of a single variable, such as the reflux ratio, is an example of simplifying a problem so that a single variable search can be carried out. Peters and Timmerhaus (1980) and Happel and Jordan (1975) include examples of such problems.

Initial design of a column involves specifying the separation of a feed of known composition and temperature. Constraints require a minimum acceptable purity of the overhead and/or bottoms product. The desired separation can be achieved with relatively low energy requirements by using a large number of trays, thus incurring larger capital costs with the reflux ratio at its minimum value. On the other hand, by increasing the reflux ratio, the overhead composition specification can be met by a fewer number of trays but with higher energy costs.

Prior to the advent of effective optimization codes, to determine the optimum reflux ratio in design, a one-dimensional search was executed via case studies using a McCabe-Thiele diagram (or a column simulator for complicated systems). Short-cut design techniques also can be employed in some instances. For example, Jafarey et al. (1979) derived an expression based on Smoker's equation that related the optimum reflux ratio to the feed mole fraction, relative volatility, and a separation factor which itself was a function of overhead and bottoms composition. The optimum value of the reflux ratio can also be determined by evaluating the ratio of annualized capital costs (column, reboiler, condenser) to operating costs (steam, cooling water).

Once a distillation column is in operation, the number of trays is fixed and there are very few degrees of freedom that can be manipulated to minimize operating costs; the reflux ratio can be used to influence the steady-state operating point. Buzzi-Ferraris and Troncani (1985) describe techniques for optimization of existing separators. Martin et al. (1981) have discussed the trade-offs between energy costs and product recovery for an existing column. Prior to the mid-1970s, reflux ratios were relatively high in order to maximize product recovery because fuel costs were low relative to product value. Since that time a more balanced viewpoint must be considered; the reflux ratio should be selected using a more general objective function based on variable costs. Figure 11.2 shows typical variable cost patterns as a function of reflux ratio.

The optimization of reflux ratio is particularly attractive for columns which operate with

1. High reflux ratio
2. High differential product values (between overhead and bottom)
3. High utility costs
4. Low relative volatility
5. Feed light key far from 50 percent

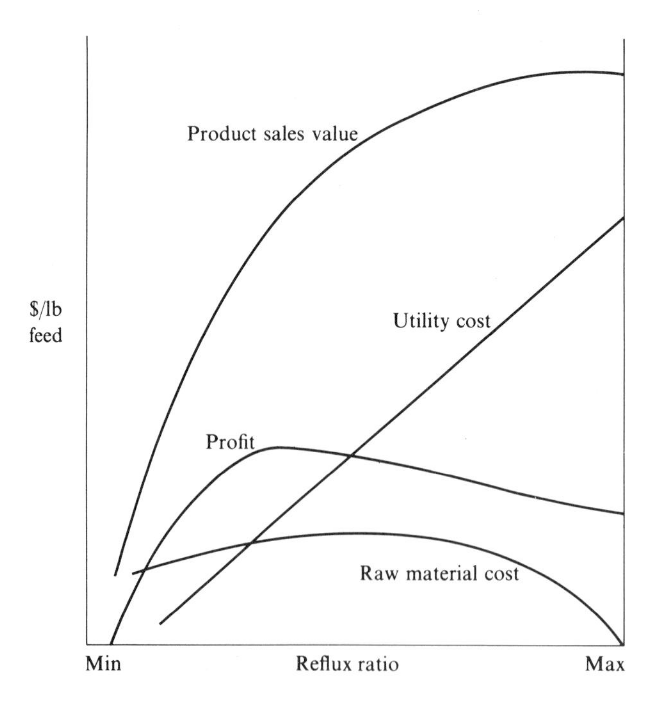

Figure 11.2 Variable cost trade-offs for distillation column.

In Example 11.4 below we illustrate the application of a one-dimensional search technique from Chap. 5 to a problem posed by Martin et al. (1981). Thereafter, in Example 11.5, we show the use of linear programming to optimize the split of heavy and light components in a train of columns.

EXAMPLE 11.4 DETERMINATION OF THE OPTIMUM REFLUX RATIO FOR A STAGED DISTILLATION COLUMN

Martin et al. (1981) described an application of optimization to an existing tower separating propane and propylene. The lighter component (propylene) is more valuable than propane. For example, propylene and propane in the overhead product were both valued at \$0.20/lb (a small amount of propane was allowable in the overhead), while propane in the bottoms was worth \$0.12/lb and propylene \$0.09/lb. The overhead stream had to be at least 95 percent propylene. Based on the data in Table E11.4*a*, we will determine the optimum reflux ratio for this column using derivations provided by McAvoy (1985). He employed correlations for column performance (operating equations) developed by Eduljee (1975), which in turn evolved from the original correlations by Gilliland.

Table E11.4*a* Notation and values for the propane-propylene splitter

Symbol	Description	Value
B	Bottoms flow rate	
C_1	Reboiler heat cost	\$3.00/10^6 Btu
C_2	Condenser cooling cost	\$0.00/10^6 Btu
C_B	Value of propylene in bottoms	
C'_B	Value of propane in bottoms	
C_F	Cost per pound of propylene	
C'_F	Cost per pound of propane	
C_D	Value of propylene in overhead	
C'_D	Value of propane in overhead	
D	Distillate flow-rate	
F	Feed rate	1,200,000 lb/day
L	Liquid flow-rate	function of R (mol/day)
N	Number of equilibrium stages	94
N_m	Minimum equilibrium stages	function of reflux ratio, R
Q_C	Condenser load requirement	$Q_C \cong \lambda V$
Q_R	Reboiler heat requirement	$Q_R \cong \lambda V$
R	Reflux ratio	(To be optimized)
R_m	Minimum reflux ratio	11.17
U	Heavy key differential value	−\$0.08/lb
V	Vapor flow-rate	function of R (mol/day)
W	Light key differential value	\$0.11/lb
X_B	Bottom light key mole fraction	(To be optimized)
X_D	Overhead light key mole fraction	0.95
X_F	Feed light key mole fraction	0.70
α	Relative volatility	1.105
λ	Latent heat	130 Btu/lb (avg. mixture)

1 Equality constraints. The Eduljee correlation involves two parameters: R_m, the minimum reflux ratio, and N_m, the equivalent number of stages to accomplish the separation at total reflux. His operating equations relate N, α, X_F, X_D, and X_B (see Table E11.4*a* for notation) all of which have known values except X_B as listed in Table E11.4*a*. Once R is specified, you can find X_B by sequential solution of the three equations given below.

First, calculate R_m

$$R_m = \frac{1}{(\alpha - 1)}\left[\frac{X_D}{X_F} - \alpha\frac{(1 - X_D)}{(1 - X_F)}\right] \tag{a}$$

Substitute the value of R_m in Eq. (*b*) to find N_m

$$\left(\frac{N - N_m}{N + 1}\right) = 0.75\left[1 - \left(\frac{R - R_m}{R + 1}\right)^{0.5668}\right] \tag{b}$$

Note that the Eduljee correlation becomes inaccurate for $R \rightarrow R_m$. Lastly, compute X_B from

$$N_m = \frac{\ln\{[X_D/(1 - X_D)]\cdot[(1 - X_B)/X_B]\}}{\ln \alpha} \tag{c}$$

Equations (*a*) to (*c*) comprise equality constraints relating X_B and R.

Once X_B is calculated, the overall material balance for the column shown in Fig. E11.4 can be computed. The pertinent equations are (the units are moles):

$$F = D + B \tag{d}$$

$$X_F F = X_D D + X_B B \tag{e}$$

Equations (*d*) and (*e*) contain two unknowns: D and B, which can be determined once F, X_F, X_B, and X_D are specified. In addition, if the assumption of constant molal overflow is made, then the liquid (L) and vapor flows (V) are

$$L = RD \tag{f}$$

$$V = (R + 1)D \tag{g}$$

2 Objective function. Next we develop expressions for the income and operating costs. The operating profit f is given by

$$f = \text{propylene sales} + \text{propane sales} - \text{utility costs} - \text{raw material costs} \tag{h}$$

$$f = \{C_D X_D D + C_B X_B B\} + \{C'_D(1 - X_D)D + C'_B(1 - X_B)B\} - \{C_1 Q_R + C_2 Q_C\} - \{C_F X_F F + C'_F(1 - X_F)F\} \tag{i}$$

The brackets indicate the correspondence between the words in Eq. (*h*) and the symbols in Eq. (*i*). Q_R is the reboiler heat requirement and Q_C is the cooling load.

Equation (*i*) can be rearranged by substituting for DX_D in the propylene sales and for BX_B in the propane sales using Eq. (*e*) and defining $-W = C_B - C_D$ and $-U = C'_D - C'_B$ as follows

$$f = C_D X_F F + C'_B(1 - X_F)F - C_F X_F F - C'_F(1 - X_F)F - C_1 Q_R - C_2 Q_C - W X_B B - U(1 - X_D)D \tag{j}$$

Note that the first four terms of f are fixed values, hence these terms can be deleted from the expression for f in the optimization. In addition, it is reasonable

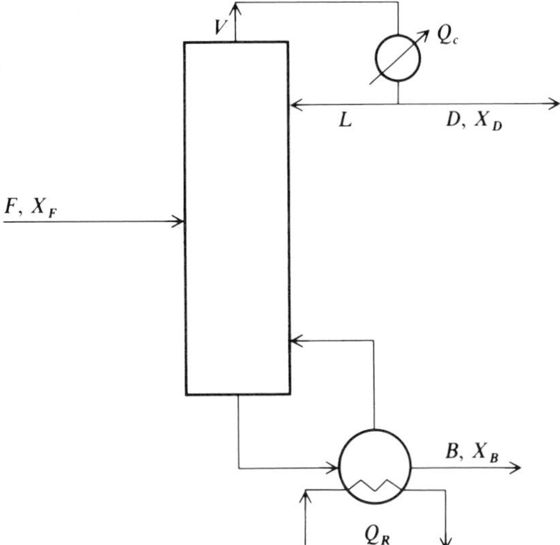

Figure E11.4 Distillation column flowsheet.

to assume $Q_R \approx Q_C \approx \lambda V$. Lastly, the right-hand side of Eq. (j) can be multiplied by -1 to give the final form of the objective function (to be minimized):

$$f_1 = (C_1 + C_2)\lambda V + WX_B B + U(1 - X_D)D \tag{k}$$

Note: λ must be converted to Btu/mol, and the costs to \$/mol.

3 Solution. Based on the data in Table E11.4a we minimized f_1 with respect to R using a quadratic interpolation one-dimensional search (see Chap. 5). The value of R_m from Eq. (a) was 11.338. The initial bracket was $12 \leq R \leq 20$, and $R = 16$, 18, and 20 were selected for the initial three points. The convergence tolerance on the optimum required that f_1 should not change by more than 0.01 from one iteration to the next.

The iterative program incorporating the quadratic interpolation search yielded the results in Table E11.4b. The optimum reflux ratio of 17.06 and the cost, f_1, was \$3870/day. Table E11.4c shows the variation in f_1 for ± 10 percent change in R. The profit function changes \$100/day or more.

Table E11.4b Iterations in quadratic interpolation test problem

	Left bracket		Center point		Right bracket		Interpolated values	
Iteration	x	f	x	f	x	f	x	f
1	16.00	3967.13	18.00	3922.14	20.00	4256.45	17.24	3872.22
2	18.00	3922.14	17.24	3872.22	16.00	3967.13	17.16	3870.79
3	17.24	3872.22	17.16	3870.79	16.00	3967.13	17.09	3870.21
4	17.16	3870.79	17.09	3870.21	16.00	3967.13	17.06	3870.18
5	17.09	3870.21	17.06	3870.18	16.00	3967.13	17.06	3870.17

Final solution
$x = 17.06$
$f = 3870.17$

Table E11.4c Sensitivity study of reflux ratio optimum

Reflux ratio (R)	X_B (mol §)	Costs ($)/day
17.07	0.0432	3870
18.77†	0.0303	4024
15.36†	0.0683	4159

† Indicates 17.07 ± 10%

Linear programming is a particularly appropriate optimization technique to apply to staged separation processes because such process models are represented by sparse matrices for the constraints. Nonlinear constraints can be accommodated within the LP framework by linearization about an operating point or by piecewise linear approximations if more accuracy is needed. Even with hundreds of linear constraints, each variable occurs only in about four to six constraints altogether so that LP codes that exploit sparseness should be used if available. Example 11.5 below illustrates the use of linear programming for a very small problem in which sparseness is not a significant factor.

EXAMPLE 11.5 USE OF LINEAR PROGRAMMING TO OPTIMIZE A SEPARATION TRAIN

In this example we illustrate use of linear programming to optimize the operation of the solvent splitter column shown in Fig. E11.5. The feed is naphtha which has a value of $42/bbl in its alternate use as a gasoline blending stock. The light ends sell at $53/bbl while the bottoms are passed through a second distillation column to yield two solvents. A medium solvent comprising 50 to 70 percent of the bottoms can be sold for $68/bbl, while the remaining heavy solvent (30 to 50 percent of the bottoms) can be sold for $42/bbl.

Another part of the plant requires 200 bbl/day of medium solvent; an additional 200 bbl/day can be sold to an external market. The maximum feed that can be processed in column 1 is 2000 bbl/day. The operational costs (i.e., utilities) associated with each distillation column are $1.25/bbl feed. The operating range for column 2 is given as the percentage split of medium and heavy solvent.

We formulate the problem as a linear programming problem by developing the objective function and constraints. Five variables are identified in Fig. E11.5, and we simplify the equations so that they are expressed in terms of the two independent variables, x_4 and x_5. Then we solve the linear programming problem to get the maximum revenue and percentages of the output streams in column 2.

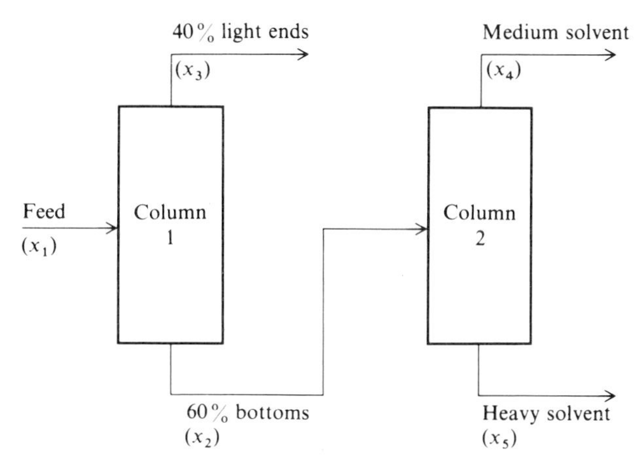

Figure E11.5 Train of distillation columns, Example 11.5.

Let us define the operating variables as follows (all are in units of bbl/day):

$$x_1 = \text{naphtha feed rate}$$

$$x_2 = \text{bottoms production rate}$$

$$x_3 = \text{light ends production rate}$$

$$x_4 = \text{medium solvent production rate}$$

$$x_5 = \text{heavy solvent production rate}$$

The objective function in terms of profit is

$$f = 68x_4 + 42x_5 + 53x_3 - 42x_1 - 1.25x_1 - 1.25x_2$$

To maximize profit, we will minimize $-f$ as is the convention in Chap. 7.

The equality constraints arise from material balances and the inequality constraints from restrictions. The equality constraints are as follows:

Equality constraints.

$$x_1 = x_3 + x_4 + x_5$$

$$0.6x_1 = x_4 + x_5 \Rightarrow x_1 = 1{,}667(x_4 + x_5)$$

$$x_3 = 0.4x_1 \Rightarrow x_3 = 0.667(x_4 + x_5)$$

$$x_2 = 0.6x_1 \Rightarrow x_2 = 0.6(1.667)(x_4 + x_5) = x_4 + x_5$$

Inequality constraints. The inequality constraints for the ranges of production are

$$\frac{x_4}{x_2} \geq 0.50 \qquad \text{and} \qquad \frac{x_5}{x_2} \geq 0.3$$

Rearranging and substituting for x_2 yields

$$x_4 - 0.5x_2 \geq 0 \Rightarrow x_4 - 0.5(x_4 + x_5) \geq 0$$

$$x_5 - 0.3x_2 \geq 0 \Rightarrow x_5 - 0.3(x_4 + x_5) \geq 0$$

$$\frac{x_4}{x_2} \geq 0.7 \qquad \frac{x_5}{x_2} \geq 0.5 \qquad \begin{array}{l} x_4 - 0.7x_2 \geq 0 \Rightarrow x_4 - 0.7(x_4 + x_5) \geq 0 \\ x_5 - 0.5x_2 \geq 0 \Rightarrow x_5 - 0.5(x_4 + x_5) \geq 0 \end{array}$$

Summarizing, we have four inequality constraints arising from production ranges

$$0.5x_4 - 0.5x_5 \geq 0 \qquad x_4 - x_5 \geq 0 \qquad (a)$$

$$0.7x_5 - 0.3x_4 \geq 0 \qquad (b)$$

$$0.3x_4 - 0.7x_5 \leq 0 \qquad (c)$$

$$0.5x_5 - 0.5x_4 \leq 0 \qquad x_5 - x_4 \leq 0 \qquad (d)$$

However, only two of the constraints are independent.

(*a*) and (*d*) are the same

(*b*) and (*c*) are the same

Other inequality constraints (supply and operating constraints) are

$$x_4 \geq 200$$

$$x_4 \leq 400$$

$$x_1 \leq 2000 \Rightarrow 1.667(x_4 + x_5) \leq 2000 \quad \text{or} \quad x_4 + x_5 \leq 1200$$

Also

$$x_4, x_5 \geq 0$$

In conclusion the inequality constraints are

$$\begin{array}{lr} (1) & x_4 - x_5 \geq 0 \\ (2) & 0.7x_5 - 0.3x_4 \geq 0 \\ (3) & x_4 \geq 200 \\ (4) & x_4 \leq 400 \\ (5) & x_4 + x_5 \leq 1200 \\ (6) & x_4 \geq 0 \\ (7) & x_5 \geq 0 \end{array}$$

Substitution of the equality constraints into the profit function gives

$$\begin{aligned} \text{Profit} &= 53(0.667)(x_4 + x_5) + 68x_4 + 42x_5 \\ &\quad -[42(1.667)(x_4 + x_5) + 1.25(1.667)(x_4 + x_5) + 1.25(x_4 + x_5)] \\ &= x_4(35.33 + 68 - 70 - 2.083 - 1.25) + x_5(35.33 + 42 - 70 - 2.083 - 1.25) \\ &= 30x_4 + 4x_5 \text{ so } f = -30x_4 - 4x_5 \qquad \text{(to be minimized)} \end{aligned}$$

Because we have reduced the problem to only two independent variables, we can solve the problem either via a graph or with the Simplex method (note that the

Simplex method could be applied to the original five-variable problem). The optimal values of x_4 and x_5 are

$$x_4 = 400 \text{ bbl/day}$$

$$x_5 = 1333 \text{ bbl/day}$$

The optimal profit is \$13,600/day.

REFERENCES

Al-Hag-Ali, N. S., and C. D. Holland, "Way to Find Distillation Optimum," *Hydrocarbon Process.*, (July, 1979), p. 165.

Buzzi-Ferraris, G., and E. Troncani, "Operational Optimization and Sensitivity Analysis of Multistage Separators", *Ind. Eng. Chem. Process Res. Dev.*, **24**, 112 (1985).

Eduljee, H. E., *Hydrocarbon Process.*, 120 (September 1975).

Gmehling, T., U. Onken, and W. Arlt, *Vapor-Liquid Equilibrium Data Collection*, DECHEMA VI, Part 1A (1981).

Happel, J., and D. G. Jordan, *Chemical Process Economics*, 2d ed., Marcel Dekker, New York (1975) p. 169.

Hartland, S., and J. C. Mecklenburgh, *The Theory of Backmixing*, Wiley, New York (1975) chap. 10.

Holland, C. D., *Fundamentals and Modeling of Separation Processes*, Prentice-Hall, Englewood Cliffs, New Jersey (1975) chap. 3.

Holland, C. D., *Multicomponent Distillation*, Prentice-Hall, Englewood Cliffs, New Jersey (1963) p. 494.

Hororka, F., R. A. Schaefer, and D. Dreisbach, *J. Am. Chem. Soc.*, **58**: 2264 (1936).

Jackson, P. J., and J. B. Agnew, "A Model Based Scheme for the On-Line Optimization of a Liquid Extraction Process," *Comput. Chem. Eng.*, **4**: 241 (1980).

Jafarey, A., J. M. Douglas, and T. M. McAvoy, "Shortcut Techniques for Distillation Column Design and Control—1. Column Design," *Ind. Eng. Chem. Process Des. Dev.*, **18**: 197 (1979).

Martin, G. D., P. R. Latour, and L. A. Richard, "Closed-Loop Optimization of Distillation Energy," *Chem. Eng. Prog.* (September, 1981) p. 33.

McAvoy, T. J., personal communication (1985).

Peters, M. S., and K. Timmerhaus, *Plant Design and Economics for Chemical Engineers*, 3d ed., McGraw-Hill, New York (1980) p. 387.

Prausnitz, J. M., T. F. Anderson, E. A. Grens, C. A. Eckert, R. Hsieh, and J. P. O'Connell, *Computer Calculations for Multicomponent Vapor-Liquid and Liquid-Liquid Equilibria*, Prentice-Hall, Englewood Cliffs, New Jersey (1980).

Sargent, R. W. H., and K. Gaminibandara, "Optimum Design of Plate Distillation Columns," in *Optimization in Action* (ed. L. D. W. Dixon), Academic Press, New York (1976)

Treybal, R. E., *Liquid Extraction*, McGraw-Hill, New York (1963)

Van Winkle, M., and W. G. Todd, "Optimum Fractionation Design by Simple Graphical Methods," *Chem. Eng.* (Sept. 20 1971) p. 136.

Wang, J. C., and Y. L. Wang, "Review on the Modeling and Simulation of Multistage Separations Processes," in *Foundations of Computer-Aided Process Design* (eds. W. D. Seider and R. S. H. Mah), AIChE (1981) p. 121.

CHAPTER 12

FLUID FLOW SYSTEMS

Optimization of fluid flow systems encompasses a wide-ranging scope of problems. In water resources planning the objective is to decide what systems to improve or build over a long time-frame. In water distribution networks and sewage systems, the time-frame may be quite long but the water and sewage flows have to balance at the network nodes. In pipeline design for bulk carriers such as oil, gas, and petroleum products, specifications on flow-rates and pressures (including storage) must be met by suitable operating strategies in the face of unusual demands. Simpler optimization problems exist in which the process models represent flow through a single pipe, flow in parallel pipes, compressors, heat exchangers, and so on. Other flow optimization problems occur in chemical reactors, where various types of process models have been proposed for the flow behavior, including well-mixed tanks, tanks with dead space and bypassing, plug-flow vessels, dispersion models, and so on. This subject is treated in Chap. 13.

Optimization (and modeling) of fluid flow systems can be put into three general classes of problems; (1) the modeling and optimization under steady state conditions, (2) the modeling and optimization under dynamic (unsteady state) conditions, and (3) stochastic modeling and optimization. All of these three classes of problems are complicated problems for large systems. Under steady state conditions, the principal difficulties in obtaining the optimum for a large system are the complexity of the topological structure, the nonlinearity of the objective function, the presence of a large number of possibly nonlinear inequality constraints, and the large number of variables. We do not consider optimization of dynamic or stochastic processes in this chapter. Instead, we focus on relatively simple steady state fluid flow processes using the following examples:

1. Optimal pipe diameter for an incompressible fluid (Example 12.1)
2. Minimum work of gas compression (Example 12.2)
3. Economic operation of a fixed-bed filter (Example 12.3)
4. Optimal design of a gas transmission line (Example 12.4)

EXAMPLE 12.1 OPTIMAL PIPE DIAMETER

The trade-off between the energy costs for transport and the investment charges for flow in a pipe determines the optimum diameter of a pipeline. With a few simplifying assumptions, you can derive an analytical formula for the optimum pipe diameter and the optimum velocity for an incompressible fluid with density ρ and viscosity μ. In developing this formula the investment charges for the pump itself are neglected as they are small compared to the pump operating costs, although these could be readily incorporated in the analysis if desired. The mass flow rate (m) of the fluid and the distance L the pipeline is to traverse are presumed known, as are ρ and μ. The variables whose values are unknown are D (pipe diameter), Δp (fluid pressure drop), and v (fluid velocity); the optimal values of the three variables are to be determined so as to minimize total annual costs. Not all of the variables are independent, as you will see below.

Total annual costs are comprised of the sum of the pipe investment charges and the operating costs for running the pump. Let C_{inv} be the annualized charges for the pipe and C_{op} be the pump operating costs. We propose that

$$C_{\text{inv}} = C_1 D^n L \tag{a}$$

$$C_{\text{op}} = \frac{C_0 m \Delta p}{\rho \eta} \tag{b}$$

where n is an exponent from a cost correlation (assumed to be 1.3), η is the pump efficiency, and C_0 and C_1 are cost coefficients. C_1 includes the capitalization charge for the pipe per unit length while C_0 corresponds to the power cost (\$/kWh) due to the pressure drop. The objective function becomes

$$C = C_{\text{inv}} + C_{\text{op}} = C_1 D^n L + \frac{C_0 m \Delta p}{\rho \eta} \tag{c}$$

Note that there are two variables in Eq. (c), D and Δp. However, they are related through a fluid flow correlation as follows (part of the process model):

$$\Delta p = \frac{2 f \rho v^2 L}{D} \tag{d}$$

where f is the friction factor. Two additional unspecified variables exist in Eq. (d), namely v and f. Both m and f are related to v as follows:

$$m = \left(\frac{\rho \pi D^2}{4}\right) v \tag{e}$$

$$f = 0.046\ \text{Re}^{-0.2} = \frac{0.046 \mu^{0.2}}{D^{0.2} v^{0.2} \rho^{0.2}} \tag{f}$$

Equation (e) merely is a definition of the mass flow-rate. Equation (f) is a standard correlation for the friction factor for turbulent flow. (Note that the correlation between f and the Reynold's number (Re) is also available as a graph, but use of data from a graph would require trial and error calculations and rule out an analytical solution.)

To this point we have isolated four variables, D, v, Δp, and f, and have introduced three equality constraints, Eqs. (d), (e), and (f), leaving one degree of freedom (one independent variable). In order to facilitate the solution of the optimization problem, we shall eliminate three of the four unknown variables (Δp, v, and f) from the objective function using the three equality constraints, leaving D as the single independent variable. Direct substitution yields the cost equation

$$C = C_1 D^{1.3} L + 0.142 \frac{C_0}{\eta} m^{2.8} \mu^{0.2} \rho^{-2.0} D^{-4.8} L \tag{g}$$

(In the American Engineering system of units C_0 will be divided by g_c to keep the units consistent). Here, C_0 will be selected with units {(\$/year)/[(lb$_m$)(ft^2)/s^3]}. We can now differentiate C with respect to D and set the resulting derivative to zero

$$\frac{dC}{dD} = 0 = 1.3 C_1 D^{0.3} L - 0.682 \frac{C_0}{\eta g_c} m^{2.8} \mu^{0.2} \rho^{-2.0} D^{-5.8} \tag{h}$$

and solve for D^{opt}:

$$D^{opt} = 0.900\left(\frac{C_0}{C_1 \eta g_c}\right)^{0.164} m^{0.459} \mu^{0.033} \rho^{-0.328} \tag{i}$$

Note that L does not appear in the result.

Equation (i) permits a quick analysis of the optimum diameter as a function of a variety of physical properties. From the exponents in Eq. (i), the density and mass flow-rate seem to be fairly important in determining D^{opt} while the ratio of the cost factors is less important. A doubling of m changes the optimum diameter by a factor of 1.4, while a doubling of the density decreases D^{opt} by a factor of 1.25. The viscosity is also not too important. For very viscous fluids, larger diameters resulting in lower velocities are indicated, while gases (low density) will give smaller diameters and higher velocities. The validity of Eq. (i) for gases is questionable, since the variation of gas velocity with pressure must be taken into account.

By the use of Eq. (e)

$$v = \frac{4m}{\pi \rho D^2} \tag{j}$$

we can discover how the optimum velocity varies as a function of m, ρ, and μ by substituting Eq. (i) for D^{opt} into (j):

$$v^{opt} = C_2 m^{0.082} \mu^{-0.066} \rho^{-0.344} \tag{k}$$

where C_2 is a consolidated constant. Consider the effect of ρ on the optimum velocity. Generally optimum velocities for liquids vary from 3 to 8 ft/s while for gases the range is from 30 to 60 ft/s. While D^{opt} is influenced noticeably by changes in m, v^{opt} is very insensitive to changes in m.

Suppose a flow problem with the following specifications is posed:

$m = 50$ lb/s

$\rho = 60$ lb/ft^3

$\mu = 6.72 \times 10^{-4}$ lb/(ft)(s)

$\eta = 0.6$ (60% pump efficiency)

Purchased cost of electricity = \$0.05/kWh

8760 h/year of operation (100% stream factor)

$C_1 = \$5.7$ (D in ft); $C_1 D^n$ is an annualized cost expressed as \$/(ft)(year)

L is immaterial as mentioned earlier

The units in Eq. (g) must be made consistent so that C is in dollars per year. For \$0.05/kWh, $C_0 = \$0.5938$ $\{(\$/\text{year})/[(\text{lb}_m)(\text{ft}^2)/\text{s}^3)]\}$. Substitution of the values specified into Eq. (i) gives $D^{opt} = 0.473$ ft = 5.7 in. The standard pipe schedule 40 size which is closest to D^{opt} is 6 in. For this pipe size (ID = 6.065 in) the optimum velocity is 4.2 ft/s. (A Schedule 80 pipe has an ID of 5.7561 in.) Figure E12.1 shows the respective contributions of operating and investment costs to the total value of C.

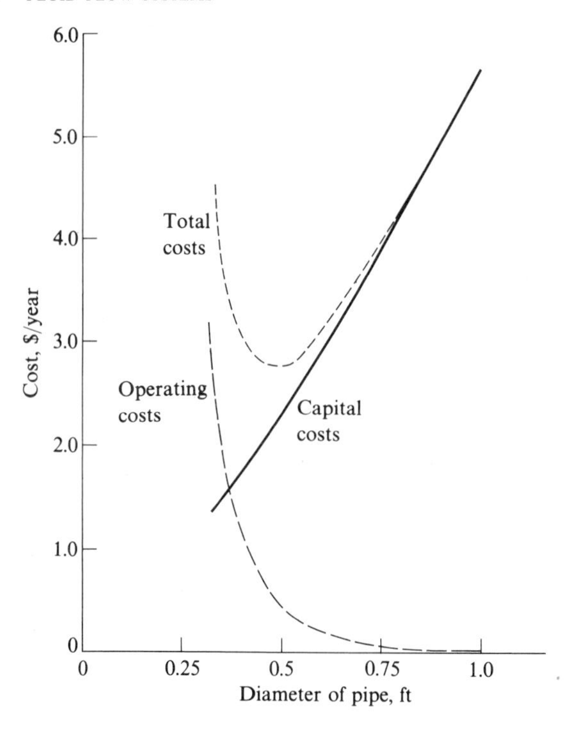

Figure E12.1 Investment, operating, and total costs for pipeline example ($L = 1$ ft).

As the process model is made more accurate and complicated, you can lose the possibility of obtaining an analytical solution of the optimization problem. For example, if (1) the pressure losses through the pipe fittings and valves are included in the model [see Happel (1975) for a typical example], (2) the pump investment costs are included as a separate term with a cost exponent ($\bar{n}$) which is not equal to 1.0, (3) elevation changes must be taken into account, (4) contained solids are present in the flow, or (5) significant changes in density occur, the optimum diameter cannot be calculated analytically.

EXAMPLE 12.2 MINIMUM WORK OF COMPRESSION

In this example we describe the calculation of the minimum work for ideal compressible adiabatic flow using three different optimization techniques, (a) analytical, (b) numerical, and (c) dynamic programming. Most real flows lie somewhere between adiabatic and isothermal flow. For adiabatic flow, the case examined here, you cannot establish a priori the relationship between pressure and density of the gas because the temperature is not known as a function of pressure or density, hence the relation between pressure and density is derived using the mechanical energy balance. If the gas is assumed to be ideal, and $k = C_p/C_v$ is assumed to be constant in the range of interest from p_1 to p_2, you can make use of the well-known relation

$$pV^k = \text{const} \qquad (a)$$

in getting the theoretical work per mole (or mass) of gas compressed for a single stage compressor (McCabe et al., 1985, p. 188)

$$W = \frac{kRT_1}{k-1}\left[\left(\frac{p_2}{p_1}\right)^{(k-1)/k} - 1\right] \tag{b}$$

where T_1 is the inlet gas temperature and R the ideal gas constant ($p_1\hat{V}_1 = RT_1$). For a three-stage compressor with intercooling back to T_1 between stages as shown in Fig. E12.2*a*, the work of compression from p_1 to p_4 is

$$\hat{W} = \frac{kRT_1}{k-1}\left[\left(\frac{p_2}{p_1}\right)^{(k-1)/k} + \left(\frac{p_3}{p_2}\right)^{(k-1)/k} + \left(\frac{p_4}{p_3}\right)^{(k-1)/k} - 3\right] \tag{c}$$

We want to determine the optimal interstage pressures p_2 and p_3 to minimize $\hat{W}$ keeping p_1 and p_4 fixed.

Analytical solution. We set up the necessary conditions using calculus and also test to make sure that the extremum found is indeed a minimum.

$$\frac{\partial \hat{W}}{\partial p_2} = 0 = RT_1[(p_1)^{(1-k)/k}(p_2)^{1/k} - (p_3)^{(k-1)/k}(p_2)^{(1-2k)/k}] \tag{d}$$

$$\frac{\partial \hat{W}}{\partial p_3} = 0 = RT_1[(p_2)^{(1-k)/k}(p_3)^{1/k} - (p_4)^{(k-1)/k}(p_3)^{(1-2k)/k}] \tag{e}$$

The simultaneous solution of Eqs. (*d*) and (*e*) yields the desired results

$$p_2^2 = p_1 p_3 \qquad \text{and} \qquad p_3^2 = p_2 p_4$$

so that the optimal values of p_2 and p_3 in terms of p_1 and p_4 are

$$p_2^* = (p_1^2 p_4)^{1/3} \tag{f}$$

$$p_3^* = (p_4^2 p_1)^{1/3} \tag{g}$$

With these conditions for pressure, the work for each stage will be the same.

To check the sufficiency conditions, we examine the Hessian matrix of $\hat{W}$ (after substituting p_2^* and p_3^*) to see if it is positive definite.

$$\nabla^2 W = RT_1\left(\frac{k-1}{k}\right)$$

$$\cdot\begin{bmatrix} 2[(p_1^*)^{(1-5k)/3k}][(p_4^*)^{-(1+k)/3k}] & [(p_1^*)^{(1-4k)3k}][(p_4^*)^{-(1+2k)/3k}] \\ [(p_1^*)^{(1-4)/3k}][(p_2^*)^{-(1+2k)/3k}] & 2[(p_1^*)^{(1-3k)/3k}][(p_4^*)^{-(1+2k)/3k}] \end{bmatrix}$$

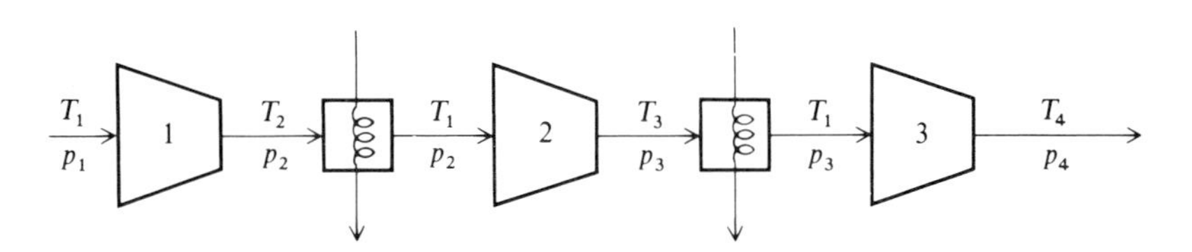

Figure E12.2*a*

The two principal minors (the two diagonal elements) must be positive because p_1^* and p_4^* are both positive, and the determinant of $\nabla^2 \hat{W}$

$$4\left[\frac{RT_1(k-1)}{k}\right]^2 [(p_1^*)^{(2-8k)/3k}(p_4^*)^{-(2+4k)/3k}]$$

$$-\left[\frac{RT_1(k-1)}{k}\right]^2 [(p_1^*)^{(2-8k)/3k}(p_4^*)^{-(2+4k)/3k}] > 0$$

is also positive, hence $\nabla^2 \hat{W}$ is positive definite (refer to Sec. 4.3).

Numerical solution. Numerical methods of solution do not produce the general solution given by Eqs. (f) and (g) but require that specific numerical values be provided for the parameters, and give specific results. Suppose that $p_1 = 100$ kPa and $p_4 = 1000$ kPa. Let the gas be air so that $k = 1.4$. Then $(k-1)/k = 0.286$. Application of the BFGS algorithm to minimize $\hat{W}$ in Eq. (c) as a function of p_2 and p_3 starting with $p_2 = p_3 = 500$ yields

$p_2^* = 215.44$ compared with $p_2^* = 215.44$ from Eq. (f)

$p_3^* = 464.17$ compared with $p_3^* = 464.16$ from Eq. (g)

Solution via dynamic programming. To use dynamic programming we also require specific numerical values for the given parameters. We start with the work for the last stage in Fig. E12.2a.

$$\hat{W}_3 = 0.286RT_1\left[\left(\frac{1000}{p_3}\right)^{0.286} - 1\right] \tag{h}$$

Because we do not know p_3, we prepare a graph or table of $\hat{W}_3(p_3)$ such as Fig. E12.2b. $\hat{W}_3 = 0$ for $p_3 = 1000$ and $\hat{W}_3 = \infty$ for $p_3 = 0$, but we only need to examine the function in the range of, say, 300 to 900 kPa.

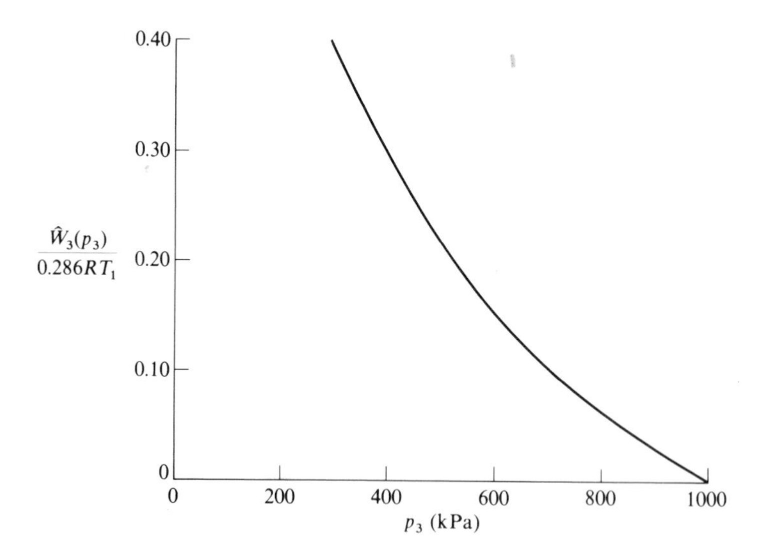

Figure E12.2b

Table E12.2*a* Values of $\hat{W}_2\ (p_2, p_3)/0.286RT_1$

p_2	**300**	**400**	**500**	**600**	**700**	**800**
200	0.126	0.219	0.300	0.309	0.431	0.487
300	......	0.086	0.157	0.219	0.274	0.324
400	......	......	0.066	0.126	0.174	0.219
500	......	......	......	0.054	0.101	0.144
600	......	......	......	......	0.045	0.086
700	......	......	......	......	......	0.039

Next we calculate $\hat{W}_2$ for the second stage. $\hat{W}_2$ is a function of both p_2 and p_3

$$\hat{W}_2 = 0.286RT_1\left[\left(\frac{p_3}{p_2}\right)^{0.286} - 1\right] \qquad (i)$$

Inasmuch as we would need a three-dimensional plot for $\hat{W}_2(p_2, p_3)$, we will make up a table instead for discretized values of p_2 and p_3 to get values of $\hat{W}_2$. Examine Table E12.2*a*.

Finally, we compute $\hat{W}_1$ which is a function only of p_2, and make a graph, Fig. E12.2*c*. The objective is to make the sum of the work from the three stages a minimum

$$\text{Minimize:} \quad \hat{W} = \hat{W}_3 + \hat{W}_2 + \hat{W}_1 \qquad (j)$$

Various values of p_3 are selected with subcases of p_2 for a certain level of discretization of p_3 and p_2. Examine Fig. E12.2*d*. The values on the arcs in Fig. E12.2*d* represent the values of $\hat{W}_i/0.286RT_1$. To get more precise values of p_2 and p_3, finer

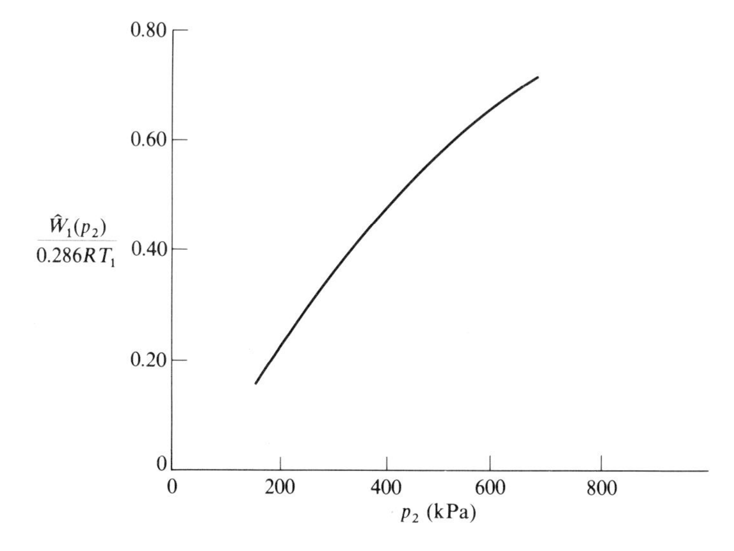

Figure E12.2*c*

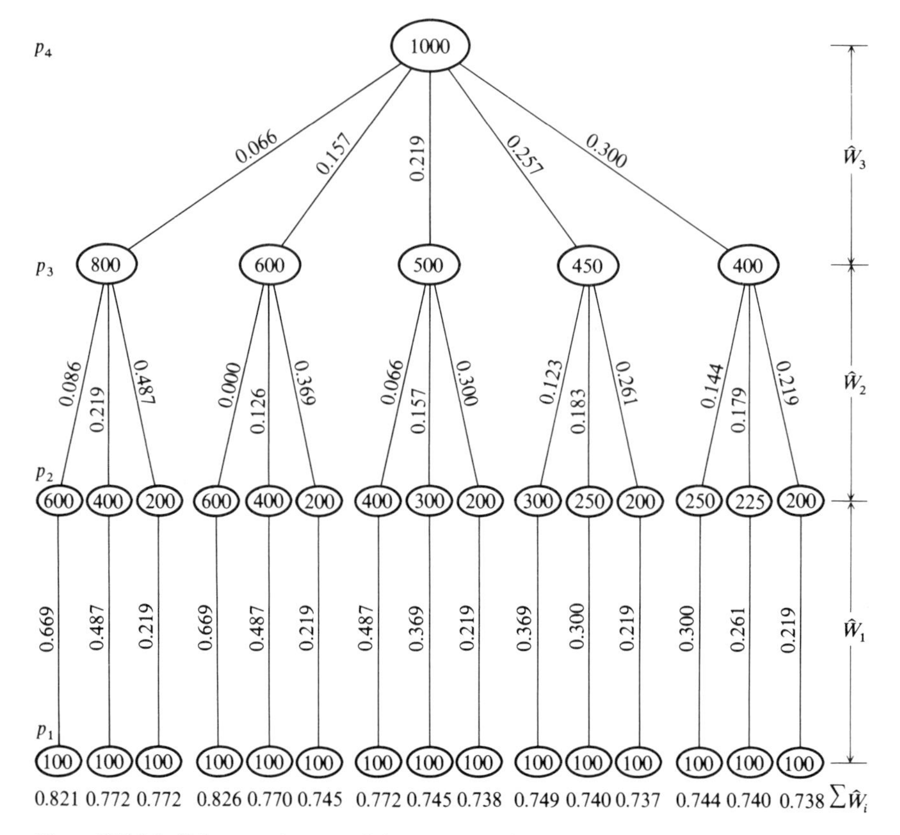

Figure E12.2*d* Values on the arcs of the tree are $W_i/0.286RT_1$.

subdivisions in the appropriate pressure regions can be employed along with interpolation. From Fig. E12.2*d* along you can note that the minimum $\hat{W}$ is not very sensitive to changes in p_2 and p_3, hence the number of computations would increase substantially as the precision of the solution is improved.

EXAMPLE 12.3 ECONOMIC OPERATION OF A FIXED-BED FILTER

Various rules of thumb exist for standard water filtration rates and cycle time before backwashing. However, higher filtration rates may appear to be economically justified when the filter loading is within conventional limits. In this example, we examine the issues involved for constant rate filtration for a dual media bed. Dual and mixed media beds result in increased production of water in a filter for two reasons. First, the larger grains (say charcoal approximately 1-mm size) as a top layer help reduce cake formation and deposition within the small (150-mm) top layer of the bed. Second, the head loss in the region of significant filtration is reduced.

With respect to the objective function for a filter, the total annual cost of filtration f is assumed to be the sum of the annualized capital costs f_c and the annual operating costs f_0. The annualized capital cost is related to the cross-sectional area of the filter by the relation

$$f_c = rbA^z \tag{a}$$

where r is the capital recovery factor involving the discount rate and economic life of the filter
b is an empirical constant
z is an empirical exponent
A is the cross-sectional area of the filter

The cross-sectional area can be calculated by dividing the design flow rate by a quantity that is equal to the number of filter runs per day times the net water production per run per cross-sectional area:

$$A = \frac{q}{1440/[(V_f/Q) + t_b]\cdot(V_f - V_b)} \tag{b}$$

where q is the design flow rate in gal/day; L/day (dual units given here)
V_f is the volume of water filtered per unit area of bed per filter run in gal/ft^2; L/m^2
V_b is the volume of filtered water used for backwash per unit area of bed in gal/ft^2; L/m^2
Q is the filtration rate in gal/(min)(ft^2); L/(min)(m^2)
t_b is the filter down time for backwash, min
1440 is the number of minutes/day

For a constant filtration rate, the length of the filter run is given by $t_f = V_f/Q$.

The water production per filter run V_f is based on a relation proposed by Letterman (1980) that assumes minimal surface cake formation by the time filtration is stopped because of head loss:

$$V_f = \frac{K_\rho}{\beta C_0}\frac{D}{n}\sum_{i=1}^{n} \log \frac{n\Delta H}{k_i DQ} \tag{c}$$

where K_ρ is a constant related to the density of the deposit within the bed
D is the overall depth of the bed, ft.
β is the overall fraction of the influent suspended solids removed during the entire filter run
C_0 is suspended solids concentration in the filter influent
n is the number of layers $i = 1, \ldots, n$ into which the filter is divided for use of Eq. (c)
ΔH is the terminal pressure (head) loss for the bed, ft.
k_i is a function of the geometric mean grain diameter d_{gi} in layer i. For rounded grains, the Kozeny–Carmen equation can be used to estimate k_i: $k_i = 0.081 d_{gi}^{-2}$, where d_{gi} is in millimeters.

Typical values are $n = 1$, $d_g = 1$ mm, $\Delta H = 10$ ft, $D = 3$ ft $(K_\rho/\beta C_0) = 700$.

The backwash flow rate is calculated from

$$q_b = \left(\frac{V_f}{V_f - V_b} - 1\right)q \tag{d}$$

We assume the backwash water is not recycled.

We next summarize the annual operating costs of the filter since they are equal to the energy costs for pumping

$$f_0 = q_b\left[1.146 \times 10^{-3} C_E\left(\frac{h}{\eta}\right)\right] \tag{e}$$

where f_0 is in dollars per year
h is the backwash pumping head in feet of water
C_E is the cost of electricity in dollars per kilowatt-hour
η is the pump efficiency
1.146×10^{-3} is the conversion factor

Let us now carry out a numerical calculation based on the following values for the filter parameters

$$h = 110 \text{ ft of water (33.5 m)}$$
$$\eta = 0.8$$
$$b = \frac{\$870}{(\text{ft}^2)^{0.86}}; \frac{\$6715}{(\text{m}^2)^{0.86}}$$
$$z = 0.86$$
$$r = 0.134 \text{ (12.5\% for 20 years) (year}^{-1})$$
$$C_E = \$0.03/\text{kWh}$$

Substitution of these values into Eqs. (a) and (e) together with Eqs. (b) and (d) yields the total cost function

$$f\left(\frac{\$}{\text{year}}\right) = 116\left[\frac{10^6 q}{1440/[(V_f/Q) + t_b](V_f - V_b)}\right]_{0 \le q \le 10}^{0.86} + 4.73 \times 10^{-3}\left[\frac{V_f}{V_f - V_b} - 1\right]q \tag{f}$$

If the values of q, t_b, and V_b are specified, and Eq. (c) is ignored, the total annual cost can be determined as a function of the water production V_f per bed area and the filtration rate Q.

Figure E12.3 shows f vs. V_f, the water filtered per run, for q (in 10^6 units) = 10 Mgal/day (3.79×10 ML/day), $t_b = 10$ min, and $V_b = 200$ gal/ft² ($8.15 \times$

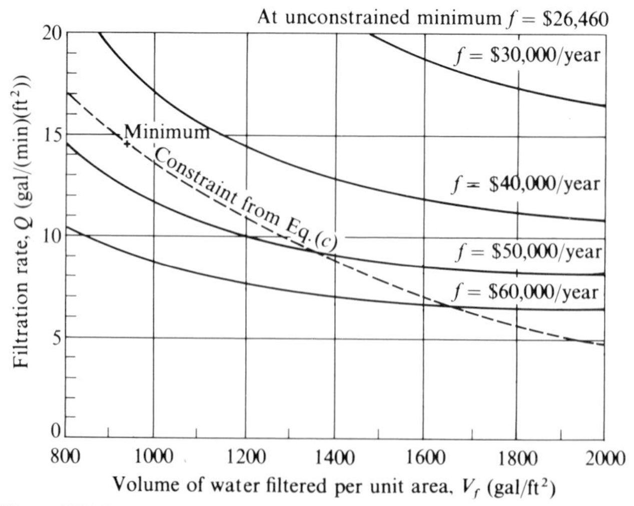

Figure E12.3

10^3 L/m^2) with Q gal/(min)(ft^2) as a parameter. The unconstrained solution is at the upper bounds on Q and V_f. Notice the flatness of f as V_f increases.

Equation (c) would be used in the design of the filter, hence Eq. (c) imposes a constraint that must be taken into account. The optimal solution becomes $V_f = 940$ gal/ft^2 and $Q = 14.2$ gal/(min)(ft^2) with Eq. (c) included in the problem (see Fig. E12.3). A rule of thumb is 2 gal/(min)(ft^2) (Letterman, 1980), as compared with the optimal value of Q.

EXAMPLE 12.4 OPTIMAL DESIGN OF A GAS TRANSMISSION NETWORK

A gas gathering and transmission system consists of sources of gas, arcs comprised of pipeline segments, compressor stations, and delivery sites. The design or expansion of a gas pipeline transmission system involves capital expenditures as well as the continuing cost of operation and maintenance. Substantial savings can be expected by improving the system design for a given delivery rate (Flanigan (1972), Graham et al. (1971)). Mah and Shacham (1978) discuss the problem of configuring and routing for a transmission network and Edgar et al. (1978) examine how to determine the optimal number of compressor stations in a system.

Larson and Wong (1968) determined the steady state optimal operating conditions of a straight (unbranched) natural gas pipeline with compressors in series. Their solution technique was to use dynamic programming to find the optimal suction and discharge pressures of a fixed number of compressor stations. The length and diameter of the pipeline segments were considered as known values because the dynamic programming technique cannot accommodate many decision variables. Martch and McCall (1972) expanded the pipeline configuration by adding branches to form a network, and posed the problem as one of capacity expansion rather than initial design. Nevertheless, the transmission network configuration was predetermined because the optimization technique employed was dynamic programming.

In 1971 Cheesman of the International Management and Engineering Group (1971a, b) introduced a computer optimization code for pipeline design. While the pipeline segment lengths and diameters were variables, the pipeline itself had to be unbranched, i.e., complicated network systems could not be handled. Olorunniwo (1981) and Olorunniwo and Jensen (1982) optimized a gas transmission network including the following features:

1. The maximum number of compressor stations that would ever be required during a specified time horizon
2. The optimal locations of these compressor stations
3. The initial construction dates of the stations
4. The optimal solution of expansion for the compressor stations
5. The optimal diameter sizes of the main pipes for each arc of the network
6. The minimum recommended thickness of the main pipes
7. The optimal diameter sizes, thicknesses, and lengths of any required parallel pipe loops on each arc of the network
8. The timing of constructions of the parallel pipe loops
9. The operating pressures of the compressors and the gas in the pipelines

They used a combination of dynamic programming and finding the shortest route through a network.

In the example here we will describe the solution of a simplified problem so that the various factors involved are clear. Suppose that a gas pipeline is to be designed so that it will transport a prespecified quantity of gas per time from point A to other points. Both the initial state (pressure, temperature, composition) at A and final states of the gas are known. What is to be determined for the design is:

1. The number of compressor stations
2. The lengths of pipeline segments between compressor stations
3. The diameters of the pipeline segments
4. The suction and discharge pressures at each station.

The criterion for design will be to achieve the minimum total cost of operation per year including capital, operation, and maintenance costs. Note that the problem considered here does *not* fix the number of compressor stations, the pipeline lengths, the diameters of pipe between stations, the location of branching points, nor limit the configuration (branches) of the system so that the design problem has to be formulated as a nonlinear integer programming problem. Figure E12.4*a*

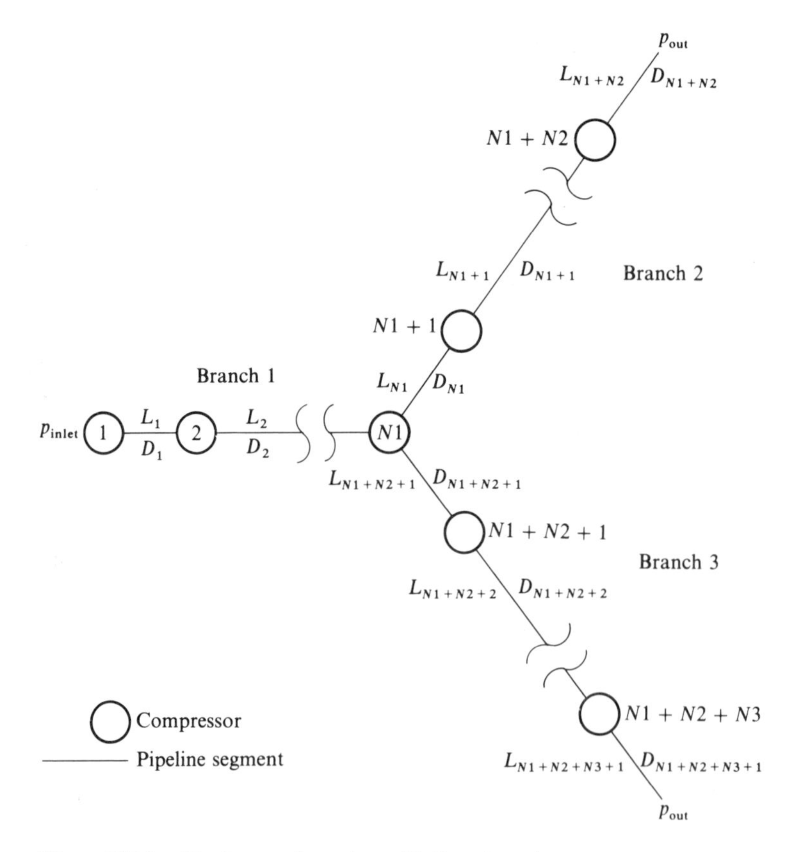

Figure E12.4*a* Pipeline configuration with three branches.

illustrates a simplified pipeline that we will use as an example in defining and solving the problem.

Before presenting the details of the design problem, we need to distinguish between two related problems, one being of a higher degree of difficulty than the other. If the capital costs of the compressors are a linear function of horsepower as shown in line *A* in Figure E12.4*b*, the transmission line problem can be solved as a nonlinear programming problem by one of the methods discussed in Chap. 8. On the other hand, if the capital costs are a linear function of horsepower with a fixed capital outlay for zero horsepower as indicated by line *B* in Fig. E12.4*b*, a condition that more properly reflects the real world, then the design problem becomes more difficult to solve and must be solved by a branch-and-bound algorithm combined with a nonlinear programming algorithm as discussed below. The reason why branch and bound is avoided for the case involving line *A* is best examined after the mathematical formulation of the objective function (cost function) has been completed. We split the discussion of the transmission line problem into five parts: (1) the pipeline configuration, (2) the variables, (3) the objective function and costs, (4) the inequality constraints, and (5) the equality constraints.

The pipeline configuration. Figure E12.4*a* shows the configuration of the pipeline we are using in this example and the notation employed for the numbering system for the compressor stations and the pipeline segments. Each compressor station is represented by a node and each pipeline segment by an arc. Pressure

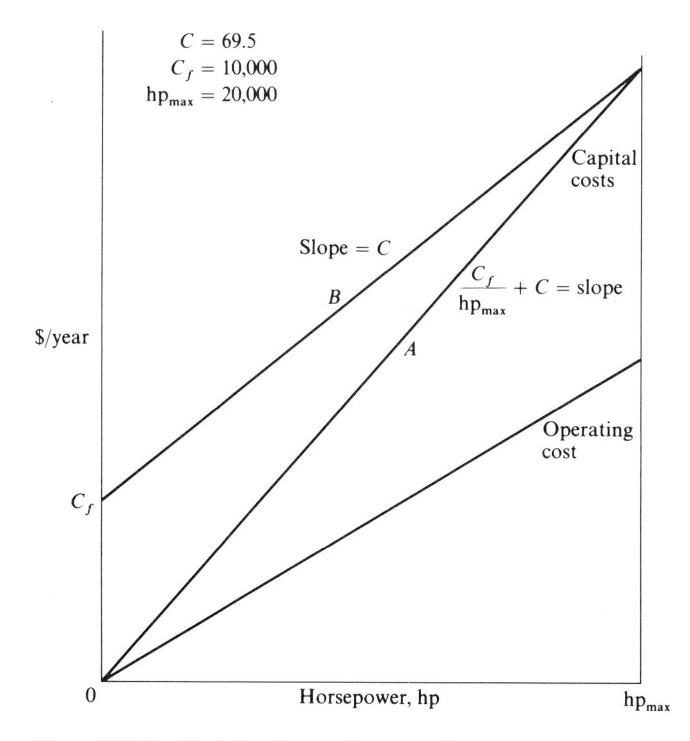

Figure E12.4*b* Capital and operating costs of compressors.

increases at a compressor and decreases along the pipeline segment. The transmission system is presumed to be horizontal. Although a simple example has been selected to illustrate a transmission system, a much more complicated network can be accommodated including various branches and loops, at the cost of additional computation time. For a given pipeline configuration each node and each arc are labeled separately. In total there are

n total compressors $\left[n = \sum (N_i)\right]$
$n - 1$ suction pressures (the initial entering pressure is known)
n discharge pressures
$n + 1$ pipeline segment lengths and diameters (note there are two segments issuing at the branch)

The variables. Each pipeline segment has associated with it five variables: (1) the flow rate, Q; (2) the inlet pressure, p_d (discharge pressure from the upstream compressor); (3) the outlet pressure, p_s (suction pressure of the downstream compressor), (4) the pipe diameter, D, and (5) the pipeline segment length, L. Inasmuch as the mass flow rate is fixed, and each compressor is assumed to have gas losses of one-half of one percent of the gas transmitted, only the last four variables need to be determined for each segment.

The objective function. Because the problem is posed as a minimum cost problem, the objective function is comprised of the sum of the yearly operating and maintenance costs of the compressors plus the sum of the discounted (over 10 years) capital costs of the pipeline segments and compressors. Each compressor is assumed to be adiabatic with an inlet temperature equal to that of the surroundings. A long pipeline segment is assumed so that by the time gas reaches the next compressor it returns to the ambient temperature. The annualized capital costs for each pipeline segment depend on pipe diameter and length, and have been estimated (Martch and McCall, 1972) at \$870/(inch)(mile)(year). The rate of work of one compressor is:

$$W = (0.08531)Q\frac{k}{k-1}T_1\left[\left(\frac{p_d}{p_s}\right)^{z(k-1)/k} - 1\right] \qquad (a)$$

where $k = C_p/C_v$ for gas at suction conditions [assumed to be 1.26 (Katz, et al., 1959)]
z = compressibility factor of gas at suction conditions
p_s = suction pressure, psia
p_d = discharge pressure, psia
T_1 = suction temperature, °R (assumed 520°R)
Q = flow rate into the compressor, MMCFD (million cubic feet per day)
W = rate of work, horsepower.

Operation and maintenance charges per year can be related directly to horsepower (Cheesman, 1971b) and have been estimated to be between 8.00 and 14.0 \$/(hp)(year) (Martch and McCall, 1972), hence the total operating costs are assumed to be a linear function of compressor horsepower.

Figure E12.4*b* shows two different forms for the annualized capital cost of the compressors. Line *A* indicates the cost is a linear function of horsepower (\$70.00/(hp)(year)) with the line passing through the origin whereas line *B* assumes a linear function of horsepower with a fixed initial capital outlay ($70.00/(hp)(year) + $10,000) to take into account installation costs, foundation, etc. For line A, the objective function in dollars per year for the example problem is

$$f = \sum_{i=1}^{n} (C_0 + C_c)Q_i(0.08531)T_1\left(\frac{k}{k-1}\right)\left[\left(\frac{p_{d_i}}{p_{s_i}}\right)^{z(k-1/k)} - 1\right]$$

$$+ \sum_{j=1}^{m} C_s L_j D_j \qquad (b)$$

where n = number of compressors in the system
m = number of pipeline segments in the system ($= n + 1$)
C_0 = yearly operating cost \$/(hp)(year)
C_c = compressor capital cost \$/(hp)(year)
C_s = pipe capital cost \$/(in)(mile)(year)
L_j = length of pipeline segment j, mile
D_j = diameter of pipeline segment j, inch

You can now see why for line A a branch and bound technique is not required to solve the design problem. Because of the way the objective function is formulated, if the ratio $(p_d/p_s) = 1$, the term involving compressor i vanishes from the first summation in the objective function. This outcome is equivalent to the deletion of compressor i in the execution of a branch and bound strategy. (Of course the pipeline segments joined at node i may be of different diameters.) But when line B represents the compressor costs, the fixed incremental cost for each compressor in the system at zero horsepower (C_f) would *not* be multiplied by the term in the square brackets of Eq. (b). Instead, C_f would be added in the sum of the costs whether or not compressor i is in the system, and a nonlinear programming technique could not be used alone. Hence, if line B applies, a different solution procedure is required.

The inequality constraints. The operation of each compressor is constrained so that the discharge pressure is greater than or equal to the suction pressure

$$\frac{p_{d_i}}{p_{s_i}} \geq 1 \qquad i = 1, 2, \ldots, n \qquad (c)$$

and the compression ratio does not exceed some prespecified maximum limit K

$$\frac{p_{d_i}}{p_{s_i}} \leq K_i \qquad i = 1, 2, \ldots, n \qquad (d)$$

In addition, upper and lower bounds are placed on each of the four variables

$$p_{d_i}^{\min} \leq p_{d_i} \leq p_{d_i}^{\max} \qquad (e)$$

$$p_{s_i}^{\min} \leq p_{s_i} \leq p_{s_i}^{\max} \qquad (f)$$

$$L_i^{\min} \leq L_i \leq L_i^{\max} \qquad (g)$$

$$D_i^{\min} \leq D_i \leq D_i^{\max} \qquad (h)$$

The equality constraints. Two classes of equality constraints exist for the transmission system. First, the length of the system is fixed. With two branches, there are two constraints

$$\sum_{j=1}^{N1-1} L_j + \sum_{j=N1}^{N1+N2} L_j = L_1^*$$
$$\sum_{j=1}^{N1-1} L_j + \sum_{j=N1+N2+1}^{N1+N2+N3+1} = L_2^* \tag{i}$$

where L_k^* represents the length of a branch. Second, the flow equation, the Weymouth relation (GPSA handbook, 1972), must hold in each pipeline segment

$$Q_j = 871 D_j^{8/3} \left[\frac{p_d^2 - p_s^2}{L_j} \right]^{1/2} \tag{j}$$

where Q_j is a fixed number, p_d is the discharge pressure at the entrance of the segment, and p_s is the suction pressure at the exit of the segment. To avoid problems in taking square roots, Eq. (j) is squared to yield

$$(871)^2 D_j^{16/3}(p_d^2 - p_s^2) - L_j Q_j^2 = 0 \tag{k}$$

Solution strategy. As mentioned previously, if the capital costs in the problem are described by line A in Fig. E12.4b, then the problem can be solved directly by a nonlinear programming algorithm. If the capital costs are represented by line B in Fig. E12.4b, then nonlinear programming in conjunction with branch and bound enumeration must be used to be able to accommodate the integer variable of a compressor being in place or not. Edgar et al. (1978) used both techniques to solve the example design problem posed here.

As explained in Sec. 9.2.2, a branch and bound enumeration is nothing more than an organized search, organized so that certain portions of the possible solution set are deleted from consideration. A tree is formed of nodes and branches (arcs). Each branch in the tree represents an added or modified inequality constraint to the problem defined for the prior node. Each node of the tree itself represents a nonlinear optimization problem without integer variables.

With respect to the example we are considering, in Fig. E12.4c, node 1 in the tree represents the original problem as posed by Eqs. (b) through (k), that is the problem in which the capital costs are represented by line A in Fig. E12.4b. When the problem at node 1 is solved, it provides a lower bound on the solution of the problem involving the cost function represented by line B in Fig. E12.4b. Note that line A always lies below line B. (If the problem at node 1 using line A had no feasible solution, the more complex problem involving line B also has no feasible solution.) Although the solution of the problem at node 1 is feasible, the solution may not be feasible for the problem defined by line B because line B involves an initial fixed capital cost at zero horsepower.

After solving the problem at node 1, a decision is made to partition on one of the three integer variables; $N1$, $N2$, or $N3$. The partition variable is determined by the following heuristic rule.

> The smallest average compression ratio of all the branches in the transmission system is calculated by adding all the compression ratios in each branch

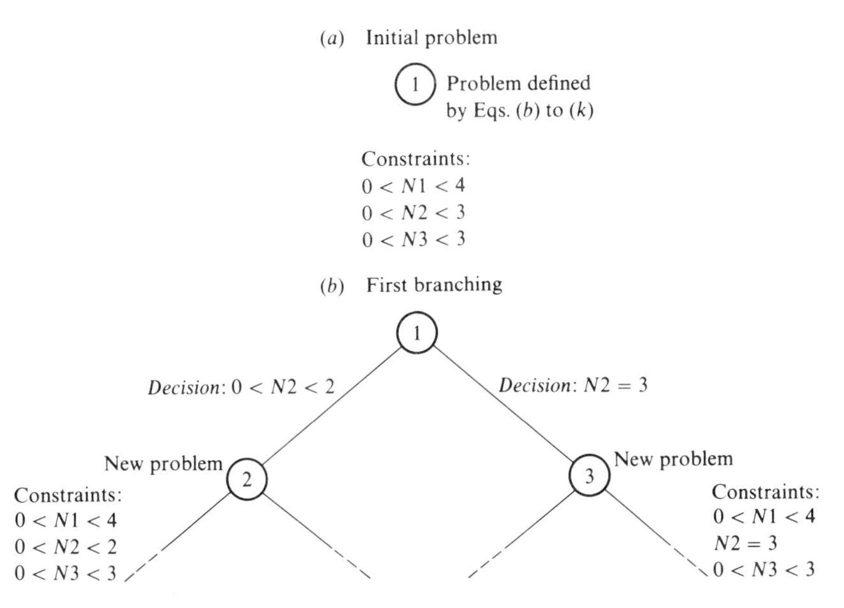

Figure E12.4c Partial tree and branches for the example design problem.

and dividing by the number of compressors in the branch. The number of compressors in the branch that has smallest ratio becomes the partition variable.

Based on this rule, the partition variable was calculated to be $N2$.

After selection of the partition variable, the next step is to determine how the variable should be partitioned. It was decided to check each compressor in the branch of the transmission line associated with the partition variable, and if any compressor operated at less than 10 percent of capacity, it was assumed the compressor was not necessary in the line. (If all operate at greater than 10 percent capacity, the compressor with the smallest compression ratio was deleted.) For example, with $N2$ selected as the partition variable, and one of the three possible compressors in branch 2 of the gas transmission network operating at less than 10 percent of capacity, the first partition would lead to the tree shown in Fig. E12.4*c*; $N2$ would either be 3 or would be $0 \leq N2 \leq 2$. Thus at each node in the tree, the upper or lower bound on the number of compressors in each branch of the pipeline is readjusted to be tighter.

The nonlinear problem at node 2 will be the same as at node 1 with two exceptions. First, the maximum number of compressors permitted in branch 2 of the transmission line is now two. Second, the objective function is changed. From the lower bounds, we know the minimum number of compressors in each branch of the pipeline. For the lower bound, the costs related to line B in Fig. E12.4*b* apply; for compressors in excess of the lower bound and up to the upper bound, the costs are represented by line A.

As the decision tree descends, the solution at each node becomes more and more constrained, until node r is reached in which the upper bound and the lower bound for the number of compressors in each pipeline branch are the same. The solution at node r will be feasible for the general problem but not necessarily optimal. Nevertheless, the important point is that the solution at node r is an upper bound on the solution of the general problem.

As the search continues through the rest of the tree, if the value of the objective function at a node is greater than that of the best feasible solution found to that stage in the search, then it is not necessary to continue down that branch of the tree. The objective function of any solution subsequently found in the branch would be larger than the solution already found. Thus, we can fathom the node, i.e., terminate the search down that branch of the tree.

The next step is to backtrack up the tree, and continue searching through other branches until all nodes in the tree have been fathomed. Another reason to fathom a particular node occurs when no feasible solution exists to the nonlinear problem at node r; then all subsequent nodes below node r will also be infeasible.

At the end of the search, the best solution found is the solution to the general problem.

Computational results. Figure E12.4*d* and Table E12.4*a* show the solution to the example design problem outlined in Fig. E12.4*a* using the cost relation of line *A* in Fig. E12.4*b*. The maximum number of compressors in branches 1, 2, and 3 were set at 4, 3, and 3, respectively. The input pressure was fixed at 500 psia at a flow rate of 600 MMCFD, and the two output pressures were set at 600 psia and 300 psia, respectively, for branches 2 and 3. The total length of branches 1 plus 2 was constrained to be 175 miles, whereas the total length of branches 1 plus 3 was constrained at 200 miles. The upper bound on the diameter of the pipeline segments in branch 1 was set at 36 inches, the upper bound on the diameters of the

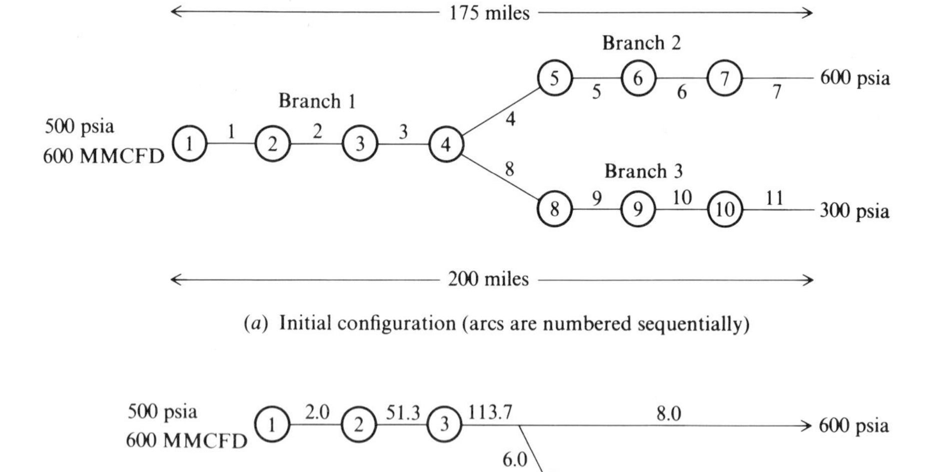

(*a*) Initial configuration (arcs are numbered sequentially)

(*b*) Optimal configuration with optimal pipeline lengths (in miles) shown on the arcs

Figure E12.4*d* Initial gas transmission system and final optimal system using the costs of line *A*, Fig. E12.4*b*.

Table E12.4*a* Values of operating variables for the optimal network configuration using the costs of line *A*, Fig. E12.4*b*

Pipeline segment	Discharge pressure (psia)	Suction pressure (psia)	Pipe diameter (inch)	Length (mile)	Flow rate (MMCFD)
1	719.1	715.4	35.0	2.0	597.0
2	1000.0	889.3	32.4	51.3	594.0
3	1000.0	735.8	32.4	113.7	591.0
4	735.7	703.8	18.0	2.0	294.0
5	703.8	670.6	18.0	2.0	292.6
6	670.6	636.1	18.0	2.0	291.1
7	636.1	600.0	18.0	2.0	289.7
8	735.8	703.8	18.0	2.0	294.0
9	685.2	859.1	18.0	2.0	292.6
10	859.1	832.5	18.0	2.0	291.1
11	832.5	300.0	18.0	27.0	289.7

Compressor station	Compression ratio	Capital cost ($/year)
1	1.44	70.00
2	1.40	70.00
3	1.00	70.00
4	1.00	70.00
5	1.00	70.00
6	1.00	70.00
7	1.00	70.00
8	1.26	70.00
9	1.00	70.00
10	1.00	70.00

pipeline segments in branches 2 and 3 at 18 inches, and the lower bound on the diameters of all pipeline segments at 4 inches. A lower bound of 2 miles was placed on each pipeline segment to ensure that the natural gas was at ambient conditions when it entered a subsequent compressor in the pipeline.

Figure E12.4*d* compares the optimal gas transmission network with the original network. From a nonfeasible starting configuration with 10-mile-long pipeline

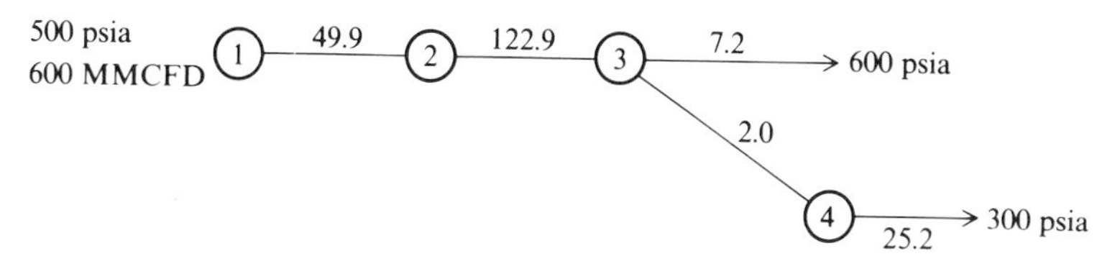

Figure E12.4*e* Optimal configuration using the costs of line *B* in Fig. E12.4*b*.

Table E12.4*b* Values of operating variables for the optimal network configuration using the costs of line *B*, Fig. E12.4*b*

Pipeline segment	Discharge pressure (psia)	Suction pressure (psia)	Pipe diameter (inch)	Length (mile)	Flow rate (MMCFD)
1	954.5	837.2	32.3	49.9	597.0
2	1000.0	699.7	32.3	122.9	594.0
3	699.7	600.0	15.2	2.2	295.5
4	699.7	665.7	18.0	2.0	295.5
5	952.2	300.0	16.9	25.2	294.0

Compressor station	Compression ratio	Capital cost ($/year)
1	1.91	69.50
2	1.19	69.50
3	1.00	69.50
4	1.43	69.50

segments, the nonlinear optimization algorithm reduced the objective function from the first feasible state of 1.399×10^7 dollars/year to 7.289×10^6 dollars/year, a savings of close to seven million dollars. Of the ten possible compressor stations, only four remained in the final optimal network. Table E12.4*a* lists the final state of the network. Note that because the suction and discharge pressures for the pipeline segments in branch 2 are identical, compressors 4, 5, 6, and 7 do not exist in the optimal configuration, nor do 9 and 10 in branch 3.

The same problem represented by Fig. E12.4*a* was solved again but using the costs represented by line *B* instead of line *A* in Fig. E12.4*b*. Figure E12.4*e* and Table E12.4*b* present the results of the computations. It is interesting to note that compressor 3 remains in the final configuration but with a compression ratio of 1, that is, compressor 3 is not doing any work. What this means is that it is cheaper to have two pipeline segments in branch 1 and two compressors each operating at about one-half capacity, plus a penalty of $10,000, than to have one pipeline segment and one compressor operating at full capacity. Compressor 3 doing no work represents just a branch in the line plus a cost penalty.

REFERENCES

Cheesman, A. P., "How to Optimize Gas Pipeline Design by Computer," *Oil and Gas J.*, *69* (No. 51), Dec. 20, 64 (1971*a*).

Cheesman, A. P., "Understanding Origin of Pressure is a Key to Better Well Planning," *Oil and Gas J.*, *69* (No. 46), Nov. 15, 146 (1971*b*).

Edgar, T. F., D. M. Himmelblau, and T. C. Bickel, "Optimal Design of Gas Transmission Network," *Soc. Petrol. Eng. J.*, ***30***: ***96*** (1978).

Flanigan, O., "Constrained Derivatives in Natural Gas Pipeline System Optimization," *J.Pet. Tech.*, 549 (May, 1972).

Gas Processor Suppliers Association, *Engineering Data Book*, 1972.

Graham, G. E., D. A. Maxwell, and A. Vallone, "How to Optimize Gas Pipeline Networks," *Pipeline Ind.*, 41–43 (June, 1971).

Happel, J., and D. G. Jordan, *Chemical Process Economics*, Second Edition, Marcel Dekker, New York, 1975.

Katz, D. L., *Handbook of Natural Gas Engineering*, McGraw-Hill, New York, 1959.

Larson, R. E., and P. J. Wong, "Optimization of Natural Gas Systems via Dynamic Programming," *Industrial and Engineering Chemistry*, *AC-12*, (No. 5), 475–481 (1968).

Letterman, R. D., "Economic Analysis of Granular Bed Filtration," *Trans. Amer. Soc. Civil Engrs.* (J. of Environmental Engr. Div.), *106*, 279 (1980).

Mah, R. S. H., and M. Shacham, "Pipeline Network Design and Synthesis," *Advances in Chem. Engr.*, *10*, (1978).

Martch, H. B., and N. J. McCall, "Optimization of the Design and Operation of Natural Gas Pipeline Systems," Paper No. SPE 4006, Society of Petroleum Engineers of AIME, 1972.

McCabe, W. L., J. C. Smith, and P. Harriott. *Unit Operations of Chemical Engineering*, 4th ed., McGraw-Hill, (1985).

Olorunniwo, F. O., "A Methodology for Optimal Design and Capacity Expansion Planning of Natural Gas Transmission Networks," Ph.D. Dissertation, The University of Texas at Austin, May, 1981.

Olorunniwo, F. O., and P. A. Jensen, "Optimal Capacity Expansion Policy for Natural Gas Transmission Networks—A Decomposition Approach," *Engr. Opt.*, *6*, 13 (1982); "Dynamic Sizing and Locationing of Facilities of Natural Gas Transmission Networks," *Engr. Opt.*, *6*, 95 (1982).

CHAPTER

13

CHEMICAL REACTOR DESIGN AND OPERATION

In practice, every chemical reaction carried out on a commercial scale involves the transfer of reactants and products of reaction, and the absorption or evolution of heat. Physical design of the reactor depends on the required structure and dimensions of the reactor, which involve the temperature and pressure distribution and the rate of chemical reaction. In this chapter, after describing the methods of formulating optimization problems for reactors and the tools for their solution, we will illustrate the techniques involved for several different reactor types.

13.1 FORMULATION OF CHEMICAL REACTOR OPTIMIZATION PROBLEMS

13.1.1 Modeling of Chemical Reactors

Optimization in the design and operation of a reactor focuses on formulating a suitable objective function plus a mathematical description of the reactor; the latter forms a set of constraints. Ideal reactors in chemical engineering are usually, but not always, represented by one or a combination of

1. Algebraic equations
2. Ordinary differential equations
3. Partial differential equations

One extreme of representation of reactor operation is complete mixing in a continuous stirred tank reactor (CSTR); the other extreme is no mixing whatsoever (plug flow). In between occur various degrees of mixing in dispersion reactors. Single ideal reactor types can be combined in various configurations to represent intermediate types of mixing as well as nonideal mixing and short-circuiting.

Ideal reactors can be classified in various ways, but for our purposes here the most convenient classification is according to the mathematical description of the reactor, as listed in Table 13.1. Each of the reactor types in Table 13.1 can be expressed in terms of integral equations, or difference equations, as well. However, not all real reactors can neatly fit into the classification in Table 13.1. The accuracy and precision of the mathematical description rest not only on the character of the mixing and the heat and mass transfer coefficients in the reactor, but also on the validity and analysis of the experimental data used to model the chemical reactions involved.

Other factors that must be considered in the modeling of reactors, factors that influence the number of equations and their degree of nonlinearity but not their form, are

1. The number and nature of the phases present in the reactor (gas, liquid, solid, and combinations thereof)

Table 13.1 Classification of reactors

Reactor type	Mathematical description (continuous variables)
Batch (well-mixed (CSTR), closed system)	Ordinary differential equations (unsteady state) Algebraic equation (steady state)
Semibatch (well-mixed (CSTR), open system)	Ordinary differential equations (unsteady state) Algebraic equations (steady state)
Continuous stirred tank reactors, individual or in series	Ordinary differential equations (unsteady state) Algebraic equations (steady state)
Plug flow reactor	Partial differential equations in one spatial variable (unsteady state) Ordinary differential equations in the spatial variable (steady state)
Dispersion reactor	Partial differential equations (unsteady state and steady state) Ordinary differential equations in one spatial variable (steady state)

2. The way of supplying and removal of heat (adiabatic, heat exchange mechanism, etc.)
3. The geometric configuration (empty cylinder, packed bed, sphere, etc.)
4. Reaction features (exothermic, endothermic, reversible, irreversible, number of species, parallel, consecutive, chain, selectivity)
5. Stability
6. The catalyst characteristics

Some references for the modeling of chemical reactors include Carberry (1976), Holland and Anthony (1979), Hill (1977), Wen and Fan (1975), and Lapidus and Amundson (1977).

13.1.2 Objective Functions for Reactors

Various questions can be posed concerning reactors that lead directly to the formulation of an objective function. Typical objective funtions in terms of the adjustable variables are:

1. Maximize conversion (yield) per volume with respect to time.
2. Maximize production per batch.
3. Minimize production time for a fixed yield.

4. Minimize total production costs/average production costs with respect to time/fraction conversion.
5. Maximize yield/number of moles of component/concentration with respect to time and/or operating conditions.
6. Design the optimal temperature sequence with respect to time/reactor length to obtain (*a*) a given fraction conversion, (*b*) a maximum rate of reaction, or (*c*) the minimum residence time.
7. Adjust the temperature profile to specifications (via sum of squares) with respect to the independent variables.
8. Minimize volume of the reactor(s) with respect to certain concentration(s).
9. Change the temperature from T_0 to T_f in minimum time subject to heat transfer rate constraints.
10. Maximize profit with respect to volume.
11. Maximize profit with respect to fraction conversion to get optimal recycle.
12. Optimize profit/volume/yield with respect to boundary/initial conditions in time.
13. Minimize consumption of energy with respect to operating conditions.

In some cases a variable can be independent and in another case the same variable can be dependent, but the usual independent variables are pressure, temperature, and flow rate or concentration of a feed. We cannot provide examples for all of these criteria, but have selected a few to show how they mesh with the optimization methods described in earlier chapters and mathematical models listed in Table 13.1.

In considering a reactor by itself, as we do in this chapter, you must keep in mind that a reactor no doubt will be only one unit in a complete process, and that at least a separator must be included in any economic analysis. Figure 13.1 depicts figuratively the relation between the yield or selectivity of a reactor and costs.

All of the various optimization techniques described in previous chapters can be applied to one or more types of reactor models. The reactor model forms a set of constraints so that most optimization problems involving reactors must accommodate steady-state algebraic equations or dynamic differential equations as well as inequality constraints. The following are the most commonly used optimization techniques reported in the literature:

1. Differential calculus after converting the constrained problem to an unconstrained one
2. Linear and nonlinear programming
3. Maximum/minimum principle
4. Dynamic programming

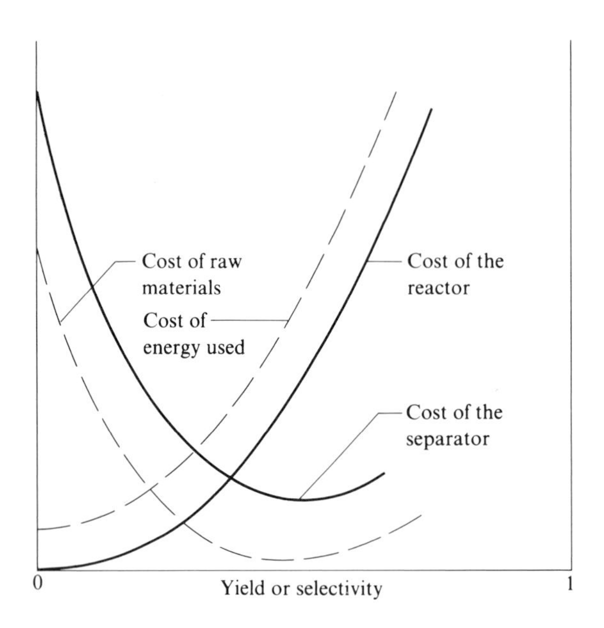

Figure 13.1 Costs of energy and raw materials for a reactor as a function of yield and selectivity [adapted and modified from P. LeGoff, "The Energetic and Economic Optimization of Heterogeneous Reactors," *Chem. Eng. Sci.*, **35**: 2089 (1980)].

Tables surveying the application of various classes of techniques are included in each section that follows. We focus mainly on categories 1 and 2; the maximum principle and dynamic programming have been used infrequently (see Aris (1961) and Lee and Aris (1963) for examples). Examples applying these techniques to reactor optimization are also presented.

13.2 USE OF DIFFERENTIAL CALCULUS IN REACTOR OPTIMIZATION

Table 13.2 lists applications in which the final formulation of the problem is in the form of an unconstrained function (usually of a single variable) that is differentiated, and the derivative(s) equated to zero. The optimum is obtained either analytically or by a one-dimensional search. Integration of differential equations is usually required and some algebraic manipulations are needed to obtain a suitable function that can be differentiated. In some cases, conclusions can be reached by direct inspection of the intermediate or final mathematical relations by looking at limits, etc. In other cases, sets of nonlinear equations must be solved, one for each independent variable. Rippin (1983), in his review, includes a tabulation of references for batch and CSTR reactor optimization. Chen (1983), in each of six chapters in the second half of his book, describes and illustrates via many examples the use of calculus in the optimization of reactors.

Table 13.2 Ordinary calculus applied to the optimization of chemical reactors

Objective function		Type of constraints				
Specific objective	Independent variable(s)	Algebraic equations	Ordinary differential equations	Type of reactor	Remarks	Reference
Mean production rate (max)	Time		√ (integrated)	CSTR + catalyst aging		[A]
Catalyst activity (max)	Time		√ (integrated)			
Cycle period (max)	Catalyst activity		√ (integrated)			
Unit cost of production (min)	Time		√ (integrated)			
Reaction time (min)	Temperature, initial concentration		√ (integrated)	Batch		[B]
Conversion rate (max) Volume gradient (min)	Temperature		√	Plug flow		[C]
Residence time (min)	Exit concentration		√ (integrated)	Plug flow CSTR in combinations		[D]
Total volume (min)	Conversion	√		Two CSTRs	Uses maximum reaction rate	[E]
Yield (max)	Molar flow rate	√		CSTR		[F]
Present value of net profit (max)	Time of cycle, feed ratio		√ (integrated)	CSTR, plug flow		[G]
Profit (max)	Reactor lengths (or catalyst weights), temperatures after preheating, bypass flows		√ (integrated)	Plug flow	Transform $2n$ variable problem into $2n$ single variable problems	[H]

(*Continued*)

Table 13.2 *(continued)*

Objective function		Type of constraints				
Specific objective	**Independent variable(s)**	**Algebraic equations**	**Ordinary differential equations**	**Type of reactor**	**Remarks**	**Reference**
Production (max)	Conversion	√		CSTR		[I]
Rate of reaction (max)	Fractional usage		√	CSTR	Uses similitude transformation	[J]
Volume of catalyst mass (min)	Molar flow rate, fraction conversion		√ (integrated)	Plug flow	*n* stages	[K]
Concentrations	Temperature		√	CSTR	Graphical analysis	[L]
Volume and number of tanks	Residence time	√		Staged CSTRs	Root finding	[M]

[A] LeGoff, P., "Optimization of the Regenerative and Replacement Cycles of a Catalyst," *Int. Chem. Eng.*, **23**: 225 (1983).
[B] Quicker, G., A. Schumpe, and W. D. Deckwer, "Minimum Endtime Policies for Batchwise Radical Chain Polymerization IV," *Chem. Eng. Sci.*, **39**: 177 (1984).
[C] Hills, B. A., "Optimum Temperature Conditions for Programmed Gas Reactors," *Br. Chem. Eng.*, **10**: 680 (1965).
[D] Bischoff, K. B., "Optimal Continuous Fermentation Reactor Design," *Can. J. Chem. Eng.*, **44**: 281 (1966).
[E] Lin, K. F., and J. T. Chen, "Optimization of Nonisothermal Adiabatic Backmix Reactors in Series for a Single Reaction," *J. Chin. Inst. Chem. Eng.*, **10**: 107 (1979).
[F] Suciu, G. D., and H. G. Zengel, "Maximum Yield of a Liquid Product Resulting from a Gas Phase Reaction," *Chem. Eng. Sci.*, **31**: 322 (1976).
[G] Barreto, G. F., O. A. Ferretti, I. H. Farina, and N. O. Lemcoff, "Optimization of the Operating Conditions of CO Converters," *Ind. Eng. Process Des. Dev.*, **20**: 594 (1981).
[H] Malengé, J. P., and L. M. Vincent, "Optimal Design of a Sequence of Adiabatic Reactors with Cold-Shot Cooling," *Ind. Eng. Chem. Process Des. Dev.*, **11**: 465 (1972).
[I] Burghardt, A., and N. Bartelmus, "Optimization of a Cascade of Reactors for an Irreversible Chemical Reaction of the *n*th Order," *Int. Chem. Eng.*, **11**: 604 (1971).
[J] Fainzil'ber, A. M., and N. A. Fridlenden, "Application of Similitude Integrals to Chemical Processes," *Br. Chem. Eng.*, **10**: 462 (1965).
[K] Dyson, D. C., and F. J. M. Horn, "Optimum Adiabatic Cascade Reactor with Direct Intercooling," *Ind. Eng. Chem. Fundam.*, **8**: 49 (1969).
[L] Mah, R. S. H., and R. Aris, "Optimal Policies for First Order Consecutive Reversible Reactions," *Chem. Eng. Sci.*, **19**: 541 (1964).
[M] Oguztorelli, M. N., W. J. Gibb, and B. Ogum, "Optimal Reactor Size Distribution with Fixed Terminal Constraints," Math. & Comput. in Simulation XXVI, Elsevier Scientific Publications, New York (1984) p. 497.

EXAMPLE 13.1 OPTIMAL RESIDENCE TIME FOR MAXIMUM YIELD IN AN IDEAL ISOTHERMAL BATCH REACTOR

The following reaction scheme takes place in a well-mixed batch (closed) reactor

$$A \underset{k_2}{\overset{k_1}{\rightleftarrows}} B \quad \begin{matrix} \overset{k_3}{\nearrow} C \\ \underset{k_4}{\searrow} D \end{matrix}$$

For the following initial concentrations and values of reaction rates

$$c_{AO} = 50 \text{ g mol/L} \qquad k_1 = 2.00 \text{ h}^{-1}$$

$$c_{BO} = 5.0 \text{ g mol/L} \qquad k_2 = 1.00 \text{ h}^{-1}$$

$$c_{CO} = 0 \text{ g mol/L} \qquad k_3 = 0.20 \text{ h}^{-1}$$

$$c_{DO} = 0 \text{ g mol/L} \qquad k_4 = 0.60 \text{ h}^{-1}$$

calculate the optimal residence time to yield a maximum value of c_B.

Solution. From the reaction scheme, we can presume the rate equations to be first-order:

$$\frac{dc_A}{dt} = -k_1 c_A + k_2 c_B \tag{a}$$

$$\frac{dc_B}{dt} = k_1 c_A - (k_2 + k_3 + k_4) c_B \tag{b}$$

$$\frac{dc_C}{dt} = k_3 c_B \tag{c}$$

$$\frac{dc_D}{dt} = k_4 c_B \tag{d}$$

and the overall material balance at any time is

$$c_A + c_B + c_C + c_D = c_{AO} + c_{BO} + c_{CO} + c_{DO} \tag{e}$$

We want to solve for c_B solely as a function of time.

Solve Eq. (b) for c_A

$$c_A = \frac{1}{k_1}\left[\frac{dc_B}{dt} + (k_2 + k_3 + k_4)c_B\right]$$

$$= \frac{1}{2.00}\left(\frac{dc_B}{dt} + 1.80 c_B\right) \tag{f}$$

Differentiate Eq. (f) with respect to time to get

$$\frac{dc_A}{dt} = \frac{1}{2.00}\left(\frac{d^2 c_B}{dt^2} + 1.80\frac{dc_B}{dt}\right) \tag{g}$$

Substitution of Eqs. (f) and (g) into Eq. (a) yields a second-order linear homogeneous differential equation in c_B:

$$\frac{d^2c_B}{dt^2} + 3.80\frac{dc_B}{dt} + 1.60c_B = 0 \tag{h}$$

which has the solution

$$c_B = c_1e^{-0.482t} + c_2e^{-3.318t} \tag{i}$$

For $c_B(0) = 5.0$ Eq. (i) gives

$$5.0 = c_1 + c_2 \tag{j}$$

From Eq. (f) and (i)

$$c_A = \frac{1}{2.00}[(-0.482)c_1e^{-0.482t} - 3.318c_2e^{-3.318t} + 1.80c_B]$$

For $c_A(0) = 50$ and $c_B(0) = 5.0$

$$50 = -0.241c_1 - 1.659c_2 + 4.5 \tag{k}$$

Thus, after solving for c_1 and c_2

$$c_B = 37.94e^{-0.482t} - 32.94e^{-3.318t} \tag{l}$$

Finally, we differentiate c_B in Eq. (l) with respect to t and equate the derivative to zero to get the maximum c_B and the optimal t^*:

$$\frac{dc_B}{dt} = 0 = 18.29e^{-0.482t} + 109.28e^{-3.318t} \tag{m}$$

The root of Eq. (m) is $t^* = 0.63$ h; $c_B^* = 23.04$.

EXAMPLE 13.2 ONE-DIMENSIONAL SEARCH FOR THE OPTIMUM RESIDENCE TIME OF A CHEMOSTAT

Jeffreson and Smith (1973) proposed the following dynamic model for a continuous flow biological chemostat (CSTR):

$$\begin{aligned}\dot{e} &= rk_2c - De \\ \dot{c} &= k_1s(e - c) - (k_2 + k_3)c - Dc \\ \dot{s} &= -k_1s(e - c) + k_3c + D(s^0 - s)\end{aligned} \tag{a}$$

where D is the independent variable and is called the dilution rate (flow volume per unit time/chemostat volume, h^{-1}), e is the biomass concentration (mol/L), c is a metabolic intermediate concentration (mol/L), s is the limiting substrate concentration (mol/L), and s^0 is the limiting feed substrate concentration (10.0 mol/L). The parameter and rate constant values are: $r = 0.09$ (mole enzyme formed/mole substrate consumed), $k_1 = 0.9$ L/(mol) (h), $k_2 = 0.7\ h^{-1}$, and $k_3 = 0.0$. One objective of such a reactor is to maximize the steady state production of biomass, which is given by

$$f = De$$

using the dilution rate as the independent variable. The steady state values of e, c, and s for the reactor model can be obtained by solving the set of equations (a) with $\dot{e} = \dot{c} = \dot{s} = 0$. Plot the behavior of f as D varies, indicating the optimum value of D.

Solution. Although the variation of the steady-state values of e, c, and s as a function of D could be found by integrating the three ordinary differential equations until steady state is reached, it is much easier to solve the three algebraic equations in (a) with $\dot{e} = \dot{c} = \dot{s} = 0$ by the Newton-Raphson or secant methods. To find the optimum value of f, you can use a one-dimensional search procedure, employing the equation-solving routine for each value of D. Figure E13.2 shows a plot of e, c, s and f ($= De$) obtained in this way. The objective function is clearly unimodal between the values of $0 \le D \le 0.6$. The objective function is relatively flat for $0.2 \le D \le 0.5$, but precipitously drops off as D approaches 0.55. This behavior is known as the "washout" conditions in a chemostat; the dilution rate simply overwhelms the reaction of biomass, causing $e = c = 0$ and $s = s^0$ (at $D = 0.553$).

For this particular reaction system, the equations can also be manipulated algebraically to obtain an analytical formula for f as a function of D. You can also show that the washout conditions can be found by solving the equation

$$s = s^0 = \frac{k_2 + k_3 + D}{k_1[(rk_2/D) - 1]}$$

This calculation is left to the reader as an exercise.

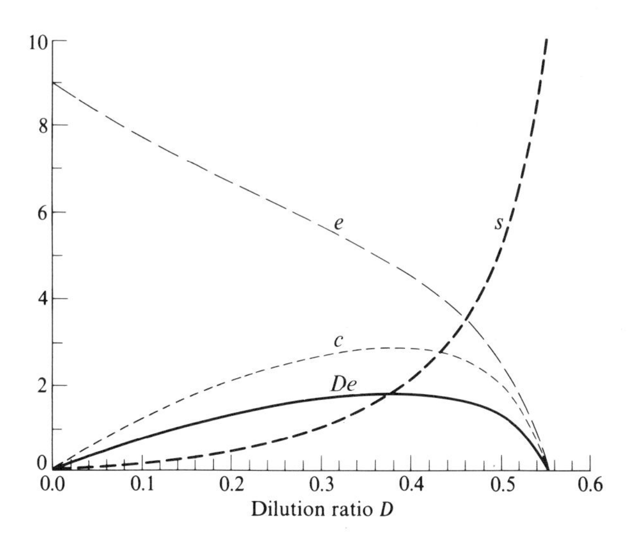

Figure E13.2 Plot of steady state values of e, c, s, and De ($=f$) as a function of dilution ratio D.

13.3 USE OF LINEAR PROGRAMMING TO OPTIMIZE REACTOR OPERATIONS

Reactor systems that can be described by a "yield matrix" are potential candidates for the application of linear programming. In these situations, each reactant is known to produce a certain distribution of products. When multiple reactants are employed, then it is desirable to optimize the amounts of each reactant so that the products satisfy flow and demand constraints. Linear programming has become widely adopted in scheduling production in olefin units and catalytic crackers (see Sourander, et al. (1984)). Below we illustrate the use of linear programming to optimize the operation of a thermal cracker.

EXAMPLE 13.3 OPTIMIZATION OF A THERMAL CRACKER

Figure E13.3 lists various feeds and the corresponding product distribution for a thermal cracker which produces olefins. The possible feeds include ethane, propane, debutanized natural gasoline (DNG), and gas oil, some of which may be fed simultaneously. Based on plant data, eight products are produced in varying proportions according to the following matrix.

Yield structure: (wt. fraction)

	Feed			
Product	**Ethane**	**Propane**	**Gas oil**	**DNG**
Methane	0.07	0.25	0.10	0.15
Ethane	0.40	0.06	0.04	0.05
Ethylene	0.50	0.35	0.20	0.25
Propane	....	0.10	0.01	0.01
Propylene	0.01	0.15	0.15	0.18
Butadiene	0.01	0.02	0.04	0.05
Gasoline	0.01	0.07	0.25	0.30
Fuel oil	....	....	0.21	0.01

The capacity to run gas feeds through the cracker is 200,000 lb/stream hour (total flow based on an average mixture). Ethane uses the equivalent of 1.1 lb of capacity per pound of ethane; propane uses 0.9 lb of capacity per pound of propane; gas oil uses 0.9 lb/lb; and DNG has a ratio of 1.0.

Downstream processing limits exist of 50,000 lb/stream hour on the ethylene and 20,000 lb/stream hour on the propylene. The fuel requirements to run the cracking system for each feedstock type are as follows:

Ethane	8364 Btu/lb
Propane	5016 Btu/lb
Gas oil	3900 Btu/lb
DNG	4553 Btu/lb

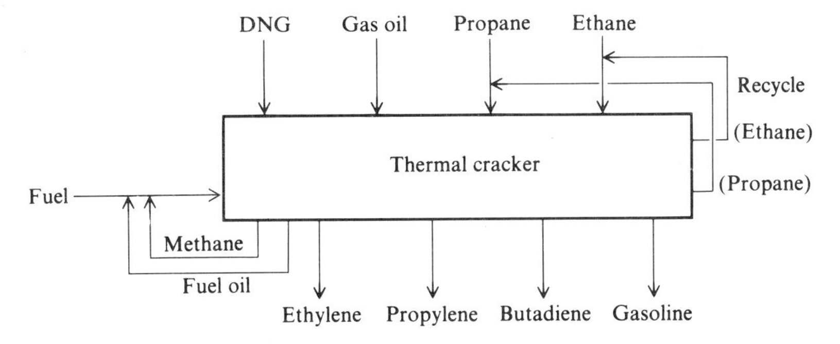

Figure E13.3 Flow diagram of thermal cracker.

Methane and fuel oil produced by the cracker are recycled as fuel. All the ethane and propane produced is recycled as feed. Heating values are as follows:

Natural gas	21,520 Btu/lb
Methane	21,520 Btu/lb
Fuel oil	18,000 Btu/lb

Because of heat losses and the energy requirements for pyrolysis, a fixed fuel requirement of 20.0×10^6 Btu/stream hour occurs. The price structure on the feeds and products and fuel costs is (all values are in cents per pound):

Feeds:	Ethane	6.55
	Propane	9.73
	Gas oil	12.50
	DNG	10.14
Products:	Methane	5.38 (fuel value)
	Ethylene	17.75
	Propylene	13.79
	Butadiene	26.64
	Gasoline	9.93
	Fuel oil	4.50 (fuel value)

Assume an energy (fuel) cost of $\$2.50/10^6$ Btu.

(*a*) Set up the objective function and constraints to maximize profit while operating within furnace and downstream process equipment constraints. The variables to be optimized are the amounts of the four feeds. Solve using the LP-Simplex method.

(*b*) Examine the sensitivity of profits to increases in the ethylene production rate.

Solution. We define the following variables for the flow rates to and from the furnace (in lb/h):

x_1 = fresh ethane feed

x_2 = fresh propane feed

x_3 = gas oil feed

x_4 = DNG feed

x_5 = ethane recycle

x_6 = propane recycle

x_7 = fuel added

Assumptions used in formulating the objective function and constraints are:

1. 20×10^6 Btu/h fixed fuel requirement (methane) to compensate for the heat loss.
2. All propane and ethane are recycled with the feed, and all methane and fuel oil will be recycled as fuel.

A basis of 1 hour will be used and all costs will be on a basis of cents per hour.

Objective function (profit). In words the profit f is

$$f = \text{product value} - \text{feed cost} - \text{energy cost}$$

Product value. The value for each product (in cents per pound) is as follows:

$$\text{Ethylene:} \quad 17.75(0.5x_1 + 0.5x_5 + 0.35x_2 + 0.35x_6 + 0.20x_3 + 0.25x_4) \qquad (a)$$

$$\text{Propylene:} \quad 13.79(0.01x_1 + 0.01x_5 + 0.15x_2 + 0.15x_6 + 0.15x_3 + 0.18x_4) \qquad (b)$$

$$\text{Butadiene:} \quad 26.64(0.01x_1 + 0.01x_5 + 0.02x_2 + 0.02x_6 + 0.04x_3 + 0.05x_4) \qquad (c)$$

$$\text{Gasoline:} \quad 9.93(0.01x_1 + 0.01x_5 + 0.07x_2 + 0.07x_6 + 0.25x_3 + 0.30x_4) \qquad (d)$$

$$\text{Total product sales} = 9.39x_1 + 9.51x_2 + 9.17x_3 + 11.23x_4 + 9.39x_5 + 9.51x_6 \qquad (e)$$

Feed cost.

$$\text{feed cost (cents/hour)} = 6.55x_1 + 9.73x_2 + 12.50x_3 + 10.14x_4 \qquad (f)$$

Energy cost. The fixed heat loss of 20×10^6 Btu/h can be expressed in terms of methane cost (5.38¢/lb) using a heating value of 21,520 Btu/lb for methane. However, the fixed heat loss represents a constant cost which is independent of the variables x_i, hence in optimization we can ignore this factor, but in evaluating the final costs this term must be taken into account. The value for x_7 will depend on the amount of fuel oil and methane produced in the cracker (x_7 will provide for any deficit in products recycled as fuel).

We combine (e) and (f) to get the objective function (cents per hour)

$$f = 2.84x_1 - 0.22x_2 - 3.33x_3 + 1.09x_4 + 9.39x_5 + 9.51x_6 \qquad (g)$$

Constraints.

(1) Cracker capacity of 200,000 lb/h

$$1.1(x_1 + x_5) + 0.9(x_2 + x_6) + 0.9x_3 + 1.0x_4 \leq 200{,}000 \tag{h}$$

or

$$1.1x_1 + 0.9x_2 + 0.9x_3 + 1.0x_4 + 1.1x_5 + 0.9x_6 \leq 200{,}000$$

(2) Ethylene processing limitation of 100,000 lb/h

$$0.5x_1 + 0.35x_2 + 0.25x_3 + 0.25x_4 + 0.5x_5 + 0.35x_6 \leq 100{,}000 \tag{i}$$

(3) Propylene processing limitation of 20,000 lb/h

$$0.01x_1 + 0.15x_2 + 0.15x_3 + 0.18x_4 + 0.01x_5 + 0.15x_6 \leq 20{,}000 \tag{j}$$

(4) Ethane recycle

$$x_5 = 0.4x_1 + 0.4x_5 + 0.06x_2 + 0.06x_6 + 0.04x_3 + 0.05x_4 \tag{k}$$

Rearranging, (*j*) becomes

$$0.4x_1 + 0.06x_2 + 0.04x_3 + 0.05x_4 - 0.6x_5 + 0.06x_6 = 0 \tag{l}$$

(5) Propane recycle

$$x_6 = 0.1x_2 + 0.1x_6 + 0.01x_3 + 0.01x_4 \tag{m}$$

Rearranging Eq. (*m*),

$$0.1x_2 + 0.01x_3 + 0.01x_4 - 0.9x_6 = 0 \tag{n}$$

(6) Heat constraint

The total fuel heating value (THV) in Btu per hour is given by

fuel — methane from cracker

$$\text{THV} = 21{,}520x_7 + 21{,}520\,(0.07x_1 + 0.25x_2 + 0.10x_3 + 0.15x_4 - 0.07x_5 + 0.25x_6)$$

fuel oil from cracker

$$+\ 18{,}000(0.21x_3 + 0.01x_4)$$

$$= 1506.4x_1 + 5380x_2 + 5932x_3 + 3408x_4 + 1506.4x_5 + 5380x_6 + 21{,}520x_7 \tag{o}$$

The required fuel for cracking (Btu/hour) is

ethane — propane — gas oil — DNG

$$8364(x_1 + x_5) + 5016(x_2 + x_6) + 3900x_3 + 4553x_4$$

$$= 8364x_1 + 5016x_2 + 3900x_3 + 4553x_4 + 8364x_5 + 5016x_6 \tag{p}$$

Therefore the sum of Eq. (*p*) + 20,000,000 Btu/h is equal to the total fuel heating value from Eq. (*o*), which gives the constraint

$$-6857.6x_1 + 364x_2 + 2032x_3 - 1145x_4 - 6857.6x_5 + 364x_6 + 21{,}520x_7 = 20{,}000{,}000 \tag{q}$$

Table E13.3 Optimal flowrates for cracking furnace for different restrictions on ethylene and propylene production

	Flow rate (lb/h)	
Stream	Case 1	Case 2
x_1 (ethane feed)	60,000	21,770
x_2 (propane feed)	0	0
x_3 (gas oil feed)	0	0
x_4 (DNG feed)	0	107,600
x_5 (ethane recycle)	40,000	23,600
x_6 (propane recycle)	0	1,195
x_7 (fuel added)	32,800	21,090
Ethylene	50,000	50,000
Propylene	1,000	20,000
Butadiene	1,000	5,857
Gasoline	1,000	32,820
Methane (recycled to fuel)	7,000	19,610
Fuel oil	0	1,076
Objective function (¢/h)	369,560	298,590

Table E13.3 lists the optimal solution of this problem obtained using the Simplex algorithm (case 1). Note that the maximum amount of ethylene is produced. As the ethylene production constraint is relaxed, the objective function value increases. Once the constraint is raised above 90,909 lb/h, the objective function remains constant.

Suppose the inequality constraints on ethylene and propylene production were changed to equality constraints (ethylene = 50,000; propylene = 20,000). The optimal solution for these conditions is shown as case 2 in Table E13.3. This specification forces the use of DNG as well as ethane.

13.4 NONLINEAR PROGRAMMING APPLIED TO CHEMICAL REACTOR OPTIMIZATION

If nonlinear algebraic equations and nonlinear inequalities comprise the reactor model, then the algorithms described in Chap. 8 can be used for optimizing reactor design and operation. One of the advantages of nonlinear programming is that both equality and inequality constraints can be accommodated directly. However, if differential equations are involved in the reactor model, you must either solve the differential equations within each cycle of optimization, or else represent the differential equations as a set of finite difference equations, i.e., use discrete approximations for the derivatives (see Sec. 6.5). Table 13.3 lists several examples of the application of nonlinear programming to reactor optimization that have appeared in the literature.

Table 13.3 Nonlinear programming applied to optimization of chemical reactors

Objective function		Type of constraints					
Specific objective	Independent variable(s)	Algebraic equations	Ordinary differential equations	Inequalities	Type of reactor	Remarks	Reference
Maximize yield of tubular catalytic reactor	Time	√	√	√	Plug flow	Deactivating catalyst; gradient method used.	[A]
Minimize costs of autothermal reactor with heat exchanger	Pressure drop, flow rate, temperature, configuration of units	√	√		Plug flow	Used finite difference approximation to differential equations. Pattern search of Hooke and Jeeves used. Yield fixed.	[B]
Minimize sums of squares related to operation of ammonia reactor	Heat transfer coefficients, catalyst activity factor	√	√		Plug flow (plus heat exchanger)	Powell's method used.	[C]
Maximize exit concentration of product of CSTR's	Inlet concentration in feed of raw material		√	√(bounds on variables)	Two CSTR's in series	Successive linear programming used.	[D]
Maximize yield of series of reactors	Concentration, recycle ratio		√		CSTR, plug flow, and combinations	Integrated differential equations. Used Hooke and Jeeves' pattern search.	[E]
Maximize profit of hydrogenation reactor	Composition, steam use, by-product loss	√	√		Plug flow	Solved differential equations numerically, Used Rosenbrock's search.	[F]

(*Continued*)

Table 13.3 *(continued)*

Objective function		Type of constraints					
Specific objective	**Independent variable(s)**	**Algebraic equations**	**Ordinary differential equations**	**Inequalities**	**Type of reactor**	**Remarks**	**Reference**
Maximize rate of production of one component in catalyst	Reactor length		√	√(bounds)	Plug flow	Analytical integration of differential equations. Used Rosenbrock's method. Also used maximum principle.	[G]
Optimize weighted sum of polydispersity, conversion, chain length in polymerization reactor	Initiator concentration, temperature, holding time	√	√	√	Well-mixed batch	Set up difference equations. Used method of Nelder and Mead.	[H]

[A] Buzzi-Ferraris, G., E. Facchi, P. Forzatti, and E. Troncani, "Control Optimization of Tubular Catalytic Reactors With Catayst Decay," *Ind. Eng. Chem. Process Des. Dev.*, **23:** 126 (1984).

[B] Zabar, E., and M. Sheintuch, "Optimization of an Autothermal Monolithic Reactor—Heat Exchanger for SO_2 Oxidization over Platinum," *Chem. Eng. Commun.*, **16**: 313 (1982).

[C] Patnaik, L. J., N. Viswanadham, and I. G. Sarma, "Steady State Optimization of an Ammonia Reactor," *Comput. Chem. Eng.*, **7**: 217 (1980).

[D] Bhattacharya, A., and B. Joseph, "Online Optimization of Chemical Processes," 1982 ACC, p. 334.

[E] Chitra, S. P., and R. Govind, "Yield Optimization for Complex Reactor Systems," *Chem. Eng. Sci.*, **36**: 1219 (1981).

[F] Sheel, J. G. P., and C. M. Crowe, "Simulation and Optimization of an Existing Ethylbenzene Dehydrogenation Reactor," *Can. J. Chem. Eng.*, **47**: 183 (1969).

[G] Gunn, D. J., "The Optimization of Bifunctional Catalyst Systems," *Chem. Eng. Sci.*, **22**: 963 (1967).

[H] Ray, W. H., "Modeling Polymerization Reactors with Applications to Optimal Design," *Can. J. Chem. Eng.*, **45**: 356 (1967).

Example 13.4 shows how to apply nonlinear programming to optimize yield in a series of reactors.

EXAMPLE 13.4 MAXIMUM YIELD WITH RESPECT TO REACTOR VOLUME

A series of four well-mixed reactors operate isothermally as illustrated in Fig. E13.4. The compound whose concentration is designated by c reacts according to the following mechanism: $r = -kc^n$ in each tank. Assume steady state operation and a fixed fluid flow rate of q.

Determine the values of the tank volumes (in effect the residence times of the compound for a constant volumetric flow rate) in each of the four tanks so as to maximize the yield of product that arises from the reaction. Note $(V_i/q) = \theta_i$, the residence time. Use the following data for coefficients in the problem

$$n = 2.5 \qquad k = 0.00625[\text{m}^3/\text{kg mol}]^{1.5}(\text{s})^{-1}$$

$$c_0 = 20 \text{ kg mol/m}^3 \qquad q = 71 \text{ m}^3/\text{h}$$

The total volume of all the tanks is fixed at 20 m^3.

Solution. The material balances for each reactor are

$$\frac{d(V_i c_i)}{dt} = q_{i-1}c_{i-1} - qc_i - V_i k c_i^n \tag{a}$$

In the steady state the derivatives vanish so that the material balances are (in terms of θ_i)

$$c_0 = c_1 + \theta_1 k c_1^n \tag{b}$$

$$c_1 = c_2 + \theta_2 k c_2^n \tag{c}$$

$$c_2 = c_3 + \theta_3 k c_3^n \tag{d}$$

$$c_3 = c_4 + \theta_4 k c_4^n \tag{e}$$

An additional constraint from the volume specification is

$$\theta_1 + \theta_2 + \theta_3 + \theta_4 = \tfrac{20}{71} \tag{f}$$

Murase, et al. (1970) proposed a method where the constrained optimization problem could be solved as a series of unconstrained problems. Instead of maximizing composition for the objective function, minimize the value of c_4 for a fixed

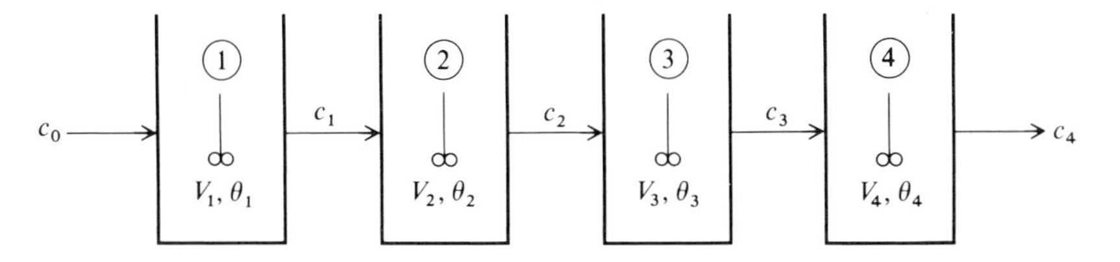

Figure E13.4 Series of four well-mixed reactors.

c_0. Unfortunately, it is not possible to solve Eqs. (*b*) to (*e*) and obtain an explicit expression for c_4. An alternative strategy is to fix c_4 and maximize c_0 because we can explicitly solve the above equations for c_0. Because for this problem c_0 is specified, we must make several trials until we match the optimal c_0 with the specified value. Thus we will solve for c_0 in terms of c_4:

$$c_2 = [c_4 + \theta_4 k c_4^n] + \theta_3 k [c_4 + \theta_4 k c_4^n]^n$$

$$c_1 = [c_4 + \theta_4 k c_4^n] + \theta_3 k [c_4 + \theta_4 k c_4^n]^n + \theta_2 k \{[c_4 + \theta_4 k c_4^n] + \theta_3 k [c_4 + \theta_4 k c_4^n]^n\}^n$$

$$c_0 = [c_4 + \theta_4 k c_4^n] + \theta_3 k [c_4 + \theta_4 k c_4^n]^n + \theta_2 k \{[c_4 + \theta_4 k c_4^n] + \theta_3 k [c_4 + \theta_4 k c_4^n]^n\}^n + \theta_1 k_1 ([c_4 + \theta_4 + k c_4^n] + \theta_3 k [c_4 + \theta_4 k c_4^n]^n + \theta_2 k \{[c_4 + \theta_4 k c_4^n] + \theta_3 k [c_4 + \theta_4 k c_4^n]^n\}^n)^n$$

Equation (*f*) can be used to eliminate θ_1.

Keep in mind that the θ_i's have to be positive, but we can use an unconstrained optimization code first and assume that only positive θ's result. (If not, an optimization code for constrained problems would have to be used.) The procedure is to specify a value for c_4 and maximize c_0 with respect to θ_2, θ_3, and θ_4. After each cycle of optimization has converged, we adjust c_4 using a secant method (linear interpolation or extrapolation), and the procedure is repeated until $c_0 = 20$ kg mol/m^3.

The Broyden-Davidon-Fletcher-Shanno code (see Chap. 6) yields the following results:

$$c_4 = 0.3961$$

$$\theta_2 = 196.9 \text{ s} \qquad V_2 = 3.884 \text{ m}^3$$

$$\theta_3 = 296.6 \text{ s} \qquad V_3 = 5.849 \text{ m}^3$$

$$\theta_4 = 406.9 \text{ s} \qquad V_4 = 8.025 \text{ m}^3$$

and consequently

$$\theta_1 = 113.9 \text{ s} \qquad V_1 = 2.242 \text{ m}^3$$

References to other versions and treatments of this type of problem via Lagrange multipliers, dynamic programming, plotting, and calculus are due to Wood and Stevens (1964), Luss (1965), Denbigh (1944), Crooks (1966), and Szepe and Levenspiel (1964).

EXAMPLE 13.5 OPTIMAL DESIGN OF AN AMMONIA REACTOR

This example based on the reactor described by Murase et al. (1970) shows one way to mesh the numerical solution of the differential equations in the process model with an optimization code. The reactor, illustrated in Fig. E13.5*a*, is based on the Haber process

$$N_2 + 3H_2 \leftrightarrow 2NH_3$$

Figure E13.5*b* illustrates suboptimal concentration and temperature profiles experienced. The temperature at which the reaction rate is a maximum decreases as the conversion increases.

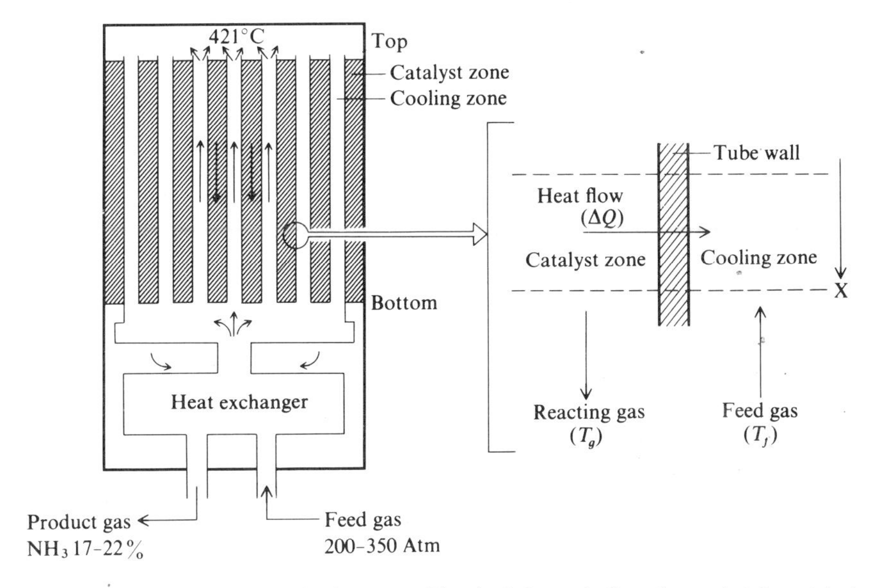

Figure E13.5*a* Ammonia synthesis reactor. The shaded area indicated contains the catalyst. [*Adapted from Murase, Roberts, and Converse (1970), p. 504.*]

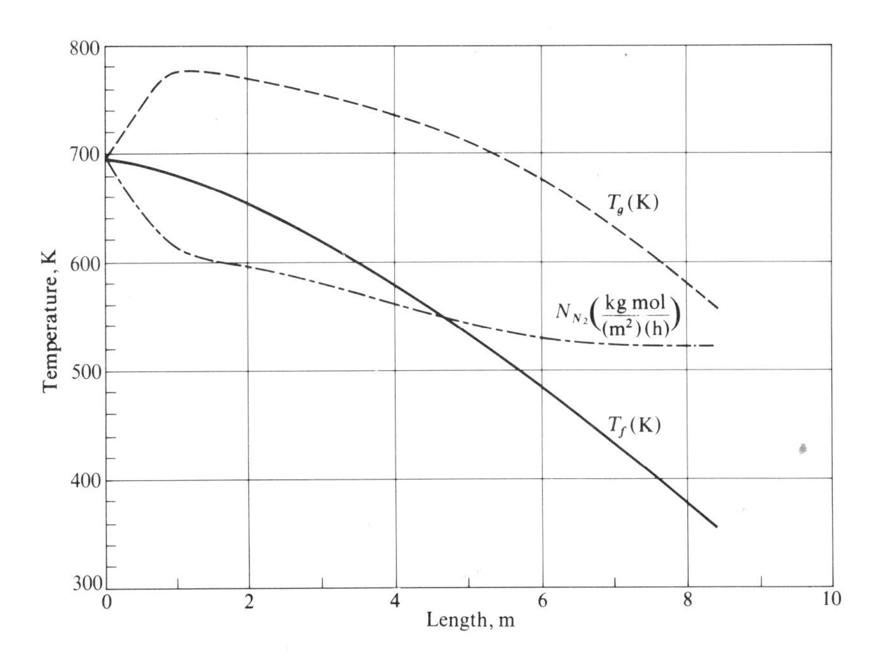

Figure E13.5*b* Temperature and concentration profiles of the NH_3 synthesis reactor.

Assumptions made in developing the model are

1. The rate expression is valid
2. Longitudinal heat and mass transfer can be ignored
3. The gas temperature in the catalytic zone is also the catalyst particle temperature
4. The heat capacities of the reacting gas and feed gas are constant
5. The catalytic activity is uniform along the reactor and equal to unity
6. The pressure drop across the reactor is negligible compared to the total pressure in the system

The notation and data to be used are listed in Table E13.5.

Objective function. The objective function for the reactor optimization is based on the difference between the value of the product gas (heating value and ammonia value) and the value of the feed gas (as a source of heat only) less the amortization of reactor capital costs. Other operating costs are omitted. As shown in Murase et al. (1970), the final consolidation of the objective function terms (corrected here) is

$$f(x, N_{N_2}, T_f, T_g) = 11.9877 \times 10^6 - 1.710 \times 10^4 N_{N_2} + 704.04 T_g - 699.3T_f - [3.4566 \times 10^7 + 2.101 \times 10^9 x]^{1/2} \quad (a)$$

Equality constraints. Only one degree of freedom exists in the problem because there are three constraints; x is designated to be the independent variable.

Table E13.5 Notation and data for Example E13.5

	Independent and dependent variables
x	Reactor length, m
N_{N_2}	Mole flow rate of N_2 per area catalyst, kg mol/(m^2)(h)
T_f	Temperature of feed gas, K
T_g	Temperature of reacting gas, K
	Parameters
C_{pf}	Heat capacity of the feed gas = 0.707 kcal/(kg)(K)
C_{pg}	Heat capacity of reacting gas = 0.719 kcal/(kg)(K)
$f(\ )$	Objective function, \$/year
f	Catalyst activity = 1.0
ΔH	Heat of reaction = −26,000 kcal/kg mol N_2
N	Mass flow of component designed by subscript through catalyst zone, kg mol/(m^2)(h)
N_1	Hours of operation per year = 8330
p	Partial pressure of component designated by subscript, psia; reactor pressure is 286 psia
R	Ideal gas constant, 1.987 kcal/(kg mol)(K)
S_1	Surface area of catalyst tubes per unit length of reactor = 10 m
S_2	Cross-sectional area of catalyst zone = 0.78 m^2
T_0	Reference temperature = 421 °C (694/K)
U	Overall heat transfer coefficient = 500 kcal/(h)(m^2)(K)
W	Total mass transfer flow rate = 26,400 kg/h

Energy balance, feed gas

$$\frac{dT_f}{dx} = -\frac{US_1}{WC_{pf}}(T_g - T_f) \tag{b}$$

Energy balance, reacting gas

$$\frac{dT_g}{dx} = -\frac{US_1}{WC_{pg}}(T_g - T_f) + \frac{(-\Delta H)S_2}{WC_{pg}}(f)\left[K_1\frac{(1.5)p_{N_2}p_{H_2}}{p_{NH_3}} - K_2\frac{p_{NH_3}}{(1.5)p_{H_2}}\right] \tag{c}$$

where $K_1 = 1.78954 \times 10^4 \exp(-20{,}800/RT_g)$

$K_2 = 2.5714 \times 10^{16} \exp(-47{,}400/RT_g)$

Mass balance, N_2

$$\frac{dN_{N_2}}{dx} = -f\left[K_1\frac{(1.5)p_{N_2}p_{H_2}}{p_{NH_3}} - K_2\frac{p_{NH_3}}{(1.5)p_{H_2}}\right] \tag{d}$$

The boundary conditions are

$$T_f(x = L) = 421\ ^\circ\mathrm{C}\ (694\ \mathrm{K}) \tag{e}$$

$$T_g(x = 0) = 421\ ^\circ\mathrm{C}\ (694\ \mathrm{K}) \tag{f}$$

$$N_{N_2}(x = 0) = 701.2\ \mathrm{kg\ mol/(h)(m^2)} \tag{g}$$

For the reaction, in terms of N_{N_2}, the partial pressures are

$$p_{N_2} = 286\left[\frac{N_{N_2}}{1 - 2(N^0_{N_2} - N_{N_2})}\right]$$

$$p_{N_2} = 286\left[\frac{3N_{N_2}}{1 - 2(N^0_{N_2} - N_{N_2})}\right]$$

$$p_{NH_3} = 286\left[\frac{2(N^0_{N_2} - N_{N_2})}{1 - 2(N^0_{N_2} - N_{N_2})}\right]$$

Inequality constraints

$$0 \le N_{N_2} \le 3220$$

$$400 \le T_f \le 800$$

$$x \ge 0$$

Feed gas composition (mole %).

N_2: 21.75; H_2: 65.25; NH_3: 5; CH_4: 4; Ar: 4

Solution procedure. Because the differential equations have to be solved numerically, a two-stage flow of information is needed in the computer program used to solve the problem. Examine Fig. E13.5c. Lasdon's generalized reduced-gradient

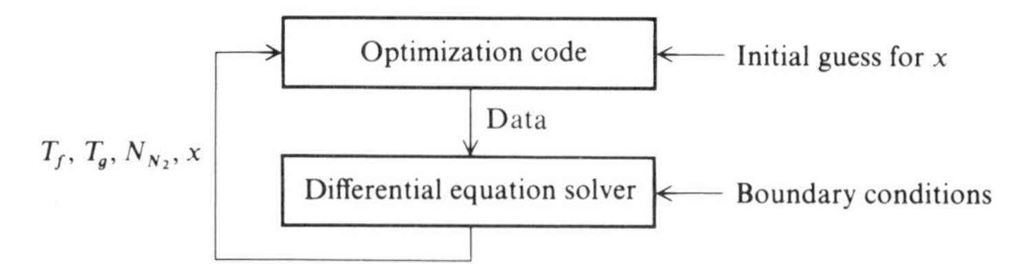

Figure E13.5c Flow diagram for solution procedure, Example 13.5.

code GRG2 (refer to Chap. 8) was coupled with the differential equation solver LSODE, yielding the following results for the exit conditions:

	Initial guesses	**Optimal solution**
N_{N2}	646 kg mol/(m^2)(h)	625 kg mol/(m^2)(h)
Mole fraction N_2	20.06%	19.4%
T_g	710 K	563 K
T_f	650 K	478 K
x	10.0 m	2.58 m
$f(x)$	8.451×10^5 \$/year	1.288×10^6 \$/year

In all, 10 one-dimensional searches were carried out, 54 objective function calls and 111 gradient calls (numerical differences were used) were made by the code, and the execution time (exclusive of compiling and printing) on a Cyber 170 was 4.8 seconds.

EXAMPLE 13.6 SOLUTION OF AN ALKYLATION PROCESS BY SEQUENTIAL QUADRATIC PROGRAMMING

A problem of long standing (Sauer, Colville, and Burwick, 1964) is to determine the optimal operating conditions for the simplified alkylation process shown in Fig. E13.6. Sauer, Colville, and Burwick solved this problem using a form of successive

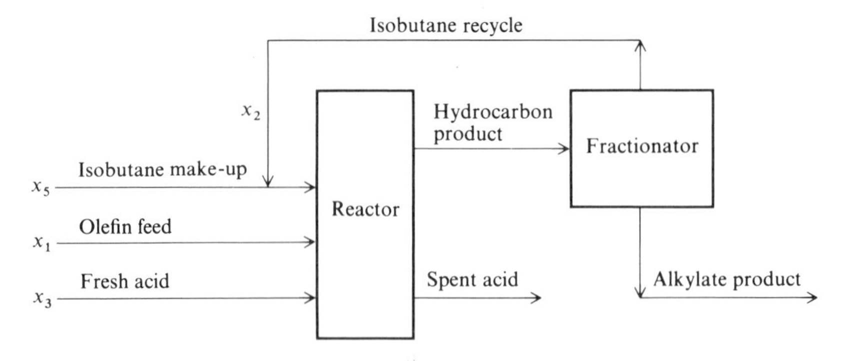

Figure E13.6

Table E13.6*a*

Symbol	Variable	Lower† bound	Upper bound	Starting value
x_1	Olefin feed (barrels per day)	0	2,000	1,745
x_2	Isobutane recycle (barrels per day)	0	16,000	12,000
x_3	Acid addition rate (thousands of pounds per day)	0	120	110
x_4	Alkylate yield (barrels per day)	0	5,000	3,048
x_5	Isobutane makeup (barrels per day)	0	2,000	1,974
x_6	Acid strength (weight percent)	85	93	89.2
x_7	Motor octane number	90	95	92.8
x_8	External isobutane-to-olefin ratio	3	12	8
x_9	Acid dilution factor	1.2	4	3.6
x_{10}	F-4 performance number	145	162	145

† Instead of 0, 10^{-6} was used.

linear programming. We first formulate the problem and then solve it by sequential quadratic programming. The notation to be used is listed in Table E13.6*a* which includes the units, upper and lower bounds, and the starting values for each x_i (a nonfeasible point). All the bounds represent economic, physical, or performance constraints.

The objective function was defined in terms of alkylate product, or output value minus feed and recycle costs; operating costs were not reflected in the function. The total profit per day, to be maximized, is:

$$f(x) = C_1 x_4 x_7 - C_2 x_1 - C_3 x_2 - C_4 x_3 - C_5 x_5 \tag{a}$$

where: C_1 = alkylate product value ($0.063 per octane-barrel)
C_2 = olefin feed cost ($5.04 per barrel)
C_3 = isobutane recycle costs ($0.035 per barrel)
C_4 = acid addition cost ($10.00 per thousand pounds)
C_5 = isobutane makeup cost ($3.36 per barrel)

To form the process model, regression analysis was carried out. The alkylate yield, x_4, was a function of the olefin feed, x_1, and the external isobutane-to-olefin ratio, x_8. The relationship determined by nonlinear regression holding the reactor temperatures between 80 to 90°F and the reactor acid strength by weight percent at 85 to 93 was

$$x_4 = x_1(1.12 + 0.13167x_8 - 0.00667x_8^2) \tag{b}$$

The isobutane makeup, x_5, was determined by a volumetric reactor balance. The alkylate yield, x_4, equals the olefin feed, x_1, plus the isobutane makeup, x_5, less shrinkage. The volumetric shrinkage can be expressed as 0.22 volume per volume of alkylate yield so that

$$x_4 = x_1 + x_5 - 0.22x_4$$

or

$$x_5 = 1.22x_4 - x_1 \tag{c}$$

The acid strength by weight percent, x_6, could be derived from an equation that expressed the acid addition rate, x_3, as a function of the alkylate yield, x_4, the acid dilution factor, x_9, and the acid strength by weight percent, x_6 (the addition acid was assumed to have acid strength of 98%)

$$1000x_3 = \frac{(x_4)(x_9)(x_6)}{(98 - x_6)}$$

or

$$x_6 = \frac{98{,}000x_3}{x_4x_9 + 1000x_3} \tag{d}$$

The motor octane number, x_7, was a function of the external isobutane-to-olefin ratio, x_8, and the acid strength by weight percent, x_6 (for the same reactor temperatures and acid strengths as for the alkylate yield, x_4)

$$x_7 = 86.35 + 1.098x_8 - 0.038x_8^2 + 0.325(x_6 - 89) \tag{e}$$

The external isobutane-to-olefin ratio, x_8, was equal to the sum of the isobutane recycle, x_2, and the isobutane makeup, x_5, divided by the olefin feed, x_1

$$x_8 = \frac{x_2 + x_5}{x_1} \tag{f}$$

The acid dilution factor, x_9, could be expressed as a linear function of the F-4 performance number, x_{10}

$$x_9 = 35.82 - 0.222x_{10} \tag{g}$$

The last dependent variable is the F-4 performance number, x_{10}, which was expressed as a linear function of the motor octane number, x_7

$$x_{10} = -133 + x_7 \tag{h}$$

The above relationships give the dependent variables in terms of the independent variables and the other dependent variables.

Equations (*c*), (*d*), and (*f*) were used as equality constraints. The other relations were modified to form two inequality constraints each so as to take account of the uncertainty that existed in their formulation. The d_l and d_u values listed in Table E13.6*b* allow for deviations from the expected values of the associated variables.

Table E13.6*b*

Deviation parameter	Value
d_{4_l}	99/100
d_{4_u}	100/99
d_{7_l}	99/100
d_{7_u}	100/99
d_{9_l}	9/10
d_{9_u}	10/9
d_{10_l}	99/100
d_{10_u}	100/99

Table E13.6*c*

Variable	Optimal value	Variable	Optimal value
x_1	1698.1	x_6	90.115
x_2	15819	x_7	95.000†
x_3	54.107	x_8	10.493
x_4	3031.2	x_9	1.5618
x_5	2000.0†	x_{10}	153.54

† At bound.

Thus there were eight inequality constraints in the model in addition to the three equality constraints and the upper and lower bounds on all of the variables.

$$[x_1(1.12 + 0.13167x_8 - 0.00667x_8^2)] - d_{4_l}x_4 \geq 0 \qquad (i)$$

$$-[x_1(1.12 + 0.13167x_8 - 0.00667x_8^2)] + d_{4_u}x_4 \geq 0 \qquad (j)$$

$$[86.35 + 1.098x_8 - 0.038x_8^2 + 0.325(x_6 - 89)] - d_{7_l}x_7 \geq 0 \qquad (k)$$

$$-[86.35 + 1.098x_8 - 0.038x_8^2 + 0.325(x_6 - 89)] + d_{7_l}x_7 \geq 0 \qquad (l)$$

$$[35.82 - 0.222x_{10}] - d_{9_l}x_9 \geq 0 \qquad (m)$$

$$-[35.82 - 0.222x_{10}] + d_{9_u}x_9 \geq 0 \qquad (n)$$

$$[-133 + 3x_7] - d_{10_l}x_{10} \geq 0 \qquad (o)$$

$$-[-133 + 3x_7] + d_{10_u}x_{10} \geq 0 \qquad (p)$$

To solve the alkylation process problem, the code NPSOL was employed (refer to Chap. 8), a successive quadratic programming code.

The values of the objective function found were:

$$f(x^0) = 872.3 \text{ initial guess}$$

$$f(x^*) = 1768.75$$

Tables E13.6*c* and E13.6*d* list values of the variables at x^* (rounded to five significant figures) and the constraints, respectively, at the optimal solution.

Note that the value of the isobutane recycle, x_5, is at its upper bound.

Table E13.6*d*

Constraint	Value at x^*	Constraint	Value at x^*
1(*i*)	0.33	7(*o*)	60.9
2(*j*)	0.18×10^{-11}	8(*p*)	1.91
3(*k*)	-0.22×10^{-12}	9(c)	0.29×10^{-10}
4(*l*)	573	10(*d*)	0.45×10^{-12}
5(*m*)	0	11(*f*)	-0.57×10^{-13}
6(*n*)	0.45×10^{-12}		

REFERENCES

Aris, R., *Optimal Design of Chemical Reactors*, Academic Press, New York (1961).

Carberry, J. J., *Chemical and Catalytic Reaction Engineering*, McGraw-Hill, New York (1976).

Chen, N. H., *Process Reactor Design*, Allyn & Bacon, Boston (1983).

Crooks, W. M., "Denbigh's 2-Tank CSTR System," *Br. Chem. Eng.*, **11**: 710 (1966).

Denbigh, K. G., *Trans. Faraday Soc.*, **40**: 352 (1944).

Hill, C. G., *An Introduction to Chemical Engineering Kinetics and Reactor Design*, John Wiley, New York (1977).

Holland, C. D., and R. G. Anthony, *Fundamentals of Chemical Reaction Engineering*, Prentice-Hall, Englewood Cliffs, New Jersey (1979).

Jeffreson, C. D., and J. M. Smith, "Stationary and Non-Stationary Models of Bacterial Kinetics in Well-Mixed Flow Reactors," *Chem. Eng. Sci.*, **28**: 629 (1973).

Lapidus, L., and N. R. Amundson, (eds.), *Chemical Reactor Theory—A Review*, Prentice-Hall, Englewood Cliffs, New Jersey, (1977).

Lee, K. Y., and R. Aris, "Optimal Adiabatic Bed Reactors for Sulfur Dioxide with Cold Shot Cooling," *Ind. Eng. Chem. Process Des. Dev.*, **2**, 200 (1963).

Luss, D., "Optimum Volume Ratios for Residence Time in Stirred Tank Reactor Sequences," *Chem. Eng. Sci.*, **20**: 171 (1965).

Murase, A., H. L. Roberts, and A. O. Converse, "Optimal Thermal Design of an Autothermal Ammonia Synthesis Reactor," *Ind. Eng. Chem. Process Des. Dev.*, **9**: 503 (1970).

Rippin, D. W. T., "Simulation of Single- and Multiproduct Batch Chemical Plants for Optimal Design and Operation," *Comput. Chem. Eng.*, **7**: 137 (1983).

Sauer, R. N., A. R. Coville, and C. W. Burwick, "Computer Points Way to More Profits," *Hydrocarbon Process Petrol. Ref.* **43**, 84 (1964).

Sourander, M. L., M. Kolari, J. C. Cugini, J. B. Poje, and D. C. White, "Control and Optimization of Olefin-Cracking Heaters," *Hydro. Proc.*, 63 (June, 1984).

Szepe, S., and O. Levenspiel, "Optimization of Back Mix Reactors in Series for a Single Reaction," *Ind. Eng. Chem. Process Des. Dev.*, **3**, 214 (1964).

Wen, C. Y., and L. T. Fan, *Models for Flow Systems and Chemical Reactors*, Marcel Dekker, New York (1975).

Wood, R. K., and W. F. Stevens, "Optimum Volume Ratios for Minimum Residence Time in Stirred-Tank Reactor Sequences," *Chem. Eng. Sci.*, **19**: 4216 (1964).

CHAPTER 14

OPTIMIZATION IN LARGE-SCALE PLANT DESIGN AND OPERATION

As discussed in Chap. 1, optimization of a large configuration of plant components can involve several levels of detail ranging from the most minute features of equipment design to the grand scale of international company operations. Considerable progress has been made in the last decade in understanding and developing more comprehensive procedures of optimization so that more complex problems can be solved. No matter what procedure is used, optimization is an integral feature in plant design and operation. Blau (1981) pointed out some of the reasons why optimization in the design of plants is just now gaining a foothold:

1. Optimization codes suitable to solve large-scale complex problems have only recently existed;
2. Uncertainty in the data and the coefficients used in plant equipment models lead to considerable ambiguity in the results of optimization;
3. Merit functions usually are not especially sensitive to changes in the variables from a local optimum, and multiple merit functions may be involved, some of a nonquantitative nature;
4. Computer and personnel costs of executing an optimization study are just now becoming reasonable.

On the other hand, optimization of operating conditions in an operating plant does not involve several of the barriers cited above. Objective functions are easy to define, values of products and costs of feeds and utilities are known, and the equipment sizes are fairly well fixed. Benefits to be achieved come from appropriate settings of the operating variables.

Synthesis of the optimal plant configuration has long been a goal of numerous investigators. By *synthesis* we mean the designation of the structure of the plant elements that will meet the designer's goals. Figure 14.1 shows the relation of synthesis to design and operation. You can see that even if no new technology is to be used, the problem is combinatorial in nature, and the number of alternatives increases astronomically. Furthermore, synthesis is really a multiobjective function problem. In his survey, Stephanopoulos (1981) stated that 250–300 research articles, 10 monographs, 3 books, and 7 review articles were published and 15 workshops and symposia on the subject of process synthesis occurred in the period 1968–1980, but only recently have there been reports of the use of this work in industrial practice. We have chosen not to discuss the general problem of synthesis in this chapter, but instead we focus on examples of optimization applied to a specified configuration or flowsheet. Some of the supplementary references give examples of simple synthesis problems.

Once the flowsheet is specified, the solution of the appropriate steady-state process material and energy balances is referred to as "flowsheeting." The essential problem in flowsheeting with or without associated optimization is to solve (satisfy) a large set of linear and nonlinear equations to an acceptable degree of precision, normally by an iterative procedure. You will recall that optimization

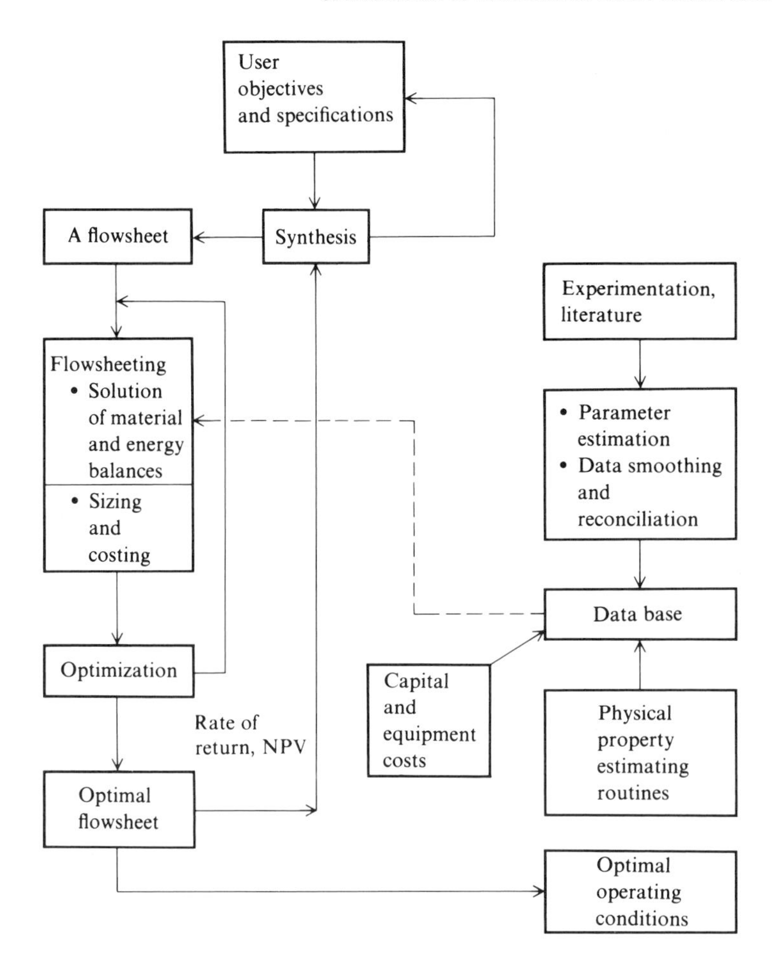

Figure 14.1 Information flow in the design process.

can be employed if the number of variables whose values are unknown exceeds the number of binding constraints. In flowsheeting without optimization, you must make sufficient specifications to take up all the degrees of freedom. Table 14.1 lists some typical codes used to execute flowsheeting. Several have superimposed proprietary optimization features. Sargent (1981) and Westerberg et al. (1979) have reviewed various methods of flowsheeting, pointing out that you frequently may encounter specialized methods in flowsheeting because they prove to be faster. We should point out that use of optimization codes to solve flowsheeting problems has the advantage that the variables can be bounded by inequality constraints so that the solution of the equations can be prevented from diverging excessively at any point.

Early attempts at optimization in conjunction with a flowsheeting program used the solution of balance equations as an inner loop with the optimization

Table 14.1 Flowsheeting codes

Name	Author	Ref.	Includes some optimization features	No. of components in data bank	Availability
ASPENPLUS	Aspen Technology	[L]	Yes	Large	Aspen Technology Corp., Cambridge, Massachusetts; original ASPEN from Nat. Tech. Info. Service, Springfield, Virginia
CAPES	Chiyoda Ltd.	[B]			Chiyoda Ltd., Yokohama, Japan
CHESS	R. L. Motard	[A]	No	98	Dept. Chem. Eng., Washington University, St. Louis, Missouri; microcomputer version from COADE, Houston, Texas
CONCEPT	Computer Aided Design Center	[E]		105	Computer Aided Design Center, Cambridge, England
DESIGN/2000	Chem Share	[M]			Chem Share, Houston, Texas
FLOWPACK II	Berger	[N]			
FLOWTRAN	Various at Monsanto Co.	[C]		>200	Monsanto Co., St. Louis, Missouri
GEMCS	A. I. Johnson	[D]	No		Dept. Chem. Eng., McMaster University, Hamilton, Ontario, Canada
MOSES	U. of Western Ontario	[I]	Yes		
MPBII	Sood and Reklaitis	[O]			Dept. Chem. Eng., Purdue University
PROCESS	Simulation Sciences	[F]	Yes	>600	Simulation Sciences, Fullerton, California
SPAD	Biegler and Hughes	[Q]	Yes	61	Chem. Eng. Dept., University of Wisconsin, Madison, Wisconsin

PROPS	Gains and Gaddy	[H]	Yes		Dept. Chem. Eng., Univ. Arkansas
SIMMOD	Chen and Stadtherr	[P]	Yes		Dept. Chem. Eng., University of Illinois
SPEEDUP	R. Sargent	[K]	Yes	100	British Technology Group, London
SYMBOL	Cambridge University	[G]	Yes	No	
TISFLO	Dutch State Mines, Central Laboratory	[J]	Yes	120	

[A] Motard, R. L., Dept. Chem. Eng., Washington University, St. Louis, Missouri.
[B] Chiyoda Ltd., Systems Eng., Section, *Outline of Capes*, Chiyoda, Ltd., Yokohama, Japan, 1979.
[C] Rosen, E. M., and A. C. Pauls, "Computer Aided Chemical Process Design: The FLOWTRAN System," *Comp. & Chem. Eng.*, **1**: 11 (1977).
[D] Johnson, A. I., *GEMCS: Users Manual*, McMaster University., Hamilton, Ontario, 1970.
[E] Thambynayagam, R. K. M., S. J. Banch, and P. Winter, "CONCEPT MK IV: A New Approach to Process Flowsheeting," paper presented at 227th Event of the European Federation of Chem. Eng. (CHEMPLANT '80), Sept. 3–5, 1980, Heviz, Hungary.
[F] Brannock, N. F., V. S. Verneuil, and Y. L. Wang, "PROCESS Simulation Program—A Comprehensive Flowsheeting Tool for Chemical Engineers," *Comput. & Chem. Eng.*, **3**: 329 (1979).
[G] Computer Aided Design Center, *SYMBOL*, Cambridge, England.
[H] Gains, L. D., and Gaddy, J. L., *Indus. Eng. Chem. Proc. Des. Dev.*, **15**: 206 (1976).
[I] Dept. Chem. Eng., University of Western Ontario, London, Ontario, Canada.
[J] de Leeuw den Bouter, J. A., and A. G. Swenker, "TISFLO, A Flowsheet Simulation Program Based on New Principles," paper presented at EFCE Conf. "Comput. Appl. in Proc. Dev.," April, 1974, Erlangen, Germany.
[K] Perkins, J. D., R. W. H. Sargent, and S. Thomas in *Computer Aided Process Plant Design*, M. E. Leesley (Ed.), p. 566, Gulf Publ., Houston (1982).
[L] Evans, L. B., et al., "ASPEN: An Advanced System for Process Engineering," *Comput. Chem. Eng.*, **3**: 319 (1979).
[M] Chem. Share Corp., *Guide to Solving Process Engineering Problems by Simulation* (Users Manual), Chem. Share Corp., Houston, Texas, 1979.
[N] Berger, F., and F. A. Perris, "FLOWPACK II—A New Generation of System for Steady-State Flowsheeting," *Comput. Chem. Eng.*, **3**: 309 (1979).
[O] Sood, M. K., and G. V. Reklaitis, "Material Balance Program—II," School of Chemical Engineering, Purdue University, W. Lafayette, Indiana, 1977.
[P] Chen, H. S., and M. A. Stadtherr, "A Simultaneous-Modular Approach to Process Flowsheeting and Optimization: I Theory and Implementation." *AIChE J*, **32**: 184 (1984).
[Q] Biegler, L. T., and R. R. Hughes, "Optimization of Propylene Chlorination Process: A Case Study Comparison of Four Algorithms," *Comput. Chem. Eng.*, **7**: 645 (1983).

subroutine choosing values of certain of the parameters in an outer loop. The optimizer usually adjusted the values of the unspecified variables or parameters and introduced revised values back into the flowsheeting program which would then perform a complete flowsheeting simulation based on those values. Revised values would be returned to the optimization routine which would again re-adjust the decision variables, and so on. Usually simple search methods were used to find the optimal values of the unspecified parameters and variables so that large numbers of iterations (as many as 1000) through the flowsheeting routines would result, at considerable cost. Thus, optimization was expensive. To make matters worse, the user might make some slight error and find that the lowest cost for a flowsheet did not correspond to a feasible operating point.

Recent optimization techniques have focused on better ways of meshing the optimization procedure with the process flowsheet calculations right from the start. These techniques progress toward the optimal values of the flowsheet parameters and variables while simultaneously moving toward the solution of the set of algebraic equations defining the flowsheet. Thus, they are far more efficient than older methods.

14.1 GENERAL METHODS OF MESHING OPTIMIZATION PROCEDURES WITH PROCESS MODELS/SIMULATORS

By the term *process model* we include all the mathematical relations that comprise the material and energy balances, the rate equations, the controls, connecting variables, and methods of computing the physical properties used in any of the relations in the model. A module is a model of an individual element in a flowsheet (such as a reactor) that can be coded, analyzed, debugged, and interpreted by itself.

Two extremes are encountered in flowsheeting software. At one extreme the entire set of equations (and inequalities) representing the process is employed so that the process model equations form the constraints for optimization, exactly the same as described in previous chapters in this book. This representation is known as the *equation-oriented* method of flowsheeting. The equations can be solved in a sequential fashion analogous to the modular representation described in the next section or simultaneously by Newton's method, Broyden's (1965) method, or by employing sparse matrix techniques to reduce the extent of matrix manipulations; references can be found in the review by Rosen (1980).

At the other extreme, the process can be represented by a collection of modules (the *modular* method) in which the equations (and other information) representing each subsystem or piece of equipment are collected together and coded so that the module may be used in isolation from the rest of the flowsheet and hence is portable from one flowsheet to another. Examine Fig. 14.2. Each module contains the equipment sizes, the material and energy balance relations, the component flowrates, and the temperatures, pressures, and phase conditions

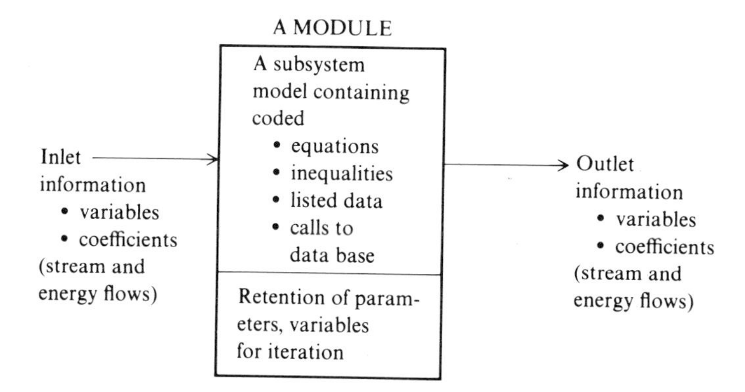

Figure 14.2 A typical process module showing the necessary interconnections of information.

of each stream that enters and leaves the physical equipment represented by the module. Values of certain of these parameters and variables determine the capital and operating costs for the units. Of course, the interconnections set up for the modules must be such that information can be transferred from module to module concerning the streams, compositions, flow rates, coefficients, and so on. In other words, the modules comprise a set of building blocks that can be arranged in general ways to represent any process. An executive routine calls the modules in the proper order, transmits information from a library of calculational subroutines, and picks out information on physical properties from an associated data base. Both sequential and simultaneous calculational sequences have been proposed for the modular approach as well as the equation-oriented approach (and intermediate mixtures of the two are possible as well). Either the program and/or the user must select the decision variables for recycle and provide estimates of certain stream values to make sure that convergence of the calculations occurs, especially in a process with many recycle streams. Reviews by Evans (1982), Rosen (1980), and Westerberg (1981) point out many of the problems and practices pertaining to flowsheeting.

An intermediate approach to combining optimization and flowsheeting is that proposed by Parker and Hughes (1981). They used each module in a flowsheeting code to develop polynomial functions (quadratic being the simplest) to approximate the module by least squares. Then, the equations could be used as constraints in an optimization code. Parker and Hughes used direct search for their optimization strategy. Some disadvantages of such an approximation strategy are that (1) adequate approximation of the module may not be possible with simple relations, and (2) the optimum of the approximate model may not lie near the optimum of the rigorous model as ascertained via a more rigorous solution. Nevertheless, approximation schemes avoid some of the difficulties encountered in closure and convergence of the recycle loops each time the process simulator is called. If the optimization algorithm always maintains feasibility from stage to stage of the iterations, then considerable computer time will be needed. Also, the number of variables in the optimization problem is increased by the number of equality constraints so developed. Perhaps most importantly, the process simulator presumably selects appropriate tear variables for iteration in the recycle loops whereas most optimization codes can not.

Another variation in the modular approach to flowsheeting, termed "two-tier" by Rosen[1] (1980) (also see Jirapongphan (1980)) and simultaneous modular by others, that might be combined with optimization uses successive "coarse tuning" and "fine tuning" for the model. In one sense this procedure is related to the approximation methods described above. Two types of models are used to represent the same unit operation: (1) a simple approximate model and (2) a more complex rigorous model. The simple models may be locally linearized rigorous models or approximate engineering models such as the Kremser equation for an absorber. Calculations using the rigorous model are used to determine the coefficients in the simple model, and the simple model is solved for the stream variables that are then fed into the rigorous model, and so on. Some optimization codes that do not yield feasible solutions on each stage of iteration can take advantage of this approach because in most of the preliminary stages of optimization the constraints do not have to be satisfied exactly.

A final procedure that we should mention is the use of decomposition in optimization. The essential idea is to create a set of smaller size or less complex problems that individually may be easy to solve, and then iterate using a coordination routine to generate an optimal solution from among the suboptimal solutions generated by the subproblems; see Brosilow and Lasdon (1965); Findeisen et al. (1970); Kulikowski (1970); Tazaki et al. (1972); Liu and Brosilow (1983). Schock and Luus (1972) described the relationship between the discrete maximum principle and two-level hierarchial optimization, and concluded that the two-level procedure has all of the shortcomings of the discrete maximum principle.

In current practice, optimization combined with modularly organized simulators seems to prevail because (1) the modules are easier to construct and understand, (2) addition and deletion of modules for a flowsheet is easily accomplished without changing the solution strategy, (3) modules are easier to program and debug than sets of equations, and diagnostics for them easier to analyze, and (4) the modules already exist and work whereas equation blocks for equipment are not prevalent. It seems appropriate to segregate the process modelling from the numerical strategy as far as possible so that the improvements resulting in changes in either portion of the computer code do not require wholesale rewriting. Difficulties encountered in equation-based codes usually

[1] Not to be confused with the two-level optimization strategy proposed in the 1960s (now obsolete) in which sub-Lagrangian functions are written for each module at the lower level of optimization, and the coupling between the lower-level and upper-level optimization is via adjoint variables which are Lagrange multipliers on the connectivity equations that link the modules. Both methods solve the constraints in the outer loop.

stem from sets of equations associated with particular units. Embedding these equations as subproblems with special routines often alleviates the difficulties. By permitting the user to mix equations and subroutines, use can be made of "macros" in the general network representing a process.

We will first discuss the equation-oriented approach to large-scale optimization in Sec. 14.2, and then in Sec. 14.3 treat the modular approach.

14.2 EQUATION-BASED LARGE-SCALE OPTIMIZATION

Optimization based on equations representing the process models can be carried out in principle by the constrained optimization methods discussed in Chap. 8. However, in practice there are a number of features that must be included in a computer code so that it can be applied to general problems rather than a particular type of plant. Some developers of software prefer a specialized program for a particular type of plant, such as the Kellogg ammonia plant, because the program can use system-dependent but highly efficient calculational procedures, making the most of special features of the process model. On the other hand, the disadvantage of such a modus operandi is that the codes are difficult to update or modify even for minor flowsheet changes, and are certainly not useful for other processes.

Four codes under development for equation-based flowsheeting are ASCEND (Locke and Westerberg, 1983), QUASILIN (Hutchinson et al., 1983, 1986), SPEEDUP (Sargent et al., 1982), and TISFLO-II (de Leeuw den Bouter, 1980). An equation-based code for plant optimization should include the following characteristics:

1. A method of stating the equations and inequality constraints. Slack variables can be used to transform the inequality constraints into equality constraints.
2. A possibility of using both continuous and discrete variables, the latter being particularly necessary to accommodate changes in phase or changes from one correlation to another (see Perkins, 1983).
3. The option to use alternate forms of a function depending on the value of logical variables that outline the state of the process. Typical examples are the shift in the relation used to calculate the friction factor from laminar to turbulent flow, or the calculation of P-V-T relations as the phase changes from gas to liquid. To fix on the appropriate form, the problem must be treated as an optimization problem in which feasibility is required on each cycle for each state with the logical variables treated as constraints for a state. Alternatively, the code can guess the state and if the logical constraints are violated, another guess can be made.
4. A method of solving submodels that defy normal equation-solving procedures or are specially suited to ad hoc methods of solution.

5. The ability to build *macros*, that is create complex models comprised of standard subelements. For example, a distillation module might be composed of trays, flash units, splitters, mixers, heat exchangers, etc.
6. Provision for equivalence variables, that is using the same name for two different variables if they are equal, such as the mass flowrate through a pump.
7. A method for solving individual models. Westerberg (1983) suggests using Newton or quasi-Newton methods combined with sparse matrix methods to convert the nonlinear algebraic equations in the model to linearized approximates. Then the linearized equations can be solved iteratively taking advantage of their structure. The alternative is to use *tearing*. By tearing we mean selecting certain output variables from a set of equations as known values so

$$
\begin{aligned}
&h_1:\ x_1^2 x_2 - 2x_3^{1.5} + 4 = 0\\
&h_2:\ x_2 + 2x_5 - 8 = 0\\
&h_3:\ x_1 x_4 x_5^2 - 2x_3 - 7 = 0\\
&h_4:\ -2x_2 + x_5 + 5 = 0\\
&h_5:\ x_2 x_4^2 x_5 + x_2 x_4 - 6 = 0
\end{aligned}
$$

(*a*) The n independent equations involving n variables ($n = 5$).

	x_1	x_2	x_3	x_4	x_5
h_1	1	1	1		
h_2		1			1
h_3	1		1	1	1
h_4		1			1
h_5		1		1	1

(*b*) The occurrence matrix (the 1's represent the occurrence of a variable in an equation).

	x_2	x_5	x_4	x_1	x_3
h_2	1	1 (I)			
h_4	1	1	(II)		
h_5	1	1	1	(III)	
h_3		1	1	1	1
h_1	1			1	1

(*c*) The rearranged (partitioned) occurrence matrix with groups of equations (sets I, II, and III) that have to be solved simultaneously collected together in a precedence order for solution.

Figure 14.3 Partitioning and tearing. The equations are partitioned as in (*c*). Equations h_2 and h_4 (set I) are solved for x_2 and x_5 first, then h_5 (set II) is solved for x_4, and lastly h_1 and h_3 are solved by tearing. Assume a value for x_3; solve h_1 for x_1; check to see if h_3 is satisfied. If not, adjust x_3, recheck h_3, and so on until both h_1 and h_3 are satisfied.

that the remaining variables can be solved by serial substitution. A residual set of equations equal to the number of tear variables will remain, and if these are not satisfied, new guesses are made for the values of the tear variables, and the sequence repeated. Examine Fig. 14.3. (Sequential modular flowsheeting codes also use tearing in solving problems involving recycle streams.) Furthermore, the sensitivity of functions with respect to variables (the first partial derivatives) developed in Newton-like methods is valuable for implementing the optimization itself.

8. A method for *precedence ordering* so as to partition a model into a sequence of smaller models containing sets of irreducible equations (equations that have to be solved simultaneously) as illustrated in Fig. 14.3.

9. A method to select initial guesses for the Newton-like solution procedure for the algebraic equations. Poor choices lead to unsatisfactory results. You want the initial guesses to be as close to the correct answer as possible so that the procedure will converge. You can sequence through a flowsheet and guess only the recycle streams, or perhaps solve approximate models, and then pass to more complex ones.

10. Provision for *scaling* of the variables and equations. By scaling of variables we mean introducing transformations that make all the variables have ranges in the same order of magnitude. By scaling of equations we mean multiplying each equation by a factor that causes the value of the deviation of each equation from zero to be of the same order of magnitude. Perkins (1983) discusses solution techniques that are theoretically scale-invariant but in practice may not be satisfactory. User interaction and analysis for a specific problem is usually the best way to introduce scaling.

Example 14.1 below illustrates an equation-oriented approach to the optimization of a small scale problem so that the details involved can be illustrated.

EXAMPLE 14.1 EQUATION-BASED OPTIMIZATION FOR A REFRIGERATION PROCESS

Figure E14.1 shows the flowsheet for the process. Feed (stream 1) is a vapor mixture of ethane, propane, and butane (in the proportions shown in the figure) at 200°F and 500 psia. The product stream (stream 8) is a liquid at ≤ -20°F having the same composition but a reduced pressure. The notation for this example is defined in Table E14.1*a*.

1 Objective function. A simple objective function will be used, namely, the minimization of the instantaneous cost of the work done by the three recycle compressors:

$$\text{Minimize} \quad f = C\left[F_3\frac{(H_9 - H_3)}{0.65} + F_5\frac{(H_{11} - H_5)}{0.65} + F_7\frac{(H_{13} - H_7)}{0.65}\right] \qquad (a)$$

The value 0.65 is the efficiency factor.

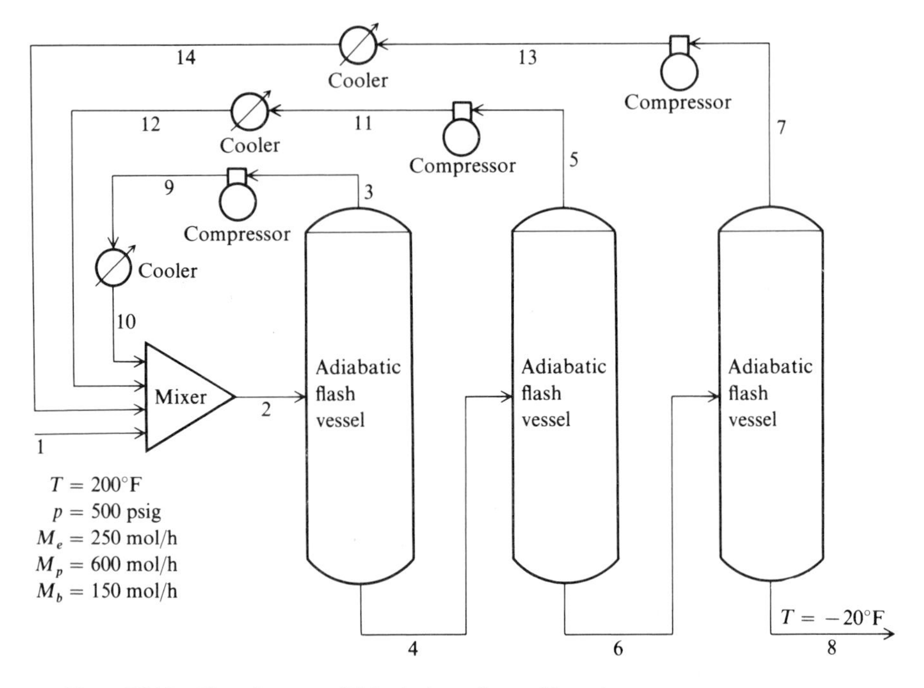

Figure E14.1 Flow diagram of light hydrocarbon refrigeration process.

2 Inequality constraints. Three inequality constraints are involved, two relating pressures and one product temperature specification:

$$p_5 - p_3 \leq 0 \qquad (b\text{-}1)$$

$$p_7 - p_5 \leq 0 \qquad (b\text{-}2)$$

$$T_7 + 20 \leq 0 \qquad (b\text{-}3)$$

In addition, all 34 values of T_j, p_j, $x_{i,j}$, and $y_{i,j}$ have lower and upper bounds.

Table E14.1*a* Notation for Example 14.1

C	A constant denoting the cost of work per unit energy
F_j	Total molar flowrate of process stream j
H_j	Molar enthalpy of process stream j
K_i	Vapor-liquid equilibrium constant for component i
L_j	Liquid molar flowrate of process stream j
p_i	Pressure of process stream identified by subscript ($p_4 = p_3$, $p_6 = p_5$, and $p_8 = p_7$)
T_i	Temperature of process stream identified by subscript ($T_4 = T_3$, $T_6 = T_5$, and $T_8 = T_7$)
V_j	Vapor molar flowrate of process stream j
$x_{i,j}$	Liquid molar flowrate of component i in process stream j [i = 1 (ethane), 2 (propane), 3 (n-butane)]
$y_{i,j}$	Vapor molar flowrate of component i in process stream j [i = 1 (ethane), 2 (propane), 3 (n-butane)]

3 Equality constraints. The equality constraints (30 in all) are the linear and nonlinear material and energy balances and the phase relations.

3.1 *Mixer.* Material balances:

$$\frac{0.5(y_{i,1} + x_{i,10} + x_{i,12} + x_{i,14} - y_{i,2} - x_{i,2})}{\max\{1, (y_{i,1} + x_{i,10} + x_{i,12} + x_{i,14} + y_{i,2} + x_{i,2})/2\}} = 0 \qquad (c)$$

$$i = 1, \ldots, 3$$

Energy balance:

$$\frac{0.05(F_1 H_1 + F_{10} H_{10} + F_{12} H_{12} + F_{14} H_{14} - F_2 H_2)}{\max\{1, (F_1 H_1 + F_{10} H_{10} + F_{12} H_{12} + F_{14} H_{14})/2\}} = 0 \qquad (d)$$

Note that the terms in the denominators of Eqs. (*c*) and (*d*) representing the average of the mass and energy, respectively, in and out, as well as the denominators of the equations below, are not needed for the balances—they are scaling factors (as is the multiplier 0.5) that are introduced to improve the conditioning of the matrices of partial derivatives of the constraints.[2] Without such scaling, the nonlinear programming code may not reach the optimal solution but instead terminate prematurely.

3.2 *Adiabatic flash vessels.*

Material balances:

$$\frac{0.5(y_{i,2} + x_{i,2} - y_{i,3} - x_{i,4})}{\max\{1, (y_{i,2} + x_{i,2} + y_{i,3} + x_{i,4})/2\}} = 0 \qquad i = 1, \ldots, 3 \qquad (e)$$

$$\frac{0.5(x_{i,4} - y_{i,5} - x_{i,6})}{\max\{1, (x_{i,4} + y_{i,5} + x_{i,6})/2\}} = 0 \qquad i = 1, \ldots, 3 \qquad (f)$$

$$\frac{0.5(x_{i,6} - y_{i,7} - x_{i,8})}{\max\{1, (x_{i,6} + y_{i,7} + x_{i,8})/2\}} = 0 \qquad i = 1, \ldots, 3 \qquad (g)$$

Energy balances: In the energy balances the multiplier 0.05 is used to assist in scaling.

$$\frac{0.05(F_2 H_2 - F_3 H_3 - F_4 H_4)}{\max\{1, (F_2 H_2 + F_3 H_3 + F_4 H_4)/2\}} = 0$$

$$\frac{0.05(F_4 H_4 - F_5 H_5 - F_6 H_6)}{\max\{1, (F_4 H_4 + F_5 H_5 + F_6 H_6)/2\}} = 0$$

$$\frac{0.05(F_6 H_6 - F_7 H_7 - F_8 H_8)}{\max\{1, (F_6 H_6 + F_7 H_7 + F_8 H_8)/2\}} = 0$$

The values for the enthalpies of the streams in the data base were based on the Curl-Pitzer correlations. The enthalpies are calculated from correlations at zero pressure (functions of temperature and composition only) and then corrected via the enthalpy deviation:

$$H = H^0 - \left(\frac{H^0 - H}{T_c}\right) T_c \qquad (h)$$

[2] The denominators in this example are not functions containing variables but simply represent scalars in the respective constraints evaluated using the values of the variables in the numerator; 1 or the other term is picked, whichever is bigger.

where H^0 is the stream molar enthalpy and the superscript 0 designates zero pressure, and T_c is the critical temperature. The enthalpy deviation term itself, $\Delta H/T_c$, is a function of the mole weighted average of the three critical properties: temperature, pressure, and compressibility.

3.3 *Energy balances for compressors.* For isentropic compression

$$\frac{0.05[T_9 - (T_3 + 459.69)(500.0/P_3)^{0.200} - 459.69]}{\max\{1, (T_3 + T_9)/2\}} = 0 \tag{i}$$

$$\frac{0.05[T_{11} - (T_5 + 459.69)(500.0/P_5)^{0.200} - 459.69]}{\max\{1, (T_5 + T_{11})/2\}} = 0 \tag{j}$$

$$\frac{0.05[T_{13} - (T_7 + 459.69)(500.0/P_7)^{0.200} - 459.69]}{\max\{1, (T_7 + T_{13})/2\}} = 0 \tag{k}$$

3.4 *Phase equilibria relations.* Evaluation of the K values for phase equilibria was based on the relation

$$K_i = \frac{\gamma_{il} v_i}{\phi_i} \tag{l}$$

where γ_{il} = activity coefficient in the liquid phase of component i evaluated from Hildebrand (1950)

v_i = fugacity coefficient of component i in the liquid phase evaluated from Chao-Seader (1961)

ϕ_i = fugacity coefficient of component i in the vapor phase evaluated from Redlich-Kwong (1949).

Based on the notation of Table E14.1*a*, in stream j

$$K_i = \frac{y_{i,j-1}/V_j}{x_{i,j}/L_j} \tag{m}$$

To assist in scaling, Eq. (m) is rearranged as follows:

$$x_{i,j} K_i \left(\frac{V_j}{L_j}\right) + x_{i,j} = y_{i,j-1} + x_{i,j}$$

$$x_{i,j} = \frac{(x_{i,j} + y_{i,j-1})L_j}{K_i V_j + L_j}$$

or

$$x_{i,j} - \frac{(x_{i,j} + y_{i,j-1})L_j}{K_i V_j + L_j} = 0$$

and divided by

$$\max\left\{1, x_{i,j} + \frac{(x_{i,j} + y_{i,j-1})L_j}{K_i V_j + L_j}\right\}$$

and multiplied by the factor 0.01:

$$\frac{0.01\left[x_{i,2} - \dfrac{(x_{i,2} + y_{i,2})L_2}{K_i V_2 + L_2}\right]}{\max\left\{1, x_{i,2} + \dfrac{(x_{i,2} + y_{i,2})L_2}{K_i V_2 + L_2}\right\}} = 0 \qquad i = 1, \ldots, 3 \qquad (n)$$

$$\frac{0.01\left[x_{i,4} - \dfrac{(x_{i,4} + y_{i,3})L_4}{K_i V_3 + L_4}\right]}{\max\left\{1, x_{i,4} + \dfrac{(x_{i,4} + y_{i,3})L_4}{K_i V_3 + L_4}\right\}} = 0 \qquad i = 1, \ldots, 3 \qquad (o)$$

$$\frac{0.01\left[x_{i,6} - \dfrac{(x_{i,6} + y_{i,5})L_6}{K_i V_5 + L_6}\right]}{\max\left\{1, x_{i,6} + \dfrac{(x_{i,6} + y_{i,5})L_6}{K_i V_5 + L_6}\right\}} = 0 \qquad i = 1, \ldots, 3 \qquad (p)$$

$$\frac{0.01\left[x_{i,8} - \dfrac{(x_{i,8} + y_{i,7})L_8}{K_i V_7 + L_8}\right]}{\max\left\{1, x_{i,8} + \dfrac{(x_{i,8} + y_{i,7})L_8}{K_i V_7 + L_8}\right\}} = 0 \qquad i = 1, \ldots, 3 \qquad (q)$$

In summary, the problem consists of 34 bounded variables (both upper bound and lower bounds) associated with the process, 12 linear equality constraints, 18 nonlinear equality constraints, and 3 linear inequality constraints.

4 Solution of the problem. Several different nonlinear programming codes including an augmented Lagrange function technique, a successive linear programming technique, and a combined penalty function and gradient projection code were used to solve the above problem, but only the Generalized Reduced Gradient Method of Abadie and Guigou (see Sec. 8.4) was effective in reaching a solution. It was not possible to use analytical derivatives in the nonlinear programming code because the energy balance equality constraints and the process stream phase equilibria constraints involve the stream molar enthalpy H_j and the phase equilibrium constant K_{ij}, respectively. H_j was calculated at zero pressure and then corrected using the Watson acentric factor. The correction for nonideality was based on correlated experimental data which cannot be differentiated analytically. The component phase equilibrium constant K_{ij} was calculated via the Redlich–Kwong equation of state; the vapor phase mixture compressibility factor z^v was determined as the largest of the three real roots from the virial equation:

$$z^3 - z^2 + C_1 z + C_2 = 0$$

where C_1 and C_2 are functions of the critical properties of the mixture. An analytical derivative of the vapor phase mixture compressibility with respect to the stream variables cannot be determined explicitly, and therefore, the derivative of the component phase equilibrium constant K_{ij} cannot be determined analytically.

As a consequence, the gradient of the objective function and the Jacobian matrix of the constraints in the nonlinear programming problem cannot be determined analytically. Finite difference substitutes as discussed in Sec. 6.5 had to be

Table E14.1*b* Starting point 1 of light hydrocarbon refrigeration process

Stream	$F/1000$ (lb mol/h)	T (°F)	p (psia)	$H/1000$ (Btu/lb mol)	Molar flow rates: Liquid/1000 C_2H_6	Liquid/1000 C_3H_8	Liquid/1000 nC_4H_{10}	Vapor/100 C_2H_6	Vapor/100 C_3H_8	Vapor/100 nC_4H_{10}
1	1.00	200	500	6.90	0.00	0.00	0.00	2.50	6.00	1.50
2	3.50	103	500	1.79	1.48	0.952	0.167	6.85	1.95	0.149
3	1.47	68.3	300	4.46	0.00	0.00	0.00	1.14	3.12	0.208
4	2.02	68.3	300	−0.156	1.03	0.834	0.161	0.00	0.00	0.00
5	0.372	34.3	175	4.33	0.00	0.00	0.00	2.88	0.793	0.0471
6	1.65	34.3	175	−1.16	0.741	0.755	0.156	0.00	0.00	0.00
7	0.652	−8.17	150	3.46	0.00	0.00	0.00	4.91	1.55	0.0607
8	1.00	−81.7	150	−4.18	0.250	0.600	0.150	0.00	0.00	0.00
9	1.47	125	500	4.70	0.00	0.00	0.00	11.4	3.12	0.208
10	1.47	50.0	500	−0.260	1.14	0.312	0.0208	0.00	0.00	0.00
11	0.372	150	500	5.49	0.00	0.00	0.00	2.88	0.793	0.0471
12	0.372	50.0	500	−0.259	2.88	0.793	0.0471	0.00	0.00	0.00
13	0.652	303	500	8.55	0.00	0.00	0.00	4.91	1.55	0.0607
14	0.652	500	500	−0.290	0.491	0.155	0.0607	0.00	0.00	0.00

Table E14.1*c*   Final solution of light hydrocarbon refrigeration process

Stream	$F/1000$ $\left(\frac{\text{lb mol}}{\text{h}}\right)$	T (°F)	p (psia)	$H/1000$ $\left(\frac{\text{Btu}}{\text{lb mol}}\right)$	Moiar flow rates: Liquid/100			Moiar flow rates: Vapor/100		
					C_2H_6	C_3H_8	nC_4H_{10}	C_2H_6	C_3H_8	nC_4H_{10}
1	1.00	200	500	6.90	0.00	0.00	0.00	2.50	6.00	1.50
2	2.97	115	500	2.07	10.6	9.15	1.66	5.63	2.46	0.221
3	1.29	80.6	306	4.69	0.00	0.00	0.00	9.04	3.58	0.259
4	1.68	80.6	306	0.0651	7.19	8.03	1.62	0.00	0.00	0.00
5	0.412	33.7	143	4.48	0.00	0.00	0.00	2.86	1.89	0.0749
6	1.27	33.7	143	−1.36	4.34	6.84	1.55	0.00	0.00	0.00
7	0.272	−20.0	511	4.09	0.00	0.00	0.00	1.84	0.843	0.0451
8	1.00	−20.0	511	−2.85	2.50	6.00	1.50	0.00	0.00	0.00
9	1.29	136	500	4.72	0.00	0.00	0.00	9.04	3.58	0.259
10	1.29	50.0	500	−0.373	9.04	3.58	0.259	0.00	0.00	0.00
11	0.412	174	500	6.06	0.00	0.00	0.00	2.86	1.19	0.0749
12	0.412	50.0	500	−0.386	2.86	1.19	0.0749	0.00	0.00	0.00
13	0.272	234	500	7.32	0.00	0.00	0.00	1.84	0.843	0.0451
14	0.272	50.0	500	−0.415	1.84	0.843	0.0451	0.00	0.00	0.00

used. To be conservative, substitutes for derivatives were computed as suggested by Curtis and Reid (1974). They estimated the ratio μ_j of the truncation error to the roundoff error in the central difference formula

$$\frac{\partial f}{\partial x_j} = \frac{f(x + d_j) - f(x - d_j)}{2d_j}$$

where d_j is the step size, as follows:

$$\mu_j = \frac{\dfrac{-(d_j/2)[f(x + d_j) - 2f(x) + f(x - d_j)]}{d_j^2}}{p\left|\dfrac{\partial f}{\partial x_j} \cdot x_j\right|}$$

where p is the magnitude of the error incurred in the storage of a number in the computer.

The Curtis–Reid method updates d_j on each calculation of a partial derivative from the relation

$$d_j^{k+1} = (d_j^k) \min \left\{1000, \sqrt{\frac{\mu_j^*}{\max \{u_j, 1\}}}\right\}$$

where u_j^* is the target value of the error ratio. To ensure that the truncation error calculation was not dominated by round-off error, Curtis and Reid suggested a value for u_j^* of 100 with an acceptable range of 10 to 1000.

The generalized reduced gradient code of Abadie and Guigou reached the solution shown in Table E14.1*c* from the several nonfeasible starting points, one of which is shown in Table E14.1*b*.

An example of an equation-oriented flowsheeting code that includes optimization is ASCEND (Locke and Westerberg, 1983). Figure 14.4 shows the configuration of the various subroutines involved. Phasing of inputs, solving a problem, and generating outputs can be ordered by the user. Several independent programs operate on a file of the values of the variables and pointers that represent a flowsheet. A user supplies or makes use of existing subroutines containing equations that model a component in the flowsheet plus an input file that defines how these submodels should be assembled. ASCEND maintains a file that includes the user-supplied definition of the configuration plus definitions of variable packets (groups of associated variables), equation packets, types of variable and equation packets associated with an element in the flowsheet, a procedure to allow the user to create complex models (called "macros") from a simple set of inputs (such as expand a distillation column from 7 to 15 trays), and default values so that the user has only to input the important parameters of interest or the exceptions to the default values. From the viewpoint of optimization, for each variable in the variable packet, in addition to the variable name, dimensions, and default values, an upper and lower bound is stored. Physical

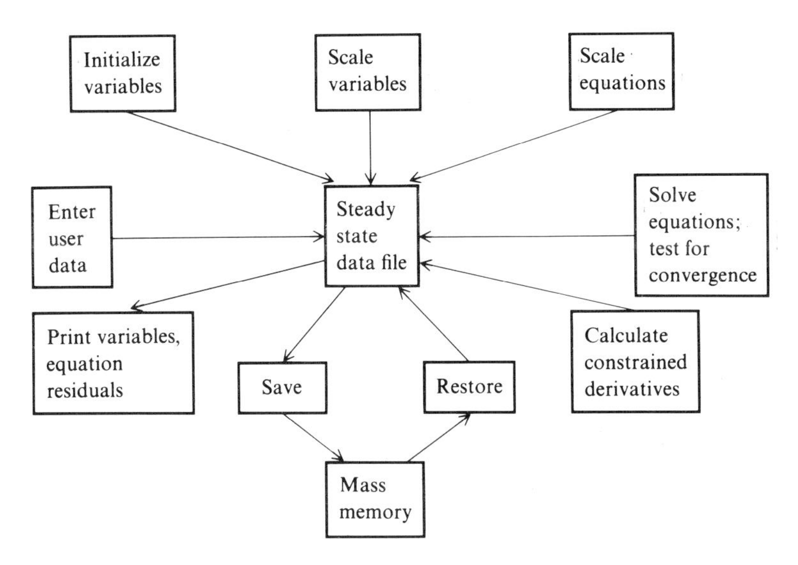

Figure 14.4 Information flow in ASCEND-II.

property options permit the user some choice in the calculation of fugacity, enthalpy, entropy, activity, and so on.

For the optimization algorithm, ASCEND-II makes use of a decomposition scheme in a modified Han–Powell successive quadratic programming method (Locke, Edahl, and Westerberg, 1981). Only the elements of the Hessian matrix that pertain to the independent variables (refer to Sec. 8.6) have to be estimated. For a large-scale flowsheet, the number of variables might be on the order of 10,000 but the number of independent variables may be only of the order of 10. Inequality constraints are converted to equality constraints through the use of bounded slack variables. At stage k the quadratic programming problem is solved (refer to Sec. 8.6) only in the space of the independent variables, and the equality constraints are linearized and solved for the dependent variables.

Internal representation and storage in ASCEND-II of a flowsheet which may contain hundreds or even thousands of variables is carried out by the GEV (Generator, Equation packet, Variable packet) method proposed by Berna, et al. (1980). Figure 14.5 shows (1) a flowsheet element, (2) the internal formulation of the element, and (3) the Jacobian representation of the element. A generator is a subroutine. It calculates the necessary partial derivatives and equation residuals for the Newton-Raphson portion of a solution. Equation packets are groups of equations grouped together because they represent the equations of the Jacobian matrix reserved for a single generator. In Fig. 14.5*b*, the corresponding equation packet would contain the equations for the material balance, equilibrium, and enthalpy balance for the flash unit. Variable packets are groups of associated variables grouped together for convenience. One example of a variable packet

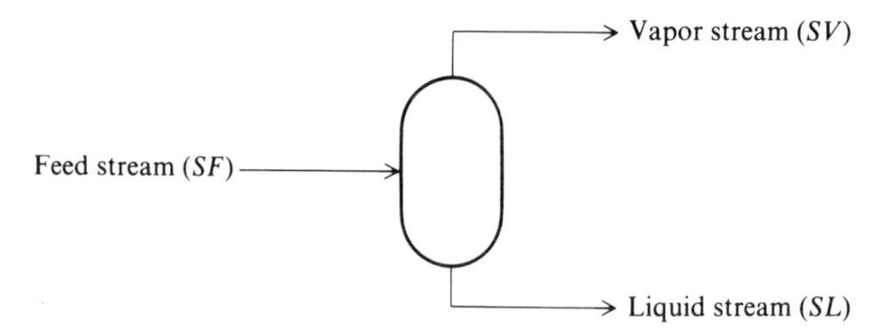

(*a*) The flowsheet element—a flash unit

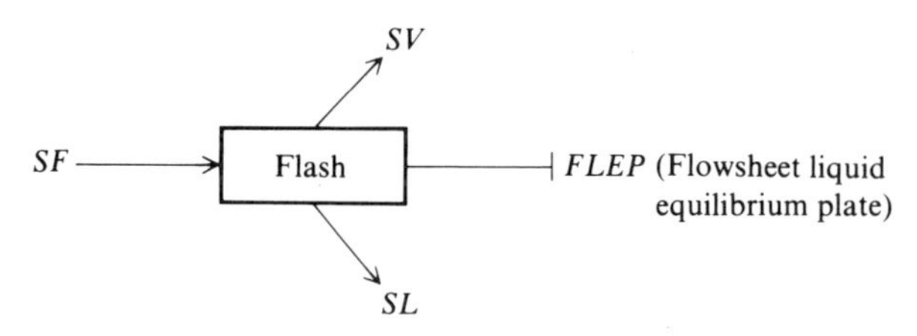

(*b*) Internal representation of the flowsheet element via GEV

	SF	*SV*	*SL*
Material balance	1	1	1
Equilibrium relation	0	1	1
Enthalpy balance	1	1	1

(*c*) Jacobian representation of the flowsheet element

Figure 14.5 Three representations of an element in a flowsheet in ASCEND-II.

would be a stream variable packet containing the flowrate, temperature, pressure, enthalpy, entropy, and component mole fractions of a stream. By specifying the elements in Fig. 14.5*c*, the user designates the generator to be used to calculate the partial derivatives and equation residuals. Thus, an entire flowsheet can be characterized by a set of generators with associated variable packets and equation packets.

EXAMPLE 14.2 APPLICATION OF ASCEND-II TO THE OPTIMIZATION OF A DISTILLATION COLUMN

This example taken from Locke and Westerberg (1983) summarizes their proposed evolutionary strategy to optimize a distillation column such as the one shown in

Fig. E14.2. The column contains five stages in each of the two-column sections. The feed is a 3-component mixture with relative volatilities for components A, B, and C of 3, 2, and 1, respectively. The feed is set at 100 mol/h with a composition of 30 percent A, 30 percent B, and 40 percent C. The objective is to minimize the composition of the heaviest component, component C, in the distillate, stream $S2$, using a total condenser. Constraints, other than requiring all flowrates, mole fractions, and split fractions to be nonnegative, are discussed below. The column was modeled by ASCEND-II with 131 equations, most of them nonlinear.

Locke and Westerberg proposed to find the optimal operating conditions of the column by first solving the column equations with zero degrees of freedom, and then carrying out the optimization calculations by removing some of the specifications of variables. Thus, to start with, the feed is fixed, the reflux rate is set to 1.0, the number of trays in the top and bottom sections are fixed, and the flowrate of the distillate stream ($S2$) is fixed at 50.0 mol/h. With these specifications, ASCEND-II was always able to find a solution of set of equations modeling the column. (With these specifications, the composition of component C in stream $S2$ was 0.0627.)

From this base point, the user may decide to run several other simulations such as heating the feed or lowering the reflux ratio to assist in getting a feeling for how the column performs. Suppose, for example, the specification of the reflux rate

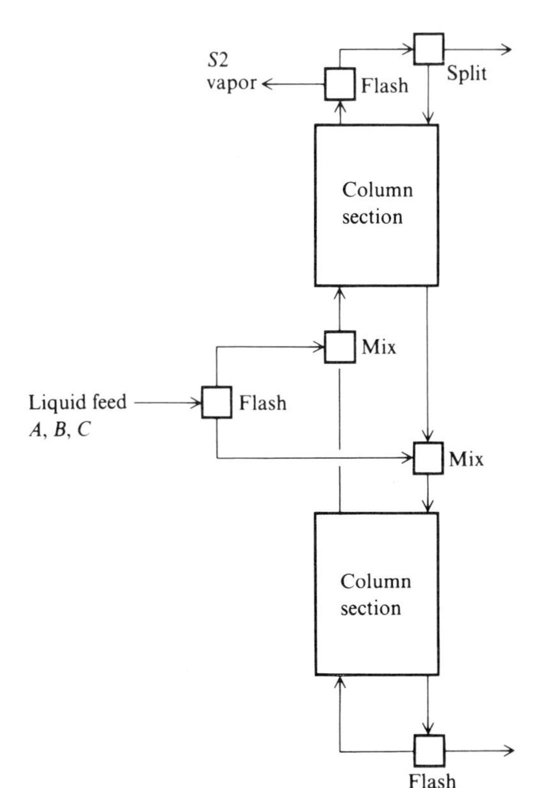

Figure E14.2 Units in ASCEND-II grouped to form a distillation column.

is relaxed and the composition of C in $S2$ is fixed instead. The user can reduce the mole fraction of C until it is so low that the iterative solution of the column equations does not converge, or negative flows are found. From such simulations practical bounds on column behavior can be established.

Several optimization calculations were carried out starting from a base case. In the first run, the reflux rate was fixed at 1.0, the tray efficiency was fixed at 1.0, and the specification on the flowrate of stream $S2$ was released. ASCEND-II found an optimum with respect to the flow of $S2$ at a value of 26.000 mol/h. The mole fraction of component C in $S2$ was 0.0518. To demonstrate that 26.000 was really the optimal flow rate, additional calculations were made specifying the flow of $S2$ at 26.2 mol/h which gave a mole fraction of C at a very slightly larger amount. Lowering $S2$ to 25.8 also gave a slightly higher mole fraction.

If the user makes an error, such as releasing the specifications on both flowrate $S2$ and the tray efficiencies in the top section of the column, resulting in two degrees of freedom for optimization, he or she finds that the optimizer "does not work." With both variables free, the optimizer does not seem to converge. Tray efficiency increases without bound! The user discovers that greater efficiency will always improve the objective function.

If the user releases the specification on the reflux rate while fixing the tray efficiency at 1.0, but puts an upper bound of 9.0 on the reflux rate, ASCEND-II would find an optimum with respect to the flowrate of $S2$ of 23.965 mol/h with the reflux rate at its upper bound. The target mole fraction would be lowered to 0.00105, an optimum that might seem counterintuitive.

In the numerical experiments reported, with flowsheets of less than 1000 equations, and with 1 to 5 degrees of freedom, typically 10 to 20 total iterations were required for convergence. When not optimizing, the number of iterations was typically 4 to 6, so it appeared that the optimization algorithm took only about three times the number of iterations needed for a design calculation with all the variables specified.

Although equation-based optimization methods offer a natural formulation of the design problem as an optimization problem, at present most industrial practice employs modular-oriented flowsheeting codes plus an optimization strategy for design and simulation.

14.3 LARGE-SCALE OPTIMIZATION USING SEQUENTIAL MODULAR FLOWSHEETING

The general concept of meshing a flowsheeting program or process simulator with an optimization code is shown in Fig. 14.6. Note that the traditional information flow sequence for connecting the codes uses the least amount of intercommunication. In the hierarchial structure, the optimization code forms an outer loop about the flowsheeting code which itself may contain nested loops. The obvious advantage of an arrangement such as in Fig. 14.6 is that existing computer codes can be employed. The number of independent variables in the

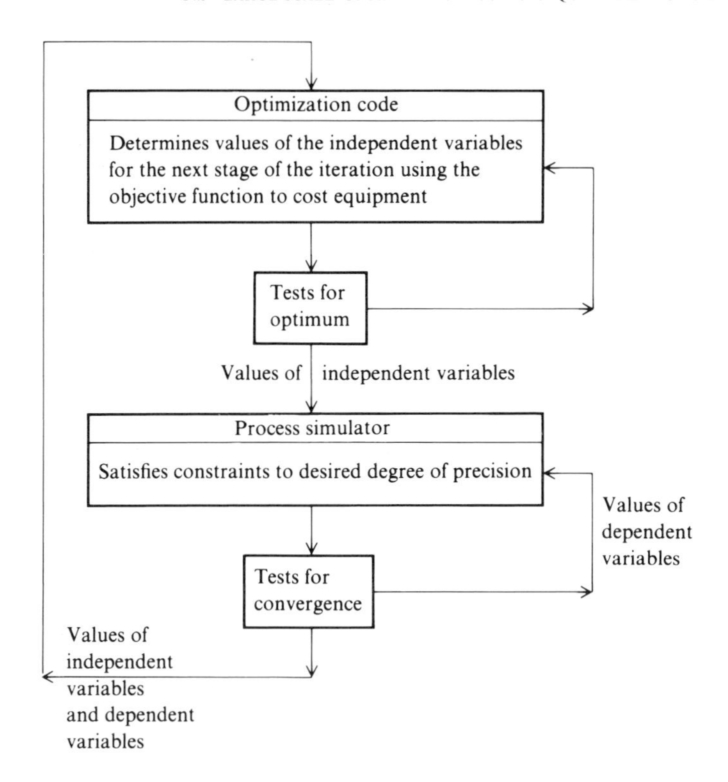

Figure 14.6 A two-level structure to combine an optimization code with a modular process simulator code.

optimization routine will be equal to the number of degrees of freedom at the point at which information is transferred from the flowsheeting routines to the optimization routine. However, if you want to superimpose or mesh an optimization routine with a modular-oriented flowsheeting code, certain special difficulties occur relative to an equation-based code:

(*a*) The output of one module is the input to another. The input and output variables in a computer module are fixed so that you cannot arbitrarily introduce an output and generate an input as can be done with an equation-based code.

(*b*) The modules make it difficult to generate reasonably accurate derivatives or their substitutes, especially if a module contains tables, functions with discrete variables, discontinuities, etc. Perturbation of the input to a module is the primary way in which a finite-difference substitute for a derivative can be generated.

(*c*) The modules may require a fixed precedence order of solution, i.e., the output of one module must become the input of another, hence convergence may be slower than in an equation-solving method, and the computational costs may be high.

(*d*) To specify a parameter in a module as a design variable, you have to place a control block around the module and adjust the parameter such that design specifications are met. This arrangement creates a loop. If the values of many design variables are to be determined, you might end up with several nested loops of calculation (which do, however, enhance stability). A similar arrangement must be used if you want to impose constraints on the simulation.

(*e*) Conditions imposed on a process (or a set of equations for that matter) may cause the unit physical states to move from a two-phase to a single-phase operation, or the reverse. As the code shifts from one module to another to represent the process properly, a severe discontinuity occurs in the objective function surface (and perhaps a constraint surface). Derivatives or their substitutes may not change smoothly and physical property values might jump about.

The most efficient way to combine a flowsheeting code with optimization is to directly mesh the optimization algorithm with the flowsheeting code. Two extreme classes of strategies exist if a sequential solution of the modules occurs in the flowsheeting code:

1. Feasible path strategies
2. Infeasible path strategies

Combinations are also possible such as proposed by Lang and Biegler (1987). In the general nonlinear programming problem in Chap. 8 we had two sets of constraints, equality and inequality constraints, and two vectors for the variables, independent and dependent. Here we have three classes of constraints:

(*a*) Inequality constraints

(*b*) Equality constraints that involve the design variables

(*c*) Equality constraints that represent the stream interconnections between modules

In addition, the vector $\mathbf{x}$ is split into vectors for the

1. Design variables
2. Stream variables
3. Dependent variables—those calculated from (*a*) and (*b*) above.

Feasible path strategies, as the name implies, on each iteration solve the equality constraints (i.e., seek convergence for each module) for fixed values of the design variables, and then adjust the design variables via the optimization procedure. The results of each iteration therefore provide a candidate design for the plant although the design may be suboptimal. Infeasible path strategies, on the other hand, do not require exact solution of the modules on each pass through the simulator. Thus, if an infeasible path method fails, the concluding solution is of little value. We will describe feasible path strategies first.

14.3.1 Feasible Path Strategies

Early attempts to formulate feasible path sequential modular optimization codes used direct-search or random-search algorithms that did not require derivatives for optimization (Friedman and Pinder, 1972; Gaines and Gaddy, 1976; Ballman and Gaddy, 1977). Most investigators have concluded that such search methods are too expensive for routine use. Biegler and Hughes (1981, 1982, 1985) developed a method of meshing process simulators with successive quadratic programming. They replaced the convergence acceleration code in the simulator with an optimization subroutine that selected new values for the decision variables at the same time that new values of the torn variables were selected.

Tearing in connection with modular simulators involves decoupling the interconnections between the modules so that sequential information flow takes place. Tearing is required because of loops of information created by recycle streams. What you do in tearing is to provide guesses for values of some of the unknowns (the tear variables), usually the recycle streams, and then calculate the values of the tear variables from the simulator subroutines. These calculated values form new guesses, and so on until the differences between the guessed values and the calculated values are sufficiently small.

Physical insight and experience in numerical analysis are important in selecting which variables to tear. For example, Fig. 14.7 illustrates an equilibrium vapor–liquid separator for which the combined material and equilibrium equations give the relation

$$\sum_{j=1}^{C} \frac{z_j(1 - K_j)}{1 - (V/F) + (VK_j/F)} = 0 \qquad (14.1)$$

where z_j is the mole fraction of species j out of C components in the feed stream, $K_j = y_j/x_j$ is the vapor–liquid equilibrium coefficient as a function of temperature, and the stream flowrates are noted in the figure. For narrow-boiling systems, you can guess V/F, $\mathbf{y}$, and $\mathbf{x}$, and use Eq. (14.1) to calculate K_j and hence the temperature. This scheme works well because T lies within a narrow range. For wide-boiling materials, the scheme does not converge well. It is better to

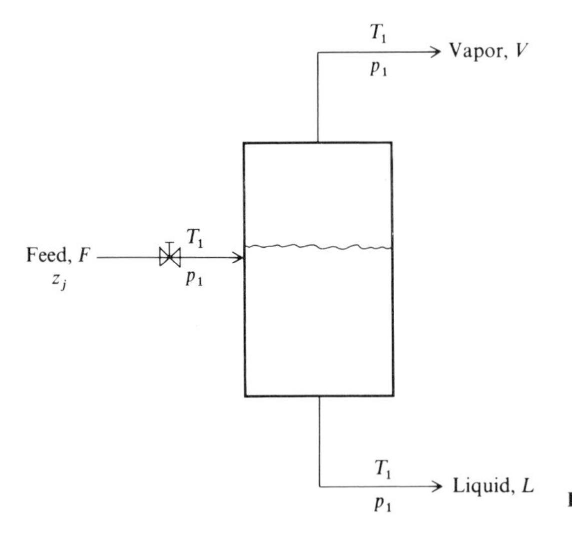

Figure 14.7 Vapor-liquid separator.

solve Eq. (14.1) for V/F by guessing T, $\mathbf{y}$, and $\mathbf{x}$, because V/F lies within a narrow range even for large changes in T.

If the convergence routines for the simulator comprise a separate module whose variables are connected to the other modules via the tear variables, an optimization code can be interfaced to the modules by acting as another convergence module or substituting for the convergence module. The optimization module would receive values of the functions and gradients (analytic or approximate), and adjust both the design and tear variables on each pass through the simulation. The system executive would not be affected nor would be the tear sets and calculation order.

Biegler and Hughes (1985) proposed two variants of a feasible path algorithm:

1. CFV—Complete Feasible Variant
2. RFV—Reduced Feasible Variant

Both employ successive quadratic programming. They use the same gradient calculations and require convergence to a solution of the flowsheet for each objective function evaluation. All the design and dependent variables are included in the convergence loop of the calculations. The perturbations used to obtain derivative information use the same calculation sequence as in the flowsheet program because only tear and design variables are perturbed. Convergence to a solution for the flowsheet is not required when making the perturbations, a major difference between the strategies in Fig. 14.6 and the feasible variant procedure. CFV includes the tear equations in the quadratic programming subproblem whereas RFV calculates reduced gradients and solves a smaller quadratic programming problem. Convergence is accelerated because the quadratic programming problem yields good starting values for the tear variables, directly in CFV and indirectly through the reduced gradients in RFV.

For sequential modular algorithms the optimization problem can be posed as (in the same notation as in Chap. 8)

$$\underset{\mathbf{x}_I}{\text{Minimize}} \quad f(\mathbf{x}_I, \mathbf{x}_D) \tag{14.2a}$$

$$\text{Subject to} \quad \mathbf{g}(\mathbf{x}_I, \mathbf{x}_D) \geq 0 \quad \text{inequality constraints} \tag{14.2b}$$

$$\mathbf{h}(\mathbf{x}_I, \mathbf{x}_D) = 0 \quad \text{design constraints} \tag{14.2c}$$

$$\bar{\mathbf{h}}(\mathbf{x}_I, \mathbf{x}_S) = 0 \quad \text{stream interconnection constraints (recycle tear equations)} \tag{14.2d}$$

$$\mathbf{x}_D = \psi(\mathbf{x}_I, \mathbf{x}_S) \tag{14.2e}$$

where $\mathbf{x}_I$ is the vector of independent variables, $\mathbf{x}_D$ is the vector of dependent variables (those used to directly calculate f, g_j, and h_j), and $\mathbf{x}_S$ is the vector of stream variables between modules. In the algorithms, Eq. (14.2*d*) is replaced by linear approximations to the tear equations, and $\mathbf{x}_D$ is calculated in each module as indicated by Eq. (14.2*e*).

In the RFV version, once the finite difference substitutes for gradients are calculated by perturbation (with respect to $\mathbf{x}_I$ only) thus getting reduced gradients, the quadratic programming subproblem becomes

$$\underset{\mathbf{s}}{\text{Minimize}} \quad \nabla^T_{\mathbf{x}_I} f(\mathbf{x}_I, \mathbf{x}_D)\mathbf{s} + \tfrac{1}{2}\mathbf{s}^T\mathbf{B}\mathbf{s} \tag{14.3}$$

$$\text{Subject to} \quad h_j(\mathbf{x}_I) + [\nabla_{\mathbf{x}_I} h_j(\mathbf{x}_I)]^T\mathbf{s} = 0 \qquad j = 1, \ldots, m$$

$$g_j(\mathbf{x}_I) + [\nabla_{\mathbf{x}_I} g_j(\mathbf{x}_I)]^T\mathbf{s} \geq 0 \qquad j = m + 1, \ldots, p$$

In the CFV version, both $\mathbf{x}_I$ and $\mathbf{x}_S$ and the recycle tear equations are included in the quadratic programming problem

$$\underset{\mathbf{s}}{\text{Minimize}} \quad \nabla^T_{\mathbf{x}_I} f(\mathbf{x}_I, \mathbf{x}_D)\mathbf{s} + \tfrac{1}{2}\mathbf{s}^T\mathbf{B}\mathbf{s} \tag{14.4}$$

$$\text{Subject to} \quad h_j(\mathbf{x}_I, \mathbf{x}_S) + [\nabla_{\mathbf{x}_I, \mathbf{x}_S} h_j(\mathbf{x}_I, \mathbf{x}_S)]^T\mathbf{s} = 0 \qquad j = 1, \ldots, m_1$$

$$\bar{h}_j(\mathbf{x}_I, \mathbf{x}_S) + [\nabla_{\mathbf{x}_I, \mathbf{x}_S} \bar{h}_j(\mathbf{x}_I, \mathbf{x}_S)]^T\mathbf{s} = 0 \qquad j = m_1, \ldots, m + 1$$

$$g_j(\mathbf{x}_I, \mathbf{x}_S) + [\nabla_{\mathbf{x}_I, \mathbf{x}_S} g_j(\mathbf{x}_I, \mathbf{x}_S)]^T\mathbf{s} \geq 0 \qquad j = m + 1, \ldots, p$$

In both algorithms $\mathbf{B}$ is updated by the BFGS method. Table 14.2 lists the steps in the feasible path algorithm.

Table 14.2 Steps for the feasible path sequential modular Algorithm (Biegler and Hughes, 1985)

Step 1
Identify $\mathbf{x}_I$ and $\mathbf{x}_D$. Set up a sequence for calculations involving $\mathbf{x}_I$ and $\mathbf{x}_D$ as well as all of the recycle loops as an outer calculation loop.

Step 2
Tear the outer calculation loop and subrecycle loops by choosing the tear variables $\mathbf{x}_S$ (usually the component flows, pressure, and specific enthalpy for each tear stream). For $k = 1$, pick the initial values of the tear variables.

Step 3
Execute a complete flowsheet simulation, and then calculate $f(\mathbf{x}_I^k, \mathbf{x}_D^k)$, $\mathbf{g}(\mathbf{x}_I^k, \mathbf{x}_D^k)$, $\mathbf{h}(\mathbf{x}_I^k, \mathbf{x}_D^k)$; $\mathbf{x}_D^k$ comes from Eq. (14.2*e*), and $\mathbf{x}_S^k$ satisfies (to within a specified tolerance) Eq. (14.2*c*).

Step 4
Start with a tear stream. Back up along the calculation loop until an unperturbed design variable, $x_{I,i}$, in a module is encountered. Perturb the design variable, and calculate the resulting dependent and tear variables in that module and all downstream modules in the calculation loop. (Dependent variables upstream are not affected.) Evaluate the gradients of f, $\mathbf{g}$, $\mathbf{h}$, and $\bar{\mathbf{h}}$ with respect to each $\mathbf{x}_{I,i}$ by using a forward difference formula in which the values of $\mathbf{x}_D$ are those from the perturbed calculations and the values of $\mathbf{x}_I$, except $x_{I,i}$ are perturbed values.

Step 5
One at a time, perturb the tear variable $\mathbf{x}_{S,i}$. Calculate the dependent variables from Eq. (14.2*e*) and evaluate the tear equations (14.2*d*). Calculate the gradients of f, $\mathbf{g}$, $\mathbf{h}$, and $\bar{\mathbf{h}}$ with respect to each $x_{S,i}$ by a forward difference equation in which the $\mathbf{x}_D$ are the perturbed values and $\mathbf{x}_I$ are the unperturbed values.

Step 5*a*
For the RFV algorithm only, calculate the reduced gradient [Eq. (8.30)].

Step 6
Update $\mathbf{B}^k$ by the BFGS method [Eq. (6.32)]. Let $\mathbf{B}^0 = \mathbf{I}$ or some other positive definite matrix.

Step 7
Solve the quadratic programming subproblem in Eq. (14.3) for RFV, or (14.4) for CFV, as described in Sec. 8.6.2.

(Continued)

Biegler and Hughes (1985) recommend values of several parameters for the algorithm in Table 14.2 to improve efficiency, including scaling of the design variables, what size perturbations δ to make, and the value of the convergence tolerance ε. Example 14.3 illustrates an application of RFV and CFV.

EXAMPLE 14.3 APPLICATION OF THE FEASIBLE PATH METHOD TO A PROCESS FOR THE CHLORINATION OF PROPYLENE

This example is taken from Biegler and Hughes (1985); numerical results have been provided by Professor Biegler whose assistance we gratefully acknowledge. Figure E14.3*a* is a flowsheet for the synthesis of allyl chloride. All of the process data and references for the data can be found in Biegler and Hughes who used a process

Step 8
Test for convergence.
For RFV, stop if

$$|\nabla^T_{\mathbf{x}_I} f(\mathbf{x}_I^k)\mathbf{s}| + |u_j g_j(\mathbf{x}_I^k)| + |w_j h_j(\mathbf{x}_I^k)| \leq \varepsilon$$

where u_j and w_j are the Lagrange multipliers associated with g_j and h_j, respectively, that can be calculated from the results of step 7.
For CFV, stop if

$$|\nabla^T_{\mathbf{x}_I} f(\mathbf{x}_I^k, \mathbf{x}_S^k)\mathbf{s}| + |u_j g_j(\mathbf{x}_I^k, \mathbf{x}_S^k)| + |w_j h_j(\mathbf{x}_I^k, \mathbf{x}_S^k)| + |v_j \bar{h}_j(\mathbf{x}_I^k, \mathbf{x}_S^k)| \leq \varepsilon$$

where v_j is the Lagrange multiplier associated with $\bar{h}_j$.

Step 9
Calculate in step length λ is search direction $\mathbf{s}$ that decreases an exact penalty function $P(\mathbf{x}_I)$, or, if the exact penatly function has decreased monotonically in previous iterations, decreases a modified Lagrangian function. Refer to Biegler (1981) for details. Each point along $\mathbf{s}$ requires a pass through step 3. The tear values are calculated:
For the RFV algorithm

$$\mathbf{x}_S^{k+1} = \mathbf{x}_S^k + \lambda^k \mathbf{w}^k$$

where

$$\mathbf{w}^k = \left[\left[\frac{\partial \bar{\mathbf{h}}^k}{\partial \mathbf{x}_S}\right]^T\right]^{-1} \left[\frac{\partial \bar{\mathbf{h}}^k}{\partial \mathbf{x}_I}\right]^T \mathbf{s}^k$$

For the CFV algorithm

$$x_{S,i}^{k+1} = x_{S,i}^k + \lambda s_i^k$$

Step 10
If the line search succeeds, let $k \to k + 1$, and return to step 4. If the line search fails, or the result of step 7 does not generate a direction of descent, reset $\mathbf{B}^k = \mathbf{B}^0$, and return to step 7. If reinitialization of $\mathbf{B}$ fails, stop the calculations.

---

simulator called SPAD and the standard modules in SPAD to represent the process plus a module they developed for the reactor and one for the scrubber/dryer. Figure E14.3*b* shows the block diagram of the modules employed.

In this example, propylene is fed at 60°F and 200 psia and mixed with a recycle stream containing mostly unreacted propylene. Rate expressions for the three reactions involved were of the form

$$r_k = k_k p_{jx} p_{Cl_2} \tag{a}$$

where r_k is the reaction rate for reaction k, k_k is the rate constant, p_{jx} is the partial pressure of the reactant, p_{Cl_2} is the partial pressure of the Cl_2 in the reacting mixture, and $k_k = A_k \exp\left[-B_k/(T + 460)\right]$, A_k and B_k being constant.

Because the plant was fairly simple, the objective function used was the net value added (sales returns less feed costs) in dollars per hour:

$$f = 22.17x_{D6} + 12.48x_{D7} + 4.91x_{D8} \tag{b}$$

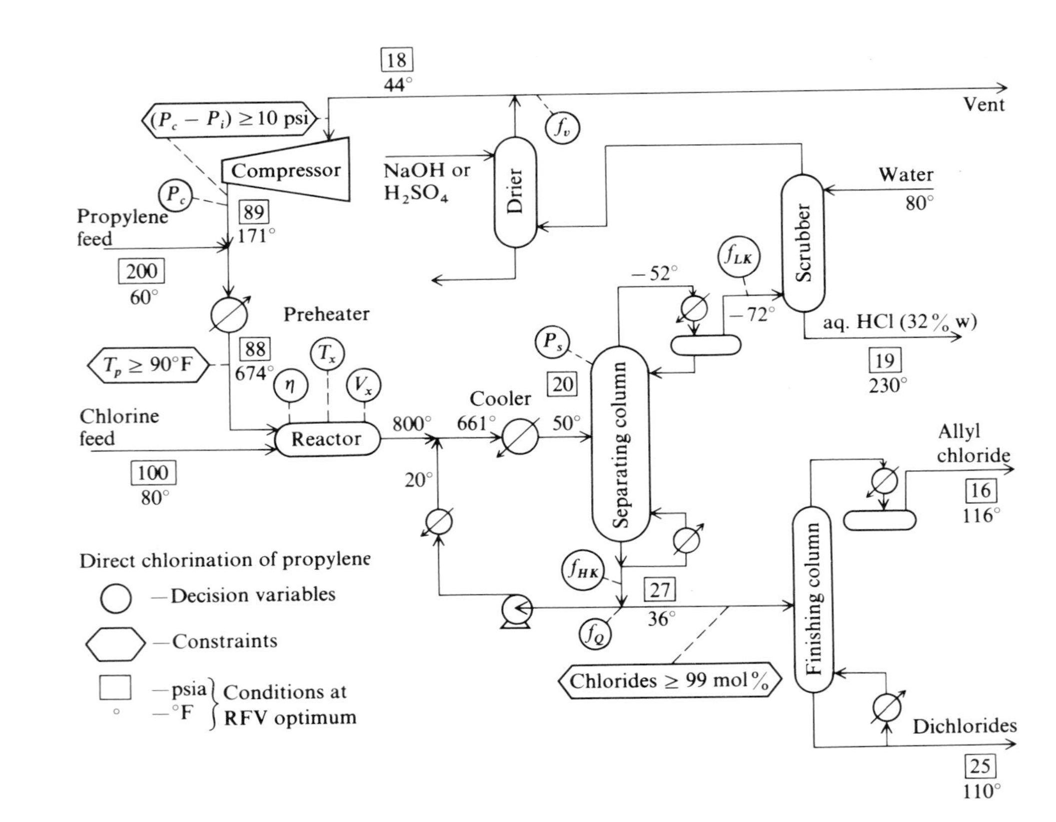

Figure E14.3a Allyl chloride process flowsheet.

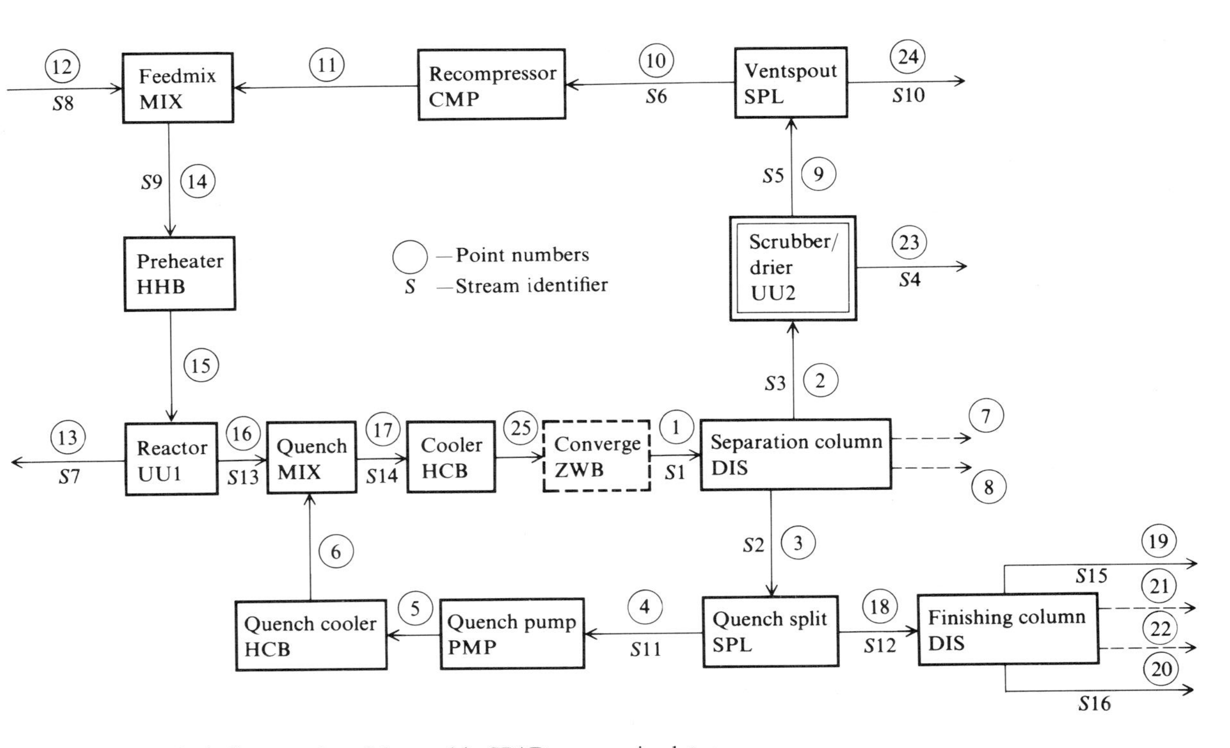

Figure E14.3*b* Block diagram of modules used in SPAD process simulator.

The inequality constraints were

$$x_{D2} - x_{D3} \geq 10 \tag{c}$$

$$\frac{x_{D6} + x_{D7} + x_{D8}}{x_{D4} + x_{D5} + x_{D6} + x_{D7} + x_{D8}} \leq 0.99 \tag{d}$$

$$x_{D1} \geq 90 \tag{e}$$

Scaling was carried out so that the objective function derivatives had absolute values between 10 and 1000, a perturbation of 10^{-2} was used, and the convergence tolerance was 10^{-3}.

Table E14.3*a* lists all of the independent, tear stream, and dependent variables. Table E14.3*b* lists the bounds, starting guesses, and optimal results for the independent (decision) variables. Table E14.3*c* summarizes the stream flows and

Table E14.3*a* List of variables and notation

	Independent variables
x_{I1}	C_3H_6/Cl_2 ratio to reactor
x_{I2}	Reactor temperature, °F
x_{I3}	Reactor volume, ft^3
x_{I4}	Separator column pressure, psia
x_{I5}	Split fraction to quench
x_{I6}	Split fraction to vent
x_{I7}	Outlet compressor pressure, psia
x_{I8}	Fraction recovery Cl_2 in separator tops
x_{I9}	Fraction recovery allyl chloride in separator bottoms
	Tear stream variables (used in Example 14.4)
x_{S1}	HCl, lb mol/h
x_{S2}	C_2, lb mol/h
x_{S3}	C_3H_6, lb mol/h
x_{S4}	Allyl chloride, lb mol/h
x_{S5}	Dichloropropane, lb mol/h
x_{S6}	Dichloropropene, lb mol/h
x_{S7}	Enthalpy, Btu/(lb)(mol)
	Dependent variables
x_{D1}	Reactor inlet temperature, psia
x_{D2}	Outlet compressor pressure, psia
x_{D3}	Inlet compressor pressure, psia
x_{D4}	Cl_2 in separator bottoms, lb mol/h
x_{D5}	C_3H_6 in separator bottoms, lb mol/h
x_{D6}	Allyl chloride in separator bottoms, lb mol/h
x_{D7}	Dichloropropane in separator bottoms, lb mol/h
x_{D8}	Dichloropropene in separator bottoms, lb mol/h

Table E14.3b Initial values and optimal results

Basis. 100 lb mol/h propylene feed	Notation	Specifications: Lower bound	Upper bound	Start	Optimal solution for three algorithms: IPOSEQ§	CFV¶	RFV††
Objective function Net value, \$/h	....	....	....	1010.13	1609.84	1609.92	1609.97
Variables							
1. C_3H_6/Cl_2 ratio to reactor	x_{I1}	2	10	10‡	4.35	5.42	5.86
2. Reactor temperature, °F	x_{I2}	800	1100	1000	800†	800†	800†
3. Reactor volume, ft^3	x_{I3}	1	15	10	15.0‡	8.58	13.04
4. Separation column pressure, psia	x_{I4}	20	50	28	20†	20†	20†
5. Split fraction to quench	x_{I5}	0.25	0.75	0.6	0.739	0.725	0.717
6. Split fraction to vent	x_{I6}	0.01	0.10	0.05	0.01†	0.01†	0.01†
7. Compressor out. pressure, psia	x_{I7}	20	100	100‡	61.60	100‡	88.78
8. Fraction recovery Cl_2 in separator tops	x_{I8}	0.95	0.99	0.99‡	0.99‡	0.99‡	0.99‡
9. Fraction recovery AC in separator bottoms	x_{I9}	0.95	0.99	0.99‡	0.99‡	0.99‡	0.99‡
Constraints							
1. $(p_c - p_{in})_{comp}$, psi	Eq. (*c*)	10	None	74	43.6	82	70.78
2. Product purity, mol %	Eq. (*d*)	99.00	None	99.02	99.25	99.36	99.38
3. Reactor preheat, °F	x_{D1} [Eq. (*e*)]	90	None	913.5	676.2	673.8	674.1
	Product yields, mol % on basis of propylene feed						
C_3H_5Cl Allyl chloride	x_{D6}	....	....	....	53.51	53.44	53.49
$C_3H_6Cl_2$ Dichloropropane	x_{D7}	....	....	....	13.92	13.93	13.93
$C_3H_4Cl_2$ Dichloropropene	x_{D8}	....	....	....	22.70	22.82	22.74

† At lower bound.
‡ At upper bound
§ Infeasible path sequential algorithm. See Example 14.4.
¶ Complete feasible path algorithm.
†† Reduced feasible path algorithm.

Table E14.3*c* Stream flows and operating parameter at the optimal solution

	Propylene feed	Chlorine feed	Recycle to feed	Reactor effluent	Product to fin. col.	Vent gas	Aqueous HCl
Stream No.	*S*8	*S*7	*S*6	*S*13	*S*12	*S*10	*S*4
Stream flows, lb mol/h							
HCl	....	....	0.10	99.24	......	0.001	99.137
Chlorine	....	113.88	58.66	59.48	0.17	0.593	0.051
Propylene	100.00	....	910.42	920.10	0.39	0.196	0.086
Allyl Chloride	....	....	1.75	55.40	53.50	0.018	0.142
Dichloropropane	....	....	....	13.93	13.93	......	....
Dichloropropene	....	....	....	22.74	22.74	......	....
Total	100.00	113.88	970.94	1170.89	90.73	9.807	99.417†
Total lb/h	4208	8075	42,609	54,892	8220	430.4	3633†
Point No.	12	13	11	16	18	24	23
Temp., °F	60.0	80.0	171.2	800.0	36.3	44.2	230.0
Press., psia	200.0	100.0	88.8	86.7	27.0	18.0	19.0
Btu/lb mol	2026.8	9105.7	10,701.7	32,328.1	1077.8	8077.4	6973.6†
ft^3/s gas	....	1.65	19.42	50.60	....	0.80	....
GPM liq	13.72	....	....	....	15.65	....	20.60 (aq.)

† Dry basis—accompanied by 428.24 lb mol/h (7708.3 lb/h) of water.

Table E14.3*d* Trajectory of the variables in optimization by the CFV algorithm (feasible path)

	Objective function	Decision variables									Constraints		
Iteration	Net val. \$/h	x_{11}, ratio	x_{12}, °F	x_{13}, ft^3	x_{14}, psia	x_{15}, frac.	x_{16}, frac.	x_{17}, psia	x_{18}, frac.	x_{19} frac.	Δp, psi	Purity, mol %	x_{D1}, °F
1	1010.13	10.00	1000	10.00	28	0.600	0.05	100	0.99	0.99	74	99.01	913.5
2	1245.63	7.85	800	15.00	20	0.573	0.038	100	0.99	0.99	82	99.15	689.4
3	1521.07	2.00	800	6.77	20	0.539	0.01	100	0.99	0.99	82	99.37	365.6
4	1580.63	3.17	800	7.18	20	0.541	0.01	100	0.99	0.99	82	99.23	559.5
5	1600.40	3.92	800	7.086	20	0.545	0.01	100	0.99	0.99	82	99.08	623.1
6	1606.87	4.48	800	7.183	20	0.551	0.01	100	0.99	0.99	82	99.00	652.7
7	1608.21	4.84	800	7.633	20	0.558	0.01	100	0.99	0.99	82	99.00	662.6
12	1609.10	5.64	800	8.85	20	0.605	0.01	100	0.99	0.99	82	99.08	678.2
14	1609.74	5.91	800	9.61	20	0.705	0.01	100	0.99	0.99	82	99.32	680.3
20	1609.92	5.42	800	8.58	20	0.725	0.01	100	0.99	0.99	82	99.36	673.8

Table E14.3*e* Trajectory of the variables in optimization by the RFV algorithm (feasible path)

	Objective function	Decision variables									Constraints		
Iteration	Net val. $/h	x_{I1}, ratio	x_{I2}, °F	x_{I3}, ft^3	x_{I4}, psia	x_{I5}, frac.	x_{I6}, frac.	x_{I7}, psia	x_{I8}, frac.	x_{I9} frac.	Δp, psi	Purity, mol %	x_{D1}, °F
1	1010.13	10.00	1000	10	28	0.6	0.05	100	0.99	0.99	74	99.01	913.5
2	1479.76	2.00	800	15	2	0.265	0.01	100	0.99	0.99	82	99.48	610.5
3	1588.84	8.85	800	15	20	0.25	0.01	100	0.99	0.99	82	99.85	713.7
4	1589.70	8.97	823.3	15	20	0.30	0.01	100	0.99	0.99	82	99.62	734.4
5	1609.03	7.23	800	15	20	0.75	0.01	100	0.99	0.99	82	99.46	687.2
6	1609.86	6.62	800	14.88	20	0.718	0.01	94.35	0.99	0.99	76.35	99.41	679.4
12	1609.97	5.86	800	13.04	20	0.717	0.01	88.78	0.99	0.99	70.78	99.38	674.1

operating parameters at the optimal solution. Tables E14.3*d* and E14.3*e*, respectively, show the progress of the execution of the optimization procedure for CFV and RFV. Each code required essentially the same computational effort on the process simulator.

Example 14.4 in the next section shows the results of an infeasible path algorithm for the same example.

14.3.2 Infeasible Path Strategies

As implied by the name, infeasible path strategies do not require the exact solution of each module on each pass through the simulation code. The optimization problem formulated in Eq. (14.2) still applies; the stream (tear) variables are designated by $\mathbf{x}_S$. The vector of stream constraints $\bar{\mathbf{h}}$ is rewritten as

$$\bar{\mathbf{h}} = \mathbf{x}_S - \boldsymbol{\phi}(\mathbf{x}_I, \mathbf{x}_D, \mathbf{x}_S) = \mathbf{0} \tag{14.5}$$

where $\bar{\mathbf{h}}(\mathbf{x}_S)$ now represents the vector of the differences between the guessed values of the tear variables and the calculated values of the tear streams, $\boldsymbol{\phi}(\mathbf{x}_S)$, calculated from the process loop. Convergence does not take place on every pass through the simulator. Instead $\bar{\mathbf{h}}(\mathbf{x}_S) \to \mathbf{0}$ as the objective function and other constraints converge to their limits during the optimization.

Biegler and Hughes (1982) proposed an infeasible path algorithm, IPOSEQ, the steps of which are listed in Table 14.3. They employed a modified successive quadratic programming algorithm to adjust both the design and tear variables, and calculated gradients via perturbation of the variables although they suggested that the chain rule could be applied to evaluate gradients after each model is perturbed individually. This latter method is more efficient for flowsheeting codes in which the design variables occur at the beginning of the calculation sequence. But special provisions must be made for information flow reversal. Without constraints, the algorithm listed in Table 14.3 reduces to the BFGS method. With zero degrees of freedom, the method becomes Newton's method for solving a square system of equations in the space of the tear variables.

Biegler (personal communication) has run the same problem (with a few different coefficients so the results are not directly comparable with those shown here) using an improved quadratic programming algorithm and the chain rule to calculate derivatives (Biegler, 1985; Biegler and Cuthrell, 1985). The results indicate a reduction in CPU time of 75 percent and a reduction in the number of iterations of 77 percent.

The algorithm contains the same type of user-selected parameters as mentioned in connection with the feasible path algorithms in Sec. 14.3.1.

Table 14.3 Steps for infeasible path sequential modular optimization algorithm [Biegler and Hughes (1982)]

Step 1
Select the design variables $\mathbf{x}_I$ and identify the variables in $\mathbf{x}_D$ that must be returned from the simulator modules to calculate values of $f(\mathbf{x})$ and the constraints.
Step 2
Choose a calculational sequence that includes all the $\mathbf{x}_I$ and $\mathbf{x}_D$ elements in a single calculational loop (so that the perturbations can be accomplished in the proper sequence).
Step 3
Identify the tear streams and choose the tear variables $\mathbf{x}_S$.
Step 4
Initialize. Let $\mathbf{x}_I = \mathbf{x}_I^1$ and $\mathbf{x}_S = \mathbf{x}_S^1$, that is, set the iteration index k to 1.
Step 5
On the (k-1) stage of iteration sequentially go through the process module calculations to update $\mathbf{x}_D$ and $\boldsymbol{\phi}(\mathbf{x}_I, \mathbf{x}_S)$ to the kth stage.
Step 6
Calculate $f(\mathbf{x}_I^k, \mathbf{x}_D^k)$, $\mathbf{g}(\mathbf{x}_I^k, \mathbf{x}_D^k)$, $\mathbf{h}(\mathbf{x}_I^k, \mathbf{x}_D^k)$, and $\bar{\mathbf{h}}(\mathbf{x}_I^k, \mathbf{x}_S^k) \equiv \mathbf{x}_S^k - \boldsymbol{\phi}(\mathbf{x}_I^k, \mathbf{x}_S^k)$.
Step 7
Start with the last module in the calculational sequence. Search backward in the sequence until an unperturbed design variable $x_{I,i}^k$ is identified as the module input. Perturb $x_{I,i}^k$. Recalculate the output for the module and all subsequent downstream modules.
Step 8
Evaluate approximates to the gradients of f, $\mathbf{g}$, $\mathbf{h}$, and $\bar{\mathbf{h}}$ with respect to $x_{I,i}$ from forward difference substitutes.
Step 9
Restore $x_{I,i}^k$ to its unperturbed value. Repeat step 7 for all the elements in $\mathbf{x}_I^k$.
Step 10
Perturb each torn variable in sequence, and recalculate the output for the modules in sequence beginning with the module for which the associated tear stream is an input.
Step 11
Evaluate approximates to the gradients of f, $\mathbf{g}$, $\mathbf{h}$, and $\bar{\mathbf{h}}$ with respect to $\mathbf{x}_{S,i}$ from forward finite difference substitutes using the reference values of $\mathbf{x}_I^k$ and the perturbed values of $\mathbf{x}_D^k$.

(*Continued*)

EXAMPLE 14.4 APPLICATION OF THE NONFEASIBLE PATH METHOD TO A PROCESS FOR THE CHLORINATION OF PROPYLENE

This example is taken from Biegler and Hughes (1985), and the process is explained in Example 14.3. The computer results have been provided by Professor Biegler. Table 14.3*b* lists the results obtained by Code IPOSEQ while Table E14.4 on page 590 lists the progress of the optimization iteration by iteration.

14.4 LARGE-SCALE OPTIMIZATION INCORPORATING SIMULTANEOUS MODULAR FLOWSHEETING STRATEGIES

Because the sequential modular strategy combined with optimization often does not prove efficient when a number of design constraints are imposed, nor is it

Step 12
If $k = 1$, set $\mathbf{B} = \mathbf{I}$ or $\mathbf{B}^0$. If $k > 1$, update $\mathbf{B}^k$ using the Broyden-Fletcher-Goldfarb-Shanno (BFGS) method [see Eq. (6.32)].

Step 13
Solve the following quadratic programming problem to get the search direction vector $\mathbf{s}^k$

$$\text{Minimize} \quad \nabla^T f(\mathbf{x}_I^k, \mathbf{x}_D^k)\mathbf{s} + \tfrac{1}{2}\mathbf{s}^T\mathbf{B}^k\mathbf{s}$$

$$\mathbf{g}(\mathbf{x}_I^k, \mathbf{x}_D^k) + [\nabla \mathbf{g}(\mathbf{x}_I^k, \mathbf{x}_D^k)]^T\mathbf{s} \geq 0$$

$$\mathbf{h}(\mathbf{x}_I^k, \mathbf{x}_D^k) + [\nabla \mathbf{h}(\mathbf{x}_I^k, \mathbf{x}_D^k)]^T\mathbf{s} = 0$$

$$\bar{\mathbf{h}}(\mathbf{x}_I^k, \mathbf{x}_S^k) + [\nabla \bar{\mathbf{h}}(\mathbf{x}_I^k, \mathbf{x}_S^k)]^T\mathbf{s} = 0$$

The Lagrange multipliers $\mathbf{u}$, $\mathbf{v}$, and $\mathbf{w}$ for $\mathbf{g}$, $\bar{\mathbf{h}}$, and $\mathbf{h}$, respectively, where $L = f(\mathbf{x}_I^k, \mathbf{x}_D^k) + \mathbf{u}^T\mathbf{g}(\mathbf{x}_I^k, \mathbf{x}_D^k) + \mathbf{v}^T\bar{\mathbf{h}}(\mathbf{x}_I^k, \mathbf{x}_D^k) + \mathbf{w}^T\mathbf{h}(\mathbf{x}_I^k, \mathbf{x}_D^k)$, come from the solution of the quadratic programming problem.

Step 14
If $|\nabla^T f(\mathbf{x}_I^k, \mathbf{x}_D^k)\mathbf{s}| + |\mathbf{u}^T\mathbf{g}(\mathbf{x}_I^k, \mathbf{x}_D^k)| + |\mathbf{v}^T\bar{\mathbf{h}}(\mathbf{x}_D^k, \mathbf{x}_D^k)| + |\mathbf{w}^T\mathbf{h}(\mathbf{x}_I^k, \mathbf{x}_D^k)| \leq \varepsilon$ where ε is the prespecific tolerance, stop. Otherwise, carry out a line search (see Biegler and Cuthrell, 1985) with the Lagrangian and/or an exact penalty function to calculate λ^k, the stepsize along $\mathbf{s}^k$ that yields an improved point. Note each iteration in the line search requires a pass through the process modules calculational sequence.

Step 15
If no value of λ can be found in the line search that improves the value of $f(\mathbf{x})$, set $\mathbf{B}^k = \mathbf{I}$ and return to step 13. Otherwise, update $\mathbf{x}_I$ and $\mathbf{x}_S$

$$\mathbf{x}_I^{k+1} = \mathbf{x}_I^k + \lambda^k\mathbf{s}_I^k$$

$$\mathbf{x}_S^{k+1} = \mathbf{x}_S^k + \lambda^k\mathbf{s}_s^k$$

Step 16
Calculate the function values $f(\mathbf{x}_I^{k+1}, \mathbf{x}_S^{k+1})$, $\mathbf{g}(\mathbf{x}_I^{k+1}, \mathbf{x}_S^{k+1})$, $\bar{\mathbf{h}}(\mathbf{x}_I^{k+1}, \mathbf{x}_S^{k+1})$, $\mathbf{h}(\mathbf{x}_I^{k+1}, \mathbf{x}_S^{k+1})$.

Step 17
Increment $k \rightarrow k + 1$ and return to step 7.

particularly efficient for optimization in general, several investigators have pursued a "simultaneous modular" strategy (Mahelec, Kluzik, and Evans, 1979; Biegler, 1983; Chen and Stadtherr, 1984). As implied by the name, a simultaneous modular strategy is designed to solve all the modules (in fact, equations representing the modules) simultaneously. How can this step be accomplished?

What has been suggested is to represent the modules by simple, usually linear, equations, the coefficients of which are determined by perturbation of inputs and states of individual modules. Only the decision variables are independent variables. From stage to stage in the calculations the coefficients in the approximating model equations are reformulated. Thus, the simultaneous modular strategy leads to a two-phase implementation:

1. Simulation using the modules themselves perhaps together with some of the connecting equations leads to a Jacobian matrix that (approximately) represents the process locally

Table E14.4 Trajectory of the variables in optimization by the IPOSEQ algorithm (infeasible path)

	Objective function	Decision variables									Constraints		
Iteration	Net val. \$/h	x_{I1}, ratio	x_{I2}, °F	x_{I3}, ft^3	x_{I4}, psia	x_{I5}, frac.	x_{I6}, frac.	x_{I7}, psia	x_{I8}, frac.	x_{I9}, frac.	Δp, psi	Purity, mol %	T_p, °F
1	604.20	10	1000	10.00	28.0	0.60	0.05	100.00	0.99	0.99	74.0	99.76	907.1
2	1041.27	2	800	15.00	33.1	0.597	0.01	100.00	0.99	0.99	68.9	99.23	292.1
3	1764.05	3.03	800	15.00	20.0	0.582	0.01	100.00	0.99	0.99	82.0	99.62	472.5
4	1582.78	3.73	800	15.00	20.0	0.582	0.01	100.00	0.99	0.99	82.0	99.52	533.7
11	1567.36	3.80	814	8.87	23.4	0.545	0.01	91.88	0.99	0.99	70.4	99.23	610.5
13	1593.07	3.91	800	10.46	20.0	0.545	0.01	91.86	0.99	0.99	73.8	99.25	596.7
15	1611.83	5.13	800	12.43	20.0	0.555	0.01	84.97	0.99	0.99	67.0	99.08	659.3
20	1593.24	5.12	800	15.00	20.0	0.617	0.01	71.56	0.99	0.99	53.6	99.08	673.4
25	1609.39	4.38	800	14.67	20.0	0.736	0.01	62.52	0.99	0.99	44.5	99.24	676.2
39	1609.84	4.35	800	15.00	20.0	0.739	0.01	61.60	0.99	0.99	43.6	99.25	676.2

2. The equations comprising the Jacobian matrix rows plus the rest of the connecting equations and the design specification equations are solved in a flowsheeting code much as with the equation-oriented strategy described in Sec. 14.2

With these two phases identified, the considerations for optimization should be familiar: (*a*) how to solve the set of nonlinear equations in phase 2, (*b*) how to efficiently compute the elements in the Jacobian matrix in phase 1, (*c*) what algorithm to use to carry out the optimization, and (*d*) how to select initial values of the variables involved in phases 1, and 2.

14.4.1 Calculation of the Elements in the Jacobian Matrix that Represents the Process

Chen and Stadtherr classified simultaneous modular strategies into three specific categories:

1. Tear all connecting streams so that each connecting stream is treated as two streams, i.e., one input and one output stream [see Mahalec, et al. (1979) or Jirapongphan (1980)]. The process model is then represented by $2(C + 2)n + m$ (perhaps nonlinear) equations where C = the number of chemical species (the $+2$ is for enthalpy and pressure), n = the number of connecting streams, and m = the number of free design specifications.
2. Each connecting stream is treated as one input stream and the model equations are substituted into the stream connection equations to reduce the number of equations representing the process. The process model then contains $(C + 2)n + m$ equations, a reduction of almost 50 percent of category 1.
3. To reduce the number of equations that must be solved simultaneously, an appropriate set of connecting streams can be torn instead of all of the connecting streams. Streams that are not torn can be eliminated by using the associated stream connection equations leading to a much smaller (but essentially full of nonzero elements) Jacobian matrix involving only tear stream variables (see Mahalec et al., 1979; McLane, et al., 1979; Perkins, 1979; Chen and Stadtherr, 1984). The process model equations then total $(C + 2)n_t + m$ where n_t is the number of torn streams. The resulting Jacobian in this category is analogous to that used in the sequential modular strategy described in Sec. 14.3.

For each of the three classes there are various ways to calculate the partial derivatives that comprise the elements of the Jacobian matrix. In Sec. 14.3 we mentioned perturbation of the modules in sequence to calculate finite difference substitutes for derivatives for the torn variables. To calculate the Jacobian matrix for the simultaneous modular strategy, one has to simulate each module $(C + 2)n_t + m + 1$ times in sequence (sequential perturbation).

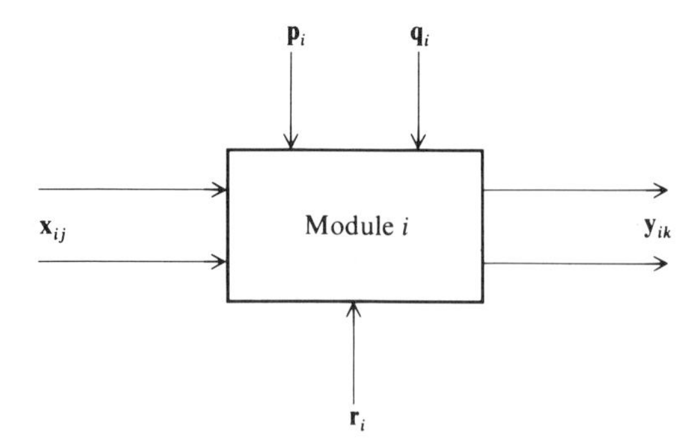

Figure 14.8 A typical module showing the input stream vectors $\mathbf{x}_{ij}$, output stream vectors $\mathbf{y}_{ik}$, specified equipment parameter vector $\mathbf{p}_i$, unspecified equipment parameter vector $\mathbf{g}_i$, and the retention (dependent) variable vector $\mathbf{r}_i$.

Another way to calculate the partial derivatives is possible. Figure 14.8 represents a typical module. If a module is simulated individually rather than in sequence after each unknown input variable is perturbed by a small amount, to calculate the Jacobian matrix $(C + 2)n_{ci} + n_{di} + 1$ simulations will be required for the ith module, where n_{ci} = number of interconnecting streams to module i and n_{di} = number of unspecified equipment parameters for module i. This method of calculation of the Jacobian matrix is usually referred as full-block perturbation.

Because of the large number of modular simulations required for full-block perturbation, Mahalec, et al. (1979) suggested using a diagonal approximation to the Jacobian matrix in which only the jth element of the input stream vector affected the corresponding element of the output stream vector. Only $n_{ci} + n_{di} +$ 1 simulations would have to be carried out for module i for this so-called diagonal-block perturbation approximation of the Jacobian matrix. As might be expected, this approach sometimes leads to an exceedingly fast solution of a problem and at other times leads to failure, and is not recommended.

Chen and Stadtherr evaluated the relative effectiveness of the sequential perturbation method versus full-block perturbation in calculating the category 3 Jacobian matrix. When full-block perturbations are used, the internal variables for the module, r_{ik}, can be temporarily saved, thus reducing the CPU time. Sequential calculations lose this advantage. For example, by means of the chain rule with full-block perturbation, one can calculate the first element of the gradient of g_i with respect to the independent variable x_{I1} from the following relation:

$$\left(\frac{\partial g_i}{\partial x_{I1}}\right)_{x_{I2}, x_{I3}, \ldots} = \left(\frac{\partial g_i}{\partial r_{i1}}\right)_{r_{i2}, r_{i3}, \ldots} \left(\frac{\partial r_{i1}}{\partial x_{I1}}\right)_{x_{I2}, x_{I3}, \ldots} + \left(\frac{\partial g_i}{\partial r_{i2}}\right)_{r_{i1}, r_{i3} \ldots} \left(\frac{\partial r_{i2}}{\partial x_{I1}}\right)_{x_{I2}, x_{I3}, \ldots} + \cdots$$

Since $(\partial g_i/\partial r_{ik})$ often can be evaluated explicitly, only $(\partial r_{ik}/\partial x_{Ii})$ then needs to be evaluated by perturbation. Similar relations for other elements of the Jacobian matrix can be derived via the chain rule.

As mentioned before, Chen and Stadtherr evaluated the relative effectiveness of sequential perturbations versus full-block perturbation in calculating the elements of the Jacobian matrix for strategy 3, that is, using torn streams, on nine different problems using their code SIMMOD. Full-block perturbation plus Euler's chain rule to calculate elements in the Jacobian proved to be twice as fast as sequential perturbation and achieved equal success in converging to the problem solutions.

14.4.2 Nonlinear Programming Algorithm

SIMMOD used the infeasible path strategy (see Sec. 14.3) together with the Han-Powell successive quadratic programming algorithm (refer to Sec. 8.6) to solve the overall nonlinear programming problem. However, the particular code used, SQPHP (Chen and Stadtherr, 1984), included modifications with respect to the line search and the scaling of the objective function and variables. Recall from Sec. 8.6 that the Han-Powell algorithm determined the best search direction for optimization at iteration k by solving a quadratic programming problem. Because the Hessian matrix of the objective function in the quadratic programming problem is a positive definite approximate of the Hessian matrix of the Lagrange function itself, the search direction is unique. However, taking full steps along this direction does not guarantee global convergence of the nonlinear problem. To improve convergence, various modifications were proposed to select the step size in the search direction in such a way as to take into account both the magnitude of the objective function and the degree of infeasibility of the constraints (Chamberlain, et al., 1982; Schittkowski, 1982a, 1982b; Mayne, 1980; Fletcher, 1982; Yamashita, 1982; Biegler and Cuthrell, 1985).

SQPHP employed the watchdog procedure of Chamberlain, et al., in which the primary search direction is obtained by the Han-Powell procedure, but a secondary search direction is calculated from the Lagrangian function itself and used as the main search direction if an improved rate of convergence can be achieved with it. (Biegler and Hughes also used this technique of line search in their work.) Chen and Stadtherr found in their test problems (only four) that this "watchdog" method reduced the number of iterations by half and the number of line searches by substantially more relative to the original Han-Powell algorithm.

14.4.3 Scaling of the Objective Function and Variables

In SIMMOD, Chen and Stadtherr scaled the objective function and variables by setting the initial approximate Hessian matrix to be a nonidentity diagonal matrix. The results of this type of scaling slightly reduce the number of line searches and iterations on some problems but increase them on others; however, the overall reliability of the code was improved. Appropriate scaling of the variables and initial choice for the Hessian matrix are equivalent factors in achieving reliable performance of a code.

14.5 CONCLUSIONS REGARDING COMBINING OPTIMIZATION WITH FLOWSHEETING CODES

There is no question that effective optimization algorithms can be meshed with flowsheeting programs. At this time it is not possible to predict which of the methods described in this chapter will prove to be superior after extensive testing. Practitioners prefer robustness far more than efficiency. Probably the codes used in industry in the future will be a hybrid of existing proposals comprised of partly modular and partly equation-based components. New hardware and the feasibility of parallel processing probably will generate new opportunities for improvement both in flowsheeting and optimization that will lead to improved performance and new types of computer codes.

14.6 TREATMENT OF LARGE-SCALE PROBLEMS WITH INTEGER-VALUED VARIABLES

As discussed in chapter 9 on integer programming, in numerous applications of optimization, the optimization problem is naturally formulated with some variables taking on only discrete values. In addition to the examples of staged equipment or units used in earlier chapters, integer-valued variables are found in problems involving facilities location, investment analysis, scheduling, transportation, and engineering design. Some problems with as many as 1000 integer variables are tractable whereas others with as few as 50 integer variables become unmanageable. For such problems, approximate methods have to be employed. As mentioned in Chap. 9, the procedure used most often for mixed integer and continuous variable problems is branch-and-bound, a method that uses partial enumeration and relies on heuristics to choose variables on which to branch and to guide the search.

In a process-optimization problem in which both continuous and integer variables occur, the continuous variables represent flowrates, pressures, temperatures, some equipment sizes, etc., whereas the integer variables represent certain equipment sizes and the existence, or not, of units. The optimization problem is formulated as

$$
\begin{aligned}
&\text{Minimize} && f(\mathbf{x}, \mathbf{y}) && && (14.6)\\
&\text{Subject to} && \mathbf{h}(\mathbf{x}, \mathbf{y}) = \mathbf{0} \\
& && \mathbf{g}(\mathbf{x}, \mathbf{y}) \geq \mathbf{0} \\
& && x_i \geq 0 && i = 1, \ldots, n_1 \quad \text{(continuous)} \\
& && y_i = \begin{Bmatrix} 0 \\ 1 \end{Bmatrix} && i = n_1 + 1, \ldots, n \quad \text{(integer)}
\end{aligned}
$$

where **h** and **g** represent the material and energy balances, design equations, design specifications, physical constraints, etc., that must be satisfied according to the flowsheet. Effective codes to solve (14.6) in its general form do not exist.

However, codes do exist to solve simplified problems such as occur if (14.6) can be converted to a mixed-integer linear-programming problem, or if

(14.6) can be formulated as a mixed-integer nonlinear-programming problem in which the binary variables appear linearly and separable from the continuous variables. Refer to Table 9.1 and also to Grossmann (1983) and Duran and Grossmann (1984) for applications.

To check the effectiveness of a particular strategy for computation, it is necessary to use problems of small dimension for which you know or can calculate an exact solution. It is extremely difficult to evaluate strategies for high-dimensional problems directly, hence inference is based on the outcome of investigations on small-dimensional problems. Thus, to sum up, the solution of large-scale nonlinear optimization problems involving several integer variables is extremely challenging, and may not be practical to obtain without various simplifications of the problem statement.

REFERENCES

Ballman, S. H., and J. L. Gaddy, "Optimization of Methanol Process by Flowsheet Simulation," *IEC Proc. Des. Dev.*, **16**: 337 (1977).

Berna, T. J., M. H. Locke, and A. W. Westerberg, "A New Approach to Optimization of Chemical Processes," *AIChE J.*, **26**: 37 (1980).

Biegler, L. T., Ph.D. Thesis, University of Wisconsin, Madison (1981).

Biegler, L. T., "Improved Infeasible Path Optimization for Sequential Modular Simulators—I: The Interface," *Comput. Chem. Eng.*, **9**, 245 (1985).

Biegler, L. T., "Simultaneous Modular Simulation and Optimization," Proceedings FOCAPD-II, Snowmass, CO 1983, (eds. A. W. Westerberg and H. H. Chien), CACHE Corp., Austin, Texas (1983).

Biegler, L. T., and J. E. Cuthrell, "Improved Feasible Path Optimization for Sequential Modular Simulators—II: The Optimization Algorithm," *Comput. Chem. Eng.*, **9**: 257 (1985).

Biegler, L. T., and R. R. Hughes, "Approximation Programming of Chemical Processes with Q/LAP," *Chem. Eng. Prog.*, (April 1981).

Biegler, L. T., and R. R. Hughes, "Infeasible Path Optimization with Sequential Modular Simulators," *AIChE J.*, **28**: 994 (1982).

Biegler, L. T., and R. R. Hughes, "Feasible Path Optimization with Sequential Modular Simulators," *Comput. Chem. Eng.*, **9**: 375 (1985).

Blau, G. E., "Session Summary—Nonlinear Programming," *Foundations of Computer-Aided Design, Vol. I* (eds. R. S. H. Mah and W. D. Seider), AIChE, New York (1981) p. 219.

Brosilow, C., and L. S. Lasdon, *AIChE Symp. Ser. No. 4*, 75 (1965).

Broyden, C. G., "A Class of Methods for Solving Nonlinear Simultaneous Equations," *Math. Comput.*, **19**: 577 (1965).

Chao, K. C., and J. D. Seader, *AIChE J.*, **7**: 598 (1961).

Chamberlain, R. M., M. J. D. Powell, C. Lemarechal, and H. C. Pedersen, "The Watchdog Method for Forcing Convergence in Algorithms for Constrained Optimization," *Math. Program. Study*, **16**: 1 (1982).

Chen, H. S., and M. A. Stadtherr, "Enhancements of the Han-Powell Method for Successive Quadratic Programming," private communication (1984).

Chen, H. S., and M. A. Stadtherr, "A Simultaneous-Modular Approach to Process Flowsheeting and Optimization: I. Theory and Implementation," *AIChE J.*, **30** (1984).

Curtis, A. R., and J. K. Reid, "The Choice of Step Lengths When Using Differences to Approximate Jacobian Matrices," *J. Inst. Math. Its Appl.*, **13**: 121 (1974).

de Leeuw den Bouter, J. A., A. G. Swenker, and M. G. G. Van Meulebrouk, "Simulation and Optimization with TISFL—II," *Proc. CHEMPLANT 80*, Heviz, Hungary (September 1980).

Duran, M. A., and I. E. Grossmann, "A Mixed Integer Nonlinear Programming Algorithm for Process Synthesis," *AIChE J.*, **32**: 592 (1986).

Evans, L. B., "Flowsheeting: A State of the Art Review," *Proc. Chemcomp. 1982* (ed. G. F. Froment, KVI, Antwerp, Belgium, 1982, p. 1.

Findeisen, W., J. Pulaczewski, and A. Maritius, "Multilevel Optimization and Dynamic Coordination of Mass Flows in a Beet Sugar Plant," *Automatica*, **6**: 581 (1970).

Fletcher, R., "Second-Order Corrections in Non-Differential Optimization," *Numerical Analysis*, Dundee, 1981, Lecture Notes in Mathematics, Vol. 912, Springer-Verlag (1982).

Friedman, P., and K. L. Pinder, "Optimization of a Simulation Model of Chemical Plants, *IEC Proc. Des. Dev.*, **11**: 512 (1972).

Gaines, L. D., and J. L. Gaddy, "Process Optimization by Flowsheet Simulation," *IEC Proc. Des. Dev.*, **15**: 206 (1976).

Grossmann, I. E., "Mixed-Integer Programming: Is It a Useful Tool in Process Synthesis," paper presented at the AIChE Diamond Jubilee Meeting, Washington, D.C., November 1983.

Hildebrand, J. H., and R. L. Scott, *The Solubility of Nonelectrolytes*, Reinhold, 1950.

Hutchinson, H. P., D. J. Jackson, and W. Morton, "Equation Oriented Flowsheet Simulation, Design and Optimization," Proc. Europ. Fed. Chem. Eng. Conf. Comput. Appl. in Chem. Eng., Paris (April 1983). "The Development of an Equation-Oriented Flowsheet Simulation and Optimization Package," *Comput. Chem. Eng.*, **10**: 19 (1986).

Jirapongphan, S., "Simultaneous Modular Convergence Concept in Process Flowsheet Optimization," Ph.D. Thesis, MIT (1980).

Kulikowski, R., "Optimization of Large-Scale Systems," *Automation*, **6**: 315 (1970).

Lang, Y. D., and L. T. Biegler, "A Unified Algorithm for Flowsheet Optimization," *Comput. Chem. Eng.*, **11:** 143 (1987).

Liu, Y. C., and C. B. Brosilow, "Modular Integration Methods for Simulation of Large Dynamic Systems," paper presented at the Washington, D.C. Meeting, AIChE, Nov. 1, 1983.

Locke, M. H., and A. W. Westerberg, "The ASCEND-II System—A Flowsheeting Application of Successive Quadratic Programming Methodology," *Comput. Chem. Eng.*, **7**: 615 (1983).

Locke, M. H., R. H. Edahl, and W. A. Westerberg, "An Improved Successive Quadratic Programming Optimization Algorithm for Engineering Design Problems," *AIChE J.*, **21**; 871 (1981).

Mahalec, V., H. Kluzik, and L. B. Evans, "Simultaneous Modular Algorithm for Steady State Flowsheet Simulation and Design," paper presented at the 12th European Symposium on Computer in Chemical Engineering, Montreux, Switzerland (1979).

McLane, M., M. K. Sood, and G. V. Reklaitis, "A Hierarchial Strategy for Large-Scale Process Calculations," *Comput. Chem. Eng.*, **3**: 383 (1979).

Mayne, D. Q., "On the Use of Exact Penalty Functions to Determine Step Length in Optimization Algorithms," *Numerical Analysis*, Dundee, 1979, Lecture Notes in Mathematics, Vol. 773, Springer-Verlag, Berlin (1980).

Perkins, J. D., "Equation Oriented Flowsheeting," in *Foundations of Computer-Aided Design*" (eds. A. W. Westerberg and H. H. Chien), CACHE Corp., Austin, Texas (1983).

Perkins, J. D., "Efficient Solution of Design Problems Using a Sequential-Modular Flowsheeting Program," *Comput. Chem. Eng.*, **3**: 375 (1979).

Parker, A. P., and R. R. Hughes, "Approximate Programming in Chemical Processes—1," *Comput. Chem. Eng.*, **5**: 123 (1981).

Redlich, O., and J. N. S. Kwong, *Chem. Rev.*, **44**: 233 (1949).

Rosen, E., "Steady State Chemical Process Simulation—State of the Art Review," *Computer Applications to Chemical Engineering*, (eds. R. G. Squires and G. V. Reklaitis), ACS Symp. Series No. 124 (1980) p. 3.

Sargent, R. W. H., "A Review of Methods for Solving Nonlinear Algebraic Equation," *Foundations of Computer-Aided Design, Vol. I* (eds. R. S. H. Mah and W. D. Seider), AIChE, New York (1981) p. 27.

Sargent, R. W. H., J. D. Perkins, and S. Thomas, "SPEEDUP: Simulation Program for Economic Evaluation and Design of Unified Processes," in *Computer-Aided Process Plant Design* (ed. M. E. Leesley), Gulf Pub. Co., Houston (1982).

Schittkowski, K., "The Nonlinear Programming Method of Wilson, Han, and Powell with an Augmented Lagrange-Type Line Search, Part I," *Numer. Math.*, **38**: 83 (1982*a*), "... Part II," **38**: 115 (1982*b*).

Schock, A. V., and R. Luus, "Relationship of the Two-Level Optimization Procedure to the Discrete Maximum Principle," *AIChE J.*, **18**: 659 (1972).

Stephanopoulos, G., "Synthesis of Process Flowsheets: An Adventure in Heuristic Design or a Utopia of Mathematical Programming," in *Foundations of Computer-Aided Design, Vol. II* (eds. R. S. H. Mah and W. D. Seider), Engineering Foundation (1981) p. 439.

Tazaki, E., A. Shindo, and T. Umeda, "Decentralized Optimization of a Chemical Process by a Feasible Method," *Automation*, **8**, 543 (1972).

Westerberg, A. W., "Optimization in Computer-Aided Design," *Foundations of Computer-Aided Chemical Process Design, Vol. I*, ed. R. S. H. Mah and W. D. Seider, AIChE, New York (1981) p. 149.

Westerberg, A. W., "A Review of the Synthesis of Distillation-Based Separation Systems," paper presented at 25th Anniversary Int. Conf. on New Developments Toward Technologies with Low Energy Requirements, INTEC, Santa Fe, Argentina, August 1983.

Westerberg, A. W., H. P. Hutchinson, R. L. Motard, and P. Winter, *Process Flowsheeting*, Cambridge University Press, Cambridge (1979).

Yamashita, H., "A Globally Convergent Constrained Quasi-Newton Method with an Augmented Lagrangian-Type Penalty Function," *Math. Program*, **23**: 75 (1982).

SUPPLEMENTARY REFERENCES

Achenie, L. E. K. and L. T. Biegler, "Algorithmic Synthesis of Chemical Reactor Networks Using Mathematical Programming", *Indus. Engr. Chem. Fundam.* **25**, 621 (1986).

Andrecovish, M. J., Synthesis of Integrated Distillation Sequences, Ph.D. Dissertation, Carnegie Mellon University (1983).

Aspen Technology Inc., "Aspen Plus Optimizer Plant", *The Aspen Leaf*, 251 Vassar St., Cambridge, Massachusetts 02139, **IV**, 1 (1986).

Assad, A., and B. Golden, "A Categorized Bibliography of Survey Articles on Management Science and Operations Research," *Manage. Sci.*, **28**: 425 (1982).

Biegler, L. T., I. E. Grossman, and A. W. Westerberg, "A Note on Approximation Techniques Used for Process Optimization," *Comput. Chem. Eng.*, **9**: 201 (1985).

Biegler, L. T., I. E. Grossman, and A. W. Westerberg, "Strategy for the Optimization of Chemical Processes," *Rev. in Chem. Eng.*, **3**: 1 (1985).

Boland, D., and E. Hindmarsh, "Heat Exchanger Network Improvements," *Chem. Eng., Prog.* **80**(7): 47 (1984).

Chan, W. K., and R. G. H. Prince, "Application of the Chain Rule of Differentiation to Sequential Modular Flowsheet Optimization," *Compute & Chem. Eng.*, **10**, 223 (1986).

Chitra, S. P., Synthesis of Optimal Serial Reactor Structures for Homogeneous Reactions, Ph.D. Dissertation, University of Cincinnati (1983).

Clark, S. M., and G. V. Reklaitis, "Investigation of Strategies for Executing Sequential Modular Simulations," *Comput. Chem. Eng.*, **8**: 205 (1984).

Cuthrell, J. E., and L. T. Biegler, "Simultaneous Solution and Optimization of Process Flowsheets with Differential Equation Models", *Chem. Eng. Res. Des*; **64**, 341 (1986).

Design Optimization Lab, Mech. Eng. Dept., Brigham Young University, Provo UT, OPTDES.-BYU-3 (4 codes).

Duran, M. A., and I. E. Grossmann, "Simultaneous Optimization and Heat Integration of Chemical Processes," *AIChE J.*, **32**: 123 (1986).

Durrer, E. J., and G. E. Slater, "Optimization of Petroleum and Natural Gas Production—A Survey," *Manage. Sci.*, **24**: 35 (1977).

Edgar, T. F., D. M. Himmelblau, and T. C. Bickel, "Optimal Design of Gas Transmission Networks," *Soc. Pet. Eng. J*, **30**: 96, April, (1978).

Hartman, K., "Experience in the Synthesis of Optimal Chemical Process Systems," Proc. 12 Symp. Comput. Appl. in Chem. Eng., Montreux, Switzerland (1979).

Hillestad, M., and T. Hertzberg, "Dynamic Simulation of Chemical Engineering Systems by the Sequential Modular Approach", *Comput. Chem. Eng.*, **10**, 377 (1986).

Hutchinson, H. P., and C. F. Shewchule, "A Computational Method for Multiple Distillation Towers," *Trans. Inst. Chem. Eng.*, **52**: 325 (1974).

Institute of Management Sciences, Chemical Production Optimization at Monsanto (A video tape), IMS, Providence, Rhode Island (1984).

Jirapongphan, S., J. F. Boston, H. I. Britt, and L. B. Evans, "A Nonlinear Simultaneous Modular Algorithm for Process Flowsheet Optimization," paper presented at the 80th Annual Meeting of the AIChE, Chicago (1980).

Joulia, X., and B. Koehret, "Simultaneous Modular Approach to Process Flowsheeting," EFCE Pub. Ser. 1983, 27 Cong. Int. "Inf. Genie Chem.," Vol. 2, C70/1-C70/7, 1983.

Lasdon, L. S., *Optimization Theory for Large Systems*, Macmillan, New York (1970).

Mahalec, V., and R. L. Motard, "Evolutionary Search for an Optimal Limiting Process Flowsheet," *Comput. Chem. Eng.*, **1**: 149 (1976).

Muraki, M., and T. Hayakawa, "Practical Synthesis Method for Heat Exchanger Network," *J. Chem. Eng. Jpn.*, **15**: 136 (1982).

Nishida, N., G. Stephanopoulos, and A. W. Westerberg, "A Review of Process Synthesis," *AIChE J.*, **27**: 321 (1981).

Nye, W. T., and A. L. Tits, "An Application-Oriented, Optimization-Based Methodology for Interactive Design of Engineering Systems," *Int. J. Control*, **43**: 1693 (1986).

Palmer, K. H., N. K. Boudwin, A. Patton, A. J. Rowland, D. Sammes, and D. M. Smith, *A Model-Management Framework For Mathematical Programming*, John Wiley & Sons, Inc., New York (1984).

Papoulias, S. A., *Studies in the Optimal Synthesis of Chemical Processing and Energy Systems*, Ph.D. Thesis, Carnegie-Mellon University, Pittsburgh, Philadelphia (1982).

Rangaiah, G. P., "Studies in Constrained Optimization of Chemical Process Problems," *Comput. Chem. Eng.*, **9**: 395 (1985).

Shacham, M., S. Macchetto, L. F. Stutzman, and P. Babcock, "Equation-Oriented Approach to Flowsheeting," *Comput. Chem. Eng.*, **6**: 79 (1982).

Shah, J. V., and A. W. Westerberg, "EROS: A Program for Quick Evaluation of Energy Recovery Systems," *Comput. Chem. Eng.*, **4**: 231 (1980).

Shindo, A., G. Nagasawox, and T. Maejima, "Computer-Aided Heat Exchange System Design," *Proc. CHEMCOMP 1982*, Antwerp, Belgium, 1982, K. V. I., 58, Jan van Rijswijcklaan, Antwerp, 1982.

Simulation Sciences, Inc., *PROCESS Simulation Program Input Manual*, Fullerton, California (1984).

Stadtherr, M. A., and H. S. Chen, "Numerical Techniques for Process Optimization by Successive Quadratic Programming," 3d. Int. Cong. on Comput. in Chem. Eng., Paris, April 1983.

Stadtherr, M. A., and H. S. Chen, and C. M. Hilton, "Development of a New Equation-Based Process Flowsheeting System: Numerical Studies," *Selected Topics on Computer-Aided Process Design and Analysis* (eds. R. S. H. Mah and G. V. Reklaitis), AIChE Symp. Ser. 78 (214), 12 (1982).

Stephanopoulos, G., "Synthesis of Process Flowsheets: An Adventure in Heuristic Design or a Utopia of Mathematical Programming?" in *Foundations of Computer-Aided Process Design*, Amer. Inst. Chem. Eng., New York (1981).

Umeda, T., and A. Ichikawa, "A Unified Approach to Processing System Design," in U.S.-Japan Seminar on the Synthesis of Optimal Heat Exchanger Systems—An Approach by the Optimal Assigment Problem in Linear Programming, Kyoto, 1975.

Westerberg, A. W., and Chien, H. H. (eds.), *Proc. 2d Int. Conf. on Foundations of Computer-Aided Process Design*, CACHE Corp., Austin, Texas (1984).

Westerberg, A. W., and Chien, H. H. (eds.), "Thoughts on a Future Equation-Oriented Flowsheeting System," *Comput. Chem. Eng.*, **9**: 517 (1985).

APPENDIX A

NOMENCLATURE

a	boundary of interval in one-dimensional search
a_i	parameters to be optimized in optimal control problem
A	annual revenue
A_i^l	lower bound on analysis for component i
A_i^u	upper bound on analysis for component i
$\mathbf{A}$	coefficient matrix in linear inequality constraint
$\mathbf{A}$	Jacobian matrix of constraints
$\bar{\mathbf{A}}$	adjoint matrix of $\mathbf{A}$
b	coefficient in quadratic function
b	boundary of interval in one-dimensional search
b_{ij}	coefficients in quadratic function
b_i	ith parameter estimate
$\mathbf{b}$	vector of coefficients in inequality constraints
$\mathbf{B}$	approximation to the Hessian matrix used in sequential quadratic programming
c	coefficient in quadratic function
$\mathbf{c}$	vector of cost coefficients
$\mathbf{c}_B$	cost coefficient vector for basic variables
$\mathbf{c}_N$	cost coefficient vector for nonbasic variables
C	cost
d_j	depreciation taken in year j
d_j	decision variable at stage j
d_{ij}	coefficient in quadratic function
D	number of units produced in manufacturing
D	diameter
D_j	cumulative depreciation in year j
$\mathbf{D}$	diagonal matrix
e	constant in equality constraint
$\mathbf{e}_j$	vector used in finite difference approximation
E	error measure
E	annual expense
f	objective function
f_{low}	lowest estimated value of f
f_n	objective estimated after n stages
F	dual of objective function in Chap. 7
F	objective function
F_i	future worth after i years
g_j	inequality constraint
$\mathbf{g}$	vector of inequality constraints
$\mathbf{g}_R$	reduced gradient
h	step size in discretization
h_j	equality constraint
$\mathbf{h}$	vector of equality constraints
$\mathbf{H}$	Hessian matrix
$\tilde{\mathbf{H}}$	modified Hessian matrix in Eq. (6.17)

$\hat{\mathbf{H}}$	modified Hessian matrix in Sec. 6.4.1 and Chap. 8
$\mathbf{H}^{-1}$	inverse Hessian matrix
i	interest rate
I_j	investment in jth year
$\mathbf{I}$	identity matrix
J	mathematical operator
$\mathbf{J}$	Jacobian matrix
k	iteration number
k_i	cost coefficients
L	Lagrangian
L	size of interval of uncertainty
L_i	lower bound
$\mathbf{L}$	lower triangular matrix
m	slope
m	number of equality constraints
m_e	number of independent equality constraints
m_i	number of independent inequality constraints
M	penalty coefficient (big M method)
M_i	ith minor of matrix
n	number of time periods in investment project
n	stage number
n	number of variables
p	number of data points
p	total number of constraints
P	principal
P	present worth
P	penalty function
$\mathbf{P}$	projection matrix
q	number of times per period interest is compounded
q	number of nonnegative dual variables
Q	annual production level
$\mathbf{Q}$	weighting matrix in quadratic programming
$\mathbf{Q}$	positive definite matrix
r	repayment multiplier
r	number of nonnegative variables in linear programming
r	penalty function weighting coefficient
r_A	penalty function weighting coefficient in augmented Lagrange method
r_j	number of levels of a variable in dynamic programming
s_i	slack or surplus variable
s_i	component of a search direction
$\mathbf{s}$	search direction
S	size parameter
S_v	salvage value
$S_c^{K_i}$	relative sensitivity of cost to coefficient K_i
t	time

t	tax rate
t	number of nonbasic variables
t_c	tax credit
t_f	final time
t_{ij}	time to complete job i on machine j
t_0	initial time
$\mathbf{u}$	vector of Lagrange multipliers
U_i	upper bound
$\mathbf{U}$	partition of Jacobian matrix
$\mathbf{v}$	vector of Lagrange multipliers
$\mathbf{v}_i$	ith eigenvector
$\mathbf{v}_i$	vector defined in necessary conditions
V_j	book value in year j
V_j	physical property value
$\mathbf{V}$	eigenvector matrix
$\mathbf{V}$	partition of Jacobian matrix
w	factor in Eq. (5.18)
w	objective function for artificial variables
$\mathbf{W}$	Hessian matrix in constrained optimization in Sec. 8.5
x_{ij}	variable in assignment problem
$\mathbf{x}$	vector of n variables
$\mathbf{x}$	model input vector
$\mathbf{x}_B$	vector of basic variables
$\mathbf{x}_N$	vector of nonbasic variables
$\mathbf{x}^p$	reference point
$\mathbf{x}^*$	optimal value of $\mathbf{x}$
$\mathbf{X}$	data matrix
y	model output
y_i	dual problem slack variables
Y	operating hours per year
Y_j	observed data point
$\mathbf{Y}$	basis in range space
z	distance variable
$\mathbf{z}$	vector in quadratic programming
$\mathbf{Z}$	basis in null space

Greek Symbols

α	eigenvalue
β	positive weighting factor
β_i	true model parameter
β^k	step size adjustment in conjugate gradient method
γ	positive weighting factor
δ	bound on step size
∇	gradient operator ("del")

Δ	difference in general
Δ	determinant
$\Delta \mathbf{g}$	$\nabla f(\mathbf{x}^{k+1}) - \nabla f(\mathbf{x}^k)$
Δt	discretization in time for optimal control problem
Δx	deviation of input variable from base point
$\Delta \mathbf{x}^k$	$\mathbf{x}^{k+1} - \mathbf{x}^k$
Δy	deviation of output variable from base point
Δ_i	determinant of ith principal minor
$\mathbf{\Delta}_I$	search direction for independent variables, GRG method
$\mathbf{\Delta}_D$	search direction for dependent variables, GRG method
ε	convergence criterion
ε	roundoff error
ε_j	error between jth data point and model prediction
λ	distance moved along a search vector (step length)
θ	component of objective function in optimal control
Θ_j	parameter in penalty function method
ρ	a scalar between 0 and 1
ρ	dimension of trust region
σ	slack variable
τ	weighting factor in Eq. (6.25)
$\boldsymbol{\phi}$	vector function in set of ordinary differential equations
Φ	component of objective function in optimal control in Sec. 8.10
Ψ_i	approximating function
$\boldsymbol{\omega}$	vector of Lagrange multipliers

Superscripts

k	stage in search
T	transpose
*	at optimal solution
$'$	first derivative

APPENDIX B

MATHEMATICAL SUMMARY

This appendix summarizes certain essential background material concerning matrices and vectors. It is by no means a complete exposition of the subject (see, for example, Amundson (1966); Fadeeva (1959); Pipes (1963), and Campbell (1968)), but concentrates mainly on those features useful in optimization.

B.1 DEFINITIONS

A matrix is an array of numbers, symbols, functions, etc.

$$\mathbf{A} = \begin{bmatrix} a_{11} & a_{12} & \cdots & a_{1m} \\ a_{21} & a_{22} & \cdots & a_{2m} \\ \vdots & & & \vdots \\ a_{n1} & a_{n2} & \cdots & a_{nm} \end{bmatrix} \tag{B.1}$$

An element of the matrix $\mathbf{A}$ is denoted by a_{ij}, where the subscript i corresponds to the row number and subscript j corresponds to the column number. Thus $\mathbf{A}$ in (B.1) has a total of n rows and m columns, and the dimensions of $\mathbf{A}$ are n by m ($n \times m$). If $m = n$, $\mathbf{A}$ is called a "square" matrix. If all elements of $\mathbf{A}$ are zero except the main diagonal (a_{ii}, $i = 1, \ldots, n$), $\mathbf{A}$ is called a diagonal matrix. A diagonal matrix with each $a_{ii} = 1$ is called the identity matrix, abbreviated $\mathbf{I}$.

Vectors are a special type of matrix, defined as having one column and n rows. For example in (B.2) $\mathbf{x}$ has n components

$$\mathbf{x} = \begin{bmatrix} x_1 \\ x_2 \\ \vdots \\ x_n \end{bmatrix} \qquad n \times 1 \text{ matrix, a vector} \tag{B.2}$$

A vector can be thought of as a point in n-dimensional space, although the graphical representation of such a point, when the dimension of the vector is greater than three, is not feasible. The general rules for matrix addition, subtraction, and multiplication described below in Sec. B.2 apply also to vectors.

The transpose of a matrix or a vector is formed by assembling the elements of the first row of the matrix as the elements of the first column of the transposed matrix, the second row into the second column, and so on. In other words, a_{ij} in the original matrix $\mathbf{A}$ becomes the component a_{ji} in the transpose $\mathbf{A}^T$. Note that the position of the diagonal components (a_{ij}) are unchanged by transposition. If the dimension of $\mathbf{A}$ is $n \times m$, the dimension of $\mathbf{A}^T$ is $m \times n$ (m rows and n columns). If square matrices $\mathbf{A}$ and $\mathbf{A}^T$ are identical, $\mathbf{A}$ is called a symmetric matrix. The transpose of a vector $\mathbf{x}$ is a row

$$\mathbf{x}^T = [x_1 \quad x_2 \quad \cdots \quad x_n] \tag{B.3}$$

B.2 BASIC MATRIX OPERATIONS

First we present the rules for equality, addition, and multiplication of matrices.

Equality: $\mathbf{A} = \mathbf{B}$ if and only if $a_{ij} = b_{ij}$ for all i and j.

Furthermore, both **A** and **B** must have the same dimensions (**A** and **B** are "conformable").

Addition: $\mathbf{A} + \mathbf{B} = \mathbf{C}$ requires that the element $c_{ij} = a_{ij} + b_{ij}$ for all i and j. **A**, **B**, and **C** must all have the same dimensions.

Multiplication: $\mathbf{AB} = \mathbf{C}$

If the matrix **A** has dimensions $n \times m$ and **B** has dimensions $q \times r$, then to obtain the product **AB** requires that $m = q$ (the number of columns of **A** equals the number of rows of **B**). The resulting matrix **C** is of dimension $n \times r$ and thus depends upon the dimensions of both **A** and **B**. An element c_{ij} of **C** is obtained by summing the products of the elements of the ith row of **A** times the corresponding elements of the jth column of **B**:

$$c_{ij} = \sum_{k=1}^{m} a_{ik} b_{kj} \tag{B.4}$$

Note that the number of terms in the summation is m, corresponding to the number of columns of **A** and the rows of **B**. Matrix multiplication in general is not commutative as is the case with scalars, i.e.,

$$\mathbf{AB} \neq \mathbf{BA}$$

Often the validity of this rule is obvious because the matrix dimensions are not conformable but even for square matrices commutation is not allowed.

Multiplication of a matrix by a scalar:

Each component of the matrix is multiplied by the scalar.

$$\begin{aligned} s\mathbf{A} &= \mathbf{B} \qquad \text{is obtained by} \\ s(a_{ij}) &= b_{ij} \end{aligned} \tag{B.5}$$

Transpose of a product of matrices

The transpose of a matrix product $(\mathbf{AB})^T$ is $(\mathbf{AB})^T = \mathbf{B}^T\mathbf{A}^T$. Likewise, $(\mathbf{ABC})^T = \mathbf{C}^T(\mathbf{AB})^T = \mathbf{C}^T\mathbf{B}^T\mathbf{A}^T$.

EXAMPLE B.1 MATRIX OPERATIONS

Consider a number of simple examples of these operations.

Multiplication:

For

$$\mathbf{A} = \underset{(3 \times 3)}{\begin{bmatrix} 1 & 0 & 0 \\ 1 & 1 & 0 \\ 1 & 1 & 1 \end{bmatrix}} \qquad \mathbf{B} = \underset{(3 \times 2)}{\begin{bmatrix} 1 & 2 \\ 3 & 4 \\ 5 & 6 \end{bmatrix}}$$

find **AB**.

Solution.

$$\underset{(3 \times 2)}{\mathbf{AB}} = \begin{bmatrix} 1(1) + 0(3) + 0(5) & 1(2) + 0(4) + 0(6) \\ 1(1) + 1(3) + 0(5) & 1(2) + 1(4) + 0(6) \\ 1(1) + 1(3) + 1(5) & 1(2) + 1(4) + 1(6) \end{bmatrix} = \begin{bmatrix} 1 & 2 \\ 4 & 6 \\ 9 & 12 \end{bmatrix}$$

Addition:

For

$$\mathbf{A} = \begin{bmatrix} 1 & 0 & 2 \\ 1 & -1 & 0 \\ 0 & 0 & 0 \end{bmatrix} \qquad \mathbf{B} = \begin{bmatrix} 1 & 5 & 1 \\ 4 & 4 & 4 \\ 1 & 0 & 1 \end{bmatrix}$$

Find **A** + **B**.

Solution.

$$\mathbf{A} + \mathbf{B} = \begin{bmatrix} 1+1 & 0+5 & 2+1 \\ 1+4 & -1+4 & 0+4 \\ 0+1 & 0+0 & 0+1 \end{bmatrix} = \begin{bmatrix} 2 & 5 & 3 \\ 5 & 3 & 4 \\ 1 & 0 & 1 \end{bmatrix}$$

Subtraction:

For

$$\mathbf{A} = \begin{bmatrix} 1 & 1 \\ 1 & 1 \end{bmatrix} \qquad \mathbf{B} = \begin{bmatrix} 2 & 6 \\ 1 & 3 \end{bmatrix}$$

find **A** − **B**.

Solution.

$$\mathbf{A} - \mathbf{B} = \begin{bmatrix} 1-2 & 1-6 \\ 1-1 & 1-3 \end{bmatrix} = \begin{bmatrix} -1 & -5 \\ 0 & -2 \end{bmatrix}$$

Transpose:

For

$$\mathbf{A} = \begin{bmatrix} 1 & 1 \\ 1 & 0 \end{bmatrix} \quad \text{and} \quad \mathbf{B} = \begin{bmatrix} 2 & 1 \\ 3 & 4 \end{bmatrix}$$

find $(\mathbf{AB})^T$.

Solution.

$$(\mathbf{AB})^T = \mathbf{B}^T\mathbf{A}^T = \begin{bmatrix} 2 & 3 \\ 1 & 4 \end{bmatrix}\begin{bmatrix} 1 & 1 \\ 1 & 0 \end{bmatrix} = \begin{bmatrix} 5 & 2 \\ 5 & 1 \end{bmatrix}$$

Multiplication of matrices by vectors:

A coordinate transformation can be performed by multiplying a matrix times a vector. If

$$\mathbf{A} = \begin{bmatrix} 1 & 1 & 2 \\ 2 & 0 & 3 \\ 4 & 8 & 4 \end{bmatrix} \quad \text{and} \quad \mathbf{x} = \begin{bmatrix} 1 \\ 1 \\ 1 \end{bmatrix}$$

find $\mathbf{y} = \mathbf{Ax}$,

$$\mathbf{y} = \begin{bmatrix} 1(1) + 1(1) + 2(1) \\ 2(1) + 0(1) + 3(1) \\ 4(1) + 8(1) + 4(1) \end{bmatrix} = \begin{bmatrix} 4 \\ 5 \\ 16 \end{bmatrix}$$

Note $\mathbf{y}$ has the same dimension as $\mathbf{x}$. We have transformed a point in three-dimensional space to another point in that same space.

Other commonly encountered vector-matrix products ($\mathbf{x}$ and $\mathbf{y}$ are n-component vectors) include

$$(a) \quad \mathbf{x}^T\mathbf{x} = \sum_{i=1}^{n} x_i^2 \qquad \text{(a scalar)} \tag{B.6}$$

$$(b) \quad \mathbf{x}^T\mathbf{y} = \langle \mathbf{x}, \mathbf{y} \rangle = \sum_{i=1}^{n} x_i y_i \tag{B.7}$$

Equation (B.7) is referred to as the inner product or dot product of two vectors. If the two vectors are *orthogonal*, then $\mathbf{x}^T\mathbf{y} = 0$. In two or three dimensions, this means that the vectors $\mathbf{x}$ and $\mathbf{y}$ are perpendicular to each other.

(c) $\mathbf{x}^T\mathbf{Ax}$ Here $\mathbf{A}$ is a square matrix of dimension $n \times n$ and the product is a scalar. If $\mathbf{A}$ is a diagonal matrix, then

$$\mathbf{x}^T\mathbf{Ax} = \sum_{i=1}^{n} a_{ii} x_i^2 \tag{B.8}$$

(*d*)

$$\mathbf{x}\mathbf{x}^T = \begin{bmatrix} x_1x_1 & x_1x_2 & \cdots & x_1x_n \\ \vdots & \vdots & & \vdots \\ x_nx_1 & & \cdots & x_nx_n \end{bmatrix} \tag{B.9}$$

Each vector has the dimensions $(n \times 1)$ and the matrix is square $(n \times n)$. Note that $\mathbf{x}\mathbf{x}^T$ is a matrix rather than a scalar (as with $\mathbf{x}^T\mathbf{x}$).

There is no matrix version of simple division, as with scalar quantities. Rather, the inverse of a matrix $(\mathbf{A}^{-1})$, which exists only for square matrices, is the closest analog to a divisor. An inverse matrix is defined such that $\mathbf{A}\mathbf{A}^{-1} = \mathbf{A}^{-1}\mathbf{A} = \mathbf{I}$ (all three matrices are $n \times n$). In scalar algebra, the equation $a \cdot b = c$ can be solved for b by simply multiplying both sides of the equation by $1/a$. For a matrix equation, the analog of solving

$$\mathbf{A}\mathbf{B} = \mathbf{C} \tag{B.10}$$

is to premultiply both sides by $\mathbf{A}^{-1}$:

$$\mathbf{A}^{-1}\mathbf{A}\mathbf{B} = \mathbf{A}^{-1}\mathbf{C}$$

$$\mathbf{I}\mathbf{B} = \mathbf{A}^{-1}\mathbf{C} \tag{B.11}$$

Since $\mathbf{I}\mathbf{B} = \mathbf{B}$, an explicit solution for $\mathbf{B}$ results. Note that the order of multiplication is critical because of the lack of commutation. Postmultiplication of both sides of Eq. (B.10) by $\mathbf{A}^{-1}$ is allowable but does not lead to a solution for $\mathbf{B}$.

How can the inverse of a matrix be calculated? To get the inverse of a diagonal matrix, you just assemble the inverse of each element on the main diagonal. If

$$\mathbf{A} = \begin{bmatrix} a_{11} & 0 & 0 \\ 0 & a_{22} & 0 \\ 0 & 0 & a_{33} \end{bmatrix}$$

then

$$\mathbf{A}^{-1} = \begin{bmatrix} 1/a_{11} & 0 & 0 \\ 0 & 1/a_{22} & 0 \\ 0 & 0 & 1/a_{33} \end{bmatrix}$$

The proof is evident by multiplication: $\mathbf{A}\mathbf{A}^{-1} = \mathbf{I}$.

For a general square matrix of size 2×2 or 3×3, the procedure is more involved and is discussed later in Examples B.3 and B.7.

The *determinant* (denoted by det $[\mathbf{A}]$ or $|\mathbf{A}|$) is reasonably easy to calculate by hand for matrices up to size 3×3:

$$\det \begin{bmatrix} a_{11} & a_{12} & a_{13} \\ a_{21} & a_{22} & a_{23} \\ a_{31} & a_{32} & a_{33} \end{bmatrix} = a_{11}a_{22}a_{33} + a_{12}a_{23}a_{31} + a_{13}a_{32}a_{21} - a_{31}a_{22}a_{13} - a_{32}a_{23}a_{11} - a_{33}a_{21}a_{12} \tag{B.12}$$

Another way to calculate the value of a determinant is to evaluate the cofactors of the determinant. The cofactor of an element a_{ij} of the matrix is found by first deleting from the original matrix the ith row and jth column corresponding to that element; the resulting array is the minor (M_{ij}) for that element and has dimension $(n - 1) \times (n - 1)$. The cofactor is defined as

$$c_{ij} = (-1)^{i+j} \det M_{ij} \tag{B.13}$$

The determinant of the original matrix is calculated by either

$$(a) \quad \sum_{j=1}^{n} a_{ij}c_{ij} \qquad (i \text{ fixed arbitrarily; row expansion}) \tag{B.14}$$

or

$$(b) \quad \sum_{i=1}^{n} a_{ij}c_{ij} \qquad (j \text{ fixed arbitrarily; column expansion}) \tag{B.15}$$

For example, if

$$\mathbf{A} = \begin{bmatrix} a_{11} & a_{12} \\ a_{21} & a_{22} \end{bmatrix}$$

an expansion of the first row gives

$$\det[\mathbf{A}] = a_{11}c_{11} + a_{12}c_{12}$$

$$c_{11} = (-1)^{1+1}a_{22} = a_{22}$$

$$c_{12} = (-1)^{1+2}a_{21} = -a_{21}$$

so that

$$\det[\mathbf{A}] = a_{11}a_{22} - a_{12}a_{21}$$

EXAMPLE B.2 CALCULATE THE VALUE OF A DETERMINANT USING COFACTORS

Calculate the det $\begin{bmatrix} 1 & 2 & 1 \\ 2 & 1 & 1 \\ 0 & 0 & 1 \end{bmatrix}$ using the first row as the expansion.

Solution.

$$\det[\mathbf{A}] = c_{11} + 2c_{12} + c_{13} = \det\begin{bmatrix} 1 & 1 \\ 0 & 1 \end{bmatrix} - 2\det\begin{bmatrix} 2 & 1 \\ 0 & 1 \end{bmatrix} + \det\begin{bmatrix} 2 & 1 \\ 0 & 0 \end{bmatrix}$$

$$1 = 1 - 4 + 0 = -3$$

It is actually easier to use the third row because of the two zeros that exist in the row.

$$\det \mathbf{A} = c_{33} = (-1)^{3+3}\begin{bmatrix} 1 & 2 \\ 2 & 1 \end{bmatrix} = -3$$

The *adjoint* of a matrix is constructed using the cofactors defined above. The elements $\bar{a}_{ij}$ of the adjoint matrix $\bar{\mathbf{A}}$ are defined as

$$\bar{a}_{ij} = c_{ji} \tag{B.16}$$

In other words, the adjoint matrix is the array composed of the transpose of the cofactors.

The adjoint of **A** can be used to directly calculate the inverse, $\mathbf{A}^{-1}$.

$$\mathbf{A}^{-1} = \frac{\text{adj}\,[\mathbf{A}]}{|\mathbf{A}|} \tag{B.17}$$

Note that the denominator of (B.17), the determinant of $\mathbf{A} \equiv |\mathbf{A}|$, is a scalar. If $|\mathbf{A}| = 0$, the inverse does not exist. A square matrix with determinant equal to zero is called a *singular* matrix. Conversely, for a nonsingular matrix **A**, $\det \mathbf{A} \neq 0$.

EXAMPLE B.3 CALCULATION OF THE INVERSE OF A MATRIX

Consider the following matrix and find its inverse.

$$\mathbf{A} = \begin{bmatrix} 1 & 4 \\ 2 & 1 \end{bmatrix} \qquad |\mathbf{A}| = 1 - 8 = -7$$

Solution. The cofactors are

$$c_{11} = 1 \qquad c_{12} = -2 \qquad c_{21} = -4 \qquad c_{22} = 1$$

$$\text{adj}\,\mathbf{A} = \begin{bmatrix} 1 & -4 \\ -2 & 1 \end{bmatrix}$$

$$\mathbf{A}^{-1} = \frac{1}{-7}\begin{bmatrix} 1 & -4 \\ -2 & 1 \end{bmatrix} = \begin{bmatrix} \dfrac{-1}{7} & \dfrac{4}{7} \\ \dfrac{2}{7} & \dfrac{-1}{7} \end{bmatrix}$$

The use of Eq. (B.17) for inversion is conceptually simple, but it is not a very efficient method for calculating the inverse matrix. A method based on use of row operations is discussed in Sec. B.3. For matrices of size larger than 3×3, we recommend that you use a computer library code to find $\mathbf{A}^{-1}$.

Another use for the matrix inverse is to express one set of variables in terms of another, an important operation in constrained optimization (see Chap. 8). For example, suppose **x** and **z** are two n-vectors which are related by

$$\mathbf{z} = \mathbf{A}\mathbf{x} \tag{B.18}$$

Then, to express **x** in terms of **z**, merely multiply both sides of (B.18) by $\mathbf{A}^{-1}$ (note that **A** must be $n \times n$):

$$\mathbf{A}^{-1}\mathbf{z} = \mathbf{x} \tag{B.19}$$

EXAMPLE B.4 RELATION OF VARIABLES

Suppose that

$$z_1 = x_1 + x_2$$

and

$$z_2 = 2x_1 + x_2$$

What are x_1 and x_2 in terms of z_1 and z_2?

Solution. Let

$$\mathbf{z} = \begin{bmatrix} z_1 \\ z_2 \end{bmatrix} \quad \text{and} \quad \mathbf{x} = \begin{bmatrix} x_1 \\ x_2 \end{bmatrix}$$

Therefore $\mathbf{z} = \mathbf{A}\mathbf{x}$, where

$$\mathbf{A} = \begin{bmatrix} 1 & 1 \\ 2 & 1 \end{bmatrix}$$

The inverse of $\mathbf{A}$ is

$$\mathbf{A}^{-1} = \begin{bmatrix} -1 & 1 \\ 2 & -1 \end{bmatrix}$$

hence $\mathbf{x} = \mathbf{A}^{-1}\mathbf{z}$ or

$$x_1 = -z_1 + z_2$$
$$x_2 = 2z_1 - z_2$$

The inverse matrix also can be employed in the solution of linear algebraic equations,

$$\mathbf{A}\mathbf{x} = \mathbf{b} \tag{B.20}$$

which arise in many applications of engineering as well as in optimization theory. In order to have a unique solution to Eq. (B.20), there must be the same number of independent equations as unknown variables. Note that the number of equations is equal to the number of rows of $\mathbf{A}$, while the number of unknowns is equal to the number of columns of $\mathbf{A}$.

With the inverse matrix, you can solve directly for $\mathbf{x}$:

$$\mathbf{x} = \mathbf{A}^{-1}\mathbf{b} \tag{B.21}$$

While this is a conceptually convenient way to solve for $\mathbf{x}$, it is not necessarily the most efficient method for doing so. We shall return to the matter of solving linear equations in Sec. B.4.

The final matrix characteristics covered here involve differentiation of function of a vector with respect to a vector. Suppose $f(\mathbf{x})$ is a scalar function of n variables $(x_1, x_2, \ldots, x_n)$. The first partial derivative of $f(\mathbf{x})$ with respect to $\mathbf{x}$ is

$$\frac{\partial f}{\partial \mathbf{x}} = \nabla_{\mathbf{x}} f = \left[\frac{\partial f}{\partial x_1} \quad \frac{\partial f}{\partial x_2} \quad \cdots \quad \frac{\partial f}{\partial x_n} \right]^T$$

For a vector function $\mathbf{h}(\mathbf{x})$, such as occurs in a series of nonlinear multivariable constraints

$$\begin{aligned} h_1(x_1, x_2, \ldots, x_n) &= 0 \\ h_2(x_1, x_2, \ldots, x_n) &= 0 \\ &\vdots \\ h_m(x_1, x_2, \ldots, x_n) &= 0 \end{aligned}$$

the matrix of first partial derivatives, called the Jacobian matrix, is

$$\mathbf{J} = \frac{\partial \mathbf{h}}{\partial \mathbf{x}} = \begin{bmatrix} \dfrac{\partial h_1}{\partial x_1} & \dfrac{\partial h_1}{\partial x_2} & \cdots & \dfrac{\partial h_1}{\partial x_n} \\ \vdots & \vdots & & \vdots \\ \dfrac{\partial h_m}{\partial x_1} & \dfrac{\partial h_m}{\partial x_2} & \cdots & \dfrac{\partial h_m}{\partial x_n} \end{bmatrix}$$

For a scalar function, the matrix of second derivatives, called the Hessian matrix, is

$$\mathbf{H}(\mathbf{x}) \equiv \nabla^2 f = \begin{bmatrix} \dfrac{\partial^2 f}{\partial x_1^2} & \dfrac{\partial^2 f}{\partial x_1 \, \partial x_2} & \cdots & \dfrac{\partial^2 f}{\partial x_1 \, \partial x_n} \\ \dfrac{\partial^2 f}{\partial x_2 \, \partial x_1} & \dfrac{\partial^2 f}{\partial x_2^2} & \cdots & \dfrac{\partial^2 f}{\partial x_2 \, \partial x_n} \\ \vdots & & & \vdots \\ \dfrac{\partial^2 f}{\partial x_n \, \partial x_1} & \cdots & \cdots & \dfrac{\partial^2 f}{\partial x_n^2} \end{bmatrix}$$

The use of this matrix and its eigenvalue properties is discussed in several chapters. For continuously differentiable functions, $\nabla^2 f$ is symmetric.

B.3 LINEAR INDEPENDENCE AND ROW OPERATIONS

As mentioned above, singular matrices have a determinant of zero value. This outcome occurs when a row or column contains all zeros or when a row (or column) in the matrix is linearly dependent on one or more of the other rows (or columns). It can be shown that for a square matrix, row dependence implies column dependence. By definition the columns of $\mathbf{A}$, $\mathbf{a}_j$, are linearly independent if

$$\sum_{j=1}^{n} d_j \mathbf{a}_j = \mathbf{0} \qquad \text{only if } d_j = 0 \text{ for all } j \tag{B.22}$$

Conversely, linear dependence occurs when some nonzero set of values for d_j satisfies Eq. (B.22). The *rank* of a matrix is defined as the number of linearly independent columns ($\leq n$).

EXAMPLE B.5 LINEAR INDEPENDENCE AND THE RANK OF A MATRIX

Calculate the rank of

$$\mathbf{A} = \begin{bmatrix} 1 & -1 & 1 \\ 1 & 0 & 1 \\ 2 & -2 & 2 \end{bmatrix}$$

Solution. Note that columns 1 and 3 are identical. Likewise the third row can be formed by multiplying the first row by 2. Equation (B.22) is

$$d_1 \begin{bmatrix} 1 \\ 1 \\ 2 \end{bmatrix} + d_2 \begin{bmatrix} -1 \\ 0 \\ -2 \end{bmatrix} + d_3 \begin{bmatrix} 1 \\ 1 \\ 2 \end{bmatrix} = \mathbf{0}$$

One solution of (B.22) that exists is $d_1 = 1$, $d_2 = 0$, $d_3 = -1$. Since a nontrivial (nonzero) solution exists, then the matrix has one dependent and two independent columns, and the rank ≤ 2 (here 2). The determinant is zero, as can be readily verified using Eq. (B.12).

In general for a matrix, the determination of linear independence cannot be performed by inspection. For large matrices, rather than solving the set of linear equations (B.22), elementary row or column operations can be used to demonstrate linear independence. These operations involve adding some multiple of one row to another row, analogous to the types of algebraic operations (discussed later) which are used to solve simultaneous equations. The transformed matrix is desired to be of upper triangular form (zeros below the main diagonal). The value of the determinant of $\mathbf{A}$ is invariant under these row (or column) operations. Implications with respect to linear independence and the use of determinants for equation-solving are discussed in Sec. B.4.

EXAMPLE B.6 USE OF ROW OPERATIONS

Use row operations to determine if the matrix

$$\mathbf{A} = \begin{bmatrix} 1 & 0 & 1 \\ 1 & 2 & 2 \\ 3 & 4 & 5 \end{bmatrix}$$

is nonsingular, i.e., composed of linearly independent columns.

Solution. First create zeros in the a_{21} and a_{31} position by multiplication/addition. The necessary transformations are

1. Multiply row 1 by (-1); add to row 2

$$\mathbf{C}_1 = \begin{bmatrix} 1 & 0 & 1 \\ 0 & 2 & 1 \\ 3 & 4 & 5 \end{bmatrix}$$

2. Multiply row 1 by (-3); add to row 3

$$\mathbf{C}_2 = \begin{bmatrix} 1 & 0 & 1 \\ 0 & 2 & 1 \\ 0 & 4 & 2 \end{bmatrix}$$

Next use row 2 to create a zero in a_{32}.

3. Multiply row 2 by (-2); add to row 3

$$\mathbf{C}_3 = \begin{bmatrix} 1 & 0 & 1 \\ 0 & 2 & 1 \\ 0 & 0 & 0 \end{bmatrix}$$

Note that neither rows 1 or 2 are changed in this step. The appearance of a row with all zero elements indicates that the matrix is singular (det $[\mathbf{A}] = 0$).

Row operations can also be used to obtain an inverse matrix. Suppose we augment $\mathbf{A}$ with an identity matrix $\mathbf{I}$ of the same dimension; then multiply the augmented matrix by $\mathbf{A}^{-1}$:

$$\mathbf{A}^{-1}[\mathbf{A} \mid \mathbf{I}] = [\mathbf{I} \mid \mathbf{A}^{-1}] \tag{B.23}$$

If $\mathbf{A}$ is transformed by row operations to obtain $\mathbf{I}$, $\mathbf{A}^{-1}$ occurs in the augmented part of the matrix.

EXAMPLE B.7 CALCULATION OF INVERSE MATRIX

Verify the results of Example B.3 using row operations.

Solution. Form the augmented matrix

$$\mathbf{C}_0 = \begin{bmatrix} 1 & 4 & 1 & 0 \\ 2 & 1 & 0 & 1 \end{bmatrix}$$

Successive transformations would be

$$\mathbf{C}_1 = \begin{bmatrix} 1 & 4 & 1 & 0 \\ 0 & -7 & -2 & 1 \end{bmatrix} \quad \mathbf{C}_2 = \begin{bmatrix} 1 & 4 & 1 & 0 \\ 0 & 1 & 2/7 & -1/7 \end{bmatrix}$$

$$\mathbf{C}_3 = \begin{bmatrix} 1 & 0 & -1/7 & 4/7 \\ 0 & 1 & 2/7 & -1/7 \end{bmatrix}$$

Therefore the inverse of $\mathbf{A}$ is

$$\mathbf{A}^{-1} = \begin{bmatrix} -\frac{1}{7} & \frac{4}{7} \\ \frac{2}{7} & -\frac{1}{7} \end{bmatrix}$$

B.4 SOLUTION OF LINEAR EQUATIONS

The need to solve sets of linear equations arises in many optimization applications. Consider Eq. (B.20) where $\mathbf{A}$ is an $n \times n$ matrix corresponding to the coefficients in n equations in n unknowns. Since $\mathbf{x} = \mathbf{A}^{-1}\mathbf{b}$, then from (B.17) $|\mathbf{A}|$ must be nonzero; $\mathbf{A}$ must have rank n, that is, no linearly dependent rows or columns exist, for a unique solution. Let us illustrate two cases where $|\mathbf{A}| = 0$:

$$2x_1 + 2x_2 = 6$$
$$x_1 + x_2 = 5$$

or

$$\begin{bmatrix} 2 & 2 \\ 1 & 1 \end{bmatrix} \begin{bmatrix} x_1 \\ x_2 \end{bmatrix} = \begin{bmatrix} 6 \\ 5 \end{bmatrix}$$

It is obvious that only one linearly independent column or row exists, and $|\mathbf{A}|$ is zero. Note that there is no solution to this set of equations. As a second case, suppose $\mathbf{b}$ were changed to $\begin{bmatrix} 6 \\ 3 \end{bmatrix}$. Here an infinite number of solutions can be obtained, but no unique solution exists.

Degenerate cases such as those above are not frequently encountered. More often, $|\mathbf{A}| \neq 0$. Let

$$\mathbf{A} = \begin{bmatrix} 2 & 1 \\ 1 & 2 \end{bmatrix} \quad \text{and} \quad \mathbf{b} = \begin{bmatrix} 1 \\ 0 \end{bmatrix}$$

or

$$2x_1 + x_2 = 1 \tag{B.24a}$$

$$x_1 + 2x_2 = 0 \tag{B.24b}$$

By algebraic substitution, x_1 and x_2 can be found. Multiply (B.24*a*) by (-0.5) and add this equation to (B.24*b*),

$$2x_1 + x_2 = 1 \tag{B.24c}$$

$$0 + 1.5x_2 = -0.5 \tag{B.24d}$$

Solve (B.24*d*) for $x_2 = -0.333$. This result can be substituted into (B.24*c*) to obtain $x_1 = 0.667$.

The steps employed in Eqs. (B.24) are equivalent to row operations. The use of row operations to simplify linear algebraic equations is the basis for Gaussian elimination. Gaussian elimination transforms the original matrix into upper triangular form, i.e., all components of the matrix below the main diagonal are zero. Let us illustrate the process by solving a set of three equations in three unknowns for $\mathbf{x}$:

EXAMPLE B.8 SOLUTION OF SIMULTANEOUS LINEAR EQUATIONS

Solve for $\mathbf{x}$ given $\mathbf{A}$ and $\mathbf{b}$.

$$\mathbf{A} = \begin{bmatrix} 1 & 0 & 1 \\ 1 & 2 & 2 \\ 2 & 1 & 1 \end{bmatrix} \qquad \mathbf{b} = \begin{bmatrix} 1 \\ 0 \\ 2 \end{bmatrix} \qquad \mathbf{x} = \begin{bmatrix} x_1 \\ x_2 \\ x_3 \end{bmatrix}$$

Solution. First a composite matrix from $\mathbf{A}$ and $\mathbf{b}$ is constructed:

$$\mathbf{C}_0 = [\mathbf{A} \mid \mathbf{b}] = \begin{bmatrix} 1 & 0 & 1 & 1 \\ 1 & 2 & 2 & 0 \\ 2 & 1 & 1 & 2 \end{bmatrix}$$

Carry out row operations, keeping the first row intact; successive matrices are

$$\mathbf{C}_1 = \begin{bmatrix} 1 & 0 & 1 & 1 \\ 0 & 2 & 1 & -1 \\ 2 & 1 & 1 & 2 \end{bmatrix} \qquad \mathbf{C}_2 = \begin{bmatrix} 1 & 0 & 1 & 1 \\ 0 & 2 & 1 & -1 \\ 0 & 1 & -1 & 0 \end{bmatrix}$$

Next, with the second row in $\mathbf{C}_2$ kept intact, the upper triangular form is achieved by operating on the third row:

$$\mathbf{C}_3 = \begin{bmatrix} 1 & 0 & 1 & 1 \\ 0 & 2 & 1 & -1 \\ 0 & 0 & -1.5 & 0.5 \end{bmatrix}$$

$\mathbf{C}_3$ can now be converted to the form of algebraic equations:

$$x_1 + x_3 = 1$$

$$2x_2 + x_3 = -1$$

$$-1.5x_3 = 0.5$$

which can be solved stage by stage starting with the last row to get $x_3 = -0.333$, $x_2 = -0.333$, $x_1 = 1.333$.

Gaussian elimination is a very efficient method for solving n equations in n unknowns, and this algorithm is readily available through most computer centers and/or services. For solution of linear equations, this method is preferred computationally over the use of the matrix inverse. For hand calculations, Cramer's rule is also popular (see Fadeeva, 1959).

The determinant of $\mathbf{A}$ is unchanged by the row operations used in Gaussian elimination. Take the first three columns of $\mathbf{C}_3$ above. The determinant is simply the product of the diagonal terms. If none of the diagonal terms are zero when the matrix is reformulated as upper triangular, then $|\mathbf{A}| \neq 0$ and a solution exists. If $|\mathbf{A}| = 0$, there is no solution to the original set of equations.

A set of nonlinear equations can be solved by combining a Taylor series linearization with the linear equation-solving approach discussed above. For solving a single nonlinear equation, $h(x) = 0$, Newton's method applied to a function of a single variable is the well-known iterative procedure

$$x^{k+1} - x^k \equiv \Delta x^k = -\frac{h(x^k)}{dh(x^k)/dx} \tag{B.25}$$

or

$$\left[\frac{dh(x^k)}{dx}\right](\Delta x^k) = -h(x^k)$$

where k is the iteration number and Δx^k is the correction to the previous value, x^k. Similarly, a set of nonlinear equations, $\mathbf{h}(\mathbf{x}) = \mathbf{0}$, can be solved iteratively using Newton's method, by solving a set of linearized equations of the form $\mathbf{Ax} = \mathbf{b}$:

$$\frac{\partial \mathbf{h}(\mathbf{x}^k)}{\partial \mathbf{x}} \cdot \Delta \mathbf{x}^k = -\mathbf{h}(\mathbf{x}^k) \tag{B.26}$$

Note that the Jacobian matrix $\partial \mathbf{h}(\mathbf{x}^k)/\partial \mathbf{x}$ on the left-hand side of (B.26) is analogous to $\mathbf{A}$ in Eq. (B.20) and $\Delta \mathbf{x}^k$ is analogous to $\mathbf{x}$. In order to compute the correction vector $\Delta \mathbf{x}$, $\partial \mathbf{h}(\mathbf{x}^k)/\partial \mathbf{x}$ must be nonsingular. However, there is no guarantee even then that Newton's method will converge to an $\mathbf{x}$ which satisfies $\mathbf{h}(\mathbf{x}) = \mathbf{0}$.

In solving sets of simultaneous linear equations, the "condition" of the matrix is quite important. If some elements are quite large and some are quite small (but nonzero), numerical roundoff and/or truncation in a computer can have a significant effect on accuracy of the solution. A type of matrix is referred to as "ill-conditioned" if it is nearly singular (equivalent to the scalar division by 0). A common measure of the degree of ill-conditioning is the condition number, namely the ratio of the eigenvalues with largest (α_h) and smallest (α_l) modulus:

$$\text{Condition number} = \frac{|\alpha_h|}{|\alpha_l|} \tag{B.27}$$

The bigger the ratio, the worse the conditioning; 1 is best. The calculation of eigenvalues are discussed in the next section. In general, as the dimension of the matrix increases, numerical accuracy of the elements is diminished. One technique to solve ill-conditioned sets of equations that has some advantages in speed and accuracy over Gaussian elimination is called "L-U decomposition" (Forsythe and Moler, 1967), in which the original matrix is decomposed into upper and lower triangular forms.

B.5 EIGENVALUES, EIGENVECTORS

An $n \times n$ matrix has n eigenvalues. We define an n-vector, $\mathbf{v}$, the eigenvector, which is associated with an eigenvalue α such that

$$\mathbf{A}\mathbf{v} = \alpha\mathbf{v} \tag{B.28}$$

Hence the product of the matrix $\mathbf{A}$ multiplying the eigenvector $\mathbf{v}$ is the same as the product obtained by multiplying the vector $\mathbf{v}$ by the scalar eigenvalue, α. One eigenvector exists for each of the n eigenvalues. Eigenvalues and eigenvectors provide unambiguous information about the nature of functions used in optimization. If all eigenvalues of $\mathbf{A}$ are positive, then $\mathbf{A}$ is positive definite. If all $\alpha_i < 0$, then $\mathbf{A}$ is negative definite. See Chap. 4 for a more complete discussion of definiteness and how it relates to convexity and concavity.

If we rearrange (B.28) (note that the identity matrix must be introduced to maintain conformable matrices),

$$(\mathbf{A} - \alpha\mathbf{I})\mathbf{v} = \mathbf{0} \tag{B.29}$$

$(\mathbf{A} - \alpha\mathbf{I})$ in (B.29) has the unknown variable α subtracted from each diagonal element of $\mathbf{A}$. Equation (B.29) is a set of linear algebraic equations where $\mathbf{v}$ is the unknown vector. However, because the right-hand side of (B.29) is zero, either $\mathbf{v} = \mathbf{0}$ (the trivial solution), or a nonunique solution exists. For example in

$$\begin{aligned} 2v_1 + 2v_2 &= 0 \\ v_1 + v_2 &= 0 \end{aligned}$$

the det $[\mathbf{A}] = 0$, and the solution is nonunique, that is, $v_1 = -v_2$. The equations are redundant. However, if one of the coefficients of v_1 or v_2 in (B.29) changes, then the only solution is $v_1 = v_2 = 0$ (the trivial solution).

The determinant of $(\mathbf{A} - \alpha\mathbf{I})$ must be zero in order for a nontrivial solution ($\mathbf{v} \neq \mathbf{0}$) to exist. Let us illustrate this idea with a (2×2) matrix:

$$\mathbf{A} = \begin{bmatrix} 1 & 2 \\ 2 & 1 \end{bmatrix} \qquad (\mathbf{A} - \alpha\mathbf{I}) = \begin{bmatrix} 1-\alpha & 2 \\ 2 & 1-\alpha \end{bmatrix}$$

$$\det \begin{bmatrix} 1-\alpha & 2 \\ 2 & 1-\alpha \end{bmatrix} = (1-\alpha)^2 - 4 = \alpha^2 - 2\alpha - 3 = 0 \tag{B.30}$$

Equation (B.30) determines values of α which yield a nontrivial solution. Factoring (B.30)

$$(\alpha - 3)(\alpha + 1) = 0 \qquad \alpha = 3, -1$$

Therefore, the eigenvalues are 3 and -1. Note that for $\alpha = 3$,

$$\mathbf{A} - \alpha\mathbf{I} = \begin{bmatrix} -2 & 2 \\ 2 & -2 \end{bmatrix}$$

and for $\alpha = -1$,

$$\mathbf{A} - \alpha\mathbf{I} = \begin{bmatrix} 2 & 2 \\ 2 & 2 \end{bmatrix}$$

both of which are singular matrices.

For each eigenvalue there exists a corresponding eigenvector. For $\alpha_1 = 3$, Eq. (B.29) becomes

$$\begin{bmatrix} (1-3) & 2 \\ 2 & (1-3) \end{bmatrix} \begin{bmatrix} v_{11} \\ v_{12} \end{bmatrix} = 0$$

$$-2v_{11} + 2v_{12} = 0$$

$$2v_{11} - 2v_{12} = 0$$

Note that these equations are equivalent and cannot be solved uniquely; the solution to both equations is $v_{11} = v_{12}$. Thus, the eigenvector has direction but not length. The direction of the eigenvector can be specified by choosing v_{11} and calculating v_{12}. For example, let $v_{11} = 1$. Then $\mathbf{v}_1 = \begin{bmatrix} 1 \\ 1 \end{bmatrix}$. The magnitude of $\mathbf{v}_1$ cannot be determined uniquely. Similarly, for $\alpha_2 = -1$, $\mathbf{v}_2 = \begin{bmatrix} 1 \\ -1 \end{bmatrix}$ is a solution of (B.29).

For a general $n \times n$ matrix, an nth order polynomial results from solving $\det(\mathbf{A} - \alpha\mathbf{I}) = 0$. This polynomial will have n roots, and some of the roots may be imaginary numbers. A computer program can be used to generate the polynomial and factor it using a root-finding technique, such as Newton's method. However, more efficient iterative techniques can be found in computer libraries to calculate both α and $\mathbf{v}$.

Principal Minors

In Chap. 4 we discuss the definitions of convexity and concavity in terms of eigenvalues; an equivalent definition using determinants of principal minors is also provided. A principal minor of $\mathbf{A}$ of order k is a submatrix found by deleting any $n - k$ columns (and their corresponding rows) from the matrix. The leading principal minor of order k is found by deleting the last $n - k$ columns and rows. In Example B.2, the leading principal minor (order 1) is 1; the leading principal minor (order 2) is $\begin{bmatrix} 1 & 2 \\ 2 & 1 \end{bmatrix}$, and for order 3 the minor is the 3×3 matrix itself.

REFERENCES

Amundson, N. R., *Mathematic Methods in Chemical Engineering: Matrices and Their Application*, Prentice-Hall, Englewood Cliffs, New Jersey (1966).
Campbell, H. G., *Matrices with Applications*, Prentice-Hall, Englewood Cliffs, New Jersey (1968).
Fadeeva, N. V., *Computational Methods of Linear Algebra*, Dover Publications, New York (1959).
Forsythe, G., and C. B. Moler, *Computer Solution of Linear Algebraic Equations*, Prentice-Hall, Englewood Cliffs, New Jersey (1967).
Pipes, L. A., *Matrix Methods for Engineers*, Prentice-Hall, Englewood Cliffs, New Jersey (1963).

SUPPLEMENTARY REFERENCES

Graybill, F. A., *Introduction to Matrices*, Wadsworth, Belmont, California (1969).
Stewart, G. W., *Introduction to Matrix Computations*, Academic Press, New York (1973).

PROBLEMS

B.1. For

$$\mathbf{A} = \begin{bmatrix} 1 & 1 \\ 2 & 1 \end{bmatrix} \qquad \mathbf{B} = \begin{bmatrix} 0 & 4 \\ 1 & 3 \end{bmatrix}$$

Find
(*a*) **AB** and **BA** (compare)
(*b*) $\mathbf{A}^T\mathbf{B}$
(*c*) $\mathbf{A} + \mathbf{B}$
(*d*) $\mathbf{A} - \mathbf{B}$
(*e*) det **A**, det **B**
(*f*) Adj **A**, Adj **B**
(*g*) $\mathbf{A}^{-1}$, $\mathbf{B}^{-1}$ (verify the answer)

B.2. Solve $\mathbf{Ax} = \mathbf{b}$, for **x**, where

$$\mathbf{A} = \begin{bmatrix} 1 & 2 & 3 \\ 3 & 2 & 1 \\ 1 & 0 & 1 \end{bmatrix} \qquad \mathbf{b} = \begin{bmatrix} 1 \\ 2 \\ 1 \end{bmatrix}$$

Use
(*a*) Gaussian elimination and demonstrate **A** is nonsingular. Check to see that the determinant does not change after each row operation.
(*b*) Use $\mathbf{x} = \mathbf{A}^{-1}\mathbf{b}$.
(*c*) Use Cramer's rule.

B.3. Suppose

$$z_1 = 3x_1 + x_3$$
$$z_2 = x_1 + x_2 + x_3 \qquad \mathbf{z} = \mathbf{Ax}$$
$$z_3 = 2x_2 + x_3$$

Find equations for x_1, x_2, and x_3 in terms of z_1, z_2, z_3. Use an algebraic method first; check the result using $\mathbf{A}^{-1}$.

B.4. For

$$\mathbf{x}_1 = \begin{bmatrix} 1 \\ 1 \\ 2 \end{bmatrix} \qquad \mathbf{x}_2 = \begin{bmatrix} 1 \\ 0 \\ 2 \end{bmatrix} \qquad \mathbf{A} = \begin{bmatrix} 1 & 0 & 0 \\ 0 & 2 & 0 \\ 0 & 0 & 4 \end{bmatrix}$$

find the magnitude (norm) of each vector.
What is $\mathbf{x}_1^T\mathbf{x}_2$? $\mathbf{x}_1\mathbf{x}_2^T$? $\mathbf{x}_1^T\mathbf{A}\mathbf{x}_1$?
Find a vector $\mathbf{x}_3$ which is orthogonal to $\mathbf{x}_1(\mathbf{x}_1^T\mathbf{x}_3 = 0)$. Are $\mathbf{x}_1$, $\mathbf{x}_2$, and $\mathbf{x}_3$ linearly independent?

B.5. For

$$\mathbf{A} = \begin{bmatrix} 1 & 1 & 2 \\ -1 & 0 & 1 \\ -2 & -3 & 1 \end{bmatrix}$$

calculate det $\mathbf{A}$ using expansion by minors of the second row. Repeat with the third column.

B.6. Calculate the eigenvalues and eigenvectors of $\begin{bmatrix} 0 & 1 \\ 1 & 4 \end{bmatrix}$. Repeat for

$$\begin{bmatrix} 2 & 0 & 0 \\ 0 & 3 & 0 \\ 0 & 0 & 2 \end{bmatrix}.$$

B.7. Show that for a 2×2 symmetric matrix, the eigenvalues must be real (do not contain imaginary components). Develop a 2×2 nonsymmetric matrix which has complex eigenvalues.

B.8. A technique called LU decomposition can be used to solve sets of linear algebraic equations. **L** and **U** are lower and upper triangular matrices, respectively. A lower triangular matrix has zeros above the main diagonal; an upper triangular matrix has zeros below the main diagonal. Any matrix **A** can be formed by the product of $\mathbf{L} \cdot \mathbf{U}$.
(*a*) For

$$\mathbf{A} = \begin{bmatrix} 1 & 1 & 0 \\ 2 & 3 & 1 \\ 1 & 0 & 1 \end{bmatrix}$$

find some **L** and **U** which satisfy $\mathbf{L} \cdot \mathbf{U} = \mathbf{A}$.
(*b*) If $\mathbf{A}\mathbf{x} = \mathbf{b}$, $\mathbf{L}\mathbf{U}\mathbf{x} = \mathbf{b}$ or $\mathbf{U}\mathbf{x} = \mathbf{L}^{-1}\mathbf{b} = \hat{\mathbf{b}}$.
Let

$$\mathbf{b} = \begin{bmatrix} 1 \\ 1 \\ 2 \end{bmatrix}$$

Calculate $\mathbf{L}^{-1}$ and $\hat{\mathbf{b}}$. Then solve for $\mathbf{x}$ using substitution from the upper triangular matrix, **U**.

B.9. You are to solve the two nonlinear equations,

$$x_1^2 + x_2^2 = 8$$

$$x_1 x_2 = 4$$

using the Newton-Raphson method. Suggested starting points are (0, 1) and (4, 4).

APPENDIX C

RANGE SPACE AND NULL SPACE AND RELATION TO REDUCED GRADIENT AND PROJECTION METHODS

One of the important concepts in constrained optimization is that of the range and null space. Given an $n \times m$ matrix $\mathbf{B}$, we denote each of the columns of $\mathbf{B}$ as a vector $\mathbf{b}_j$ in Euclidean n dimensional space

$$\mathbf{B} = [\mathbf{b}_1 \quad \mathbf{b}_2 \quad \cdots \quad \mathbf{b}_m] \qquad \mathbf{b}_j = [b_{1j} \quad b_{2j} \quad \cdots \quad b_{nj}]^T$$

The set of all vectors that can be expressed as linear combinations of $\mathbf{b}_1, \mathbf{b}_2, \ldots, \mathbf{b}_m$ form a linear vector space spanned by the m column vectors termed the *range space* $\mathscr{R}$ of $\mathbf{B}$; it is a subspace of the n-dimensional space. The range space dimension, or rank r, equals the number of linearly independent columns of $\mathbf{B}$.

Any set of linearly independent vectors in $\mathscr{R}$ are said to form a (non-unique) *basis* for the subspace $\mathscr{R}$. By way of example, for the matrix $\mathbf{B}$

$$\mathbf{B} = [\mathbf{b}_1 \quad \mathbf{b}_2 \quad \mathbf{b}_3] = \begin{bmatrix} 1 & 1 & 1 \\ 0 & 1 & -1 \\ 2 & 0 & 4 \end{bmatrix}$$

the rank of $\mathbf{B}$ is $r = 2$, which can be shown by row operations (see Sec. B.3). The range space $\mathscr{R}$ of $\mathbf{B}$ can be generated by taking linear combinations of columns 1 and 2, which form a basis to get the vectors in the range space

$$\mathbf{x} = \sum_{j=1}^{2} u_j \mathbf{b}_j$$

or in general,

$$\mathbf{x} = \mathbf{B}\mathbf{u} \tag{C.1}$$

where $\mathbf{u} = [u_1 \quad u_2 \quad \cdots \quad u_n]^T$, and u_j is just the coordinate of $\mathbf{x}$ with respect to the basis vector $\mathbf{b}_j$. Figure C.1 illustrates the concept in two dimensions, and in particular illustrates an *orthogonal basis*, i.e., all the basis vectors are orthogonal to each other, $\mathbf{b}_2^T \mathbf{b}_1 = 0$.

For every subspace $\mathscr{R}$ in the n-dimensional space in which the vectors $\mathbf{x}$ lie, there is a complementary subspace $\mathscr{N}$ termed the *null space* whose members (vectors) $\mathbf{y}$ are defined by

$$\mathbf{x}^T \mathbf{y} = \mathbf{0} \tag{C.2}$$

In other words, the vectors $\mathbf{y}$ in the null space are orthogonal to all the vectors $\mathbf{x}$ in the range space. Often the subspace $\mathscr{N}$ is termed the orthogonal complement of $\mathscr{R}$, and the two subspaces are disjoint. If the dimension of $\mathscr{R}$ is r, the

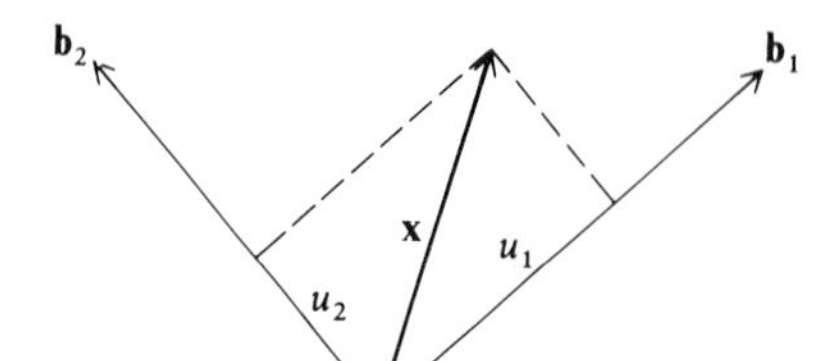

Figure C.1 Every vector $\mathbf{x}$ lying in the range space $\mathscr{R}$ can be expressed as a linear combination of the basis vectors $\mathbf{b}_j$.

dimension of $\mathcal{N}$ is $n - r$. Furthermore, $\mathcal{R}$ and $\mathcal{N}$ span the whole space of n-dimensions so that any vector $\mathbf{x}$ can be expressed as the sum of

$$\mathbf{x}_{\mathcal{R}} + \mathbf{x}_{\mathcal{N}} = \mathbf{x} \tag{C.3}$$

and the basis of $\mathcal{R}$ plus the basis of $\mathcal{N}$ will span all of the n-dimensional space. For example, if the (nonunique) basis of $\mathcal{R}$ is (the matrix has a rank of 2)

$$\begin{matrix} \mathbf{b}_1 & \mathbf{b}_2 \end{matrix}$$
$$\begin{bmatrix} 1 & 1 \\ 0 & 1 \\ 2 & 0 \end{bmatrix}$$

then a (nonunique) basis for $\mathcal{N}$ can be determined from $\mathbf{B}^T\mathbf{y} = 0$

$$\begin{bmatrix} 1 & 0 & 2 \\ 1 & 1 & 0 \end{bmatrix} \begin{bmatrix} y_1 \\ y_2 \\ y_3 \end{bmatrix} = 0$$

such as $\mathbf{y} = \begin{bmatrix} 2 \\ -2 \\ -1 \end{bmatrix}$ which has a rank of 1.

EXAMPLE C.1 THE RANGE SPACE AND NULL SPACE

In a nonlinear programming problem, each linear (or linearized) equality constraint $\mathbf{Ax} = \mathbf{b}$ can be transformed to $\mathbf{a}_j\mathbf{x} = 0$, $j = 1, 2, \ldots, m$, where $\mathbf{x}$ is the vector of variables, x_i, $i = 1, 2, \ldots, n$, and $\mathbf{a}_j$ represents the set of coefficients in equation j, a row vector. The full set of constraints is organized as illustrated by the matrix and vectors in Fig. EC.1a so that $\mathbf{Ax} = \mathbf{0}$. Each equation defines a hyperplane passing through the origin. Figure EC.1b illustrates the case for $n = 2$ and $j = 1$. In Fig. EC.1b the equation is $\mathbf{a}_1\mathbf{x}_{\mathcal{N}} = 0$ or

$$[2 \quad 1]\begin{bmatrix} x_1 \\ x_2 \end{bmatrix} = 0$$

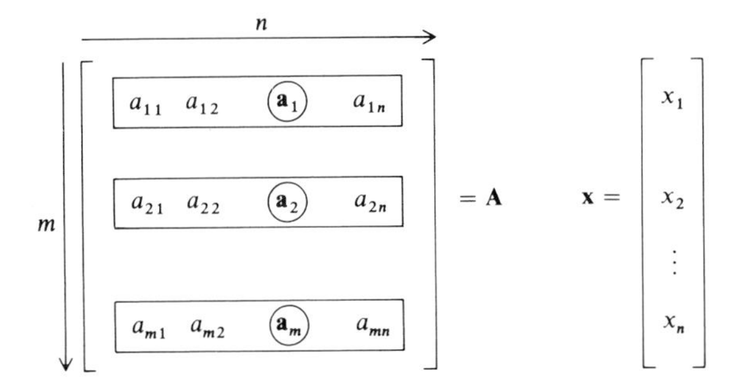

Figure EC.1*a*

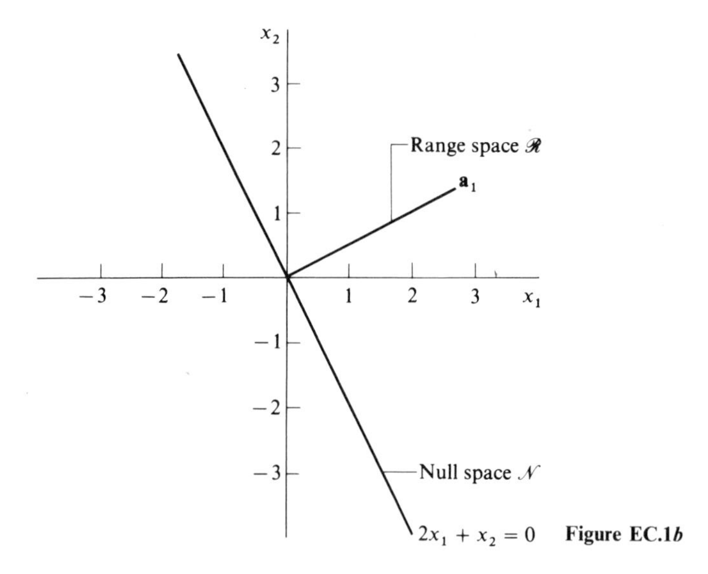

Figure EC.1*b*

The null space falls along the line $2x_1 + x_2 = 0$. A range space is the vector $\mathbf{a}_1$. Any other vector orthogonal to the line could comprise the range space.

If you had two linear equations as constraints, $\mathbf{a}_1\mathbf{x} = 0$ and $\mathbf{a}_2\mathbf{x} = 0$, then the only vector that satisfies both equations is $\mathbf{x} = [0 \quad 0]^T$, that is, the null space becomes a point.

In three-dimensional space for $\mathbf{x}$, if $j = 1$, the constraint is a plane and the null space has two dimensions; if $j = 1, 2$, the constraints intersect to form a line and the null space has one dimension; if $j = 1, 2, 3$, the constraints intersect to form a point and the null space is a point.

For more details on range and null space, see Stewart (1973) and Graybill (1969).

For any vector $\mathbf{x}$ whose elements are composed of the variables $x_1, \ldots, x_n$, we know $\mathbf{x}$ can be split among the range space and the null space

$$\mathbf{x} = \mathbf{x}_{\mathscr{R}} + \mathbf{x}_{\mathscr{N}}$$

and furthermore that $\mathbf{x}_{\mathscr{N}}$ is orthogonal to $\mathbf{x}_{\mathscr{R}}$, that is, $\mathbf{x}_{\mathscr{N}}^T\mathbf{x}_{\mathscr{R}} = 0$. Consequently $\mathbf{x}_{\mathscr{N}}$ can be interpreted as the orthogonal projection of $\mathbf{x}$ on $\mathscr{N}$. The orthogonal vector $\mathbf{x}_{\mathscr{R}}$ is the shortest vector to $\mathscr{N}$; examine Fig. C.2.

Any vector $\mathbf{x}$ with n components can be formed uniquely from a combination of a basis $\mathbf{Y}$ in the range space $\mathscr{R}$ and a basis $\mathbf{Z}$ in the null space $\mathscr{N}$

$$\mathbf{x} = \mathbf{Y}\mathbf{x}_{\mathscr{R}} + \mathbf{Z}\mathbf{x}_{\mathscr{N}} \tag{C.4}$$

where $\mathbf{Y}$ is a matrix with m independent columns, $\mathbf{Z}$ is a matrix with $(n - m)$ independent columns orthogonal to $\mathbf{Y}$, that is, $\mathbf{Y}^T\mathbf{Z} = 0$, $\mathbf{x}_{\mathscr{R}}$ is the m-dimensional component of $\mathbf{x}$ in the range space, and $\mathbf{x}_{\mathscr{N}}$ is the $(n - m)$ dimensional compo-

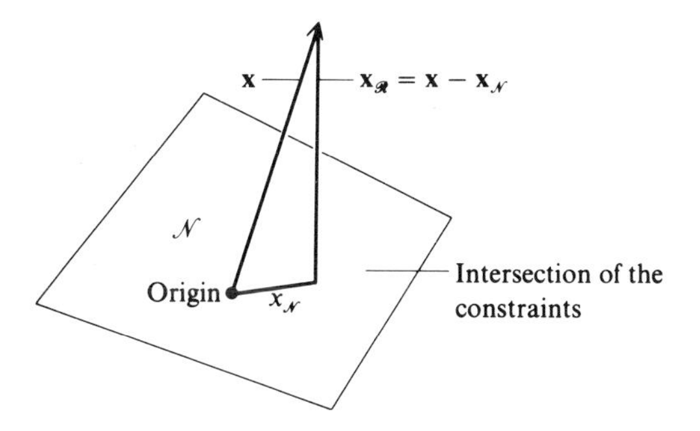

Figure C.2 Relationship of $\mathbf{x}_{\mathscr{R}}$ and $\mathbf{x}_{\mathscr{N}}$.

nent of **x** in the null space. For example, with two constraints ($m = 2$), and three variables ($n = 3$), suppose the equations for the constraints are

$$x_1 + x_3 = 2$$

$$x_2 + x_3 = 3$$

Then the Jacobian matrix of the two constraints is

$$\mathbf{A} = \overset{n \longrightarrow}{\begin{bmatrix} 1 & 0 & 1 \\ 0 & 1 & 1 \end{bmatrix}} \Big\downarrow m$$

We note by inspection that the rank of **A** is $r = 2$, and consequently the rank of the null space will be $r = 1$. Suppose we let **Y** be $\mathbf{A}^T$ as an example of a basis in $\mathscr{R}$

$$\mathbf{Y} = \begin{bmatrix} 1 & 0 \\ 0 & 1 \\ 1 & 1 \end{bmatrix}$$

Then **Z** can be any matrix (really a vector here) orthogonal to **A** such as

$$\mathbf{Z} = \begin{bmatrix} 1 \\ 1 \\ -1 \end{bmatrix}$$

Examine Fig. C.3. Figure C.4 illustrates the range space and null space for the case in which $m = 1$ and $n = 3$.

Next, suppose that $\hat{\mathbf{x}}$ is a feasible point so that $\mathbf{A}\hat{\mathbf{x}} = \mathbf{b}$. Because of Eq. (C.4)

$$\mathbf{A}\hat{\mathbf{x}} = \mathbf{A}\mathbf{Y}\hat{\mathbf{x}}_{\mathscr{R}} + \mathbf{A}\mathbf{Z}\hat{\mathbf{x}}_z = \mathbf{b} \tag{C.5}$$

but $\mathbf{AZ} = \mathbf{0}$, hence $\mathbf{A}\hat{\mathbf{x}} = \mathbf{A}\mathbf{Y}\hat{\mathbf{x}}_{\mathscr{R}} = \mathbf{b}$ or

$$\hat{\mathbf{x}}_{\mathscr{R}} = (\mathbf{A}\mathbf{Y})^{-1}\mathbf{b} \tag{C.6}$$

What this means is that all feasible points have the same range space component $\hat{\mathbf{x}}_{\mathscr{R}}$, and the solution to the optimization problem is determined solely by the null space portion of **x**.

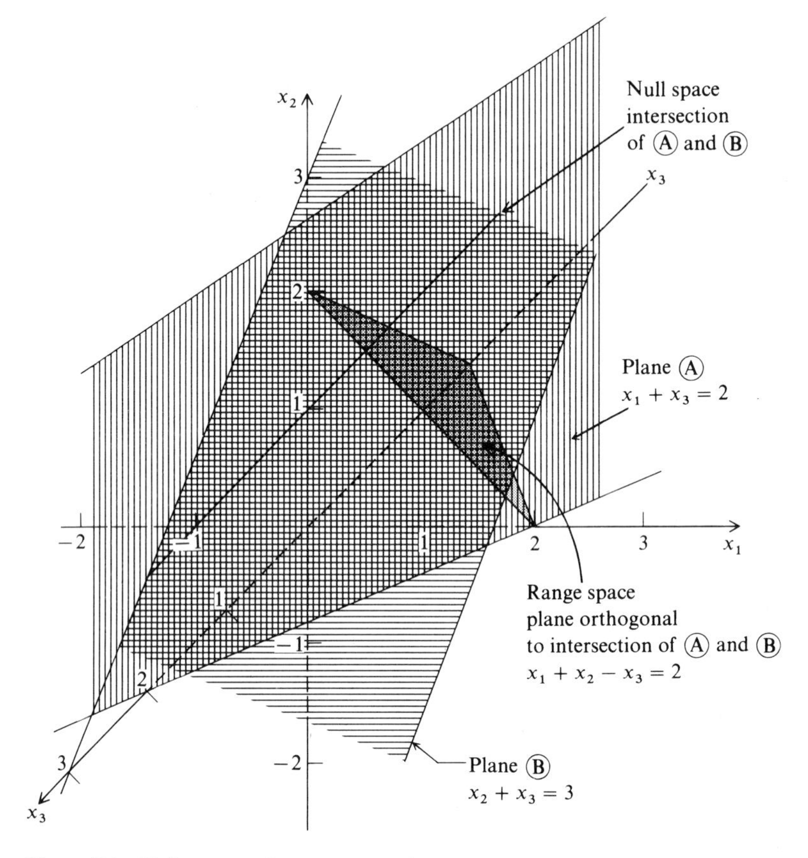

Figure C.3 Null space and range space with two constraints and three variables.

For the example above the equation $\mathbf{AYx}_{\mathscr{R}} = \mathbf{b}$ gives

$$\begin{bmatrix} 1 & 0 & 1 \\ 0 & 1 & 1 \end{bmatrix} \begin{bmatrix} 1 & 0 \\ 0 & 1 \\ 1 & 1 \end{bmatrix} \begin{bmatrix} \hat{x}_{1\mathscr{R}} \\ \hat{x}_{2\mathscr{R}} \end{bmatrix} = \begin{bmatrix} 2 \\ 3 \end{bmatrix} \quad \text{or} \quad \begin{aligned} \hat{x}_{1\mathscr{R}} &= 1 \\ \hat{x}_{2\mathscr{R}} &= 1.5 \end{aligned}$$

and Eq. (C.4) is then

$$\hat{\mathbf{x}} = \begin{bmatrix} 1 & 0 \\ 0 & 1 \\ 1 & 1 \end{bmatrix} \begin{bmatrix} 1 \\ 1.5 \end{bmatrix} + \begin{bmatrix} 1 \\ 1 \\ -1 \end{bmatrix} \hat{\mathbf{x}}_{\mathscr{N}} = \begin{bmatrix} 1 \\ 1.5 \\ 2.5 \end{bmatrix} + \begin{bmatrix} 1 \\ 1 \\ -1 \end{bmatrix} \hat{\mathbf{x}}_{\mathscr{N}}$$

Thus, a linearly constrained problem is an unconstrained problem in the null space variables.

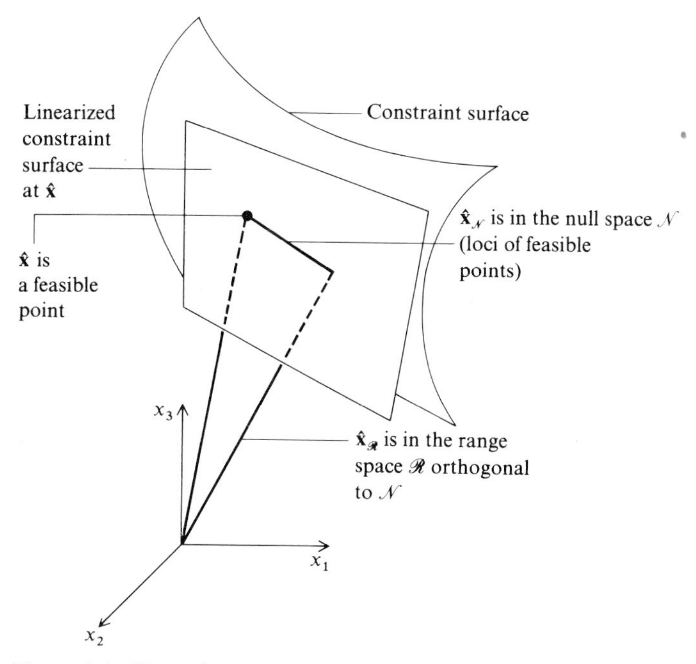

Figure C.4 Illustration of the range space and the null space for a feasible point **x**. (Constraint surface is convex, i.e., behind the plane)

In searching for the optimum of a linearly constrained NLP problem, a step $\Delta\mathbf{x}$ from one feasible point $\mathbf{x}^k$ where $\mathbf{A}\mathbf{x}^k = \mathbf{b}$ to the next feasible point $\mathbf{x}^{k+1}$ where $\mathbf{A}\mathbf{x}^{k+1} = \mathbf{b}$ must be orthogonal to the rows of **A** because

$$\mathbf{A}\Delta\mathbf{x}^k = \mathbf{0} \tag{C.7}$$

if **A** is a constant matrix. Any such step $\Delta\mathbf{x}^k$ occurs in what is termed a *feasible direction* $\mathbf{s}^k$ because $\Delta\mathbf{x}^k = \lambda^k\mathbf{s}^k$. Thus, $\mathbf{s}^k$ is orthogonal to the m rows of **A**

$$\mathbf{A}\mathbf{s}_Z^k = \mathbf{0} \tag{C.8}$$

and lies in the null space. The vector $\mathbf{s}^k$ can be expressed as a linear combination of the columns of **Z**, $\mathbf{s}^k = \mathbf{Z}\mathbf{s}_Z^k$, hence the subscript z placed on $\mathbf{s}^k$ in Eq. (C.8). How $\mathbf{s}_Z$ can be calculated is described next. The important point to keep in mind is that $\mathbf{s}_Z$ is of reduced dimension $(n - m)$. In most large-scale chemical engineering problems there are many variables and constraints but only a few degrees of freedom. If the equality constraints are linear, **A** is a constant matrix. On the other hand, if the constraints are linearized, the values of the gradients of the constraints which comprise the elements of **A**, the Jacobian matrix, change from stage to stage, hence one should use the designation $\mathbf{A}^k$.

How should the search direction $\mathbf{s}_Z^k$ in the null space be calculated? Some algorithms, particularly those pertaining to problems involving only linear constraints, use the *projected gradient* as the search direction

$$\mathbf{s}_Z^k = -\mathbf{P}^k\nabla f(\mathbf{x}^k) \tag{C.9}$$

where $\mathbf{P}$ is called the *projection matrix*. $\mathbf{P}$ might be equal to the orthogonal projection matrix[1]

$$\mathbf{P} = \mathbf{Z}^T = \mathbf{I} - \mathbf{A}_m^T(\mathbf{A}_m\mathbf{A}_m^T)^{-1}\mathbf{A}_m \tag{C.10}$$

or the generalized projection matrix

$$\mathbf{P}^* = \mathbf{I} - \mathbf{Q}\mathbf{A}_m^T(\mathbf{A}_m\mathbf{Q}\mathbf{A}_m^T)^{-1}\mathbf{A}_m \tag{C.11}$$

where $\mathbf{Q}$ is a square nonsingular matrix such as the Hessian matrix of $f(\mathbf{x})$, of the Lagrangian function $L(\mathbf{x}, \boldsymbol{\omega})$, or approximations thereto, and $\mathbf{A}_m$ is the Jacobian matrix of the constraints. The projected gradient ensures that the search direction remains feasible.

The generalized reduced-gradient method uses a special form for $\mathbf{Z}$. Suppose that $\mathbf{A}$, the Jacobian matrix of the constraints, is of full rank and is partitioned as follows

$$\mathbf{A} = [\mathbf{V} \mid \mathbf{U}] \tag{C.12}$$

Let a set of m (independent) columns of $\mathbf{A}$ form $\mathbf{V}$, an $m \times m$ nonsingular matrix. The balance of the (independent) columns of $\mathbf{A}$ form $\mathbf{U}$. The vector of the variables $\mathbf{x}$ is correspondingly partitioned into $\mathbf{x} = \begin{bmatrix} \mathbf{x_V} \\ \mathbf{x_U} \end{bmatrix}$. The (linearized) equality constraints of a nonlinear programming problem can then be expressed as

$$[\mathbf{V} \mid \mathbf{U}]\begin{bmatrix} \mathbf{x_V} \\ \mathbf{x_U} \end{bmatrix} = \mathbf{b} \tag{C.13}$$

---

[1] From another viewpoint, if we expand $f(\mathbf{x})$ about a feasible point $\hat{\mathbf{x}}$ with a search direction in the null space $\mathbf{s} = \mathbf{Z}\mathbf{s_Z}$

$$f(\hat{\mathbf{x}} + \lambda\mathbf{s}) = f(\hat{\mathbf{x}} + \lambda\mathbf{Z}\mathbf{s_Z}) = f(\hat{\mathbf{x}}) + \nabla^T f(\hat{\mathbf{x}})\mathbf{Z}\mathbf{s_Z} + \tfrac{1}{2}\mathbf{s_Z}^T\mathbf{Z}^T\mathbf{H}\mathbf{Z}\mathbf{s_Z}$$

a Newton step can be calculated from

$$(\mathbf{Z}^T\mathbf{H}\mathbf{Z})\mathbf{s_Z} = -\mathbf{Z}^T\nabla f(\mathbf{x})$$

The right-hand side is the projected gradient of $f(\mathbf{x})$ and the term in parentheses on the left-hand side is known as the *projected Hessian*.

To get the projection of a vector $\mathbf{v}$ on the constraint intersection, we want

$$\mathbf{v}_{\mathcal{N}} = \mathbf{P}\mathbf{v} \tag{a}$$

We know both that $\mathbf{v}_{\mathcal{N}} = \mathbf{v} - \mathbf{v}_{\mathcal{R}}$ and $\mathbf{v}_{\mathcal{R}} = \sum_{j=1}^{m} u_j\mathbf{a}_j = \mathbf{A}_m^T\mathbf{u}$ where the subscript on $\mathbf{A}$ denotes the dimension of $\mathbf{A}$. Therefore, $\mathbf{v}_{\mathcal{N}} = \mathbf{v} - \mathbf{A}_m^T\mathbf{u}$. We also know that

$$\mathbf{v}_{\mathcal{R}}^T\mathbf{v}_{\mathcal{N}} = 0 = (A_m^T u)(\mathbf{v} - (\mathbf{A}_m^T\mathbf{u})) \tag{b}$$

Solve (b) for $\mathbf{u}$

$$\mathbf{u} = (\mathbf{A}_m\mathbf{A}_m^T)^{-1}\mathbf{A}_m\mathbf{v}$$

hence

$$\mathbf{v}_{\mathcal{N}} = [\mathbf{v} - \mathbf{A}_m^T(\mathbf{A}_m A_m^T)^{-1}\mathbf{A}_m\mathbf{v}] = \mathbf{P}\mathbf{v} = [\mathbf{I} - \mathbf{A}_m^T(\mathbf{A}_m\mathbf{A}_m^T)^{-1}\mathbf{A}_m]\mathbf{v} \tag{c}$$

If $\mathbf{A}$ is so partitioned, then by definition $\mathbf{Z}$ has to be orthogonal to all vectors in the range space $\mathscr{R}$, hence

$$\mathbf{Z} = \begin{bmatrix} -\mathbf{V}^{-1}\mathbf{U} \\ \mathbf{I} \end{bmatrix} \tag{C.14}$$

as can be demonstrated by direct multiplication of $\mathbf{A}$ times $\mathbf{Z}$

$$\mathbf{AZ} = \mathbf{0}$$

What the partitioning of $\mathbf{A}$ means in practice is that $\mathbf{Z}$ does not have to be explicitly computed—only $\mathbf{V}$ and $\mathbf{U}$, which are easy to identify, and the search can be split into two phases, using $\mathbf{s_V}$ and $\mathbf{s_U}$.

To clarify these ideas, let us examine a numerical example. Let $n = 3$, $m = 2$, and $\mathbf{A}$ be

$$\mathbf{A} = \begin{bmatrix} 1 & 2 & 1 \\ -2 & 0 & 3 \end{bmatrix}$$

For this $\mathbf{A}$

$$\mathbf{V} = \begin{bmatrix} 1 & 2 \\ -2 & 0 \end{bmatrix} \qquad \mathbf{U} = \begin{bmatrix} 1 \\ 3 \end{bmatrix}$$

and

$$\mathbf{V}^{-1} = \begin{bmatrix} 0 & -\frac{1}{2} \\ \frac{1}{2} & \frac{1}{4} \end{bmatrix}$$

as can be verified from $\mathbf{VV}^{-1} = \mathbf{I}$. Then we can calculate from Eq. (C.14)

$$\mathbf{Z} = \begin{bmatrix} -\begin{bmatrix} 0 & -\frac{1}{2} \\ \frac{1}{2} & \frac{1}{4} \end{bmatrix}\begin{bmatrix} 1 \\ 3 \end{bmatrix} \\ 1 \end{bmatrix} = \begin{bmatrix} \frac{3}{2} \\ -\frac{5}{4} \\ 1 \end{bmatrix}$$

As a check, we can verify that $\mathbf{AZ} = \mathbf{0}$

$$\begin{bmatrix} 1 & 2 & 1 \\ -2 & 0 & 3 \end{bmatrix}\begin{bmatrix} \frac{3}{2} \\ -\frac{5}{4} \\ 1 \end{bmatrix} = \begin{bmatrix} 0 \\ 0 \end{bmatrix}$$

In the generalized reduced-gradient method, $\mathbf{x}$ is split into two components, one containing m dependent variables (also termed *basic variables* by analogy with linear programming) associated with $\mathbf{V}$, and the other containing $(n - m)$ variables termed *independent* variables (also termed *superbasic*) associated with $\mathbf{U}$. (Variables held fixed at their bounds are usually called *nonbasic* variables.) The gradient elements are similarly partitioned. Consequently, the null space search direction $\mathbf{s_Z}$ is just $\mathbf{s_U}$ with $(n - m)$ elements.

The term *reduced-gradient*, denoted by $\mathbf{g_R}$, is given to the following vector

$$\mathbf{s} = (\mathbf{Z})^T \nabla f(\mathbf{x}) = \begin{bmatrix} -\mathbf{V}^{-1}\mathbf{U} \\ \mathbf{I} \end{bmatrix}^T \begin{bmatrix} \nabla_{\mathbf{x_V}} f(\mathbf{x}) \\ \nabla_{\mathbf{x_U}} f(\mathbf{x}) \end{bmatrix}$$

$$= -\mathbf{U}^T(\mathbf{V}^{-1})^T \nabla_{\mathbf{x_V}} f(\mathbf{x}) + \nabla_{\mathbf{x_U}} f(\mathbf{x}) \equiv \mathbf{g_R} \qquad \text{(C.15)}$$

Also the transpose of $\mathbf{g_R}$ is

$$\mathbf{g_R^T} = -[\nabla_{\mathbf{x_V}} f(\mathbf{x})]^T \mathbf{V}^{-1}\mathbf{U} + [\nabla_{\mathbf{x_U}} f(\mathbf{x})]^T \qquad \text{(C.16)}$$

where the subscripts on ∇ designate the set of variables involved in the partial derivatives in the gradient elements.

We can calculate from the data in Example 8.6 the expressions

$$\nabla_{\mathbf{x_U}} f(\mathbf{x}) = \frac{\partial f(\mathbf{x})}{\partial x_2} = [4x_2] \qquad \nabla_{\mathbf{x_V}} f(\mathbf{x}) = \frac{\partial f(\mathbf{x})}{\partial x_1} = [4x_1]$$

$$\mathbf{U} = \frac{\partial h(\mathbf{x})}{\partial x_2} = [2] \qquad \mathbf{V} = \frac{\partial h(\mathbf{x})}{\partial x_1} = [1]$$

$$\mathbf{g_R^T} = -[(4x_1)(1)^{-1}(2)]^T + 4x_2 = -8x_1 + 4x_2$$

$$\mathbf{g_R} = -[2]^T[1^{-1}]^T[4x_1] + 4x_2 = -8x_1 + 4x_2$$

These last two relations are the same because $\mathbf{g_R}$ has the dimensions of $n - m = 2 - 1 = 1$ (Eqs. (C.15) and (C.16)).

REFERENCES

Graybill, F. A., *Introduction to Matrices*, Wadsworth, Belmont, California (1969).
Stewart, G. W., *Introduction to Matrix Computations*, Academic Press, New York, (1973).

NAME INDEX

(Upright numbers designate the reference location; *italic numbers designate the full citation.*)

SUBJECT INDEX